Protein Synthesis and Translational Control

A subject collection from *Cold Spring Harbor Perspectives in Biology*

OTHER SUBJECT COLLECTIONS FROM *COLD SPRING HARBOR PERSPECTIVES IN BIOLOGY*

The Synapse

Extracellular Matrix Biology

Protein Homeostasis

Calcium Signaling

The Golgi

Germ Cells

The Mammary Gland as an Experimental Model

The Biology of Lipids: Trafficking, Regulation, and Function

Auxin Signaling: From Synthesis to Systems Biology

The Nucleus

Neuronal Guidance: The Biology of Brain Wiring

Cell Biology of Bacteria

Cell–Cell Junctions

Generation and Interpretation of Morphogen Gradients

Immunoreceptor Signaling

NF-κB: A Network Hub Controlling Immunity, Inflammation, and Cancer

Symmetry Breaking in Biology

The Origins of Life

The p53 Family

SUBJECT COLLECTIONS FROM *COLD SPRING HARBOR PERSPECTIVES IN MEDICINE*

Parkinson's Disease

Type 1 Diabetes

Angiogenesis: Biology and Pathology

HIV: From Biology to Prevention and Treatment

The Biology of Alzheimer Disease

Protein Synthesis and Translational Control

A subject collection from *Cold Spring Harbor Perspectives in Biology*

EDITED BY

John W.B. Hershey
University of California, Davis

Nahum Sonenberg
McGill University

Michael B. Mathews
UMDNJ–New Jersey Medical School

COLD SPRING HARBOR LABORATORY PRESS
Cold Spring Harbor, New York • www.cshlpress.org

Protein Synthesis and Translational Control

A Subject Collection from *Cold Spring Harbor Perspectives in Biology*
Articles online at www.cshperspectives.org

Executive Editor	Richard Sever
Managing Editor	Maria Smit
Project Manager	Barbara Acosta
Permissions Administrator	Carol Brown
Production Editor	Diane Schubach
Production Manager/Cover Designer	Denise Weiss
Publisher	John Inglis

Front cover artwork: The cover art depicts the structure of the eukaryotic ribosome from the yeast *Saccharomyces cerevisiae*. The ribosome consists of four RNA chains (gray ribbons) and 79 different proteins (colored ribbons). With a total mass of 3.3 MDa, it is more intricate and ~40% larger than its bacterial counterpart. The previous edition of the translational control series, *Translational Control in Biology and Medicine* (2007), showed the structure of the bacterial ribosome on its cover. The display of the structure of the eukaryotic ribosome on the cover of this volume is a demonstration of the recent remarkable advances made in the protein synthesis field. The image was kindly provided by Marat Yusupov, Directeur de Recherche du CNRS.

Library of Congress Cataloging-in-Publication Data

Protein synthesis and translational control : a subject
collection from Cold Spring Harbor perspectives in biology/
edited by John W.B. Hershey, Nahum Sonenberg, Michael B. Mathews.
    p. cm. -- (Cold Spring Harbor perspectives in biology)
  Includes bibliographical references and index.
  ISBN 978-1-936113-46-0 (hardcover : alk. paper)
1. Proteins--Synthesis. 2. Genetic translation. 3. Genetic
regulation. I. Hershey, John W. B. II. Sonenberg, Nahum. III.
Mathews, Michael.

  QH450.5.P77 2012
  572'.6--dc23

                      2012016831

10 9 8 7 6 5 4 3 2 1

All World Wide Web addresses are accurate to the best of our knowledge at the time of printing.

For a complete catalog of all Cold Spring Harbor Laboratory Press publications, visit our website at www.cshlpress.org.

Contents

Contents

Preface

THE MECHANISM OF PROTEIN SYNTHESIS and its regulation have been studied intensively for more than a half-century, yet much remains to be learned. This is a particularly exciting time for such studies, as the role of translational control in regulating gene expression is broadly recognized as more important than previously thought. In the past, many studies focused on defining the translational machinery and how it functions. The translation of specific mRNAs suspected of being regulated was also studied, establishing a variety of mechanisms for controlling the translational efficiency of mRNAs. During the past few years, high-throughput methods have been applied to studies of translational control, resulting in the realization that such regulation is applied to the majority of mRNAs. Situated at the nexus between nucleic acids and proteins, the importance of translational control, now appreciated for its role in establishing the cell's proteome, is comparable to that of transcriptional control—a realization that makes studies of translational control even more compelling and essential.

The fact that protein synthesis is regulated broadly means that we need to understand a vast range of translational controls that operate on most mRNAs. This is an enormous challenge, as mRNAs differ in structure, in their modes of initiation, and in the assortment of *cis*-acting sequences that coordinate different regulating elements. Many mRNAs are themselves a collection of different structures due to alternative promoters, splicing, or processing. In addition, multiple regulatory mechanisms may operate on individual mRNAs, complicating their identification. To address this problem effectively, a precise knowledge of the mechanism of protein synthesis is required. Recent advances in ribosome structure, single-molecule studies, and reaction kinetics should provide the depth of understanding required to explain regulation.

While we were contemplating editing a fourth edition of *Translational Control*, John Inglis suggested that we consider creating a book for the Perspectives series for the Cold Spring Harbor Laboratory Press. Our previous editions, namely *Translational Control* (1996), *Translational Control of Gene Expression* (2000), and *Translational Control in Biology and Medicine* (2007), provided comprehensive reviews of the process and regulation of protein synthesis. For the Perspectives series, we have attempted to focus on the current status of the field, with emphasis on aspects that need further elucidation and development. We have chosen a limited number of specialized areas that we feel are particularly important for future developments in the field. The volume begins with a number of chapters that examine fundamental mechanisms of protein synthesis and continues with chapters that describe approaches or mechanisms that apply broadly to many mRNAs. A number of chapters address a specific aspect of cell metabolism where translational control plays a prominent role. The volume ends with an examination of how insights into translational control can be used to develop therapeutic agents.

We thank all of the authors for their superb efforts in generating thoughtful and exciting chapters. The quality of the book rests on their efforts. We also thank John Inglis and Richard Sever for their encouragement, project manager Barbara Acosta for her competent and tireless attention to our submissions, and the production staff of the Press.

JOHN W.B. HERSHEY
NAHUM SONENBERG
MICHAEL B. MATHEWS

Principles of Translational Control: An Overview

John W.B. Hershey[1], Nahum Sonenberg[2], and Michael B. Mathews[3]

[1]Department of Biochemistry and Molecular Medicine, School of Medicine, University of California, Davis, California 95616

[2]Department of Biochemistry and Goodman Cancer Research Center, 1160 Pine Avenue West, Montreal, Quebec H3A 1A3, Canada

[3]Department of Biochemistry and Molecular Biology, UMDNJ—New Jersey Medical School, Newark, New Jersey 07103-2714

Correspondence: jwhershey@ucdavis.edu; nahum.sonenberg@mcgill.ca; mathews@umdnj.edu

Translational control plays an essential role in the regulation of gene expression. It is especially important in defining the proteome, maintaining homeostasis, and controlling cell proliferation, growth, and development. Numerous disease states result from aberrant regulation of protein synthesis, so understanding the molecular basis and mechanisms of translational control is critical. Here we outline the pathway of protein synthesis, with special emphasis on the initiation phase, and identify areas needing further clarification. Features of translational control are described together with numerous specific examples, and we discuss prospects for future conceptual advances.

Protein synthesis is an indispensable process in the pathway of gene expression, and is a key component in its control. Regulation of translation plays a prominent role in most processes in the cell and is critical for maintaining homeostasis in the cell and the organism. The synthesis rate of a protein in general is proportional to the concentration and translational efficiency of its mRNA. Translational control governs the efficiency of mRNAs and thus plays an important role in modulating the expression of many genes that respond to endogenous or exogenous signals such as nutrient supply, hormones, or stress. Because the vast majority of eukaryotic mRNAs have quite long half-lives (>2 h) (Raghavan et al. 2002), rapid regulation of the cellular levels of the proteins they encode must be achieved by controlling their mRNA translational efficiencies and protein degradation rates. During early stages of viral infection (Walsh et al. 2012) and in cells lacking active transcription such as oocytes and reticulocytes, translational control is often the only mechanism to regulate the synthesis of proteins. Furthermore, protein synthesis accounts for a large proportion of the energy budget of a cell, especially one that is rapidly growing or biosynthetically active, and therefore requires tight regulation. Because protein synthesis is intimately integrated with cell metabolism, aberrations in its regulation contribute to a number of disease states. It is therefore apparent that a detailed knowledge of the mechanisms that contribute to translational control is essential in understanding cell homeostasis and disease.

PROTEIN SYNTHESIS PATHWAY

Protein synthesis is a highly conserved process that links amino acids together on ribosomes based on the sequence of an mRNA template. To appreciate the complex translation pathway in human cells, it is useful first to consider protein synthesis in bacteria. The bacterial pathway is coupled to transcription of the DNA into mRNA, made possible because no nuclear membrane separates these processes. It comprises four phases: initiation, elongation, termination, and recycling (Fig. 1). The initiation phase involves the binding of the small ribosomal subunit (30S) to an unstructured region in the mRNA (the Shine-Dalgarno region) that is complementary to a portion of the 16S rRNA, and is stabilized through an interaction between the ribosome-bound initiator formyl-methionyl-tRNA and the initiation codon, usually AUG. Although formation of the 30S initiation complex is promoted by three initiation factors, identification of the initiation site in the mRNA is based primarily on RNA–RNA interactions. The large ribosomal subunit (50S) then binds to form a 70S initiation complex, which contains the formyl-methionyl-tRNA in the ribosomal P-site and which is prepared to enter the elongation phase. Elongation involves three steps: the binding of an aminoacyl-tRNA whose anticodon is complementary to the mRNA codon in the ribosomal A-site; formation of a peptide bond by transfer of the amino acid or peptide attached to the tRNA in the P-site to the aminoacyl-tRNA in the A-site; and translocation of the newly formed peptidyl-tRNA from the A-site to the P-site, together with the mRNA. These reactions are promoted by a number of elongation factors and by

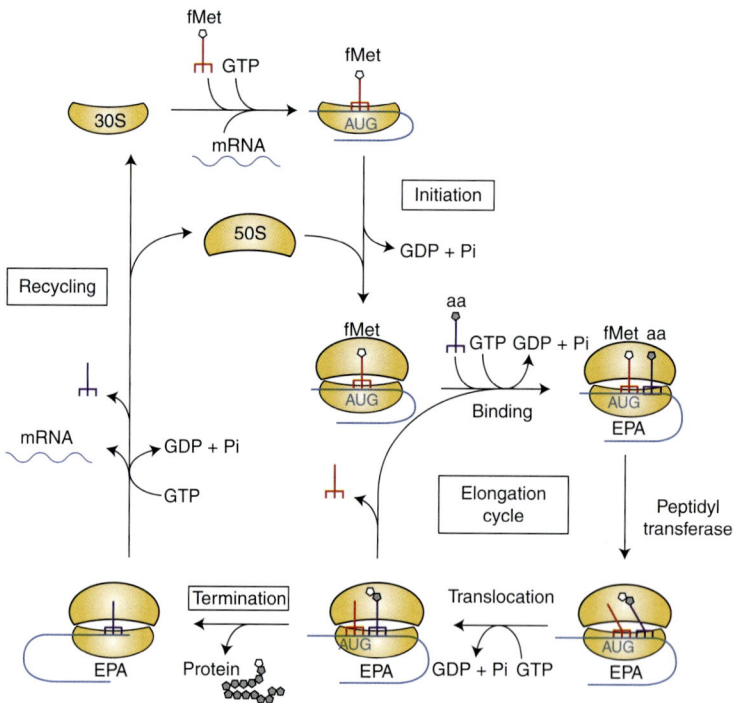

Figure 1. Pathway of protein synthesis in bacteria. The simplified cartoon shows the four phases of protein synthesis and how the ribosomes, tRNAs, mRNA, and GTP interact. Not shown are the initiation, elongation, and termination factors that promote the reactions. Following initiation, each turn of the elongation cycle results in the addition of another amino acid (gray pentagon) to the growing peptide chain (not shown). Termination occurs when a termination codon (UAA, UAG, UGA) appears in the ribosomal A-site and involves hydrolysis of the peptidy-tRNA in the P-site. (Figure constructed by Nancy Villa, University of California, Davis.)

Cite this article as Cold Spring Harb Perspect Biol doi: 10.1101/cshperspect.a011528 the ribosome itself, with rRNA playing a particularly notable part. When a termination codon (UAA, UAG, UGA) enters the A-site, termination factors bind to the ribosome and promote the hydrolysis of the peptidyl-tRNA. Ribosomes are then recycled through interactions with a number of protein factors to generate ribosomal subunits capable of undergoing another round of protein synthesis. Detailed descriptions of the bacterial pathway are found in a number of reviews (Laursen et al. 2005; Noller 2007).

Protein synthesis in higher cells shares many similarities with that in bacteria. The genetic code is identical and the aminoacyl-tRNAs and their synthetases are very similar, but eukaryotic ribosomal subunits, named 40S and 60S, are larger and richer in protein, as illustrated by recent high-resolution structures (see Wilson and Cate 2012). Whereas the elongation phase is strongly conserved, the initiation and termination/recycling phases differ substantially. A conspicuous feature of eukaryotic protein synthesis is the fact that mRNAs are translated in the cytoplasm, making translation uncoupled from transcription. Mature eukaryotic mRNAs possess a m^7G-cap at their $5'$-terminus and, in most cases, a poly (A) tail at their $3'$-terminus. They are generally monocistronic, unlike most bacterial mRNAs, and the pathway and mechanism for the formation of 40S and 80S initiation complexes differ substantially from those in bacteria. For example, a large number of initiation factors (at least 12) promote the binding of the mRNA and initiator methionyl-tRNA$_i$ (Met-tRNA$_i$)—which is not formylated—to the 40S ribosomal subunit. Therefore 40S initiation complex formation involves numerous protein–RNA and protein–protein interactions, in contrast to what occurs in bacteria. Given the preeminence of the initiation phase in the regulation of protein synthesis, we develop the mechanism of eukaryotic initiation in considerable detail in the following section. Termination and recycling resemble the reactions in prokaryotes, except that different sets of proteins promote these phases. Eukaryotic initiation pathways are outlined in Figure 2; detailed descriptions of the molecular mechanisms are found in Hinnebusch and Lorsch (2012).

MECHANISM OF EUKARYOTIC INITIATION

To elucidate translational control mechanisms, it is essential to define the detailed molecular mechanism of protein synthesis. The major initiation pathway, scanning, involves binding of a 40S–Met-tRNA$_i$ complex to the $5'$-terminus of an m^7G-capped mRNA, followed by downstream scanning along the mRNA until an AUG (or near-cognate) initiation codon is recognized. The 60S ribosomal subunit then joins the 40S initiation complex to form an 80S initiation complex capable of entry into the elongation phase. These reactions are promoted by twelve or more initiation factors comprising over 25 proteins (see Hinnebusch and Lorsch 2012). Although much has been learned about how mammalian cells initiate protein synthesis, a number of gaps and uncertainties remain. For example, identification of initiation factors has been based on their stimulation of in vitro initiation assays constructed with purified components, and verified by genetic methods. However, the recent discoveries of new proteins apparently involved in the pathway (e.g., DHX29 [Parsyan et al. 2009] and DDX3 [Lai et al. 2008]) suggest that all relevant initiation factors may not have been identified. In addition, the relevance of some identified factors is uncertain. For example, eIF2A promotes the binding of Met-tRNA$_i$ to 40S ribosomal subunits, but its role in translation initiation is not well established (Komar et al. 2005; Ventoso et al. 2006). A newly identified factor, eIF2D, promotes tRNA binding into the ribosomal P-site in the absence of GTP, but its mechanism of action and role in initiation have not been defined (Dmitriev et al. 2010). eIF5A promotes protein synthesis, but whether it is involved in the initiation or elongation phase (Saini et al. 2009), or possibly just in formation of the first peptide bond, is controversial (reviewed in Henderson and Hershey 2011). eIF6 was first identified as an antiribosome association factor but its role in initiation was then questioned (Si and Maitra 1999); however, recent structural studies show clearly how binding to the nascent premature 60S subunit prevents junction with the 40S initiation complex (Klinge et al. 2011).

Cite this article as *Cold Spring Harb Perspect Biol* doi: 10.1101/cshperspect.a011528

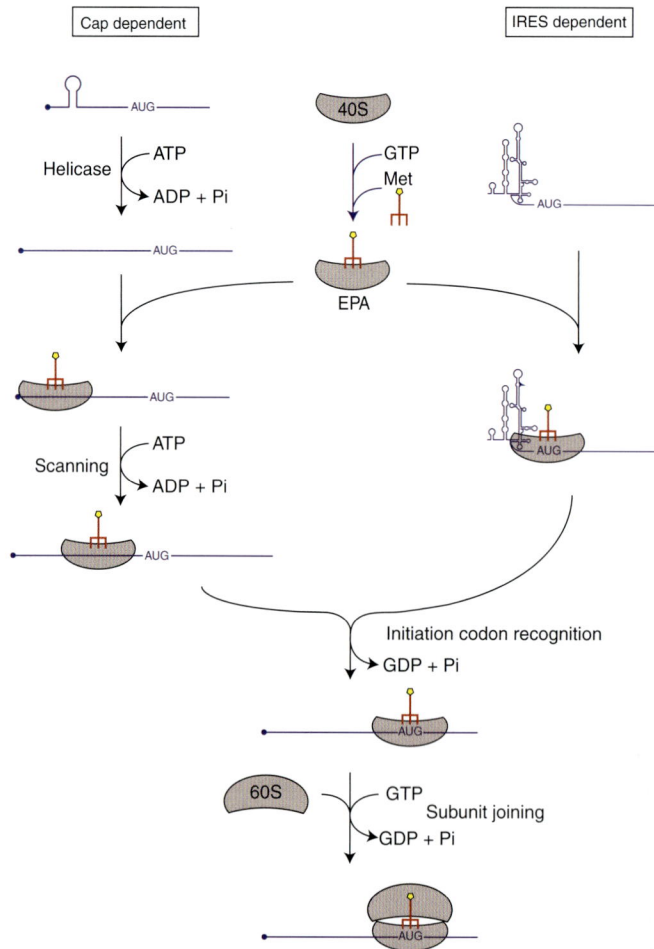

Figure 2. Pathway of eukaryotic initiation. The simplified cartoon shows two types of initiation mechanisms (m^7G-cap-dependent scanning and IRES-dependent internal), and how the ribosomes, methionyl-tRNA$_i$, mRNA, and ATP/GTP interact. Not shown are the initiation factors or the possibility that scanning follows IRES-directed binding of the 40S ribosomal subunit during internal initiation. (Figure constructed by Nancy Villa, University of California, Davis.)

Despite its centrality, aspects of the scanning mechanism are not yet well elucidated. eIF4A is a well-established RNA helicase that functions while tethered to the cap-binding complex. However, does it continue to unwind RNA following 40S ribosomal subunit recruitment to the mRNA, or do other identified helicases such as DHX29 (Parsyan et al. 2009) and DDX3 (Lai et al. 2008) provide the helicase function during scanning? Are there mechanistic clues in the unusual bidirectionality of the helicase activity of eIF4A, and the departures

from stoichiometry in the levels of some of the factors? Ribosome profiling methods detect initiation at numerous sites in a large number of mRNAs, some at non-AUG codons (Ingolia et al. 2011), but how such initiation events are regulated is unclear. Because rigorous kinetic analyses of many of the reactions in initiation have not been performed, we do not have a full description of their reaction rates, yet such information is essential for detecting and understanding regulation during initiation. In contrast, great progress has been made recently in elucidating the structure of eukaryotic ribosomes (see Ben-Shem et al. 2011; Klinge et al. 2011; Rabl et al. 2011; Wilson and Cate 2012), although atomic resolution structures of initiation complexes are still lacking.

Besides the scanning pathway, a sizable number of cellular mRNAs—generally estimated as 5%–10%—may use a different way to recruit the ribosome. Direct binding of the 40S ribosomal subunit to an internal region of the mRNA, called the internal ribosome entry site (IRES), bypasses the necessary recognition of the m^7G cap (Fig. 2). IRES-mediated initiation is often used for the translation of viral mRNAs, and is reported for some cellular mRNAs as well (see Jackson 2012 for a critical analysis of cellular mRNAs containing IRESs). Still other initiation pathways have been described, involving shunting (Yueh and Schneider 2000; Pooggin et al. 2006), tethering (Martin et al. 2011), translation enhancers (Vivinus et al. 2001), the TISU element that frequently functions with mRNAs possessing very short 5′-UTRs (Elfakess et al. 2011), and a poly-adenylate leader in the 5′-UTR that appears to function in the absence of a number of the canonically required initiation factors (Shirokikh and Spirin 2008). Although we already possess much sophisticated knowledge of how these initiation pathways proceed, there remains much to be learned that is essential for a full understanding of translational control.

FEATURES OF TRANSLATIONAL CONTROL

Regulation of protein synthesis may occur at different steps of the pathway, with the initiation phase being the most common target. Which phase of protein synthesis is affected is often identified by determining polysome profiles (Merrick and Hensold 2000) and ribosome transit times (Fan and Penman 1970; Palmiter 1972). One of the salient features of translational control involves the number of mRNAs affected. A given mechanism might affect the translation of a single mRNA, a subset of mRNAs, or most mRNAs. Global regulation often is based on the activation or inhibition of one or more components of the translational machinery, whereas specific regulation frequently occurs

through the action of *trans*-acting proteins (see Gebauer et al. 2012) or microRNAs (see Roux and Topisirovic 2012) binding to *cis*-elements in the mRNA. Some mRNAs are capable of escaping the effects of global activation or inhibition. Therefore, the response caused by a given mechanism may be complex in terms of the mRNAs affected. Methods that address this latter issue are ribosome profiling (Ingolia et al. 2009) involving high-throughput deep sequencing of ribosome-protected mRNA sequences, or DNA microarray technology, involving identification of mRNAs in size-fractionated polysomes (see Larsson et al. 2012). Such analyses of changes in ribosome profiles caused by a difference in physiological state enable identification of the mRNAs that are most affected. The polysome profiling method is particularly powerful, as it determines where ribosomes are positioned on essentially all cellular mRNAs at a specific point in time, thereby shedding light on the phase of protein synthesis that changes.

Another important feature of translational control is that a change in physiological state can activate multiple regulatory mechanisms that affect the rate of protein synthesis. Such redundancy complicates mechanistic studies, because interfering with one mechanism does not necessarily alter the overall extent of inhibition or activation. A further complication is that a given mechanism may itself cause only a minor change in protein synthesis rate. However, when multiple weak mechanisms act on the system together, significant translational control can result. Mechanisms that are modest in their action are especially difficult to elucidate, as their effects sometimes only slightly alter a specific reaction rate. To detect and assess the importance of such mechanisms, sophisticated and highly accurate kinetic analyses are required, and are increasingly being pursued.

REGULATORY MECHANISMS

Recruitment of the mRNA to the 40S ribosomal subunit is thought to be the rate-limiting step of initiation, and is often modulated. The binding of methionyl-tRNA$_i$ also is frequently regulated, and subsequent steps such as scanning and initiator codon recognition may be affected as well. How are these reactions regulated? The cellular levels of the canonical initiation factors differ in various cell or tissue types, thereby affecting initiation rates. Modulating the activities of the initiation factors by phosphorylation is often used to regulate global rates of protein synthesis. The best-studied examples are phosphorylation of eIF2α, which results in an inhibition of Met-tRNA$_i$ binding to ribosomes (see Hinnebusch and Lorsch 2012; Pavitt and Ron 2012), and phosphorylation of 4E-BPs (sequestors of the cap-binding protein, eIF4E), which relieves translational repression caused by decreased mRNA recognition and binding to ribosomes (see Hinnebusch and Lorsch 2012; Roux and Topisirovic 2012). Numerous other initiation factors are phosphorylated, often as targets of signal transduction pathways, as are ribosomes and the elongation factor eEF2, but how such events regulate protein synthesis is not yet well established. Besides phosphorylation, posttranslational modifications such as methylation, ubiquitination, and glycosylation, may affect protein synthesis, but these have not been studied extensively. One can anticipate that mass spectrometric methods will identify new modifications of importance in the near future.

mRNA levels appear not to be rate-limiting for global protein synthesis in many cells, as a substantial number of mRNAs are found as untranslated, free mRNPs rather than in active polysomes. However, mRNAs also can be sequestered in stress granules or P-bodies (see Decker and Parker 2012) or localized in specific regions of a cell's cytoplasm (see Chao et al. 2012; Lasko 2012), indicating that mRNA accessibility can influence the efficiency of translation.

Regulation of translation through the action of microRNAs is an exciting new area of study. MicroRNAs can stimulate the degradation of mRNAs or affect protein synthesis directly (see Braun et al. 2012). The mode of regulation is mRNA-specific, although a single microRNA may affect a number of different mRNAs. Precise mechanisms whereby the microRNAs affect protein synthesis are yet to be elucidated. Recent in vitro experiments detecting early microRNA-based inhibition of protein synthesis prior to

mRNA deadenylation may resolve the controversy of which effect is dominant (Fabian et al. 2009).

Trans-acting proteins affect initiation rates by binding to specific mRNAs and interfering with various steps of the pathway (see Gebauer et al. 2012). Such proteins frequently recognize a binding site in the 3′-UTR, yet affect events occurring near the 5′-m^{7}G cap. These regulatory mechanisms often function during early development (see Gebauer et al 2012; Lasko 2012) via protein-mediated crosstalk between the two ends of the mRNAs. Indeed, active mRNAs are thought to be circularized through an interaction between the poly (A)-binding protein (PABP) and eIF4G, which is a component of the m^{7}G-cap-binding complex (Wells et al. 1998). Some mRNAs, for example the histone mRNAs that lack a poly (A) tail, are circularized through specialized proteins that bind near the 3′-terminus of the mRNA and react with the cap-binding protein complex (Cakmakci et al. 2008). However, mutant yeast lacking the PABP-eIF4G interaction show normal translation rates (Park et al. 2011), which suggests that circularization does not invariably promote initiation, at least not in yeast. Another possibility, not yet established, is that circularization enhances the ability of a terminating ribosome to reinitiate on the same mRNA, perhaps by a mechanism that differs from the canonical scanning mechanism. During analyses of the generation of polysomes in vitro, the rate of addition of a new ribosome to a polysome was slower than the ribosome transit time, yet polysome size increased, suggesting that ribosomes already present on a polysome reinitiate more efficiently on the same mRNA than new ribosomes initiate (Nelson and Winkler 1987).

Although less commonly reported, the elongation and termination phases (see Dever and Green 2012) also are targets of translational control. The rate of elongation is thought to be maximal under most conditions (Ingolia et al. 2011), but can be inhibited by specific mechanisms. Whether such inhibition affects the elongation rates of all mRNAs equally is not well established. Only when elongation is slowed sufficiently, such that initiation is no longer rate-limiting, is the rate of protein synthesis affected. The occurrence of rare codons or strong secondary structure in the coding region of a mRNA is thought to slow the rate of elongation. Some mRNAs can thereby be induced to undergo a shift in reading frame at a specific region, generating a protein whose sequence and length differ from the unshifted one. For proteins that are to be inserted into the endoplasmic reticulum, their elongation is paused by the signal recognition particle, enabling the amino terminus of the nascent protein to dock onto the endoplasmic reticulum, after which elongation resumes (see Benham 2012). The rate of elongation affects the folding of proteins (see Cabrita et al. 2010); if elongation is too fast, e.g., when recombinant eukaryotic proteins are synthesized in bacteria, many proteins fail to fold properly unless the overall rate is reduced (Siller et al. 2010). Alternatively, slowing the elongation rate at specific regions of the mRNA may enhance proper folding (Zhang et al. 2009), further demonstrating the link between rates or elongation and protein folding. Furthermore, the folding of the nascent protein as it emerges from the large ribosomal subunit can affect the elongation rate, either positively or negatively (see Cabrita et al. 2010).

The termination phase also may be regulated (reviewed in Dinman and Berry 2007). Under most circumstances, termination is not rate-limiting for protein synthesis, because ribosomes are not found stacked up at the ends of mRNAs. However, termination can be suppressed, enabling either frame-shifting or read-through to occur, thereby extending the nascent protein at its carboxyl terminus. The UGA stop codon can be reprogrammed to enable the insertion of a seleno-cysteine residue rather than to terminate synthesis. Incorporated into proteins by the translational process itself, seleno-cysteine has been called the "21st amino acid," and it is now followed by the 22nd—pyrrolysine, encoded by a UAG codon in some methanogenic archaea and bacteria (Atkins and Gesteland 2002). Such "amendments" to the elongation and termination steps are influenced by the sequence context of the codon, or by other features of the mRNA.

FUTURE PROSPECTS

New discoveries of the involvement of translational control in cell metabolism, proliferation and disease are being reported constantly. The ribosome profiling method has already identified unexpected changes in the translation of numerous specific mRNAs and can be expected to generate a vast amount of new data. Handling the plethora of information requires new and sophisticated bioinformatic methods that are rapidly being developed and refined (see Larsson et al. 2012). A challenge is presented by the proliferation of gene products, including alternatively processed mRNAs and protein isoforms, produced in higher cells. This diversification introduces additional levels of complexity that need to be accommodated in these analyses. Such high-throughput approaches do not generally elucidate details of the molecular mechanisms involved, however. To understand the observed changes in mRNA translation, many of them rather modest in extent, it is necessary to have a precise knowledge of the mechanism of protein synthesis.

What are the major challenges for understanding translational control mechanisms? We already have a fairly detailed description of the process of protein synthesis during the initiation, elongation, termination, and recycling phases. With the recent ability to obtain crystals of eukaryotic ribosomes (Ben-Shem et al. 2011; Klinge et al. 2011; Rabl et al. 2011), we can anticipate atomic level structures of ribosome complexes that are essential for describing how peptide bonds are formed and how the various factors interact on the surface of the ribosome to promote initiation, elongation, and termination. However, crystals of 40S and 80S initiation complexes have eluded researchers, and even cryo-electron microscopic approaches have not yet yielded sufficiently precise structures of these important intermediates. Another area lacking structural information at high resolution pertains to mRNAs. Although computer programs can predict structural motifs in RNA (Cruz and Westhof 2011), the actual 3-dimensional structures, especially those of the 5'-UTR, are only now beginning to be determined (Steen et al.

2011). Such detailed structures of mRNAs and their native mRNP complexes are eagerly awaited, as they surely are important for mRNA binding, scanning and initiator codon recognition during initiation. mRNA structures also affect the elongation and termination rates, thereby affecting protein folding and the regulation of frameshifting. So high-resolution mRNA structural information, especially pertaining to the initiation phase, is needed.

Another area in which our knowledge is limited pertains to the kinetics of the various reactions, interactions and conformational changes involved in protein synthesis. The elongation phase is relatively well characterized, especially for prokaryotes, but there are numerous initiation steps that are yet to be studied in detail. Insights into the kinetics of initiation complex formation have been gained from studies primarily performed with yeast components (reviewed in Lorsch and Dever 2010 and Hinnebusch and Lorsch 2012), yet much is yet to be learned. Do initiation factors form subcomplexes off the ribosome, or do they first bind to the 40S subunits, and if so, in what order? Which proteins mediate the binding of Met-tRNA$_i$ to 40S ribosomes, and how is that rate affected by other initiation factors? What is the rate of ribosome scanning of the 5′-UTR, and is this rate affected by changes in the activities of associated initiation factors, e.g., those involved in RNA helicase activity? Why do ribosomes appear to idle at the initiator AUG codon? That is, what limits their rapid progression into the elongation phase? Most kinetic analyses average the effects of many molecules over time. The application of single molecule studies to the kinetics of protein synthesis (see Petrov et al. 2012) likely will generate new views of how such reactions proceed. Additional work employing both single molecule and standard kinetic methods are needed to properly recognize and evaluate many of the translational control mechanisms that operate at the initiation phase, especially those mechanisms that only marginally affect reaction rates.

Complementing our constantly increasing understanding of the molecular mechanisms of translational control is the expectation that more and more examples of regulation at the level of protein synthesis will be discovered. A number of promising areas of research are featured in this volume. We anticipate that regulation by microRNAs will prove to be important for the translation for most mRNAs (Braun et al. 2012). How does the secretion or cotranslational folding of nascent proteins affect their synthesis (Cabrita et al. 2010; Benham 2012)? Other areas in which translational control already is firmly established are described in literature dealing with cell development (Lasko 2012), cancer (Ruggero 2012), synaptic plasticity and memory (Darnell and Richter 2012), and viruses (Walsh et al. 2012). As translational control mechanisms are better understood and as high throughput methods identify when such regulation occurs, we can confidently anticipate that we will learn of additional areas of cellular metabolism that are modulated through protein synthesis. Indeed, it is becoming clear that translational control and transcriptional control are comparably important in regulating gene expression.

The relevance and importance of protein synthesis in disease and medicine is increasingly recognized and appreciated. The dysregulation of protein synthesis in a specific disease provides a target for therapeutic intervention (see Malina et al. 2012). As our knowledge of the structures and detailed mechanisms of protein synthesis improve, this information can be applied to enable more rational drug design. Therefore, research in the area of translational control will contribute to a better understanding of many disease states and to the development of novel therapeutic agents.

REFERENCES

*Reference is also in this collection.

Atkins J, Gesteland R. 2002. The 22nd amino acid. *Science* **296:** 1409–1410.
* Benham AM. 2012. Protein secretion and the endoplasmic reticulum. *Cold Spring Harb Perspect Biol* doi: 10.1101/cshperspect.a012872.
Ben-Shem A, de Loubresse N, Melnikov S, Jenner L, Yusupova G, Yusupov M. 2011. The structure of the eukaryotic ribosome at 3.0 Å resolution. *Science* **334:** 1524–1529.

Cite this article as *Cold Spring Harb Perspect Biol* doi: 10.1101/cshperspect.a011528

* Braun JE, Huntzinger E, Izaurralde E. 2012. A molecular link between miRISCs and deadenylases provides new insight into the mechanism of gene silencing by micro-RNAs. *Cold Spring Harb Perspect Biol* doi: 10.1101/cshperspect.a012328.

Cabrita DL, Dobson CM, Christodoulou J. 2010. Protein folding on the ribosome. *Curr Opin Struct Biol* **20:** 33–45.

Cakmakci N, Lerner R, Wagner E, Zheng L, Marzluff W. 2008. SLIP1, a factor required for activation of histone mRNA translation by the stem-loop binding protein. *Mol Cell Biol* **28:** 1182–1194.

* Chao JA, Yoon YJ, Singer RH. 2012. Imaging translation in single cells using fluorescent microscopy. *Cold Spring Harb Perspect Biol* doi: 10.1101/cshperspect.a012310.

Cruz J, Westhof E. 2011. Sequence-based identification of 3D structural modules in RNA with RMDetect. *Nat Methods* **8:** 513–521.

* Darnell J, Richter J. 2012. Cytoplasmic RNA binding proteins and the control of complex brain function. *Cold Spring Harb Perspect Biol* doi: 10.1101/cshperspect.a012344.

* Decker CJ, Parker R. 2012. P bodies and stress granules and their possible roles in the control of translation and mRNA degradation. *Cold Spring Harb Perspect Biol* doi: 10.1101/cshperspect.a012286.

* Dever TE, Green R. 2012. The elongation, termination, and recycling phases of translation in eukaryotes. *Cold Spring Harb Perspect Biol* doi: 10.1101/cshperspect.a013706.

Dinman J, Berry M. 2007. Regulation of Termination and Recoding. In *Translational control in biology and medicine* (ed. Mathews MB, Sonenberg N, Hershey JWB), pp. 625–654. Cold Spring Harbor Laboratory Press, Cold Spring Harbor, NY.

Dmitriev S, Terenin I, Andreev D, Ivanov P, Dunaevsky J, Merrick W, Shatsky I. 2010. GTP-independent tRNA delivery to the ribosomal P-site by a novel eukaryotic translation factor. *J Biol Chem* **285:** 26779–26787.

Elfakess R, Sinvani H, Haimov O, Svitkin Y, Sonenberg N, Dikstein R. 2011. Unique translation initiation of mRNAs containing TISU element. *Nucleic Acids Res* **39:** 7598–7609.

Fabian M, Mathonnet G, Sundermeier T, Mathys H, Zipprich J, Svitkin Y, Rivas F, Jinek M, Wohlschlegel J, Doudna J, et al. 2009. Mammalian miRNA RISC recruits CAP1 and PABP to affect PABP-dependent deadenylation. *Mol Cell* **35:** 868–880.

Fan H, Penman S. 1970. Regulation of protein synthesis in mammalian cells. II. Inhibition of protein synthesis at the level of initiation during mitosis. *J Mol Biol* **50:** 655–670.

* Gebauer F, Preiss T, Hentze MW. 2012. From *cis*-regulatory elements to complex RNPs and back. *Cold Spring Harb Perspect Biol* doi: 10.1101/cshperspect.a012245.

Henderson A, Hershey J. 2011. The role of eIF5A in protein synthesis. *Cell Cycle* **10:** 3617–3618.

* Hinnebusch AG, Lorsch JR. 2012. The mechanism of eukaryotic translation initiation: New insights and challenges. *Cold Spring Harb Perspect Biol* doi: 10.1101/cshperspect.a011544.

Ingolia N, Ghaemmaghami S, Newman J, Weissman J. 2009. Genome-wide analysis in vivo of translation with nucleotide resolution using ribosome profiling. *Science* **324:** 218–223.

Ingolia N, Lareau L, Weissman J. 2011. Ribosome profiling of mouse embryonic stem cells reveals the complexity and dynamics of mammalian proteomes. *Cell* **147:** 789–802.

* Jackson R. 2012. The current status of vertebrate cellular mRNA IRESs. *Cold Spring Harb Perspect Biol* doi: 10.1101/cshperspect.a011569.

Klinge S, Voigts-Hoffmann F, Leibundgut M, Arpagaus S, Ban N. 2011. Crystal structure of the eukaryotic 60S ribosomal subunit in complex with initiation factor 6. *Science* **334:** 941–948.

Komar A, Gross S, Barth-Baus D, Strachen R, Hensold J, Goss Kinzy T, Merrick W. 2005. Novel characteristics of the biological properties of the yeast Saccharomyces cerevisiae eukaryotic initiation factor 2A. *J Biol Chem* **280:** 15601–15611.

Lai M, Lee Y, Tarn W. 2008. The DEAD-box RNA helicase DDX3 associates with export messenger ribonucleoproteins as well as tip-associated protein and participates in translational control. *Mol Biol Cell* **19:** 3847–3858.

* Larsson O, Tian B, Sonenberg N. 2012. The genome-wide landscape of translational control. *Cold Spring Harb Perspect Biol* doi: 10.1101/cshperspect.a012302.

* Lasko P. 2012. mRNA localization and translational control in Drosophila. *Cold Spring Harb Perspect Biol* doi: 10.1101/cshperspect.a012294.

Laursen B, Sorensen H, Mortensen K, Sperling-Petersen H. 2005. Initiation of protein synthesis in bacteria. *Microbiol Mol Biol Rev* **69:** 101–123.

Lorsch J, Dever T. 2010. Molecular view of 43S complex formation and start site selection in eukaryotic translation initiation. *J Biol Chem* **285:** 21203–21207.

* Malina A, Mills JR, Pelletier J. 2012. Emerging therapeutics targeting mRNA translation. *Cold Spring Harb Perspect Biol* doi: 10.1101/cshperspect.a012377.

Martin F, Barends S, Jaeger S, Schaeffer L, Prongidi-Fix L, Eriani G. 2011. Cap-assisted internal initiation of translation of histone H4. *Mol Cell* **41:** 197–209.

Merrick W, Hensold J. 2000. The use of sucrose gradients in studies on eukaryotic translation. In *Current protocols in cell biology*, pp. 11.19.11–11.19.26. John Wiley & Sons, New York.

Nelson E, Winkler M. 1987. Regulation of mRNA entry into polysomes. Parameters affecting polysome size and the fraction of mRNA in polysomes. *J Biol Chem* **262:** 11501–11506.

Noller H. 2007. Structure of the Bacterial Ribosome and Some Implications for Translational Regulation. In *Translational control in biology and medicine* (ed. Mathews MB, Sonenberg N, Hershey JWB), pp. 41–58. Cold Spring Harbor Laboratory Press, Cold Spring Harbor, NY.

Palmiter R. 1972. Regulation of protein synthesis in chick oviduct. II. Modulation of polypeptide elongation and initiation rates by estrogen and progesterone. *J Biol Chem* **247:** 6770–6780.

Park E, Walker S, Lee J, Rothenburg S, Lorsch J, Hinnebusch A. 2011. Multiple elements in the eIF4G1 N-terminus promote assembly of eIF4G1-PABP mRNPs in vivo. *EMBO J* **30:** 302–316.

Parsyan A, Shahbazian D, Martineau Y, Petroulakis E, Alain T, Larsson O, Mathonnet G, Tettweiler G, Hellen C, Pestova T, et al. 2009. The helicase protein DHX29 promotes translation initiation, cell proliferation, and tumorigenesis. *Proc Natl Acad Sci* **106:** 22217–22222.

* Pavitt GD, Ron D. 2012. New insights into translational regulation in the endoplasmic reticulum unfolded protein response. *Cold Spring Harb Perspect Biol* doi: 10.1101/cshperspect.a012278.

* Petrov A, Chen J, O'Leary S, Tsai A, Puglisi JD. 2012. Single-molecule analysis of translational dynamics. *Cold Spring Harb Perspect Biol* doi: 10.1101/cshperspect.a011551.

Pooggin M, Ryabova L, He X, Futterer J, Hohn T. 2006. Mechanism of ribosome shunting in rice tungro bacilliform pararetrovirus. *RNA* **12:** 841–850.

Rabl J, Leibundgut M, Ataide SF, Haag A, Ban N. 2011. Crystal structure of the eukaryotic 40S ribosomal subunit in complex with initiation factor 1. *Science* **331:** 730–736.

Raghavan A, Orgilvie R, Reilly C, Abelson M, Raghavan S, Vasdewani J, Krathwohl M, Bohjanen P. 2002. Genome-wide analysis of mRNA decay in resting and activated primary human T lymphocytes. *Nucleic Acids Res* **30:** 5529–5538.

* Roux PP, Topisirovic I. 2012. Regulation of mRNA translation by signaling pathways. *Cold Spring Harb Perspect Biol* doi: 10.1101/cshperspect.a012252.

* Ruggero D. 2012. Translational control in cancer etiology. *Cold Spring Harb Perspect Biol* doi: 10.1101/cshperspect.a012336.

Saini P, Eyler D, Green R, Dever T. 2009. Hypusine-containing protein eIF5A promotes translation elongation. *Nature* **459:** 118–121.

Shirokikh N, Spirin A. 2008. Poly(A) leader of eukaryotic mRNA bypasses the dependence of translation on initiation factors. *Proc Natl Acad Sci* **105:** 10738–10743.

Si K, Maitra U. 1999. The Saccharomyces cerevisiae homologue of mammalian translation initiation factor 6 does not function as a translation initiation factor. *Mol Cell Biol* **19:** 1416–1426.

Siller E, DeZwaan D, Anderson J, Freeman B, Barral J. 2010. Slowing bacterial translation speed enhances eukaryotic protein folding efficiency. *J Mol Biol* **396:** 1310–1318.

Steen K, Siegfried N, Weeks K. 2011. Selective 2′-hydroxyl acylation analyzed by protection from exoribonuclease (RNase-detected SHAPE) for direct analysis of covalent adducts and of nucleotide flexibility in RNA. *Nat Protoc* **6:** 1683–1694.

Ventoso I, Sanz M, Molina S, Berlanga J, Carrasco L, Esteban M. 2006. Translational resistance of late alphavirus mRNA to eIF2α phosphorylation: A strategy to overcome the antiviral effect of protein kinase PKR. *Genes Dev* **20:** 87–100.

Vivinus S, Baulande S, van Zanten M, Campbell F, Topley P, Ellis J, Dessen P, Coste H. 2001. An element within the 5′ untranslated region of human Hsp70 mRNA, which acts as a general enhancer of mRNA translation. *Eur J Biochem* **268:** 1908–1917.

* Walsh D, Mathews MB, Mohr I. 2012. Tinkering with translation: Protein synthesis in virus-infected cells. *Cold Spring Harb Perspect Biol* doi: 10.1101/cshperspect.a012351.

Wells S, Hillner P, Vale R, Sachs A. 1998. Circularization of mRNA by eukaryotic translation initiation factors. *Mol Cell* **2:** 135–140.

* Wilson DN, Cate JHD. 2012. The structure and function of the eukaryotic ribosome. *Cold Spring Harb Perspect Biol* doi: 10.1101/cshperspect.a011536.

Yueh A, Schneider R. 2000. Translation by ribosome shunting on adenovirus and hsp70 mRNAs facilitated by complementarity to 18S rRNA. *Genes Dev* **14:** 414–421.

Zhang G, Hubalewska M, Ignatova Z. 2009. Transient ribosomal attenuation coordinates protein synthesis and co-translational folding. *Nat Struct Mol Biol* **16:** 274–280.

Cite this article as *Cold Spring Harb Perspect Biol* doi: 10.1101/cshperspect.a011528

The Structure and Function of the Eukaryotic Ribosome

Daniel N. Wilson[1,2] and Jamie H. Doudna Cate[3,4]

[1]Center for Integrated Protein Science Munich (CiPSM), 81377 Munich, Germany

[2]Gene Center and Department of Biochemistry, Ludwig-Maximilians-Universität München, 81377 Munich, Germany

[3]Departments of Molecular and Cell Biology and Chemistry, University of California at Berkeley, Berkeley, California 94720

[4]Physical Biosciences Division, Lawrence Berkeley National Laboratory, Berkeley, California 94720

Correspondence: wilson@lmb.uni-muenchen.de and jcate@lbl.gov

Structures of the bacterial ribosome have provided a framework for understanding universal mechanisms of protein synthesis. However, the eukaryotic ribosome is much larger than it is in bacteria, and its activity is fundamentally different in many key ways. Recent cryo-electron microscopy reconstructions and X-ray crystal structures of eukaryotic ribosomes and ribosomal subunits now provide an unprecedented opportunity to explore mechanisms of eukaryotic translation and its regulation in atomic detail. This review describes the X-ray crystal structures of the *Tetrahymena thermophila* 40S and 60S subunits and the *Saccharomyces cerevisiae* 80S ribosome, as well as cryo-electron microscopy reconstructions of translating yeast and plant 80S ribosomes. Mechanistic questions about translation in eukaryotes that will require additional structural insights to be resolved are also presented.

All ribosomes are composed of two subunits, both of which are built from RNA and protein (Figs. 1 and 2). Bacterial ribosomes, for example of *Escherichia coli*, contain a small subunit (SSU) composed of one 16S ribosomal RNA (rRNA) and 21 ribosomal proteins (r-proteins) (Figs. 1A and 1B) and a large subunit (LSU) containing 5S and 23S rRNAs and 33 r-proteins (Fig. 2A). Crystal structures of prokaryotic ribosomal particles, namely, the *Thermus thermophilus* SSU (Schluenzen et al. 2000; Wimberly et al. 2000), *Haloarcula marismortui* and *Deinococcus radiodurans* LSU (Ban et al. 2000; Harms et al. 2001), and *E. coli* and *T. thermophilus* 70S ribosomes (Yusupov et al. 2001; Schuwirth et al. 2005; Selmer et al. 2006), reveal the complex architecture that derives from the network of interactions connecting the individual r-proteins with each other and with the rRNAs (Brodersen et al. 2002; Klein et al. 2004). The 16S rRNA can be divided into four domains, which together with the r-proteins constitute the structural landmarks of the SSU (Wimberly et al. 2000) (Fig. 1A): The 5′ and 3′ minor (h44) domains with proteins S4, S5, S12, S16, S17, and S20 constitute the body

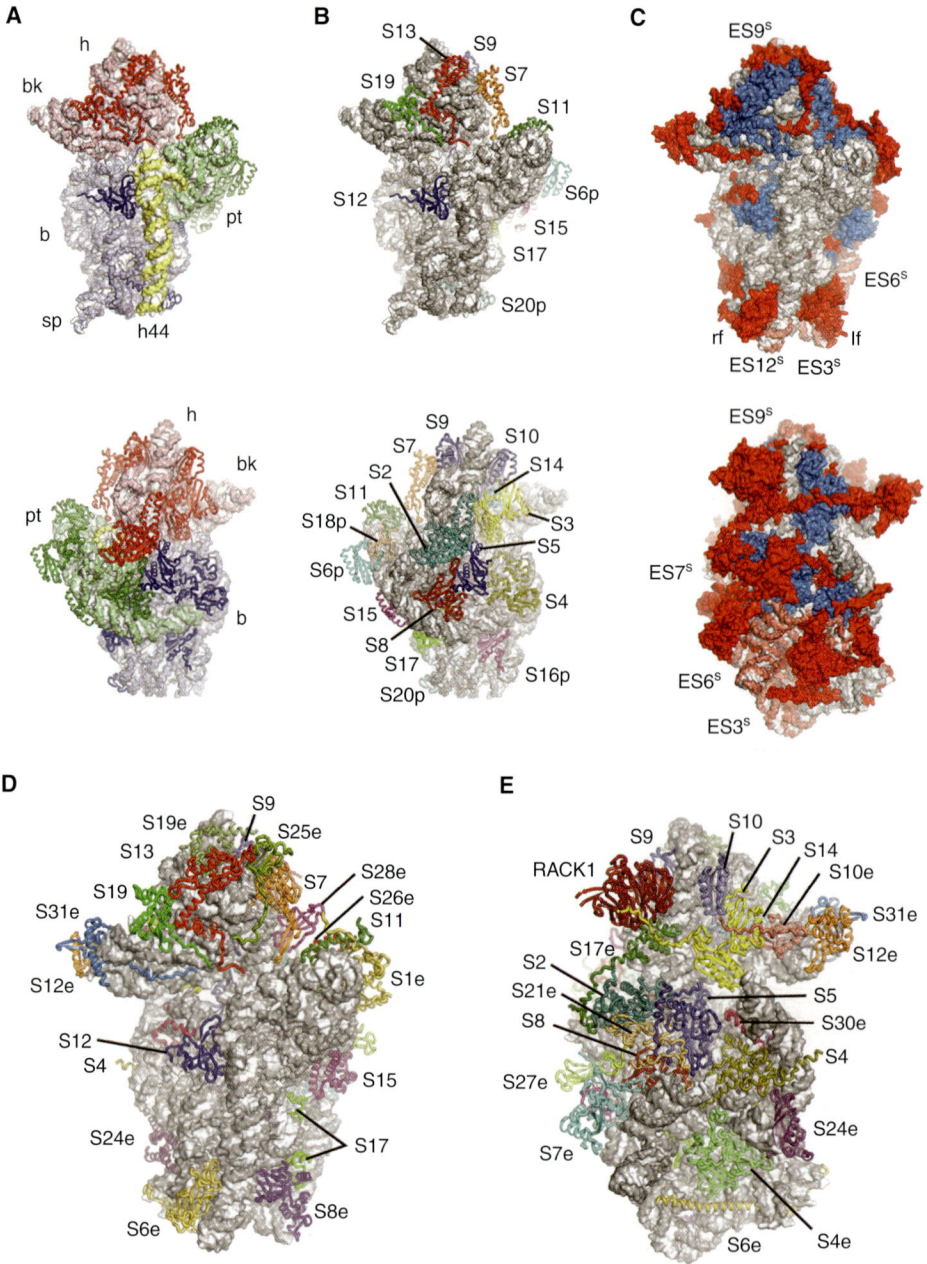

Figure 1. The bacterial and eukaryotic small ribosomal subunit. (*A,B*) Interface (*upper*) and solvent (*lower*) views of the bacterial 30S subunit (Jenner et al. 2010a). (*A*) 16S rRNA domains and associated r-proteins colored distinctly: b, body (blue); h, head (red); pt, platform (green); and h44, helix 44 (yellow). (*B*) 16S rRNA colored gray and r-proteins colored distinctly and labeled. (*C–E*) Interface and solvent views of the eukaryotic 40S subunit (Rabl et al. 2011), with (*C*) eukaryotic-specific r-proteins (red) and rRNA (pink) shown relative to conserved rRNA (gray) and r-proteins (blue), and with (*D,E*) 18S rRNA colored gray and r-proteins colored distinctly and labeled.

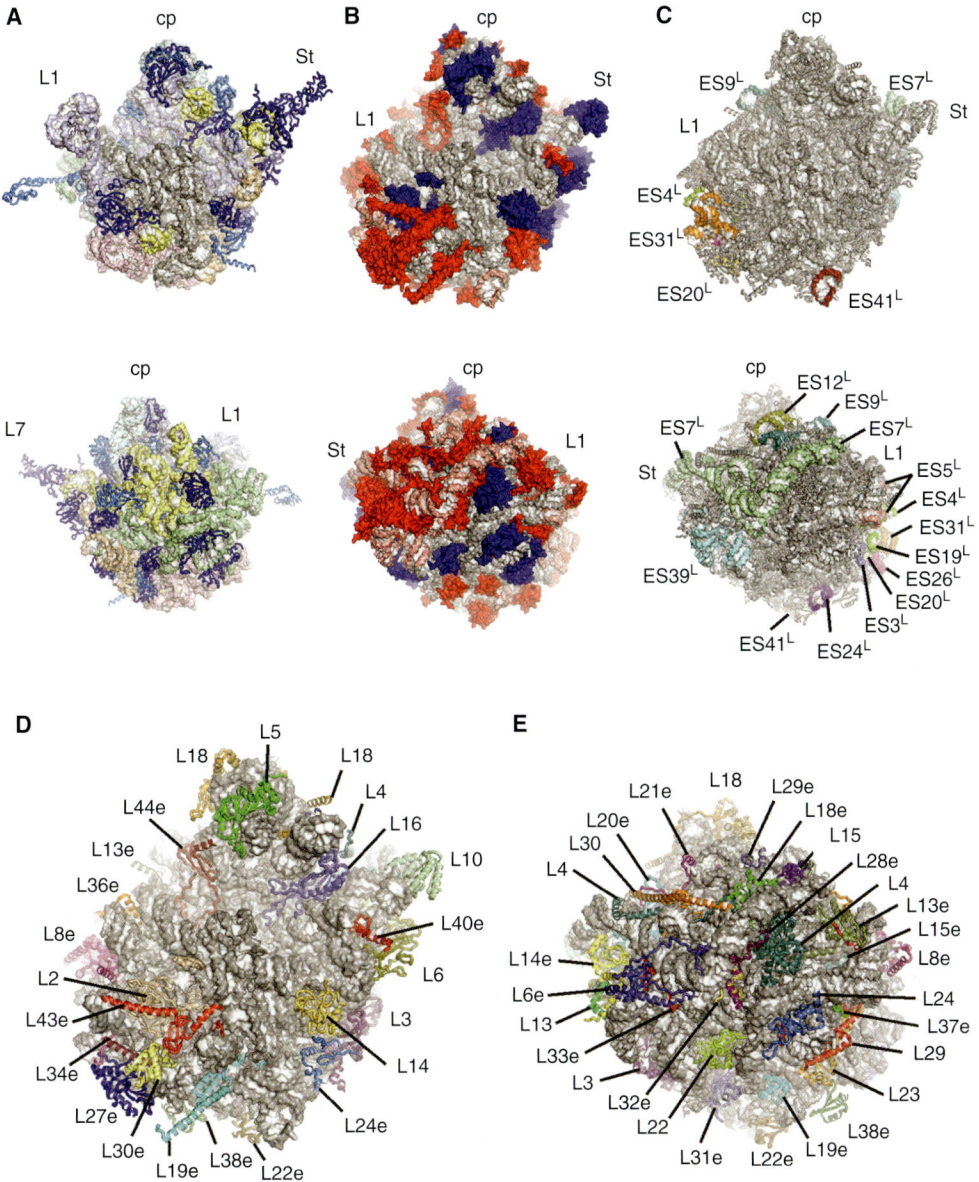

Figure 2. The bacterial and eukaryotic large ribosomal subunit. (*A*) Interface (*upper*) and solvent (*lower*) views of the bacterial 50S subunit (Jenner et al. 2010b), with 23S rRNA domains and bacterial-specific (light blue) and conserved (blue) r-proteins colored distinctly: cp, central protuberance; L1, L1 stalk; and St, L7/L12 stalk (or P-stalk in archaea/eukaryotes). (*B–E*) Interface and solvent views of the eukaryotic 60S subunit (Klinge et al. 2011), with (*B*) eukaryotic-specific r-proteins (red) and rRNA (pink) shown relative to conserved rRNA (gray) and r-proteins (blue), (*C*) eukaryotic-specific expansion segments (ES) colored distinctly, and (*D,E*) 28S rRNA colored gray and r-proteins colored distinctly and labeled.

(and spur or foot) of the SSU; the 3' major domain forms the head, which is protein rich, containing S2, S3, S7, S9, S10, S13, S14, and S19; whereas the central domain makes up the platform by interacting with proteins S1, S6, S8, S11, S15, and S18 (Fig. 1B). The rRNA of the LSU can be divided into seven domains (including the 5S rRNA as domain VII), which—in contrast to the SSU—are intricately interwoven with the r-proteins as well as each other (Ban et al. 2000; Brodersen et al. 2002) (Fig. 2A). Structural landmarks on the LSU include the central protuberance (CP) and the flexible L1 and L7/L12 stalks (Fig. 2A).

In contrast to their bacterial counterparts, eukaryotic ribosomes are much larger and more complex, containing additional rRNA in the form of so-called expansion segments (ES) as well as many additional r-proteins and r-protein extensions (Figs. 1C–E and 2C–E). Compared with the ~4500 nucleotides of rRNA and 54 r-proteins of the bacterial 70S ribosome, eukaryotic 80S ribosomes contain >5500 nucleotides of rRNA (SSU, 18S rRNA; LSU, 5S, 5.8S, and 25S rRNA) and 80 (79 in yeast) r-proteins. The first structural models for the eukaryotic (yeast) ribosome were built using 15-Å cryo-electron microscopy (cryo-EM) maps fitted with structures of the bacterial SSU (Wimberly et al. 2000) and archaeal LSU (Ban et al. 2000), thus identifying the location of a total of 46 eukaryotic r-proteins with bacterial and/or archaeal homologs as well as many ES (Spahn et al. 2001a). Subsequent cryo-EM reconstructions led to the localization of additional eukaryotic r-proteins, RACK1 (Sengupta et al. 2004) and S19e (Taylor et al. 2009) on the SSU and L30e (Halic et al. 2005) on the LSU, as well as more complete models of the rRNA derived from cryo-EM maps of canine and fungal 80S ribosomes at ~9 Å (Chandramouli et al. 2008; Taylor et al. 2009). Recent cryo-EM reconstructions of plant and yeast 80S translating ribosomes at 5.5–6.1 Å enabled the correct placement of an additional six and 10 r-proteins on the SSU and LSU, respectively, as well as the tracing of many eukaryotic-specific r-protein extensions (Armache et al. 2010a,b). The full assignment of the r-proteins in the yeast and fungal 80S ribosomes,

however, only became possible with the improved resolution (3.0–3.9 Å) resulting from the crystal structures of the SSU and LSU from *Tetrahymena thermophila* (Klinge et al. 2011; Rabl et al. 2011) and the *Saccharomyces cerevisiae* 80S ribosome (Figs. 1D,E and 2D,E) (Ben-Shem et al. 2011).

RIBOSOMAL RNA OF THE EUKARYOTIC RIBOSOME

In terms of rRNA, the major differences between bacterial and eukaryotic ribosomes is the presence in eukaryotes of five expansion segments ($ES3^S$, $ES6^S$, $ES7^S$, $ES9^S$, and $ES12^S$, following the nomenclature of Gerbi [1996]) and five variable regions (VRs) (h6, h16, h17, h33, and h41) on the SSU, as well as 16 expansion segments ($ES3^L$, $ES4^L$, $ES5^L$, $ES7^L$, $ES9^L$, $ES10^L$, $ES12^L$, $ES15^L$, $ES19^L$, $ES20^L$, $ES24^L$, $ES26^L$, $ES27^L$, $ES31^L$, $ES39^L$, and $ES41^L$) and two VRs (H16–18 and H38) on the LSU (Figs. 1C and 2C) (Cannone et al. 2002). On the LSU most ES are located on the back and sides of the particle, leaving the subunit interface and exit tunnel regions essentially unaffected (Taylor et al. 2009; Armache et al. 2010a; Ben-Shem et al. 2010; Klinge et al. 2011). The largest concentration of additional rRNA (~40%) on the yeast LSU is positioned behind the P stalk and is formed by $ES7^L$ (~200 nucleotides) and $ES39^L$ (~150 nucleotides), with a second patch (~150 nucleotides) located behind the L1 stalk formed by the clustering of $ES19^L$, $ES20^L$, $ES26^L$, and $ES31^L$ (Figs. 2C and 3A). In addition, the highly flexible $ES27^L$ (150 nucleotides), which was not observed in the crystal structures (Ben-Shem et al. 2011; Klinge et al. 2011), adopts two distinct conformations in cryo-EM reconstructions of yeast ribosomes (Beckmann et al. 2001; Armache et al. 2010a). On the yeast SSU the majority (~75%) of the additional rRNA comprises $ES3^S$ (~100 nucleotides) and $ES6^S$ (~200 nucleotides), which interact and cluster together to form the left foot of the particle (Figs. 1C and 3B) (Armache et al. 2010a; Ben-Shem et al. 2011; Rabl et al. 2011).

Comparison of rRNA sequences of diverse organisms, ranging from bacteria to mammals,

Cite this article as *Cold Spring Harb Perspect Biol* doi: 10.1101/cshperspect.a011536

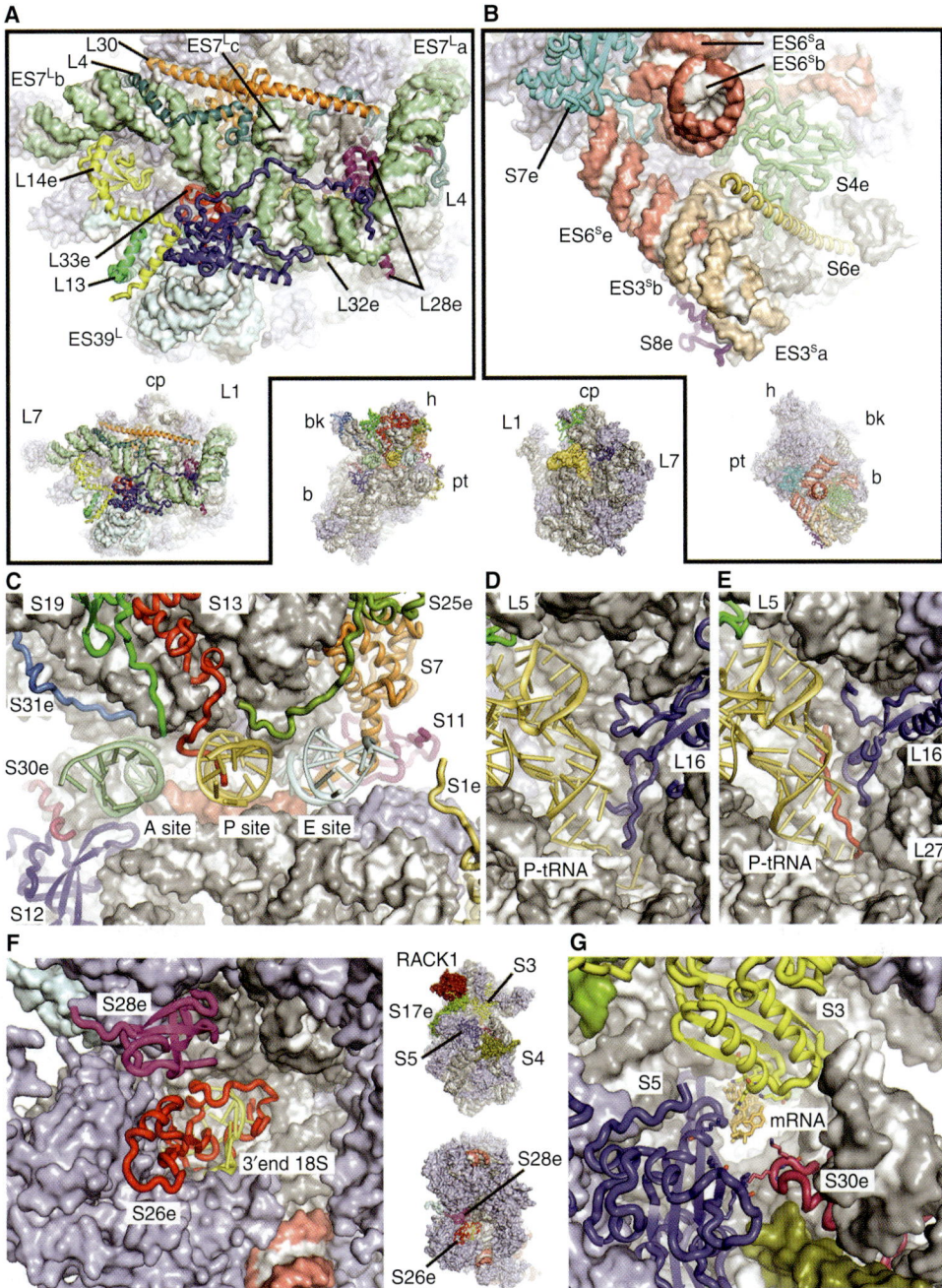

Figure 3. Structural and functional aspects of the eukaryotic ribosome. Interweaving of rRNA and r-proteins on the (A) LSU near ES7^L and ES39^L (Klinge et al. 2011), and (B) SSU near ES3 and ES6 (Rabl et al. 2011). Extension of r-proteins at the tRNA-binding sites on the (C) SSU (Armache et al. 2010b; Rabl et al. 2011), LSU of the (D) bacterial (Jenner et al. 2010b), and (E) eukaryotic (Armache et al. 2010b) peptidyltransferase centers. R-proteins located at the mRNA (F) exit, and (G) entry sites (Klinge et al. 2011).

reveals that the major differences in ES are restricted to four sites on the LSU, namely, ES7^L, ES15^L, ES27^L, and ES39^L. These ES are significantly longer ($\sim$850, $\sim$180, $\sim$700, and $\sim$220 nucleotides) in human 80S ribosomes than in yeast ($\sim$200, $\sim$20, $\sim$150, and $\sim$120 nucleotides, respectively) (Cannone et al. 2002). Moreover, cryo-EM reconstructions of mammalian ribosomes (Dube et al. 1998; Morgan et al. 2000; Spahn et al. 2004b; Chandramouli et al. 2008; Budkevich et al. 2011) reveal little to no density for the longer ES in mammalian ribosomes, indicating that they are highly mobile elements. In *Tetrahymena*, deletion of ES27^L is lethal (Sweeney et al. 1994), suggesting a functionally important role for this ES. Despite the high variability in length of ES27^L, ranging from $\sim$150 nucleotides in yeast to $\sim$700 nucleotides in mammals (Cannone et al. 2002), deletion of ES27^L can be complemented with a corresponding ES27^L from other species (Sweeney et al. 1994). ES27^L has been suggested to play a role in coordinating the access of nonribosomal proteins to the tunnel exit (Beckmann et al. 2001), but this remains to be shown. The role of other ES remains unclear. Their presence in eukaryotic ribosomes may reflect the increased complexity of translation regulation in eukaryotic cells, as evident for assembly, translation initiation, and development, as well as the phenomenon of localized translation (Sonenberg and Hinnebusch 2009; Freed et al. 2010; Wang et al. 2010).

RIBOSOMAL PROTEINS OF THE EUKARYOTIC RIBOSOME

The yeast 80S ribosome contains 79 r-proteins (SSU, 33; LSU, 46), 35 of which (SSU, 15; LSU, 20) have bacterial/archaeal homologs, whereas 32 (SSU, 12; LSU, 20) have only archaeal homologs (Lecompte et al. 2002). Thus, 12 (SSU, 6; LSU, 6) r-proteins of the yeast 80S are specific for eukaryotes. Cytoplasmic 80S ribosomes of *Tetrahymena* and higher eukaryotes, such as humans, contain an additional LSU r-protein, L28e, and thus have 13 eukaryotic-specific r-proteins and 80 (SSU, 33; LSU, 47) in total. Together with the ES, the additional r-proteins/

r-protein extensions form an intricate layer of additional RNA–protein mass that locates predominantly to the solvent surfaces of the ribosome (Figs. 1C and 2B). More than half of the conserved r-proteins contain extensions, which in some cases, such as S5, L4, L7, and L30, establish long-distance interactions far (50–100 Å) from the globular core of the protein. Interaction of eukaryotic-specific extensions with conserved core proteins using interprotein shared β-sheets has been noted, for example, between L14e and L6 (Ben-Shem et al. 2011) as well as L21e and L30 (Klinge et al. 2011).

The eukaryotic LSU contains $\sim$1 MDa of additional protein: 200 kDa of eukaryotic-specific domains or extensions and 800 kDa of r-proteins that are absent in bacteria. Most of this additional protein mass is located in a ring around the back and sides of the LSU, where it interacts with ES (Fig. 2B). Two large concentrations of additional RNA–protein mass exemplify the intertwined and coevolving nature of the ribosome (Yokoyama and Suzuki 2008). One cluster on the LSU comprises ES7^L, ES39^L, five eukaryotic r-proteins (L6e, L14e, L28e, L32e, and L33e), as well as eukaryotic-specific extensions of conserved r-proteins (L4, L13, and L30) (Fig. 3A). In this cluster yeast ES7^L comprises three helices, ES7^{L}a–c, whereas wheat germ (plant) ES7^L has five helices, ES7^{L}a–e, including a three-way junction extending from ES7^{L}c (Armache et al. 2010b). Curiously, the extension of L6e is longer in wheat germ as compared with yeast and appears to wrap around ES7^L and insert through the three-way junction of ES7^{L}a–c (Armache et al. 2010b). ES7^{L}a is stabilized by L28e in wheat germ and *Tetrahymena*, whereas this helix is more flexible in baker's yeast lacking L28e. Stabilization of ES by eukaryotic r-proteins is also evident for ES27^L, with the two different yeast conformations being stabilized by interaction with either L38e or L27e (Armache et al. 2010b). The second major ES cluster comprises ES19^L, ES20^L, ES26^L, and ES31^L, which are intimately associated with eukaryotic-specific r-proteins L27e, L30e, L34e, L43e, and the carboxy-terminal extension of L8e (Fig. 2C–E) (Ben-Shem et al. 2011). A single-stranded loop region of ES31^L provides an interaction platform for many of these r-proteins, notably the carboxy-terminal helix of L34e. Similarly, ES39^L also has many single-stranded loop regions that provide interaction sites for r-proteins, such as L20e and L14e.

The protein-to-RNA ratio of bacterial SSU is ~1:2, whereas the dramatic increase in r-protein mass for the eukaryotic SSU results in an almost 1:1 ratio. The SSU structures reveal that most of the additional eukaryotic-specific r-proteins and extensions cover the back of the SSU particle, forming a web of interactions with each other as well as with conserved r-proteins and rRNA (Fig. 1C–E) (Ben-Shem et al. 2011; Rabl et al. 2011). The beak of the eukaryotic SSU has acquired three r-proteins, S10e, S12e, and S31e, which appear to compensate for the reduced h33 compared with the bacterial SSU rRNA (Rabl et al. 2011). R-proteins are also seen to interact with the expansion segments ES3^S and ES6^S, via r-proteins S4e, S6e, S7e, and S8e (Fig. 3B). S6e has a long carboxy-terminal helix that stretches from the left to right foot, and that is phosphorylated in most eukaryotes (Meyuhas 2008). Based on the peripheral position of S6e, any regulation of translation via S6e phosphorylation is likely to be via indirect recruitment of specific regulatory factors (Rabl et al. 2011). The mRNA exit site on the eukaryotic SSU also differs from the bacterial one because of the presence of S26e and S28e surrounding the 3′ end of the 18S rRNA (Fig. 3F) (Armache et al. 2010a; Rabl et al. 2011). S26e overlaps the binding position of the *E. coli* r-protein S21p (Schuwirth et al. 2005), whereas S28e has a similar fold to the bacterial RNA-binding domain of r-protein S1p (Rabl et al. 2011). Such differences may reflect the distinct elements found in the 5′ untranslated regions of eukaryotic mRNAs, as well as the divergence in the translation initiation phase from bacteria (Sonenberg and Hinnebusch 2009). Indeed, eIF3, which is absent in bacteria, interacts with this general region of the SSU (Bommer et al. 1991; Srivastava et al. 1992; Siridechadilok et al. 2005), as do internal ribosome entry site (IRES) elements present in the 5′ untranslated region of viral mRNAs (Spahn

et al. 2001b; Schuler et al. 2006; Muhs et al. 2011). S30e replaces part of S4 at the mRNA entry site of the eukaryotic SSU and has conserved lysine residues that extend into the mRNA channel (Fig. 3G), suggesting that S30e, together with S3, plays a role in unwinding mRNA secondary structure (Rabl et al. 2011). S3 has a long carboxy-terminal extension that spans across S17e and interacts with RACK1 (Fig. 3G) (Rabl et al. 2011). RACK1 is a scaffold protein that binds to several signaling proteins, therefore connecting signaling transduction pathways with translation (Nilsson et al. 2004). Thus, in addition to stabilization of rRNA ES architecture of the ribosome, eukaryotic-specific r-proteins and extensions appear to be important for binding of eukaryotic-specific regulatory factors, particularly factors that interact with the SSU to regulate translation initiation of specific mRNAs.

THE tRNA-BINDING SITES ON THE EUKARYOTIC RIBOSOME

The binding sites for the aminoacyl-transfer RNA (tRNA) (A site), peptidyl-tRNA (P site), and deacylated tRNA (exit or E site) on the bacterial ribosome are composed predominantly of rRNA (Yusupov et al. 2001; Selmer et al. 2006). This rRNA is conserved in archaeal and eukaryotic ribosomes, suggesting that the basic mechanism by which the ribosome distinguishes the cognate tRNA from the near- or noncognate tRNAs at the A site during decoding (Ogle and Ramakrishnan 2005; Schmeing et al. 2011) is also likely to be conserved. Nevertheless, many r-proteins encroach on the tRNA-binding sites and appear to play important roles in decoding, accommodation, and stabilization of tRNAs (Fig. 3C) (Yusupov et al. 2001; Selmer et al. 2006; Jenner et al. 2010b). These r-proteins may be responsible for the slightly different positioning of tRNAs on the eukaryotic ribosome compared with the bacterial ribosome (Budkevich et al. 2011). On the SSU a conserved loop of S12 participates in monitoring of the second and third positions of the mRNA–tRNA codon–anticodon duplex (Ogle and Ramakrishnan 2005). Additionally, the carboxy-terminal extensions of r-proteins S19 and S9/S13 stretch from globular domains located on the head of the SSU to interact with anticodon stem-loop (ASL) regions of A- and P-tRNA, respectively, whereas S7, and to a lesser extent S11, interacts with the ASL of E-tRNA (Fig. 3C) (Yusupov et al. 2001; Selmer et al. 2006; Jenner et al. 2010b). Although these tRNA interactions are likely to be maintained in eukaryotic 80S ribosomes, additional interactions are probable on the SSU because of the presence of extensions of four eukaryotic r-proteins that approach the tRNA-binding sites, namely, the amino-terminal extensions of S30e and S31e that reach into the A site; S25e, which is positioned between the P and E sites; and S1e at the E site (Fig. 3C) (Armache et al. 2010b; Ben-Shem et al. 2011; Rabl et al. 2011). S31e is expressed with an amino-terminal ubiquitin fusion, suggesting that the lethality from lack of cleavage (Lacombe et al. 2009) arises because of the inability of tRNA and/or initiation factors to bind to the SSU (Rabl et al. 2011).

Additional stabilization of tRNA binding is observed via interaction between LSU r-proteins with the elbow regions of tRNAs, namely, the A- and P-tRNA, through contact with conserved r-proteins L16 and L5, respectively, as well as the E-tRNA with the L1 stalk (Yusupov et al. 2001; Selmer et al. 2006; Jenner et al. 2010b). The carboxyl terminus of the bacterial-specific r-protein L25p also interacts with the elbow region of A-tRNA (Jenner et al. 2010b). This r-protein is absent in archaeal and eukaryotic ribosomes. At the peptidyltransferase center (PTC) of the LSU, the CCA ends of the A- and P-tRNAs are stabilized through interaction with the conserved A- and P-loops of the 23S rRNA, thus positioning the α-amino group of the A-tRNA for nucleophilic attack on the carbonyl carbon of the peptidyl-tRNA (Leung et al. 2011). The high sequence and structural conservation of the PTC and of the tRNA substrates suggests that the insights into the mechanism of peptide bond formation gained from studying archaeal and bacterial ribosomes (Simonovic and Steitz 2009) are transferable to eukaryotic ribosomes. Nevertheless, the varying specificity for binding of antibiotics

to the PTC of bacterial versus eukaryotic LSU indicates that subtle differences do in fact exist (Wilson 2011). In addition to differences in the conformation of rRNA nucleotides, one of the major differences between the bacterial and eukaryotic PTC is related to r-proteins. Eukaryotic L16 contains a highly conserved loop that reaches into the PTC and contacts the CCA end of the P-tRNA (Fig. 3D) (Armache et al. 2010b; Bhushan et al. 2010b). This loop is absent in bacteria, and instead the space is occupied by the amino-terminal extension of bacterial-specific r-protein L27p (Fig. 3E) (Voorhees et al. 2009). The binding site of the CCA end of the E-tRNA on the eukaryotic LSU resembles the archaeal, rather than the bacterial, context. Whereas bacterial-specific r-protein L28p contributes to the E site of the bacterial LSU (Selmer et al. 2006), the archaeal and eukaryotic r-protein L44e contains an internal loop region (Fig. 2D) through which the CCA end of the E-tRNA inserts (Schmeing et al. 2003). Moreover, the carboxyl terminus of L44e is longer in eukaryotes, such as yeast, than in archaea, providing the potential for additional interactions with the P- and/or E-tRNA. Nevertheless, the E site restricts binding of only deacylated tRNAs via a direct interaction between the 2'OH of A76 and the base of C2394 (E. coli 23S rRNA numbering) (Schmeing et al. 2003; Selmer et al. 2006). The base equivalent to C2394 is conserved across all kingdoms (Cannone et al. 2002), suggesting a universal mechanism of deacylated-tRNA discrimination at the E site on the LSU.

BINDING SITES OF INITIATION FACTORS ON THE RIBOSOME

In bacteria, translation initiation is driven in large part by base pairing between the mRNA just 5' of the start codon and the 3' end of 16S rRNA—the Shine–Dalgarno interaction—which defines the ribosome binding site (Geissmann et al. 2009; Simonetti et al. 2009). Three proteins contribute to bacterial initiation, termed initiation factors 1, 2, and 3 (IF1, IF2, and IF3), and help to load initiator tRNA into the small-subunit P site at the correct start codon (Simonetti et al. 2009). In eukaryotes, translation initiation generally requires a scanning mechanism that starts at the $5'$-7-methyl-guanosine ($5'$-m^{7}G) cap and proceeds to the appropriate AUG start codon, often the first AUG codon encountered by the initiation machinery (Jackson et al. 2010). To accomplish scanning, a whole suite of eukaryotic translation initiation factors (eIFs) is involved, with names from eIF1 through eIF6, as described in more detail by Lorsch et al. (2012). Only two of the three bacterial proteins, IF1 and IF2, are conserved in eukaryotes, as counterparts of eIF1A and eIF5B, respectively (Benelli and Londei 2009). However, eIF1A and eIF5B have augmented or divergent roles to play in eukaryotic translation initiation (Jackson et al. 2010). IF3 is not conserved in eukaryotes, but seems to have a functional counterpart in eIF1 (Lomakin et al. 2003, 2006). Similar to what is observed for r-proteins in eukaryotes, eIF1 and eIF1A have extensions or "tails" that are important for their function (Olsen et al. 2003; Fekete et al. 2005, 2007; Cheung et al. 2007; Reibarkh et al. 2008; Saini et al. 2010). Most of the interactions between the 40S subunit and eukaryotic translation initiation factors are only known from genetic, biochemical, and low-resolution cryo-EM reconstructions and models of partial initiation complexes (Lomakin et al. 2003; Valasek et al. 2003; Fraser et al. 2004, 2007; Unbehaun et al. 2004; Siridechadilok et al. 2005; Passmore et al. 2007; Szamecz et al. 2008; Shin et al. 2009; Yu et al. 2009; Chiu et al. 2010; Kouba et al. 2011). With the determination of the recent X-ray crystal structures of the *T. thermophila* 40S and 60S subunits, in complexes with eIF1 and eIF6, respectively (Klinge et al. 2011; Rabl et al. 2011), our understanding of the structural basis for translation initiation in eukaryotes has increased greatly, but still lags behind our structural knowledge of bacterial translation initiation (Simonetti et al. 2009).

Initiation factor eIF1 promotes binding of initiator tRNA, in the form of a ternary complex of eIF2–GTP–Met–tRNA$_i^{Met}$, to preinitiation complexes of the SSU. It also serves to prevent initiation at non–start codons, likely by promoting an "open" state of the SSU (Jackson et al. 2010; Hinnebusch 2011). Consistent with this model, a cryo-EM reconstruction of the yeast 40S subunit in complex with eIF1 and eIF1A revealed that these two proteins induce an opening of the mRNA- and tRNA-binding groove in the 40S subunit that may contribute to scanning and correct start codon selection (Passmore et al. 2007). Release of eIF1 when the start codon is recognized is proposed to result in the closing of this groove, thereby locking the mRNA and initiator tRNA in place (Nanda et al. 2009). In the structure of the 40S subunit, eIF1 is bound adjacent to the SSU P site, in such a way that it would prevent full docking of the initiator tRNA ASL in the P-site cleft (Fig. 4A). Notably, the position of eIF1 is more compatible

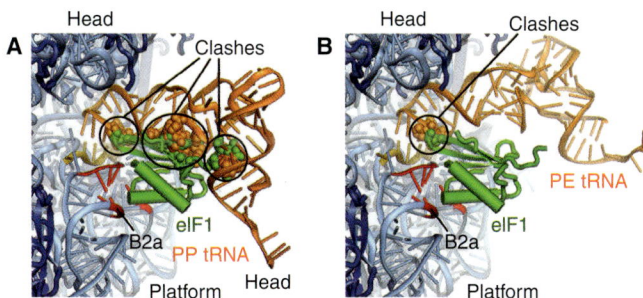

Figure 4. Positioning of eIF1 near the SSU P site. (*A*) Steric clash between eIF1 and P-site tRNA in the canonical P/P configuration. Structure of the 40S subunit–eIF1 complex superimposed with the unrotated state of the ribosome in Dunkle et al. (2011). (*B*) Binding of eIF1 is more compatible with tRNA in the P/E configuration. Structure of the 40S subunit–eIF1 complex superimposed with the rotated state of the ribosome in Dunkle et al. (2011). Nucleotides in 18S rRNA that would contribute to contacts with the LSU in bridge B2a are colored red.

with tRNA docked in a hybrid configuration seen in the bacterial ribosome, in which the tRNA is bound in the SSU P site and LSU E site (P/E-tRNA) (Fig. 4B) (Dunkle et al. 2011). As part of start codon selection, dissociation of eIF1 may allow initiator tRNA to adopt an intermediate P/I orientation, observed in bacterial initiation complexes with IF2 (Allen et al. 2005; Julian et al. 2011), or the P/P configuration, in which it could access the LSU P site upon subunit association (Jackson et al. 2010).

The binding site for eIF1 would also block the premature binding of the 60S subunit, because it is situated right where a critical contact ("bridge" B2a) forms between the two ribosomal subunits (Fig. 4) (Rabl et al. 2011). Part of eIF1 also extends into the mRNA-binding groove, adjacent to where the P-site codon would be situated. From biochemical and genetic experiments, the amino-terminal tail of eIF1 plays an important role in recruiting the eIF2–GTP–Met–tRNA$_i^{Met}$ ternary complex to preinitiation complexes (Cheung et al. 2007). However, the structure of the eIF1–40S complex provides only the first structural hints into how the ternary complex is recruited and how start codons are selected. Future structures with more of the translation initiation factors, as well as with initiator tRNA, will be needed to unravel the molecular basis for start codon selection.

The role in initiation of translation initiation factor eIF6 is not as clearly defined. It has been proposed to be an antiassociation factor that prevents premature association of the two ribosomal subunits, and it also acts in late stages of pre-60S assembly (Brina et al. 2011). In the recent X-ray crystal structure of the 60S subunit (Klinge et al. 2011), and as previously observed (Gartmann et al. 2010), eIF6 binds to the GTPase center, the region of the LSU where GTPases such as those responsible for mRNA decoding (eukaryotic elongation factor 1 [eEF1]) and mRNA and tRNA translocation (eEF2) interact with the ribosome. The location of eIF6 would sterically prevent SSU interactions with the LSU, helping to explain its antiassociative activity. Its position near the GTPase center is also highly suggestive of how it might be released in a GTPase-dependent manner during

LSU assembly (Senger et al. 2001; Menne et al. 2007; Finch et al. 2011), and also how it might be used to regulate the availability of 60S subunits as a means to control cell growth and proliferation (Gandin et al. 2008).

THE RIBOSOMAL TUNNEL OF EUKARYOTIC RIBOSOMES

As the nascent polypeptide chain (NC) is being synthesized, it passes through a tunnel within the LSU and emerges at the solvent side, where protein folding occurs. Cryo-EM reconstructions and X-ray crystallography structures of bacterial, archaeal, and eukaryotic cytoplasmic ribosomes have revealed the universality of the dimensions of the ribosomal tunnel (Frank et al. 1995; Beckmann et al. 1997; Ban et al. 2000; Ben-Shem et al. 2011; Klinge et al. 2011). The ribosomal tunnel is ~80 Å long, 10–20 Å wide, and predominantly composed of core rRNA (Nissen et al. 2000), consistent with an overall electronegative potential (Lu et al. 2007). The extensions of the r-proteins L4 and L22 contribute to formation of the tunnel wall, forming a so-called constriction where the tunnel narrows (Nissen et al. 2000). Near the tunnel exit the ribosomal protein L39e is present in eukaryotic and archaeal ribosomes (Nissen et al. 2000), whereas a bacterial-specific extension of L23 occupies an overlapping position in bacteria (Harms et al. 2001).

For many years the ribosomal tunnel was thought of only as a passive conduit for the NC. However, growing evidence indicates that the tunnel plays a more active role in regulating the rate of translation, in providing an environment for early protein folding events, and in recruiting translation factors to the tunnel exit site (Wilson and Beckmann 2011). At the simplest level, long stretches of positively charged residues, such as arginine or lysine, in an NC can reduce or halt translation, most likely through interaction with the negatively charged rRNA in the tunnel (Lu and Deutsch 2008). More specific regulatory systems also exist in bacteria and eukaryotes, in which stalling during translation of upstream open reading frames (uORFs of the cytomegalovirus [CMV] gp48 and arginine attenuator peptide [AAP] CPA1 genes) or leader peptides (TnaC, SecM) leads to modulation of expression of downstream genes (Tenson and Ehrenberg 2002). Interestingly, the translational stalling events depend critically on the sequence of the NC and the interaction of the NC with the ribosomal tunnel. Cryo-EM reconstructions of bacterial TnaC- and SecM-stalled 70S ribosomes (Seidelt et al. 2009; Bhushan et al. 2011) and eukaryotic CMV- and AAP-stalled 80S ribosomes (Bhushan et al. 2010b) reveal the distinct pathways and conformations of the NCs in the tunnel as well as the interactions between the NCs and tunnel wall components. Compared with bacteria, eukaryotic r-protein L4 has an insertion that establishes additional contacts with the CMV- and AAP-NCs (Bhushan et al. 2010b), whereas the bacterial stalling sequences interact predominantly with L22 (Seidelt et al. 2009; Bhushan et al. 2011). The dimensions of the ribosomal tunnel preclude the folding of domains as large as an IgG domain (~17 kDa) (Voss et al. 2006), whereas α-helix formation has been demonstrated biochemically (Deutsch 2003; Woolhead et al. 2004) and visualized structurally within distinct regions of the tunnel (Bhushan et al. 2010a). Folding of NCs within

the tunnel may have implications for not only protein folding, but also downstream events, such as recruitment of chaperones or targeting machinery (Bornemann et al. 2008; Berndt et al. 2009; Pool 2009).

INTERACTIONS BETWEEN THE RIBOSOMAL SUBUNITS

During translation the ribosome undergoes global conformational rearrangements that are required for mRNA decoding, mRNA and tRNA translocation, termination, and ribosome recycling. These changes involve intersubunit rotation, as well as swiveling of the head domain of the SSU (Fig. 5A). The interactions between the ribosomal subunits, or "bridges," change with each of these rearrangements, and are therefore dynamic in composition. The intersubunit bridges were originally mapped in bacteria by modeling high-resolution SSU and LSU structures into cryo-EM reconstructions and low-resolution X-ray crystal structures (Gabashvili et al. 2000; Yusupov et al. 2001; Valle et al. 2003), and in more recent high-resolution structures of the intact bacterial ribosome (Schuwirth et al. 2005; Dunkle et al. 2011). The bridges in

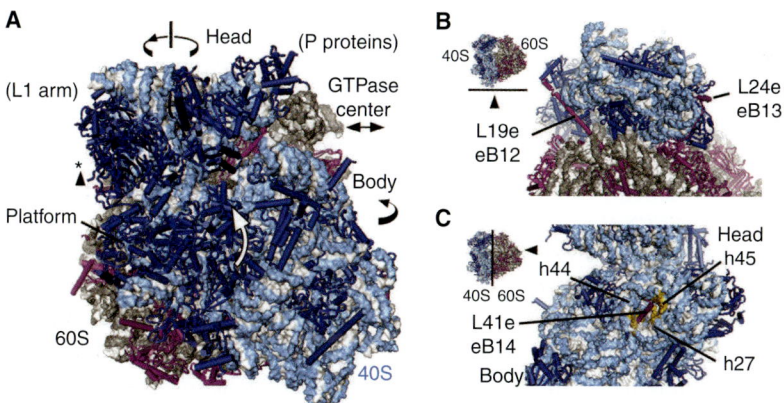

Figure 5. Intersubunit rotation required for translation. (*A*) Key conformational rearrangements in the ribosome. Rotation of the SSU body, head domain, and opening of the mRNA- and tRNA-binding groove during mRNA and tRNA translocation (asterisk) are indicated by arrows. Closing of the SSU body toward the LSU during mRNA decoding is also indicated by an arrow. Dynamic regions of the LSU (L1 arm, P proteins, and GTPase center) are labeled. (*B*) Bridges eB12 and eB13 in the yeast ribosome at the periphery of the subunits. LSU proteins contributing to the bridges are marked. The view is indicated to the left. (*C*) Bridge eB14 in the yeast ribosome, near the pivot point of intersubunit rotation. LSU protein L41e and 18S rRNA helices in the SSU contributing to the bridge (gold) are indicated.

eukaryotic ribosomes have been mapped using similar approaches. The high-resolution structures of the yeast 80S ribosome now provide an atomic-resolution view of the bridges for rotated states of the ribosome (Ben-Shem et al. 2011), and cryo-EM reconstructions of translating ribosomes at ~5- to 6-Å resolution reveal the intersubunit bridges in the unrotated state of the ribosome (Armache et al. 2010a,b).

Whereas the bacterial ribosome preferentially adopts the unrotated state of the two subunits, the eukaryotic ribosome seems to adopt rotated states more readily (Spahn et al. 2004a; Chandramouli et al. 2008; Ben-Shem et al. 2011; Budkevich et al. 2011). A possible reason for this difference in behavior is the fact that the interaction surface between the two ribosomal subunits has nearly doubled in eukaryotes compared with bacteria, primarily because of the appearance of numerous additional bridges at the periphery of the subunit interface. These new bridges are composed mainly of protein–protein and protein–rRNA contacts, some of the more notable involving long extensions from the LSU to contact the body and platform of the SSU, bridges eB12 and eB13 (Fig. 5B) (Ben-Shem et al. 2011). One striking exception to this general trend is one new bridge right at the center of the subunit interface, near the pivot point of intersubunit rotation (Ben-Shem et al. 2011). This bridge, termed eB14, is composed of a single short α-helical peptide, designated L41e, that is nearly entirely buried in a pocket composed of 18S rRNA in the SSU. Remarkably, this pocket is highly conserved in eukaryotes and in bacteria (Fig. 5C) (Schluenzen et al. 2000; Wimberly et al. 2000; Cannone et al. 2002; Ben-Shem et al. 2011), but no corresponding peptide in bacteria has been identified. The importance of this peptide in eukaryotic ribosome function remains unknown.

MECHANISMS OF mRNA DECODING, TRANSLOCATION, TERMINATION, AND RIBOSOME RECYCLING

Remarkably for processes that are functionally conserved in all domains of life, the mechanisms used by eukaryotes for mRNA decoding, mRNA

and tRNA translocation, translation termination, and ribosome recycling differ in significant ways from those in bacteria (Triana-Alonso et al. 1995; Andersen et al. 2000; Gaucher et al. 2002; Jorgensen et al. 2003; Alkalaeva et al. 2006; Khoshnevis et al. 2010; Pisarev et al. 2010). The recent breakthroughs in the structural biology of the eukaryotic ribosome provide a structural framework to unravel these differences. The large number of approximately nanometer or subnanometer cryo-EM reconstructions of eukaryotic ribosomes in different functional states (Halic et al. 2004, 2005, 2006a,b; Spahn et al. 2004a; Gao et al. 2005; Andersen et al. 2006; Schuler et al. 2006; Taylor et al. 2007, 2009; Chandramouli et al. 2008; Sengupta et al. 2008; Becker et al. 2009, 2011, 2012; Armache et al. 2010a,b; Bhushan et al. 2010a,b; Gartmann et al. 2010; Budkevich et al. 2011) now can be interpreted using high-resolution structures of the ribosome (Jarasch et al. 2011) in combination with X-ray crystal structures of the individual factors (Noble and Song 2008; Chen et al. 2010).

Although there are many differences in the translation elongation and termination factors between bacteria and eukaryotes, these factors seem to exploit common features of the ribosome conserved in all domains of life. One notable example is the mechanism for GTPase activation in mRNA decoding, in which the sarcin–ricin loop was shown to reorganize the catalytic center in bacterial EF-Tu (eukaryotic ortholog of eEF1A) during mRNA decoding (Voorhees et al. 2010). A second example is the convergent evolution of a motif in release factors that is responsible for stimulating the hydrolysis of completed proteins from peptidyl-tRNA during termination. Bacterial and eukaryotic release factors (RF1 and RF2 in bacteria, eRF1 in eukaryotes) are composed of entirely different protein topologies (Song et al. 2000; Vestergaard et al. 2001; Shin et al. 2004). Furthermore, eukaryotic RF1 requires the GTPase eRF3 and ATPase ABCE1 to stimulate termination and ribosome recycling (Khoshnevis et al. 2010; Pisarev et al. 2010; Becker et al. 2012), whereas bacterial termination and ribosome recycling use different factors (Zavialov et al. 2001; Savelsbergh et al. 2009). Strikingly, given these differences, the key residues in RFs that insert into the PTC to promote peptidyl-tRNA hydrolysis, a GGQ motif, are universally conserved. A second example occurs with the GTPases involved in elongation. Bacteria rely on the GTPases EF-Tu and EF-G, whereas eukaryotes use the GTPases eEF1A and eEF2. Eukaryotic eEF2 cannot function on the bacterial ribosome, unless the bacterial L10 and L12 proteins in the LSU are replaced by the eukaryotic acidic proteins P0 and P1/P2 (Uchiumi et al. 1999, 2002). Notably, this protein-swapping experiment also illustrates how the underlying rRNA functions are probably universal.

CONCLUSIONS

The last few years have witnessed a surge of new structures of the bacterial and eukaryotic ribosome in different steps of the translation cycle. The recent X-ray crystal structures of the *T. thermophila* 40S and 60S ribosomal subunits and yeast 80S ribosome now provide an unprecedented framework for interpreting the many cryo-EM reconstructions of the eukaryotic ribosome and biochemical insights into the eukaryotic translation mechanism. In a few years, it is not hard to imagine that many of the steps in eukaryotic translation will be understood in atomic detail based on new cryo-EM and X-ray crystal structures of the eukaryotic ribosome.

ACKNOWLEDGMENTS

This work is supported by the EMBO Young Investigator program (to D.N.W.) and by the National Institutes of Health grant R56-AI095687 (to J.H.D.C).

REFERENCES

*Reference is also in this collection.

Alkalaeva EZ, Pisarev AV, Frolova LY, Kisselev LL, Pestova TV. 2006. In vitro reconstitution of eukaryotic translation reveals cooperativity between release factors eRF1 and eRF3. *Cell* 125: 1125–1136.

Allen GS, Zavialov A, Gursky R, Ehrenberg M, Frank J. 2005. The cryo-EM structure of a translation initiation complex from *Escherichia coli*. *Cell* 121: 703–712.

Andersen GR, Pedersen L, Valente L, Chatterjee I, Kinzy TG, Kjeldgaard M, Nyborg J. 2000. Structural basis for nucleotide exchange and competition with tRNA in the yeast elongation factor complex eEF1A:eEF1Bα. *Mol Cell* 6: 1261–1266.

Andersen CB, Becker T, Blau M, Anand M, Halic M, Balar B, Mielke T, Boesen T, Pedersen JS, Spahn CM, et al. 2006. Structure of eEF3 and the mechanism of transfer RNA release from the E-site. *Nature* 443: 663–668.

Armache JP, Jarasch A, Anger AM, Villa E, Becker T, Bhushan S, Jossinet F, Habeck M, Dindar G, Franckenberg S, et al. 2010a. Cryo-EM structure and rRNA model of a translating eukaryotic 80S ribosome at 5.5-Å resolution. *Proc Natl Acad Sci* 107: 19748–19753.

Armache JP, Jarasch A, Anger AM, Villa E, Becker T, Bhushan S, Jossinet F, Habeck M, Dindar G, Franckenberg S, et al. 2010b. Localization of eukaryote-specific ribosomal proteins in a 5.5-Å cryo-EM map of the 80S eukaryotic ribosome. *Proc Natl Acad Sci* 107: 19754–19759.

Ban N, Nissen P, Hansen J, Moore PB, Steitz TA. 2000. The complete atomic structure of the large ribosomal subunit at 2.4 Å resolution. *Science* 289: 905–920.

Becker T, Bhushan S, Jarasch A, Armache JP, Funes S, Jossinet F, Gumbart J, Mielke T, Berninghausen O, Schulten K, et al. 2009. Structure of monomeric yeast and mammalian Sec61 complexes interacting with the translating ribosome. *Science* 326: 1369–1373.

Becker T, Armache JP, Jarasch A, Anger AM, Villa E, Sieber H, Motaal BA, Mielke T, Berninghausen O, Beckmann R. 2011. Structure of the no-go mRNA decay complex Dom34–Hbs1 bound to a stalled 80S ribosome. *Nat Struct Mol Biol* 18: 715–720.

Becker T, Franckenberg S, Wickles S, Shoemaker CJ, Anger AM, Armache JP, Sieber H, Ungewickell C, Berninghausen O, Daberkow I, et al. 2012. Structural basis of highly conserved ribosome recycling in eukaryotes and archaea. *Nature* 482: 501–506.

Beckmann R, Bubeck D, Grassucci R, Penczek P, Verschoor A, Blobel G, Frank J. 1997. Alignment of conduits for the nascent polypeptide chain in the ribosome—Sec61 complex. *Science* 278: 2123–2126.

Beckmann R, Spahn CM, Eswar N, Helmers J, Penczek PA, Sali A, Frank J, Blobel G. 2001. Architecture of the protein-conducting channel associated with the translating 80S ribosome. *Cell* 107: 361–372.

Ben-Shem A, Jenner L, Yusupova G, Yusupov M. 2010. Crystal structure of the eukaryotic ribosome. *Science* 330: 1203–1209.

Ben-Shem A, Garreau de Loubresse N, Melnikov S, Jenner L, Yusupova G, Yusupov M. 2011. The structure of the eukaryotic ribosome at 3.0 Å resolution. *Science* 334: 1524–1529.

Benelli D, Londei P. 2009. Begin at the beginning: Evolution of translational initiation. *Res Microbiol* 160: 493–501.

Berndt U, Oellerer S, Zhang Y, Johnson AE, Rospert S. 2009. A signal-anchor sequence stimulates signal recognition particle binding to ribosomes from inside the exit tunnel. *Proc Natl Acad Sci* 106: 1398–1403.

Bhushan S, Gartmann M, Halic M, Armache JP, Jarasch A, Mielke T, Berninghausen O, Wilson DN, Beckmann R. 2010a. α-Helical nascent polypeptide chains visualized within distinct regions of the ribosomal exit tunnel. *Nat Struct Mol Biol* 17: 313–317.

Bhushan S, Meyer H, Starosta AL, Becker T, Mielke T, Berninghausen O, Sattler M, Wilson DN, Beckmann R. 2010b. Structural basis for translational stalling by human cytomegalovirus and fungal arginine attenuator peptide. *Mol Cell* **40**: 138–146.

Bhushan S, Hoffmann T, Seidelt B, Frauenfeld J, Mielke T, Berninghausen O, Wilson DN, Beckmann R. 2011. SecM-stalled ribosomes adopt an altered geometry at the peptidyl transferase center. *PLoS Biol* **9**: e1000581.

Bommer UA, Lutsch G, Stahl J, Bielka H. 1991. Eukaryotic initiation factors eIF-2 and eIF-3: Interactions, structure and localization in ribosomal initiation complexes. *Biochimie* **73**: 1007–1019.

Bornemann T, Jockel J, Rodnina MV, Wintermeyer W. 2008. Signal sequence-independent membrane targeting of ribosomes containing short nascent peptides within the exit tunnel. *Nat Struct Mol Biol* **15**: 494–499.

Brina D, Grosso S, Miluzio A, Biffo S. 2011. Translational control by 80S formation and 60S availability: The central role of eIF6, a rate limiting factor in cell cycle progression and tumorigenesis. *Cell Cycle* **10**: 3441–3446.

Brodersen DE, Clemons WM Jr, Carter AP, Wimberly BT, Ramakrishnan V. 2002. Crystal structure of the 30 S ribosomal subunit from *Thermus thermophilus*: Structure of the proteins and their interactions with 16 S RNA. *J Mol Biol* **316**: 725–768.

Budkevich T, Giesebrecht J, Altman RB, Munro JB, Mielke T, Nierhaus KH, Blanchard SC, Spahn CM. 2011. Structure and dynamics of the mammalian ribosomal pretranslocation complex. *Mol Cell* **44**: 214–224.

Cannone JJ, Subramanian S, Schnare MN, Collett JR, D'Souza LM, Du Y, Feng B, Lin N, Madabusi LV, Muller KM, et al. 2002. The Comparative RNA Web (CRW) Site: An online database of comparative sequence and structure information for ribosomal, intron, and other RNAs. *BMC Bioinformatics* **3**: 2.

Chandramouli P, Topf M, Menetret JF, Eswar N, Cannone JJ, Gutell RR, Sali A, Akey CW. 2008. Structure of the mammalian 80S ribosome at 8.7 Å resolution. *Structure* **16**: 535–548.

Chen L, Muhlrad D, Hauryliuk V, Cheng Z, Lim MK, Shyp V, Parker R, Song H. 2010. Structure of the Dom34–Hbs1 complex and implications for no-go decay. *Nat Struct Mol Biol* **17**: 1233–1240.

Cheung YN, Maag D, Mitchell SF, Fekete CA, Algire MA, Takacs JE, Shirokikh N, Pestova T, Lorsch JR, Hinnebusch AG. 2007. Dissociation of eIF1 from the 40S ribosomal subunit is a key step in start codon selection in vivo. *Genes Dev* **21**: 1217–1230.

Chiu WL, Wagner S, Herrmannova A, Burela L, Zhang F, Saini AK, Valasek L, Hinnebusch AG. 2010. The C-terminal region of eukaryotic translation initiation factor 3a (eIF3a) promotes mRNA recruitment, scanning, and, together with eIF3j and the eIF3b RNA recognition motif, selection of AUG start codons. *Mol Cell Biol* **30**: 4415–4434.

Deutsch C. 2003. The birth of a channel. *Neuron* **40**: 265–276.

Dube P, Wieske M, Stark H, Schatz M, Stahl J, Zemlin F, Lutsch G, van Heel M. 1998. The 80S rat liver ribosome at 25 Å resolution by electron cryomicroscopy and angular reconstitution. *Structure* **6**: 389–399.

Dunkle JA, Wang L, Feldman MB, Pulk A, Chen VB, Kapral GJ, Noeske J, Richardson JS, Blanchard SC, Cate JH. 2011. Structures of the bacterial ribosome in classical and hybrid states of tRNA binding. *Science* **332**: 981–984.

Fekete CA, Applefield DJ, Blakely SA, Shirokikh N, Pestova T, Lorsch JR, Hinnebusch AG. 2005. The eIF1A C-terminal domain promotes initiation complex assembly, scanning and AUG selection in vivo. *EMBO J* **24**: 3588–3601.

Fekete CA, Mitchell SF, Cherkasova VA, Applefield D, Algire MA, Maag D, Saini AK, Lorsch JR, Hinnebusch AG. 2007. N- and C-terminal residues of eIF1A have opposing effects on the fidelity of start codon selection. *EMBO J* **26**: 1602–1614.

Finch AJ, Hilcenko C, Basse N, Drynan LF, Goyenechea B, Menne TF, Gonzalez Fernandez A, Simpson P, D'Santos CS, Arends MJ, et al. 2011. Uncoupling of GTP hydrolysis from eIF6 release on the ribosome causes Shwachman–Diamond syndrome. *Genes Dev* **25**: 917–929.

Frank J, Zhu J, Penczek P, Li Y, Srivastava S, Verschoor A, Radermacher M, Grassucci R, Lata RK, Agrawal RK. 1995. A model of protein synthesis based on cryo–electron microscopy of the *E. coli* ribosome. *Nature* **376**: 441–444.

Fraser CS, Lee JY, Mayeur GL, Bushell M, Doudna JA, Hershey JW. 2004. The j-subunit of human translation initiation factor eIF3 is required for the stable binding of eIF3 and its subcomplexes to 40 S ribosomal subunits in vitro. *J Biol Chem* **279**: 8946–8956.

Fraser CS, Berry KE, Hershey JW, Doudna JA. 2007. eIF3j is located in the decoding center of the human 40S ribosomal subunit. *Mol Cell* **26**: 811–819.

Freed EF, Bleichert F, Dutca LM, Baserga SJ. 2010. When ribosomes go bad: Diseases of ribosome biogenesis. *Mol Biosyst* **6**: 481–493.

Gabashvili IS, Agrawal RK, Spahn CM, Grassucci RA, Svergun DI, Frank J, Penczek P. 2000. Solution structure of the *E. coli* 70S ribosome at 11.5 Å resolution. *Cell* **100**: 537–549.

Gandin V, Miluzio A, Barbieri AM, Beugnet A, Kiyokawa H, Marchisio PC, Biffo S. 2008. Eukaryotic initiation factor 6 is rate-limiting in translation, growth and transformation. *Nature* **455**: 684–688.

Gao H, Ayub MJ, Levin MJ, Frank J. 2005. The structure of the 80S ribosome from *Trypanosoma cruzi* reveals unique rRNA components. *Proc Natl Acad Sci* **102**: 10206–10211.

Gartmann M, Blau M, Armache JP, Mielke T, Topf M, Beckmann R. 2010. Mechanism of eIF6-mediated inhibition of ribosomal subunit joining. *J Biol Chem* **285**: 14848–14851.

Gaucher EA, Das UK, Miyamoto MM, Benner SA. 2002. The crystal structure of eEF1A refines the functional predictions of an evolutionary analysis of rate changes among elongation factors. *Mol Biol Evol* **19**: 569–573.

Geissmann T, Marzi S, Romby P. 2009. The role of mRNA structure in translational control in bacteria. *RNA Biol* **6**: 153–160.

Gerbi SA. 1996. Expansion segments: Regions of variable size that interrupt the universal core secondary structure of ribosomal RNA. In *Ribosomal RNA—Structure, evolution, processing, and function in protein synthesis* (ed.

Cite this article as *Cold Spring Harb Perspect Biol* doi: 10.1101/cshperspect.a011536

RA Zimmermann, AE Dahlberg), pp. 71–87. CRC Press, Boca Raton, FL.

Halic M, Becker T, Pool MR, Spahn CM, Grassucci RA, Frank J, Beckmann R. 2004. Structure of the signal recognition particle interacting with the elongation-arrested ribosome. *Nature* **427:** 808–814.

Halic M, Becker T, Frank J, Spahn CM, Beckmann R. 2005. Localization and dynamic behavior of ribosomal protein L30e. *Nat Struct Mol Biol* **12:** 467–468.

Halic M, Blau M, Becker T, Mielke T, Pool MR, Wild K, Sinning I, Beckmann R. 2006a. Following the signal sequence from ribosomal tunnel exit to signal recognition particle. *Nature* **444:** 507–511.

Halic M, Gartmann M, Schlenker O, Mielke T, Pool MR, Sinning I, Beckmann R. 2006b. Signal recognition particle receptor exposes the ribosomal translocon binding site. *Science* **312:** 745–747.

Harms J, Schluenzen F, Zarivach R, Bashan A, Gat S, Agmon I, Bartels H, Franceschi F, Yonath A. 2001. High resolution structure of the large ribosomal subunit from a mesophilic eubacterium. *Cell* **107:** 679–688.

Hinnebusch AG. 2011. Molecular mechanism of scanning and start codon selection in eukaryotes. *Microbiol Mol Biol Rev* **75:** 434–467.

Jackson RJ, Hellen CU, Pestova TV. 2010. The mechanism of eukaryotic translation initiation and principles of its regulation. *Nat Rev Mol Cell Biol* **11:** 113–127.

Jarasch A, Dziuk P, Becker T, Armache JP, Hauser A, Wilson DN, Beckmann R. 2011. The DARC site: A database of aligned ribosomal complexes. *Nucleic Acids Res* **40:** D495–D500.

Jenner L, Demeshkina N, Yusupova G, Yusupov M. 2010a. Structural rearrangements of the ribosome at the tRNA proofreading step. *Nat Struct Mol Biol* **17:** 1072–1078.

Jenner LB, Demeshkina N, Yusupova G, Yusupov M. 2010b. Structural aspects of messenger RNA reading frame maintenance by the ribosome. *Nat Struct Mol Biol* **17:** 555–560.

Jorgensen R, Ortiz PA, Carr-Schmid A, Nissen P, Kinzy TG, Andersen GR. 2003. Two crystal structures demonstrate large conformational changes in the eukaryotic ribosomal translocase. *Nat Struct Biol* **10:** 379–385.

Julian P, Milon P, Agirrezabala X, Lasso G, Gil D, Rodnina MV, Valle M. 2011. The cryo-EM structure of a complete 30S translation initiation complex from *Escherichia coli*. *PLoS Biol* **9:** e1001095.

Khoshnevis S, Gross T, Rotte C, Baierlein C, Ficner R, Krebber H. 2010. The iron-sulphur protein RNase L inhibitor functions in translation termination. *EMBO Rep* **11:** 214–219.

Klein DJ, Moore PB, Steitz TA. 2004. The roles of ribosomal proteins in the structure assembly, and evolution of the large ribosomal subunit. *J Mol Biol* **340:** 141–177.

Klinge S, Voigts-Hoffmann F, Leibundgut M, Arpagaus S, Ban N. 2011. Crystal structure of the eukaryotic 60S ribosomal subunit in complex with initiation factor 6. *Science* **334:** 941–948.

Kouba T, Rutkai E, Karaskova M, Valasek LS. 2011. The eIF3c/NIP1 PCI domain interacts with RNA and RACK1/ASC1 and promotes assembly of translation pre-

initiation complexes. *Nucleic Acids Res* doi: 10.1093/nar/gkr1083.

Lacombe T, Garcia-Gomez JJ, de la Cruz J, Roser D, Hurt E, Linder P, Kressler D. 2009. Linear ubiquitin fusion to Rps31 and its subsequent cleavage are required for the efficient production and functional integrity of 40S ribosomal subunits. *Mol Microbiol* **72:** 69–84.

Lecompte O, Ripp R, Thierry JC, Moras D, Poch O. 2002. Comparative analysis of ribosomal proteins in complete genomes: An example of reductive evolution at the domain scale. *Nucleic Acids Res* **30:** 5382–5390.

Leung EK, Suslov N, Tuttle N, Sengupta R, Piccirilli JA. 2011. The mechanism of peptidyl transfer catalysis by the ribosome. *Annu Rev Biochem* **80:** 527–555.

Lomakin IB, Kolupaeva VG, Marintchev A, Wagner G, Pestova TV. 2003. Position of eukaryotic initiation factor eIF1 on the 40S ribosomal subunit determined by directed hydroxyl radical probing. *Genes Dev* **17:** 2786–2797.

Lomakin IB, Shirokikh NE, Yusupov MM, Hellen CU, Pestova TV. 2006. The fidelity of translation initiation: Reciprocal activities of eIF1, IF3 and YciH. *EMBO J* **25:** 196–210.

* Lorsch JR. 2012. Translational control: Pathway, mechanism of protein synthesis. *Cold Spring Harb Perspect Biol* doi: 10.1101/cshperspect.a011544.

Lu J, Deutsch C. 2008. Electrostatics in the ribosomal tunnel modulate chain elongation rates. *J Mol Biol* **384:** 73–86.

Lu J, Kobertz WR, Deutsch C. 2007. Mapping the electrostatic potential within the ribosomal exit tunnel. *J Mol Biol* **371:** 1378–1391.

Menne TF, Goyenechea B, Sanchez-Puig N, Wong CC, Tonkin LM, Ancliff PJ, Brost RL, Costanzo M, Boone C, Warren AJ. 2007. The Shwachman–Bodian–Diamond syndrome protein mediates translational activation of ribosomes in yeast. *Nat Genet* **39:** 486–495.

Meyuhas O. 2008. Physiological roles of ribosomal protein S6: One of its kind. *Int Rev Cell Mol Biol* **268:** 1–37.

Morgan DG, Menetret JF, Radermacher M, Neuhof A, Akey IV, Rapoport TA, Akey CW. 2000. A comparison of the yeast and rabbit 80 S ribosome reveals the topology of the nascent chain exit tunnel, inter-subunit bridges and mammalian rRNA expansion segments. *J Mol Biol* **301:** 301–321.

Muhs M, Yamamoto H, Ismer J, Takaku H, Nashimoto M, Uchiumi T, Nakashima N, Mielke T, Hildebrand PW, Nierhaus KH, et al. 2011. Structural basis for the binding of IRES RNAs to the head of the ribosomal 40S subunit. *Nucleic Acids Res* **39:** 5264–5275.

Nanda JS, Cheung YN, Takacs JE, Martin-Marcos P, Saini AK, Hinnebusch AG, Lorsch JR. 2009. eIF1 controls multiple steps in start codon recognition during eukaryotic translation initiation. *J Mol Biol* **394:** 268–285.

Nilsson J, Sengupta J, Frank J, Nissen P. 2004. Regulation of eukaryotic translation by the RACK1 protein: A platform for signalling molecules on the ribosome. *EMBO Rep* **5:** 1137–1141.

Nissen P, Hansen J, Ban N, Moore PB, Steitz TA. 2000. The structural basis of ribosome activity in peptide bond synthesis. *Science* **289:** 920–930.

Noble CG, Song H. 2008. Structural studies of elongation and release factors. *Cell Mol Life Sci* **65:** 1335–1346.

Ogle JM, Ramakrishnan V. 2005. Structural insights into translational fidelity. *Annu Rev Biochem* **74:** 129–177.

Olsen DS, Savner EM, Mathew A, Zhang F, Krishnamoorthy T, Phan L, Hinnebusch AG. 2003. Domains of eIF1A that mediate binding to eIF2, eIF3 and eIF5B and promote ternary complex recruitment in vivo. *EMBO J* **22:** 193–204.

Passmore LA, Schmeing TM, Maag D, Applefield DJ, Acker MG, Algire MA, Lorsch JR, Ramakrishnan V. 2007. The eukaryotic translation initiation factors eIF1 and eIF1A induce an open conformation of the 40S ribosome. *Mol Cell* **26:** 41–50.

Pisarev AV, Skabkin MA, Pisareva VP, Skabkina OV, Rakotondrafara AM, Hentze MW, Hellen CU, Pestova TV. 2010. The role of ABCE1 in eukaryotic posttermination ribosomal recycling. *Mol Cell* **37:** 196–210.

Pool MR. 2009. A trans-membrane segment inside the ribosome exit tunnel triggers RAMP4 recruitment to the Sec61p translocase. *J Cell Biol* **185:** 889–902.

Rabl J, Leibundgut M, Ataide SF, Haag A, Ban N. 2011. Crystal structure of the eukaryotic 40S ribosomal subunit in complex with initiation factor 1. *Science* **331:** 730–736.

Reibarkh M, Yamamoto Y, Singh CR, del Rio F, Fahmy A, Lee B, Luna RE, Ii M, Wagner G, Asano K. 2008. Eukaryotic initiation factor (eIF) 1 carries two distinct eIF5-binding faces important for multifactor assembly and AUG selection. *J Biol Chem* **283:** 1094–1103.

Saini AK, Nanda JS, Lorsch JR, Hinnebusch AG. 2010. Regulatory elements in eIF1A control the fidelity of start codon selection by modulating tRNA$_i^{Met}$ binding to the ribosome. *Genes Dev* **24:** 97–110.

Savelsbergh A, Rodnina MV, Wintermeyer W. 2009. Distinct functions of elongation factor G in ribosome recycling and translocation. *RNA* **15:** 772–780.

Schluenzen F, Tocilj A, Zarivach R, Harms J, Gluehmann M, Janell D, Bashan A, Bartels H, Agmon I, Franceschi F, et al. 2000. Structure of functionally activated small ribosomal subunit at 3.3 Å resolution. *Cell* **102:** 615–623.

Schmeing TM, Moore PB, Steitz TA. 2003. Structures of deacylated tRNA mimics bound to the E site of the large ribosomal subunit. *RNA* **9:** 1345–1352.

Schmeing TM, Voorhees RM, Kelley AC, Ramakrishnan V. 2011. How mutations in tRNA distant from the anticodon affect the fidelity of decoding. *Nat Struct Mol Biol* **18:** 432–436.

Schuler M, Connell SR, Lescoute A, Giesebrecht J, Dabrowski M, Schroeer B, Mielke T, Penczek PA, Westhof E, Spahn CM. 2006. Structure of the ribosome-bound cricket paralysis virus IRES RNA. *Nat Struct Mol Biol* **13:** 1092–1096.

Schuwirth BS, Borovinskaya MA, Hau CW, Zhang W, Vila-Sanjurjo A, Holton JM, Cate JH. 2005. Structures of the bacterial ribosome at 3.5 Å resolution. *Science* **310:** 827–834.

Seidelt B, Innis CA, Wilson DN, Gartmann M, Armache JP, Villa E, Trabuco LG, Becker T, Mielke T, Schulten K, et al. 2009. Structural insight into nascent polypeptide chain–mediated translational stalling. *Science* **326:** 1412–1415.

Selmer M, Dunham CM, Murphy FV IV, Weixlbaumer A, Petry S, Kelley AC, Weir JR, Ramakrishnan V. 2006. Structure of the 70S ribosome complexed with mRNA and tRNA. *Science* **313:** 1935–1942.

Senger B, Lafontaine DL, Graindorge JS, Gadal O, Camasses A, Sanni A, Garnier JM, Breitenbach M, Hurt E, Fasiolo F. 2001. The nucle(ol)ar Tif6p and Efl1p are required for a late cytoplasmic step of ribosome synthesis. *Mol Cell* **8:** 1363–1373.

Sengupta J, Nilsson J, Gursky R, Spahn CM, Nissen P, Frank J. 2004. Identification of the versatile scaffold protein RACK1 on the eukaryotic ribosome by cryo-EM. *Nat Struct Mol Biol* **11:** 957–962.

Sengupta J, Nilsson J, Gursky R, Kjeldgaard M, Nissen P, Frank J. 2008. Visualization of the eEF2–80S ribosome transition-state complex by cryo–electron microscopy. *J Mol Biol* **382:** 179–187.

Shin DH, Brandsen J, Jancarik J, Yokota H, Kim R, Kim SH. 2004. Structural analyses of peptide release factor 1 from *Thermotoga maritima* reveal domain flexibility required for its interaction with the ribosome. *J Mol Biol* **341:** 227–239.

Shin BS, Kim JR, Acker MG, Maher KN, Lorsch JR, Dever TE. 2009. rRNA suppressor of a eukaryotic translation initiation factor 5B/initiation factor 2 mutant reveals a binding site for translational GTPases on the small ribosomal subunit. *Mol Cell Biol* **29:** 808–821.

Simonetti A, Marzi S, Jenner L, Myasnikov A, Romby P, Yusupova G, Klaholz BP, Yusupov M. 2009. A structural view of translation initiation in bacteria. *Cell Mol Life Sci* **66:** 423–436.

Simonovic M, Steitz TA. 2009. A structural view on the mechanism of the ribosome-catalyzed peptide bond formation. *Biochim Biophys Acta* **1789:** 612–623.

Siridechadilok B, Fraser CS, Hall RJ, Doudna JA, Nogales E. 2005. Structural roles for human translation factor eIF3 in initiation of protein synthesis. *Science* **310:** 1513–1515.

Sonenberg N, Hinnebusch AG. 2009. Regulation of translation initiation in eukaryotes: Mechanisms and biological targets. *Cell* **136:** 731–745.

Song H, Mugnier P, Das AK, Webb HM, Evans DR, Tuite MF, Hemmings BA, Barford D. 2000. The crystal structure of human eukaryotic release factor eRF1—Mechanism of stop codon recognition and peptidyl-tRNA hydrolysis. *Cell* **100:** 311–321.

Spahn CM, Beckmann R, Eswar N, Penczek PA, Sali A, Blobel G, Frank J. 2001a. Structure of the 80S ribosome from *Saccharomyces cerevisiae*–tRNA-ribosome and subunit–subunit interactions. *Cell* **107:** 373–386.

Spahn CM, Kieft JS, Grassucci RA, Penczek PA, Zhou K, Doudna JA, Frank J. 2001b. Hepatitis C virus IRES RNA-induced changes in the conformation of the 40S ribosomal subunit. *Science* **291:** 1959–1962.

Spahn CM, Gomez-Lorenzo MG, Grassucci RA, Jorgensen R, Andersen GR, Beckmann R, Penczek PA, Ballesta JP, Frank J. 2004a. Domain movements of elongation factor eEF2 and the eukaryotic 80S ribosome facilitate tRNA translocation. *EMBO J* **23:** 1008–1019.

Spahn CM, Jan E, Mulder A, Grassucci RA, Sarnow P, Frank J. 2004b. Cryo-EM visualization of a viral internal ribosome entry site bound to human ribosomes: The IRES functions as an RNA-based translation factor. *Cell* **118:** 465–475.

Cite this article as *Cold Spring Harb Perspect Biol* doi: 10.1101/cshperspect.a011536

Srivastava S, Verschoor A, Frank J. 1992. Eukaryotic initiation factor 3 does not prevent association through physical blockage of the ribosomal subunit–subunit interface. *J Mol Biol* **226:** 301–304.

Sweeney R, Chen L, Yao MC. 1994. An rRNA variable region has an evolutionarily conserved essential role despite sequence divergence. *Mol Cell Biol* **14:** 4203–4215.

Szamecz B, Rutkai E, Cuchalova L, Munzarova V, Herrmannova A, Nielsen KH, Burela L, Hinnebusch AG, Valasek L. 2008. eIF3a cooperates with sequences 5′ of uORF1 to promote resumption of scanning by post-termination ribosomes for reinitiation on GCN4 mRNA. *Genes Dev* **22:** 2414–2425.

Taylor DJ, Nilsson J, Merrill AR, Andersen GR, Nissen P, Frank J. 2007. Structures of modified eEF2 80S ribosome complexes reveal the role of GTP hydrolysis in translocation. *EMBO J* **26:** 2421–2431.

Taylor DJ, Devkota B, Huang AD, Topf M, Narayanan E, Sali A, Harvey SC, Frank J. 2009. Comprehensive molecular structure of the eukaryotic ribosome. *Structure* **17:** 1591–1604.

Tenson T, Ehrenberg M. 2002. Regulatory nascent peptides in the ribosomal tunnel. *Cell* **108:** 591–594.

Triana-Alonso FJ, Chakraburtty K, Nierhaus KH. 1995. The elongation factor 3 unique in higher fungi and essential for protein biosynthesis is an E site factor. *J Biol Chem* **270:** 20473–20478.

Uchiumi T, Hori K, Nomura T, Hachimori A. 1999. Replacement of L7/L12.L10 protein complex in *Escherichia coli* ribosomes with the eukaryotic counterpart changes the specificity of elongation factor binding. *J Biol Chem* **274:** 27578–27582.

Uchiumi T, Honma S, Nomura T, Dabbs ER, Hachimori A. 2002. Translation elongation by a hybrid ribosome in which proteins at the GTPase center of the *Escherichia coli* ribosome are replaced with rat counterparts. *J Biol Chem* **277:** 3857–3862.

Unbehaun A, Borukhov SI, Hellen CU, Pestova TV. 2004. Release of initiation factors from 48S complexes during ribosomal subunit joining and the link between establishment of codon–anticodon base-pairing and hydrolysis of eIF2-bound GTP. *Genes Dev* **18:** 3078–3093.

Valasek L, Mathew AA, Shin BS, Nielsen KH, Szamecz B, Hinnebusch AG. 2003. The yeast eIF3 subunits TIF32/a, NIP1/c, and eIF5 make critical connections with the 40S ribosome in vivo. *Genes Dev* **17:** 786–799.

Valle M, Zavialov A, Sengupta J, Rawat U, Ehrenberg M, Frank J. 2003. Locking and unlocking of ribosomal motions. *Cell* **114:** 123–134.

Vestergaard B, Van LB, Andersen GR, Nyborg J, Buckingham RH, Kjeldgaard M. 2001. Bacterial polypeptide release factor RF2 is structurally distinct from eukaryotic eRF1. *Mol Cell* **8:** 1375–1382.

Voorhees RM, Weixlbaumer A, Loakes D, Kelley AC, Ramakrishnan V. 2009. Insights into substrate stabilization from snapshots of the peptidyl transferase center of the intact 70S ribosome. *Nat Struct Mol Biol* **16:** 528–533.

Voorhees RM, Schmeing TM, Kelley AC, Ramakrishnan V. 2010. The mechanism for activation of GTP hydrolysis on the ribosome. *Science* **330:** 835–838.

Voss NR, Gerstein M, Steitz TA, Moore PB. 2006. The geometry of the ribosomal polypeptide exit tunnel. *J Mol Biol* **360:** 893–906.

Wang DO, Martin KC, Zukin RS. 2010. Spatially restricting gene expression by local translation at synapses. *Trends Neurosci* **33:** 173–182.

Wilson DN. 2011. On the specificity of antibiotics targeting the large ribosomal subunit. *Ann NY Acad Sci* **1241:** 1–16.

Wilson DN, Beckmann R. 2011. The ribosomal tunnel as a functional environment for nascent polypeptide folding and translational stalling. *Curr Opin Struct Biol* **21:** 274–282.

Wimberly BT, Brodersen DE, Clemons WM Jr, Morgan-Warren RJ, Carter AP, Vonrhein C, Hartsch T, Ramakrishnan V. 2000. Structure of the 30S ribosomal subunit. *Nature* **407:** 327–339.

Woolhead CA, McCormick PJ, Johnson AE. 2004. Nascent membrane and secretory proteins differ in FRET-detected folding far inside the ribosome and in their exposure to ribosomal proteins. *Cell* **116:** 725–736.

Yokoyama T, Suzuki T. 2008. Ribosomal RNAs are tolerant toward genetic insertions: Evolutionary origin of the expansion segments. *Nucleic Acids Res* **36:** 3539–3551.

Yu Y, Marintchev A, Kolupaeva VG, Unbehaun A, Veryasova T, Lai SC, Hong P, Wagner G, Hellen CU, Pestova TV. 2009. Position of eukaryotic translation initiation factor eIF1A on the 40S ribosomal subunit mapped by directed hydroxyl radical probing. *Nucleic Acids Res* **37:** 5167–5182.

Yusupov MM, Yusupova GZ, Baucom A, Lieberman K, Earnest TN, Cate JH, Noller HF. 2001. Crystal structure of the ribosome at 5.5 Å resolution. *Science* **292:** 883–896.

Zavialov AV, Buckingham RH, Ehrenberg M. 2001. A post-termination ribosomal complex is the guanine nucleotide exchange factor for peptide release factor RF3. *Cell* **107:** 115–124.

The Mechanism of Eukaryotic Translation Initiation: New Insights and Challenges

Alan G. Hinnebusch[1] and Jon R. Lorsch[2]

[1]Laboratory of Gene Regulation and Development, Eunice Kennedy Shriver National Institute of Child Health and Human Development, National Institutes of Health, Bethesda, Maryland 20892

[2]Department of Biophysics and Biophysical Chemistry, Johns Hopkins University School of Medicine, Baltimore, Maryland 21205

Correspondence: ahinnebusch@nih.gov; jlorsch@jhmi.edu

Translation initiation in eukaryotes is a highly regulated and complex stage of gene expression. It requires the action of at least 12 initiation factors, many of which are known to be the targets of regulatory pathways. Here we review our current understanding of the molecular mechanics of eukaryotic translation initiation, focusing on recent breakthroughs from in vitro and in vivo studies. We also identify important unanswered questions that will require new ideas and techniques to solve.

This work aims to present the current state of our knowledge of the molecular mechanics of translation initiation in eukaryotes. We focus on advances that have taken place over the last few years and, because of space limitations, assume readers will be able to find references to the foundational literature for the field (published before 2000) in the more recent works that are cited here. As always, we apologize for not having the space to cite many important works. Please view this as merely an introduction to the field rather than a complete summary.

OVERVIEW OF THE INITIATION PATHWAY

Figure 1 presents the basic outline of the eukaryotic cap-dependent initiation pathway, and the reader is referred to a number of recent reviews that summarize the evidence supporting the current paradigm outlined below (Hinnebusch et al. 2007; Pestova et al. 2007; Lorsch and Dever 2010; Hinnebusch 2011; Parsyan et al. 2011). Identification of the initiation codon by the eukaryotic translational machinery begins with binding of the ternary complex (TC) consisting of initiator methionyl-tRNA (Met-tRNA$_i$) and the GTP-bound form of eukaryotic initiation factor 2 (eIF2) to the small (40S) ribosomal subunit to form the 43S preinitiation complex (PIC). Binding of the TC to the 40S subunit is promoted by eIFs 1, 1A, 5, and the eIF3 complex (Fig. 1). A network of physical interactions links eIFs 1, 3, 5, and TC in a multifactor complex (MFC) in yeast (Asano et al. 2000), plants (Dennis et al. 2009), and mammals (Sokabe et al. 2011). In budding yeast, the MFC enhances the formation or stability of the 43S PIC in vivo (reviewed in Hinnebusch et al. 2007).

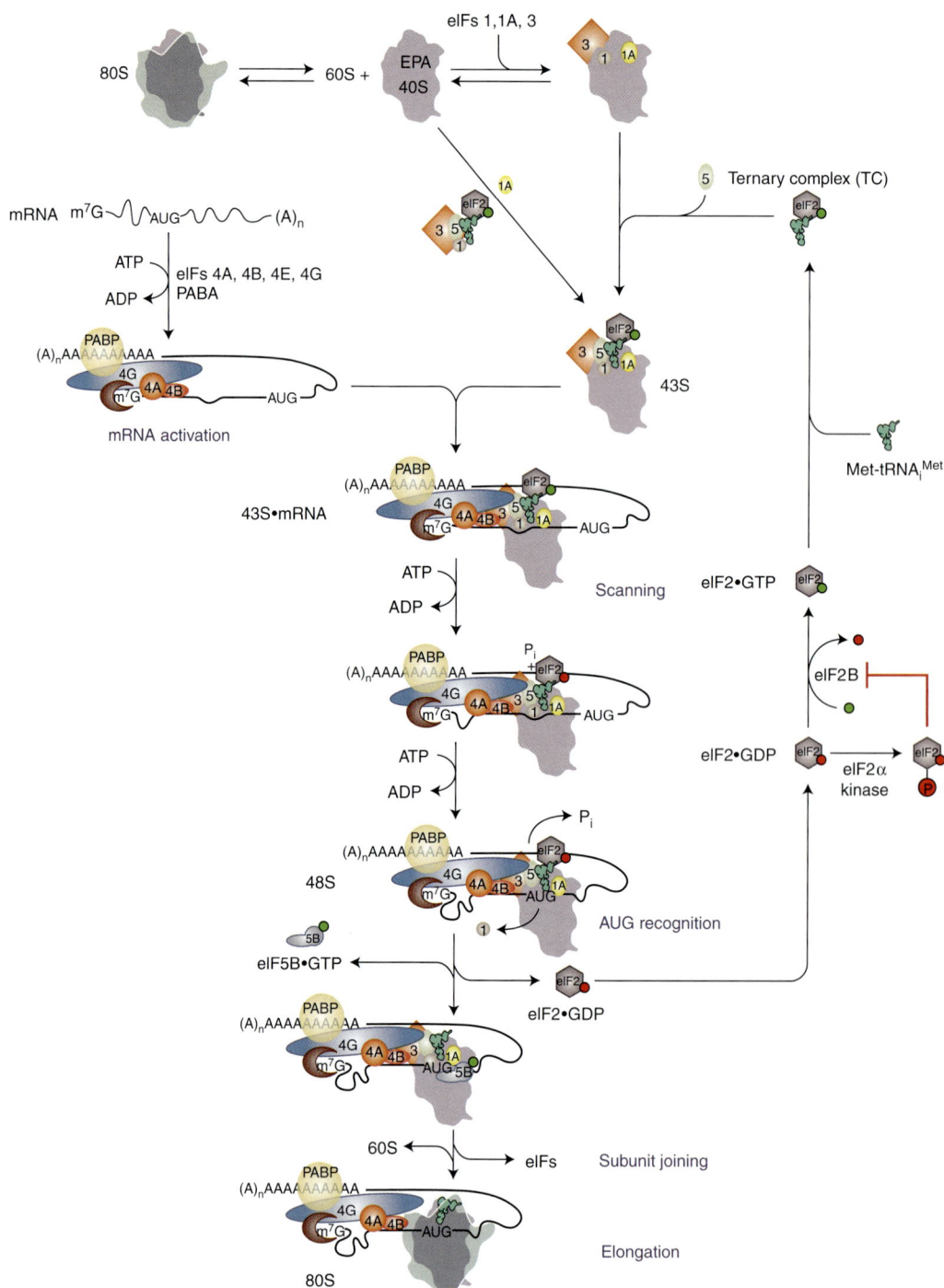

Figure 1. Model of canonical eukaryotic translation initiation pathway. The pathway is shown as a series of discrete steps starting with dissociation of 80S ribosomes into subunits. Binding of factors is depicted both as a single step via the multifactor complex and as two separate steps, with eIFs 1, 1A, and 3 binding first followed by binding of ternary complex and eIF5. The resulting 43S preinitiation complex (PIC) is then loaded onto an activated mRNP near the 5′ cap. (*Legend continues on facing page.*)

Cite this article as *Cold Spring Harb Perspect Biol* doi: 10.1101/cshperspect.a011544

The 43S PIC binds to the messenger RNA (mRNA) near the 5′-7-methylguanosine cap in a process facilitated by eIF3, the poly(A)-binding protein (PABP), and eIFs 4B, 4H (in mammals), and 4F. The eIF4F complex is comprised of the cap-binding protein eIF4E, eIF4G, and the RNA helicase eIF4A. eIF4G is a scaffold protein that harbors binding domains for PABP, eIF4E, eIF4A, and (in mammals) eIF3. Both yeast and human eIF4G also bind RNA. The binding domains for eIF4E and PABP in eIF4G, along with its RNA-binding activity, enable eIF4G to coordinate independent interactions with mRNA via the cap, poly(A) tail, and sequences in the mRNA body to assemble a stable, circular messenger ribonucleoprotein (mRNP), referred to as the "closed-loop" structure. The eIF4G–eIF3 interaction is expected to establish a protein bridge between this "activated mRNP" and the 43S PIC to stimulate 43S attachment to the mRNA, and the helicase activity of eIF4A is thought to generate a single-stranded landing pad in the mRNA on which the 43S PIC can load (reviewed in Hinnebusch et al. 2007; Pestova et al. 2007; Lorsch and Dever 2010; Hinnebusch 2011).

Once bound near the cap, the 43S PIC scans the mRNA leader for an AUG codon in a suitable sequence context. Base-pairing between the anticodon of Met-tRNA$_i$ and the AUG in the peptidyl-tRNA (P) site of the 40S subunit is the initial event in start codon recognition (Lomakin et al. 2006; Kolitz et al. 2009; Hinnebusch 2011). AUG recognition causes arrest of the scanning PIC and triggers conversion of eIF2 in the TC to its GDP-bound state via gated phosphate (P$_i$) release and the action of the GTPase-activating (GAP) factor eIF5. Following release of eIF2·GDP and several other eIFs present in the PIC, joining of the large (60S) subunit is catalyzed by eIF5B to produce an 80S initiation complex (IC) containing Met-tRNA$_i$ base-paired to AUG in the P site and ready to begin the elongation phase of protein synthesis (Fig. 1) (reviewed in Hinnebusch et al. 2007; Pestova et al. 2007; Hinnebusch 2011).

RECRUITMENT OF Met-tRNA$_i$ TO THE 40S RIBOSOMAL SUBUNIT

eIF2 Is a G-Protein Switch that Carries Met-tRNA$_i$ onto the Ribosome

The Met-tRNA$_i$ is delivered to the 40S subunit in the TC with eIF2·GTP. The affinity of Met-tRNA$_i$ is greater for eIF2·GTP than for eIF2·GDP, and this affinity switch depends on the methionine moiety on the Met-tRNA$_i$ (Kapp and Lorsch 2004). This contribution of methionine, plus the stimulatory role of the unique A1:U72 base pair (bp) in the acceptor stem of tRNA$_i$ in binding eIF2 (Kapp and Lorsch 2004; Pestova et al. 2007), presumably act to prevent binding of elongator tRNAs to the factor. This specificity, coupled with the requirement for eIF2 to load tRNA onto the 40S subunit, is thought to eliminate the need for a mechanism to reject elongator tRNAs during PIC assembly, a process in bacteria that relies heavily on IF3 (Hershey and Merrick 2000). (As described below, a structural homolog of IF3 is absent in eukaryotes, but eIF1 acts similarly to ensure selection of AUG as a start codon.) Understanding the structural basis for the stimulatory effects of methionine, the A1:U72 bp, and GTP versus GDP on initiator tRNA binding to eIF2 would be advanced by high-resolution structural analysis of the complete TC. Whereas the crystal structure of the archaeal ortholog (aIF2) has been solved, as well as various aIF2 subcomplexes bound to GDP or GTP analogs (reviewed in Schmitt et al. 2010), no crystal structures or cryo-EM

Figure 1. (*Continued*) Subsequent scanning of the mRNA allows recognition of the start codon, which triggers downstream steps in the pathway including eIF1 release from the PIC, P$_i$ release from eIF2, and conversion to the closed, scanning-arrested state of the complex. eIF2·GDP released after subunit joining is recycled back to eIF2·GTP by the exchange factor eIF2B. eIF5B in its GTP-bound form promotes joining of the 60S subunit to the preinitiation complex, which triggers release of eIF5B·GDP and eIF1A to form the final 80S initiation complex, which can begin the elongation phase of protein synthesis. Throughout, GTP is depicted as a green ball and GDP as a red ball. (Modified from Hinnebusch 2011; reproduced, with permission, from the author.)

(electron microscopy) models of heterotrimeric eIF2 have been described.

eIF2γ binds directly to both GTP and Met-tRNA$_i$ and it appears that the α and β subunits each increase the affinity of the eIF2 complex for Met-tRNA$_i$ by ∼100-fold (Naveau et al. 2010), but it is unknown whether this stimulatory effect involves direct contacts between Met-tRNA$_i$ and eIF2α or eIF2β. Based on the crystal structure of a heterotrimer of aIF2β, aIF2γ, and a portion of aIF2α (Yatime et al. 2007) it has been proposed that the α and β subunits allosterically induce a conformation in aIF2γ with high affinity for Met-tRNA$_i$ (Naveau et al. 2010). Evidence consistent with an allosteric mechanism, at least for eIF2α, comes from directed hydroxyl radical cleavage mapping of Met-tRNA$_i$ binding to yeast eIF2 in reconstituted PICs. Met-tRNA$_i$ was cleaved by free radicals generated from particular positions in eIF2γ or eIF2β, but not from eIF2α, suggesting the latter does not make direct contact with the tRNA. Interestingly, the patterns of cleavage imply a mode of initiator binding to eIF2γ dramatically different from that seen in crystal structures of the EF-Tu·GDPNP·Phe-tRNA^{Phe} TC, which delivers aminoacylated tRNAs to the A site during elongation. In contrast to the latter complex, domain III of eIF2γ, the subunit homologous to EF-Tu, does not contact the T stem of Met-tRNA$_i$; instead the sole contact is with the methionylated acceptor end of the tRNA in a pocket in eIF2γ formed between the G domain and domain II (Shin et al. 2011). A recent crystal structure of the TC formed by an archaeal aIF2, GDPNP, and E. coli Met-tRNA(fMet) also demonstrated that the tRNA is bound by aIF2 in a manner dramatically distinct from that of elongator tRNA binding to EF-Tu (Schmitt et al. 2012). Consistent with previous models (Schmitt et al. 2010; Shin et al. 2011), the acceptor end of the tRNA binds to aIF2γ according to the EF-Tu paradigm; however, the T-stem minor groove does not contact aIF2 and, instead, the T-loop in the tRNA "elbow" interacts with regions of the aIF2α subunit. As these last contacts were not detected in the hydroxyl radical probing of the eIF2 TC (Shin et al. 2011), it remains to be seen

whether they are important in solution and conserved in eukaryotic TC.

Importantly, the patterns of free radical-induced cleavages of 18S rRNA observed in this last study suggested that eIF2γ domain III interacts with h44 of 18S rRNA, but no other contacts between eIF2 and 18S rRNA were detected. Using the cleavage data to dock eIF2γ onto h44 and the 3' end of Met-tRNA$_i$, making use of high-resolution structures of a bacterial 70S·tRNA·mRNA complex (Selmer et al. 2006), the 40S·eIF1 complex (Rabl et al. 2011), and aIF2αγ and aIF2βγ heterodimers (Yatime et al. 2006, 2007) (among others), a structural model of the 43S PIC was constructed (Shin et al. 2011).

Although this model represents an important step, high-resolution crystal structures and cryo-EM reconstructions of free TC and TC bound to the 43S PIC remain critical goals. In addition, the model does not include known interactions of the eIF2β N-terminal domain (NTD) (lacking in aIF2β) with eIFs 1 and 5 in the MFC (Asano et al. 2000; Singh et al. 2004), and there might be contacts between eIF2α or eIF2β with 40S ribosomal proteins not detected by the hydroxyl radical mapping. Identifying mutations in yeast eIF2 subunits, ribosomal proteins, and 18S rRNA that reduce TC binding to the PIC should help identify the eIF2·40S contacts most critical in vivo. Only one such mutation has been identified in domain III of eIF2γ (R439A) and it produces a synergistic reduction in TC binding to reconstituted 43S·mRNA complexes when combined with an 18S rRNA substitution in helix 28 (A1152U) (Shin et al. 2011) that likely weakens interaction of the anticodon stem loop (ASL) of Met-tRNA$_i$ with the 40S P site (Dong et al. 2008). Identifying h44 mutations with these phenotypes would provide valuable support for the Shin et al. model of the 43S PIC.

Binding of TC to the 40S Subunit Is Promoted by Other Factors

TC does not bind to the 40S subunit on its own, and instead requires the assistance of eIFs 1, 1A, 5, and the eIF3 complex (Asano et al. 2001; Algire et al. 2002; Majumdar et al. 2003; Kolupaeva et al. 2005; Pestova et al. 2007). All of these factors, except for eIF1A, are components of the MFC. As all MFC components can bind directly to the 40S subunit (Hinnebusch et al. 2007; Pestova et al. 2007; Sokabe et al. 2011), they would be expected to bind cooperatively in the context of the MFC. Indeed, there is considerable evidence that disrupting particular contacts between MFC components reduces the rate or stability of TC binding to 40S subunits in yeast cells (Valášek et al. 2002, 2004; Nielsen et al. 2004; Singh et al. 2005, 2006; reviewed in Hinnebusch et al. 2007). Recent studies on reconstituted mammalian MFC indicate that the rate of MFC binding to 40S·eIF1A complexes is indistinguishable from TC binding to 40S·eIF1A complexes preloaded with eIFs 1, 3, and 5 (Sokabe et al. 2011), suggesting that the stimulatory effects of other MFC components on TC binding can be exerted outside of the preformed MFC. Thus, it is important to determine in vivo whether TC generally binds to the 40S subunit in the context of the MFC or, rather, the MFC represents only one possible pathway for TC recruitment (as depicted in Fig. 1), or serves another function. It is intriguing that dissociation of Met-tRNA₁ from eIF2·GDP is enhanced when eIF2 resides in the mammalian MFC, and eIF5 figures prominently in this activity (Sokabe et al. 2011). This might implicate eIF5 in the final step of AUG selection, release of Met-tRNA₁ from eIF2·GDP, in addition to its function in promoting GTP hydrolysis in the TC. As discussed below, the physical connections among MFC components also function in AUG recognition in yeast cells. Interestingly, these connections might be regulated in plants by phosphorylation of the interacting segments in eIF3c, eIF5, and eIF2β by casein kinase 2 (Dennis et al. 2009).

The unstructured amino-terminal tail (NTT) of yeast eIF1A interacts with eIF2 and eIF3 (Olsen et al. 2003), which could also help stabilize TC binding to the 40S. In addition to physically contacting eIF2, however, eIF1A and eIF1 stimulate TC binding indirectly by stabilizing an "open" conformation of the 40S subunit that is permissive for rapid TC loading (Passmore et al. 2007). As discussed below, it is likely that Met-tRNA₁ binds differently to the PIC in this open conformation than in the "closed," scanning-arrested state that prevails after AUG recognition. Presumably, TC binding to the PIC in the model produced by Shin et al. (2011) represents the closed conformation because it was generated using reconstituted 43S·mRNA complexes in which start codon recognition has already taken place (Kolitz et al. 2009). Hence, determining the locations of TC in the free 43S PIC and open, scanning conformation of the 43S·mRNA complex should also be important goals for future research.

Recycling of eIF2·GDP

In the course of initiation, the GTP in TC is hydrolyzed to GDP, and eIF2·GDP must be recycled to eIF2·GTP for renewed TC assembly, a reaction catalyzed by the heteropentameric eIF2B complex. The essential exchange reaction is catalyzed by the carboxy-terminal segment of eIF2Bε, which interacts directly with the G domain of eIF2γ and with lysine-rich regions of eIF2β (Gomez and Pavitt 2000; Gomez et al. 2002; Alone and Dever 2006; Hinnebusch et al. 2007). The other eIF2B subunits, notably the α-β-δ "regulatory" subcomplex also contribute to eIF2·GDP binding through interactions with eIF2α (Dev et al. 2010), and this latter interaction is enhanced by phosphorylation of Ser51 by one of the eIF2α kinases, which are activated in stress conditions to down-regulate general initiation (Krishnamoorthy et al. 2001; Hinnebusch et al. 2007). It was proposed that tighter binding of eIF2α-P to the eIF2B regulatory subcomplex disrupts productive interaction of the catalytic (ε) subunit with eIF2γ, rendering phosphorylated eIF2α·GDP a competitive inhibitor of eIF2B that impedes recycling of unphosphorylated eIF2. Although this model is consistent with a large body of biochemical and genetic data, it should be tested further by structural analysis of eIF2B in complexes with phosphorylated and unphosphorylated eIF2·GDP. There is in vivo evidence in yeast that the recycling of eIF2·GDP by eIF2B is negatively regulated by formation of a competing eIF5·eIF2·GDP complex (Singh et al. 2006). Moreover, eIF5 contains a segment in the linker region connecting its amino- and carboxy-terminal domains that interacts with eIF2γ and inhibits GDP release from eIF2, serving as a GDP dissociation inhibitor (GDI) (Jennings and Pavitt 2010).

eIF2-Independent Met-tRNA$_i$ Recruitment

Recent studies indicate that in mammalian reconstituted systems, the protein Ligatin/eIF2D can deliver Met-tRNA$_i$ to the 40S subunit independently of eIF2·GTP in the case of certain specialized mRNAs (internal ribosome entry sites [IRES] containing, leaderless, or with A-rich 5′ untranslated regions [UTRs]) in which the AUG can be placed directly in the P site independently of scanning (Dmitriev et al. 2010; Skabkin et al. 2010). This could explain the ability of certain viral mRNAs containing IRESs to maintain translation in the face of eIF2α phosphorylation, a host defense mechanism triggered by many viruses (Dever et al. 2007). It was proposed that Ligatin/eIF2D can increase both on and off rates of tRNA binding in the P site, which could explain its other known activity of dissociating deacylated elongator tRNAs from recycled ribosomes after termination (Skabkin et al. 2010). There is also evidence that the protein eIF2A can bind to the IRES of hepatitis C virus (HCV) and enhance Met-tRNA$_i$ loading to the 40S subunit when eIF2 is phosphorylated, and knockdown of eIF2A reduces HCV proliferation in cells (Kim et al. 2011).

mRNA RECRUITMENT TO THE 43S PIC

eIF4F Actively Promotes Loading of mRNA onto the PIC

A critical aspect of the scanning mechanism concerns the reactions involved in directing the 43S PIC to the 5′ end of the mRNA. eIF4F stimulates this step through interaction of eIF4E with the cap structure, recruiting eIF4A to the 5′ UTR (Pestova et al. 2007). eIF4G holds eIF4A in its active conformation (Oberer et al. 2005; Schutz et al. 2008; Hilbert et al. 2011; Nielsen et al. 2011; Ozes et al. 2011), enabling it to unwind the mRNA and produce a single-stranded

binding site for the 43S PIC near the 5′ cap. It is believed that eIF4G also helps to recruit the 43S PIC directly, via physical interactions with eIF3 or eIF5 in the PIC (Asano et al. 2001; Pestova et al. 2007). There is genetic and biochemical evidence implicating eIF4A and eIF4F in promoting 43S attachment to mRNAs, in some cases even if they contain relatively short 5′ UTRs without obvious secondary structures. As might be expected, a greater requirement for these factors has been observed for mRNAs with more structured 5′ UTRs (Svitkin et al. 2001; Pestova and Kolupaeva 2002; Mitchell et al. 2010; Hinnebusch 2011). In addition, 43S attachment to model mRNAs expected to lack any structure in the 5′ UTR can occur in reconstituted systems without eIF4F (Pestova and Kolupaeva 2002; Mitchell et al. 2010).

Simultaneous binding of eIF4E to the cap, PABP to the poly(A) tail, and eIF4E and PABP to their separate binding sites in the eIF4G NTD enables circularization of the mRNA (Pestova et al. 2007), and it is frequently assumed that this "closed-loop" conformation is crucial for efficient recruitment of the 43S PIC. However, the importance of the PABP–eIF4G interaction seems to vary with the cell type. Eliminating the PABP–eIF4G interaction by deleting or mutating the PABP-binding domain in eIF4G is not lethal in yeast (Tarun et al. 1997); and even if the eIF4G–eIF4E interaction is impaired, deleting the PABP-binding domain has no effect on yeast cell growth provided that the RNA-binding region in the amino terminus of eIF4G1 (RNA1) is intact (Park et al. 2011a). It appears that RNA1 and the PABP- and eIF4E-binding domains in yeast eIF4G collaborate to promote stable association of eIF4G with mRNA near the cap, and formation of a closed loop may be incidental to the efficiency of 43S attachment. Impairing the PABP–eIF4G interaction had only a modest effect on translation in rabbit reticulocyte lysates (RRLs) (Hinton et al. 2007) but a dramatic effect was seen in Krebs-2 cell extracts, in which it reduced eIF4E binding to the cap, 48S assembly, and 60S subunit joining (Kahvejian et al. 2005). The PABP independence of RRLs likely results from the high ratio of eIF4F to general RNA-binding proteins, as addition of the RNA-binding protein YB-1 to RRLs confers PABP dependence (Svitkin et al. 2009). In addition, tight binding of mammalian eIF4F to the capped 5' end of mRNA requires the RNA-binding domain in the middle of eIF4G (Yanagiya et al. 2009). Thus, interaction of eIF4G with PABP bound to the poly(A) tail might be critical only when YB-1 or other general RNA-binding proteins effectively compete with eIF4G for direct binding to the mRNA—a situation that apparently does not exist in yeast cells under favorable culture conditions. Interestingly, although the eIF4E–cap interaction adds little to the binding affinity of eIF4F for mRNA in vitro (Kaye et al. 2009), the eIF4E–cap interaction with eIF4G should provide eIF4F with yet another way to circumvent competition with general RNA-binding proteins. In general, it appears that a number of the interactions among the components of the mRNP are redundant and may serve to safeguard the system against failure at a single point and to give the mRNA recruitment machinery an advantage over competing RNA–protein and protein–protein interactions rather than playing central mechanistic roles.

In yeast eIF4G, there are two other RNA-binding domains in the middle and carboxyl terminus (RNA2 and RNA3, respectively) (Berset et al. 2003), which appear to perform critical functions downstream from eIF4F mRNA·PABP assembly (Park et al. 2011a). Interestingly, the RNA3 domain contains a binding site for the DEAD-box RNA helicase Ded1/Ddx3 (Hilliker et al. 2011), an essential protein in yeast implicated in ribosomal scanning (Berthelot et al. 2004; Abaeva et al. 2011; Hinnebusch 2011). Although eliminating the eIF4G-binding domain in the carboxyl terminus of Ded1 impairs translation in vitro, it does not affect cell growth (Hilliker et al. 2011), implying either that the RNA3 domain has another critical function in vivo besides Ded1 recruitment or that Ded1 can be recruited by a redundant pathway. It is clearly important to identify the molecular functions of RNA2 and RNA3 in mRNA recruitment and/or ribosomal scanning. In this regard, a recent study has shown that the three RNA-binding sites in yeast eIF4G work together

to impart a strong preference on the eIF4F complex for unwinding RNA duplexes with 5'-single-stranded overhangs over duplexes with 3'-overhangs (Rajagopal et al. 2012). This polarity may be important for establishing the 5'-3' directionality of scanning by the PIC.

eIF4A is not a processive helicase and is thought to melt short helices in the mRNA by binding in its ATP-bound form to an unpaired RNA strand, with ATP hydrolysis serving either to disrupt the neighboring duplex or to release eIF4A for subsequent rounds of RNA binding and melting (Sengoku et al. 2006; Liu et al. 2008; Bulygin et al. 2010; Parsyan et al. 2011). In the crystal structure of free eIF4A, its amino- and carboxy-terminal RecA-like domains are widely separated and a functional active site does not exist (Caruthers et al. 2000). Interaction with the "HEAT" domains of eIF4G holds the RecA-like domains of eIF4A near each other in a conformation that may be poised to interact with substrates and release products (Oberer et al. 2005; Schutz et al. 2008; Hilbert et al. 2011). It seems clear that eIF4A undergoes a cycle of conformational and ligand-affinity changes driven by ATP hydrolysis and/or nucleotide binding and release and that the conformation of the enzyme is modulated by interactions with other proteins (Oberer et al. 2005; Pestova et al. 2007; Schutz et al. 2008; Marintchev et al. 2009; Hilbert et al. 2011). Exactly how these changes result in RNA unwinding is not yet clear, nor is the stoichiometry of events. It is noteworthy that eIF4A is the most abundant initiation factor; at a concentration of $50~\mu M$ in yeast it exists in fivefold excess over ribosomes (von der Haar and McCarthy 2002) and at a concentration similar to that of actin. Thus, it is possible that multiple eIF4A molecules act during recruitment of an individual mRNA to the PIC, both within eIF4F and outside of it. A full understanding of the mechanism of action of eIF4A will require additional structural and biophysical studies, including use of ensemble and single-molecule kinetics approaches.

In addition to recruiting and activating eIF4A, there is evidence that a segment of mammalian eIF4G helps to recruit the 43S PIC to the mRNA 5' end by its interactions with the e subunit of eIF3 (Korneeva et al. 2000; LeFebvre et al. 2006). This conclusion is based on the inhibitory effects of overexpressing eIF3e (presumably to out-compete eIF3 binding to eIF4G) on translation initiation and on eIF4G and eIF2 association with native PICs. It would be valuable to extend the analysis to include cells depleted of eIF3e, as it is possible that the e subunit does not make the sole (or even most critical) contact between eIF3 and eIF4G. Neither eIF3e nor the eIF3-binding segment of eIF4G is present in yeast (Marintchev et al. 2009), and yeast eIF3 and eIF4G do not directly interact (Asano et al. 2001; Mitchell et al. 2010). However, yeast eIF4G and eIF5 interact directly (Mitchell et al. 2010), and the carboxy-terminal domain (CTD) of eIF5 can bridge interaction between eIF4G2 and the eIF3c-NTD (eIF5's direct partner in yeast eIF3) and stimulate eIF4G–eIF3 association in yeast cell extracts (Asano et al. 2001). Although a mutation in the eIF5-CTD that disrupts its interaction with eIF4G impaired 43S binding to mRNA in extracts, this defect was not seen in living cells, possibly because the eIF5-CTD mutation also reduces eIF5 GAP function and blocks the downstream conversion of PICs to 80S ICs (Asano et al. 2001). A stimulatory function of eIF5 on mRNA recruitment to the PIC was not observed in a reconstituted yeast system, however (Mitchell et al. 2010). Hence, more work is required to determine whether the eIF4G–eIF5 interaction significantly enhances 43S binding to mRNAs in yeast cells, and if a similar interaction in mammalian cells is redundant with the eIF3–eIF4G interaction.

The Mysterious eIF4B

Mammalian eIF4B binds in vitro to eIF3a through its internal "DRYG" repeats and thus could potentially form a protein bridge between the eIF4F·mRNP and 43S PIC, functioning redundantly with the eIF3–eIF4G interaction. It has been proposed that mammalian eIF4B can also stimulate 43S binding to mRNA more directly by binding simultaneously to mRNA, through its carboxy-terminal, arginine-rich RNA-binding domain, and to 18S rRNA

through its amino-terminal RNA recognition motif (RRM). Yeast eIF4B (Tif3) also appears to possess a single-stranded RNA (ssRNA)-binding domain located carboxy-terminal to the conserved amino-terminal RRM, and it was concluded that both halves of Tif3 are required for its ability to stimulate translation in vitro and in vivo; hence, the putative mRNA–rRNA bridging mechanism could apply to yeast eIF4B as well (reviewed in Pestova et al. 2007; Hinnebusch 2011). It is important to test the effects of disrupting the eIF3a- and RNA-binding domains of eIF4B or Tif3 on the efficiency of 43S binding to mRNA both in vitro and in vivo.

Mammalian eIF4B is best known for its function in stimulating the helicase activity of eIF4A—an activity it shares with a homolog, eIF4H (Pestova et al. 2007; Rozovsky et al. 2008; Parsyan et al. 2011). Consistent with this, PIC binding and scanning of structured mRNAs in an in vitro mammalian system was shown to be highly dependent on eIF4B (Dmitriev et al. 2003). A recent study suggests that eIF4B increases the efficiency with which eIF4G-stimulated ATP hydrolysis is coupled to RNA duplex unwinding by eIF4A, and that eIF4H is less efficient than eIF4B in this respect (Ozes et al. 2011). This is consistent with the finding that the carboxy-terminal, RNA-binding region of mammalian eIF4B is required for stimulation of helicase activity (Rozovsky et al. 2008) and the fact that eIF4H is shorter and lacks most of the carboxy-terminal region found in eIF4B.

The mechanism by which eIF4B stimulates eIF4A helicase activity remains unclear. There is evidence that eIF4B stimulates binding of both ATP and RNA by eIF4A (Bi et al. 2000; Rozovsky et al. 2008; Marintchev et al. 2009; Nielsen et al. 2011), possibly by enhancing interdomain closure in the manner described for eIF4G. eIF4B could also load onto single-stranded RNA extensions to stabilize eIF4A binding to the duplex-containing substrate (Rozovsky et al. 2008), it could capture the single-stranded RNA products of the helicase reaction to prevent reannealing, or it could stabilize a conformation of the eIF4A-RNA complex incompatible with duplex formation. Presumably, eIF4H is incapable of one or more of these activities, rendering it less effective than eIF4B in stimulating eIF4A helicase activity (Ozes et al. 2011).

Cross-linking studies have indicated that eIF4B, eIF4H, and eIF4A are bound to mRNA from 12 nucleotides to at least 52 nucleotides from the cap and suggest that multiple molecules of each factor interact with a single mRNA both near the eIF4F-cap complex and further downstream from it (Lindqvist et al. 2008). It is currently unclear, however, whether eIF4A and eIF4B directly interact with each other and whether eIF4B can bind to eIF4G (Marintchev et al. 2009; Nielsen et al. 2011).

As described above, the current model for mRNA recruitment posits that the eIF4 factors and PABP cooperatively assemble on an mRNA and mediate unwinding to produce the activated mRNP. The activated mRNP then binds to the 43S PIC via interactions between the eIF4 factors and factors associated with the PIC, and the 5′ end of the mRNA is loaded into the mRNA-binding channel of the 40S subunit. One alternative to this prevailing model is that the eIF4 factors assemble on the PIC to form a "holoPIC," which then directly recruits an mRNA. In this model unwinding of the mRNA actually takes place on the holoPIC, allowing the unwound segments of the 5′ UTR to be directly fed into the mRNA-binding channel of the 40S subunit. This model is appealing because it is unclear how the unwound 5′ end of the activated mRNP can be handed off to the PIC without refolding occurring first. Distinguishing between the activated mRNP and holoPIC mechanisms will require the development of new, quantitative assays to directly measure unwinding and mRNA loading in the presence of different combinations of initiation components.

eIF3, a Central Hub in mRNA Recruitment

In addition to the eIF4 group of factors, eIF3 also plays a critical role in mRNA recruitment to the PIC. eIF3 is a large complex of 13 nonidentical subunits (a−m) in mammals, and only 6 subunits (a, b, c, g, h, and j) in budding yeast. There is accumulating evidence that eIF3 interacts primarily with the solvent-exposed, "back-side" of the 40S, that it spans the entry and exit pores of the mRNA-binding channel, and that it likely interacts with the mRNA itself at these locations to stabilize 43S attachment or to regulate scanning (Pestova et al. 2007; Hinnebusch 2011). Recent findings from Cate et al. indicate that the bulk of the density visible in cryo-EM models of mammalian eIF3·40S complexes is contributed by the so-called PCI/MPN octamer, which represents only ∼1/2 of the mass of holo-eIF3 and lacks homologs of the essential yeast eIF3 subunits b, g, and i. The PCI/MPN octamer can bind the HCV IRES, 40S subunits, eIF1 and eIF1A, but cannot stimulate 48S PIC assembly, additionally requiring the b-g-i subcomplex for this key activity (Sun et al. 2011). Clearly, a high-resolution model of eIF3 binding to the 40S subunit is an important goal for future research. It has long been known that both yeast and mammalian eIF3 promote 43S binding to mRNA, but because eIF3 also stimulates 43S assembly, it was unclear if it acts directly in 43S attachment to mRNA. Consistent with a direct role are findings that 40S binding of mammalian eIF3 is stimulated by ssRNAs that can likely occupy the mRNA-binding channel of the 40S, although the stabilizing effect of mRNA on eIF3·40S interaction might play a greater role following AUG recognition and release of eIF2·GDP than in 43S attachment to the 5′ UTR (Unbehaun et al. 2004; Kolupaeva et al. 2005). However, recent findings indicate a direct role for yeast eIF3 in 43S binding to capped, native mRNA in vitro, even more critical than that of eIF4F and eIF4B (Mitchell et al. 2010), and conserved residues in the carboxyl terminus of eIF3a have been implicated in this function both in vivo (Chiu et al. 2010) and in vitro (AG Hinnebusch and JR Lorsch, unpubl. observations).

UV-cross-linking data indicate direct interactions of mammalian eIF3a and eIF3d at the mRNA exit channel (Pisarev et al. 2008), which is consistent with the role of yeast eIF3a in reinitiation on *GCN4* mRNA (Szamecz et al. 2008). It is thought that these eIF3 subunits comprise an extension of the mRNA exit channel. Consistent with this, yeast eIF3 more strongly enhanced 43S binding to a model mRNA with a long leader upstream of AUG (that would protrude from the exit channel) than one containing a short leader but a long 3′ extension (that would protrude from the entry channel) (Mitchell et al. 2010). Yeast eIF3a substitutions that impair 43S attachment to mRNA also produce phenotypes in vivo indicating defects in scanning and AUG recognition (Chiu et al. 2010). Considering evidence that the yeast eIF3a (Tif32) CTD interacts directly with 40S structural elements (h16 and Rps3) (Valášek et al. 2003; Chiu et al. 2010) that promote the open conformation of the mRNA channel latch (Passmore et al. 2007), it was suggested that the eIF3a CTD facilitates opening of the latch, although it could also help to recruit a helicase that functions at the entry channel to remove secondary structure. Clearly, more detailed structural information about interactions of eIF3 subunits with the ribosome, mRNA, eIF4G, and eIF4B are required to develop a molecular picture of its manifold roles in 43S attachment to mRNA.

Knocking out the Model?

As might be expected from their established biochemical functions, eIF4E, eIF4G, and eIF4A are all essential proteins in yeast. Moreover, mutational analysis indicates that the eIF4E and eIF4A interactions with their respective domains in eIF4G are also essential for yeast cell viability (Hinnebusch et al. 2007). However, depletion of eIF4G1 to undetectable levels in a yeast strain lacking the other isoform (eIF4G2) does not abolish translation initiation, reducing it by only ~75% and leaving a considerable fraction of polysomes intact (Jivotovskaya et al. 2006), even though cell division is blocked. In contrast, a similar depletion of eIF3 subunits virtually eliminates polysomes and detectable translation. In addition, microarray analysis of polyribosomal mRNAs after eIF4G depletion revealed substantial alterations in translational efficiencies for only a fraction of cellular mRNAs (Park et al. 2011b). These results suggest that eIF4G is rate enhancing, rather than fundamentally crucial, for translation initiation on the large majority of yeast mRNAs in cells, and raise

the question of how 43S PICs are directed to the 5′ ends of mRNAs in the absence of cap–eIF4F interaction. It is possible that the RNA-binding sites in yeast eIF4G, which impart a 5′ end dependence on eIF4F, might provide an additional means for directing PICs to the 5′ ends of mRNAs (Rajagopal et al. 2012). Additionally, perhaps the A + U bias and lack of strong secondary structure for the majority of yeast 5′ UTRs (Shabalina et al. 2004; Lawless et al. 2009) renders them intrinsically permissive for 43S attachment, albeit at rates significantly below those possible with eIF4F present. This conclusion is consonant with findings from the yeast reconstituted system, in which omitting eIF4G from reactions containing eIF3, eIF4A, and eIF4B (in addition to eIFs 1, 1A, and TC) reduced the rate of mRNA recruitment by 20-fold, but did not alter the end point of 48S PIC assembly, at least for one native mRNA tested (*RPL41A*), whereas no mRNA recruitment was observed without eIF3 (Mitchell et al. 2010). Interestingly, the group of mRNAs displaying the largest reductions in translational efficiencies in eIF4G-depleted yeast cells was among the most efficiently translated in wild-type cells and displayed shorter than average 5′-UTR lengths (Park et al. 2011b). Furthermore, none of the yeast mRNAs predicted to contain strong secondary structures in their 5′-UTRs (Lawless et al. 2009) were found to be unusually dependent on eIF4G, suggesting that another factor(s), possibly DEAD-box helicases Ded1 or Dbp1, can substitute for eIF4G to enable 43S attachment or scanning on structured 5′ UTRs in yeast.

The significant G + C bias of mammalian mRNAs (Shabalina et al. 2004) might be expected to impart a much stronger requirement for eIF4F for translation initiation. However, substantial siRNA-mediated depletion of both eIF4GI and eIF4GII simultaneously in mammalian cells had only a moderate effect on translation rates (Ramirez-Valle et al. 2008), and simultaneous depletion of eIF4GI and the eIF4G-like protein DAP5 left >30% of translation intact. The possibility that the residual pool of eIF4G was rendered more active by a compensatory reduction in a negative regulator of eIF4F seems unlikely because assembly of eIF4F was strongly reduced; however, it remains a possibility that depletion of eIF4GI, eIF4GII, and DAP5 simultaneously would more severely impact translation. These studies underscore the importance of examining the in vivo consequences of depleting initiation factors that are deemed to be essential for translation initiation purely on the basis of work performed using in vitro systems. They may also point again to a redundancy of function in the translational machinery, ensuring system failure cannot come from disruption at a single point.

SCANNING AND AUG RECOGNITION

Once the 43S PIC has been loaded onto the 5′ end of an mRNA it scans the 5′ UTR for the start codon, using complementarity with the anticodon of Met-tRNA$_i$ to identify the AUG. There are two key aspects of this process that, to some extent, are mechanistically distinct. The first concerns the factors that promote a conformation of the 43S PIC that is competent for threading along the mRNA with base-by-base inspection of the nucleotide sequence for an AUG in suitable context, and which trigger irreversible hydrolysis of GTP in the TC on AUG recognition (refer to Figs. 2 and 3). The second aspect, discussed further below, concerns the requirement to unwind secondary structure in the mRNA 5′ UTR to enable the mRNA to pass through the 40S mRNA entry channel in single-stranded form for base-by-base inspection in the P site. There is also the issue of how the 5′–3′ directionality of the scanning process is established.

An Open and Shut Case: eIFs 1, 1A, and 5 Mediate Conformational Changes Required for Start Codon Recognition

Toe-printing experiments in the mammalian reconstituted system suggested that eIF1 and eIF1A stabilize an "open" conformation of the 43S PIC conducive to scanning, and that eIF1 impedes formation of a closed state required for progression to downstream steps in the pathway in a manner that is overcome efficiently only when an AUG in preferred sequence context occupies the P site (Pestova and Kolupaeva 2002). Subsequently, it was found that eIF1 is ejected from the PIC on start codon recognition (Maag et al. 2005), consistent with this proposal (Fig. 2). The structure of the *Tetrahymena* 40S subunit bound to eIF1 appears to explain the mechanism of eIF1 release on start codon recognition, as modeling of tRNA into the P site indicates that it sterically clashes with eIF1 (Rabl et al. 2011). These results support the notion that the anticodon end of the tRNA is not deeply bound in the P site during scanning (P$_{out}$ state) and only fully engages in the site (P$_{in}$ state) on codon:anticodon pairing (Yu et al. 2009; Saini et al. 2010), which in turn drives eIF1 out of the site owing to the steric clash.

Ejection of eIF1 triggers release of P$_i$ from eIF2 in the PIC. GTP hydrolysis by eIF2 occurs nearly as fast before start codon recognition as after it, but P$_i$ is only released rapidly once eIF1 has been ejected from the complex (Algire et al. 2005). The connection between eIF1 release and P$_i$ release was shown by the fact that mutations in eIF1 that slow or speed up release of the factor from the complex correspondingly slow or speed up P$_i$ release, which occurs at the same rate as eIF1 release in all cases (Algire et al. 2005; Cheung et al. 2007; Nanda et al. 2009). Supporting the central role of eIF1 as a gatekeeper in start codon recognition, substitutions in the factor that increase initiation at near-cognate UUG codons in vivo (Sui$^-$ phenotype) generally weaken eIF1 binding to 40S subunits and accelerate release of eIF1 and P$_i$ from reconstituted PICs, whereas a substitution in eIF1A that suppresses UUG initiation (Ssu$^-$ phenotype) retards eIF1 dissociation in vitro (Cheung et al. 2007). In addition, overexpressing wild-type eIF1 suppresses UUG initiation in Sui$^-$ mutants (Valasek et al. 2004), consistent with a requirement for its release to trigger downstream events following initial start codon:anticodon pairing in the P site.

eIF1 and eIF1A bind directly and cooperatively to the 40S subunit (Maag and Lorsch 2003), with eIF1 occupying the platform near the P site (Lomakin et al. 2003; Rabl et al. 2011), and the globular (OB-fold) domain of eIF1A most likely occupying the A site (Yu et al. 2011) in the manner observed for its bacterial

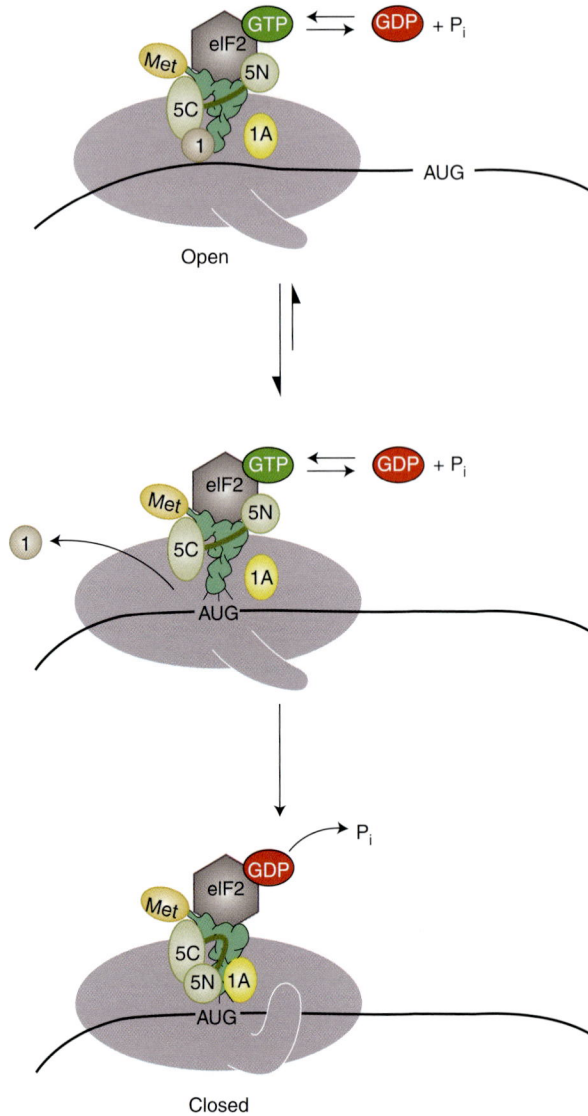

Figure 2. Model of structural rearrangements in the PIC accompanying start codon recognition. (*Top*) Before start codon recognition, the PIC exists in an open conformation, promoted by eIF1 and eIF1A, which is capable of scanning the mRNA. (*Middle*) Base pairing between the anticodon of the initiator tRNA and the start codon promotes movement of the tRNA from the P_{out} to P_{in} states and release of eIF1 from the complex. (*Bottom*) Ejection of eIF1, in turn, triggers release of P_i from eIF2, converting it to its GDP-bound form. Because eIF1 stabilizes the open state of the PIC, its departure also results in conversion of the complex to the closed, scanning-arrested conformation (shown as the closure of a latch on the mRNA entry site). Release of eIF1 is promoted by eIF5, possibly by competition between one of eIF5's domains (depicted here as the amino-terminal domain; 5N) and eIF1 for the same binding site in the PIC. Start codon recognition also induces an interaction between eIF1A and eIF5, which further stabilizes the closed state of the complex. (Modified from Hinnebusch 2011; reproduced, with permission.)

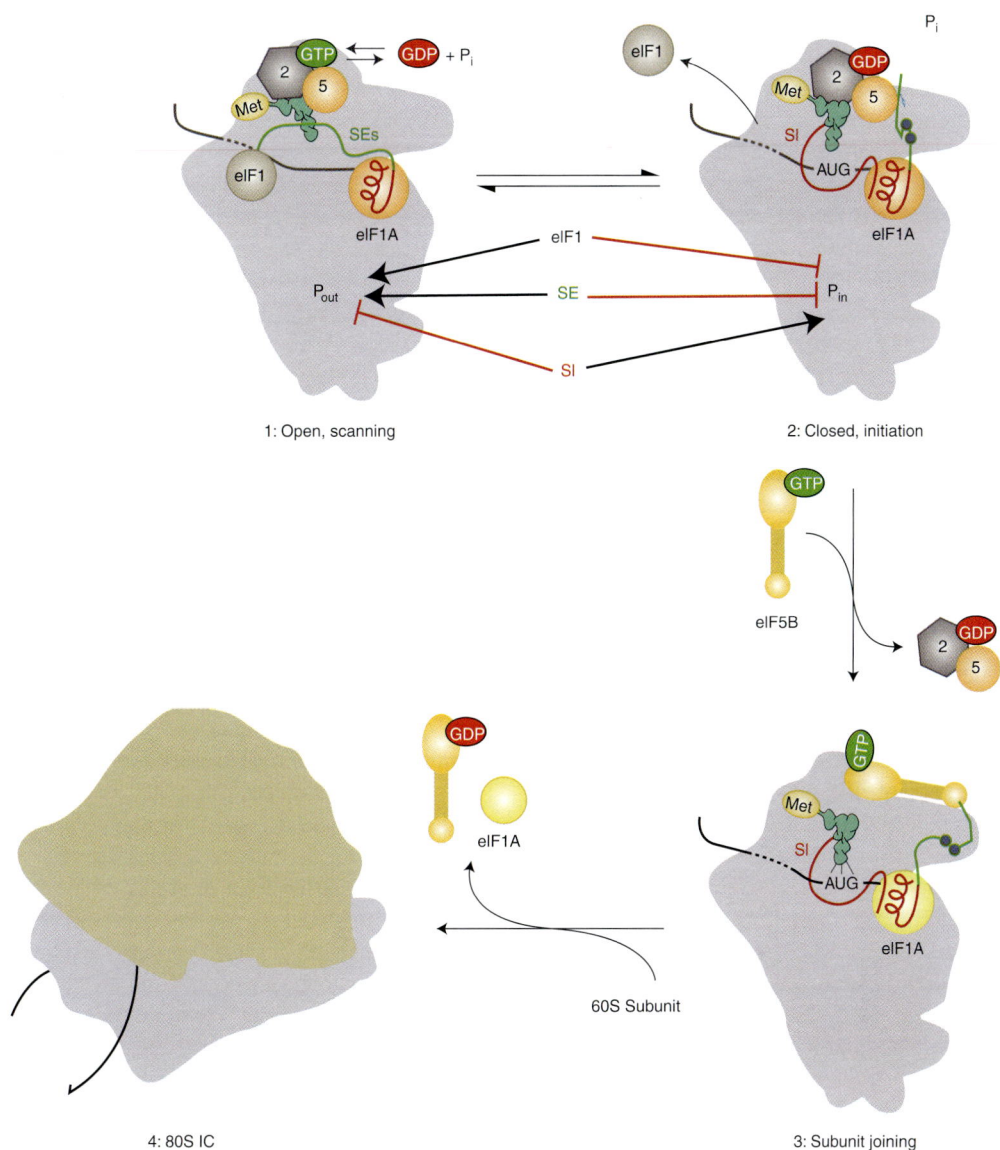

Figure 3. Model of the roles of eIF1A's amino- and carboxy-terminal tails in mediating start codon recognition and later steps in eukaryotic translation initiation. Before start codon recognition (complex 1) the amino- and carboxy-terminal tails (CTTs; shown in red and green, respectively) are both in the P site of the 40S subunit. On start codon recognition (complex 2), the initiator tRNA moves from the P_{out} to P_{in} state, causing both eIF1 and the CTT of eIF1A to be ejected from the P site. eIF1 stabilizes the open conformation of the PIC and the P_{out} state of the initiator tRNA. The scanning enhancer (SE) elements in the CTT of eIF1A (shown as blue balls) stabilize the open state of the PIC relative to the closed state. Conversely, the scanning inhibitor (SI) element in eIF1A's NTT destabilizes the open state, thus promoting closed complex formation. The CTT of eIF1A may interact directly with eIF5 after start codon recognition, and it is hypothesized that this interaction triggers P_i release from eIF2. After P_i release and dissociation of eIF2·GDP and eIF5 from the PIC (complex 3), the CTT of eIF1A is free to interact with eIF5B·GTP, recruiting it to the complex and promoting subunit joining. Release of eIF5B·GDP and eIF1A from the resulting 80S IC produces the final translation-competent ribosome, poised at the start codon to commence decoding of the mRNA. (Modified from Hinnebusch 2011; reproduced, with permission.)

ortholog (IF1) (Carter et al. 2001). Cryo-EM analysis shows that when eIF1 and eIF1A are bound simultaneously to the 40S, they provoke a structural rearrangement that involves an open conformation of the "latch" of the mRNA entry channel, which is thought to render the PIC competent for scanning. In contrast, the 40S·eIF1A complex, which might be expected to resemble the PIC following eIF1 release, is in a closed state with the latch in a locked conformation that might impede scanning (Passmore et al. 2007). Importantly, eIF1 and eIF1A together stimulate the rate of TC binding to the 40S, but TC binds more tightly to the PIC in the absence of eIF1, leading to the deduction that TC binds to the "open" conformation of the PIC in a metastable state (P_{out}) conducive for scanning but incompatible with start codon recognition. On AUG recognition, the TC achieves a more stable interaction with the P site (P_{in}) through an isomerization that requires eIF1 dissociation and rearrangement to the closed 40S conformation (Passmore et al. 2007; Kolitz et al. 2009). The physical basis for the stimulatory effects of eIF1 and eIF1A and the open conformation of the 40S subunit on the rate of TC binding are not understood, although for eIF1A this function has been localized to its unstructured carboxy-terminal tail (CTT) (Saini et al. 2010). As described above, the crystal structure of the *Tetrahymena* 40S·eIF1 complex suggests that eIF1 may physically block full entry of the anticodon end of the tRNA into the P site (Rabl et al. 2011), preventing conversion to the P_{in} state before start codon recognition. Additional structural studies will be required to fully elucidate the modes of action of eIF1 and eIF1A in the conformational transitions associated with scanning and start codon recognition.

As mentioned above, the HEAT domain in the eIF5-CTD interacts with the eIF2β-NTD, which could help stabilize TC binding to the 43S PIC in the open (P_{out}) conformation. It appears that eIF5 also promotes rearrangement to the more stable P_{in} conformation on AUG recognition by enhancing the dissociation of eIF1 (Nanda et al. 2009). One possibility is that the eIF5-CTD stimulates eIF1 eviction by weakening one of the contacts that anchors

eIF1 to the MFC. At least in yeast, however, it appears that the interactions of eIF1 and the eIF5-CTD with the eIF2β-NTD (Singh et al. 2004) and the eIF3c-NTD (Singh et al. 2004) are not mutually exclusive. Another possibility is that the eIF5 NTD facilitates ejection of eIF1 by competing with it for binding to a common site in the PIC—an idea prompted by eIF5-NTD's structural similarity with eIF1 (Conte et al. 2006). The NTD of eIF5 interacts with eIF2γ via its eIF1-like folded region (Alone and Dever 2006) and also contains a long, unstructured amino-terminal tail (NTT) that includes the key arginine residue required for GTPase activation (Das et al. 2001; Paulin et al. 2001). Nanda et al. (2009) speculated that either the NTD or CTD of eIF5 might move into eIF1's binding site in the PIC after the latter is released following start codon recognition. This movement could in turn affect the position of the NTT of eIF5 on eIF2γ, allowing P_i release from the TC (Figs. 2 and 3). An attractive feature of this model is that it provides a mechanism for coupling P_i release to eIF1 dissociation, as there is currently no evidence that eIF1 can interact with the eIF2γ G domain and block P_i release directly. Testing the models will require determination of which domain of eIF5 is responsible for promoting eIF1 release and data concerning the positions of eIF5's domains in the PIC before and after start codon recognition.

Consistent with the role of eIF5 in promoting eIF1 dissociation from the PIC (Nanda et al. 2009), in both mammals and yeast, overexpressing eIF5 reduces the requirement for optimal context and AUG start codon for efficient initiation, whereas overexpressing eIF1 has the opposite effect (Nanda et al. 2009; Ivanov et al. 2010, 2011; Martin-Marcos et al. 2011). These activities are used to negatively autoregulate translation of eIF1 in yeast (Martin-Marcos et al. 2011) and both eIF1 and eIF5 in mammals (Ivanov et al. 2010; Loughran et al. 2012). High-level eIF1 decreases recognition of its own start codon, which has a poor context. High-level eIF5 expression increases initiation at uORFs in its mRNA leader (whose AUGs are in poor context), which then blocks initiation at the main eIF5 ORF. Overexpressing eIF1 has the opposite effect on eIF5 production by promoting leaky scanning of the upstream open reading frames (uORFs) and attendant increased recognition of the main ORF AUG, which has a favorable context. Thus, the opposing effects of eIF5 and eIF1 on accuracy of start codon selection underlie a network of auto- and cross-regulatory interactions that should stabilize the stringency of start codon selection.

At least in yeast, the NTT and CTT of eIF1A display opposing activities that enable the PIC to toggle between the open and closed 40S conformations (Fekete et al. 2007), and also the P_{in} and P_{out} modes of Met-tRNA$_i$ binding to the 40S. Two 10-amino acid repeats in the CTT, dubbed SE elements, cooperate with eIF1 to promote the open 40S conformation and accelerate TC loading in the P_{out} state (Saini et al. 2010). Based on directed hydroxyl radical cleavage-mapping, the (mammalian) eIF1A CTT occupies the P site in a manner incompatible with canonical P-site binding of Met-tRNA$_i$ (Yu et al. 2011). Hence, the CTT (with its SE elements) must presumably be ejected from the P site for isomerization to the stable P_{in} state upon AUG recognition. In this view, the SE elements resemble eIF1 in having to be removed from their binding site in the open 40S conformation to achieve the closed state. Consistent with this, mutations inactivating the SE elements show the same Sui$^-$ phenotype (elevated UUG:AUG initiation ratio) (Saini et al. 2010) produced by eIF1 mutations that provoke abnormally rapid eIF1 dissociation from the PIC (Cheung et al. 2007). As the eIF1A CTT functionally interacts with the eIF5 GAP domain on AUG recognition (Maag et al. 2006), it is attractive to envision that ejection of the CTT from the P site is stabilized by physical interaction with the eIF5-NTD (Fig. 3). The yeast eIF1A NTT and a helical domain adjacent to the CTT contain SI elements that appear to oppose SE function by destabilizing the open, P_{out} conformation, thereby promoting the closed, P_{in} configuration for AUG recognition. Accordingly, SI mutations suppress UUG initiation in Sui$^-$ mutants (Fekete et al. 2007; Saini et al. 2010). The eIF1A NTT also seems to occupy the P site, but should not clash with Met-tRNA$_i$ in the P_{in} state (Yu et al. 2011).

Genetics Brings It All Together: Other Factors Participate in Start Codon Recognition

In addition to eIF1 and eIF1A, the subunits of eIF2 and eIF5 have been implicated in stringent AUG recognition by the isolation of Sui$^-$ mutations (Donahue 2000). Biochemical analysis suggested that certain Sui$^-$ mutations in eIF2γ weaken Met-tRNA$_i$ binding to eIF2·GTP, which might allow inappropriate release of initiator tRNA into the P site at near-cognate start codons. More recently, however, other mutations in eIF2γ that weaken binding of Met-tRNA$_i$ have been found that increase rather than decrease the stringency of start codon recognition, suggesting that the orientation of tRNA binding to eIF2, rather than simply its affinity, may be crucial for proper start codon recognition (Alone et al. 2008). Sui$^-$ mutations in eIF2β, in contrast, appear to primarily elevate eIF5-independent GTP hydrolysis by the TC, and the *SUI5* mutation was reported to increase the GAP function of eIF5 (Donahue 2000). Both of the latter effects might accelerate P$_i$ release, and thereby enhance initiation, at near-cognate triplets. As the eIF2β Sui$^-$ mutations map to the zinc-binding domain (ZBD), and the ZBD likely interacts directly with the eIF2γ G domain (Yatime et al. 2007), the affected ZBD residues could control access of the GAP domain of eIF5 to the eIF2γ GTP-binding pocket or prevent P$_i$ release at non-AUG codons. As noted above, in addition to promoting start codon selection by its GAP function, there is evidence that eIF5 also functions to stabilize the closed conformation of the PIC through functional interaction with eIF1A (Maag et al. 2006) and by accelerating eIF1 release (Nanda et al. 2009); however, the physical basis of these non-GAP activities is unknown. Their importance is indicated, however, by the fact that the G31R Sui$^-$ mutation (SUI5) in eIF5 completely inverts the preference for AUG and UUG codons in an assay measuring the effect of the factor on the open:closed complex equilibrium (Maag et al. 2006).

There is also genetic evidence that the interactions between eIF3c, eIF1, and eIF5 promote the opposing functions of the latter two factors in start codon recognition, as eIF3c-NTD mutations that weaken its binding to eIF1 or eIF5 confer Sui⁻ or Ssu⁻ phenotypes, respectively (Valasek et al. 2004). More work is needed to determine whether eIF1 binds simultaneously to the 40S subunit and the eIF3c-NTD to stabilize the open conformation, as both interactions seem to require helix α1 in eIF1 (Reibarkh et al. 2008; Rabl et al. 2011), or whether eIF1 toggles between these alternative binding sites at different stages of the initiation pathway (i.e., before and after start codon recognition). There is also evidence that a patch of basic residues on the surface of eIF1 that encompasses helix α2 (the "KH" region) interacts with the eIF2β-NTD and eIF5-CTD to promote the open conformation of the PIC (Cheung et al. 2007; Reibarkh et al. 2008); however, the relative importance of these interactions in maintaining 40S association of eIF1 in the scanning complex, or in stimulating eIF1 release at the start codon, remains to be determined. In addition, a mutation in eIF4G that reduces its interaction with eIF1 has been found to produce a weak Sui⁻ phenotype in yeast, suggesting that eIF4G might also act to stabilize binding of eIF1 to the scanning PIC (He et al. 2003). Use of FRET probes on eIF1 and each of its potential binding partners might reveal the dynamics of its myriad interactions during the transition from the open to closed PIC conformations (Maag et al. 2005).

How Does the Sequence Context around the Start Codon Influence Initiation Site Recognition?

There is evidence that eIF2α mediates the stimulatory effect of a permissive sequence context surrounding the AUG codon and interacts with the −3 nucleotide of the mRNA, the key contextual determinant of AUG recognition (Pisarev et al. 2006). Interaction of eIF2α with the 40S E site, in which the −3 nucleotide would reside when AUG is in the P site, would require a substantial reorientation of eIF2α from its location in the 43S model of Shin et al. (Shin et al. 2011); however, crystal structures reveal that domains 1–2 of the aIF2α are highly flexible (Stolboushkina et al. 2008), so

perhaps this is possible. eIF2β and eIFs 1 and 1A also function in discriminating against poor AUG context, and the same domains/residues in these factors involved in this activity also discriminate against non-AUG start codons, supporting the model that the AUG and context nucleotides act as a unit to stabilize the closed PIC conformation (Pisarev et al. 2006; Ivanov et al. 2010; Martin-Marcos et al. 2011). Although sequence context regulates AUG recognition in animals and yeast, with the exception of A at −3, the preferred nucleotides differ (Shabalina et al. 2004). In addition to residues immediately upstream of the AUG, it seems that a 5′ UTR of sufficient length to occupy the mRNA exit channel (∼12 nt) and extend some distance from the backside of the 40S is also required to stabilize the closed PIC conformation (Pestova and Kolupaeva 2002). It will be important to determine whether segments of 18S rRNA or ribosomal proteins in the 40S E site or mRNA exit channel (Rabl et al. 2011) interact directly with the context nucleotides and 5′-UTR sequences in a manner that regulates initiation efficiency or accuracy.

What Roles Do rRNA and tRNA Play in Start Codon Recognition and Scanning?

In addition to the various eIFs that regulate scanning and start codon recognition, there is evidence that specific residues in the 18S rRNA and in Met-tRNA$_i$ (beyond the anticodon) function in TC binding and AUG recognition. Substitutions that perturb the location or identity of the "bulge" residue in h28 of 18S rRNA decrease the rate and stability of TC binding to mutant 40S subunits in vitro. The bulge G contacts the +1 position of the P-site codon (A of AUG) in crystal structures of bacterial 70S·mRNA·tRNA complexes. Substitutions in other 18S residues predicted to contact the AUG or the ASL of Met-tRNA$_i$ also confer in vivo phenotypes indicating unstable TC binding to the scanning PIC or the PIC arrested at AUG (Dong et al. 2008). Thus, the contacts with the codon or ASL of P-site tRNA seen in bacterial elongation complexes are likely also critical for stabilizing Met-tRNA$_i$ binding to eukaryotic PICs. Genetic analysis also implicated A1193 in the h31 loop, located directly below the codon–anticodon duplex in bacterial 70S complexes, and G875 in h22, in stabilizing Met-tRNA$_i$ binding at the AUG in the closed complex (P_{in}), either directly (A1193U) or indirectly by reducing the 40S association of eIFs that promote TC binding (G875A) (Nemoto et al. 2010). Further structural studies are required to determine whether bases or proteins in the 40S subunit itself play a direct role in responding to codon–anticodon pairing in the P site, as bacterial bases A1492, A1493, and G530 do in the A site during tRNA selection in the elongation phase of protein synthesis.

tRNA$_i^{Met}$ contains highly conserved bases (identity elements) that are not present in elongator tRNAs (Marck and Grosjean 2002). These identity elements appear to play important roles in allowing specific binding of the initiator tRNA by initiation factors (e.g., eIF2) and preventing its binding to elongation factors. They may also be involved in transmitting the signal that the start codon has been located from initial codon:anticodon pairing to the rest of the PIC, promoting downstream events. The ASL of tRNA$_i$ contains three consecutive G:C pairs that are conserved from bacteria to eukaryotes. In yeast, changing the first and third of these G:C pairs to their identities in elongator tRNA$_e^{Met}$, eliminates the strong thermodynamic coupling between TC binding to the PIC and an AUG codon in the mRNA (Kapp et al. 2006), suggesting a role for these bases in start codon recognition. Base-pair substitutions at these positions also destabilized mammalian 48S PICs, at least following eIF5-stimulated GTP hydrolysis in the TC (Lomakin et al. 2006).

Strikingly, substitution of two other identity bases in eukaryotic tRNAMet$_i^{Met}$, A54, and A60 in the T loop, with the cognate elongator tRNA residues was found to suppress the loss of coupling between the AUG codon and TC in binding to the PIC caused by the elongator/initiator base swap at the third G:C pair in the ASL (G31:C39; Kapp et al. 2006). It was suggested that the A54,A60 replacements might lower the energy barrier to a structural rearrangement of Met-tRNA$_i$ necessary for its full accommoda-

tion in the P site (i.e., P_{in} state), and thereby compensate for the effect of the G31:C39 replacement, which may decrease the ability of the tRNA to fully engage in the P site (Kapp et al. 2006). In this view, structural features of the tRNA$_i^{Met}$ contribute with eIF1 and the SE elements in the eIF1A CTT to block the rearrangement from the P_{out} to P_{in} state in the absence of a perfect AUG duplex in the P site. Testing the hypothesis that the tRNA$_i^{Met}$ body plays an active role in the proper response to start codon recognition will require additional mutagenesis studies to probe the functions of the unique structural elements in the initiator tRNA.

N6-threonylcarbamoyl modification of A37 (t6A37) of tRNA$_i^{Met}$, immediately adjacent to the anticodon triplet, is thought to stabilize the first base pair of the codon:anticodon duplex for ANN or UNN codons (Agris 2008), which includes decoding of AUG by tRNA$_i^{Met}$. Consistent with this, inactivation or elimination of yeast proteins that catalyze t6A37 formation, including Sua5 (El Yacoubi et al. 2009; Lin et al. 2009) and subunits of the EKC/KEOPS complex Kae1 and Pcc1 (Daugeron et al. 2011; El Yacoubi et al. 2011; Srinivasan et al. 2011) evokes phenotypes indicating impaired recognition of AUG codons by the scanning 43S PIC.

Untangling the RNA Helicases Involved in Translation Initiation

As mentioned above, in addition to achieving a ribosomal conformation conducive to scanning, there is also a need to remove secondary structures to allow the mRNA to pass through the 40S mRNA entry or exit channels and permit its base-by-base inspection in the P site. In addition, assuming scanning is a directional process (5' to 3') one or more factors must transduce the energy required to impart this directionality, presumably using ATP hydrolysis. By analyzing the effect of increasing the 5' UTR length on the time required to complete the first round of translation of a reporter mRNA in cell extracts, it was concluded that scanning occurs at a frequency of 6–8 bases/sec in wheat germ and mammalian cell extracts, even in the presence of a strong secondary structure in the 5'

UTR (Vassilenko et al. 2011), which is similar to the figure of ~10/sec calculated for yeast extracts (Berthelot et al. 2004). The linear relationship between the translation time and 5′-UTR length implies a substantial bias toward 5′–3′ movement at each base; although a certain amount of backward scanning cannot be excluded (Berthelot et al. 2004; Vassilenko et al. 2011). The fact that increasing leader length does not reduce translational efficiency in yeast cells was taken to indicate that multiple PICs can load and scan simultaneously on long 5′ UTRs to compensate for the increased time required for each PIC to reach the AUG codon in extended leaders (Berthelot et al. 2004).

Multiple DEAD-box RNA helicases have been implicated in translation initiation, including eIF4A, Dhx29, and Ded1/Ddx3. Although it is not fully understood which ones participate in scanning, there is increasing evidence that eIF4A stimulates both ribosome attachment at the 5′ end and scanning through structured 5′ UTRs, whereas Dhx29 and Ded1 act primarily in scanning and are particularly important for negotiating long or highly structured 5′ UTRs.

A model of the scanning complex that takes into account much of the accumulated data regarding the role and position of eIF4F has been presented by Marintchev and colleagues. In this model, eIF4G spans the mRNA exit and entry channels on the 40S, positioning eIF4A and eIF4B at the entry channel for unwinding structure ahead of the ribosome. eIF4F can remain engaged with the cap as scanning proceeds, with the scanned nucleotides forming a loop between the cap-eIF4F assembly and the ribosome (Marintchev et al. 2009). The location of the HEAT-1 domain of eIF4G in this model, below the 40S platform, differs from that predicted from a cryo-EM reconstruction (Siridechadilok et al. 2005), but it is consistent with hydroxyl radical cleavage mapping of eIF4G HEAT-1 in reconstituted mammalian 43S PICs (Yu et al. 2011).

The notion that the eIF4 factors are positioned on the downstream side of scanning PICs needs to take into account the lack of evidence for factors stably associated with the mRNA 3′ of native 48S complexes, whereas residues 5′ of the ribosome are protected from RNAse digestion (Spirin 2009). The latter is more compatible with eIF4G working exclusively at the mRNA exit channel to "pull" the mRNA through the 40S subunit (Siridechadilok et al. 2005). Spirin has suggested that, from this location, eIF4G could deliver eIF4B–eIF4A·ATP complexes to single-stranded mRNA as it emerges from the exit channel, and the eIF4B would remain bound to the mRNA and block backward movement by the PIC. When the ribosome moves again, the cycle would repeat itself, constituting a "Brownian ratchet" with eIF4B as the pawl (Spirin 2009). It is also possible that unwinding takes place at the entrance channel and RNA structures reannealing after emerging from the exit channel serve to prevent backward motion of the PIC and confer 5′–3′ directionality. There is not enough experimental evidence currently available to reject or select any one of these models of the scanning PIC.

Genetic evidence from yeast indicates that the essential Ded1/Ddx3 protein cooperates with eIF4F in translation initiation (Hinnebusch 2011), and suggests that Ded1 and Dbp1 are more critical than eIF4A for efficient scanning through a long 5′ UTR in vivo (Berthelot et al. 2004). Using the reconstituted mammalian system, it was found that Dhx29, and to a lesser extent yeast Ded1, enable scanning through highly stable stem loops in a manner that is not possible with eIF4F alone. Dhx29 and Ded1 could not replace eIF4F for initiation on the β-globin 5′ UTR, which was attributed to their inability to stimulate 43S PIC attachment to an mRNA with cap-proximal structure. Thus, Dhx29 and Ded1 might specifically stimulate scanning through secondary structures, whereas eIF4F would enhance both 43S attachment and scanning through secondary structures of moderate stability (Pisareva et al. 2008; Abaeva et al. 2011). Considering the substantial reduction in total protein synthesis evoked by knockdown of Dhx29 in mammalian cells (Parsyan et al. 2009), it seems likely that Dhx29 promotes translation of a large fraction of the cell's mRNAs and not only those burdened with highly stable 5′-UTR structures.

SUBUNIT JOINING

After the start codon has been recognized, the remaining factors must dissociate from the complex and the 60S subunit must join with the 40S subunit to form the final 80S IC with Met-tRNA$_i$ and the start codon in the P site. Subunit joining is facilitated by a second GTPase initiation factor, eIF5B (Pestova et al. 2000), which is the ortholog of the bacterial translation initiation factor IF2. Below we describe the current state of knowledge of these final stages of eukaryotic translation initiation.

As discussed above, start codon recognition results in conversion of eIF2 to its GDP-bound state, which has a low affinity for Met-tRNA$_i$ (Kapp and Lorsch 2004). eIF2·GDP must dissociate from the tRNA and 40S subunit before 60S subunit joining. Experiments in the reconstituted mammalian system showed that the presence of eIF5B and 60S subunits enhanced release of eIF2·GDP from PICs following start codon recognition, most likely by driving the equilibrium toward dissociation by capturing the vacated 40S subunits as 80S complexes (Unbehaun et al. 2004). Whether eIF5B or the 60S subunit actively accelerates release of eIF2·GDP is not yet clear. It seems likely that eIF5, which binds tightly to eIF2 in both its GTP- and GDP-bound forms (Algire et al. 2005; Singh et al. 2007), leaves the PIC along with eIF2. Determining whether eIF2·GDP dissociates on its own from the 40S complex after start codon recognition or if its release is actively promoted by an initiation factor or the 60S subunit will require the development of new kinetic assays for this step in the initiation pathway. Mutations in eIF2 that reduce subunit joining, which should produce constitutive derepression of *GCN4* translation (Martin-Marcos et al. 2007) without decreasing TC occupancy of 43S PICs, would also be useful to probe factor release from the PIC.

After eIF2·GDP dissociation, eIF5B·GTP binds to the complex and accelerates the rate of 60S subunit joining (Acker et al. 2006, 2009). The ability of eIF5B to facilitate this latter step requires an interaction between the extreme carboxyl terminus of eIF1A (DIDDI) and the CTD of eIF5B (Marintchev et al. 2003; Olsen

et al. 2003; Acker et al. 2006; Fringer et al. 2007). The requirement for this interaction underscores the importance of eIF1A and, in particular, its CTT, in the initiation pathway. Before start codon recognition, the CTT is bound in the P site, which presumably sequesters the DIDDI sequence, preventing it from recruiting eIF5B to the complex prematurely. When start codon recognition occurs, the CTT is ejected from the P site. As outlined above, it may then be involved in an interaction with eIF5 that serves to promote P$_i$ release from eIF2. Once eIF2·GDP and eIF5 dissociate from the PIC, the DIDDI end of the CTT is freed to recruit eIF5B·GTP to the complex and promote joining of the 60S subunit. This series of events suggests that the CTT of eIF1A is a key controller of the timing of the steps in the initiation pathway (Fig. 3).

After subunit joining, eIF5B hydrolyzes its bound GTP, which lowers its affinity for the 80S IC and triggers its release (Lee et al. 2002; Shin et al. 2002, 2007). Release of eIF1A occurs only after dissociation of eIF5B from the 80S IC, making eIF1A the only factor to remain on the ribosome during the entire initiation pathway, with the possible exception of eIF3 (Acker et al. 2009). GTP hydrolysis by eIF5B appears to alter the conformation of the 80S IC, as release of eIF1A occurs slowly in the presence of a mutant form of eIF5B that does not hydrolyze GTP but has weakened affinity for the ribosome and thus can still dissociate after subunit joining (eIF5B-T439A,H505Y). A role for eIF5B GTP hydrolysis in altering the final conformation of the IC is consistent with recent single-molecule FRET studies in the bacterial system that indicated a similar role for GTP hydrolysis by the orthologous factor IF2 (Marshall et al. 2009), and with the leaky scanning and Gcn⁻ phenotypes produced in yeast expressing the eIF5B-T439A,H505Y mutant (Shin et al. 2002).

eIF1A has been reported to be the lowest abundance initiation factor in yeast cells (von der Haar and McCarthy 2002). This observation might be explained by the fact that eIF1A still binds with reasonable affinity to the final 80S IC (Acker et al. 2009) and presumably could compete with incoming eEF1A·GTP·aa-tRNA complexes for binding to the A site if present at high concentrations. In addition, the central role played by the CTT of eIF1A in all stages of initiation might also have put selective pressure on cells to keep the factor's concentration low to prevent the formation of spurious interactions that could prematurely trigger transition from one stage of initiation to another.

Although we now know a considerable amount about these late events in translation initiation, a number of important mysteries remain. We do not know when eIF5B interacts with the 60S subunit. Does it bind to it before it interacts with the PIC and carry it along like a tugboat pulling a ship or does it bind first to the PIC and then wait for an encounter with the 60S to dock the two together? We also know very little about how eIF5B and its interaction with eIF1A accelerate the rate of subunit joining. Understanding the molecular mechanics of this process will require additional structural studies of the 80S IC bound to eIF5B before and after GTP hydrolysis as well as additional mutagenesis and kinetics experiments to elucidate the key regions of each component required to accelerate subunit joining.

FUTURE PROSPECTS

The combination of in vitro and in vivo studies used by the field over the past 10 years has allowed dramatic advances in our understanding of the roles of the individual components of the translation initiation machinery. We have also begun to gain an understanding of what different regions of each of these components do. Obtaining a clearer picture of the molecular mechanics of this critical stage of gene expression will require the continued coupling of increasingly sophisticated in vitro and in vivo approaches, including additional advances in structural biology (Ben-Shem et al. 2011) and kinetic and thermodynamic measurements, forays into single-molecule studies, and genomewide analyses such as ribosome profiling (Ingolia et al. 2009). The focus will be on understanding how the parts work together to coordinate and facilitate the complex interactions and rearrangements required for the assembly of the 80S

IC on an mRNA, poised to enter the elongation phase of protein synthesis.

REFERENCES

Abaeva IS, Marintchev A, Pisareva VP, Hellen CU, Pestova TV. 2011. Bypassing of stems versus linear base-by-base inspection of mammalian mRNAs during ribosomal scanning. *EMBO J* 30: 115–129.

Acker MG, Shin BS, Dever TE, Lorsch JR. 2006. Interaction between eukaryotic initiation factors 1A and 5B is required for efficient ribosomal subunit joining. *J Biol Chem* 281: 8469–8475.

Acker MG, Shin BS, Nanda JS, Saini AK, Dever TE, Lorsch JR. 2009. Kinetic analysis of late steps of eukaryotic translation initiation. *J Mol Biol* 385: 491–506.

Agris PF. 2008. Bringing order to translation: The contributions of transfer RNA anticodon-domain modifications. *EMBO Rep* 9: 629–635.

Algire MA, Maag D, Savio P, Acker MG, Tarun SZ Jr, Sachs AB, Asano K, Nielsen KH, Olsen DS, Phan L, et al. 2002. Development and characterization of a reconstituted yeast translation initiation system. *RNA* 8: 382–397.

Algire MA, Maag D, Lorsch JR. 2005. Pi release from eIF2, not GTP hydrolysis, is the step controlled by start-site selection during eukaryotic translation initiation. *Mol Cell* 20: 251–262.

Alone PV, Dever TE. 2006. Direct binding of translation initiation factor eIF2γ-G domain to its GTPase-activating and GDP-GTP exchange factors eIF5 and eIF2B ε. *J Biol Chem* 281: 12636–12644.

Alone PV, Cao C, Dever TE. 2008. Translation initiation factor 2γ mutant alters start codon selection independent of Met-tRNA binding. *Mol Cell Biol* 28: 6877–6888.

Asano K, Clayton J, Shalev A, Hinnebusch AG. 2000. A multifactor complex of eukaryotic initiation factors eIF1, eIF2, eIF3, eIF5, and initiator tRNA^{Met} is an important translation initiation intermediate in vivo. *Genes Dev* 14: 2534–2546.

Asano K, Shalev A, Phan L, Nielsen K, Clayton J, Valášek L, Donahue TF, Hinnebusch AG. 2001. Multiple roles for the carboxyl terminal domain of eIF5 in translation initiation complex assembly and GTPase activation. *EMBO J* 20: 2326–2337.

Ben-Shem A, Garreau de Loubresse N, Melnikov S, Jenner L, Yusupova G, Yusupov M. 2011. The structure of the eukaryotic ribosome at 3.0 Å resolution. *Science* 334: 1524–1529.

Berset C, Zurbriggen A, Djafarzadeh S, Altmann M, Trachsel H. 2003. RNA-binding activity of translation initiation factor eIF4G1 from *Saccharomyces cerevisiae*. *RNA* 9: 871–880.

Berthelot K, Muldoon M, Rajkowitsch L, Hughes J, McCarthy JE. 2004. Dynamics and processivity of 40S ribosome scanning on mRNA in yeast. *Mol Microbiol* 51: 987–1001.

Bi X, Ren J, Goss DJ. 2000. Wheat germ translation initiation factor eIF4B affects eIF4A and eIFiso4F helicase activity by increasing the ATP binding affinity of eIF4A. *Biochemistry* 16: 5758–5765.

Cite this article as *Cold Spring Harb Perspect Biol* doi: 10.1101/cshperspect.a011544

Bulygin KN, Khairulina YS, Kolosov PM, Ven'yaminova AG, Graifer DM, Vorobjev YN, Frolova LY, Kisselev LL, Karpova GG. 2010. Three distinct peptides from the N domain of translation termination factor eRF1 surround stop codon in the ribosome. *RNA* **16:** 1902–1914.

Carter AP, Clemons WM Jr, Brodersen DE, Morgan-Warren RJ, Hartsch T, Wimberly BT, Ramakrishnan V. 2001. Crystal structure of an initiation factor bound to the 30S ribosomal subunit. *Science* **291:** 498–501.

Caruthers JM, Johnson ER, McKay DB. 2000. Crystal structure of yeast initiation factor 4A, a DEAD-box RNA helicase. *Proc Natl Acad Sci* **97:** 13080–13085.

Cheung YN, Maag D, Mitchell SF, Fekete CA, Algire MA, Takacs JE, Shirokikh N, Pestova T, Lorsch JR, Hinnebusch AG. 2007. Dissociation of eIF1 from the 40S ribosomal subunit is a key step in start codon selection in vivo. *Genes Dev* **21:** 1217–1230.

Chiu WL, Wagner S, Herrmannova A, Burela L, Zhang F, Saini AK, Valasek L, Hinnebusch AG. 2010. The C-terminal region of eukaryotic translation initiation factor 3a (eIF3a) promotes mRNA recruitment, scanning, and, together with eIF3j and the eIF3b RNA recognition motif, selection of AUG start codons. *Mol Cell Biol* **30:** 4415–4434.

Conte MR, Kelly G, Babon J, Sanfelice D, Youell J, Smerdon SJ, Proud CG. 2006. Structure of the eukaryotic initiation factor (eIF) 5 reveals a fold common to several translation factors. *Biochemistry* **45:** 4550–4558.

Das S, Ghosh R, Maitra U. 2001. Eukaryotic translation initiation factor 5 functions as a GTPase-activating protein. *J Biol Chem* **276:** 6720–6726.

Daugeron MC, Lenstra TL, Frizzarin M, El Yacoubi B, Liu X, Baudin-Baillieu A, Lijnzaad P, Decourty L, Saveanu C, Jacquier A, et al. 2011. Gcn4 misregulation reveals a direct role for the evolutionary conserved EKC/KEOPS in the t6A modification of tRNAs. *Nucleic Acids Res* **39:** 6148–6160.

Dennis MD, Person MD, Browning KS. 2009. Phosphorylation of plant translation initiation factors by CK2 enhances the in vitro interaction of multifactor complex components. *J Biol Chem* **284:** 20615–20628.

Dev K, Qiu H, Dong J, Zhang F, Barthlme D, Hinnebusch AG. 2010. The β/Gcd7 subunit of eukaryotic translation initiation factor 2B (eIF2B), a guanine nucleotide exchange factor, is crucial for binding eIF2 in vivo. *Mol Cell Biol* **30:** 5218–5233.

Dever TE, Dar AC, Sicheri F. 2007. The eIF2a kinases. In *Translational control in biology and medicine* (ed. Mathews M, Sonenberg N, Hershey JWB), pp. 319–344. Cold Spring Harbor Laboratory Press, Cold Spring Harbor, NY.

Dmitriev SE, Terenin IM, Dunaevsky YE, Merrick WC, Shatsky IN. 2003. Assembly of 48S translation initiation complexes from purified components with mRNAs that have some base pairing within their 5′ untranslated regions. *Mol Cell Biol* **23:** 8925–8933.

Dmitriev SE, Terenin IM, Andreev DE, Ivanov PA, Dunaevsky JE, Merrick WC, Shatsky IN. 2010. GTP-independent tRNA delivery to the ribosomal P-site by a novel eukaryotic translation factor. *J Biol Chem* **285:** 26779–26787.

Donahue T. 2000. Genetic approaches to translation initiation in *Saccharomyces cerevisiae*. In *Translational control of gene expression* (ed. Sonenberg N, Hershey JWB, Math-

ews MB), pp. 487–502. Cold Spring Harbor Laboratory Press, Cold Spring Harbor, NY.

Dong J, Nanda JS, Rahman H, Pruitt MR, Shin BS, Wong CM, Lorsch JR, Hinnebusch AG. 2008. Genetic identification of yeast 18S rRNA residues required for efficient recruitment of initiator tRNA(Met) and AUG selection. *Genes Dev* **22:** 2242–2255.

El Yacoubi B, Lyons B, Cruz Y, Reddy R, Nordin B, Agnelli F, Williamson JR, Schimmel P, Swairjo MA, de Crecy-Lagard V. 2009. The universal YrdC/Sua5 family is required for the formation of threonylcarbamoyladenosine in tRNA. *Nucleic Acids Res* **37:** 2894–2909.

El Yacoubi B, Hatin I, Deutsch C, Kahveci T, Rousset JP, Iwata-Reuyl D, Murzin AG, de Crecy-Lagard V. 2011. A role for the universal Kae1/Qri7/YgjD (COG0533) family in tRNA modification. *EMBO J* **30:** 882–893.

Fekete CA, Mitchell SF, Cherkasova VA, Applefield D, Algire MA, Maag D, Saini A, Lorsch JR, Hinnebusch AG. 2007. N- and C-terminal residues of eIF1A have opposing effects on the fidelity of start codon selection. *EMBO J* **26:** 1602–1614.

Fringer JM, Acker MG, Fekete CA, Lorsch JR, Dever TE. 2007. Coupled release of eukaryotic translation initiation factors 5B and 1A from 80S ribosomes following subunit joining. *Mol Cell Biol* **27:** 2384–2397.

Gomez E, Pavitt GD. 2000. Identification of domains and residues within the ε subunit of eukaryotic translation initiation factor 2B (eIF2b) required for guanine nucleotide exchange reveals a novel activation function promoted by eIF2B complex formation. *Mol Cell Biol* **20:** 3965–3976.

Gomez E, Mohammad SS, Pavitt GD. 2002. Characterization of the minimal catalytic domain within eIF2B: The guanine-nucleotide exchange factor for translation initiation. *EMBO J* **21:** 5292–5301.

He H, von der Haar T, Singh CR, Ii M, Li B, Hinnebusch AG, McCarthy JE, Asano K. 2003. The yeast eukaryotic initiation factor 4G (eIF4G) HEAT domain interacts with eIF1 and eIF5 and is involved in stringent AUG selection. *Mol Cell Biol* **23:** 5431–5445.

Hershey JWB, Merrick WC. 2000. Pathway and mechanism of initiation of protein synthesis. In *Translational control of gene expression* (ed. Sonenberg N, Hershey JWB, Mathews MB), pp. 33–88. Cold Spring Harbor Laboratory Press, Cold Spring Harbor, NY.

Hilbert M, Kebbel F, Gubaev A, Klostermeier D. 2011. eIF4G stimulates the activity of the DEAD box protein eIF4A by a conformational guidance mechanism. *Nucleic Acids Res* **39:** 2260–2270.

Hilliker A, Gao Z, Jankowsky E, Parker R. 2011. The DEAD-box protein Ded1 modulates translation by the formation and resolution of an eIF4F-mRNA complex. *Mol Cell* **43:** 962–972.

Hinnebusch AG. 2011. Molecular mechanism of scanning and start codon selection in eukaryotes. *Microbiol Mol Biol Rev* **75:** 434–467.

Hinnebusch AG, Dever TE, Asano K. 2007. Mechanism of translation initiation in the yeast *Saccharomyces cerevisiae*. In *Translational control in biology and medicine* (ed. Mathews M, Sonenberg N, Hershey JWB), pp. 225–268. Cold Spring Harbor Laboratory Press, Cold Spring Harbor, NY.

Hinton TM, Coldwell MJ, Carpenter GA, Morley SJ, Pain VM. 2007. Functional analysis of individual binding activities of the scaffold protein eIF4G. *J Biol Chem* **282:** 1695–1708.

Ingolia NT, Ghaemmaghami S, Newman JR, Weissman JS. 2009. Genome-wide analysis in vivo of translation with nucleotide resolution using ribosome profiling. *Science* **324:** 218–223.

Ivanov IP, Loughran G, Sachs MS, Atkins JF. 2010. Initiation context modulates autoregulation of eukaryotic translation initiation factor 1 (eIF1). *Proc Natl Acad Sci* **107:** 18056–18060.

Jennings MD, Pavitt GD. 2010. eIF5 has GDI activity necessary for translational control by eIF2 phosphorylation. *Nature* **465:** 378–381.

Jivotovskaya AV, Valasek L, Hinnebusch AG, Nielsen KH. 2006. Eukaryotic translation initiation factor 3 (eIF3) and eIF2 can promote mRNA binding to 40S subunits independently of eIF4G in yeast. *Mol Cell Biol* **26:** 1355–1372.

Kahvejian A, Svitkin YV, Sukarieh R, M'Boutchou MN, Sonenberg N. 2005. Mammalian poly(A)-binding protein is a eukaryotic translation initiation factor, which acts via multiple mechanisms. *Genes Dev* **19:** 104–113.

Kapp LD, Lorsch JR. 2004. GTP-dependent recognition of the methionine moiety on initiator tRNA by translation factor eIF2. *J Mol Biol* **335:** 923–936.

Kapp LD, Kolitz SE, Lorsch JR. 2006. Yeast initiator tRNA identity elements cooperate to influence multiple steps of translation initiation. *RNA* **12:** 751–764.

Kaye NM, Emmett KJ, Merrick WC, Jankowsky E. 2009. Intrinsic RNA binding by the eukaryotic initiation factor 4F depends on a minimal RNA length but not on the m7G cap. *J Biol Chem* **284:** 17742–17750.

Kim JH, Park SM, Park JH, Keum SJ, Jang SK. 2011. eIF2A mediates translation of hepatitis C viral mRNA under stress conditions. *EMBO J* **30:** 2454–2464.

Kolitz SE, Takacs JE, Lorsch JR. 2009. Kinetic and thermodynamic analysis of the role of start codon/anticodon base pairing during eukaryotic translation initiation. *RNA* **15:** 138–152.

Kolupaeva VG, Unbehaun A, Lomakin IB, Hellen CU, Pestova TV. 2005. Binding of eukaryotic initiation factor 3 to ribosomal 40S subunits and its role in ribosomal dissociation and anti-association. *RNA* **11:** 470–486.

Korneeva NL, Lamphear BJ, Hennigan FL, Rhoads RE. 2000. Mutually cooperative binding of eukaryotic translation initiation factor (eIF) 3 and eIF4A to human eIF4G-1. *J Biol Chem* **275:** 41369–41376.

Krishnamoorthy T, Pavitt GD, Zhang F, Dever TE, Hinnebusch AG. 2001. Tight binding of the phosphorylated α subunit of initiation factor 2 (eIF2α) to the regulatory subunits of guanine nucleotide exchange factor eIF2B is required for inhibition of translation initiation. *Mol Cell Biol* **21:** 5018–5030.

Lawless C, Pearson RD, Selley JN, Smirnova JB, Grant CM, Ashe MP, Pavitt GD, Hubbard SJ. 2009. Upstream sequence elements direct post-transcriptional regulation of gene expression under stress conditions in yeast. *BMC Genomics* **10:** 7–26.

Lee JH, Pestova TV, Shin BS, Cao C, Choi SK, Dever TE. 2002. Initiation factor eIF5B catalyzes second GTP-dependent step in eukaryotic translation initiation. *Proc Natl Acad Sci* **99:** 16689–16694.

LeFebvre AK, Korneeva NL, Trutschl M, Cvek U, Duzan RD, Bradley CA, Hershey JW, Rhoads RE. 2006. Translation initiation factor eIF4G-1 binds to eIF3 through the eIF3e subunit. *J Biol Chem* **281:** 22917–22932.

Lin CA, Ellis SR, True HL. 2009. The Sua5 protein is essential for normal translational regulation in yeast. *Mol Cell Biol* **30:** 354–363.

Lindqvist L, Imataka H, Pelletier J. 2008. Cap-dependent eukaryotic initiation factor-mRNA interactions probed by cross-linking. *RNA* **14:** 960–969.

Liu F, Putnam A, Jankowsky E. 2008. ATP hydrolysis is required for DEAD-box protein recycling but not for duplex unwinding. *Proc Natl Acad Sci* **105:** 20209–20214.

Lomakin IB, Kolupaeva VG, Marintchev A, Wagner G, Pestova TV. 2003. Position of eukaryotic initiation factor eIF1 on the 40S ribosomal subunit determined by directed hydroxyl radical probing. *Genes Dev* **17:** 2786–2797.

Lomakin IB, Shirokikh NE, Yusupov MM, Hellen CU, Pestova TV. 2006. The fidelity of translation initiation: Reciprocal activities of eIF1, IF3 and YciH. *EMBO J* **25:** 196–210.

Lorsch JR, Dever TE. 2010. Molecular view of 43 S complex formation and start site selection in eukaryotic translation initiation. *J Biol Chem* **285:** 21203–21207.

Loughran G, Sachs MS, Atkins JF, Ivanov IP. 2012. Stringency of start codon selection modulates autoregulation of translation initiation factor eIF5. *Nucleic Acids Res* **40:** 2898–2906.

Maag D, Lorsch JR. 2003. Communication between eukaryotic translation initiation factors 1 and 1A on the yeast small ribosomal subunit. *J Mol Biol* **330:** 917–924.

Maag D, Fekete CA, Gryczynski Z, Lorsch JR. 2005. A conformational change in the eukaryotic translation preinitiation complex and release of eIF1 signal recognition of the start codon. *Mol Cell* **17:** 265–275.

Maag D, Algire MA, Lorsch JR. 2006. Communication between eukaryotic translation initiation factors 5 and 1A within the ribosomal pre-initiation complex plays a role in start site selection. *J Mol Biol* **356:** 724–737.

Majumdar R, Bandyopadhyay A, Maitra U. 2003. Mammalian translation initiation factor eIF1 functions with eIF1A and eIF3 in the formation of a stable 40 S preinitiation complex. *J Biol Chem* **278:** 6580–6587.

Marck C, Grosjean H. 2002. tRNomics: Analysis of tRNA genes from 50 genomes of Eukarya, Archaea, and bacteria reveals anticodon-sparing strategies and domain-specific features. *RNA* **8:** 1189–1232.

Marintchev A, Kolupaeva VG, Pestova TV, Wagner G. 2003. Mapping the binding interface between human eukaryotic initiation factors 1A and 5B: A new interaction between old partners. *Proc Natl Acad Sci* **100:** 1535–1540.

Marintchev A, Edmonds K, Marintcheva B, Hendrickson E, Oberer M, Suzuki C, Herdy B, Sonenberg N, Wagner G. 2009. Topology and regulation of the human eIF4A/4G/4H helicase complex in translation initiation. *Cell* **136:** 447–460.

Cite this article as *Cold Spring Harb Perspect Biol* doi: 10.1101/cshperspect.a011544

Marshall RA, Aitken CE, Puglisi JD. 2009. GTP hydrolysis by IF2 guides progression of the ribosome into elongation. *Mol Cell* **35**: 37–47.

Martin-Marcos P, Hinnebusch AG, Tamame M. 2007. Ribosomal protein L33 is required for ribosome biogenesis, subunit joining and repression of GCN4 translation. *Mol Cell Biol* **27**: 5968–5985.

Martin-Marcos P, Cheung YN, Hinnebusch AG. 2011. Functional elements in initiation factors 1, 1A and 2β discriminate against poor AUG context and non-AUG start codons. *Mol Cell Biol* **31**: 4814–4831.

Mitchell SF, Walker SE, Algire MA, Park EH, Hinnebusch AG, Lorsch JR. 2010. The 5′-7-methylguanosine cap on eukaryotic mRNAs serves both to stimulate canonical translation initiation and block an alternative pathway. *Mol Cell* **39**: 950–962.

Nanda JS, Cheung YN, Takacs JE, Martin-Marcos P, Saini AK, Hinnebusch AG, Lorsch JR. 2009. eIF1 controls multiple steps in start codon recognition during eukaryotic translation initiation. *J Mol Biol* **394**: 268–285.

Naveau M, Lazennec-Schurdevin C, Panvert M, Mechulam Y, Schmitt E. 2010. tRNA binding properties of eukaryotic translation initiation factor 2 from *Encephalitozoon cuniculi*. *Biochemistry* **49**: 8680–8688.

Nemoto N, Singh CR, Udagawa T, Wang S, Thorson E, Winter Z, Ohira T, Ii M, Valasek L, Brown SJ, et al. 2010. Yeast 18 S rRNA is directly involved in the ribosomal response to stringent AUG selection during translation initiation. *J Biol Chem* **285**: 32200–32212.

Nielsen KH, Szamecz B, Valasek L, Jivotovskaya A, Shin BS, Hinnebusch AG. 2004. Functions of eIF3 downstream of 48S assembly impact AUG recognition and GCN4 translational control. *EMBO J* **23**: 1166–1177.

Nielsen KH, Behrens MA, He Y, Oliveira CL, Sottrup Jensen L, Hoffmann SV, Pedersen JS, Andersen GR. 2011. Synergistic activation of eIF4A by eIF4B and eIF4G. *Nucleic Acids Res* **39**: 2678–2689.

Oberer M, Marintchev A, Wagner G. 2005. Structural basis for the enhancement of eIF4A helicase activity by eIF4G. *Genes Dev* **19**: 2212–2223.

Olsen DS, EM S, Mathew A, Zhang F, Krishnamoorthy T, Phan L, Hinnebusch AG. 2003. Domains of eIF1A that mediate binding to eIF2, eIF3 and eIF5B and promote ternary complex recruitment *in vivo*. *EMBO J* **22**: 193–204.

Ozes AR, Feoktistova K, Avanzino BC, Fraser CS. 2011. Duplex unwinding and ATPase activities of the DEAD-box helicase eIF4A are coupled by eIF4G and eIF4B. *J Mol Biol* **412**: 674–687.

Park E, Walker S, Lee J, Rothenburg S, Lorsch J, Hinnebusch A. 2011a. Multiple elements in the eIF4G1 N-terminus promote assembly of eIF4G1 PABP mRNPs in vivo. *EMBO J* **30**: 302–316.

Park EH, Zhang F, Warringer J, Sunnerhagen P, Hinnebusch AG. 2011b. Depletion of eIF4G from yeast cells narrows the range of translational efficiencies genome-wide. *BMC Genomics* **12**: 1–18.

Parsyan A, Shahbazian D, Martineau Y, Petroulakis E, Alain T, Larsson O, Mathonnet G, Tettweiler G, Hellen CU, Pestova TV, et al. 2009. The helicase protein DHX29 promotes translation initiation, cell proliferation, and tumorigenesis. *Proc Natl Acad Sci* **106**: 22217–22222.

Parsyan A, Svitkin Y, Shahbazian D, Gkogkas C, Lasko P, Merrick WC, Sonenberg N. 2011. mRNA helicases: The tacticians of translational control. *Nat Rev Mol Cell Biol* **12**: 235–245.

Passmore LA, Schmeing TM, Maag D, Applefield DJ, Acker MG, Algire MA, Lorsch JR, Ramakrishnan V. 2007. The eukaryotic translation initiation factors eIF1 and eIF1A induce an open conformation of the 40S ribosome. *Mol Cell* **26**: 41–50.

Paulin FE, Campbell LE, O'Brien K, Loughlin J, Proud CG. 2001. Eukaryotic translation initiation factor 5 (eIF5) acts as a classical GTPase-activator protein. *Curr Biol* **11**: 55–59.

Pestova TV, Kolupaeva VG. 2002. The roles of individual eukaryotic translation initiation factors in ribosomal scanning and initiation codon selection. *Genes Dev* **16**: 2906–2922.

Pestova TV, Lomakin IB, Lee JH, Choi SK, Dever TE, Hellen CUT. 2000. The joining of ribosomal subunits in eukaryotes requires eIF5B. *Nature* **403**: 332–335.

Pestova TV, Lorsch JR, Hellen CUT. 2007. The mechanism of translation initiation in eukaryotes. In *Translational control in biology and medicine* (ed. Mathews M, Sonenberg N, Hershey JWB), pp. 87–128. Cold Spring Harbor Laboratory Press, Cold Spring Harbor, NY.

Pisarev AV, Kolupaeva VG, Pisareva VP, Merrick WC, Hellen CU, Pestova TV. 2006. Specific functional interactions of nucleotides at key −3 and +4 positions flanking the initiation codon with components of the mammalian 48S translation initiation complex. *Genes Dev* **20**: 624–636.

Pisarev AV, Kolupaeva VG, Yusupov MM, Hellen CU, Pestova TV. 2008. Ribosomal position and contacts of mRNA in eukaryotic translation initiation complexes. *EMBO J* **27**: 1609–1621.

Pisareva VP, Pisarev AV, Komar AA, Hellen CU, Pestova TV. 2008. Translation initiation on mammalian mRNAs with structured 5′UTRs requires DExH-box protein DHX29. *Cell* **135**: 1237–1250.

Rabl J, Leibundgut M, Ataide SF, Haag A, Ban N. 2011. Crystal structure of the eukaryotic 40S ribosomal subunit in complex with initiation factor 1. *Science* **331**: 730–736.

Rajagopal V, Park EH, Hinnebusch AG, Lorsch JR. 2012. Specific domains in yeast eIF4G strongly bias the RNA unwinding activity of the eIF4F complex towards duplexes with 5′-overhangs. *J Biol Chem* doi: 10.1074/jbc.M112.347278.

Ramirez-Valle F, Braunstein S, Zavadil J, Formenti SC, Schneider RJ. 2008. eIF4GI links nutrient sensing by mTOR to cell proliferation and inhibition of autophagy. *J Cell Biol* **181**: 293–307.

Reibarkh M, Yamamoto Y, Singh CR, del Rio F, Fahmy A, Lee B, Luna RE, Ii M, Wagner G, Asano K. 2008. Eukaryotic initiation factor (eIF) 1 carries two distinct eIF5-binding faces important for multifactor assembly and AUG selection. *J Biol Chem* **283**: 1094–1103.

Rozovsky N, Butterworth AC, Moore MJ. 2008. Interactions between eIF4AI and its accessory factors eIF4B and eIF4H. *RNA* **14**: 2136–2148.

Saini AK, Nanda JS, Lorsch JR, Hinnebusch AG. 2010. Regulatory elements in eIF1A control the fidelity of start codon selection by modulating tRNA(i)(Met) binding to the ribosome. *Genes Dev* **24:** 97–110.

Schmitt E, Naveau M, Mechulam Y. 2010. Eukaryotic and archaeal translation initiation factor 2: A heterotrimeric tRNA carrier. *FEBS Lett* **584:** 405–412.

Schmitt E, Panvert M, Lazennec-Schurdevin C, Coureux PD, Perez J, Thompson A, Mechulam Y. 2012. Structure of the ternary initiation complex aIF2-GDPNP-methionylated initiator tRNA. *Nat Struct Mol Biol* **19:** 450–454.

Schutz P, Bumann M, Oberholzer AE, Bieniossek C, Trachsel H, Altmann M, Baumann U. 2008. Crystal structure of the yeast eIF4A-eIF4G complex: An RNA-helicase controlled by protein-protein interactions. *Proc Natl Acad Sci* **105:** 9564–9569.

Selmer M, Dunham CM, Murphy FVt, Weixlbaumer A, Petry S, Kelley AC, Weir JR, Ramakrishnan V. 2006. Structure of the 70S ribosome complexed with mRNA and tRNA. *Science* **313:** 1935–1942.

Sengoku T, Nureki O, Nakamura A, Kobayashi S, Yokoyama S. 2006. Structural basis for RNA unwinding by the DEAD-box protein *Drosophila Vasa*. *Cell* **125:** 287–300.

Shabalina SA, Ogurtsov AY, Rogozin IB, Koonin EV, Lipman DJ. 2004. Comparative analysis of orthologous eukaryotic mRNAs: Potential hidden functional signals. *Nucleic Acids Res* **32:** 1774–1782.

Shin BS, Maag D, Roll-Mecak A, Arefin MS, Burley SK, Lorsch JR, Dever TE. 2002. Uncoupling of initiation factor eIF5B/IF2 GTPase and translational activities by mutations that lower ribosome affinity. *Cell* **111:** 1015–1025.

Shin BS, Acker MG, Maag D, Kim JR, Lorsch JR, Dever TE. 2007. Intragenic suppressor mutations restore GTPase and translation functions of a eukaryotic initiation factor 5B switch II mutant. *Mol Cell Biol* **27:** 1677–1685.

Shin BS, Kim JR, Walker SE, Dong J, Lorsch JR, Dever TE. 2011. Initiation factor eIF2γ promotes eIF2-GTP-Met-tRNA(i)(Met) ternary complex binding to the 40S ribosome. *Nat Struct Mol Biol* **18:** 1227–1234.

Singh CR, He H, Ii M, Yamamoto Y, Asano K. 2004. Efficient incorporation of eukaryotic initiation factor 1 into the multifactor complex is critical for formation of functional ribosomal preinitiation complexes in vivo. *J Biol Chem* **279:** 31910–31920.

Singh CR, Curtis C, Yamamoto Y, Hall NS, Kruse DS, He H, Hannig EM, Asano K. 2005. Eukaryotic translation initiation factor 5 is critical for integrity of the scanning preinitiation complex and accurate control of GCN4 translation. *Mol Cell Biol* **25:** 5480–5491.

Singh CR, Lee B, Udagawa T, Mohammad-Qureshi SS, Yamamoto Y, Pavitt GD, Asano K. 2006. An eIF5/eIF2 complex antagonizes guanine nucleotide exchange by eIF2B during translation initiation. *EMBO J* **25:** 4537–4546.

Singh CR, Udagawa T, Lee B, Wassink S, He H, Yamamoto Y, Anderson JT, Pavitt GD, Asano K. 2007. Change in nutritional status modulates the abundance of critical preinitiation intermediate complexes during translation initiation in vivo. *J Mol Biol* **370:** 315–330.

Siridechadilok B, Fraser CS, Hall RJ, Doudna JA, Nogales E. 2005. Structural roles for human translation factor eIF3 in initiation of protein synthesis. *Science* **310:** 1513–1515.

Skabkin MA, Skabkina OV, Dhote V, Komar AA, Hellen CU, Pestova TV. 2010. Activities of Ligatin and MCT-1/DENR in eukaryotic translation initiation and ribosomal recycling. *Genes Dev* **24:** 1787–1801.

Sokabe M, Fraser CS, Hershey JW. 2011. The human translation initiation multi-factor complex promotes methionyl-tRNAi binding to the 40S ribosomal subunit. *Nucleic Acids Res* **40:** 905–913.

Spirin AS. 2009. How does a scanning ribosomal particle move along the 5'-untranslated region of eukaryotic mRNA? Brownian ratchet model. *Biochemistry* **48:** 10688–10692.

Srinivasan M, Mehta P, Yu Y, Prugar E, Koonin EV, Karzai AW, Sternglanz R. 2011. The highly conserved KEOPS/EKC complex is essential for a universal tRNA modification, t6A. *EMBO J* **30:** 873–881.

Stolboushkina E, Nikonov S, Nikulin A, Blasi U, Manstein DJ, Fedorov R, Garber M, Nikonov O. 2008. Crystal structure of the intact archaeal translation initiation factor 2 demonstrates very high conformational flexibility in the α- and β-subunits. *J Mol Biol* **382:** 680–691.

Sun C, Todorovic A, Querol-Audí J, Bai Y, Villa N, Snyder M, Ashchyan J, Lewis CS, Hartland A, Gradia S, et al. 2011. Functional reconstitution of human eukaryotic translation initiation factor 3 (eIF3). *Proc Natl Acad Sci* **108:** 20473–20478.

Svitkin Y, Pause A, Haghighat A, Pyronnet S, Witherell GW, Belsham G, Sonenberg N. 2001. The requirement for eukaryotic initiation factor 4A (eIF4A) in translation is in direct proportion to the degree of mRNA 5' secondary structure. *RNA* **7:** 382–394.

Svitkin YV, Evdokimova VM, Brasey A, Pestova TV, Fantus D, Yanagiya A, Imataka H, Skabkin MA, Ovchinnikov LP, Merrick WC, et al. 2009. General RNA-binding proteins have a function in poly(A)-binding protein-dependent translation. *EMBO J* **28:** 58–68.

Szamecz B, Rutkai E, Cuchalova L, Munzarova V, Herrmannova A, Nielsen KH, Burela L, Hinnebusch AG, Valasek L. 2008. eIF3a cooperates with sequences 5' of uORF1 to promote resumption of scanning by post-termination ribosomes for reinitiation on GCN4 mRNA. *Genes Dev* **22:** 2414–2425.

Tarun SZ, Wells SE, Deardorff JA, Sachs AB. 1997. Translation initiation factor eIF4G mediates in vitro poly (A) tail-dependent translation. *Proc Natl Acad Sci* **94:** 9046–9051.

Unbehaun A, Borukhov SI, Hellen CU, Pestova TV. 2004. Release of initiation factors from 48S complexes during ribosomal subunit joining and the link between establishment of codon-anticodon base-pairing and hydrolysis of eIF2-bound GTP. *Genes Dev* **18:** 3078–3093.

Valášek L, Nielsen KH, Hinnebusch AG. 2002. Direct eIF2-eIF3 contact in the multifactor complex is important for translation initiation in vivo. *EMBO J* **21:** 5886–5898.

Valášek L, Mathew A, Shin BS, Nielsen KH, Szamecz B, Hinnebusch AG. 2003. The Yeast eIF3 subunits TIF32/a and NIP1/c and eIF5 make critical connections with the 40S ribosome in vivo. *Genes Dev* **17:** 786–799.

Valášek L, Nielsen KH, Zhang F, Fekete CA, Hinnebusch AG. 2004. Interactions of eukaryotic translation initiation factor 3 (eIF3) subunit NIP1/c with eIF1 and eIF5 promote preinitiation complex assembly and regulate start codon selection. *Mol Cell Biol* **24:** 9437–9455.

Vassilenko KS, Alekhina OM, Dmitriev SE, Shatsky IN, Spirin AS. 2011. Unidirectional constant rate motion of the ribosomal scanning particle during eukaryotic translation initiation. *Nucleic Acids Res* **39:** 5555–5567.

von der Haar T, McCarthy JE. 2002. Intracellular translation initiation factor levels in *Saccharomyces cerevisiae* and their role in cap-complex function. *Mol Microbiol* **46:** 531–544.

Yanagiya A, Svitkin YV, Shibata S, Mikami S, Imataka H, Sonenberg N. 2009. Requirement of RNA binding of mammalian eukaryotic translation initiation factor 4GI (eIF4GI) for efficient interaction of eIF4E with the mRNA cap. *Mol Cell Biol* **29:** 1661–1669.

Yatime L, Mechulam Y, Blanquet S, Schmitt E. 2006. Structural switch of the γ subunit in an archaeal aIF2αγ heterodimer. *Structure* **14:** 119–128.

Yatime L, Mechulam Y, Blanquet S, Schmitt E. 2007. Structure of an archaeal heterotrimeric initiation factor 2 reveals a nucleotide state between the GTP and the GDP states. *Proc Natl Acad Sci* **104:** 18445–18450.

Yu Y, Marintchev A, Kolupaeva VG, Unbehaun A, Veryasova T, Lai SC, Hong P, Wagner G, Hellen CU, Pestova TV. 2009. Position of eukaryotic translation initiation factor eIF1A on the 40S ribosomal subunit mapped by directed hydroxyl radical probing. *Nucleic Acids Res* **37:** 5167–5182.

Yu Y, Abaeva IS, Marintchev A, Pestova TV, Hellen CU. 2011. Common conformational changes induced in type 2 picornavirus IRESs by cognate *trans*-acting factors. *Nucleic Acids Res* **39:** 4851–4865.

The Elongation, Termination, and Recycling Phases of Translation in Eukaryotes

Thomas E. Dever[1] and Rachel Green[2]

[1]Laboratory of Gene Regulation and Development, *Eunice Kennedy Shriver* National Institute of Child Health and Human Development, National Institutes of Health, Bethesda, Maryland 20892

[2]Department of Molecular Biology and Genetics, Johns Hopkins University School of Medicine, Baltimore, Maryland 21205

Correspondence: tdever@nih.gov; ragreen@jhmi.edu

This work summarizes our current understanding of the elongation and termination/recycling phases of eukaryotic protein synthesis. We focus here on recent advances in the field. In addition to an overview of translation elongation, we discuss unique aspects of eukaryotic translation elongation including eEF1 recycling, eEF2 modification, and eEF3 and eIF5A function. Likewise, we highlight the function of the eukaryotic release factors eRF1 and eRF3 in translation termination, and the functions of ABCE1/Rli1, the Dom34:Hbs1 complex, and Ligatin (eIF2D) in ribosome recycling. Finally, we present some of the key questions in translation elongation, termination, and recycling that remain to be answered.

The mechanism of translation elongation is well conserved between eukaryotes and bacteria (Rodnina and Wintermeyer 2009), and, in general, studies on the mechanism of translation elongation have focused on bacterial systems. Following translation initiation, an 80S ribosome is poised on a messenger RNA (mRNA) with the anticodon of Met-tRNA$_i$ in the P site base-paired with the start codon. The second codon of the open reading frame (ORF) is present in the A (acceptor) site of the ribosome awaiting binding of the cognate aminoacyl-tRNA. The eukaryotic elongation factor eEF1A, the ortholog of bacterial EF-Tu, binds aminoacyl-tRNA in a GTP-dependent manner and then directs the tRNA to the A site of the ribosome (Fig. 1). Codon recognition by the tRNA triggers GTP hydrolysis by eEF1A, releasing the factor and enabling the aminoacyl-tRNA to be accommodated into the A site. Recent high-resolution structures of the bacterial ribosome bound to EF-Tu and aminoacyl-tRNA revealed distortion of the anticodon stem and at the junction between the acceptor and D stems that enables the aminoacyl-tRNA to interact with both the decoding site on the small subunit and with EF-Tu. It is thought that the energetic penalty for this distortion is paid for by the perfect codon–anticodon match and the attendant stabilizing interactions that occur between the A site and cognate tRNA to promote high-fidelity decoding (Schmeing et al. 2009, 2011). These interactions might exceed those involving 16S rRNA bases A1492, A1493, and G530 with the

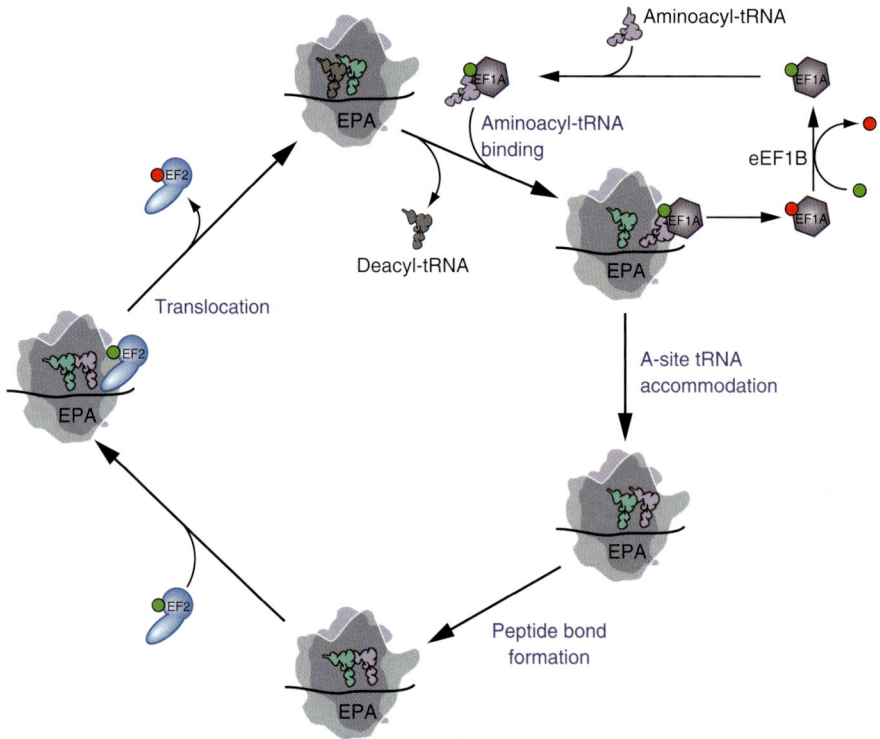

Figure 1. Model of the eukaryotic translation elongation pathway. In this model the large ribosomal subunit is drawn transparent to visualize tRNAs, factors, and mRNA binding to the decoding center at the interface between the large and small subunits and tRNAs interacting with the peptidyl transferase center in the large subunit. Starting at the top, an eEF1A·GTP·aminoacyl-tRNA ternary complex binds the aminoacyl-tRNA to the 80S ribosome with the anticodon loop of the tRNA in contact with the mRNA in the A site of the small subunit. Following release of eEF1A·GDP, the aminoacyl-tRNA is accommodated into the A site, and the eEF1A·GDP is recycled to eEF1A·GTP by the exchange factor eEF1B. Peptide bond formation is accompanied by transition of the A- and P-site tRNAs into hybrid states with the acceptors ends of the tRNAs moving to the P and E sites, respectively. Binding of eEF2·GTP promotes translocation of the tRNAs into the canonical P and E sites, and is followed by release of eEF2·GDP, which unlike eEF1A does not require an exchange factor. The ribosome is now ready for the next cycle of elongation with release of the deacylated tRNA from the E site and binding of the appropriate eEF1A·GTP·aminoacyl-tRNA to the A site. Throughout, GTP is depicted as a green ball and GDP as a red ball; also, the positions of the mRNA, tRNAs, and factors are drawn for clarity and are not meant to specify their exact places on the ribosome.

minor groove of the codon–anticodon helix (Ogle et al. 2001) to include residues in ribosomal proteins and other regions of the tRNA (Jenner et al. 2010). The recent structures of the ribosome bound to EF-Tu and aminoacyl-tRNA also revealed that the conserved nucleotide A2662 (*Thermus thermophilus* numbering) in the sarcin–ricin loop of 23S rRNA in the large subunit interacts with the conserved catalytic His residue in the G domain enabling the His residue to coordinate and position the water molecule required for GTP hydrolysis (Voorhees et al. 2010). It is expected that these mechanisms of initial aminoacyl-tRNA binding, codon recognition, and GTPase activation will be shared between bacteria and eukaryotes.

Following accommodation of the aminoacyl-tRNA into the A site, peptide bond formation with the P-site peptidyl-tRNA occurs rapidly. The peptidyl transferase center (PTC), consisting primarily of conserved ribosomal RNA (rRNA) elements on the large ribosomal subunit, positions the substrates for catalysis. Recent crystal structures of the *Saccharomyces cerevisiae* 80S ribosome and the *T. thermophila* 60S subunit revealed that the rRNA structure of the PTC is nearly superimposable between the eukaryotic and bacterial ribosomes (Ben-Shem et al. 2010, 2011; Klinge et al. 2011), supporting the idea that the mechanism of peptide bond formation, the heart of protein synthesis, is universally conserved.

Following peptide bond formation, ratcheting of the ribosomal subunits triggers movement of the tRNAs into so-called hybrid P/E and A/P states with the acceptor ends of the tRNAs in the E and P-sites and the anticodon loops remaining in the P and A sites, respectively. Translocation of the tRNAs to the canonical E and P sites requires the elongation factor eEF2 in eukaryotes, which is the ortholog of bacterial EF-G. Binding of the GTPase eEF2 or EF-G in complex with GTP is thought to stabilize the hybrid state and promote rapid hydrolysis of GTP. Conformational changes in eEF2/EF-G accompanying GTP hydrolysis and P_i release are thought to alternatively unlock the ribosome allowing tRNA and mRNA movement and then lock the subunits in the posttranslocation state. P_i release is also coupled to release of the factor from the ribosome. A structure of EF-G bound to a posttranslocation bacterial ribosome revealed the interaction of EF-G domain IV with the mRNA, P-site tRNA, and decoding center on the small ribosomal subunit (Gao et al. 2009), consistent with the notion that EF-G and eEF2 function, at least in part, to prevent backward movement of the tRNAs in the unlocked state of the ribosome.

In the posttranslocation state of the ribosome, a deacylated tRNA occupies the E site and the peptidyl-tRNA is in the P site. The A site is vacant and available for binding of the next aminoacyl-tRNA in complex with eEF1A. Although it has been proposed that E-site tRNA release is coupled to binding of aminoacyl-tRNA in the A site (Nierhaus 1990), recent single molecule and ensemble kinetic analyses indicate that release of the E-site tRNA is not strictly coupled to binding of aminoacyl-tRNA in the A site (Semenkov et al. 1996; Uemura et al. 2010; Chen et al. 2011).

Although these basic mechanisms of translation elongation and peptide bond formation are conserved between bacteria and eukaryotes, several features of translation elongation are unique in eukaryotes. Moreover, recent studies have characterized additional factors that may function in translation elongation.

eEF1 RECYCLING

Following GTP hydrolysis both EF-Tu and eEF1A are released from the ribosome in complex with GDP. The spontaneous rate of GDP dissociation from these factors is slow, and a guanine nucleotide exchange factor is required to recycle the inactive GDP-bound factor to its active GTP-bound state. In the case of EF-Tu, the single polypeptide factor EF-Ts promotes nucleotide exchange. In contrast, the eukaryotic factor eEF1B, composed of two to four subunits, catalyzes guanine nucleotide exchange on eEF1A. Interestingly, despite the strong homology between eEF1A and EF-Tu, the catalytic eEF1Bα subunit does not resemble EF-Ts and the two proteins promote guanine nucleotide exchange by distinct mechanisms (Rodnina and Wintermeyer 2009). Whereas EF-Ts binds to the EF-Tu G domain and indirectly destabilizes Mg^{2+} binding leading to GDP release, eEF1Bα inserts an essential Lys residue into the Mg^{2+} and γ-phosphate binding site to directly destabilize Mg^{2+} binding. It is unclear why eukaryotes use a more elaborate exchange factor, although it might provide a means to regulate translation elongation (Sivan et al. 2011).

eEF2 MODIFICATION

As noted above, eEF2 and EF-G promote translocation by binding to the ribosome and inserting domain IV of the factor into the decoding center of the small subunit. Interestingly, a conserved His residue (His699 in yeast eEF2) at the tip of domain IV of eEF2 is posttranslationally modified to diphthamide. The diphthamide modification is formed in two steps and requires the action of five proteins, DPH1-5. Despite the universal conservation of the diphthamide modification in eukaryotes and archaea, diphthamide is nonessential for cell viability. The DPH1-5 genes can be deleted in yeast, and derivatives of CHO cells that fail to make diphthamide grow normally (Liu et al. 2004). However, knockout mice lacking DPH1(Ovca1), DPH3, or DPH4 were either embryonic lethal or showed severe developmental defects (Chen and Behringer 2004; Liu et al. 2006; Webb et al. 2008), perhaps suggesting a critical role for diphthamide at a specific time during development. Currently, the only known function of diphthamide is to serve as a site of ADP-ribosylation by diphtheria toxin and related toxins. Modification of eEF2 inactivates the factor and blocks translation. As it seems improbable that cells would retain diphthamide solely to enable pathogen inhibition of protein synthesis, it is presumed that the diphthamide modification somehow enhances eEF2 function. Consistent with this idea, amino acid substitutions at His699 in yeast eEF2 have been reported to block or impair yeast cell growth (Kimata and Kohno 1994). Moreover, a yeast strain expressing the eEF2-H699N mutant displayed reduced translation and increased sensitivity to translational inhibitors; but, as expected, the mutant was resistant to diphtheria toxin (Ortiz et al. 2006). Interestingly, both the eEF2-H699N mutant and a *dph5* mutant that fails to make diphthamide displayed enhanced -1 ribosomal frameshifting in vivo (Ortiz et al. 2006). This latter finding indicates a positive role for diphthamide in translation; however, additional studies are needed to define the function of this modification. Given its location at the tip of eEF2 domain IV, it is appealing to speculate that diphthamide may contact the mRNA, tRNA, or rRNA in the decoding center of the ribosome to promote or enhance the fidelity of translocation.

In addition to the constitutive modification of eEF2 by diphthamide, the factor is also phosphorylated in mammalian cells by a novel Ca^{2+}-activated protein kinase eEF2K. Phosphorylation of eEF2 is thought to block translation by impairing factor binding to the ribosome (Carl-berg et al. 1990). Interestingly, despite its apparent role in blocking total protein synthesis, eEF2 phosphorylation in neurons has been linked to enhanced localized translation of activity-regulated cytoskeleton-associated protein Arc/Arg3.1, an immediate early gene induced by sensory inputs and learning that functions in postsynaptic endocytosis (Park et al. 2008; Waung et al. 2008). Although mRNA specificity in this regulation might reflect localized phosphorylation of eEF2 in the cell, this result bears resemblance to previous reports of enhanced translation of a subset of mRNAs in cells treated with the translation inhibitor cycloheximide (Walden and Thach 1986). In this latter case it was proposed that inhibition of general translation enhances the translation of the normally uncompetitive mRNAs by freeing up a limiting factor required for their translation. Accordingly, it will be interesting to further examine the factor requirements for Arc/Arg3.1 mRNA translation. Moreover, the recently described ribosomal profiling technique (Ingolia et al. 2009) may provide a further test to this model and identify additional mRNAs whose translation is enhanced under conditions where global translation elongation is inhibited.

FUNGAL SPECIFIC FACTOR eEF3

In addition to the canonical eEF1 and eEF2, translation elongation in all yeast and higher fungi that have been examined requires an additional factor eEF3. eEF3 is an ATPase and contains two ATP-binding cassettes (ABCs). Thus, whereas yeast eEF1 and eEF2 can functionally replace their mammalian counterparts to promote translation with mammalian ribosomes in vitro, mammalian eEF1 and eEF2 must be supplemented with eEF3 to promote protein synthesis with yeast ribosomes (Skogerson and Engelhardt 1977). Consistent with this ribosome specificity for eEF3, the factor binds to ribosomes and a cryo-EM structure of eEF3 bound to posttranslocation yeast 80S ribosomes has been reported (Andersen et al. 2006). Ribosome binding by eEF3 was enhanced in the presence of the nonhydrolyzable ATP analog ADPNP, consistent with the idea that ATP hydrolysis is required for eEF3 dissociation from the ribosome.

In contrast to eEF1A and eEF2, which bind to the ribosomal A site, eEF3 spans across the top of the two subunits contacting the central protuberance of the 60S subunit and the head of the 40S subunit (Andersen et al. 2006). Moreover, a chromodomain inserted within the second ABC domain of eEF3 was found to bind near the ribosomal E site. It has been proposed that eEF3 may facilitate release of deacylated tRNA from the E site following translocation (Triana-Alonso et al. 1995; Andersen et al. 2006). Both genetic and physical interactions have been reported between eEF1A and eEF3 (Anand et al. 2003, 2006); however, further study is needed to determine whether this interaction is associated with the proposed coupling between the ribosomal A and E sites.

It is unclear why translation with yeast ribosomes requires eEF3 but translation with other eukaryotic or bacterial ribosomes does not. No close eEF3 homologs can be found in the genome sequences of other organisms, so it is unlikely that an eEF3-like protein has been overlooked in studies of bacterial or mammalian translation. Although there have been reports of ribosome-associated ATPase activities in various eukaryotic and bacterial systems, it is not clear whether these activities are associated with protein synthesis, and the lack of eEF3-like proteins in other organisms suggests that these reported activities are not related to eEF3 function. Comparison of the crystal structures of the yeast 80S ribosome with the structures of bacterial ribosomes does not reveal a unique feature that would indicate the need for an additional elongation factor. It is noteworthy that the yeast 80S ribosome structure contains an additional nonribosomal protein Stm1p. The Stm1p was bound to the head of the 40S subunit and then snaked through the mRNA entry tunnel of the ribosome (Ben-Shem et al. 2011). Interestingly, Stm1p appears to inhibit the function of 80S ribosomes (Balagopal and Parker 2011) and to oppose eEF3 function (Van Dyke et al. 2009). In yeast lacking Stm1p, eEF3 binding to ribosomes is enhanced; and overexpression of eEF3 impairs the growth of the cells lacking Stm1p

(Van Dyke et al. 2009). Further genetic, biochemical, and structural studies are needed to determine the function of eEF3 in translation elongation and to resolve its unique requirement in fungi.

eIF5A/EF-P

Recent studies have revealed an additional factor requirement for translation elongation. The factor eIF5A was originally characterized for its ability to stimulate the transfer of methionine from Met-tRNA$_i$ in the 80S initiation complex to the aminoacyl-tRNA analog puromycin (Kemper et al. 1976). Because this methionyl-puromycin synthesis assay monitors formation of the first peptide bond, the factor that stimulated the assay was denoted as an initiation factor. Interestingly, the structurally related bacterial protein EF-P was likewise identified by its ability to stimulate the synthesis of methionyl-puromycin using a reconstituted in vitro translation system from *E. coli* (Glick and Ganoza 1975). Further examination of the function of eIF5A in globin mRNA translation revealed that eIF5A lowered the Mg^{2+} optimum for protein synthesis in assays lacking spermidine; however, this stimulatory effect of eIF5A was not seen in assays containing spermidine where the optimum Mg^{2+} concentration was already low (Schreier et al. 1977).

This impact of spermidine on detection of eIF5A activity is noteworthy given the unique posttranslational modification of eIF5A. In all eukaryotes and archaea, a conserved Lys residue in eIF5A is posttranslationally modified in two steps to hypusine (Park et al. 2010). In the first step, an N-butylamine moiety is transferred from spermidine to the ε-amino group of the Lys side chain to form deoxyhypusine. A subsequent hydroxylation reaction completes the modification. Unmodified eIF5A fails to stimulate the formation of methionyl-puromycin (Park et al. 1991); consistent with this, the deoxyhypusine synthase gene that catalyzes the biosynthesis of the hypusine is essential in yeast. The Lys residue in eIF5A that is modified to hypusine is located in a loop at the top of domain I of the protein (Kim et al. 1998), and the corresponding residue in EF-P is either an Arg or Lys (Bailly and de Crecy-Lagard 2010; Navarre et al. 2010; Yanagisawa et al. 2010). Interestingly, in at least some bacteria expressing the Lys variant of EF-P, this residue is modified by the addition of a β-lysine (Bailly and de Crecy-Lagard 2010; Navarre et al. 2010; Yanagisawa et al. 2010), which resembles the hypusine side chain. Although this posttranslational modification is required for EF-P stimulation of methionyl-puromycin synthesis by bacterial ribosomes (Park et al. 2012), further studies are needed to define the mechanistic role of the EF-P modification.

In attempts to resolve the function of eIF5A in yeast, various labs have depleted or inactivated the factor and examined the impact on general protein synthesis using polysome analyses. Depletion of eIF5A using transcriptional shutoff or degron approaches has indicated defects either in translation initiation or elongation, perhaps reflecting differences in assay conditions or in the efficiency of eIF5A depletion (Zanelli et al. 2006; Gregio et al. 2009; Saini et al. 2009; Henderson and Hershey 2011). Inactivation of an eIF5A temperature-sensitive mutant resulted in polysome retention in the absence of the elongation inhibitor cycloheximide (Saini et al. 2009), indicative of a role for eIF5A throughout translation elongation. As eIF5A has been suggested to stimulate the translation of only a subset of mRNAs in the cell (Kang and Hershey 1994), it will be valuable to examine the impact of eIF5A inactivation on genome-wide translation using ribosomal profiling (Ingolia et al. 2009).

Further supporting a role for eIF5A in general translation elongation, inactivation of eIF5A in yeast caused increased ribosomal transit times (Gregio et al. 2009; Saini et al. 2009), and genetic analyses revealed functional interactions between eIF5A and eEF2 (Saini et al. 2009; Dias et al. 2012). Although eEF2 was found to cosediment with eIF5A in pull-down assays (Jao and Chen 2006; Zanelli et al. 2006), this interaction appears to be bridged by ribosomes and so further experiments will be needed to determine whether eEF2 and eIF5A can simultaneously bind to the same 80S complex.

Finally, using a reconstituted in vitro translation initiation and elongation system from

yeast, addition of eIF5A was found to stimulate the rate of methionyl-puromycin synthesis and tripeptide synthesis by twofold (Saini et al. 2009). Interestingly, addition of eIF5A also stimulated the rate of peptide release in assays containing release factors eRF1 and eRF3 (Saini et al. 2009). In all of these assays eIF5A activity was fully dependent on its hypusine modification. These studies suggest that eIF5A stimulates the reactivity of peptidyl-tRNA in the ribosomal P site with either aminoacyl-tRNA or protein ligands that enter the A site. Consistently, a crystal structure of the bacterial 70S ribosome revealed EF-P binding in a site adjacent to the P-site bound Met-tRNAᵢ (Blaha et al. 2009). Taken together, the unique spermidine-derived posttranslational modification of eIF5A, the impact of eIF5A and spermidine on the optimal Mg^{2+} concentration for peptide bond synthesis, and the binding site of EF-P adjacent to the P-site tRNA suggest that eIF5A/EF-P serves as an efficient polyamine delivery system to promote reactivity of the P-site tRNA. To further define the function of eIF5A in translation, it will be necessary to accurately map the binding site of eIF5A on the 80S ribosome and to determine the timing of E-site tRNA release and eIF5A binding during the translation elongation cycle. In addition, more in-depth kinetic studies to determine the steps in elongation, termination, and/or initiation affected by the factor would be valuable in elucidating its mechanism of action.

EUKARYOTIC TRANSLATION TERMINATION

Translation termination takes place when the end of the coding sequence is reached by the ribosome and a stop codon (UAA, UGA, or UAG) enters the A site. Termination in eukaryotes is catalyzed by two protein factors, eRF1 and eRF3, that appear to collaborate in the process (Stansfield et al. 1995; Zhouravleva et al. 1995; Alkalaeva et al. 2006). The class I factor, eRF1, is responsible for high-fidelity stop codon recognition and peptidyl-tRNA hydrolysis. The class II factor, eRF3, is a translational GTPase that is more closely related to EF-Tu than to EF-G (Atkinson et al. 2008). Although bacteria also possess both class I (RF1 and RF2) and class II (RF3) release factors with similar nomenclature, there are striking structural and mechanistic differences between the classes in eukaryotes and bacteria. Most importantly, the class I release factors are wholly different proteins with no apparent evolutionary relationship. These factors appear to have evolved after the divergence of the bacterial and eukaryotic lineages and are different evolutionary solutions to the problem of termination (and as we shall see, recycling).

Like RF1 and RF2, eRF1 is, broadly speaking, a tRNA-shaped protein factor composed of three domains (Song et al. 2000). The amino-terminal domain is responsible for codon recognition and contains a distal loop with a highly conserved NIKS motif that has been proposed to decode stop codons through codon:anticodon-like interactions. Chemical crosslinking experiments suggest that this loop is indeed in close proximity to the stop codon nucleotides (Chavatte et al. 2002). Other regions of eRF1 also appear to contribute to stop codon recognition including the YxCxxxF motif (Kolosov et al. 2005; Fan-Minogue et al. 2008; Bulygin et al. 2010). Overall, the findings in eukaryotes suggest that stop codon recognition is more complex than in the bacterial system.

The middle (M) domain of eRF1 is functionally analogous to the tRNA acceptor stem and as such extends into the PTC to promote peptide release (Song et al. 2000). Like bacterial RF1 and RF2, this domain contains a universally conserved Gly-Gly-Gln (GGQ) motif that appears to be essential in promoting the chemistry of peptide hydrolysis as detailed in the bacterial system (Frolova et al. 1999; Laurberg et al. 2008; Weixlbaumer et al. 2008). It is particularly interesting to note that these tripeptide motifs are an example of convergent evolution in the otherwise unrelated class I release factors. GGQ is clearly a successful chemical solution for catalyzing peptidyl-tRNA hydrolysis in the highly conserved, RNA-rich PTC of the ribosome.

The carboxyl terminus of eRF1 is involved in facilitating interactions with the class II release factor eRF3 (Merkulova et al. 1999; Kono-

nenko et al. 2008; Cheng et al. 2009). eRF3 itself has a variable amino terminus (Ter-Avanesyan et al. 1993) and a more conserved carboxyl terminus that is directly involved in interactions with the M and C domains of eRF1. Although eRF3 is an essential gene, the carboxy-terminal fragment is sufficient in yeast to complement the deletion of eRF3 (Ter-Avanesyan et al. 1993; Kononenko et al. 2008; Cheng et al. 2009). The nonessential amino terminus has been implicated in binding interactions with PABP and Upf1 and in the prion properties of the factor [PSI$^+$] (Paushkin et al. 1996; Hoshino et al. 1999; Cosson et al. 2002; Ivanov et al. 2008).

In vitro, eRF3 both accelerates peptide release and increases termination efficiency at stop codons in a manner that depends on GTP hydrolysis (Alkalaeva et al. 2006; Eyler and Green 2011). Dissociation of GTP from eRF3 is slowed by eRF1 binding off the ribosome and, as such, eRF1 has been proposed to play the role of a GTP dissociation inhibitor (TDI, for GTP dissociation inhibitor) (Pisareva et al. 2006). The ternary complex, eRF1:eRF3:GTP, next engages the ribosome, triggering GTP hydrolysis (Frolova et al. 1996) ultimately leading to the deposition of the M domain of eRF1 in the PTC. In this scenario, eRF3 is playing a role similar to that of EF-Tu (to which it is closely related) in controlling delivery of a tRNA-like molecule into the PTC. During the delivery process, the presence of a stop codon in the A site can be evaluated by eRF1 to achieve the reported high levels of discrimination (Salas-Marco and Bedwell 2005). These roles for eRF1/eRF3 are incorporated into a model in Figure 2. Interestingly, there is no eRF3 homolog in archaea, where instead aEF1α (the elongation factor equivalent to EF-Tu in bacteria and eEF1α in eukaryotes) is thought to take its role in the termination reaction mechanism (Saito et al. 2010). Finally, as previously alluded to, bacteria have no eRF3 homolog, but instead have a different class 2 release factor, RF3; this factor is more closely related to EF-G than to EF-Tu and does not appear to play an equivalent role to eRF3 in promoting the release reaction (Freistroffer et al. 1997; Zaher and Green 2011) or as discussed next, in the downstream events of recycling.

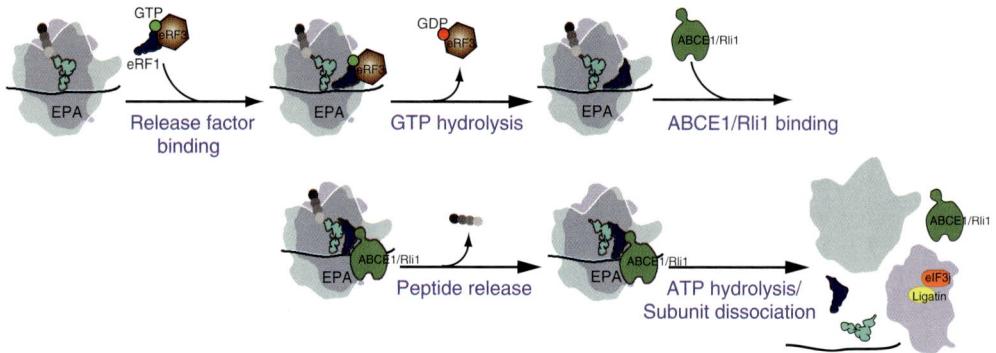

Figure 2. Model of the eukaryotic translation termination and recycling pathways. In this model the large ribosomal subunit is drawn as transparent to visualize tRNAs, factors, and mRNA binding to the decoding center at the interface between the large and small subunit. Throughout, GTP is depicted as a green ball and GDP as a red ball; also, the positions of the mRNA, tRNAs, and factors are drawn for clarity and are not meant to specify their exact places on the ribosome. On recognition of a stop codon, the eRF1:eRF3:GTP ternary complex binds to the A site of the ribosome in a preaccommodated state, GTP hydrolysis occurs, and eRF3 is released. ABCE1/Rli1 binds and facilitates the accommodation of eRF1 into an optimally active configuration.

EUKARYOTIC TRANSLATION RECYCLING (AND SOME CONNECTIONS TO REINITIATION)

Recycling is the process that takes place once the completed polypeptide chain has been released. At this stage, the 80S ribosome still is bound to the mRNA, the now deacylated tRNA, and likely the class I release factor eRF1. X-ray structures from the bacterial system indicate that both pre- and posttermination complexes are found in a preratcheted (or classical) configuration (Korostelev et al. 2008; Laurberg et al. 2008; Weixlbaumer et al. 2008; Jin et al. 2010) and so this seems likely also to be the case in the eukaryotic system. At this stage the ribosomal subunits must be dissociated and the mRNA and deacylated tRNA released to regenerate the necessary components for subsequent rounds of translation. In this section we describe recent studies that have greatly increased our understanding of these events in eukaryotes.

It is worth noting that in some cases full dissociation of the ribosomal complex will occur following termination, whereas in other cases partial dissolution of the complex will allow for a class of events that is loosely termed "reinitiation." Historically, reinitiation is a term used to describe a process wherein ribosomes translate two or more ORFs in a transcript without undergoing complete recycling between these events (Kozak 1984). Incomplete recycling could potentially also take place at the termination codon of an mRNA containing a single ORF, allowing scanning along the 3' untranslated region (UTR) and facilitating transfer of the 40S subunit to the 5' UTR and a subsequent round of translation of the same ORF. What seems to be most firmly established is that PABP, eIF4G, and eIF4E interact with one another specifically (Tarun and Sachs 1996), thus potentially bringing into close proximity the 5' and 3' ends of an mRNA. As outlined by Hinnebusch and Lorsch (2012), however, the mechanistic implications of these interactions are not yet fully understood. Such binding interactions could serve mostly to protect the mRNA from degradation or could additionally be important in promoting translation, decay, or other processes (Bernstein et al. 1989; Sachs and Davis 1989; Jacobson 1996). Although the field has long discussed the potentiating role of these interactions in promoting translation, more recent studies (also discussed by Hinnebusch and Lorsch 2012) have provided evidence that closed-loop mRNA formation via the PABP-eIF4G interaction is nonessential in vivo (Tarun et al. 1997), and may serve a redundant function in recruiting eIF4F to mRNA during initiation (Park et al. 2011). It will be important moving forward to determine the particular biochemical benefits that are specified by communication between the 5′ and 3′ ends of an mRNA.

THE ELUSIVE RECYCLING FACTOR IN EUKARYOTES

Recycling is reasonably well defined in bacterial systems and involves a specialized factor, ribosome recycling factor (RRF), that interacts with the posttermination ribosome complex following the stimulated dissociation of the class I release factor (RF1 or RF2) by RF3. RRF interacts with a ratcheted state of the ribosome (with a deacylated tRNA bound in the P/E state) and destabilizes intersubunit bridging interactions (Gao et al. 2005; Dunkle et al. 2011). EF-G:GTP promotes subunit dissociation and IF3 binds to the resulting small ribosomal subunit to stabilize the dissociation event and promote release of the deacylated tRNA and the mRNA (Hirokawa et al. 2005; Peske et al. 2005; Zavialov et al. 2005). At this stage, the dissociated ribosomal subunits are ready for the next round of initiation.

In eukaryotes, there is no homolog of RRF, and, as discussed above, the termination factors are both structurally and mechanistically distinct from their equivalents in bacteria. Unlike in bacteria, eRF3 does not appear to promote the departure of the class I release factor eRF1 (a known biochemical role for RF3 [Freistroffer et al. 1997], and indeed, current evidence suggests that eRF1 remains associated with the ribosomal complex following termination (Pisarev et al. 2007). This posttermination complex containing bound eRF1 and a deacylated tRNA, potentially in an unratcheted state, is what must be targeted by the recycling machinery in eukaryotes. Although initial reports argued that eIF3 might play an active role in recycling in higher eukaryotes (Pisarev et al. 2007), the steady-state nature of these studies left open the possibility that eIF3 merely functioned to stabilize dissociated subunits by directly binding to the subunit interface. Such a view is consistent with a similar role for IF3 in bacteria and with the fact that eIF3 does not possess any intrinsic capacity for coupling energy to the process of subunit dissociation. Subsequent studies by several groups identified the multifunctional ABC-family protein ABCE1 found in eukaryotes and archaea as a likely candidate for promoting ribosomal recycling (Pisarev et al. 2010; Barthelme et al. 2011). This cytosolic ATPase is highly conserved throughout the eukaryotic kingdom (Kerr 2004; Dean and Annilo 2005) and is essential in all organisms tested (Dong et al. 2004; Zhao et al. 2004; Andersen and Leevers 2007).

HOW DOES RECYCLING ACTUALLY WORK?

Mechanistic insights into ribosome recycling recently came unexpectedly from studies of several proteins implicated in the no-go decay (NGD) pathway, through which mRNAs with translating ribosomes stalled on them are degraded. These two factors, Dom34 and Hbs1, are related to the eukaryotic termination factors eRF1 and eRF3, respectively, and appear to be important in triggering the NGD response (Doma and Parker 2006). Biochemical studies in reconstituted systems (Shoemaker et al. 2010; Pisareva et al. 2011) established that these factors bind to the A site of ribosomal complexes to promote subunit dissociation. Although Dom34 is related to eRF1, it lacks the codon recognition motif (NIKS) that is responsible for discriminating between sense and stop codons, and lacks the full extension of the M domain (with the GGQ motif) that promotes peptide release (Lee et al. 2007; Graille et al. 2008). Consistent with this structural view, the Dom34:Hbs1 complex promotes subunit dissociation (and not peptide release) in a codon-independent manner (Shoemaker et al. 2010). These ideas subsequently led to the demonstration that the termination factors themselves (eRF1 and eRF3) trigger slow rates of subunit dissociation (Shoemaker et al. 2010). These data make some sense given that earlier studies had shown that eRF1 is retained following the termination reaction (Pisarev et al. 2007) and is required for eukaryotic recycling (Pisarev et al. 2010).

What then does ABCE1 (Rli1 in yeast) do to promote ribosome recycling? Although the canonical release factors do appear to possess some intrinsic ribosome recycling activity, the addition of Rli1 to this reaction substantially increases the rate of the observed reaction in vitro (Pisarev et al. 2010; Shoemaker and Green 2011) and this activity depends on ATP hydrolysis. As for related ABC-family ATPases, ABCE1/Rli1 is proposed to somehow convert the chemical energy from ATP hydrolysis into mechanical motions that can separate the subunits. Similarly, the Dom34-dependent subunit dissociation activity is also substantially promoted (~20-fold) by the presence of Rli1 (Shoemaker and Green 2011). Biochemical studies on the equivalent complex (Pelota:ABCE1) in a mammalian in vitro reconstituted system (Pisareva et al. 2011) are markedly consistent with these studies in yeast with Dom34:Rli1.

In addition to a role in recycling, Rli1 has also been shown to directly promote the rate of peptide release by eRF1:eRF3, in a manner that does not depend on ATP hydrolysis (Khoshnevis et al. 2010; Shoemaker and Green 2011). It seems reasonable to speculate that by promoting the release activity, Rli1 can help in staging the sequential events of termination and recycling. With both sets of factors, eRF1:eRF3 and Dom34:Hbs1, there are data to indicate that eRF3 or Hbs1 must be able to hydrolyze GTP in order for recycling to occur (Pisarev et al. 2010, Shoemaker et al. 2010; Pisareva et al. 2011; Shoemaker and Green 2011). Moreover, following GTP hydrolysis by these factors, their affinity for ribosomes is decreased and the factors are readily chased from the complex (Shoemaker and Green 2011). These data can be put together in a model (Fig. 2) in which under normal conditions, the eRF1:eRF3 complex recognizes stop codons and GTP hydrolysis by eRF3 permits departure of the GDP form of the factor (in a fashion akin to tRNA delivery by EF-Tu). Some form of accommodation takes place wherein the GGQ end of the release factor swings into the catalytic center of the large subunit. Peptide release is then catalyzed, stimulated by an ATP-independent activity of Rli1. Finally, ATP hydrolysis on Rli1 is coupled to subunit dissociation. Separated subunits then are bound by available initiation factors that prepare them for subsequent rounds of initiation or reinitiation (Pisarev et al. 2007). Deacylated tRNA and mRNA are likely dissociated from the isolated small subunits following recycling, with their departure enhanced by Ligatin (also known as eIF2D) and, to a lesser extent, by the pair of proteins MCT-1/DENR that are related in sequence to the different halves of Ligatin (Dmitriev et al. 2010; Skabkin et al. 2010). These factors may function by stabilizing the open state of the 40S subunit, from which tRNA would be expected to dissociate more rapidly. Release of tRNA and mRNA from recycled 40S subunits can also be stimulated in vitro by eIF1, eIF1A, and the j-subunit of eIF3 (Pisarev et al. 2007). It is currently unclear whether this mechanism involving conventional initiation factors or that involving Ligatin operates in vivo to complete the recycling process.

We note that the overall scheme we outline is similar in many ways to that for tRNA selection in bacteria, which is also facilitated by a G-protein factor, EF-Tu (Pape et al. 1998). The Dom34:Hbs1:Rli1 system shares many similarities with the tRNA selection pathway, although substrate recognition is less well understood. The most significant insights into substrate selection by these proteins came from studies showing a rather strict length dependence for the mRNA extending 3' of the stall site on the ribosome; the less mRNA present in this position, the more potent the recycling by Pelota:Hbs1:ABCE1 in the mammalian system (Pisareva et al. 2011). These observations were broadly corroborated in the yeast system where it was further established that the length dependence was specifically conferred by the presence of Hbs1 (Shoemaker and Green 2011). The idea that NGD depends on recognition of truncated mRNAs is easily reconciled by literature indicating that the stalling of ribosomes during NGD (and also in nonstop decay [NSD]) eventually leads to endonucleolytic cleavage (Doma and Parker 2006). Once cleavage has occurred, 80S ribosomal complexes attached to truncated mRNAs are readily identified as targets for downstream events in NGD or NSD. Genome-wide approaches are likely to increase further our understanding of the structural features of ribosomal complexes that lead to the initiation of NGD.

STRUCTURAL INSIGHTS INTO EUKARYOTIC TERMINATION AND RECYCLING

Structural insights on termination and recycling have been significant during the past few years. Isolated partial structures of the complexed termination factors eRF1:eRF3 and their homologs Dom34:Hbs1 have both been determined recently (Cheng et al. 2009; Chen et al. 2010). The eRF1:eRF3 structure includes the full-length eRF1 species, but eRF3 lacks the nonessential amino-terminal domain as well as the GTPase domain; the Dom34:Hbs1 structure includes all domains of each protein. For both *Schizosaccharomyces pombe* and human eRF1:eRF3 complexes and for the Dom34:Hbs1 complex, the factors associate with each other via their carboxy-terminal domains. Moreover, in both cases, binding of the GTPase factors (eRF3/Hbs1) resulted in gross conformational changes in eRF1/Dom34, resulting in the latter more closely resembling the shape of a tRNA molecule.

Consistently, cryo-EM reconstructions show that eRF1:eRF3 and Dom34:Hbs1 bind to eukaryotic ribosomes in a manner similar to tRNA:eEF1A (Becker et al. 2011). The same mode of binding is also observed in archaea (Kobayashi et al. 2010), indicating conservation of these processes. Following GTP hydrolysis, either eRF3 or Hbs1 dissociates and Rli1/ABCE1 binds. The binding of this factor seems to facilitate the positioning of the central domain of Dom34 in the PTC (and thus by analogy, eRF1) (Becker et al. 2012); such positioning could readily explain the acceleration of peptide release promoted by Rli1 (Shoemaker and Green 2011). Subsequent ATP hydrolysis by Rli1 likely drives subunit dissociation, although a full molecular understanding of this process remains undefined. Beckmann and colleagues have further proposed that Rli1 drives Dom34 and/or eRF1 through the subunit interface region in a fashion similar to the movements of RRF promoted by EF-G in bacteria. If true, this motion could disrupt critical subunit bridges and lead directly to subunit splitting (Gao et al. 2005).

Interestingly, ABCE1/Rli1 was first studied in yeast as a factor involved in initiation. Rli1 is stably associated with the multifactor complex (MFC) free of ribosomes and with native 40S subunits in extracts. Depletion of the protein from cells leads to polysome runoff, characteristic of an initiation defect, and reduces association of MFC components with native 40S subunits without affecting MFC integrity. In addition, Rli1 promotes assembly of the 43S PIC (Dong et al. 2004), and results consistent with this conclusion were reported for mammalian ABCE1 (Chen et al. 2006). The defect in PIC formation might be an indirect consequence of defective recycling, although enrichment of larger rather than smaller polysomes would be expected from the failure to dissociate 80S posttermination complexes at stop codons, and the free 40S subunits in such cells should not be defective for MFC binding. Hence, an alternative view is that ABCE1/Rli1 operates at the interface between termination, recycling, and initiation by helping to recruit the MFC to the free 40S subunits generated by the recycling reaction. Rli1 also has been reported to function in ribosome biogenesis (Dong et al. 2004; Yarunin et al. 2005). It will be important to determine which of ABCE1/Rli1's functions makes it an essential protein in vivo.

FUTURE DIRECTIONS

Many mechanistic questions remain to be addressed concerning elongation, termination, and recycling. For elongation, more in-depth kinetic analyses are needed to elucidate the function of eIF5A and its hypusine modification in translation initiation, elongation, and/or termination. To help define the role of eIF5A in elongation, it will be helpful to obtain additional insights regarding the timing of E-site tRNA release and eIF5A binding during the elongation cycle. Also, clarification of the role of eEF3 in fungal translation elongation may provide insights into bacterial and mammalian translation elongation that apparently lack a requirement for a comparable ATPase.

For termination, we lack substantial information on the communication between stop codon decoding and peptide release. Despite significant efforts, it remains to be fully determined how stop codons are recognized by eRF1 in the A site. Minimal binding energy is readily attributed to the relatively limited contacts thought to exist between the A-site codon and eRF1 in the decoding center, and yet these subtle differences in binding are somehow communicated to stimulate peptide release at an active site >75 Å away. Although direct comparison to the bacterial process is not warranted given the lack of conservation of the involved factors, the induced-fit mechanisms documented there (Youngman et al. 2007; He and Green 2010) likely play a similar role in the eukaryotic process. The effect of GTP hydrolysis on termination is also poorly understood. eRF1 alone can catalyze termination, but the rates of both termination and recycling are increased in the presence of eRF3. What conformational changes does eRF3 facilitate (either in eRF1 or the ribosome) that allow for more efficient catalysis? Furthermore, how does GTP hydrolysis by eRF3 (or Hbs1) commit or stage the ribosome for termination and coupling to recycling? These are some of the questions that will need to be addressed to decipher the mechanism of signal transduction between codon recognition, GTP hydrolysis, and peptide release in eukaryotes.

Finally, for recycling, how does Rli1/ABCE1 facilitate subunit dissociation? Rli1/ABCE1 contains two asymmetric ATPase sites and activity in at least one of these sites is required for cooperative function between Rli1 and eRF1 during recycling (Barthelme et al. 2011). The differential activity of these sites could be key in regulating the distinct functions of Rli1 in termination, recycling, and initiation. In vitro systems are well poised to address these questions, as many active site mutations in Rli1 are incompatible with in vivo analysis in this essential protein. It is worth noting that ABCE1/Rli1 is one of the few examples of an ATPase that directly engages the ribosome to promote a core translational event (the other being the fungal-specific elongation factor eEF3). Which other ATPases (e.g., Upf1 [Ghosh et al. 2010]) or factors (e.g.,

PABP [Hoshino et al. 1999; Cosson et al. 2002] or Tpa1 [Keeling et al. 2006]) in the cell might also directly engage the ribosome and translational GTPases to promote the core mechanisms of termination and recycling? Such insights will expand our view of the extensive network of interactions that dictate the translational output of the cell.

REFERENCES

*Reference is also in this collection.

Alkalaeva EZ, Pisarev AV, Frolova LY, Kisselev LL, Pestova TV. 2006. In vitro reconstitution of eukaryotic translation reveals cooperativity between release factors eRF1 and eRF3. *Cell* **125:** 1125–1136.

Anand M, Chakraburtty K, Marton MJ, Hinnebusch AG, Kinzy TG. 2003. Functional interactions between yeast translation eukaryotic elongation factor (eEF) 1A and eEF3. *J Biol Chem* **278:** 6985–6991.

Anand M, Balar B, Ulloque R, Gross SR, Kinzy TG. 2006. Domain and nucleotide dependence of the interaction between *Saccharomyces cerevisiae* translation elongation factors 3 and 1A. *J Biol Chem* **281:** 32318–32326.

Andersen DS, Leevers SJ. 2007. The essential *Drosophila* ATP-binding cassette domain protein, pixie, binds the 40 S ribosome in an ATP-dependent manner and is required for translation initiation. *J Biol Chem* **282:** 14752–14760.

Andersen CB, Becker T, Blau M, Anand M, Halic M, Balar B, Mielke T, Boesen T, Pedersen JS, Spahn CM, et al. 2006. Structure of eEF3 and the mechanism of transfer RNA release from the E-site. *Nature* **443:** 663–668.

Atkinson GC, Baldauf SL, Hauryliuk V. 2008. Evolution of nonstop, no-go and nonsense-mediated mRNA decay and their termination factor-derived components. *BMC Evol Biol* **8:** 290.

Bailly M, de Crecy-Lagard V. 2010. Predicting the pathway involved in post-translational modification of elongation factor P in a subset of bacterial species. *Biol Direct* **5:** 3.

Balagopal V, Parker R. 2011. Stm1 modulates translation after 80S formation in *Saccharomyces cerevisiae*. *RNA* **17:** 835–842.

Barthelme D, Dinkelaker S, Albers SV, Londei P, Ermler U, Tampe R. 2011. Ribosome recycling depends on a mechanistic link between the FeS cluster domain and a conformational switch of the twin-ATPase ABCE1. *Proc Natl Acad Sci* **108:** 3228–3233.

Becker T, Armache JP, Jarasch A, Anger AM, Villa E, Sieber H, Motaal BA, Mielke T, Berninghausen O, Beckmann R. 2011. Structure of the no-go mRNA decay complex Dom34-Hbs1 bound to a stalled 80S ribosome. *Nat Struct Mol Biol* **18:** 715–720.

Becker T, Franckenberg S, Wickles S, Shoemaker CJ, Anger AM, Armache JP, Sieber H, Ungewickell C, Berninghausen O, Daberkow I, et al. 2012. Structural basis of highly conserved ribosome recycling in eukaryotes and archaea. *Nature* **482:** 501–506.

Ben-Shem A, Jenner L, Yusupova G, Yusupov M. 2010. Crystal structure of the eukaryotic ribosome. *Science* **330**: 1203–1209.

Ben-Shem A, Garreau de Loubresse N, Melnikov S, Jenner L, Yusupova G, Yusupov M. 2011. The structure of the eukaryotic ribosome at 3.0 A resolution. *Science* **334**: 1524–1529.

Bernstein P, Peltz SW, Ross J. 1989. The poly(A)-poly(A)-binding protein complex is a major determinant of mRNA stability in vitro. *Mol Cell Biol* **9**: 659–670.

Blaha G, Stanley RE, Steitz TA. 2009. Formation of the first peptide bond: The structure of EF-P bound to the 70S ribosome. *Science* **325**: 966–970.

Bulygin KN, Khairulina YS, Kolosov PM, Ven'yaminova AG, Graifer DM, Vorobjev YN, Frolova LY, Kisselev LL, Karpova GG. 2010. Three distinct peptides from the N domain of translation termination factor eRF1 surround stop codon in the ribosome. *RNA* **16**: 1902–1914.

Carlberg U, Nilsson A, Nygard O. 1990. Functional properties of phosphorylated elongation factor 2. *Eur J Biochem* **191**: 639–645.

Chavatte L, Seit-Nebi A, Dubovaya V, Favre A. 2002. The invariant uridine of stop codons contacts the conserved NIKSR loop of human eRF1 in the ribosome. *EMBO J* **21**: 5302–5311.

Chen CM, Behringer RR. 2004. Ovca1 regulates cell proliferation, embryonic development, and tumorigenesis. *Genes Dev* **18**: 320–332.

Chen ZQ, Dong J, Ishimura A, Daar I, Hinnebusch AG, Dean M. 2006. The essential vertebrate ABCE1 protein interacts with eukaryotic initiation factors. *J Biol Chem* **281**: 7452–7457.

Chen L, Muhlrad D, Hauryliuk V, Cheng Z, Lim MK, Shyp V, Parker R, Song H. 2010. Structure of the Dom34-Hbs1 complex and implications for no-go decay. *Nat Struct Mol Biol* **17**: 1233–1240.

Chen C, Stevens B, Kaur J, Smilansky Z, Cooperman BS, Goldman YE. 2011. Allosteric vs. spontaneous exit-site (E-site) tRNA dissociation early in protein synthesis. *Proc Natl Acad Sci* **108**: 16980–16985.

Cheng Z, Saito K, Pisarev AV, Wada M, Pisareva VP, Pestova TV, Gajda M, Round A, Kong C, Lim M, et al. 2009. Structural insights into eRF3 and stop codon recognition by eRF1. *Genes Dev* **23**: 1106–1118.

Cosson B, Couturier A, Chabelskaya S, Kiktev D, Inge-Vechtomov S, Philippe M, Zhouravleva G. 2002. Poly(A)-binding protein acts in translation termination via eukaryotic release factor 3 interaction and does not influence [PSI+] propagation. *Mol Cell Biol* **22**: 3301–3315.

Dean M, Annilo T. 2005. Evolution of the ATP-binding cassette (ABC) transporter superfamily in vertebrates. *Ann Rev Genomics Hum Genet* **6**: 123–142.

Dias CA, Gregio AP, Rossi D, Galvao FC, Watanabe TF, Park MH, Valentini SR, Zanelli CF. 2012. eIF5A interacts functionally with eEF2. *Amino Acids* **42**: 697–702.

Dmitriev SE, Terenin IM, Andreev DE, Ivanov PA, Dunaevsky JE, Merrick WC, Shatsky IN. 2010. GTP-independent tRNA delivery to the ribosomal P-site by a novel eukaryotic translation factor. *J Biol Chem* **285**: 26779–26787.

Doma MK, Parker R. 2006. Endonucleolytic cleavage of eukaryotic mRNAs with stalls in translation elongation. *Nature* **440**: 561–564.

Dong J, Lai R, Nielsen K, Fekete CA, Qiu H, Hinnebusch AG. 2004. The essential ATP-binding cassette protein RLI1 functions in translation by promoting preinitiation complex assembly. *J Biol Chem* **279**: 42157–42168.

Dunkle JA, Wang L, Feldman MB, Pulk A, Chen VB, Kapral GJ, Noeske J, Richardson JS, Blanchard SC, Cate JH. 2011. Structures of the bacterial ribosome in classical and hybrid states of tRNA binding. *Science* **332**: 981–984.

Eyler DE, Green R. 2011. Distinct response of yeast ribosomes to a miscoding event during translation. *RNA* **17**: 925–932.

Fan-Minogue H, Du M, Pisarev AV, Kallmeyer AK, Salas-Marco J, Keeling KM, Thompson SR, Pestova TV, Bedwell DM. 2008. Distinct eRF3 requirements suggest alternate eRF1 conformations mediate peptide release during eukaryotic translation termination. *Mol Cell* **30**: 599–609.

Freistroffer DV, Pavlov MY, MacDougall J, Buckingham RH, Ehrenberg M. 1997. Release factor RF3 in *E. coli* accelerates the dissociation of release factors RF1 and RF2 from the ribosome in a GTP-dependent manner. *EMBO J* **16**: 4126–4133.

Frolova L, Le Goff X, Zhouravleva G, Davydova E, Philippe M, Kisselev L. 1996. Eukaryotic polypeptide chain release factor eRF3 is an eRF1- and ribosome-dependent guanosine triphosphatase. *RNA* **2**: 334–341.

Frolova LY, Tsivkovskii RY, Sivolobova GF, Oparina NY, Serpinsky OI, Blinov VM, Tatkov SI, Kisselev LL. 1999. Mutations in the highly conserved GGQ motif of class 1 polypeptide release factors abolish ability of human eRF1 to trigger peptidyl-tRNA hydrolysis. *RNA* **5**: 1014–1020.

Gao N, Zavialov AV, Li W, Sengupta J, Valle M, Gursky RP, Ehrenberg M, Frank J. 2005. Mechanism for the disassembly of the posttermination complex inferred from cryo-EM studies. *Mol Cell* **18**: 663–674.

Gao YG, Selmer M, Dunham CM, Weixlbaumer A, Kelley AC, Ramakrishnan V. 2009. The structure of the ribosome with elongation factor G trapped in the posttranslocational state. *Science* **326**: 694–699.

Ghosh S, Ganesan R, Amrani N, Jacobson A. 2010. Translational competence of ribosomes released from a premature termination codon is modulated by NMD factors. *RNA* **16**: 1832–1847.

Glick BR, Ganoza MC. 1975. Identification of a soluble protein that stimulates peptide bond synthesis. *Proc Natl Acad Sci* **72**: 4257–4260.

Graille M, Chaillet M, van Tilbeurgh H. 2008. Structure of yeast Dom34: A protein related to translation termination factor Erf1 and involved in No-Go decay. *J Biol Chem* **283**: 7145–7154.

Gregio AP, Cano VP, Avaca JS, Valentini SR, Zanelli CF. 2009. eIF5A has a function in the elongation step of translation in yeast. *Biochem Biophys Res Commun* **380**: 785–790.

He SL, Green R. 2010. Visualization of codon-dependent conformational rearrangements during translation termination. *Nat Struct Mol Biol* **17**: 465–470.

Henderson A, Hershey JW. 2011. Eukaryotic translation initiation factor (eIF) 5A stimulates protein synthesis in *Saccharomyces cerevisiae. Proc Natl Acad Sci* **108:** 6415–6419.

* Hinnebusch AG, Lorsch JR. 2012. The mechanism of eukaryotic translation initiation: New insights and challenges. *Cold Spring Harb Perspect Biol* doi: 10.1101/ cshperspect.a011544.

Hirokawa G, Nijman RM, Raj VS, Kaji H, Igarashi K, Kaji A. 2005. The role of ribosome recycling factor in dissociation of 70S ribosomes into subunits. *RNA* **11:** 1317– 1328.

Hoshino S, Imai M, Kobayashi T, Uchida N, Katada T. 1999. The eukaryotic polypeptide chain releasing factor (eRF3/ GSPT) carrying the translation termination signal to the 3′-Poly(A) tail of mRNA. Direct association of eRF3/ GSPT with polyadenylate-binding protein. *J Biol Chem* **274:** 16677–16680.

Ingolia NT, Ghaemmaghami S, Newman JR, Weissman JS. 2009. Genome-wide analysis in vivo of translation with nucleotide resolution using ribosome profiling. *Science* **324:** 218–223.

Ivanov PV, Gehring NH, Kunz JB, Hentze MW, Kulozik AE. 2008. Interactions between UPF1, eRFs, PABP and the exon junction complex suggest an integrated model for mammalian NMD pathways. *EMBO J* **27:** 736–747.

Jacobson A. 1996. Poly(A) metabolism and translation: The closed-looped model. In *Translational control* (ed. Hershey JWB, Mathews M, Sonenberg N), pp. 505– 548. Cold Spring Harbor Laboratory Press, Cold Spring Harbor, NY.

Jao DL, Chen KY. 2006. Tandem affinity purification revealed the hypusine-dependent binding of eukaryotic initiation factor 5A to the translating 80S ribosomal complex. *J Cell Biochem* **97:** 583–598.

Jenner L, Demeshkina N, Yusupova G, Yusupov M. 2010. Structural rearrangements of the ribosome at the tRNA proofreading step. *Nat Struct Mol Biol* **17:** 1072–1078.

Jin H, Kelley AC, Loakes D, Ramakrishnan V. 2010. Structure of the 70S ribosome bound to release factor 2 and a substrate analog provides insights into catalysis of peptide release. *Proc Natl Acad Sci* **107:** 8593–8598.

Kang HA, Hershey JW. 1994. Effect of initiation factor eIF-5A depletion on protein synthesis and proliferation of *Saccharomyces cerevisiae. J Biol Chem* **269:** 3934–3940.

Keeling KM, Salas-Marco J, Osherovich LZ, Bedwell DM. 2006. Tpa1p is part of an mRNP complex that influences translation termination, mRNA deadenylation, and mRNA turnover in *Saccharomyces cerevisiae. Mol Cell Biol* **26:** 5237–5248.

Kemper WM, Berry KW, Merrick WC. 1976. Purification and properties of rabbit reticulocyte protein synthesis initiation factors M2Bα and M2Bβ. *J Biol Chem* **251:** 5551–5557.

Kerr ID. 2004. Sequence analysis of twin ATP binding cassette proteins involved in translational control, antibiotic resistance, and ribonuclease L inhibition. *Biochem Biophys Res Commun* **315:** 166–173.

Khoshnevis S, Gross T, Rotte C, Baierlein C, Ficner R, Krebber H. 2010. The iron-sulphur protein RNase L inhibitor functions in translation termination. *EMBO Rep* **11:** 214–219.

Kim KK, Hung LW, Yokota H, Kim R, Kim SH. 1998. Crystal structures of eukaryotic translation initiation factor 5A from *Methanococcus jannaschii* at 1.8 Å resolution. *Proc Natl Acad Sci* **95:** 10419–10424.

Kimata Y, Kohno K. 1994. Elongation factor 2 mutants deficient in diphthamide formation show temperature-sensitive cell growth. *J Biol Chem* **269:** 13497–13501.

Klinge S, Voigts-Hoffmann F, Leibundgut M, Arpagaus S, Ban N. 2011. Crystal structure of the eukaryotic 60S ribosomal subunit in complex with initiation factor 6. *Science* **334:** 941–948.

Kobayashi K, Kikuno I, Kuroha K, Saito K, Ito K, Ishitani R, Inada T, Nureki O. 2010. Structural basis for mRNA surveillance by archaeal Pelota and GTP-bound EF1α complex. *Proc Natl Acad Sci* **107:** 17575–17579.

Kolosov P, Frolova L, Seit-Nebi A, Dubovaya V, Kononenko A, Oparina N, Justesen J, Efimov A, Kisselev L. 2005. Invariant amino acids essential for decoding function of polypeptide release factor eRF1. *Nucleic Acids Res* **33:** 6418–6425.

Kononenko AV, Mitkevich VA, Dubovaya VI, Kolosov PM, Makarov AA, Kisselev LL. 2008. Role of the individual domains of translation termination factor eRF1 in GTP binding to eRF3. *Proteins* **70:** 388–393.

Korostelev A, Asahara H, Lancaster L, Laurberg M, Hirschi A, Zhu J, Trakhanov S, Scott WG, Noller HF. 2008. Crystal structure of a translation termination complex formed with release factor RF2. *Proc Natl Acad Sci* **105:** 19684– 19689.

Kozak M. 1984. Selection of initiation sites by eucaryotic ribosomes: Effect of inserting AUG triplets upstream from the coding sequence for preproinsulin. *Nucleic Acids Res* **12:** 3873–3893.

Laurberg M, Asahara H, Korostelev A, Zhu J, Trakhanov S, Noller HF. 2008. Structural basis for translation termination on the 70S ribosome. *Nature* **454:** 852–857.

Lee HH, Kim YS, Kim KH, Heo I, Kim SK, Kim O, Kim HK, Yoon JY, Kim HS, Kim do J, et al. 2007. Structural and functional insights into Dom34, a key component of no-go mRNA decay. *Mol Cell* **27:** 938–950.

Liu S, Milne GT, Kuremsky JG, Fink GR, Leppla SH. 2004. Identification of the proteins required for biosynthesis of diphthamide, the target of bacterial ADP-ribosylating toxins on translation elongation factor 2. *Mol Cell Biol* **24:** 9487–9497.

Liu S, Wiggins JF, Sreenath T, Kulkarni AB, Ward JM, Leppla SH. 2006. Dph3, a small protein required for diphthamide biosynthesis, is essential in mouse development. *Mol Cell Biol* **26:** 3835–3841.

Merkulova TI, Frolova LY, Lazar M, Camonis J, Kisselev LL. 1999. C-terminal domains of human translation termination factors eRF1 and eRF3 mediate their in vivo interaction. *FEBS Lett* **443:** 41–47.

Navarre WW, Zou SB, Roy H, Xie JL, Savchenko A, Singer A, Edvokimova E, Prost LR, Kumar R, Ibba M, et al. 2010. PoxA, yjeK, and elongation factor P coordinately modulate virulence and drug resistance in *Salmonella enterica. Mol Cell* **39:** 209–221.

Nierhaus KH. 1990. The allosteric three-site model for the ribosomal elongation cycle: Features and future. *Biochemistry* **29:** 4997–5008.

Ogle JM, Brodersen DE, Clemons WM Jr, Tarry MJ, Carter AP, Ramakrishnan V. 2001. Recognition of cognate transfer RNA by the 30S ribosomal subunit. *Science* **292:** 897–902.

Ortiz PA, Ulloque R, Kihara GK, Zheng H, Kinzy TG. 2006. Translation elongation factor 2 anticodon mimicry domain mutants affect fidelity and diphtheria toxin resistance. *J Biol Chem* **281:** 32639–32648.

Pape T, Wintermeyer W, Rodnina MV. 1998. Complete kinetic mechanism of elongation factor Tu-dependent binding of aminoacyl-tRNA to the A site of the *E. coli* ribosome. *EMBO J* **17:** 7490–7497.

Park MH, Wolff EC, Smit-McBride Z, Hershey JW, Folk JE. 1991. Comparison of the activities of variant forms of eIF-4D. The requirement for hypusine or deoxyhypusine. *J Biol Chem* **266:** 7988–7994.

Park S, Park JM, Kim S, Kim JA, Shepherd JD, Smith-Hicks CL, Chowdhury S, Kaufmann W, Kuhl D, Ryazanov AG, et al. 2008. Elongation factor 2 and fragile X mental retardation protein control the dynamic translation of Arc/Arg3.1 essential for mGluR-LTD. *Neuron* **59:** 70–83.

Park MH, Nishimura K, Zanelli CF, Valentini SR. 2010. Functional significance of eIF5A and its hypusine modification in eukaryotes. *Amino Acids* **38:** 491–500.

Park EH, Zhang F, Warringer J, Sunnerhagen P, Hinnebusch AG. 2011. Depletion of eIF4G from yeast cells narrows the range of translational efficiencies genome-wide. *BMC Genomics* **12:** 1–18.

Park JH, Johansson HE, Aoki H, Huang B, Kim HY, Ganoza MC, Park MH. 2012. Post-translational modification by β-lysylation is required for the activity of *E. coli* elongation factor P (EF-P). *J Biol Chem* **287:** 2579–2590.

Paushkin SV, Kushnirov VV, Smirnov VN, Ter-Avanesyan MD. 1996. Propagation of the yeast prion-like [psi+] determinant is mediated by oligomerization of the SUP35-encoded polypeptide chain release factor. *EMBO J* **15:** 3127–3134.

Peske F, Rodnina MV, Wintermeyer W. 2005. Sequence of steps in ribosome recycling as defined by kinetic analysis. *Mol Cell* **18:** 403–412.

Pisarev AV, Hellen CU, Pestova TV. 2007. Recycling of eukaryotic posttermination ribosomal complexes. *Cell* **131:** 286–299.

Pisarev AV, Skabkin MA, Pisareva VP, Skabkina OV, Rakotondrafara AM, Hentze MW, Hellen CU, Pestova TV. 2010. The role of ABCE1 in eukaryotic posttermination ribosomal recycling. *Mol Cell* **37:** 196–210.

Pisareva VP, Pisarev AV, Hellen CU, Rodnina MV, Pestova TV. 2006. Kinetic analysis of interaction of eukaryotic release factor 3 with guanine nucleotides. *J Biol Chem* **281:** 40224–40235.

Pisareva VP, Skabkin MA, Hellen CU, Pestova TV, Pisarev AV. 2011. Dissociation by Pelota, Hbs1 and ABCE1 of mammalian vacant 80S ribosomes and stalled elongation complexes. *EMBO J* **30:** 1804–1817.

Rodnina MV, Wintermeyer W. 2009. Recent mechanistic insights into eukaryotic ribosomes. *Curr Opin Cell Biol* **21:** 435–443.

Sachs AB, Davis RW. 1989. The Poly(A) binding protein is required for Poly(A) shortening and 60S ribosomal subunit-dependent translation initiation. *Cell* **58:** 857–867.

Saini P, Eyler DE, Green R, Dever TE. 2009. Hypusine-containing protein eIF5A promotes translation elongation. *Nature* **459:** 118–121.

Saito K, Kobayashi K, Wada M, Kikuno I, Takusagawa A, Mochizuki M, Uchiumi T, Ishitani R, Nureki O, Ito K. 2010. Omnipotent role of archaeal elongation factor 1α (EF1α) in translational elongation and termination, and quality control of protein synthesis. *Proc Natl Acad Sci* **107:** 19242–19247.

Salas-Marco J, Bedwell DM. 2005. Discrimination between defects in elongation fidelity and termination efficiency provides mechanistic insights into translational readthrough. *J Mol Biol* **348:** 801–815.

Schmeing TM, Voorhees RM, Kelley AC, Gao YG, Murphy FVt, Weir JR, Ramakrishnan V. 2009. The crystal structure of the ribosome bound to EF-Tu and aminoacyl-tRNA. *Science* **326:** 688–694.

Schmeing TM, Voorhees RM, Kelley AC, Ramakrishnan V. 2011. How mutations in tRNA distant from the anticodon affect the fidelity of decoding. *Nat Struct Mol Biol* **18:** 432–436.

Schreier MH, Erni B, Staehelin T. 1977. Initiation of mammalian protein synthesis: Purification and characterization of seven initiation factors. *J Mol Biol* **116:** 727–753.

Semenkov YP, Rodnina MV, Wintermeyer W. 1996. The "allosteric three-site model" of elongation cannot be confirmed in a well-defined ribosome system from *Escherichia coli*. *Proc Natl Acad Sci* **93:** 12183–12188.

Shoemaker CJ, Green R. 2011. Kinetic analysis reveals the ordered coupling of translation termination and ribosome recycling in yeast. *Proc Natl Acad Sci* **108:** E1392–E1398.

Shoemaker CJ, Eyler DE, Green R. 2010. Dom34:Hbs1 promotes subunit dissociation and peptidyl-tRNA drop-off to initiate no-go decay. *Science* **330:** 369–372.

Sivan G, Aviner R, Elroy-Stein O. 2011. Mitotic modulation of translation elongation factor 1 leads to hindered tRNA delivery to ribosomes. *J Biol Chem* **286:** 27927–27935.

Skabkin MA, Skabkina OV, Dhote V, Komar AA, Hellen CU, Pestova TV. 2010. Activities of Ligatin and MCT-1/DENR in eukaryotic translation initiation and ribosomal recycling. *Genes Dev* **24:** 1787–1801.

Skogerson L, Engelhardt D. 1977. Dissimilarity in protein chain elongation factor requirements between yeast and rat liver ribosomes. *J Biol Chem* **252:** 1471–1475.

Song H, Mugnier P, Das AK, Webb HM, Evans DR, Tuite MF, Hemmings BA, Barford D. 2000. The crystal structure of human eukaryotic release factor eRF1-mechanism of stop codon recognition and peptidyl-tRNA hydrolysis. *Cell* **100:** 311–321.

Stansfield I, Jones KM, Kushnirov VV, Dagkesamanskaya AR, Poznyakovski AI, Paushkin SV, Nierras CR, Cox BS, Ter-Avanesyan MD, Tuite MF. 1995. The products of the SUP45 (eRF1) and SUP35 genes interact to mediate translation termination in *Saccharomyces cerevisiae*. *EMBO J* **14:** 4365–4373.

Tarun SZ, Sachs AB. 1996. Association of the yeast poly(A) tail binding protein with translation initiation factor eIF-4G. *EMBO J* **15:** 7168–7177.

Tarun SZ, Wells SE, Deardorff JA, Sachs AB. 1997. Translation initiation factor eIF4G mediates in vitro poly

(A) tail-dependent translation. *Proc Natl Acad Sci* **94:** 9046–9051.

Ter-Avanesyan MD, Kushnirov VV, Dagkesamanskaya AR, Didichenko SA, Chernoff YO, Inge-Vechtomov SG, Smirnov VN. 1993. Deletion analysis of the SUP35 gene of the yeast *Saccharomyces cerevisiae* reveals two non-overlapping functional regions in the encoded protein. *Mol Microbiol* **7:** 683–692.

Triana-Alonso FJ, Chakraburtty K, Nierhaus KH. 1995. The elongation factor unique in higher fungi and essential for protein biosynthesis is an E site factor. *J Biol Chem* **270:** 20473–20478.

Uemura S, Aitken CE, Korlach J, Flusberg BA, Turner SW, Puglisi JD. 2010. Real-time tRNA transit on single translating ribosomes at codon resolution. *Nature* **464:** 1012–1017.

Van Dyke N, Pickering BF, Van Dyke MW. 2009. Stm1p alters the ribosome association of eukaryotic elongation factor 3 and affects translation elongation. *Nucleic Acids Res* **37:** 6116–6125.

Voorhees RM, Schmeing TM, Kelley AC, Ramakrishnan V. 2010. The mechanism for activation of GTP hydrolysis on the ribosome. *Science* **330:** 835–838.

Walden WE, Thach RE. 1986. Translational control of gene expression in a normal fibroblast. Characterization of a subclass of mRNAs with unusual kinetic properties. *Biochemistry* **25:** 2033–2041.

Waung MW, Pfeiffer BE, Nosyreva ED, Ronesi JA, Huber KM. 2008. Rapid translation of Arc/Arg3.1 selectively mediates mGluR-dependent LTD through persistent increases in AMPAR endocytosis rate. *Neuron* **59:** 84–97.

Webb TR, Cross SH, McKie L, Edgar R, Vizor L, Harrison J, Peters J, Jackson IJ. 2008. Diphthamide modification of eEF2 requires a J-domain protein and is essential for normal development. *J Cell Sci* **121:** 3140–3145.

Weixlbaumer A, Jin H, Neubauer C, Voorhees RM, Petry S, Kelley AC, Ramakrishnan V. 2008. Insights into translational termination from the structure of RF2 bound to the ribosome. *Science* **322:** 953–956.

Yanagisawa T, Sumida T, Ishii R, Takemoto C, Yokoyama S. 2010. A paralog of lysyl-tRNA synthetase aminoacylates a conserved lysine residue in translation elongation factor P. *Nat Struct Mol Biol* **17:** 1136–1143.

Yarunin A, Panse VG, Petfalski E, Dez C, Tollervey D, Hurt EC. 2005. Functional link between ribosome formation and biogenesis of iron-sulfur proteins. *EMBO J* **24:** 580–588.

Youngman EM, He SL, Nikstad LJ, Green R. 2007. Stop codon recognition by release factors induces structural rearrangement of the ribosomal decoding center that is productive for peptide release. *Mol Cell* **28:** 533–543.

Zaher HS, Green R. 2011. A primary role for release factor 3 in quality control during translation elongation in *Escherichia coli*. *Cell* **147:** 396–408.

Zanelli CF, Maragno AL, Gregio AP, Komili S, Pandolfi JR, Mestriner CA, Lustri WR, Valentini SR. 2006. eIF5A binds to translational machinery components and affects translation in yeast. *Biochem Biophys Res Commun* **348:** 1358–1366.

Zavialov AV, Hauryliuk VV, Ehrenberg M. 2005. Splitting of the posttermination ribosome into subunits by the concerted action of RRF and EF-G. *Mol Cell* **18:** 675–686.

Zhao Z, Fang LL, Johnsen R, Baillie DL. 2004. ATP-binding cassette protein E is involved in gene transcription and translation in *Caenorhabditis elegans*. *Biochem Biophys Res Commun* **323:** 104–111.

Zhouravleva G, Frolova L, Le Goff X, Le Guellec R, Inge-Vechtomov S, Kisselev L, Philippe M. 1995. Termination of translation in eukaryotes is governed by two interacting polypeptide chain release factors, eRF1 and eRF3. *EMBO J* **14:** 4065–4072.

Single-Molecule Analysis of Translational Dynamics

Alexey Petrov[1], Jin Chen[1,2], Seán O'Leary[1], Albert Tsai[1,2], and Joseph D. Puglisi[1]

[1]Department of Structural Biology, Stanford University School of Medicine, Stanford, California 94305-5126

[2]Department of Applied Physics, Stanford University, Stanford, California 94305-4090

Correspondence: puglisi@stanford.edu

Decades of extensive biochemical and biophysical research have outlined the mechanism of translation. Rich structural studies have provided detailed snapshots of the translational machinery at all phases of the translation cycle. However, the relationship between structural dynamics, composition, and function remains unknown. The multistep nature of each stage of the translation cycle results in rapid desynchronization of individual ribosomes, thus hindering elucidation of the underlying mechanisms by conventional bulk biophysical and biochemical methods. Single-molecule approaches unsusceptible to these complications have led to the first glances at both compositional and conformational dynamics on the ribosome and their impact on translational control. These experiments provide the necessary link between static structure and mechanism, often providing new perspectives. Here we review recent advances in the field and their relationship to structural and biochemical data.

Translation and its regulation are intrinsically dynamic processes. In all organisms, to initiate translation, ribosomes must assemble from isolated subunits and an initiator transfer RNA (tRNA) on a messenger RNA (mRNA) at a specific start codon to establish a reading frame; protein factors guide this process. Elongation occurs through selection by the ribosome of cognate aminoacyl tRNAs, subsequent positioning of tRNAs for peptide bond formation chemistry, and movements of the tRNAs and mRNAs with respect to the codon (translocation). The directional process is iterative until termination at a stop codon, where the protein chain is released, and the ribosomal particle disassembled and recycled. Multiple ribosomes form higher-order polysomes on a single mRNA, with their own intrinsic dynamics.

Dynamics are central to the mechanism and control of translation. Here we explicitly define dynamics as time-dependent changes in either composition or conformation of the translational machinery. Conformational dynamics in chemical systems are governed by an array of processes with vastly different timescales. Generally, dynamic processes become slower as they involve larger numbers of atoms. These range from electronic motions (timescale 10^{-14} sec), bond vibrations (10^{-13}–10^{-12} sec), through protein side chain or nucleic acid base/sugar local conformational changes (10^{-11}–10^{-6} sec), to larger conformational rearrangements (domain movements, etc.; $10^{-6}-10^2$ sec) that are often functionally cooperative. Compositional dynamics are determined by bimolecular association and dissociation rate constants: bimolecular arrival rates for ligands are governed by intermolecular collision frequencies, electrostatic interactions, and proper binding orientations for productive binding events, whereas dissociation rates are governed by energy barriers for dissociation of noncovalent intermolecular interactions.

Fluctuations in molecular conformation and composition must be harnessed by the ribosome for accurate and rapid translation. The timescales of these dynamic changes dictate the overall rates of translation initiation and elongation: 0.2–0.5 initiation events/sec and elongation rates of 20–40 amino acids/sec in vivo, 1–5 amino acids/sec in vitro (Dennis and Bremer 1974a,b; Underwood et al. 2005). The ribosome uses external sources of free energy during translation—ATP hydrolysis during eukaryotic scanning, GTP hydrolysis by initiation, elongation and termination factors, and peptide bond formation. The free energy released by these irreversible reactions is used to drive the fidelity of initiation and elongation and the directional movement of the ribosome during both processes. The ribosome is thus a molecular motor.

The link between ribosome and ligand dynamics and the control of protein synthesis remains a key mystery of translation. The past decade has witnessed the three-dimensional structures of prokaryotic, archaeal, and eukaryotic ribosomes at atomic resolution. How factors, tRNA, and ligands interact with the ribosome has been revealed by cryo-electron microscopy (cryo-EM) (at lower 6–12 Å resolution) and X-ray diffraction studies (as low as 2.5 Å for 30S from *Thermus thermophilus* [Kurata et al. 2008] and 2.4 Å for 50S from *Haloarcula marismortui* [Ban et al. 2000]). These structures have shown how GTPase factors engage with the 70S ribosomes at a conserved factor-binding site on the large subunit to mediate GTPase activity and subsequent conformational changes. Another key observation of early cryo-EM and more recent structural studies is

that the ribosome adopts two general intersubunit conformations, related by a 6° rotation of the two subunits (Valle et al. 2003; Schuwirth et al. 2005; Agirrezabala et al. 2008; Zhang et al. 2009; Fischer et al. 2010; Dunkle et al. 2011). Peptide bond formation leads to a counterclockwise rotation of the small subunit with respect to the large subunit, and EF-G in the GTP form binds to this state. The peripheral domain L1 region of the ribosome was observed structurally to change its state in correlation with the two ribosomal conformations (Valle et al. 2003; Schuwirth et al. 2005; Agirrezabala et al. 2008), suggesting a coupling of domains within the ribosome. The intersubunit conformation of the ribosome was also correlated by EM to the relative orientations of tRNAs: In the nonrotated state (locked conformation), tRNAs are observed in the classical P site and A site, whereas in the rotated state (unlocked conformation) the tRNAs are in the Noller hybrid states with the 3' ends of the tRNAs moved to the E and P sites and their respective anticodons in the P and A sites (Agirrezabala et al. 2008). These static structural views suggested a further correlation of tRNA and ribosome conformation during translation.

The structural snapshots and prior biochemical studies are suggestive of dynamics during translation, yet experimental methods with resolution in real time are required to observe them directly. Here we focus on single-molecule methodologies that have provided an unprecedented view into the dynamics of prokaryotic translation. In the future, the same techniques can be applied to the study of eukaryotic translation dynamics.

WHY SINGLE-MOLECULE APPROACHES?

Dynamics have been traditionally measured using bulk methods, with signals sensitive to dynamic changes for a large collection of molecules. These signals, such as emission from a fluorescent dye, must be sensitive to conformational or compositional changes of the system as it evolves in time. In the case of translating ribosomes, problems arise in synchronizing a large collection of molecules to detect a change

Cite this article as *Cold Spring Harb Perspect Biol* doi: 10.1101/cshperspect.a011551

in the bulk signal for a specific process. Imagine, for example, a signal that changes on tRNA binding to the A site. In order to detect a change in this signal, the system must be synchronized such that all tRNAs are unbound at the start of the measurement, at which point reaction usually started through rapid mixing. This ensures that the observed time-dependent fluorescence change reports only on the approach to equilibrium from the unbound state, and from this signal kinetic information can be extracted. However, if we want to look at a subsequent tRNA-binding event, we would have to pause the evolution of the system, remix the reagents, and repeat the measurement. In short, dynamics cannot be measured in real time during multiple rounds of elongation. This need for synchronization is a fatal limitation of bulk kinetic investigations to probe multistep dynamic systems.

Single-molecule experiments allow direct measurement of dynamics without the need for synchronization. Commonly used organic fluorophores such as rhodamine derivatives and cyanine dyes emit sufficient numbers of photons to be readily detected with modern cameras at millisecond time resolution. To distinguish weak single-fluorophore fluorescence from background illumination and noise, various techniques such as total internal reflection fluorescence microscopy (TIRF) and zero-mode waveguides (ZMWs) are utilized (Ha 2001; Levene et al. 2003). The rapid diffusion of fluorophores in solution limits observation of the free molecules, because on the timescale of data acquisition a labeled molecule leaves the illumination volume before emitting enough photons to be observed.

The problem of diffusion can be turned into an advantage by spatially constraining the system through surface immobilization—in translation experiments this is most often accomplished using biotin-streptavidin interactions to bind mRNAs or ribosomes to an optically transparent surface. Let us take a simple example of the bimolecular binding event of a dye-labeled tRNA to an immobilized ribosome. In the single-molecule fluorescence experiment, the freely diffusing unbound tRNA is invisible and binding of the tRNA to an immobilized

ribosome leads to a burst of observed fluorescence, as the fluorophore, emitting a large number of photons, is now localized within a small observation volume, as opposed to freely diffusing in solution. Binding of multiple tRNAs is manifested at the single-molecule level as a series of fluorescence bursts and interburst delays. The observed burst lifetimes and delay times yield time constants that are reciprocal to the rate constants for dissociation and association, respectively.

The power of this approach is revealed when we observe binding of a second tRNA labeled with a differently colored dye. The single-molecule analysis of this multicolor experiment would reveal the relative arrival time for the two different tRNAs, by the time interval between the fluorescence bursts, as well as how long the two bound tRNAs overlapped on the same ribosome, as a duration of the bursts overlap. These results can be obtained directly from the data without the need to synchronize the ribosomes during the experiment. "Postsynchronization" in silico yields the relative timing of two or more events. This procedure amounts to aligning all single-molecule traces from an experiment using a single common event that defines the new zero point along the time axis of each trace. Further, in the case of a heterogeneous system, single-molecule traces may be sorted and each subset analyzed independently. Therefore the order of binding events, kinetics of the subsequent binding and dissociation events, and average overlap time of occupancy by multiple ligands can be measured independent of system complexity, and correlations between dynamic events in multistep, heterogeneous systems can be observed directly with a single-molecule approach.

Conformational dynamics are also readily investigated using single-molecule fluorescence. The main tool for this application is Förster resonance energy transfer (FRET), which involves energy transfer between a donor and acceptor dye through coupling of transition dipoles. The efficiency of energy transfer depends on $1/R^6$ in which R is the interdye distance, as well as on the spectral overlap of the two dyes and on dye orientation terms. For Cy3 and Cy5 dyes, commonly used dyes used in these experiments, the efficiency of FRET varies from 1 (e.g., all emissive energy from Cy3 is transferred to Cy5) at distances below about 20 Å, to 0 at distances above 80 Å. Thus FRET is highly sensitive to distance changes in the region of 30–60 Å, well suited for investigation of the ribosome (250 Å in diameter).

FRET represents a high time-resolution probe of biomolecular conformation. In single-molecule FRET (smFRET), donor and acceptor dye emissions are measured simultaneously for individual molecules, and these intensities are converted to FRET through the equation $E_{obs} = I_{acceptor}/(I_{donor} + I_{acceptor})$, in which I_{donor} and $I_{acceptor}$ are fluorescence intensities of the donor and acceptor dyes, correspondingly. Changes in interdye distance are revealed by anticorrelated changes in donor and acceptor intensities. Molecular conformation can be monitored by smFRET with millisecond time resolution, limited by fluorophore brightness and camera sensitivity. smFRET has been a powerful tool to explore ribosomal and ligand dynamics during translation, as outlined below.

In addition to dynamics, single-molecule methods can directly measure forces generated by molecular motors and mechanical stability of molecular complexes. Optical traps are instrumental in revealing the mechanism of motor proteins and in mechanistic studies of translation. Optical traps can apply constantly increasing force to the point of complex rupture to test mechanical stability of the complexes, thus directly reporting on the tensile strength of the intermolecular interactions, such as those between mRNA and the ribosome. Alternatively, the trap can be employed as a molecular ruler to track relative movement of the two components in the complex, for example, traveling of the ribosome along mRNA. Optical tweezers also permit application of an intermediate assisting or hindering force to the molecular motor, thus allowing elucidation of the mechanism of the motor mobility. These experiments are only possible at the single-molecule level. Modern optical traps allow distance measurement at the angstrom level of resolution and application of forces in the range of tens of piconewtons,

with subpiconewton precision. The ribosome reads mRNA in 1.3-nm-long triplets and requires up to 20 pN to dislocate from mRNA, thus it is suitable for study with single-molecule force methods.

Single-molecule experiments must be tackled with care and diligence. The approach, by its very sensitivity, is fraught with potential artifacts. Surface immobilization can perturb behavior of biological systems, and nonspecific surface interactions can interfere with single-molecule measurements. Biologically relevant signals are convoluted with photophysical and photochemical artifacts due to high-intensity illumination—blinking, photobleaching, and photodamage limit single-molecule measurements and must be addressed by use of specific dyes, removal of molecular oxygen, and addition of chemical agents to improve dye behavior. The challenging and time-consuming nature of the single-molecule approach limits throughput. As a result, single-molecule experiments should be used in conjunction with conventional kinetic and molecular biology methods to address biological questions. Despite these limitations, the past decade has seen single-molecule data contributing deeply to our understanding of translation.

TRANSLATION INITIATION

The goal of translation initiation is to select the mRNA, recognize the correct start codon, and assemble an elongation-competent 70S particle with an initiator tRNA in the P site. In prokaryotes initiation is guided by three initiation factors: IF1, the GTPase IF2, and IF3. The mRNA is directly recruited to the 30S subunit via interactions between the mRNA Shine-Dalgarno sequence and a complementary sequence in the 16S ribosomal RNA (rRNA). The Shine-Dalgarno sequence is located 5–9 bases upstream of the AUG start codon, causing the start codon to be positioned correctly into the P site. IF2 and initiator tRNA are recruited to the 30S subunit to form a 30S preinitiation complex. IF2 then promotes joining of the 50S ribosomal subunit. Formation of the 70S initiation complex triggers rapid GTP hydrolysis by IF2. The GDP-bound form of IF2 has lesser affinity for the 70S ribosome and rapidly dissociates from the ribosome leaving the 70S particle ready for the first round of elongation (Antoun et al. 2003).

ORDER OF IF2 AND INITIATOR tRNA ARRIVAL

The timing of key initiation events is crucial for translational control, although much of it remains unclear. It is a common prejudice in the literature that IF2 recruits initiator tRNA to the ribosome, but the experimental evidence is slim. The original biochemical studies showed that IF2 stabilizes tRNA in the 30S preinitiation complex, but did not elucidate the order of ligand recruitment, leading to conflicting hypotheses (Lockwood et al. 1971; Wu et al. 1996; Wu and RajBhandary 1997). There are multiple possible arrival mechanisms: One of the ligands may arrive first, recruiting or permitting binding of the second one; both ligands may arrive simultaneously; or the order of arrival may be stochastic.

Although recent experiments suggested that IF2 and fMet-tRNA^{fMet} bind sequentially to the 30S subunit (Milon et al. 2010), IF2(GTP) also forms a weak complex with the tRNA ($K_D \sim$ 1 μM) (Lockwood et al. 1971; Petersen et al. 1979; Wu and RajBhandary 1997; Spurio et al. 2000; Milon et al. 2010), potentially allowing simultaneous binding of the two molecules. Tsai et al. (2012) used a ZMW-based single-molecule approach utilizing fluorescently labeled IF2 and tRNA^{fMet} to determine whether IF2 and tRNA binding is simultaneous, sequential, or random. The mixture of fMet-(Cy3)-tRNA^{fMet}, Cy5-IF2, and Cy3.5-50S was delivered to immobilized Alexa488-30S. The appearance of a stable 50S signal was used to identify productive tRNA- and IF2-binding events. The relative timing of IF2 and tRNA^{fMet} arrival to the ribosome was directly observed by single-molecule analysis (Fig. 1).

Without IF1 and IF3 at low concentrations of IF2 and the initiator tRNA (20 nM each) the tRNA arrives first in 65% of the initiation events, IF2 arrives first in 30%, and in only 5% of cases do both IF2 and tRNA arrive simultaneously. The presence of IF1 and IF3 shifts the arrival order, with 50% of ribosomes having IF2 arriving before tRNA^{fMet}, 40% having tRNA^{fMet} arriving before IF2, and 10% showing simultaneous arrival. This is consistent with IF1 and IF3 destabilizing tRNA^{fMet} in the 30S PIC (Antoun et al. 2006) and increasing the affinity of IF2 to the 30S ribosomal subunit in the absence of initiator tRNA, leading to a higher ratio of molecules in which IF2 arrives first in their presence (Lockwood et al. 1972; Caserta et al. 2006). Increasing IF2 and tRNA^{fMet} concentrations to 1 μM raised the fraction of ribosomes showing simultaneous arrival to 45%, whereas lowering the fraction of IF2 arriving first to 35% and the fraction of tRNA^{fMet} arriving first to 10%. The increase in simultaneous arrival with concentration of ligands suggests that at higher concentrations, a significant fraction of the tRNA and IF2 arrive as IF2-tRNA complexes. Thus, the order of IF2 and initiator tRNA arrival does not strictly follow a defined sequence, and is greatly affected by ligand concentrations and the presence of other initiation factors. Although at low concentrations the order of arrival is stochastic, in the presence of IF1 and IF3 and at high ligand concentrations simultaneous arrival may be a more common mechanism.

The observed dependence of the IF2 and tRNA recruitment on the presence of initiation factors and reaction conditions may explain the disagreement in results obtained by different groups. The recent observation by Milon et al. (Milon et al. 2010) suggests that IF2 binds first to the 30S subunit and then recruits tRNA. These experiments were conducted in 20 mM MgCl$_2$, and 0.25 mM GTP, whereas single-molecule measurements were performed at 5 mM MgCl$_2$, and 4 mM GTP ($\sim$1–2 mM free Mg^{2+}). The IF2-tRNA complex is destabilized by high Mg^{2+} (Majumdar et al. 1976; Sundari et al. 1976; Spurio et al. 2000). Therefore the difference in Mg^{2+} concentrations could be a reason why Milon et al. have not observed simultaneous arrival of tRNA and IF2, whereas low magnesium conditions in single-molecule experiments facilitated formation of the IF2 and tRNA complex. The total concentration of magnesium in *Escherichia coli* is in the range of 50–200 mM with most of this being bound to

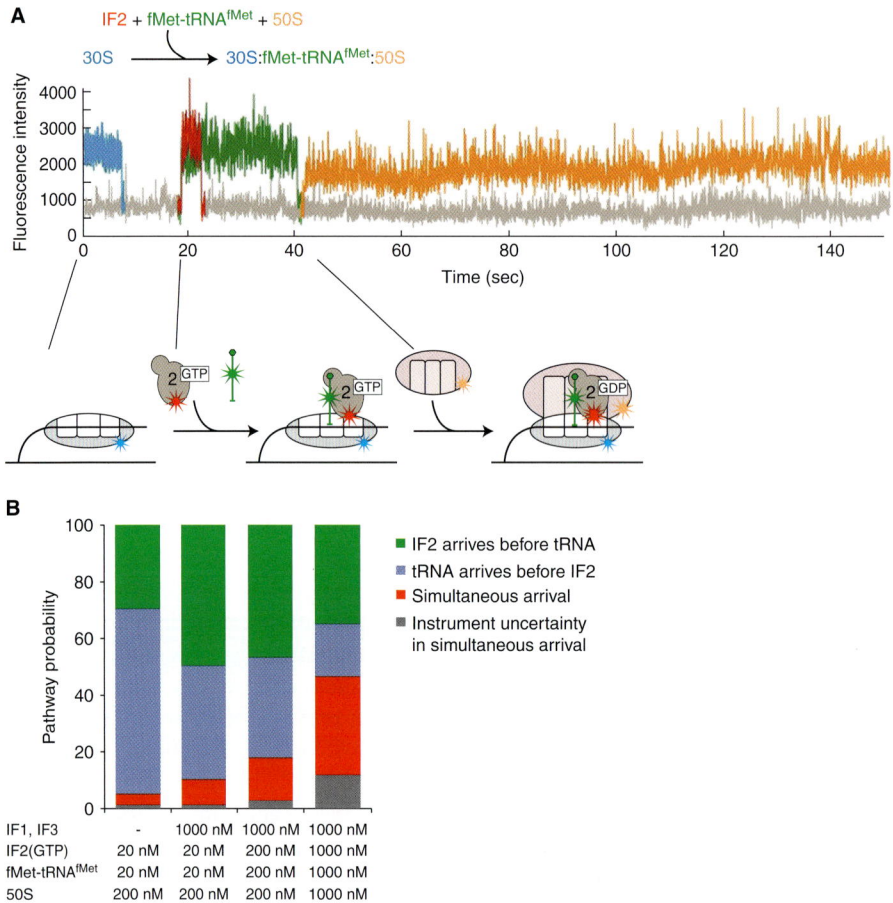

Figure 1. Timing of IF2 and tRNA arrival. (*A*) Example trace of timing of IF2 and initiator tRNA binding. 30S-mRNA initiation complexes were immobilized in zero-mode waveguide (ZMW) wells. Fluorescent dyes are shown as stars. The presence of single 30S complexes was identified by Alexa488 fluorescence of labeled small ribosomal subunits, which photobleaches rapidly on start of observation due to the short photobleaching lifetime of the Alexa488 dye. Cy5-labeled IF2, Cy3-labeled fMet-tRNA^{fMet}, and Cy3.5 50S subunits allow direct observation of molecular events during initiation. In this example trace, IF2 and tRNA arrive simultaneously and then IF2 and tRNA signals disappear due to photobleaching as both ligands are required for subsequent stable 70S complex formation. The initiation is finalized by stable arrival of 50S at an ~42 sec time point. The order of arrival was directly determined to a single frame precision (33 msec) by the sequence of the fluorescent pulses. (*B*) Observed ratios of 30S preinitiation complex (PIC) formation pathways at different ligand concentrations. The order of arrival of labeled IF2 and initiator tRNA for each ribosome was observed and the ratios of possible arrival pathways at different ligand concentrations are plotted. (Modified, with permission, from Tsai et al. 2012.)

proteins or in chelates with anionic metabolites. The free Mg^{2+} concentration was estimated to be between 1 and 2 mM (Lusk et al. 1968; Alatossava et al. 1985). It is possible that magnesium concentration plays a role in the global regulation of the initiation mechanism at translational expression by fine tuning the order of events during initiation.

TRANSITION TO ELONGATION

The final key event that marks the end of initiation is the release of initiation factors and arrival of the first elongator tRNA encoded by the second codon on the mRNA. Yet the relative timing of IF2 release, 50S subunit joining, and elongator tRNA binding are not known. To

Cite this article as *Cold Spring Harb Perspect Biol* doi: 10.1101/cshperspect.a011551

monitor these events in real time, Tsai et al. (2012) delivered Cy5-IF2, Cy3.5-50S, and Phe-(Cy2)tRNA^{Phe} in a ternary complex with EF-Tu and GTP to 30S PICs loaded with fMet-(Cy3)tRNA^{fMet}, simultaneously tracking four different labeled components. An IF2 signal was followed by rapid and stable 50S joining. Only IF2 with GTP yielded stable tRNA-bind-

ing post 50S subunit joining, whereas IF2 with GDPNP, a nonhydrolyzable analog of GTP, only results in brief tRNA sampling with no stable binding. Postsynchronizing to the departure of IF2 revealed a noticeable overlap between the IF2 and 50S signals (Fig. 2). This 2 sec overlap time of IF2 and 50S occupancy on the 30S PIC was independent of 50S concentration,

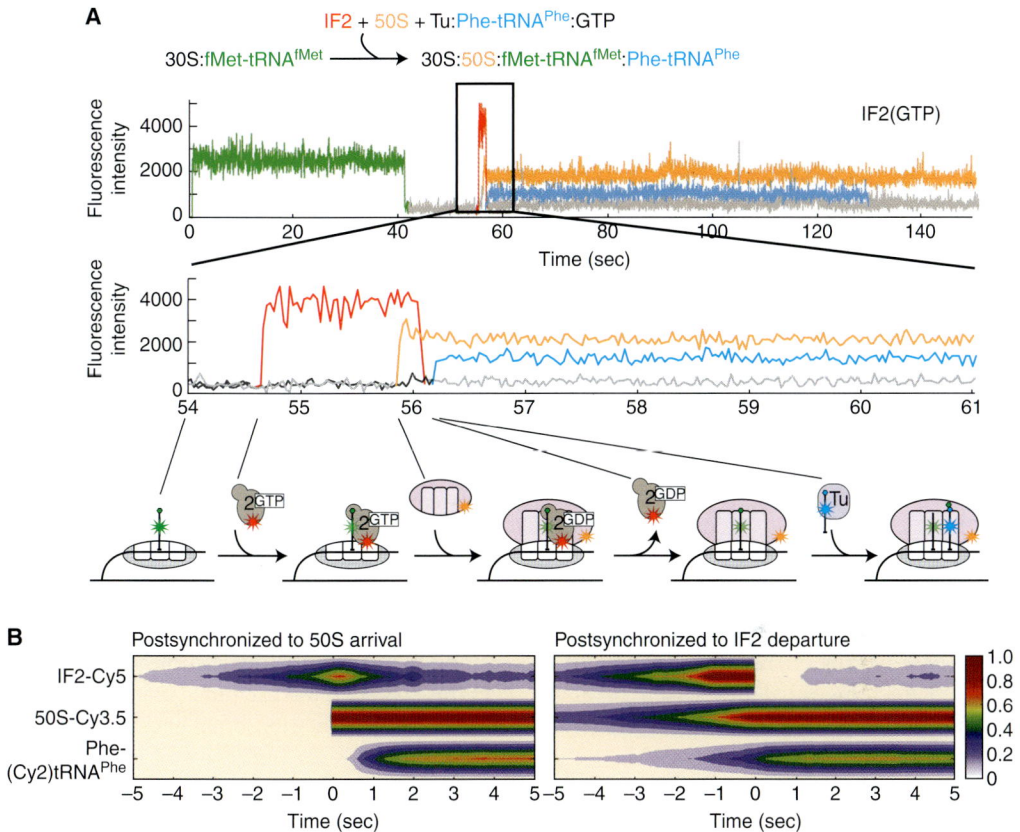

Figure 2. Transition into elongation. (A) Example trace showing the observed order of the late events during initiation. 30S PICs were immobilized via the mRNA in zero-mode waveguide (ZMW) wells and scored by Cy3-labeled fMet-tRNA^{fMet}. Cy3.5-labeled 50S subunit, Cy2-labeled Phe-tRNA^{Phe} in a ternary complex with EF-Tu(GTP), and Cy5-labeled IF2 were then delivered to the 30S PIC. The disappearance of the IF2 signal corresponds to IF2 dissociation from the 70S complex, as evidenced by the IF2 residence time that is much shorter than the photobleaching lifetime of the Cy5 dye. The disappearance of all other signals is due to dye photobleaching as indicated by their long photobleaching-limited lifetimes. (B) Postsynchronized plots of IF2, 50S subunit, and elongator tRNA. In the left panel, all single-molecule traces were postsynchronized to the arrival of 50S, with arrival events defining a new $t = 0$ and the right panel is postsynchronized to the departure of IF2 from the 70S ribosome. The presence of a signal from the molecule of interest may be read lengthwise, with the color reflecting the mean occupational density at each time point. The left panel shows an ∼2 sec overlap between the IF2 and 50S signals along with a delay in elongator tRNA signal, with most tRNAs arriving predominantly after IF2 release. The right panel shows the heterogeneity of elongator tRNA arrival. In approximately 20% of the cases, tRNA arrives before IF2 leaves the ribosome.

suggesting that a unimolecular process occurs during the overlap. During this period, IF2 rapidly hydrolyzes GTP, rearranges itself, tRNA, and ribosome conformation, and dissociates from the ribosome; consistent with this interpretation, the lifetime of IF2-GDP on 70S ribosomes was 1.2 sec. Elongator tRNA arrival required GTP hydrolysis by IF2 and was drastically decreased in the presence of GDPNP. The majority of tRNA (80%) arrived after IF2 departure. In these subsets of molecules, similar to the early kinetic studies (Tomsic et al. 2000), postsynchronization to the 50S arrival time point showed an ~2 sec lag between 50S joining and the majority of elongator tRNA arrival. The duration was independent of tRNA concentration (>200 nM), thus indicating that tRNA arrival is a gated unimolecular reaction. The similar duration of the lag and IF2 occupancy time on the ribosome suggests that IF2 release may guide tRNA recruitment. However, the observation that in 20% of 70S subunits, elongator tRNA arrives before IF2 release indicates that IF2 control over tRNA arrival is not absolute.

CLEARING THE SHINE-DALGARNO SEQUENCE

The Shine-Dalgarno mRNA sequence and 16S rRNA form between five and nine base pairs, with a base-pairing energy of 3–14 kcal/mol. These interactions must be broken to allow the transition into elongation. The recent structural and single-molecule data provide insight into the potential mechanism for this process (Korostelev et al. 2007).

The mechanical stability of the mRNA–ribosome interactions was examined by optical tweezer assays. In this setup, the mRNA in the translational complexes was immobilized on the surface of the slide and the 30S subunit was attached to the polystyrene bead held by the optical trap. The optical trap was used to exert a constantly increasing tension force, until the complexes ruptured. The rupture force required to dislocate mRNA from the ribosome directly reports on the geometry and stability of the mRNA–ribosome interactions. The force required to rupture mRNA from the 70S particle depends on the presence of the tRNA ligands and mRNA sequence. The Shine-Dalgarno interaction adds ~10 pN, the A-site tRNA 10 pN, and the P-site tRNA ~ 5 pN to the tensile strength of the initiation complexes. However, on formation of the first peptide bond the contribution of the Shine-Dalgarno sequence disappears, suggesting the release of the Shine-Dalgarno–ribosome interaction (Uemura et al. 2007).

DIRECT TRACKING OF THE RIBOSOME MOVEMENT DURING ELONGATION

The force approach has been used for direct observation of ribosome movement along mRNA. In a breakthrough study, Wen et al. (2008) used a suspended dumbbell assay in which the ends of an mRNA with a hairpin in the center of its sequence are attached to polystyrene beads (hence called "dumbbell") held by dual beam optical tweezers. The optical tweezers allow application of a stretching force sufficient to hold the mRNA in its linear form, permitting accurate measurement of RNA length, as a distance between two trap centers. The translating ribosome unwinds the RNA hairpin and the resulting increase in mRNA length reports on the ribosome position. This was the first direct dynamic observation of ribosome movement along a mRNA at the single-molecule level. Later this approach was expanded to investigate the mechanism of the ribosomal helicase (Qu et al. 2011). The results indicate that the ribosome utilizes two modes of unwinding. First, it promotes and stabilizes the open state of the RNA duplex by sequestering the unwound portion of the duplex. Second, it mechanically unwinds the duplex by pulling on mRNA. The biochemical characterization of the ribosomal helicase activity suggests that proteins S3, S4, and possibly S5 form the helicase center of the ribosome (Takyar et al. 2005) and compose a tight ring around incoming mRNA (Wimberly et al. 2000). These observations are consistent with single-molecule force data and suggest that these proteins may stabilize the open state of the mRNA duplexes or work as an "extrusion die," thus participating in the active mechanism.

Cite this article as *Cold Spring Harb Perspect Biol* doi: 10.1101/cshperspect.a011551

Future studies are needed to show the role of individual ribosomal components in helicase activity and differentiate among various helicase mechanisms.

RIBOSOME CONFORMATIONAL CHANGES DURING INITIATION AND ELONGATION

Previous cryo-EM methods and structural studies showed that the ribosome adopts two intersubunit conformations—the locked and unlocked states—that are related by a 6° ratchet-like rotation of the two subunits (Frank and Agrawal 2000; Valle et al. 2003). Peptide bond formation leads to a counterclockwise rotation of the small subunit with respect to the large subunit, from the locked to the unlocked state. Then, eventual GTP hydrolysis and translocation leads to the clockwise locking of the ribosome. These global conformational changes are correlated with the movements of the L1 protein of the ribosome and the tRNA transitions between the classical state and the hybrid state (Blanchard et al. 2004a,b; Agirrezabala et al. 2008; Fei et al. 2009; Fischer et al. 2010).

Methods to observe ribosome conformation in real time with codon resolution have revealed these dynamic changes directly during translation. Using Cy3-labeled 30S and Cy5-labeled 50S, Marshall et al. characterized an intersubunit FRET signal that reports on the global conformation of the ribosome (Dorywalska et al. 2005; Marshall et al. 2008). The 30S subunit was labeled at the terminal loop of h44 located in the spur of the small subunit, and the 50S subunit was labeled at the loop of h101 placed opposite the central protuberance. The two labeling dyes are separated by ∼45 Å according to available structural data, providing a FRET signal that reports on the conformation of the two subunits, yet is distant from the active sites of the ribosome (Fig. 3A).

During the transition from initiation to elongation, IF2 guides the appropriate assembly of the 70S initiation complex on subunit joining. Marshall et al. showed that IF2 accelerates subunit joining, with a subset of ribosomes joining in the rotated low-FRET state. After ∼30 msec, which agrees with the experimental-

ly determined rates for GTP hydrolysis by IF2, the ribosome undergoes a quick transition to the high-FRET state (Marshall et al. 2009). No such transitions were observed with GDPNP. Thus, IF2 GTP hydrolysis guides the ribosome joined in an unproductive low-FRET state into an elongation-competent high-FRET state. However, the sequence of events was not universally observed for every initiating complex. A significant number of ribosomes initiated in the high-FRET (nonrotated) state. The rates of GTP hydrolysis and subsequent intersubunit rotation are comparable with the data acquisition rates, thus it is unclear whether in those molecules conformational changes occurred too fast to be observed, or that the 50S subunits joined in the nonrotated state, resulting in an alternative initiation pathway.

After initiation, the intersubunit FRET signal alternates between a high-FRET state and a low-FRET state. By employing fluorescently labeled tRNA^{Phe}, Aitken et al. (Aitken and Puglisi 2010) directly showed that the transition from high to low FRET occurs concurrently with the arrival of the first elongator tRNA, whereas in the presence of EF-G(GTP), the ribosome then rapidly returns back to the high-FRET (nonrotated) state (Marshall et al. 2008; Aitken and Puglisi 2010). This repeating cycle of high–low–high FRET transitions was observed over multiple rounds of elongation (Aitken and Puglisi 2010). By employing mRNAs of various lengths and withholding a specific tRNA, Aitken et al. observed that the maximum number of high–low–high FRET cycles produced by an elongating ribosome corresponds to the number of codons of the mRNA translated. The addition of the antibiotic erythromycin, which blocks the peptide exit tunnel at a position in which a polypeptide of seven amino acids would reach (Schlunzen et al. 2001), resulted in a significant reduction in the number of ribosomes undergoing translation for more than six FRET cycles (Fig. 3). Thus, a cycle of high–low–high FRET transitions corresponds to one full cycle of elongation. The intersubunit FRET signal provides a method to track multiple elongation cycles and to monitor global conformational dynamics of the ribosome in real time.

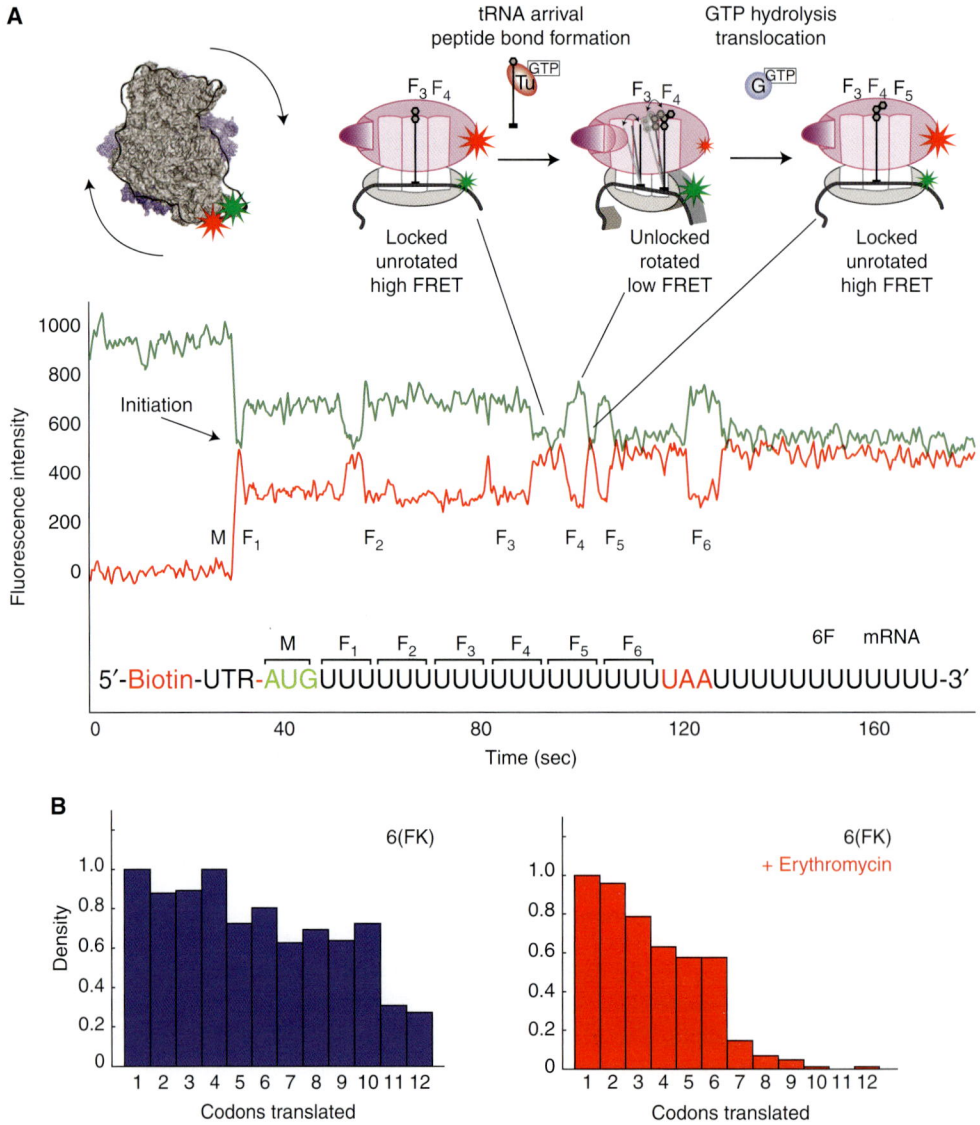

Figure 3. Following global conformation of ribosome during elongation. (*A*) An intersubunit FRET signal reports on two different rotational states of the ribosome. These states, called "unrotated" and "rotated," have high and low FRET efficiency, respectively. The intersubunit FRET signal allows the observation of changes in global ribosome conformation through multiple rounds of elongation. Cy5-labeled 50S subunits, ternary complexes, and EF-G are delivered to surface-immobilized Cy3-labeled 30S PICs assembled on an mRNA encoding six phenylalanines (designated 6F). The appearance of FRET corresponds to the 50S subunit joining during initiation. During elongation, multiple cycles of high–low–high FRET are observed, each corresponding to the ribosome unlocking and locking during one round of elongation. The elongation cycle at the fourth elongator codon is used as an example of high–low–high FRET cycle. (*B*) Number of FRET cycles observed on mRNAs (designated 6FK) encoding six alternating Phe and Lys pairs. The maximum number of elongation cycles is consistent with the number of codons of the mRNA. Erythromycin stalls ribosomes by blocking the nascent chain at codon 7, which is exactly what is observed. (Modified, with permission, from Aitken and Puglisi 2010.)

Cite this article as *Cold Spring Harb Perspect Biol* doi: 10.1101/cshperspect.a011551

What drives the ribosomal FRET transition? The observed timing of FRET transitions correlates with ample cryo-EM and X-ray data that show intersubunit rotation, suggesting that it reflects a ratcheting motion of the ribosome (Marshall et al. 2008). Because no spontaneous transitions between two states were observed, it was concluded that they are separated by a large energy barrier. Because transition between the two states is dependent on the arrival of tRNA and EF-G, it is possible that the irreversible transition between the two states is induced by peptide bond formation and GTP hydrolysis by EF-G, indicating that the free energies of peptidyl transfer (~ -8 kcal mol^{-1}) and GTP hydrolysis (~ -10 kcal mol^{-1}) are required for the rearrangement of the two subunits.

CORRELATIONS BETWEEN tRNA AND RIBOSOME CONFORMATIONS DURING TRANSLATION

Translation is a dynamic process that requires the intricate interplay between the ribosome, tRNAs, and multiple factors. The major problem for multicomponent experiments is that the small number of dyes suitable for single-molecule fluorescence experiments limits the number of components that can be observed simultaneously. Chen et al. (2012) used the same labeling strategy as Marshall et al. (2008), but replaced Cy5 as the FRET acceptor with a non-fluorescent, black hole quencher (BHQ). The use of BHQ frees the spectral region of the acceptor dye for labeling other components of the system in multiplexed experiments, so it is possible to use Cy5 to label other translation factors, such as tRNA or elongation factors, whereas fluctuating Cy3 intensity can still be utilized to determine the global conformational dynamics of the ribosome.

Using this approach the authors followed changes in ribosome conformation with Cy3/BHQ-labeled ribosomes and correlated them to tRNA dynamics by employing Cy5-labeled tRNAs. Arrival of Cy5-tRNA, shown as a red fluorescent pulse, was concomitant with the transition from high FRET to low FRET of the ribosome. Thus, the arrival of tRNA occurs simultaneously with ribosome unlocking, within the time resolution of 100 msec (Chen et al. 2012). Similarly, the departure of tRNA occurs simultaneously with the ribosome locking and translocation, also within the time resolution of 100 msec. This shows that tRNA arrival and departure are correlated with the ribosome conformational changes. This method can be further extended to reveal the correlation between the ribosomal conformational changes and factor and tRNA dynamics.

tRNA TRANSIT THROUGH THE RIBOSOME

tRNAs must be correctly selected and then must transit through the ribosome during translation elongation. Aminoacylated tRNA first arrives in the A site as a ternary complex with EF-Tu·GTP. On tRNA selection, the ribosome catalyzes the formation of the peptide bond between the aminoacylated tRNA and the nascent peptide chain attached to the tRNA in the P site. After the peptidyl-transfer reaction takes place, the ribosome changes to a conformation in which the tRNAs and mRNA are conformationally mobile. EF-G catalyzes translocation, moving the A-site tRNA into the P site and also clearing the A site for the next tRNA. At this stage, the original A-site tRNA is now stably bound in the P site and this completes one round of elongation. After another round of elongation, the P-site tRNA is moved into the E site, where tRNA eventually dissociates from the E site, completing its life cycle on the ribosome. There are two proposed mechanisms of how tRNA dissociates from the ribosome: either allosterically with the arrival of the next tRNA to the A site while potentially regulating tRNA arrival and selection, or spontaneously as soon as it reaches the E site. Uemura et al. (2010) directly tracked tRNAs labeled with fluorescent dyes in elongation experiments, and observed rapid and spontaneous release of tRNA from the E site. However, in Chen et al. (2011), the investigators reported that both pathways of tRNA release occur. These experiments were conducted at 15 mM magnesium, significantly higher than the physiological concentration, potentially allowing for overstabilizing the E-site tRNA and factor-independent spontaneous transloca-tion of the ribosome. Moreover, the partition between the two mechanisms of tRNA arrival remained constant in the range of tRNA con-centrations tested, despite an expected increase in the E-site departure rate for the allosteric pathway with increasing tRNA concentration. On the other hand, at a lower (5 mM) Mg^{2+} concentration, Uemura et al. (2010) did not observe a detectable overlap between the P- and E-site tRNA signals even when the ribo-some is translocating quickly at 1 μM tRNA and elongation factor concentrations, suggest-ing that tRNAs rapidly and (within ∼50 msec) spontaneously depart from the E site (Fig. 4).

tRNA DYNAMICS IN THE RIBOSOME

The binding dynamics and conformation of the tRNAs on the ribosome play important roles in the tRNA selection and translocation steps of elongation (Korostelev et al. 2006). At the begin-ning of an elongation cycle (Fig. 5), the ribo-some is in the posttranslocation state, with the P-site tRNA stably bound to the ribosome with little conformational fluctuation. In this state, the entire tRNA is located within the P site of both the small and large subunits. The A site is empty at this stage, awaiting the arrival of the next tRNA. When a tRNA arrives in a ternary complex (TC) with EF-Tu·GTP, the anticodon loop on the tRNA comes into contact with the codon on the mRNA in the small subunit A site.

Here, the tRNA is in the A/T state and initial selection occurs in which the codon–anticodon interaction is checked to determine if the tRNA is cognate to the codon. In the A-site tRNA to P-site tRNA (tRNA–tRNA) FRET experiments conducted by Blanchard et al. (Blanchard et al. 2004a,b), this is observed as a low-FRET efficien-cy state. tRNAs that are not cognate to the next codon on the mRNA only very briefly sample this state (lifetime less than 50 msec) and then dissociate from the A site.

In the case of a cognate tRNA, the correct codon–anticodon interactions induce a local conformational change in the A site of the small subunit, destacking two adenines (A1492 and A1493) in the decoding site within helix 44 of the 16S rRNA so that the bases are in an orien-tation to interact with the anticodon arm of the tRNA (Ogle et al. 2001). These interactions stabilize the tRNA in the A site to allow time for further contact between EF-Tu and the GTPase activation center on the large subunit. On GTPase activation, tRNA is further inserted into the A site, resulting in a higher medium-FRET efficiency (0.50) to the P-site tRNA. Sub-sequent GTP hydrolysis by EF-Tu destabilizes the EF-Tu on the ribosome and sets the stage for the full accommodation of the tRNA. Sub-sequently, accommodation offers a final chance to reject the tRNA if it is incorrect. After clear-ing the accommodation step, the tRNA is fully in the A site of both subunits, detected as a high-FRET efficiency (0.75) state. This completes a

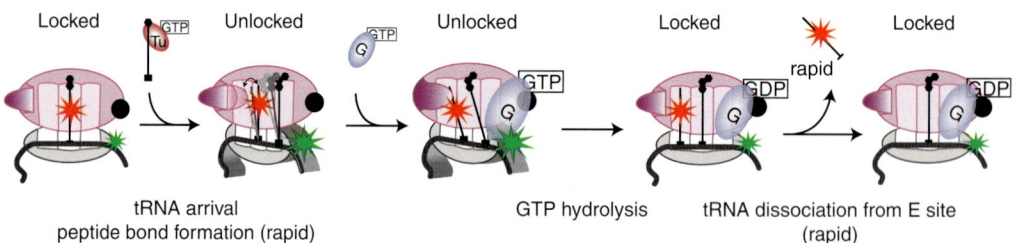

Figure 4. Correlating tRNA occupancy and conformation of the ribosome. tRNA arrives in the A site (corre-sponds to the appearance of the red fluorescence in the trace above). Peptide bond formation occurs rapidly and the ribosome adopts rotated conformation (low to high FRET transition). Subsequent EF-G catalyzed trans-location results in reverse intersubunit rotation (high to low FRET transition) and tRNA movement into the E and P sites, which is followed by rapid dissociation of Cy5 E-site tRNA (disappearance of the red tRNA signal). (Modified, with permission, from Chen et al. 2012.)

    Cite this article as *Cold Spring Harb Perspect Biol* doi: 10.1101/cshperspect.a011551

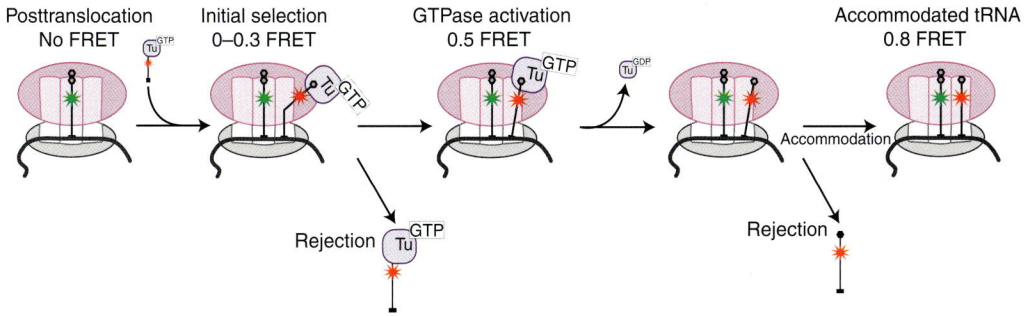

Figure 5. tRNA selection. tRNA arrival to the A site of the small subunit is a multistep process that involves two stages of error checking. FRET value indicated for each stage is from Blanchard et al. (2004b). The tRNA arrives at the A site as a ternary complex with EF-Tu(GTP) and adopts a bent A/T conformation. Initial selection takes place through the decoding site of the small subunit A site checking the codon–anticodon interaction and stabilizing the tRNA if the correct interaction is formed. If the tRNA is not rejected, the next stage, GTPase activation, follows. EF-Tu rapidly hydrolyzes GTP, and dissociates from the ribosome. The released acceptor end of tRNA spontaneously moves into the peptidyl transferase center. During this process ribosome has one last chance to check the tRNA for correctness and reject it if it is incorrect. After successful accommodation ribosome is ready to catalyze peptide bond formation between the polypeptide on the P-site tRNA and the amino acid on the A-site tRNA.

tRNA selection, which occurs within 100 msec after initial binding of a cognate tRNA. This two-stage selection mechanism improves selectivity by amplifying the limited increase in tRNA stability from correct codon–anticodon interactions compared with incorrect codon–anticodon pairs (Thompson and Stone 1977; Ruusala et al. 1982).

Immediately after tRNA accommodation, both the P-site and the A-site tRNAs are completely bound in their respective ribosomal sites in the classical conformation state (denoted as A/A-P/P, indicating that the tRNAs are completely in their respective sites), with very little tRNA conformation fluctuations (Fig. 6). On peptide bond formation that transfers the polypeptide chain from the P-site to the A-site tRNA, the ribosome unlocks and allows the portions of the tRNA in the large subunit to fluctuate. At this stage, the tRNAs can adopt hybrid conformations in which their anticodon loop is still in its original site in the small subunit but the other parts of the tRNA have moved into the next site in the large subunit (denoted as A/P-P/E). Fluctuations in tRNA conformations between the classical state and hybrid states are seen as frequent FRET efficiency fluctuations ($2-5$ sec^{-1}) between the high- and medium-FRET states. The medium-FRET states likely represent a collection of different tRNA conformations, as Munro et al. (2007) identified at least two distinct hybrid states in subsequent tRNA–tRNA FRET studies. This second identified hybrid state may represent tRNA conformations in which only the P-site tRNA has partially fluctuated into the next site (A/A-P/E).

The dynamic nature of the tRNAs when the ribosome is in the pretranslocation state plays an important role in finding the right conformation for the tRNAs to adopt before translocation mediated by EF-G occurs. As translocation involves shifting the P- and A-site tRNAs into the next site, having the tRNAs in the A/P-P/E hybrid state, in which the tRNAs have already partially moved into the next site, likely presents a lower energy barrier. Therefore, the ability for tRNAs in the ribosome to fluctuate and freely explore the hybrid states may have a direct impact on how efficiently translocation occurs. Accordingly, Feldman et al. (2010) reported that addition of antibiotics from the aminoglycoside family that stabilize the classical (A/A-P/P) state results in commensurate reduction in EF-G-catalyzed translocation rates. Members of this family bind to the A site of the small subunit in the decoding site and perturb

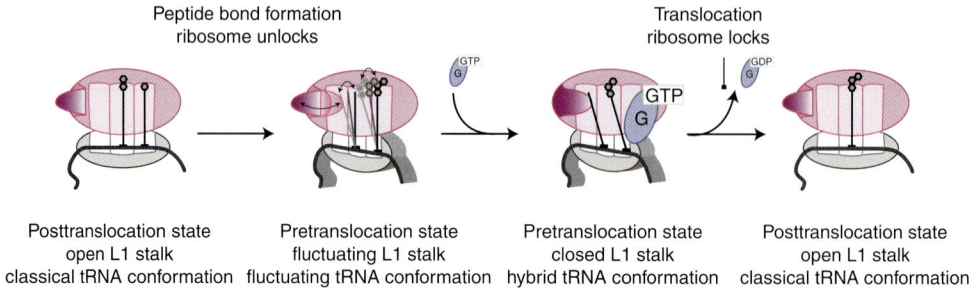

Figure 6. Translocation. The process of ribosome translocation in relation to tRNA dynamics is depicted in the diagram. The flap to the left of the large subunit represents the L1 stalk. The process begins immediately after tRNA accommodation in the A site (as detailed in Fig. 5) in the posttranslocation ribosome state, in which the L1 stalk is open and the tRNAs are in the classical A/A-P/P conformation. On peptide bond formation, the polypeptide chain is transferred to the A-site tRNA and the ribosome unlocks into the pretranslocation state. At this stage, the ribosome and tRNA become conformationally dynamic, with the L1 stalk fluctuating between the open and closed states and the tRNA between the classical conformation (A/A-P/P) and hybrid conformations (A/A-P/E or A/P-P/E). EF-G(GTP) binding locks the L1 stalk into the closed state and the tRNA in the hybrid state (A/P-P/E), readying the ribosome for translocation. On translocation, aided by EF-G, the ribosome locks into a conformationally static state, opening the L1 stalk, releasing the E-site tRNA, and stabilizing the P-site tRNA in the classical conformation.

its local structure, resulting, for many members of the aminoglycoside family, in A1492 and A1493 being destacked regardless if a cognate tRNA is present. This effectively freezes the small subunit A site in a conformation that further stabilizes the classical tRNA conformation. Furthermore, as EF-G interacts with A1492 and A1493 during translocation (Gao et al. 2009), the conformation of helix 44 forced by aminoglycosides may also slow translocation by mechanical opposition. Thus, these effects combine to increase significantly the activation energy barrier for EF-G-mediated translocation.

tRNA–RIBOSOME INTERACTIONS AND TRANSLOCATION

In addition to the conformational fluctuations of the tRNAs themselves in the ribosome, the P-site tRNA also interacts with the L1 stalk of the large subunit (located near the E site and composed of helices 76–78 of the 23S rRNA and ribosomal protein L1) when it is in a P/E configuration. Such an interaction could be central in moving the tRNA when the ribosome translocates. Using FRET between the P-site tRNA and ribosomal protein L1, Fei et al. (2008) reported that the L1 stalk and the P-site tRNA in

posttranslocation complexes are relatively static and are distant (50–70 Å, FRET of 0.2–0.4) from each other. After peptide bond formation unlocks the ribosome, the L1 stalk and P-site tRNA become dynamic and fluctuate at rates of $1-3 \text{ sec}^{-1}$ between the original low-FRET state and a high-FRET state (0.8, a distance of ~ 35 Å), in which the P-site tRNA in the P/E hybrid conformation is within distance to interact with the L1 stalk. The investigators further observed that EF-G binding to the ribosome shifts all molecules into the hybrid conformation (high-FRET state), maintaining contact until the tRNA is moved into the E site. Thus, tRNA and the ribosome must work in concert in their respective conformational dynamics in order to set the stage for translocation.

Furthermore, studies employing FRET between ribosomal proteins (Cornish et al. 2009; Fei et al. 2009, 2011) observed that the tRNA acylation state in the A and P sites and the translation state of the ribosome can change the opening and closing dynamics of the L1 stalk. Employing FRET between ribosomal proteins L33-L1 and L9-S6, Cornish et al. (2009) observed that with an acylated P-site tRNA, normally present in the posttranslocation state, the ribosome adopts an open L1 stalk conformation.

Presumably, this allows the E-site tRNA to dissociate freely from the ribosome after translocation has occurred. Conversely, the L1 stalk is closed when the P-site tRNA is deacylated, which normally is the condition in the pretranslocation ribosome immediately after peptide bond formation has taken place. Fei et al. (2009, 2011) also observed the opening of the L1 stalk with FRET between ribosomal proteins L1–L9 when the ribosome is in a posttranslocation state. However, the investigators also observed significantly more fluctuation in the conformation of the L1 stalk compared with the previous study. In the pretranslocation state, the L1 stalk of a majority of ribosomes ($\sim$70%) is highly dynamic and fluctuates between the open and closed states at rates of 2–4 sec^{-1}. Similar to their P-site tRNA to ribosomal protein L1 FRET study, the fluctuations in the pretranslocation complexes are locked into the closed state on EF-G binding if an elongator tRNA is in the P site. On the other hand, if an initiator tRNA occupies the P site, EF-G is unable to lock the L1 stalk and it continues to fluctuate. Mutant initiator tRNAs that mimic elongator tRNAs in flexibility restored the ability of EF-G to lock the L1 stalk before translocation, suggesting that the structural flexibility of the tRNA could play an important part in regulating translocation.

The studies by Cornish et al. (2009) and Fei et al. (2009, 2011) detected different levels of L1 stalk fluctuations above when different labeling sites were used; it is not currently clear if the fluctuations observed in these studies represent local conformational dynamics or global changes on the ribosome. As a specific FRET experiment only provides a single relative distance constraint, without further experiments, it is difficult to conclude if the L1 stalk opening and closing and its interactions with the P-site tRNA in the post- and pretranslocation state directly correlate to global conformational changes on the ribosome.

APPLICATIONS TO EUKARYOTIC TRANSLATION

The single-molecule toolkit developed for the study of prokaryotic translation is in principle immediately portable to eukaryotic translation. Although the technology itself requires little adaptation for this transition, the intrinsic complexity of the eukaryotic translation machinery makes application of single-molecule approaches considerably more difficult than in the case of prokaryotes. This complexity is principally due to (1) the increased number of protein factors associated with eukaryotic ribosomes, (2) the modulatory effects of eukaryotic mRNA structural elements, such as the 5'-cap structure and 3'-poly(A) tail, and mRNA circularization, and (3) the increased number of regulatory events in eukaryotic translation, particularly initiation. These elements greatly increase the number of dynamic events associated with each stage of the translation cycle, as well as making reconstitution of translation in vitro problematic due to the difficulty of isolating individual factors and optimizing experimental conditions.

Notwithstanding these obstacles, a variety of bulk rapid-mixing kinetic studies have been performed recently on reconstituted *Saccharomyces cerevisiae* translation systems, and these may form a basis for single-molecule experiments. In particular, the work of Lorsch and coworkers has led to the development of a robust eukaryotic in vitro translation system amenable to experiments with fluorescently labeled protein factors (Acker et al. 2007). This system has been used to study the roles of various initiation factors in subunit joining and in cap recognition (Reibarkh et al. 2008; Mitchell et al. 2010; Park et al. 2011). In parallel, novel strategies for the preparation and immobilization of site-specifically labeled yeast ribosomes have provided a platform for transitioning the bulk systems used in the rapid-mixing kinetics studies to single-molecule experiments (Petrov and Puglisi 2010).

Alongside the canonical eukaryotic initiation pathway, factor-independent initiation observed with internal ribosome entry sites (IRESs) found within many viral mRNAs is an attractive entry point into single-molecule studies of biologically and biomedically relevant eukaryotic translation initiation in vitro.

Interpretation of data reporting on conformational dynamics from both bulk and single-molecule studies had been hindered by the absence of a high-resolution structure of the 80S eukaryotic ribosome. The X-ray crystal structure of the yeast 80S ribosome was reported recently (Ben-Shem et al. 2010, 2011; Klinge et al. 2011). These structural data facilitate not only interpretation of existing biochemical results, but also will allow the design of new experiments. In the single-molecule case, such structural data allows identification of favorable positions for incorporation of fluorescent and other labels to follow ribosomal subunit and intersubunit factor interplay.

CONCLUSIONS

Single-molecule methods have provided the first direct glimpses of both compositional and conformational dynamics in the ribosome. The methods provide a necessary link between static structure and bulk mechanism, often providing new perspectives on old problems, as illustrated here by studies of initiation and tRNA binding. Continuing improvements in labeling strategies, single-molecule excitation and detection, and data analysis will further deepen the impact of these methods. The challenge will be to unravel the complexity of eukaryotic translation, providing a dynamic view of biological regulation.

ACKNOWLEDGMENTS

Single-molecule research in the Puglisi group is funded by NIH grants GM51266 and GM099687. We would like to thank Dr. Sotaro Uemura and all members of Puglisi laboratory for helpful discussions.

REFERENCES

Acker MG, Kolitz SE, Mitchell SF, Nanda JS, Lorsch JR. 2007. Reconstitution of yeast translation initiation. *Methods Enzymol* **430:** 111–145.

Agirrezabala X, Lei J, Brunelle JL, Ortiz-Meoz RF, Green R, Frank J. 2008. Visualization of the hybrid state of tRNA binding promoted by spontaneous ratcheting of the ribosome. *Molec Cell* **32:** 190–197.

Aitken CE, Puglisi JD. 2010. Following the intersubunit conformation of the ribosome during translation in real time. *Nat Struct Mol Biol* **17:** 793–800.

Alatossava T, Jutte H, Kuhn A, Kellenberger E. 1985. Manipulation of intracellular magnesium content in polymyxin B nonapeptide-sensitized *Escherichia coli* by ionophore A23187. *J Bacteriol* **162:** 413–419.

Antoun A, Pavlov MY, Andersson K, Tenson T, Ehrenberg M. 2003. The roles of initiation factor 2 and guanosine triphosphate in initiation of protein synthesis. *EMBO J* **22:** 5593–5601.

Antoun A, Pavlov MY, Lovmar M, Ehrenberg M. 2006. How initiation factors tune the rate of initiation of protein synthesis in bacteria. *EMBO J* **25:** 2539–2550.

Ban N, Nissen P, Hansen J, Moore PB, Steitz TA. 2000. The complete atomic structure of the large ribosomal subunit at 2.4 Å resolution. *Science* **289:** 905–920.

Ben-Shem A, Jenner L, Yusupova G, Yusupov M. 2010. Crystal structure of the eukaryotic ribosome. *Science* **330:** 1203–1209.

Ben-Shem A, Garreau de Loubresse N, Melnikov S, Jenner L, Yusupova G, Yusupov M. 2011. The structure of the eukaryotic ribosome at 3.0 Å resolution. *Science* **334:** 1524–1529.

Blanchard SC, Gonzalez RL, Kim HD, Chu S, Puglisi JD. 2004a. tRNA selection and kinetic proofreading in translation. *Nat Struct Mol Biol* **11:** 1008–1014.

Blanchard SC, Kim HD, Gonzalez RL Jr, Puglisi JD, Chu S. 2004b. tRNA dynamics on the ribosome during translation. *Proc Natl Acad Sci* **101:** 12893–12898.

Caserta E, Tomsic J, Spurio R, La Teana A, Pon CL, Gualerzi CO. 2006. Translation initiation factor IF2 interacts with the 30S ribosomal subunit via two separate binding sites. *J Mol Biol* **362:** 787–799.

Chen C, Stevens B, Kaur J, Smilansky Z, Cooperman BS, Goldman YE. 2011. Allosteric vs. spontaneous exit-site (E-site) tRNA dissociation early in protein synthesis. *Proc Natl Acad Sci* **108:** 16980–16985.

Chen J, Tsai A, Petrov A, Puglisi JD. 2012. Non-fluorescent quenchers to correlate single-molecule conformational and compositional dynamics. *J Am Chem Soc* **134:** 5734–5737.

Cornish PV, Ermolenko DN, Staple DW, Hoang L, Hickerson RP, Noller HF, Ha T. 2009. Following movement of the L1 stalk between three functional states in single ribosomes. *Proc Natl Acad Sci* **106:** 2571–2576.

Dennis PP, Bremer H. 1974a. Differential rate of ribosomal protein synthesis in *Escherichia coli* B/r. *J Mol Biol* **84:** 407–422.

Dennis PP, Bremer H. 1974b. Macromolecular composition during steady-state growth of *Escherichia coli* B-r. *J Bacteriol* **119:** 270–281.

Dorywalska M, Blanchard SC, Gonzalez RL, Kim HD, Chu S, Puglisi JD. 2005. Site-specific labeling of the ribosome for single-molecule spectroscopy. *Nucleic Acids Res* **33:** 182–189.

Dunkle JA, Wang L, Feldman MB, Pulk A, Chen VB, Kapral GJ, Noeske J, Richardson JS, Blanchard SC, Cate JH. 2011. Structures of the bacterial ribosome in classical and hybrid states of tRNA binding. *Science* **332:** 981–984.

Fei J, Kosuri P, MacDougall DD, Gonzalez RL Jr. 2008. Coupling of ribosomal L1 stalk and tRNA dynamics during translation elongation. *Mol Cell* **30:** 348–359.

Fei J, Bronson JE, Hofman JM, Srinivas RL, Wiggins CH, Gonzalez RL Jr. 2009. Allosteric collaboration between elongation factor G and the ribosomal L1 stalk directs tRNA movements during translation. *Proc Natl Acad Sci* **106:** 15702–15707.

Fei J, Richard AC, Bronson JE, Gonzalez RL Jr. 2011. Transfer RNA-mediated regulation of ribosome dynamics during protein synthesis. *Nat Struct Mol Biol* **18:** 1043–1051.

Feldman MB, Terry DS, Altman RB, Blanchard SC. 2010. Aminoglycoside activity observed on single pre-translocation ribosome complexes. *Nat Chem Biol* **6:** 54–62.

Fischer N, Konevega AL, Wintermeyer W, Rodnina MV, Stark H. 2010. Ribosome dynamics and tRNA movement by time-resolved electron cryomicroscopy. *Nature* **466:** 329–333.

Frank J, Agrawal RK. 2000. A ratchet-like inter-subunit reorganization of the ribosome during translocation. *Nature* **406:** 318–322.

Gao YG, Selmer M, Dunham CM, Weixlbaumer A, Kelley AC, Ramakrishnan V. 2009. The structure of the ribosome with elongation factor G trapped in the posttranslocational state. *Science* **326:** 694–699.

Ha T. 2001. Single-molecule fluorescence resonance energy transfer. *Methods* **25:** 78–86.

Klinge S, Voigts-Hoffmann F, Leibundgut M, Arpagaus S, Ban N. 2011. Crystal structure of the eukaryotic 60S ribosomal subunit in complex with initiation factor 6. *Science* **334:** 941–948.

Korostelev A, Trakhanov S, Laurberg M, Noller HF. 2006. Crystal structure of a 70S ribosome-tRNA complex reveals functional interactions and rearrangements. *Cell* **126:** 1065–1077.

Korostelev A, Trakhanov S, Asahara H, Laurberg M, Lancaster L, Noller HF. 2007. Interactions and dynamics of the Shine Dalgarno helix in the 70S ribosome. *Proc Natl Acad Sci* **104:** 16840–16843.

Kurata S, Weixlbaumer A, Ohtsuki T, Shimazaki T, Wada T, Kirino Y, Takai K, Watanabe K, Ramakrishnan V, Suzuki T. 2008. Modified uridines with C5-methylene substituents at the first position of the tRNA anticodon stabilize U.G wobble pairing during decoding. *J Biol Chem* **283:** 18801–18811.

Levene MJ, Korlach J, Turner SW, Foquet M, Craighead HG, Webb WW. 2003. Zero-mode waveguides for single-molecule analysis at high concentrations. *Science* **299:** 682–686.

Lockwood AH, Chakraborty PR, Maitra U. 1971. A complex between initiation factor IF2, guanosine triphosphate, and fMet-tRNA: An intermediate in initiation complex formation. *Proc Natl Acad Sci* **68:** 3122–3126.

Lockwood AH, Sarkar P, Maitra U. 1972. Release of polypeptide chain initiation factor IF-2 during initiation complex formation. *Proc Natl Acad Sci* **69:** 3602–3605.

Lusk JE, Williams RJ, Kennedy EP. 1968. Magnesium and the growth of *Escherichia coli*. *J Biol Chem* **243:** 2618–2624.

Majumdar A, Bose KK, Gupta NK. 1976. Specific binding of *Excherichia coli* chain initiation factor 2 to fMet-tRnaf-Met. *J Biol Chem* **251:** 137–140.

Marshall RA, Dorywalska M, Puglisi JD. 2008. Irreversible chemical steps control intersubunit dynamics during translation. *Proc Natl Acad Sci* **105:** 15364–15369.

Marshall RA, Aitken CE, Puglisi JD. 2009. GTP hydrolysis by IF2 guides progression of the ribosome into elongation. *Mol Cell* **35:** 37–47.

Milon P, Carotti M, Konevega AL, Wintermeyer W, Rodnina MV, Gualerzi CO. 2010. The ribosome-bound initiation factor 2 recruits initiator tRNA to the 30S initiation complex. *EMBO Rep* **11:** 312–316.

Mitchell SF, Walker SE, Algire MA, Park EH, Hinnebusch AG, Lorsch JR. 2010. The 5′-7-methylguanosine cap on eukaryotic mRNAs serves both to stimulate canonical translation initiation and to block an alternative pathway. *Mol Cell* **39:** 950–962.

Munro JB, Altman RB, O'Connor N, Blanchard SC. 2007. Identification of two distinct hybrid state intermediates on the ribosome. *Mol Cell* **25:** 505–517.

Ogle JM, Brodersen DE, Clemons WM Jr, Tarry MJ, Carter AP, Ramakrishnan V. 2001. Recognition of cognate transfer RNA by the 30S ribosomal subunit. *Science* **292:** 897–902.

Park EH, Walker SE, Lee JM, Rothenburg S, Lorsch JR, Hinnebusch AG. 2011. Multiple elements in the eIF4G1 N-terminus promote assembly of eIF4G1·PABP mRNPs in vivo. *EMBO J* **30:** 302–316.

Petersen HU, Roll T, Grunberg-Manago M, Clark BF. 1979. Specific interaction of initiation factor IF2 of *E. coli* with formylmethionyl-tRNA f Met. *Biochem Biophys Res Commun* **91:** 1068–1074.

Petrov A, Puglisi JD. 2010. Site-specific labeling of *Saccharomyces cerevisiae* ribosomes for single-molecule manipulations. *Nucleic Acids Res* **38:** e143.

Qu X, Wen JD, Lancaster L, Noller HF, Bustamante C, Tinoco I Jr. 2011. The ribosome uses two active mechanisms to unwind messenger RNA during translation. *Nature* **475:** 118–121.

Reibarkh M, Yamamoto Y, Singh CR, del Rio F, Fahmy A, Lee B, Luna RE, Ii M, Wagner G, Asano K. 2008. Eukaryotic initiation factor (eIF) 1 carries two distinct eIF5-binding faces important for multifactor assembly and AUG selection. *J Biol Chem* **283:** 1094–1103.

Ruusala T, Ehrenberg M, Kurland CG. 1982. Is there proofreading during polypeptide synthesis? *EMBO J* **1:** 741–745.

Schlunzen F, Zarivach R, Harms J, Bashan A, Tocilj A, Albrecht R, Yonath A, Franceschi F. 2001. Structural basis for the interaction of antibiotics with the peptidyl transferase centre in eubacteria. *Nature* **413:** 814–821.

Schuwirth BS, Borovinskaya MA, Hau CW, Zhang W, Vila-Sanjurjo A, Holton JM, Cate JH. 2005. Structures of the bacterial ribosome at 3.5 Å resolution. *Science* **310:** 827–834.

Spurio R, Brandi L, Caserta E, Pon CL, Gualerzi CO, Misselwitz R, Krafft C, Welfle K, Welfle H. 2000. The C-terminal subdomain (IF2 C-2) contains the entire fMet-tRNA binding site of initiation factor IF2. *J Biol Chem* **275:** 2447–2454.

Sundari RM, Stringer EA, Schulman LH, Maitra U. 1976. Interaction of bacterial initiation factor 2 with initiator tRNA. *J Biol Chem* **251:** 3338–3345.

Takyar S, Hickerson RP, Noller HF. 2005. mRNA helicase activity of the ribosome. *Cell* **120:** 49–58.

Thompson RC, Stone PJ. 1977. Proofreading of the codon-anticodon interaction on ribosomes. *Proc Natl Acad Sci* **74:** 198–202.

Tomsic J, Vitali LA, Daviter T, Savelsbergh A, Spurio R, Striebeck P, Wintermeyer W, Rodnina MV, Gualerzi CO. 2000. Late events of translation initiation in bacteria: A kinetic analysis. *EMBO J* **19:** 2127–2136.

Tsai A, Petrov R, Marshall A, Korlach J, Uemura S, Puglisi JD. 2012. Heterogeneous pathways and timing of factor departure during translation initiation. *Nature* (in press).

Uemura S, Dorywalska M, Lee TH, Kim HD, Puglisi JD, Chu S. 2007. Peptide bond formation destabilizes Shine-Dalgarno interaction on the ribosome. *Nature* **446:** 454–457.

Uemura S, Aitken CE, Korlach J, Flusberg BA, Turner SW, Puglisi JD. 2010. Real-time tRNA transit on single translating ribosomes at codon resolution. *Nature* **464:** 1012–1017.

Underwood KA, Swartz JR, Puglisi JD. 2005. Quantitative polysome analysis identifies limitations in bacterial cell-free protein synthesis. *Biotechnol Bioeng* **91:** 425–435.

Valle M, Zavialov A, Sengupta J, Rawat U, Ehrenberg M, Frank J. 2003. Locking and unlocking of ribosomal motions. *Cell* **114:** 123–134.

Wen JD, Lancaster L, Hodges C, Zeri AC, Yoshimura SH, Noller HF, Bustamante C, Tinoco I. 2008. Following translation by single ribosomes one codon at a time. *Nature* **452:** 598–603.

Wimberly BT, Brodersen DE, Clemons WM Jr, Morgan-Warren RJ, Carter AP, Vonrhein C, Hartsch T, Ramakrishnan V. 2000. Structure of the 30S ribosomal subunit. *Nature* **407:** 327–339.

Wu XQ, RajBhandary UL. 1997. Effect of the amino acid attached to *Escherichia coli* initiator tRNA on its affinity for the initiation factor IF2 and on the IF2 dependence of its binding to the ribosome. *J Biol Chem* **272:** 1891–1895.

Wu XQ, Iyengar P, RajBhandary UL. 1996. Ribosome-initiator tRNA complex as an intermediate in translation initiation in *Escherichia coli* revealed by use of mutant initiator tRNAs and specialized ribosomes. *EMBO J* **15:** 4734–4739.

Yusupov MM, Yusupova GZ, Baucom A, Lieberman K, Earnest TN, Cate JH, Noller HF. 2001. Crystal structure of the ribosome at 5.5 Å resolution. *Science* **292:** 883–896.

Yusupova GZ, Yusupov MM, Cate JH, Noller HF. 2001. The path of messenger RNA through the ribosome. *Cell* **106:** 233–241.

Yusupova G, Jenner L, Rees B, Moras D, Yusupov M. 2006. Structural basis for messenger RNA movement on the ribosome. *Nature* **444:** 391–394.

Zhang W, Dunkle JA, Cate JH. 2009. Structures of the ribosome in intermediate states of ratcheting. *Science* **325:** 1014–1017.

Cite this article as *Cold Spring Harb Perspect Biol* doi: 10.1101/cshperspect.a011551

The Current Status of Vertebrate Cellular mRNA IRESs

Richard J. Jackson

Department of Biochemistry, University of Cambridge, Cambridge CB2 1QW, United Kingdom
Correspondence: rjj@mole.bio.cam.ac.uk

Internal ribosome entry sites/segments (IRESs) were first discovered over 20 years ago in picornaviruses, followed by the discovery of two other types of IRES in hepatitis C virus (HCV), and the dicistroviruses, which infect invertebrates. In the meantime, reports of IRESs in eukaryotic cellular mRNAs started to appear, and the list of such putative IRESs continues to grow to the point in which it now stands at ∼100, 80% of them in vertebrate mRNAs. Despite initial skepticism from some quarters, there now seems universal agreement that there is genuine internal ribosome entry on the viral IRESs. However, the same cannot be said for cellular mRNA IRESs, which continue to be shrouded in controversy. The aim of this article is to explain why vertebrate mRNA IRESs remain controversial, and to discuss ways in which these controversies might be resolved.

The first part of this article reviews the current understanding of viral IRESs, mainly picornavirus IRESs, because they have strongly influenced thinking on putative cellular IRESs, and because picornavirus IRESs, especially the encephalomyocarditis virus (EMCV) IRES, and to a lesser extent the human rhinovirus (HRV) and poliovirus (PV) IRESs, are frequently used as positive controls in tests for putative cellular mRNA IRESs. All viral IRESs are readily classifiable into distinct families on the basis of sequence and secondary structure: (1) the intergenic IRES of invertebrate dicistroviruses, (2) the hepatitis C virus (HCV) and related animal virus IRESs, and (3) the picornavirus IRESs that can be further classified into several distinct subgroups, including one class (exemplified by porcine teschovirus 1 and simian virus 9) that is remarkably similar to the HCV-like IRESs in structure and initiation factor requirements (Hellen and de Breyne 2007; de Breyne et al. 2008). The predicted secondary structures of these RNA virus 5′-UTRs are particularly robust, because they are founded not only on direct structure probing but also on extensive phylogenetic comparisons. The very high error frequency of RNA replication results in enormous genetic drift, both within species and between species, and so there are numerous covariances validating the proposed base-pairing.

In contrast, the putative IRESs identified in cellular mRNAs defy classification because they are all different from one another in sequence and predicted secondary structures (Baird et al. 2006), which have necessarily been elucidated entirely from structure probing, as there is insufficient genetic drift, even between different animal species, to provide useful phylogenetic data. It may well be that the usual mammalian species for such comparisons (mainly primates, rodents, and ruminants) are too close in evolution, and it might be more informative to widen the comparison to include more distant vertebrates, such as birds, frogs, and marsupials, which has proved helpful in the discovery of other regulatory elements in mRNAs (Koeller et al. 1989; Sherrill and Lloyd 2008).

Putative cellular IRESs have been generally considered closest to the true picornavirus IRESs (mainly on the grounds of similar initiation factor requirements), although it will become apparent that any similarity is very remote. The dicistrovirus intergenic IRESs and the HCV-like IRESs differ from picornavirus IRESs in that they bind 40S ribosomal subunits directly in the absence of any canonical translation initiation factors. Initiation on the dicistrovirus intergenic IRESs does not even require Met-tRNA$_i$ and does not occur at an AUG codon (Sasaki and Nakashima 2000; Wilson et al. 2000). Although initiation on the HCV-like IRESs does require Met-tRNA$_i$ (usually as a ternary complex with eIF2 and GTP), the additional canonical initiation factor requirements are limited to eIF3, eIF5, and eIF5B, with no requirement for eIF4A, 4B, 4E, or 4G (Pestova et al.1998, 2008).

STRUCTURE AND FUNCTION OF PICORNAVIRUS IRESs (AND SOME COMPARISONS WITH PUTATIVE CELLULAR IRESs)

Classification of Picornavirus IRESs on the Basis of Sequence and Structure

Apart from one outlier, hepatitis A virus (HAV), every picornavirus IRES can be placed unambiguously into one of four distinct groups: the long-standing Type I IRESs, which include HRV, PV, and other enteroviruses, and Type II IRESs, which include foot-and-mouth disease virus (FMDV) and EMCV (Alexander et al. 1994; Jackson and Kaminski 1995); the more recently discovered Aichivirus (AV) group (Yu et al. 2011); and the HCV-like group already mentioned. Within each of the two major pi-

cornavirus IRES classes (Types I and II), there is quite strong conservation of primary sequence, particularly in unpaired loops or bulges, and even stronger conservation of predicted secondary structure (Jackson and Kaminski 1995). However, there is very little similarity between the different classes (including HAV and AV) apart from a ~25 nt tract at the 3′-end (as defined by deletion analysis), which is G-poor throughout, pyrimidine-rich at its 5′-end, followed by a middle section that is hypervariable even between different strains and isolates of the same virus species (Pöyry et al. 1992), and ends in an AUG triplet (Fig. 1). The deletion mapping shows that the 5′-boundary of these IRESs is slightly "fuzzy" in that progressive deletion of the first ~100 nt of the PV-2 5′-UTR, or the first ~120 nt downstream from the polyC tract of EMCV, reduces internal initiation by up to ~30% (Jang and Wimmer 1990; Nicholson et al. 1991). However, further deletion from the 5′-end results in a precipitous fall in activity, and the position in which this dramatic decrease occurs is usually taken as the 5′-boundary of the core IRES, which is typically ~450 nt in length. The 3′-boundary is very sharp and coincides with the AUG immediately downstream from the oligopyrimidine tract. Any attempt to shorten the pyrimidine-rich tract and thereby move the AUG further upstream by more than a very few residues abrogates activity (Iizuka et al. 1989; Kaminski et al. 1994). Subdomains of the core IRES show no IRES activity. In contrast, for many putative cellular mRNA IRESs, several quite short segments have been reported to promote fairly efficient internal initiation, leading to the suggestion that such IRESs are composed of multiple short modules, which promote internal initiation by acting in combination (Stoneley and Willis 2004).

Picornavirus 5′-UTRs have numerous AUG triplets at about the frequency expected for random occurrence (one per 64 nt), whereas a survey of the mammalian mRNA 5′-UTR database showed that AUG triplets occurred only slightly more frequently in the 66 mRNAs with putative IRESs (median frequency of one per 300 nt) than in the mRNA population at large (Baird et al. 2006). The AUG triplets in picornavirus

Cite this article as *Cold Spring Harb Perspect Biol* doi: 10.1101/cshperspect.a011569

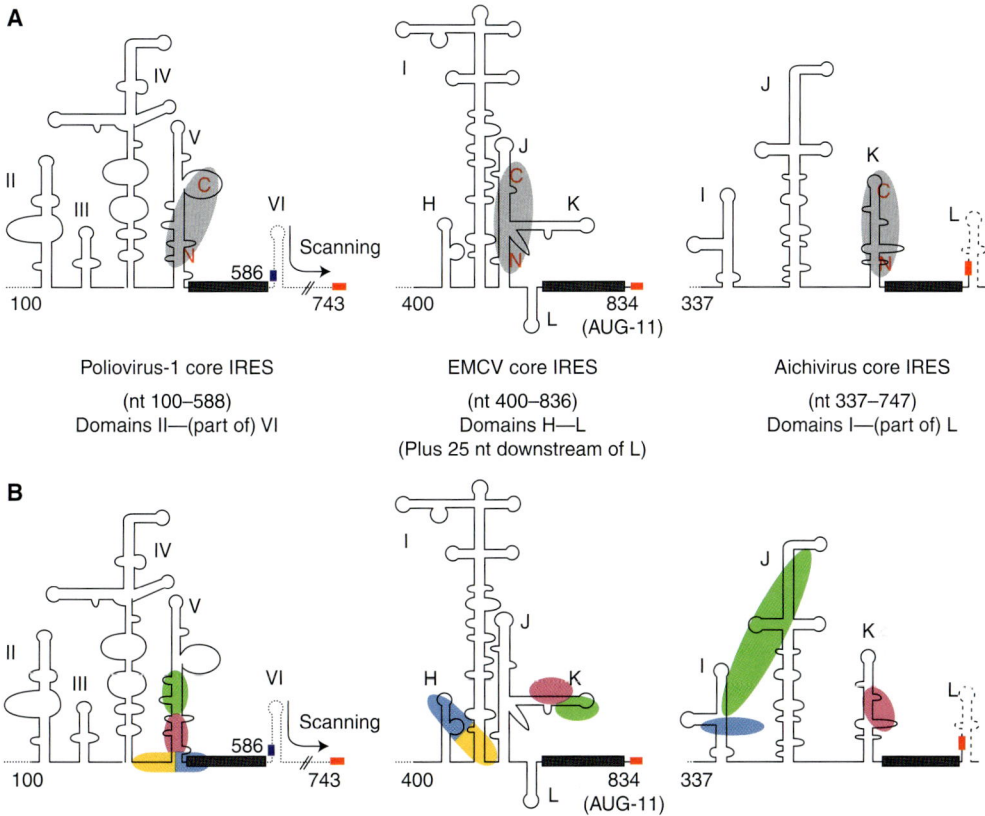

Figure 1. Secondary structure models of the three main classes of picornavirus IRESs, with the eIF4G and PTB (polypyrimidine tract binding protein) binding sites also shown. (*A*) Secondary structures of the designated core IRESs, with individual structural domains labeled according to standard nomenclature. The dotted lines represent 5′-UTR sequences outside the core IRES boundaries, and the dashed line at 3′-end of the Aichivirus structure shows 5′-proximal viral coding sequences. The dark gray rectangle represents the ∼25 nt pyrimidine-rich tract at the 3′-end of the IRESs, the small red rectangle shows the authentic initiation site AUG, and the blue rectangle represents the putative ribosome recruitment AUG of Type I IRESs (at nt 586 in poliovirus type 1). The binding site of the central domain of eIF4G (p50 fragment—see Fig. 2A) on each IRES, as determined by footprinting and tethered hydroxyl radical probing, is shown in light gray, with the amino and carboxyl termini indicated (in red) to show the orientation of binding (Kolupaeva et al. 2003; de Breyne et al. 2009; Yu et al. 2011). (*B*) Sites and orientation of PTB binding, as determined by tethered hydroxyl radical probing. The interaction sites of each RBD (RNA-binding domain) of PTB-1 are shown on the three IRES secondary structure maps, using the same color coding as in Figure 2B, namely: RBD-1 in green, RBD-2 in pink, RBD-3 in blue and RBD-4 in yellow (Kafasla et al. 2009, 2010). The Aichivirus results showed that RBD-1 interacts strongly with the apical regions of both domains I and J, suggesting that these regions are closer to each other than can be shown on a two-dimensional diagram (Yu et al. 2011), which explains the elongated depiction of RBD-1. No contacts between RBD-4 and the Aichivirus IRES were detected.

5′-UTRs are also subject to genetic drift, as shown by the fact that in 33 clinical isolates of poliovirus type 3 the number ranged from 5 to 15 (Pöyry et al. 1992); only three of them were absolutely conserved and the downstream short ORFs were not conserved in length. The only conserved AUG important for IRES activity is the one located at the 3′-end of the IRES. In the case of the EMCV IRES this AUG (AUG-11) is the authentic initiation codon, and there is very good evidence that 43S preinitiation complexes bind initially at, or extremely near, this AUG.

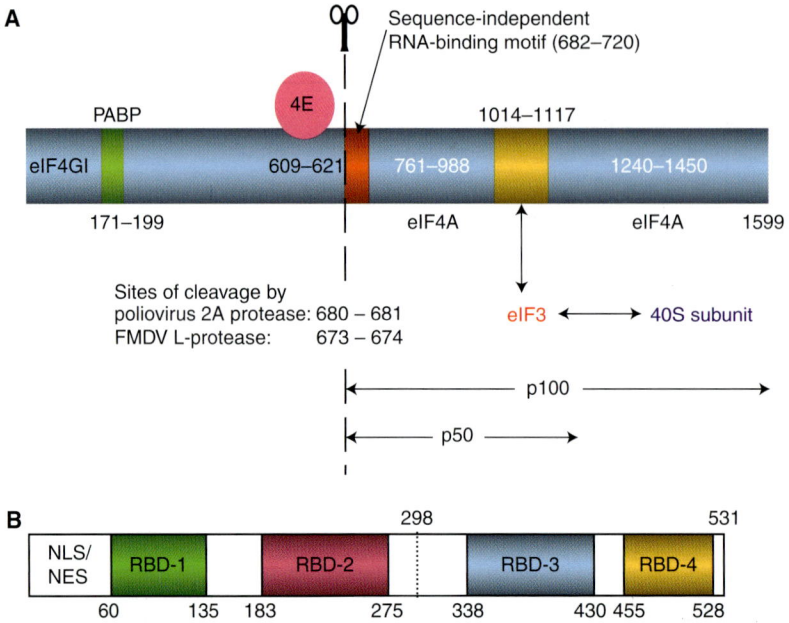

Figure 2. Domain structures of eIF4GI and polypyrimidine tract binding protein (PTB). (*A*) Domain structure of the longest isoform (1599 amino acids) of eIF4GI (blue) with associated eIF4E (magenta), showing the interaction sites of poly(A) binding protein (PABP) in green, and eIF3 (gold). The two sites of potential interaction with eIF4A are shown, although there is usually only a single bound eIF4A. The sites at which eIF4GI is cleaved by poliovirus 2A protease and FMDV L-protease are shown; the 2A cleavage site defines the amino termini of the carboxy-terminal two-thirds fragment (p100) of eIF4GI, and the central one-third domain (p50). The sequence-independent RNA-binding motif at the amino terminus of p100 (and p50) is highlighted in orange; this motif is necessary for scanning, but is not required for internal initiation on the EMCV IRES (Ali and Jackson 2001; Prévôt et al. 2003). (*B*) Domain structure of PTB-1. The amino-terminal ∼55 amino acid residues have nuclear import and export signals, but play no part in RNA-binding. The positions of the 4 RBDs (RNA-binding domains) are shown. RBDs-2 and -3 are longer than the other two RBDs because their RNA-binding surface has an additional β-strand. The linkers between the RBDs are flexible except for that between RBDs-3 and -4, which interact with each other in a back-to-back configuration, and act as a coordinated pair (Oberstrass et al. 2005). PTB-2 and PTB-4 differ from the canonical PTB-1 in having inserts (arising from alternative splicing) of 19 or 26 amino acids, respectively, at residue 298 of PTB-1.

Almost all translation initiation is at this AUG, with a little at AUG-12, situated four codons further downstream, but there is hardly any initiation at a nonconserved AUG present in some EMCV strains 8 nt upstream of AUG-11 (Kaminski et al. 1990, 1994).

In contrast, very little, if any, initiation occurs at the equivalent AUG located just downstream from the oligopyrimidine tract in Type I IRESs (Pestova et al. 1994; Kaminski et al. 2010), and the authentic initiation codon is the next AUG further downstream, at a distance of ∼35 nt in rhinoviruses and ∼160 nt in polioviruses and other enteroviruses. Despite its negligible

activity as an initiation site, this AUG at the end of the pyrimidine-rich tract is nevertheless very important for efficient initiation at the correct downstream start site. Mutation of the near-silent AUG in HRV-2 and PV-2 reduced initiation at the authentic start site by ∼70% (Meerovitch et al. 1991; Kaminski et al. 2010), and conferred a small plaque phenotype in the PV-2 background (Pelletier et al. 1988). This has led to the suggestion that 43S preinitiation complexes are first recruited at this AUG, but instead of initiating there, they are transferred to the next AUG further downstream, most probably by a linear scanning process, or by a minor variant of linear scanning in which a few residues are by-passed (Kaminski et al. 2010).

Canonical Initiation Factor Requirements for Picornavirus IRESs

Initiation on Type I, Type II, and AV IRESs requires all the canonical initiation factors except that (1) eIF4E is completely redundant, and (2) the carboxy-terminal two-thirds fragment (p100) of eIF4G, or the central one-third fragment (p50), can substitute for full-length eIF4G and is usually somewhat better than the full-length protein. Both p50 and p100 retain the eIF3 interaction site and at least one of the two eIF4A interaction sites present in intact eIF4G (Fig. 2A). These eIF4G derivatives bind directly to the basal half of Domain V of Type 1 IRESs (de Breyne et al. 2009), the three-way junction between Domains J and K of the Type II IRESs (Kolupaeva et al. 1998; Lomakin et al. 2000), and Domain K of the AV IRESs (Yu et al. 2011), and in all cases binding is strongly enhanced by inclusion of eIF4A (Fig. 1). Although the amino terminus of p100 (or p50) has general RNA binding activity with little nucleotide sequence specificity (Prévôt et al. 2003), the high affinity binding to the viral IRESs appears to involve a more carboxy-terminal domain of eIF4G p100, because deletion of the amino terminus of p100 (or p50) has little effect on its activity in driving initiation on the EMCV IRES, but abrogates its ability to support scanning-dependent initiation (Ali and Jackson 2001; Prévôt et al. 2003).

Curiously, although there is some similarity between the p100 binding sites on Type II and AV IRESs, the binding site on Type I IRESs is completely different in sequence and secondary structure. Nevertheless, the positions of these binding sites suggest a common mechanism in which a 43S preinitiation complex (a 40S ribosomal subunit with associated eIF2/GTP/Met-tRNA$_i$ ternary complex, and other initiation factors including eIF3) could be initially recruited via the interaction of the eIF3 with the IRES-associated p100 (or p50), and then delivered to the AUG located immediately downstream from the oligopyrimidine tract (Fig. 1).

Although the binding of eIF4G (or p100) to this internal site is a prerequisite for picornavirus IRES activity, it is clearly not sufficient, because the EMCV JK domain on its own (without the upstream H and I domains, and the downstream sequences extending to AUG-11) shows no significant activity in the standard bicistronic mRNA test for IRESs (described below).

The HAV IRES is exceptional in that it appears to require all the canonical initiation factors including the complete eIF4F complex (consisting of eIF4G, the RNA helicase eIF4A, and the cap-binding factor eIF4E), because it is inhibited by cap analogs, or 4E-BP, or cleavage of eIF4G by the poliovirus 2A protease or the FMDV L-protease (Ali et al. 2001; Borman et al. 2001). Although these publications proposed two alternative reasons for this unusual requirement of the HAV IRES, the true explanation remains unknown.

IRES *Trans*-Acting Factor (ITAF) Requirements

In addition to the canonical initiation factors, the activity of picornaviral IRESs is also dependent on other RNA-binding proteins, or ITAFs, but the requirements differ quite markedly for different IRESs, even between closely related IRESs. For example, although the HRV and FMDV IRESs have a very strong requirement for polypyrimidine tract binding protein (PTB) (Hunt and Jackson 1999; Pilipenko et al. 2000), the EMCV IRES generally shows high activity in the absence of any ITAFs and is only slightly stimulated by PTB (Pestova et al. 1996; Kaminski and Jackson 1998), whereas the PV and AV IRESs occupy an intermediate position between the two extremes of FMDV and EMCV (Hunt and Jackson 1999; Yu et al. 2011). The FMDV IRES also requires another RNA-binding protein of 45 kDa known as ITAF 45 (Pilipenko et al. 2000). All Type I IRESs tested so far require polyC binding protein-2 (Walter et al. 1999), which seems to act in conjunction with SRp20 (Bedard et al. 2007). In addition, the type I HRV IRES has a very strong requirement for Unr (upstream of N-ras), whereas this ITAF has little effect on the PV IRES in vitro (Hunt et al. 1999), although a strong dependency was seen in transfection assays (Boussadia et al. 2003). The PV IRES is also reported to be strongly stimulated by the autoantigen La (Svitkin et al. 1994; Costa-Mattioli et al. 2004).

The list of ITAFs reported to affect the activity of various putative cellular mRNA IRESs is even longer (reviewed in King et al. 2010); in addition to PTB (also known as hnRNP I), PCBP-2, Unr and La, it includes several other hnRNP proteins and other RNA-binding proteins that are predominantly located in the nucleus. As with the viral IRESs, PTB is the most "promiscuous" in the sense that it stimulates more putative cellular IRESs than any other ITAF (King et al. 2010).

As for the underlying mechanism, there is no evidence that any of the ITAFs play a direct role in recruiting 43S preinitiation complexes to the IRES. The idea of a more indirect effect is also consistent with the rather puzzling finding that closely related viral IRESs (e.g., EMCV and FMDV) differ quite markedly in their dependency on ITAFs. A frequent suggestion is that ITAF binding might remodel the secondary and tertiary structure of the IRES into a form that is more optimal for internal initiation, and this receives some support from the observation that ITAF binding not only protects some IRES residues against enzymatic and chemical attack (as would be expected), but also makes some residues more susceptible to these reagents (Pilipenko et al. 2000). Further clues have emerged from recent studies using tethered hydroxyl radical probing to determine which of the four RNA-binding domains (RBDs) of PTB (Fig. 2B) binds to which site on the IRES. This has shown that the core EMCV IRES binds a single PTB at widely dispersed sites (Fig. 1), such that PTB binding could constrain the three-dimensional flexibility of the IRES (Kafasla et al. 2009). In contrast, however, PTB binding to the PV-1 IRES was highly localized to the basal half of Domain V and the single-stranded flanking linkers (Fig. 1). The PTB binding site overlapped the eIF4G-binding site on the PV-1 IRES, and it was shown that PTB binding subtly repositions eIF4G (Kafasla et al. 2010), which might provide an explanation for how PTB activates Type I IRESs.

ALTERNATIVES TO THE CAP-DEPENDENT SCANNING MECHANISM OF INITIATION OF CELLULAR mRNA TRANSLATION

Because cellular mRNAs are the products of transcription by RNA Pol II, they all have m^7G-caps and so they are all potentially translatable by the cap-dependent scanning mechanism; and although AUG triplets in 5′-UTRs can attenuate scanning or even act as barriers (depending on their position, context and the length of the following ORF), such AUGs occur only slightly more frequently in 5′-UTRs with putative IRESs than in the mRNA population at large (Baird et al. 2006). The question, therefore, is whether some cellular mRNAs can be translated by alternative mechanisms, which might operate in parallel with the scanning mechanism under normal conditions, but could predominate under conditions when cap-dependent (scanning) initiation is compromised.

Although the focus of this review is IRES-dependent initiation, it is worth briefly noting that there are two other "nonstandard" initiation mechanisms (excluding special mechanisms for leaderless mRNAs or mRNAs with extremely short 5′-UTRs, which are irrelevant to this article). One is ribosome shunting, which is considered a form of discontinuous scanning that has been seen with the adenovirus tripartite leader 5′-UTR and the Hsp 70 5′-UTR, and appears to be favored over scanning when eIF4F availability or activity is reduced (Yueh and Schneider 1996, 2000).

The other is the mechanism of "cap-independent" initiation of translation of several different types of plant viral RNAs (listed in Miller et al. 2007; Mokrejs et al. 2010), which all have uncapped positive strand RNA genomes. Initiation is strictly dependent on a 3′-UTR motif known as a cap-independent translation enhancer (CITE, or simply TE), which binds eIF4F through its eIF4G and/or eIF4E subunits, and promotes initiation by a mechanism of scanning from the uncapped 5′-end. This mechanism shows that a 5′-cap (and therefore, by implication, also eIF4E) is not necessary for efficient initiation by scanning from the 5′-end, provided there is an effective means of eIF4G recruitment to compensate for absence of the eIF4E-cap interaction. Like the EMCV JK domain, these TEs do not show significant IRES activity in the standard bicistronic mRNA test (described below). Although the natural position is always in the 3′-UTR, many such TEs function equally well when repositioned to replace the endogenous 5′-UTR. In view of the similarity between plant and vertebrate translation mechanisms, it seems quite possible that this initiation mechanism could also operate in vertebrates (reviewed in Shatsky et al. 2010).

CELLULAR mRNA IRESs

Indicators of a Possible IRES in a Cellular mRNA

Because there is presently no high throughput screening method for revealing cellular mRNA IRESs, they have to be identified by testing each candidate on a case-by-case basis. Two features are often taken as preliminary indicators of candidates that would be worth testing: (1) an unusually long and GC-rich 5′-UTR that is predicted to be highly structured; and (2) persistence of translation in stress conditions, or in mitosis (Qin and Sarnow 2004), when eIF4F activity or availability is compromised. However, there are caveats associated with both these indicators. All too often, a high negative ΔG value for the predicted 5′-UTR secondary structure is taken as strong evidence that initiation via cap-dependent scanning will be inevitably inefficient. However, this overlooks the fact that the ΔG value is related to the equilibrium constant for the transition from the folded to the completely unfolded forms, and is inversely related to the probability of spontaneous and unassisted complete unwinding of the whole 5′-UTR, yet scanning requires only localized step-wise unwinding of the 5′-UTR rather than a complete unwinding of the whole sequence. Thus, it is the stability of local elements of secondary structure that determines whether they are significant barriers to scanning (Özeş et al. 2011), rather than the ΔG value of the complete 5′-UTR. Indeed it has been shown that in a Krebs-2 in vitro system, or in an RNA transfection assay, an mRNA with the ∼900 nt

(60% GC) LINE-1 5′-UTR, is translated by the cap-dependent scanning mechanism at 50% relative efficiency compared to the β-globin 5′-UTR (Dmitriev et al. 2007). It should also be noted that some long 5′-UTRs with apparent IRES activity have subsequently been shown to be incompletely spliced variants with retained introns (Baranick et al. 2008), for example the original eIF4GI 5′-UTR studied by Gan and Rhoads (1996), in contrast to the eIF4GI 5′-UTRs described by Johannes and Sarnow (1998) and Byrd et al. (2005).

As for the second indicator, the "persistence" of translation of some mRNAs under conditions of reduced eIF4F activity is often relative rather than absolute. For example, nucleophosmin mRNA translation does decrease in mitosis, but by only half as much as global mRNA translation is reduced (Qin and Sarnow 2004). Similarly, there is actually a decrease in c-myc synthesis during apoptosis, but the decrease is delayed by ∼2 h with respect to global protein synthesis (Bushell et al. 2006). Admittedly there are some intriguing cases of real increases in translation efficiency, for example vimentin mRNA translation during mitosis (Qin and Sarnow 2004). On the other hand, the increased VEGF expression in hypoxia (0.1% O_2 for 6 h) was found to be largely ascribable to increased mRNA abundance (Young et al. 2008), although different hypoxia regimes in other cell types have been reported to activate IRES-dependent translation of VEGF mRNA (Braunstein et al. 2007).

It would be naïve to assume that the affinity of eIF4F for the 5′-end of all mRNAs is the same. It is much more likely that there is differential affinity for different mRNAs, in which case those mRNAs with high affinity will continue to be translated moderately efficiently via the scanning mechanism even when eIF4F activity has been significantly reduced. A likely reflection of this can be found in the observation that addition of cap analog (m^7GTP) or especially 4E-BP1 to a Krebs-2 in vitro system resulted in less inhibition of a reporter with the Apaf-1 (especially), c-myc or Hsp70 5′-UTRs than with the actin 5′-UTR (Andreev et al. 2009). In addition, it should be appreciated that the relief of competition consequent on inhibition of bulk mRNA translation following partial eIF4F inactivation will increase the availability of other initiation factors and ribosomal subunits for those mRNAs with high functional affinity for the residual eIF4F.

It is also worth noting some other observations that are somewhat at variance (but maybe not directly contradictory) with the idea that low eIF4F activity might favor IRES-dependent translation. First, it has been found that a 90% knock-down of eIF4GI, which reduced global translation by only ~20%, preferentially decreased translation of low abundance mRNAs that have short upstream ORFs (Ramírez-Valle et al. 2008). In contrast, the translation of several mRNAs with putative IRESs, including p120 catenin, seemed to be promoted by the overexpression of eIF4GI that commonly occurs in inflammatory breast cancer (Silvera et al. 2009). Finally, the example of the picornaviral HAV IRES shows that some bona fide IRESs may actually require an intact eIF4F complex; and the existence of CITE elements shows that "cap-independent initiation" does not necessarily equate with IRES-dependent initiation.

The Standard Bicistronic Plasmid Test for Cellular mRNA IRESs

Because all cellular mRNAs have at least the potential to be translated by a cap-dependent scanning mechanism, this potential needs to be suppressed to test for possible IRES activity. The usual test follows the same approach as originally used to show the poliovirus IRES (Pelletier and Sonenberg 1988), namely to insert the candidate 5'-UTR as the intercistronic spacer of a bicistronic DNA plasmid construct, although monocistronic constructs with a very stable 5'-proximal hairpin provide a valid alternative. Although many different bicistronic expression constructs have been used (most of them listed in Kozak 2003, 2005), the majority of tests have employed a dual luciferase construct, with *Renilla* luciferase (RLuc) as the upstream cistron, and firefly (*Photinus*) luciferase (FLuc) in the downstream position (Fig. 3A). IRES activity is indicated by a significant in-

crease in the FLuc/RLuc expression ratio as compared with a control bicistronic construct lacking the putative IRES, though the raw data for RLuc and FLuc expression should also be presented so that we can see whether a decrease in RLuc expression, which is by no means unknown (van Eden et al. 2004), has contributed significantly to the IRES-dependent increase in FLuc/RLuc expression ratio. (If insertion of the test 5'-UTR does significantly decrease RLuc expression, this could be a strong warning that splicing may have removed part or all of the RLuc cistron from some transcripts.) The intercistronic spacer in the control construct is usually a short multiple cloning site, and it could be argued that a better control would have an intercistronic spacer more similar in length to the putative IRES, such as the reverse complement of the test 5'-UTR sequence, as has been used in some publications.

Many 5'-UTR sequences tested in this way promote a huge increase in the FLuc/RLuc ratio, often a few hundred-fold and significantly greater than the typical ~20-fold increase observed with the bona fide EMCV IRES. The current record appears to be the triose phosphate isomerase (TPI) 5'-UTR, which elicits almost a 1000-fold increase in the FLuc/RLuc ratio (Young et al. 2008). Impressive though these stimulations may appear at first sight, it should be remembered that FLuc expression from the control (no IRES) construct should, in principle, be zero, and so a 100-fold increase in what is probably close to zero could still be quite small in absolute terms. This highlights a major weakness in this method of determining IRES activity, because it does not give any indication of the efficiency of the apparent IRES-dependent translation in comparison with the efficiency of translation of (1) a monocistronic FLuc mRNA with the same 5'-UTR, and (2) the upstream RLuc cistron (i.e., the molar ratio of FLuc/RLuc synthesis rates). The first issue could be addressed by simply deleting the RLuc 5'-UTR and ORF sequences from the bicistronic construct depicted in Figure 3A, yet this control seems to be never included in DNA transfection assays, although it is common practice in RNA

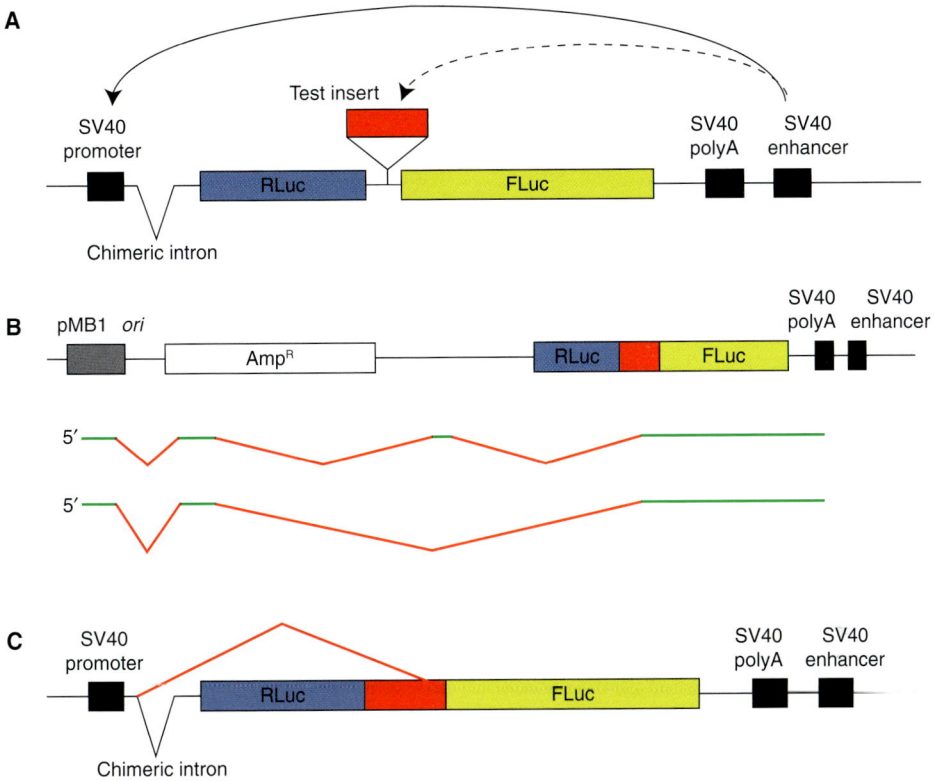

Figure 3. Mechanisms by which monocistronic mRNAs can arise from the dual luciferase plasmid construct commonly used to test for IRESs. In all three subpanels the *Renilla* luciferase (RLuc) ORF is shown in blue, the *Photinus* luciferase (FLuc) ORF in yellow, and the cellular mRNA 5′-UTR under test for IRES activity is in red. Panel (*B*) is on a reduced scale. (*A*) Configuration of the dual luciferase construct and its essential elements (promoter, chimeric intron, polyA addition signal, and enhancer). The downstream SV40 enhancer strongly activates transcription from the SV40 promoter, but can also activate transcription from any cryptic promoter element present in the putative IRES (as shown by the dashed line), thereby generating a monocistronic FLuc mRNA. (*B*) Transcription from a promoter in the vector backbone, near the pMB1 origin, also gives rise to monocistronic FLuc mRNA if the putative IRES has 3′-splice sites. The diagram shows two of the most abundant spliced RNA products found (Lemp et al. 2012), with the retained (exon) sequences shown in green, and the spliced out introns in red. About half of such monocistronic FLuc mRNAs are devoid of upstream AUG triplets, and so could give rise to high FLuc expression via cap-dependent scanning. These unanticipated mRNAs are equally abundant irrespective of whether the SV40 promoter is absent (as depicted) or present (Lemp et al. 2012). (*C*) Generation of a monocistronic FLuc mRNA by splicing from the 5′-splice donor site of the chimeric intron to 3′-splice site(s) fortuitously present in the putative IRES.

transfections (discussed below). For the second issue, there seems to be no instance where the actual molar ratio of FLuc/Ruc expression has been calculated from the luciferase activity data, yet the luciferase activities per mole of each protein under the particular assay conditions that were used must surely be known, or could be quite easily determined experimentally.

Some IRESs Are Only Active if Expressed via Transcription by RNA Polymerase II

One reason for the widespread use of DNA transfection assays is that many putative cellular mRNA IRESs do not function efficiently, if at all, in cell-free translation systems (not even in extracts from HeLa or other tissue culture cells), nor in direct RNA transfections, nor in DNA transfections in which mRNA synthesis occurs in the cytoplasm because it is driven by T7 RNA polymerase generated via infection of the cells with a recombinant vaccinia virus that expresses T7 polymerase (Stoneley et al. 2000; Shiroki et al. 2002; Holcik et al. 2003). Thus, in these cases IRES activity is manifest only if the mRNA is generated in the nucleus via Pol II. This has led to the suggestion that a "nuclear experience" is required for these IRESs to be active.

One obvious and plausible explanation for this requirement is that the apparent IRES activity might actually be attributable to the production of a monocistronic FLuc mRNA, either from a cryptic Pol II promoter sequence or through a minor splicing event. When the "nuclear experience" requirement was first observed, it was claimed that such artefacts had been ruled out, and the proposed explanation was that IRES activity is absolutely dependent on certain ITAFs being deposited on the nascent bicistronic mRNA in the nucleus, and coexported to the cytoplasm. The presumption was that such ITAFs are either (normally) restricted to the nuclear compartment, or that they can only be deposited appropriately during transcription or subsequent pre-mRNA processing. However, no direct evidence supporting this hypothesis has been forthcoming in the many years since it was first proposed. Meanwhile, in the past ∼8 years there have been so many reports of monocistronic mRNA artefacts arising from cryptic promoters or from splicing that this must now surely be taken as the default explanation for the "nuclear experience," unless or until convincing evidence to the contrary is produced.

Essential Controls to Screen for Monocistronic mRNAs Arising from Cryptic Promoters and/or Splicing

It has long been common practice to screen for possible monocistronic FLuc mRNA production by using Northern blots, or, more recently, qPCR measurement of the relative quantity of RLuc and FLuc ORF sequences. However, it is very doubtful whether these methods would detect monocistronic FLuc mRNA present at a level of less than 2%, which is the likely required threshold suggested by the relative FLuc expression from monocistronic and bicistronic mRNAs with the same test 5′-UTR in RNA transfection assays (discussed in a subsequent section). In any case, this approach is rather unsatisfactory because it is indirect and involves two controversial issues: (1) what is the required sensitivity, and (2) whether the assay achieves this level of stringency (Kozak 2003, 2005). As discussed below, better methods are now available for revealing more directly whether monocistronic mRNAs are giving rise to significant FLuc expression. Many investigators mistakenly consider the test of inserting a very stable hairpin upstream of the RLuc cistron to be a decisive control. As expected, this drastically reduces RLuc synthesis, usually with little or no effect on FLuc expression, but this outcome does not constitute proof of an IRES, because precisely the same result could be obtained if virtually all FLuc expression was from an unanticipated monocistronic mRNA.

Although some investigators screen for cryptic promoters by testing for FLuc expression when the bicistronic construct with the putative IRES sequence is transferred into a completely different plasmid background lacking known promoters (and in some such tests also lacking an enhancer), a more rigorous approach is to delete the SV40 promoter but retain the SV40 enhancer of the expression plasmid construct (Fig. 3A) actually used to assay IRES activity. This invariably results in an 80%–90% reduction in RLuc expression, and a similar percentage reduction in FLuc yield is seen in the case of the EMCV IRES and the no IRES control. However, for a great many 5′-UTRs with putative IRES activity (HIF-1α, HIF-2α, XIAP, c-myc, VEGF, VEGF receptor1, EGR-1, Glut-1, PIM-1, TPI, p27^{Kip1}, and p57^{Kip2}) there was, at best, only a slight reduction in downstream cistron translation, and in many cases a significant increase in FLuc expression was seen (Liu et al. 2005; Wang et al. 2005; Bert et al. 2006; Young et al. 2008). These results provide strong evidence that a cryptic promoter is producing FLuc mRNAs that certainly lack a complete

Cite this article as *Cold Spring Harb Perspect Biol* doi: 10.1101/cshperspect.a011569

RLuc ORF and may well be monocistronic. Although there is a report (Vopalensky et al. 2008) of weak cryptic promoter activity in the 3′-proximal part of the FLuc ORF (that would be troublesome only when FLuc is used as the upstream cistron), no such activity has been found in the RLuc ORF, and so the usual presumption is that the cryptic promoter lies in the putative IRES sequence itself. However, it has recently been shown that there is a cryptic promoter within the backbone (near the origin of replication) of virtually all expression vectors in current use (Lemp et al. 2012), and although the primary transcripts from this promoter are polycistronic, they can give rise to monocistronic FLuc mRNAs lacking upstream AUG triplets (and therefore translatable by cap-dependent scanning) if the tested 5′-UTR has potential 3′-splice sites (Fig. 3B). A detailed RNA analysis (e.g., 5′-RACE) is needed to determine whether the ultimate origin of the monocistronic FLuc mRNA is this vector backbone promoter or a cryptic promoter in the putative IRES (Lemp et al. 2012).

Even if (almost) all transcription were from the intended (SV40) promoter, splicing can still cause artefacts, as almost all expression vectors in current use have an intron in the 5′-UTR (Fig. 3), because introns have been shown to stimulate the synthesis, processing and nucleo-cyto-plasmic transport of mRNA (Le Hir et al. 2003), but an intron in the 3′-UTR located more than ∼50 nt downstream from the stop codon will trigger nonsense mediated decay (Nagy and Maquat 1998). However, the propensity for alternative splicing in vertebrates is so extremely high that the system will likely attempt all possible permutations, not just the intended splicing pattern. So this ubiquitous presence of a 5′-proximal splice donor site poses a strong risk of mis-splicing to any cryptic 3′-splice site present in the putative IRES under test (Fig. 3C), as shown by the finding that when all potential 5′-splice sites in the 5′-UTR and the upstream cistron (in this case coding for *Gaussia* luciferase) were inactivated by mutations, the BiP, NRF, VEGF, and XIAP 5′-UTRs, as well as the eIF4GI 5′-UTR reported by Gan and Rhoads (1996), failed to show significant IRES activity above the "no IRES" control, or the β-globin 5′-UTR, although the EMCV IRES positive control yielded high activity (Baranick et al. 2008). In the same publication a sensitive assay involving incorporating the IRES and a GFP reporter into a murine leukaemia virus proviral DNA (Fig. 4) revealed that the same eIF4GI 5′-UTR, as well as the NRF and XIAP 5′-UTRs all had 3′-splice sites, which accounted for the apparent IRES activity, although here again the EMCV IRES scored as a bona fide IRES. Moreover, a database

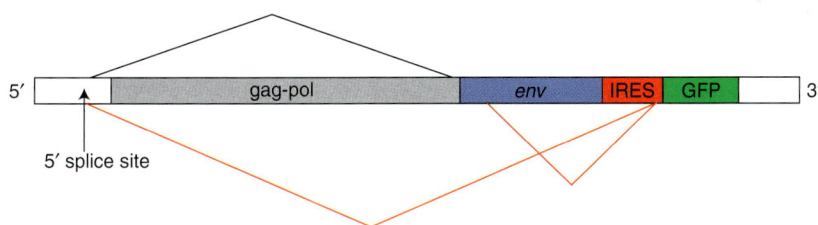

Figure 4. Highly sensitive murine leukaemia virus test for the presence of 3′-splice sites in putative IRESs (Baranick et al. 2008). The putative IRES and GFP reporter are inserted into the proviral DNA (driven by a CMV promoter) downstream from the *env* gene. The normal MLV splicing pattern (in the absence of the IRES-GFP insert) is shown in black, above the gene map. The presence of a 3′-splice site in the putative IRES promotes the alternative splicing pattern shown in red below the gene map. Consequently, there is very high GFP expression (i.e., apparent IRES activity) from the resulting capped monocistronic mRNA, but viral replication is severely impaired because of the decrease in full-length unspliced RNA for packaging, coupled with the reduction in gag, pol, and env protein synthesis. In contrast, with the EMCV IRES, GFP production is significantly lower (though nevertheless 10- to 20-fold greater than background), but there is no inhibitory effect on viral replication. The sensitivity of this assay is because of the double readout of (1) GFP expression (i.e., apparent IRES activity), and (2) viral replication.

Cite this article as *Cold Spring Harb Perspect Biol* doi: 10.1101/cshperspect.a011569

search revealed ESTs in which these 3'-splice sites in all three putative IRES sequences had been used in their endogenous cellular mRNA context. A survey of 81 cellular 5'-UTR sequences with claimed IRESs, revealed EST evidence for functional 3'-splice sites in a small minority, including BCL2, DAP5, HIAP2, MTG8a, ODC, UNR, and utrophin A (Baranick et al. 2008). Another salutary warning of how splicing-dependent artefacts can give highly misleading results is the fact that a 42 nt sequence corresponding to a 3'-splice site in the β-globin gene masquerades as an exceptionally potent IRES when tested in the dual luciferase plasmid system (Lemp et al. 2012). In view of these potential problems arising from splicing, perhaps it would be worth investing time in constructing an alternative expression vector in which the customary 5'-proximal intron (Fig. 3A) is replaced either by an intron in the FLuc ORF itself or just downstream from its stop codon, or by an intron that is spliced by the alternative U11/U12snRNA-dependent pathway.

siRNA Screen for FLuc Expression from Unanticipated mRNA Species

Although the retroviral test is undoubtedly very sensitive, a more generally accessible screen, developed by Rick Lloyd's group, has the advantage of detecting FLuc expression from unanticipated mRNA species irrespective of whether they arise from cryptic promoters or from splicing. It involves cotransfection or coexpression of an siRNA targeted against RLuc coding sequences (van Eden et al. 2004), which usually results in a 70%–80% decrease in RLuc expression, a level of reduction that is necessary if the results are to be readily interpretable. On the plausible assumption that the 3'-fragment of siRNA-mediated endonucleolytic cleavage is rapidly degraded to completion, the percentage reductions in FLuc and RLuc expression should be nearly equal if all the FLuc expression is from the anticipated bicistronic mRNA (i.e., dependent on an IRES), and this outcome is observed in tests of the EMCV and HRV picornavirus IRESs (Bert et al. 2006; Andreev et al. 2009). If there are stable degradation intermediates, the

reduction in FLuc expression could be less than that of RLuc, but decay intermediates of sufficient stability to seriously prejudice the interpretation should be detectable by Northern blotting (even though 5'-RACE would be needed to map the 5'-ends). In the one case where this Northern blotting was performed, no decay intermediates were detected (Saffran and Smiley 2009).

If the reduction in FLuc expression is significantly less than that of RLuc, and yet no stable decay intermediates can be detected, the inescapable conclusion is that some FLuc must be expressed from unanticipated mRNAs, which lack the siRNA target site. The numerous possible configurations for such aberrant mRNAs can be broadly categorized according to whether the RLuc ORF is completely missing or present only in a truncated form, and whether the whole or only part of the test 5'-UTR is retained. If the whole RLuc ORF is missing, FLuc expression is probably from a monocistronic mRNA, which would be potentially translatable by scanning. If part of the RLuc ORF is retained but most of the test 5'-UTR is eliminated in a way that places this RLuc mini-ORF in-frame with the FLuc ORF, the result would be a monocistronic mRNA encoding a fusion protein that would likely have at least some FLuc activity (van Eden et al. 2004). On the other hand, if the whole of the test 5'-UTR is retained together with the RLuc mini-ORF, the outcome would be a bicistronic mRNA which, in principle, could only express FLuc via an IRES-dependent mechanism that would be immune from siRNA-mediated knock-down. However, if IRES activity is manifest when the upstream cistron is a RLuc mini-ORF, a similar efficiency of IRES-dependent FLuc expression should also be showed by the anticipated full-length bicistronic mRNA (with a complete RLuc ORF), but in this case it would be subject to siRNA-mediated knock-down. This full-length bicistronic mRNA will certainly be present, and in most cases it will be the majority RNA species, according to typical Northern blot results.

With these various scenarios in mind, it is clear that if the siRNA causes no reduction whatsoever in FLuc expression, the most likely

Cite this article as *Cold Spring Harb Perspect Biol* doi: 10.1101/cshperspect.a011569

explanation is that it is all coming from a monocistronic mRNA rather than the intended bicistronic species. This outcome, and even siRNA-dependent stimulation of FLuc expression, has been observed with HIF-1α, VEGF, XIAP, c-myc, and EGR-1 5'-UTRs (van Eden et al. 2004; Bert et al. 2006). If the siRNA causes a reduction in FLuc expression that is statistically significant but distinctly less than the reduction in RLuc expression (or in FLuc expression when the EMCV IRES is tested), as has been reported for the Apaf-1 5'-UTR (Andreev et al. 2009), there are two alternative conclusions. One is that some FLuc expression is from the siRNA-sensitive bicistronic transcript and some from a monocistronic artefact. The alternative is that all FLuc expression is from bicistronic mRNAs, and therefore probably IRES-dependent, but some of these bicistronic mRNAs have a truncated RLuc ORF lacking the siRNA target site.

These various possibilities can only be distinguished by a thorough RNA analysis. At least in the first instance, this analysis would best be made on mRNA recovered from the siRNA-treated cells, because the reduced background of anticipated full-length transcripts should make it easier to detect the aberrant siRNA-resistant mRNA species, which could be present only at quite low abundance. One approach is to carry out RT-PCR with primers matching the very 5'-end of the anticipated transcript (upstream of the chimeric intron) and the 5'-end of the FLuc ORF (van Eden et al. 2004). However, this strategy presupposes that all aberrant transcripts originate from the intended (SV40) promoter, and so 5'-RACE would be a more open-ended and better method for RNA analysis (Lemp et al. 2012). In either case, the cDNA products should be cloned, and sufficient clones sequenced.

Direct RNA Transfections and In Vitro Translation Assays

In view of the problems arising from splicing and cryptic Pol II promoters, many investigators have turned to transfections of capped and polyadenylated bicistronic mRNAs, synthesized in vitro by a bacteriophage RNA polymerase

and usually with the same RLuc and FLuc configuration. This avoids the potential artefacts associated with DNA transfections, although it is not without its own problems, because at least some transfection protocols result in much of the input RNA remaining sequestered in intracellular vesicles, unavailable for translation and also largely immune from mRNA degradation enzymes (Barreau et al. 2006). It has been suggested that introducing the RNA by electroporation, or even microinjection, might avoid this problem, but in the meantime the existence of two intracellular pools invalidates any attempt to relate protein expression to mRNA stability or abundance, because these measurements involve recovering total RNA, and not just the pool of translated mRNAs.

Nevertheless, these considerations do not invalidate assays of expression measured during the linear phase of translation, which has been shown to persist for at least ~6 h posttransfection. Two striking results have emerged from such assays of bicistronic mRNAs. First, although the EMCV IRES elicits a 100- to 250-fold stimulation of the FLuc/RLuc expression ratio, the stimulation seen with putative cellular mRNA IRESs is seldom greater than 10-fold, even in the case of 5'-UTRs that stimulate >50-fold in DNA transfection assays. For example, in one such assay the greatest stimulation was 6.2-fold (Apaf-1 5'-UTR), almost fourfold for the β-globin 5'-UTR and only 1.9-fold for the c-myc 5'-UTR (Andreev et al. 2009). In other reports the increases seen with the HIF-1α, HIF-2α, VEGF, XIAP, PIM-1, and p27^{Kip1} 5'-UTRs were all less than ~6-fold (Liu et al. 2005; Wang et al. 2005; Bert et al. 2006; Young et al. 2008). These results raise the semantic question of whether such small stimulations, barely above that seen with the β-globin 5'-UTR and only a pale shadow of the EMCV IRES result, really warrant description as "IRES activity."

The other, even more striking outcome concerns comparison of the bicistronic mRNA with a monocistronic FLuc mRNA bearing the same test 5'-UTR, the comparison that regrettably is never performed with plasmid DNA transfections. In two independent reports, FLuc expression from the bicistronic mRNA with the Apaf-1,

HIF-1α, c-myc, VEGF, and XIAP 5'-UTRs was <2% (range 0.125%–1.66%) of the expression from the corresponding monocistronic mRNA, although the EMCV IRES was 50%–125% as active in the bicistronic background compared with the monocistronic mRNA (Bert et al. 2006; Andreev et al. 2009).

These data imply that a switch from exclusive scanning-dependent initiation to exclusive IRES-dependent initiation would result in a >50-fold decrease in the translation efficiency of these mRNAs. This seriously undermines the idea that the persistence of translation and polysome association of a given mRNA under stress conditions is caused by such a switch, unless one invokes a massive stress-induced activation of such IRESs. The few cases in which the apparent IRES activity under stress conditions has been examined by RNA transfection have given contrary results: no increase in apparent VEGF, HIF-1α, and HIF-2α IRES activity was seen in hypoxia (Young et al. 2008), but a ~5-fold increase in BCL-2 IRES activity was observed following treatment of cells with etoposide for 8 h (Sherrill et al. 2004). Proponents of the "nuclear experience" hypothesis will argue that the RNA transfection results are artefactually low because of the absence of this "experience," but until we have more evidence as to what the "nuclear experience" entails (including evidence that it cannot be entirely or even partly explained by splicing or cryptic promoters), there is no logical justification for not accepting the RNA transfection results at face value.

In vitro translation of bicistronic mRNAs with putative IRESs in a Krebs-2 cell-free system (untreated with micrococcal nuclease) gave a hierarchy of apparent IRES activity similar to that seen in RNA transfections of HEK 293 cells, but (except for the EMCV IRES) the stimulations of the FLuc/RLuc expression ratio were even smaller than the low level observed in the RNA transfection assays (Andreev et al. 2009). It was suggested that the untreated Krebs-2 extract is a more appropriate system than the usual nuclease-treated rabbit reticulocyte lysate (RRL), because it shows both a greater stimulation by poly(A) tails and also synergy between 5'-caps and poly(A) tails, which may be partly ascribable to the presence of competing endogenous mRNAs. Nevertheless, several 5'-UTRs with putative IRES activity stimulated reporter expression when tested in a nuclease-treated HeLa cell-free extract as monocistronic polyadenylated Appp-capped mRNAs in which the test 5'-UTR was inserted between a 5'-proximal stable hairpin and the FLuc reporter (Thoma et al. 2004). The c-myc 5'-UTR gave the maximum stimulation of 12-fold (Hunsdorfer et al. 2005), which is much higher than was observed in the Krebs-2 system with a bicistronic mRNA (Andreev et al. 2009), but this higher activity may conceivably be attributable to the absence of any competing translation of other mRNAs. It would be better if assays in nuclease-treated extracts included at least an equimolar quantity of a heterologous m^7Gppp-capped and polyadenylated mRNA (e.g., coding for RLuc if the reporter is FLuc), so as to reflect the competition between scanning-dependent and IRES-dependent mRNAs that will occur in intact cells.

The nuclease-treated RRL can translate EMCV RNA extremely efficiently but only if KCl is used rather than the customary KOAc (Jackson 1991), which is generally found to support higher translation of other mRNAs, including those translated by cap-dependent scanning. The RRL system is clearly deficient in ITAFs required for the Type I HRV IRES (Hunt and Jackson 1999; Hunt et al. 1999), and is also inefficient in the translation of capped polyadenylated monocistronic mRNAs with long 5'-UTRs, such as LINE-1, Apaf-1, and c-myc (Andreev et al. 2009), which is suggestive of poor processivity of scanning. Both of these defects are corrected by supplementing the system with a relatively small amount of HeLa cytoplasmic extract (Hunt and Jackson 1999; Dmitriev et al. 2007).

Further Investigations into the Properties of Cellular mRNA IRESs

It is not possible to comment sensibly on the outcome of further studies of cellular mRNA IRESs, such as deletion mapping the IRES boundaries, mapping the putative ribosome entry site, and examining the influence of potential ITAFs on IRES activity, because most of these have been based on plasmid DNA transfection assays. Consequently, until it is definitively proven that there are no monocistronic FLuc mRNAs generated from the bicistronic DNA construct with a given IRES, we cannot be sure whether the mapping experiments, for example, are showing the boundaries of a genuine IRES, or the positions of the cryptic promoter(s) or the splice sites that are the origin of the monocistronic mRNA. An example that illustrates that this a real problem and not just hypothetical conjecture, is a mapping of the c-myc IRES to a 50 nt segment in the 5′-UTR, which was further narrowed down to two 14 nt motifs, one at each end of the 50 nt segment (Cencig et al. 2004). Recent work employing 5′-RACE has shown that one of the 14 nt motifs corresponds to a functional transcription start site (of a monocistronic FLuc mRNA) and the other to an associated promoter element (Lemp et al. 2012). This finding suggests that the explanation for the modular nature reported for many putative cellular IRESs (Stoneley and Willis 2004) may well be that the short modules are actually splice sites, cryptic promoters or transcription start sites.

In the case of potential ITAFs, which have been mainly explored by studying the influence of overexpression or siRNA-mediated knockdown on apparent IRES activity in plasmid DNA transfection assays, there are, admittedly, additional cell-free translation data showing weak stimulation (less than threefold) of apparent IRES activity when some putative ITAFs were added (Cobbold et al. 2008). However, it should be appreciated that the basal level of IRES activity in such in vitro systems is exceedingly low, and thus the absolute increase in FLuc expression will be very small indeed.

CONCLUSIONS AND PERSPECTIVES

To an outsider approaching this topic with as few preconceived notions as possible, the current state of the field seems extremely confusing, because many pre-2004 claims for cellular mRNA IRESs, which appeared to be well established on the basis of plasmid transfection results, have been seriously challenged by the more stringent controls developed in the past 8 years. Future historians of science are likely to be very surprised by the way in which the mRNA translation community embraced the idea of cellular mRNA IRESs so unreservedly that the warning signs of potential artefacts associated with plasmid DNA transfections (e.g., when insertion of the test 5′-UTR strongly reduces RLuc expression) were ignored for such a long time, and sometimes continue to be ignored. There still seems to be no general agreement as to which controls are the most decisive, and the importance of thorough RNA analysis remains greatly underappreciated. The most frequently used approaches are really best suited to showing that the anticipated bicistronic transcript is the major abundant mRNA species, whereas what is required is a strategy for detecting and characterizing low abundance monocistronic FLuc mRNAs, whatever their origin.

It is hard to avoid the conclusion that plasmid DNA transfection assays for IRES activity cannot be trusted unless they meet the criterion of near-equal percentage reductions in the FLuc and RLuc activities on coexpressing an siRNA against the upstream RLuc cistron. Thus far, only the Types I and II picornavirus IRESs (exemplified by EMCV and HRV) consistently come close to meeting this criterion (Bert et al. 2006; Araud et al. 2007; Andreev et al. 2009). There are also single reports in which the Unr 5′-UTR (albeit tested with FLuc as the upstream cistron) and DAP5 5′-UTR met this criterion (Araud et al. 2007; Schepens et al. 2007), but in view of the EST evidence for 3′-splice sites in both cases (Baranick et al. 2008), independent confirmation of these findings is desirable. At the other extreme, the HIF-1α, VEGF, XIAP, and EGR-1 5′-UTRs comprehensively failed the siRNA test (Bert et al. 2006). In addition, the putative IRESs in the eIF4GI 5′-UTR of Gan and Rhoads (1996), and in the c-myc, HIF-2α, VEGF receptor1, Glut-1, TPI, NRF, PIM-1, p57^{Kip2}, and p27^{Kip1} 5′-UTRs all failed in at least one of the other recently developed sensitive controls for cryptic promoters and/or splicing artefacts. Failure in these tests suggests that the massive stimulations of the FLuc/RLuc expression ratio elicited by some of these 5'-UTRs in plasmid transfections are most unlikely to be a measure of true IRES activity. However, because the siRNA test and these other controls are essentially methods of screening for probable artefacts, rather than strictly quantitative assays, the possibility remains open that a 5'-UTR that fails these tests could nevertheless have some IRES activity, albeit an extremely weak activity probably in the same range as generally observed in RNA transfection or in vitro translation assays.

The literature has claims for ∼100 cellular mRNA IRESs (see Mokrejs et al. 2010 for the most recent, but not quite complete list), the great majority of them based primarily on plasmid DNA transfection assays, usually with only Nothern blot or RT-qPCR controls. Although a number have been examined by in vitro translation, and many, but not all, show modest IRES activity in these tests, it is difficult to evaluate these results because such a wide variety of conditions and cell-free systems have been used. Few of them (apart from those listed above) have been subjected to the siRNA test or any of the other stringent controls that have been recently developed, and comparatively few have been thoroughly tested in RNA transfections. The current status of a great many of these ∼100 putative IRESs (apart from those listed in the preceding paragraph) must therefore be regarded as "uncertain" or "not proven," until these gaps in the data are plugged.

Because the results of plasmid DNA transfections without really stringent controls cannot be trusted, increasing reliance may have to be placed on the outcome of RNA transfections. For a comprehensive test, the following polyadenylated mRNA species should be assayed: m^{7}Gppp-capped bicistronic mRNA (1) with the test 5'-UTR, and (2) with no insert and/or the reverse complement of the test 5'-UTR; Apppp-capped monocistronic mRNA with a 5'-proximal stable hairpin, and (3) the test 5'-UTR, or (4) either no test sequence or the reverse complement of the test 5'-UTR; (5) m^{7}Gppp-capped, and (6) ApppG-capped monocistronic mRNAs with the test 5'-UTR but no hairpin.

The comparisons of (1) with (2), and (3) with (4) give two independent measures of IRES activity, the latter avoiding the potential complication of cross-interference between translation of two cistrons in close mutual proximity; and the comparisons of (1) with (5), and (3) with (5) give the exceedingly important ratio of IRES-dependent versus scanning-dependent initiation efficiency. The (5) versus (6) comparison provides a measure of the "cap-dependency" of initiation via the scanning mechanism, although it should be appreciated that IRES-dependent translation could also be occurring in parallel with scanning on such mRNAs. It would be worth considering transfecting the mRNAs by electroporation rather than with chemical reagents, and in due course it could be instructive to also test the outcome of appending the natural 3'-UTR to the monocistronic reporters, rather than the 3'-UTR supplied by the expression vector. Putative IRESs that are thought to play an essential role in the persistence of translation under stress conditions should also be tested under these conditions, to reveal any stress-dependent activation.

Although it is early days, current indications suggest that the cellular mRNA IRES activity observed in such assays will usually turn out to be rather weak. So far, cellular IRES-dependent FLuc expression from a bicistronic mRNA has never exceeded 2% of the expression from a m^{7}Gppp-capped monocistronic mRNA with the same test 5'-UTR; and the stimulation of the FLuc/RLuc expression ratio following insertion of the test 5'-UTR into the bicistronic mRNA is usually in the range 5- ± 2-fold (maximum ∼10-fold) over the no IRES control, which should in principle be zero, and in practice is likely to be very small. Although there may be situations in which even such weak IRES activity is nevertheless biologically relevant, there are other situations, such as the persistence of translation of certain mRNAs under stress conditions, in which it is difficult to see how IRES activity as weak as this could possibly account for the observed outcome. Rather than jump to the presumption that IRES-dependent translation must be the explanation, it would be preferable to keep an open mind

Cite this article as *Cold Spring Harb Perspect Biol* doi: 10.1101/cshperspect.a011569

as to the other possibilities, namely: persistence of scanning-dependent translation (perhaps because the particular mRNA has an unusually high affinity for the residual eIF4F activity), ribosome shunting, or a mechanism dependent on a TE-like element in the 5′-UTR.

ACKNOWLEDGMENTS

I thank Rick Lloyd and Chris Logg for helpful comments on a draft version of this article, Chris Smith for enlightenment on the intricacies of alternative splicing, and Christopher Hellen for help with the figures.

REFERENCES

Alexander L, Lu HH, Wimmer E. 1994. Polioviruses containing picornavirus type 1 and/or type 2 internal ribosomal entry site elements: Genetic hybrids and the expression of a foreign gene. *Proc Natl Acad Sci* **91:** 1406–1410.

Ali IK, Jackson RJ. 2001. The translation of capped mRNAs has an absolute requirement for the central domain of eIF4G but not for the cap-binding initiation factor eIF4E. *Cold Spring Harb Symp Quant Biol* **66:** 377–387.

Ali IK, McKendrick L, Morley SJ, Jackson RJ. 2001. Activity of the hepatitis A virus IRES requires association between the cap-binding translation initiation factor (eIF4E) and eIF4G. *J Virol* **75:** 7854–7863.

Andreev DE, Dmitriev SE, Terenin IM, Prassolov VS, Merrick WC, Shatsky IN. 2009. Differential contribution of the m^{7}G-cap to the 5′ end dependent translation initiation of mammalian mRNAs. *Nucleic Acids Res* **37:** 6135–6147.

Araud T, Genolet R, Jaquier-Gubler P, Curran J. 2007. Alternatively spliced isoforms of the human elk-1 mRNA within the 5′-UTR: Implications for ELK-1 expression. *Nucleic Acids Res* **35:** 4649–4663.

Baird SD, Turcotte M, Korneluk RG, Holcik M. 2006. Searching for IRES. *RNA* **12:** 1755–1785.

Baranick BT, Lemp NA, Nagashima J, Hiraoka K, Kasahara N, Logg CR. 2008. Splicing mediates the activity of four putative cellular internal ribosome entry sites. *Proc Natl Acad Sci* **105:** 4733–4738.

Barreau C, Dutertre S, Paillard L, Osborne HB. 2006. Liposome-mediated RNA transfection should be used with caution. *RNA* **12:** 1790–1793.

Bedard KM, Daijogo S, Semler BL. 2007. A nucleo-cytoplasmic SR protein functions in viral IRES-mediated translation initiation. *EMBO J* **26:** 459–467.

Bert AG, Grépin R, Vadas MJ, Goodall GJ. 2006. Assessing IRES activity in the HIF-1α and other cellular 5′-UTRs. *RNA* **12:** 1074–1083.

Borman AM, Michel YM, Kean KM. 2001. Detailed analysis of the requirements of hepatitis A virus internal ribosome

entry segment for the eukaryotic initiation factor complex eIF4F. *J Virol* **75:** 7864–7871.

Boussadia O, Niepmann M, Créancier L, Prats AC, Dautry F, Jacquemin-Sablon H. 2003. Unr is required in vivo for efficient initiation of translation from the internal ribosome entry sites of both rhinovirus and poliovirus. *J Virol* **77:** 3353–3359.

Braunstein S, Karpisheva K, Pola C, Goldberg J, Hochman T, Yee H, Cangiarella J, Arju R, Formenti SC, Schneider RJ. 2007. A hypoxia-controlled cap-dependent to cap-independent translation switch in breast cancer. *Mol Cell* **28:** 501–512.

Bushell M, Stoneley M, Kong YW, Hamilton TL, Spriggs KA, Dobbyn HC, Qin X, Sarnow P, Willis AE. 2006. Polypyrimidine tract binding proteins regulates IRES-mediated gene expression during apoptosis. *Mol Cell* **23:** 401–412.

Byrd MP, Zamora M, Lloyd RE. 2005. Translation of eukaryotic translation initiation factor 4GI (eIF4GI) proceeds from multiple mRNAs containing a novel cap-dependent internal ribosome entry site (IRES) that is active during poliovirus infection. *J Biol Chem* **280:** 18610–18622.

Cencig S, Nanbru C, Le SY, Gueydan C, Huez G, Kruys V. 2004. Mapping and characterization of the minimal internal ribosome entry segment in the human c-myc mRNA 5′ untranslated region. *Oncogene* **23:** 267–277.

Cobbold LC, Spriggs KA, Haines SJ, Dobbyn HC, Hayes C, de Moor CH, Lilley KS, Bushell M, Willis AE. 2008. Identification of internal ribosome entry segment (IRES)-trans-acting factors for the Myc family of IRESs. *Mol Cell Biol* **28:** 40–49.

Costa-Mattioli M, Svitkin Y, Sonenberg N. 2004. La autoantigen is necessary for optimal function of the poliovirus and hepatitis C virus internal ribosome entry site in vivo and in vitro. *Mol Cell Biol* **24:** 6861–6870.

de Breyne S, Yu Y, Pestova TV, Hellen CU. 2008. Factor requirements for translation initiation on the Simian picornavirus internal ribosomal entry site. *RNA* **14:** 367–380.

de Breyne S, Yu Y, Unbehaun A, Pestova TV, Hellen CUT. 2009. Direct functional interaction of initiation factor eIF4G with type 1 internal ribosomal entry sites. *Proc Natl Acad Sci* **106:** 9197–91202.

Dmitriev SE, Andreev DE, Terenin IM, Olovnikov IA, Prassolov VS, Merrick WC, Shatsky IN. 2007. Efficient translation initiation directed by the 900-nucleotide-long and GC-rich 5′-untranslated region of the human retrotransposon LINE-1 mRNA is strictly cap-dependent rather than internal ribosome entry site mediated. *Mol Cell Biol* **27:** 4685–4697.

Gan W, Rhoads RE. 1996. Internal initiation of translation directed by the 5′-untranslated region of the mRNA for eIF4G, a factor involved in the picornavirus-induced switch from cap-dependent to internal initiation. *J Biol Chem* **271:** 623–626.

Hellen CUT, de Breyne S. 2007. A distinct group of hepacivirus/pestivirus-like internal ribosomal entry sites in members of diverse picornavirus genera: Evidence for modular exchange of functional noncoding RNA elements by recombination. *J Virol* **81:** 5850–5863.

Holcík M, Gordon BW, Korneluk RG. 2003. The internal ribosome entry site-mediated translation of antiapoptotic protein XIAP is modulated by the heterogeneous nuclear ribonucleoproteins C1 and C2. *Mol Cell Biol* **23:** 280–288.

Hundsdoerfer P, Thoma C, Hentze MW. 2005. Eukaryotic translation initiation factor 4GI and p97 promote cellular internal ribosome entry sequence-driven translation. *Proc Natl Acad Sci* **102:** 13421–13426.

Hunt SL, Jackson RJ. 1999. Polypyrimidine–tract binding protein (PTB) is necessary, but not sufficient, for efficient internal initiation of translation of human rhinovirus-2 RNA. *RNA* **5:** 344–359.

Hunt SL, Hsuan JJ, Totty N, Jackson RJ. 1999. Unr, a cellular cytoplasmic RNA-binding protein with five cold shock domains, is required for internal initiation of translation of human rhinovirus RNA. *Genes Dev* **13:** 437–448.

Iizuka N, Kohara M, Hagino-Yamagishi K, Abe S, Komatsu T, Tago K, Arita M, Nomoto A. 1989. Construction of less neurovirulent polioviruses by introducing deletions into the 5' noncoding sequence of the genome. *J Virol* **63:** 5354–5363.

Jackson RJ. 1991. Potassium salts influence the fidelity of mRNA translation initiation in rabbit reticulocyte lysates: Unique features of encephalomyocarditis virus RNA translation. *Biochim Biophys Acta* **1088:** 345–358.

Jackson RJ, Kaminski A. 1995. Internal initiation of translation in eukaryotes: The picornavirus paradigm and beyond. *RNA* **1:** 985–1000.

Jang SK, Wimmer E. 1990. Cap-independent translation of encephalomyocarditis virus RNA: Structural elements of the internal ribosomal entry site and involvement of a cellular 57-kD RNA-binding protein. *Genes Dev* **4:** 1560–1572.

Johannes G, Sarnow P. 1998. Cap-independent polysomal association of natural mRNAs encoding c-myc, BiP, and eIF4G conferred by internal ribosome entry site. *RNA* **4:** 1500–1513.

Kafasla P, Morgner N, Pöyry TA, Curry S, Robinson CV, Jackson RJ. 2009. Polypyrimidine tract binding protein stabilizes the encephalomyocarditis virus IRES structure via binding multiple sites in a unique orientation. *Mol Cell* **34:** 556–568.

Kafasla P, Morgner N, Robinson CV, Jackson RJ. 2010. Polypyrimidine tract binding protein stimulates the poliovirus IRES by modulating eIF4G binding. *EMBO J* **29:** 3710–3722.

Kaminski A, Jackson RJ. 1998. The polypyrimidine tract binding protein (PTB) requirement for internal initiation of translation of cardiovirus RNAs is conditional rather than absolute. *RNA* **4:** 626–638.

Kaminski A, Howell MT, Jackson RJ. 1990. Initiation of encephalomyocarditis virus RNA translation: The authentic initiation site is not selected by a scanning mechanism. *EMBO J* **9:** 3753–3759.

Kaminski A, Belsham GJ, Jackson RJ. 1994. Translation of encephalomyocarditis virus RNA: Parameters influencing the selection of the internal initiation site. *EMBO J* **13:** 1673–1681.

Kaminski A, Pöyry TAA, Skene PJ, Jackson RJ. 2010. Mechanism of initiation site selection promoted by the human

rhinovirus 2 internal ribosome entry site. *J Virol* **84:** 6578–6589.

King HA, Cobbold LC, Willis AE. 2010. The role of IRES *trans*-acting factors in regulating translation initiation. *Biochem Soc Trans* **38:** 1581–1586.

Koeller DM, Casey JL, Hentze MW, Gerhardt EM, Chan LN, Klausner RD, Harford JB. 1989. A cytosolic protein binds to structural elements within the iron regulatory region of the transferrin receptor mRNA. *Proc Natl Acad Sci* **86:** 3574–3578.

Kolupaeva VG, Pestova TV, Hellen CU, Shatsky IN. 1998. Translation eukaryotic initiation factor 4G recognizes a specific structural element within the internal ribosome entry site of encephalomyocarditis virus RNA. *J Biol Chem* **273:** 18599–18604.

Kolupaeva VG, Lomakin IB, Pestova TV, Hellen CUT. 2003. Eukaryotic initiation factors 4G and 4A mediate conformational changes downstream of the initiation codon of the encephalomyocarditis virus internal ribosomal entry site. *Mol Cell Biol* **23:** 687–698.

Kozak M. 2003. Alternative ways to think about mRNA sequences and proteins that appear to promote internal initiation of translation. *Gene* **318:** 1–23.

Kozak M. 2005. A second look at cellular mRNA sequences said to function as internal ribosome entry sites. *Nucleic Acids Res* **33:** 6593–6602.

Le Hir H, Nott A, Moore MJ. 2003. How introns influence and enhance eukaryotic gene expression. *Trends Biochem Sci* **28:** 215–220.

Lemp NA, Hiraoka K, Kasahara N, Logg CR. 2012. Cryptic transcripts from the ubiquitous pMB1 origin of replication confound functional tests for *cis*-regulatory elements. *Nucleic Acids Res* doi: 10.1093/nar/gks451.

Liu Z, Dong Z, Han B, Yang Y, Liu Y, Zhang JT. 2005. Regulation of expression by promoters versus internal ribosome entry site in the 5'-untranslated sequence of the human cyclin-dependent kinase inhibitor 27^{Kip1}. *Nucleic Acids Res* **33:** 3763–3771.

Lomakin IV, Hellen CUT, Pestova TV. 2000. Physical association of eukaryotic initiation factor 4G (eIF4G) with eIF4A strongly enhances binding of eIF4G to the internal ribosomal entry site of encephalomyocarditis virus is required for internal initiation of translation. *Mol Cell Biol* **20:** 6019–6029.

Meerovitch K, Nicholson R, Sonenberg N. 1991. In vitro mutational analysis of cis-acting RNA translational elements within the poliovirus type 2 5' untranslated region. *J Virol* **65:** 5895–5901.

Miller WA, Wang Z, Treder K. 2007. The amazing diversity of cap-independent translation elements in the 3'-untranslated regions of plant viral RNAs. *Biochem Soc Trans* **35:** 1629–1633.

Mokrejs M, Masek T, Vopalensky A, Hlubucek P, Delbos P, Pospisek M. 2010. IRESite—a tool for the examination of viral and cellular internal ribosome entry sites. *Nucleic Acids Res* **38:** D131–D136.

Nagy E, Maquat LE. 1998. A rule for termination-codon position within intron-containing genes: When nonsense affects RNA abundance. *Trends Biochem Sci* **23:** 198–199.

Nicholson R, Pelletier J, Le S-Y, Sonenberg N. 1991. Structural and functional analysis of the ribosome landing pad of poliovirus type 2: In vivo translation studies. *J Virol* **65:** 5886–5894.

Oberstrass FC, Auweter SD, Erat M, Hargous Y, Henning A, Wenter P, Reymond L, Amir-Ahmady B, Pitsch S, Black DL, et al. 2005. Structure of PTB bound to RNA: Specific binding and implications for splicing regulation. *Science* **309:** 2054–2057.

Özeş AR, Feoktistova K, Avanzino BC, Fraser CS. 2011. Duplex unwinding and ATPase activities of the DEAD-box helicase eIF4A are coupled by eIF4G and eIF4B. *J Mol Biol* **412:** 674–687.

Pelletier J, Sonenberg N. 1988. Internal initiation of translation of eukaryotic mRNA directed by a sequence derived from poliovirus RNA. *Nature* **334:** 320–325.

Pelletier J, Flynn ME, Kaplan G, Racaniello V, Sonenberg N. 1988. Mutational analysis of upstream AUG codons of poliovirus RNA. *J Virol* **62:** 4486–4492.

Pestova TV, Hellen CU, Wimmer E. 1994. A conserved AUG triplet in the 5′ nontranslated region of poliovirus can function as an initiation codon in vitro and in vivo. *Virology* **204:** 729–737.

Pestova TV, Hellen CUT, Shatsky IN. 1996. Canonical eukaryotic initiation factors determine initiation of translation by internal ribosomal entry. *Mol Cell Biol* **16:** 6859–6869.

Pestova TV, Shatsky IN, Fletcher SP, Jackson RJ, Hellen CUT. 1998. A prokaryotic-like mode of cytoplasmic eukaryotic ribosome binding to the initiation codon during internal translation initiation of hepatitis C and classical swine fever virus RNAs. *Genes Dev* **12:** 67–83.

Pestova TV, de Breyne S, Pisarev AV, Abaeva IS, Hellen CU. 2008. eIF2-dependent and eIF2-independent modes of initiation on the CSFV IRES: A common role of domain II. *EMBO J* **27:** 1060–1072.

Pilipenko EV, Pestova TV, Kolupaeva VG, Khitrina EV, Poperechnaya AN, Agol VI, Hellen CUT. 2000. A cell cycle-dependent protein serves as a template-specific translation initiation factor. *Genes Dev* **14:** 2028–2045.

Pöyry T, Kinnunen L, Hovi T. 1992. Genetic variation in vivo and proposed functional domains of the 5′ noncoding region of poliovirus RNA. *J Virol* **66:** 5313–5319.

Prévôt D, Décimo D, Herbreteau CH, Roux F, Garin J, Darlix JL, Ohlmann T. 2003. Characterization of a novel RNA-binding region of eIF4GI critical for ribosomal scanning. *EMBO J* **22:** 1909–1921.

Qin X, Sarnow P. 2004. Preferential translation of internal ribosome entry site-containing mRNAs during the mitotic cycle in mammalian cells. *J Biol Chem* **279:** 13721–13728.

Ramírez-Valle F, Braunstein S, Zavadil J, Formenti SC, Schneider RJ. 2008. eIF4GI links nutrient sensing by mTOR to cell proliferation and inhibition of autophagy. *J Cell Biol* **181:** 293–307.

Saffran HA, Smiley JR. 2009. The XIAP IRES activates 3′ cistron expression by inducing production of monocistronic mRNA in the βgal/CAT bicistronic reporter system. *RNA* **15:** 1980–1985.

Sasaki J, Nakashima N. 2000. Methionine-independent initiation of translation in the capsid protein of an insect RNA virus. *Proc Natl Acad Sci* **97:** 1512–1515.

Schepens B, Tinton SA, Bruynooghe Y, Parthoens E, Haegman M, Beyaert R, Cornelis S. 2007. A role for hnRNP C1/C2 and Unr in internal initiation of translation during mitosis. *EMBO J* **26:** 158–169.

Shatsky IN, Dmitriev SE, Terenin IM, Andreev DE. 2010. Cap- and IRES-independent scanning mechanism of translation initiation as an alternative to the concept of cellular IRESs. *Mol Cells* **30:** 285–293.

Sherrill KW, Lloyd RE. 2008. Translation of cIAP2 mRNA is mediated exclusively by a stress-modulated ribosome shunt. *Mol Cell Biol* **28:** 2011–2022.

Sherrill KW, Byrd MP, Van Eden ME, Lloyd RE. 2004. BCL-2 translation is mediated via internal ribosome entry during cell stress. *J Biol Chem* **279:** 29066–29074.

Shiroki K, Ohsawa C, Sugi N, Wakiyama M, Miura K, Watanabe M, Suzuki Y, Sugano S. 2002. Internal ribosome entry site-mediated translation of Smad5 in vivo: Requirement for a nuclear event. *Nucleic Acids Res* **30:** 2851–2861.

Silvera D, Arju R, Darvishian F, Levine PH, Zolfaghari L, Goldberg J, Hochman T, Formenti SC, Schneider RJ. 2009. Essential role for eIF4GI overexpression in the pathogenesis of inflammatory breast cancer. *Nat Cell Biol* **11:** 903–908.

Stoneley M, Willis AE. 2004. Cellular internal ribosome entry segments: Structures, *trans*-acting factors and regulation of gene expression. *Oncogene* **23:** 3200–3207.

Stoneley M, Subkhankulova T, Le Quesne JP, Coldwell MJ, Jopling CL, Belsham GJ, Willis AE. 2000. Analysis of the c-myc IRES; a potential role for cell-type specific trans-acting factors and the nuclear compartment. *Nucleic Acids Res* **28:** 687–694.

Svitkin YV, Meerovitch K, Lee HS, Dholakia JN, Kenan DJ, Agol VI, Sonenberg N. 1994. Internal translation initiation on poliovirus RNA: Further characterization of La function in poliovirus translation in vitro. *J Virol* **68:** 1544–1550.

Thoma C, Bergamini G, Galy B, Hundsdoerfer P, Hentze MW. 2004. Enhancement of IRES-mediated translation of the c-myc and BiP mRNAs by the poly(A) tail is independent of intact eIF4G and PABP. *Mol Cell* **15:** 925–935.

van Eden ME, Byrd MP, Sherrill KW, Lloyd RE. 2004. Demonstrating internal ribosome entry sites in eukaryotic mRNAs using stringent RNA test procedures. *RNA* **10:** 720–730.

Vopálenský V, Masek T, Horváth O, Vicenová B, Mokrejs M, Pospísek M. 2008. Firefly luciferase gene contains a cryptic promoter. *RNA* **14:** 1720–1729.

Walter BL, Nguyen JH, Ehrenfeld E, Semler BL. 1999. Differential utilization of poly(rC) binding protein 2 in translation directed by picornavirus IRES elements. *RNA* **5:** 1570–1585.

Wang Z, Weaver M, Magnuson N. 2005. Cryptic promoter activity in the DNA sequence corresponding to the pim-1 5′-UTR. *Nucleic Acids Res* **33:** 2248–2258.

Wilson JE, Pestova TV, Hellen CU, Sarnow P. 2000. Initiation of protein synthesis from the A site of the ribosome. *Cell* **102:** 511–520.

Young RM, Wang S-J, Gordan JD, Ji X, Liebhaber SA, Simon MC. 2008. Hypoxia-mediated selective mRNA translation by an internal ribosome entry site-independent mechanism. *J Biol Chem* **283:** 16309–16319.

Yu Y, Sweeney TR, Kafasla P, Jackson RJ, Pestova TV, Hellen CUT. 2011. The mechanism of translation initiation on Aichivirus RNA mediated by a novel type of picornavirus IRES. *EMBO J* **30:** 4423–4436.

Yueh A, Schneider RJ. 1996. Selective translation initiation by ribosome jumping in adenovirus-infected and heat-shocked cells. *Genes Dev* **10:** 1557–1567.

Yueh A, Schneider RJ. 2000. Translation by ribosome shunting on adenovirus and hsp70 mRNAs facilitated by complementarity to 18S rRNA. *Genes Dev* **14:** 414–421.

From *Cis*-Regulatory Elements to Complex RNPs and Back

Fátima Gebauer[1], Thomas Preiss[2], and Matthias W. Hentze[3]

[1]Gene Regulation Programme, Centre for Genomic Regulation (CRG) and UPF, 08003-Barcelona, Spain

[2]Genome Biology Department, The John Curtin School of Medical Research, The Australian National University, Acton (Canberra) ACT 0200, Australia

[3]European Molecular Biology Laboratory (EMBL), 69117 Heidelberg, Germany

Correspondence: fatima.gebauer@crg.eu; thomas.preiss@anu.edu.au; hentze@embl.de

Messenger RNAs (mRNAs), the templates for translation, have evolved to harbor abundant *cis*-acting sequences that affect their posttranscriptional fates. These elements are frequently located in the untranslated regions and serve as binding sites for *trans*-acting factors, RNA-binding proteins, and/or small non-coding RNAs. This article provides a systematic synopsis of *cis*-acting elements, *trans*-acting factors, and the mechanisms by which they affect translation. It also highlights recent technical advances that have ushered in the era of transcriptome-wide studies of the ribonucleoprotein complexes formed by mRNAs and their *trans*-acting factors.

Translational regulatory mechanisms are based on two key principles: signal-dependent covalent modifications of general translation (initiation) factors and *trans*-acting RNA-binding factors (RNA-binding proteins [RBPs] and miRNAs) to alter the translational fate of mRNAs. The first *cis*-regulatory elements to be found in eukaryotic mRNAs were the upstream open reading frames (uORFs) of the yeast GCN4 mRNA (Mueller and Hinnebusch 1986) and the iron-responsive elements of mammalian ferritin mRNAs (Hentze et al. 1987; Leibold and Munro 1987). Coincidentally, they provided one example of each: *cis*-acting elements that function in the context of modified initiation factor activity (GCN4), or that serve as binding sites for the first translational regulatory proteins, the iron regulatory proteins (IRPs) (Hinnebusch 2005; Hentze et al. 2010), respectively. More than two decades later, translational control is recognized as a major control point for the flux from genetic information to shaping proteomes, and is even reported to be the predominant mechanism for the control of gene expression (Schwanhausser et al. 2011).

Initially not anticipated, mRNAs now have to be seen as linear yet structured arrays of numerous *cis*-acting elements, mostly in the 5′ and 3′ untranslated regions (UTRs) but probably spreading across the whole message. This situation is mirrored with regard to mRNA engagement with *trans*-acting factors, and mRNAs must therefore be examined as messenger ribonucleoprotein particles (mRNPs).

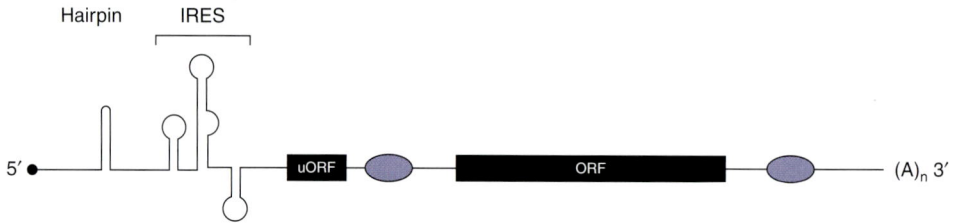

Figure 1. *Cis*-acting elements that influence mRNA translation. The nearly ubiquitous $5'$ ^{m7}GpppN cap structure (black circle) and $3'$ poly(A) tail (($A)_n$) strongly stimulate translation. Secondary structures (e.g., hairpin) and upstream open reading frames (uORFs) in the $5'$ UTR usually inhibit translation. Internal ribosome entry sequences (IRES) stimulate translation independently of the cap structure. Binding sites for regulatory RNA-binding proteins or microRNAs (ovals) can provide positive or (more frequently) negative regulation. (Adapted and modified from Gebauer and Hentze 2004.)

Cis-acting elements come in different flavors (Fig. 1). Hairpin or higher-order (e.g., pseudo-knot) intramolecular mRNA structures can influence translation in their own right (i.e., without binding factors). It has been well documented that they can affect initiation, particularly when positioned in the $5'$ UTR close to the cap structure (Pelletier and Sonenberg 1985; Kozak 1986), and elongation, being involved in numerous frameshifting events (Namy et al. 2004). Internal ribosome entry sequences (IRES) represent another important *cis*-acting element that typically occurs in $5'$ UTRs but has also been reported to occur within the coding region of mRNAs (Holcik et al. 2000). In cellular mRNAs, IRES coexist with the $5'$-cap structure and endow mRNAs with the potential to be translated under conditions in which cap-dependent translation is compromised (e.g., different forms of cell stress, apoptosis). A third category of common *cis*-acting elements comprises uORFs. They occur singularly or multiply within the $5'$ UTRs of numerous mRNAs and influence the translation of the downstream major ORF, usually negatively. This effect can be exerted by the element itself, but appears to be used also for RBP-mediated translational regulation (see below). A notable exception is the GCN4 mRNA in yeast and ATF4 mRNA in mammals, where uORFs serve to promote the translation of the downstream major ORF under conditions of increased eIF2 phosphorylation (Hinnebusch 2005). Binding sites for regulatory RBPs or miRNAs can be combined on given mRNAs

to yield translational outcomes that integrate multiple signals via the respective *trans*-acting factors. In a complementary concept, nearly all *trans*-acting factors bind to a multitude of mRNAs that frequently encode functionally related proteins, subjecting these families of mRNAs to coordinated, operon-like regulation (Keene 2007). Finally, RNA editing or modification (e.g., methylation) can provide an additional layer of regulatory intervention for *cis*-acting elements (Li et al. 2009; Squires et al. 2012; Zhang et al. 2012).

The arrival of technologies to examine mRNAs and RBPs in a highly parallel, transcriptome-wide fashion places translational control at the heart of modern systems biology. Below, we discuss how system-wide studies and reductionist mechanistic experimentation must converge and complement each other for a deeper understanding of translational control.

THE mRNP AS A TEMPLATE FOR TRANSLATION AND TRANSLATIONAL CONTROL

Eukaryotic mRNAs pass through all stages of their life cycle from transcription, to processing, transport (including specific subcellular mRNA localization), translation, and decay, as assemblies of dynamic mRNPs rather than as "naked" nucleic acids (Martin and Ephrussi 2009; Moore and Proudfoot 2009; Sonenberg and Hinnebusch 2009). The implications of this notion are particularly profound for the interaction of the mRNA with the translation apparatus (translation factors and ribosome) as a template for protein synthesis. Parallels have previously been drawn between the processes of eukaryotic transcription and translation, likening the near-ubiquitous mRNA 5′-cap structure and 3′-poly(A) tail to constitutive promoter elements, whereas *cis*-regulatory motifs in the UTRs were viewed as mRNA-specific elements that act to control translation in a combinatorial and context-specific manner akin to transcriptional regulatory elements such as silencers and enhancers (Sachs and Buratowski 1997; Gebauer and Hentze 2004). This useful concept may be extended by an analogy between DNA in chromatin as the "real" template for transcription and the mRNA interacting with the translation machinery packaged into complex mRNP particles, which may even adopt additional states of "compaction," for instance, in the form of repressive cytoplasmic foci such as stress granules and processing bodies (Anderson and Kedersha 2009; Balagopal and Parker 2009). Adoption of such an mRNP-centric view could provide similar conceptual breakthroughs for translation research as the chromatin model did for the transcription field.

At present, we are only scratching the surface in terms of understanding the composition of mRNPs, and the analogy with transcription may break down at the point of histones and nucleosomes. Even the most common RBPs (e.g., the hnRNPs) do not appear to organize mRNPs in a regular fashion as histones do with the chromatin template. There is a pressing need for identifying all cellular proteins interacting with mRNA within native mRNPs ("mRNA interactomes"), and then to study their organization and dynamic interplay. At least the determination of the first mRNA interactomes has now become feasible (see below).

NEW TECHNOLOGIES TO STUDY *CIS*-ACTING ELEMENTS AND *TRANS*-ACTING FACTORS

The complete description of an mRNA interactome in a given cellular condition requires cataloging the proteins that bind to each ex-

pressed mRNA and a high-resolution mapping of all respective binding site(s) on the mRNA. Conventional affinity copurification and low-throughput identification of binding partners have long been useful tools for characterizing individual examples and have started the journey along this road. However, such small-scale methods cannot cope with the magnitude of the new tasks at hand. Valiant attempts have been made to draw the power of bioinformatics into the challenge and to predict the occurrence of known *cis*-acting motifs across larger spectra of mRNAs. However, these approaches are limited by the degeneracy of the primary RNA sequence motifs involved (e.g., miRNA target sites) (Saito and Saetrom 2010) and/or by the important role that RNA secondary and tertiary structure plays in defining functional regulatory motifs (Parker et al. 2011; Zhao et al. 2011). The computational prediction of RBPs has helped significantly in the identification of new candidates, but it is by definition limited to proteins bearing known RNA-binding domains.

Identification of *Cis*-Acting Elements by Cross-Linking

A long existing but recently refined and popularized approach to study RBPs as well as the mRNAs and sites they bind to is to covalently cross-link the two to each other in vivo. Covalent cross-links can be induced chemically (Valasek et al. 2007) or by ultraviolet (UV) light. UV light of 254 nm cross-links the naturally photoreactive nucleotide bases, especially pyrimidines, and specific amino acids (Phe, Trp, Tyr, Cys, and Lys) (Hockensmith et al. 1986; Brimacombe et al. 1988). UV-cross-linking requires direct contact (zero distance) between protein and RNA and does not promote protein–protein cross-linking (Greenberg 1979). A recent version of in vivo UV cross-linking is called PAR-CLIP (photoactivatable-ribonucleoside-enhanced cross-linking and immunoprecipitation) (Fig. 2A) (Hafner et al. 2010). The photoactivatable nucleotide 4-thiouridine (4SU) is taken up by cultured cells and incorporated into nascent RNAs, and efficient cross-linking is induced by 365-nm UV light

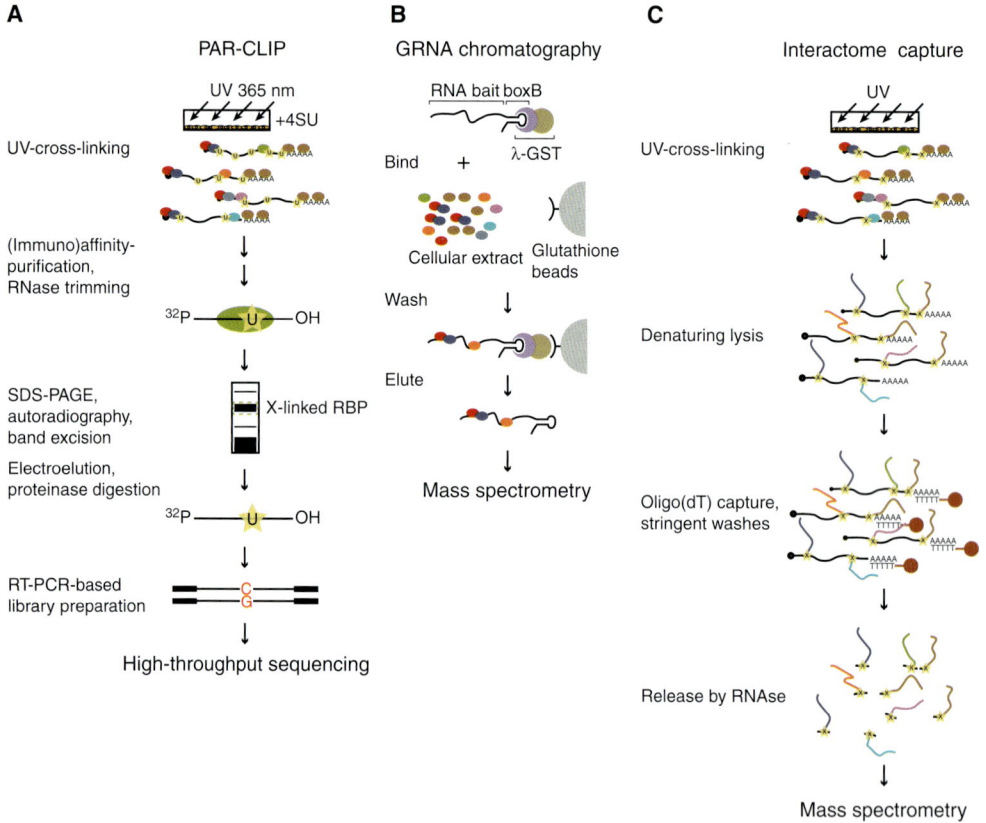

Figure 2. Methods to study mRNP composition. (*A*) Photoactivatable-ribonucleoside-enhanced cross-linking and immunoprecipitation (PAR-CLIP). Cells are cultured in media containing 4-thiouridine (4SU), leading to incorporation of the photoactivatable nucleoside into cellular RNA. Cross-linking with UV light of 365 nm leads to covalent attachment of RBPs to RNA targets that withstands partial RNAse digestion, immunoprecipitation, and purification by denaturing gel electrophoresis. Isolated RNA fragments are identified by next-generation sequencing, aided by a tendency of the cross-linked site to show thymidine (T) to cytidine (C) transitions. (*B*) GRNA chromatography using specific interaction between a 21-amino-acid peptide from the λ phage N anti-terminator protein and the boxB hairpin. A fusion of λN peptide with glutathione *S*-transferase (GST), and incorporation of the boxB hairpin into bait RNA converts glutathione Sepharose into an RNA affinity matrix (GRNA resin), which is incubated with cellular extracts. Proteins specifically bound to the matrix are eluted and identified by mass spectrometry. (*C*) Interactome capture. The procedure begins with RNP cross-linking in living cells by conventional UV 254-nm cross-linking or as in the PAR-CLIP approach. Following lysis, the complete cellular complement of (m)RNPs is purified by binding to an oligo(dT) resin and stringent washing under conditions that dissociate noncovalent RNA–protein interactions. Specifically bound proteins are released by RNase digestion and identified by mass spectrometry. (Diagrams are based on data from Czaplinski et al. 2005, Hafner et al. 2010, and Castello et al. 2012, respectively.)

irradiation, followed by affinity capture of the RBP under study, using specific antibodies against the native protein itself or a tagged version of the RBP expressed in the cells. The isolated complex is then subjected to limited RNase digestion, radioactive end labeling of the bound RNAs, and size selection by denaturing gel electrophoresis. Finally, RNA recovered from the complexes is used for reverse transcription and next-generation sequencing. The RNase trimming step ensures that transcript regions bound by the bait protein will be preferentially sequenced, whereas the propensity of the PAR-cross-linking chemistry to induce thymidine (T) to cytidine (C) transitions during reverse transcription helps with precise binding site identification. Even if PAR-CLIP is currently very popular, combinations of UV-cross-linking, RNP immunoprecipitation, isolation of cross-linked RNA segments, and cDNA sequencing have been developed before (e.g., CLIP) (Ule 2003), and related methods using microarrays or high-throughput sequencing exist to profile RNAs associated with immunopurified RNA-binding proteins (RIP-Chip, RIP-Seq) (Tenenbaum et al. 2000).

Identification of *Trans*-Acting Factors by RNA Affinity Chromatography

The converse approach to CLIP experiments, using a given RNA under study as bait to purify and identify interacting proteins, is also commonly used. Here the RNA is typically engineered to contain a small sequence or structural tag that will facilitate specific capture of the mRNP complex. GRNA affinity chromatography is an example of such an approach (Fig. 2B) (Czaplinski et al. 2005; Duncan 2006). It uses the specific binding of an RNA hairpin to a short peptide from the N anti-terminator protein of the λ phage (termed BoxB hairpin and λN peptide, respectively, originally developed as a tool for tethered function assays) (De Gregorio et al. 1999). For GRNA chromatography, native mRNP complexes are assembled on boxB-tagged RNAs in vitro, which are then purified with the help of λ-GST fusion proteins bound to a solid support. Proteins that copurify with the tagged RNA are then identified by mass spectrometry. Besides the boxB/λ tether, several other combinations of RNA aptamers (e.g., the streptomycin or tobramycin tags) or tethering proteins (e.g., MS2 coat protein) have been used successfully (Beach and Keene 2008).

Although these approaches have been quite successful in vitro, only a few success stories have been reported from in vivo settings (Hogg and Collins 2007). Recently, the MS2 coat protein fused to a tag consisting of two His6 clusters separated by a cleavage site for the TEV protease and an in vivo biotinylation site has been used to capture IRES-binding proteins from 293 cells (Tsai et al. 2011). Furthermore, the development of "designer RNA-binding domains" that can be engineered to bind any desired target RNA sequence (Filipovska et al. 2011; Mackay et al. 2011) holds further promise for the purification of specific native mRNPs from living cells.

Discovery of RBPs by Interactome Capture

Given the high sensitivity of contemporary mass spectrometric approaches to determine thousands of proteins in complex mixtures, two groups recently achieved the first step in the quest for complete mRNA interactomes (see above), the identification of "all" mRNA-binding proteins of a mammalian cell (Castello et al. 2012). This work first used efficient in vivo cross-linking (conventional UV254 cross-linking, or PAR-CL) to preserve physiologically relevant RNA–protein interactions, followed by capture of the polyadenylated (m)RNAs with their cross-linked RBPs on an oligo(dT) matrix, stringent washing, subsequent release of bound proteins by RNase digestion, and finally identification by mass spectrometry (Fig. 2C). This approach extends very early work on hnRNP proteins in vivo (Dreyfuss et al. 1984) and recent studies that used hybridization of labeled mRNA preparations to protein arrays to identify cellular RNA-binding proteins in vitro (Scherrer et al. 2010; Tsvetanova et al. 2010). It also has the advantage of sensitively and selectively detecting the near-complete array of native protein–mRNA interactions as they occur in living cells. This interactome capture approach can now be used or adapted to study the mRNA interactomes of other cells and to investigate changes in interactome composition as a function of biological conditions, such as differences in cell growth or cell cycle phase, or forms of stress (hypoxia, oxidative stress, nutrient deprivation, etc.).

FROM COMPLEX RNPs TO DEFINED *CIS/TRANS* INTERACTIONS

Proteins rarely act alone to regulate translation. Rather, multi-subunit complexes assemble on (the UTRs of) the transcript, directed by interactions between the RBP components of the complex and regulatory *cis*-acting sequences or structures of the mRNA. These complexes may also contain regulatory RNAs (e.g., mi-RISC), using nucleic acid hybridization as a principle of site-specific binding. It has now become clear that components of regulatory complexes and regulatory factors (or the translational machinery) need to interact dynamically to achieve accurate translational control. In the following, we focus on three illustrative examples of coordinated translational regulation. Two of them involve the cooperation between multiple *cis*-elements and *trans*-acting factors to ensure tight and timely control of translation. In the third example, temporal control of translation is achieved by the stepwise assembly and activation of a regulatory complex that coordinates the expression of a posttranscriptional operon.

TIGHT TRANSLATIONAL REPRESSION, A MULTIFUNCTIONAL RBP AND COMBINATORIAL REGULATION: *Msl2* mRNA

MSL2 is the limiting component of the *Drosophila* dosage compensation complex, a chromatin assembly that equalizes the expression of X-linked genes between males (XY) and females (XX) by promoting hypertranscription of the single male X chromosome (for review, see Gelbart and Kuroda 2009). Dosage compensation

must be repressed in females for viability, and this is primarily achieved by preventing MSL2 expression. Two posttranscriptional control mechanisms cooperate to inhibit *msl2*, and both are exerted by the female-specific RBP Sex-lethal (SXL). In the nucleus, SXL binds to oligo-uridine stretches adjacent to the splice sites of a small facultative intron in the *msl2* 5′ UTR to inhibit its splicing; this splicing inhibition retains the SXL-binding sites in the mature mRNA. In the cytoplasm, SXL inhibits *msl2* translation by binding to specific sites in both the retained intron and the 3′ UTR (for review, see Graindorge et al. 2011). Binding of SXL to both UTRs is necessary for tight translational repression, but partial inhibition can be achieved by each UTR alone, allowing the stepwise dissection of the mechanism (Bashaw and Baker 1997; Kelley et al. 1997; Gebauer et al. 1999). Extensive mutational and functional analyses have revealed that SXL regulates translation by a dual mechanism: SXL bound to the 3′ UTR inhibits the recruitment of the 43S ribosomal complex to the mRNA, whereas SXL bound to the 5′ UTR blocks the scanning of those complexes that have presumably escaped the 3′-UTR-mediated control (Fig. 3) (Gebauer et al. 2003; Beckmann et al. 2005). A uORF located just upstream of the SXL-binding site has been shown recently to be important for 5′-UTR-mediated repression (Medenbach et al. 2011); SXL promotes the recognition of the

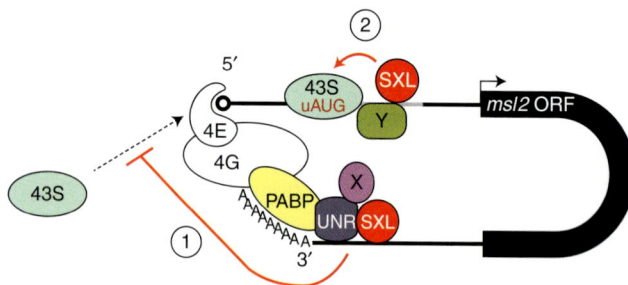

Figure 3. Mechanism of translational repression of *msl2* mRNA. SXL binds to both the 5′ and 3′ UTRs of *msl2* to achieve strong repression. SXL bound to the 3′ UTR recruits UNR to bind to the RNA in close proximity. In turn, UNR interacts with poly(A) tail-bound PABP to inhibit 43S ribosomal complex recruitment at a step downstream from closed-loop formation (1). SXL bound to the 5′ UTR inhibits ribosomal scanning by promoting recognition of an upstream AUG (uAUG), thus preventing 43S complexes from reaching the main *msl2* ORF (2). Additional unidentified factors (X, Y) are likely involved. (Adapted in modified form from Graindorge et al. 2011.)

Cite this article as *Cold Spring Harb Perspect Biol* doi: 10.1101/cshperspect.a012245

upstream initiator AUG by scanning 43S complexes, thus preventing them from reaching the main ORF. SXL does not simply act by steric hindrance, because PTB bound in the same position does not promote uAUG recognition. These data suggest that SXL establishes specific contacts with additional factors and/or the translational machinery for repression via the 5′ UTR; they also provide a first example of RBP regulation of a uORF.

Translational repression via the 3′ UTR certainly requires additional factors. Binding of SXL alone is insufficient to repress the 43S recruitment step, and the highly conserved SXL homolog from *Musca domestica* cannot inhibit translation despite binding to the same sites with similar apparent affinities, suggesting that specific contacts are made between SXL, the *msl2* 3′ UTR, and other factors necessary for repression (Gebauer et al. 2003; Grskovic et al. 2003). One of the critical factors was identified as the protein Upstream of N-ras (UNR), a conserved regulator also known for its role in IRES-mediated translation and mRNA stability control in mammals (Fig. 3) (for review, see Mihailovich et al. 2010). UNR is required for translational repression of *msl2* reporters in vitro, for repression of endogenous *msl2* in cell culture, and for inhibition of dosage compensation in female flies (Abaza et al. 2006; Duncan 2006; Patalano et al. 2009). Binding of UNR to the 3′ UTR of *msl2* depends on SXL, and therefore, even though UNR is present in males, it does not bind to *msl2* and repress its translation because of the absence of SXL. Thus, SXL confers a sex-specific function to UNR.

How does the SXL–UNR complex function to repress translation? Although a poly(A) tail is not strictly required for regulation, translational repression via the 3′ UTR is stimulated by the poly(A) tail (Duncan et al. 2009). Interactions between UNR and poly(A) tail-bound PABP are thought to underlie this stimulation by mechanisms that are yet unknown. PABP binds to the poly(A) tail and contacts the cap-binding complex, yielding a closed-loop conformation of the mRNA that is thought to be optimal for efficient ribosome recruitment (Tarun and Sachs 1996; Wells et al. 1998). Intriguingly, closed-loop formation of the *msl2* mRNA is not affected by UNR, indicating that the SXL–UNR complex inhibits ribosome recruitment by targeting a step in translation initiation that is downstream from eIF4F binding (Fig. 3) (Duncan et al. 2009). Close examination of the *msl2* 3′ UTR indicates the presence of sequences required for translational repression but dispensable for SXL–UNR binding, suggesting that the full 3′-UTR repressor complex contains additional factors (C Militti, E Szostak, and F Gebauer, unpubl.). Understanding the composition of this complex and the interactions of its components with the translational machinery may yield novel clues as to how the SXL/UNR-organized complex on the 3′ UTR of the *msl2* mRNA controls the recruitment of ribosomes.

In summary, tight translational repression of *msl2* mRNA is achieved through an elaborate combinatorial mechanism that involves targeting different steps of translation initiation from both ends of the mRNA, involving multiple RBPs and a uORF as an additional *cis*-regulatory element. How all of these elements act together to ensure efficient and coordinated repression warrants further investigation.

MORE COMPLEXITY FOR TEMPORAL AND SPATIAL CONTROL OF TRANSLATION: *nanos* mRNA

Nanos (Nos) is a posterior determinant required for the formation of abdominal segments during early *Drosophila* development. Synthesis of Nos exclusively at the posterior of the embryo is achieved by localization and translational activation of the *nos* transcript at this region, combined with translational repression elsewhere (Gavis and Lehmann 1994). *nos* mRNA is transcribed and actively translated in nurse cells, and subsequently transferred to the adjacent growing oocyte through ring canals. In the bulk cytoplasm of late oocytes and early embryos, *nos* mRNA is translationally repressed via sequences contained in a discrete region of the 3′ UTR proximal to the stop codon, the translational control element (TCE) (Dahanukar and Wharton 1996; Gavis et al. 1996; Smibert et al. 1996). The TCE consists

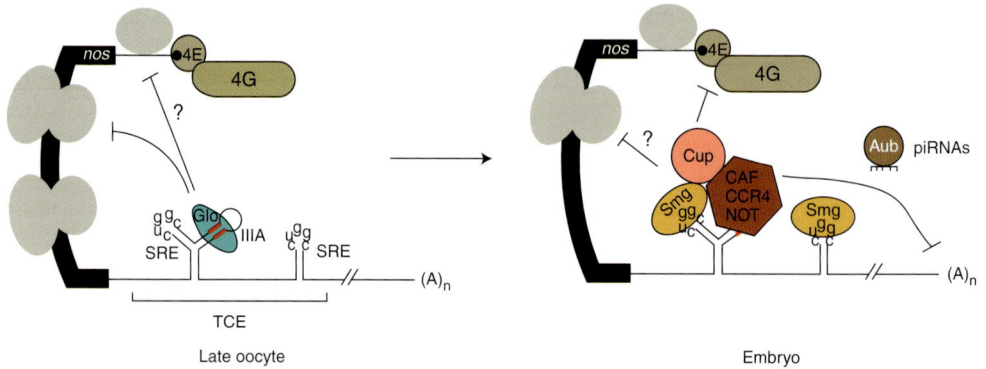

Figure 4. Mechanism of translational repression of *nanos* mRNA. *nanos* mRNA switches from a translationally active state in nurse cells to a silenced state in late oocytes and early embryos. Repression in late oocytes is driven by Glorund (Glo) binding to stem IIIA of the TCE and seems to be effected primarily at the elongation step. In embryos, Smaug (Smg) is synthesized and takes over repression by binding to the SREs within the TCE; the relative contribution of each SRE is uncertain, although SRE1 seems to contribute more to Smaug binding than SRE2. Smg recruits the eIF4E-binding protein Cup and the CAF–CCR4–NOT complex to block translation initiation and promote deadenylation and degradation of *nos* mRNA. The piwi pathway has been recently reported to participate in deadenylation. Additional steps of translation could be affected both in late oocytes and early embryos (see text for details).

of three stem–loops that are necessary for repression (Fig. 4). The AU-rich stem of one of these structures (stem IIIA, following the nomenclature of Crucs et al. 2000) is necessary for translational repression in oocytes and is thought to bind the hnRNP F/H protein Glorund (Crucs et al. 2000; Forrest et al. 2004; Kalifa et al. 2006). The other two stems carry loops consisting of CUGGC, which are recognized by the repressor Smaug and are therefore referred to as Smaug recognition elements (SREs) (Smibert et al. 1996, 1999; Dahanukar et al. 1999). Smaug is expressed exclusively in the early embryo and acts as a major regulator of maternal mRNA destabilization upon egg activation (Tadros et al. 2007). Point mutations in the SREs affect Smaug binding and lead to *nos* derepression in the embryo without effects on localization of the mRNA (Smibert et al. 1996). Smaug binds to the SRE via its SAM (sterile α motif) domain (Aviv et al. 2003; Green et al. 2003). A central guanine in the SRE appropriately oriented by the stem–loop structure is critical for SAM domain recognition (Johnson and Donaldson 2006; Oberstrass et al. 2006).

Smaug promotes deadenylation of *nos* mRNA by recruitment of the CAF–CCR4–

NOT complex (Jeske et al. 2006; Zaessinger et al. 2006). Deadenylation is necessary for *nos* repression, but deadenylated *nos* reporters can still be strongly repressed in vitro in a manner that partially depends on the SREs (Jeske et al. 2006). These data suggest that Smaug-mediated repression has two components: one that is independent of the poly(A) tail, and one that depends on deadenylation. Indeed, translational repression is often coupled to deadenylation, which contributes to maintain the repressed state (for review, see Miller and Olivas 2011). Smaug interacts with Cup, a protein that binds eIF4E and blocks the recruitment of eIF4G to the cap complex (Fig. 4) (Nelson et al. 2004; for review, see Richter and Sonenberg 2005). Mutation of the eIF4E-binding motifs of Cup partially relieves repression of *nos* transgenes, and the binding of Cup and eIF4G to *nos* mRNPs is mutually exclusive, implying a function of Cup in mediating a translation initiation block by Smaug (Nelson et al. 2004; Jeske et al. 2011). Contradictory results have been reported concerning the requirement of the cap structure for Smaug-mediated repression, however (Andrews et al. 2011; Jeske et al. 2011). Furthermore, a significant fraction of *nos* mRNA was found in association with polysomes in early embryos even under conditions in which the mRNA is completely unlocalized and Nos protein is undetectable, suggesting that a postinitiation step is affected (Clark et al. 2000). Consistently, translation mediated by the Cricket paralysis virus (CrPV) IRES, which mediates translation initiation without requirement for any of the cellular translation initiation factors, can be inhibited by the SREs (Jeske et al. 2011). Therefore, repression by Smaug might involve initiation and postinitiation events. Interestingly, recent data suggest that Cup can elicit translational repression independent of its eIF4E-binding motifs, and that Cup also promotes deadenylation directly via association with the CAF–CCR4–NOT complex (Igreja and Izaurralde 2011). This raises the possibility that Cup mediates repressor functions of Smaug beyond translation initiation. It will be interesting to analyze to what extent Smaug function is preserved in Cup-depleted extracts.

Recently, late ovary extracts that recapitulate repression mediated by the IIIA stem (the Glorund-binding site) have been developed (Andrews et al. 2011). Repression seems to be cap independent in these extracts. In addition, *nos* mRNA is found associated with polysomes, and Glorund is present in polysomes in association with the repressed mRNA, suggesting that Glorund inhibits translation at a postinitiation step. On the other hand, repression in late oocytes is poly(A) dependent, which may reflect an effect of Glorund on initiation as well. A comparison of the polysomal association of *nos* mRNA in total ovary extracts, which are enriched for early-stage oocytes, with that in late ovary and embryo extracts indicates a gradual shift to lighter fractions, consistent with the temporal acquisition of distinct mechanisms of translational repression (Andrews et al. 2011).

Altogether, currently available data can be integrated into the following model explaining the temporal switch in *nos* mRNA expression, from activation in nurse cells to repression later in development. In late oocytes, Glorund imposes a block on elongation/termination that results in quick repression; in early embryos, Smaug consolidates *nos* inhibition at the initi-

ation and possibly postinitiation steps and promotes deadenylation and degradation of the transcript by recruiting the CAF–CCR4–NOT complex (Fig. 4). Intriguingly, the piRNA pathway has recently been implicated in this mechanism. Rouget et al. (2010) found that CCR4-mediated deadenylation of *nos* depends on piRNAs complementary to a distal region of *nos* 3' UTR and that Aubergine, an Argonaute protein, interacts with Smaug and CCR4. Further experiments are required to decipher how the piRNA pathway cooperates with the Smaug–CCR4 complex.

SIGNAL-DEPENDENT TEMPORAL CONTROL OF AN (ANTI-)INFLAMMATORY RNA OPERON: THE GAIT COMPLEX

During inflammation, the synthesis of ceruloplasmin (Cp) is transiently induced by interferon (IFN)-γ in myeloid cells and ceases at ∼24 h of IFN-γ treatment by the action of a translation repressor complex termed GAIT (IFN-γ-activated inhibitor of translation) (for review, see Mukhopadhyay et al. 2009). The GAIT complex recognizes a bipartite stem–loop structure in the 3' UTR of Cp mRNA, the GAIT element (Fig. 5) (Sampath et al. 2003). The GAIT complex is composed of four proteins: ribosomal protein L13a, glutamyl-prolyl tRNA synthetase (EPRS), NS1-associated protein 1 (NSAP1), and glyceraldehyde 3-phosphate dehydrogenase (GAPDH) (Mazumder et al. 2003; Sampath et al. 2004). EPRS is the subunit that binds directly to the RNA, whereas phosphorylated L13a is responsible for interactions with the translational machinery (Sampath et al. 2004; Kapasi et al. 2007; Arif et al. 2009). Similar to 3'-UTR-bound SXL, the GAIT complex inhibits 43S recruitment without affecting closed-loop formation (Mazumder et al. 2001; Kapasi et al. 2007). Furthermore, because repression requires PABP and the poly(A) tail, it was proposed that mRNA circularization actually favors the correct positioning of GAIT close to the 5' UTR (Mazumder et al. 2001). L13a binds eIF4G at its eIF3-binding site and blocks the eIF4G–eIF3 interaction, a step required for 43S complex recruitment (Kapasi et al. 2007; Arif et al. 2009).

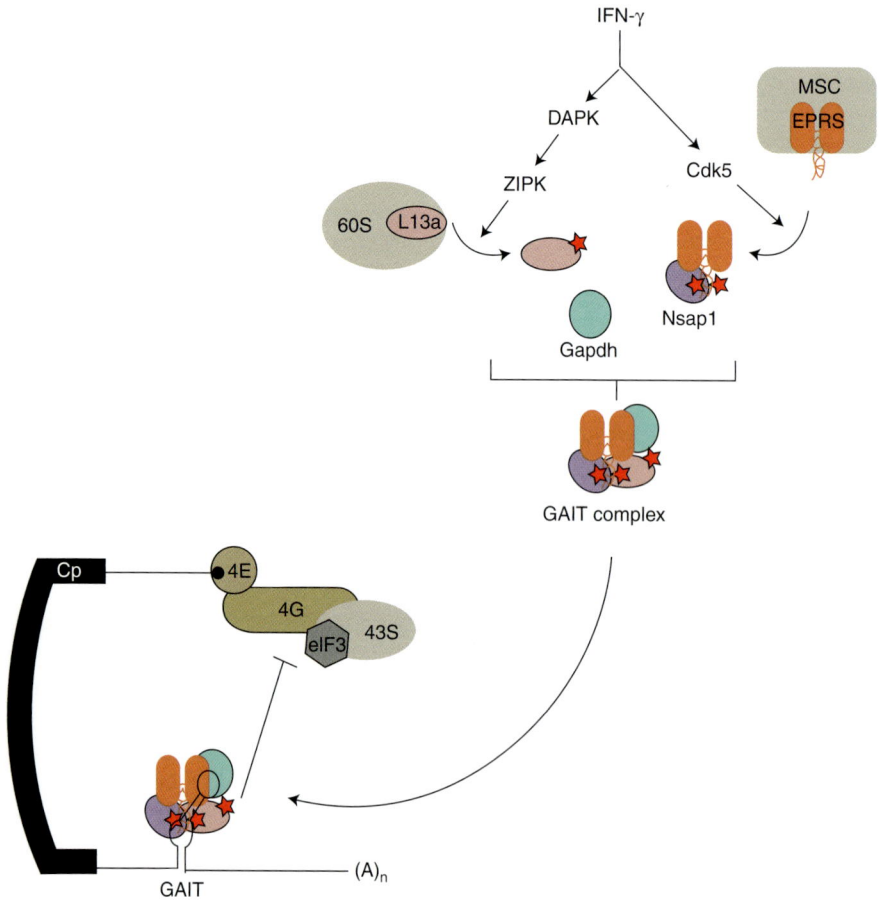

Figure 5. Translational repression by the GAIT complex. During inflammation, EPRS and L13a are phosphorylated and released from the multisynthetase complex (MSC) and the 60S ribosomal subunit, respectively. These proteins bind to NSAP1 and GAPDH to form the heterotetrameric GAIT complex, which binds to a split stem structure present in the 3′ UTR of target mRNAs. L13a then inhibits the recruitment of the 43S ribosomal complex by blocking the interaction between 43S-associated eIF3 and eIF4G.

Several lessons can be learned from this instructive example of translational control. First, temporal control of translation is achieved by the regulated, ordered assembly of the GAIT complex. Under control conditions, EPRS resides in the cytosolic tRNA multisynthetase complex (MSC) and acts as an enzyme catalyzing the addition of amino acids to tRNA. EPRS contains two catalytic domains (ERS and PRS) linked by a region that is not required for its enzymatic activity. This region contains three helix–turn–helix structures termed WHEP domains (named after the three tRNA synthetases

that contain them) that serve RNA binding (Jia et al. 2008). Phosphorylation of EPRS by Cdk5 during the early phase of the IFN-γ response induces its release from the MSC (Sampath et al. 2004; Arif et al. 2009, 2011). Free, phosphorylated EPRS interacts with NSAP1 through one of the WHEP domains, resulting in a complex with no RNA-binding activity (Fig. 5). Later on, L13a—which resides in the large ribosomal subunit—is phosphorylated and released from the ribosome (Mazumder et al. 2003). Phosphorylated L13a and GAPDH then interact with the phosphorylated EPRS–NSAP1 dimer, resulting in a heterotetrameric complex that exposes the WHEP domains for interactions with the GAIT element (Jia et al. 2008).

The second lesson concerns the functional versatility of RBPs. As mentioned above, EPRS and GAPDH are polypeptides that can function as enzymes or contribute to translational control depending on cellular conditions. The founding example of an enzyme that also functions as an RBP is IRP1. In iron-replete cells, IRP1 bears an iron–sulfur cluster necessary for cytosolic aconitase activity. In iron-starved cells, the iron–sulfur cluster is lost, and a pocket is uncovered that binds to mRNAs encoding factors involved in iron metabolism, resulting in stabilization or translational repression of its target mRNAs (for review, see Pantopoulos 2004). Interconnectivity between cellular metabolism and RNA regulation might be an emerging theme in gene expression (Hentze and Preiss 2010).

The third lesson pertains to the observation that L13a phosphorylation leads to L13a-depleted ribosomes that are functionally normal. Practically the entire cellular complement of L13a is released upon phosphorylation, yielding no defects in general translation (Mazumder et al. 2003). This result suggests that the ribosome also serves as a repository of regulatory molecules for release from its surface when their functions are required. The concept of large macromolecular complexes as depots for regulatory proteins has been discussed elsewhere (Ray et al. 2007).

Finally, the fourth lesson relates to the existence of an RNA regulon for GAIT-mediated translational control during the inflammatory response. L13a phosphorylation is the culminating event of a kinase cascade in which IFN-γ activates DAPK, which, in turn, activates ZIPK (Fig. 5). Both the DAPK and ZIPK mRNAs contain GAIT elements in their 3′ UTRs and are repressed by the GAIT complex, establishing a negative-feedback loop that contributes to the temporal limitation of the inflammatory response (Mukhopadhyay et al. 2008). Furthermore, in silico searches and genome-wide polysome profiling have provided a list of candidate targets containing GAIT elements, many of which encode proteins involved in inflammation (Ray and Fox 2007; Vyas et al. 2009). These results imply that the GAIT complex appears to coordinate a posttranscriptional operon involved in the resolution of the inflammatory response.

CONCLUSIONS AND PERSPECTIVES

Starting with the mapping of *cis*-acting elements and the identification of *trans*-acting factors mostly by biochemical and genetic approaches during the past decades, the investigation of mRNA translation has now entered into a phase of transcriptome-wide, highly parallel analyses. These approaches have begun to help define RBP-binding sites across the transcriptome and paint a picture of dense mRNP assemblies involving a multitude of *trans*-acting factors. Several concepts have emerged along the way: (1) The binding sites of RBPs, and RBPs themselves, influence each other on the mRNAs, giving rise to combinatorial outcomes. (2) Essentially all RBPs have numerous target mRNAs, further driving combinatorial modes of translational control. (3) The resulting mRNPs are highly dynamic structures that undergo rearrangements in response to biological signaling. (4) DExH/D RNA helicases not only remodel RNA structure by unwinding RNA–RNA duplex structures, but can directly remodel RNPs in different biological settings (Jankowsky and Bowers 2006). Reductionist biochemical work on model systems revealed that the same RBPs can influence translation by more than one mechanism, often influenced by other RBPs in a combinatorial way. We expect that reductionist mechanistic investigations by biochemical approaches and transcriptome-wide, time-resolved in vivo analyses including ribosome profiling (Ingolia et al. 2011) will converge to yield unprecedented insights into translation and translational control, both at the level of individual mRNAs and whole transcriptomes.

ACKNOWLEDGMENTS

Work in F.G.'s laboratory is supported by grants BFU2009-08243 and Consolider CSD2009-00080 from MICINN; T.P. is supported by grants from the National Health and Medical Research Council of Australia and the Australian Research Council; M.W.H. acknowledges continuous support from the Deutsche Forschungsgemeinschaft. We are grateful to laboratory members and collaborators for their many critical contributions to cited work from our laboratories, and apologize to those colleagues whose work we could not cite for reasons of focus.

REFERENCES

Abaza I, Coll O, Patalano S, Gebauer F. 2006. *Drosophila* UNR is required for translational repression of male-specific lethal 2 mRNA during regulation of X-chromosome dosage compensation. *Genes Dev* **20**: 380–389.

Anderson P, Kedersha N. 2009. RNA granules: Post-transcriptional and epigenetic modulators of gene expression. *Nat Rev Mol Cell Biol* **10**: 430–436.

Andrews S, Snowflack DR, Clark IE, Gavis ER. 2011. Multiple mechanisms collaborate to repress *nanos* translation in the *Drosophila* ovary and embryo. *RNA* **17**: 967–977.

Arif A, Jia J, Mukhopadhyay R, Willard B, Kinter M, Fox PL. 2009. Two-site phosphorylation of EPRS coordinates multimodal regulation of noncanonical translational control activity. *Mol Cell* **35**: 164–180.

Arif A, Jia J, Moodt RA, DiCorleto PE, Fox PL. 2011. Phosphorylation of glutamyl-prolyl tRNA synthetase by cyclin-dependent kinase 5 dictates transcript-selective translational control. *Proc Natl Acad Sci* **108**: 1415–1420.

Aviv T, Lin Z, Lau S, Rendl LM, Sicheri F, Smibert CA. 2003. The RNA-binding SAM domain of Smaug defines a new family of post-transcriptional regulators. *Nat Struct Biol* **10**: 614–621.

Balagopal V, Parker R. 2009. Polysomes, P bodies and stress granules: States and fates of eukaryotic mRNAs. *Curr Opin Cell Biol* **21**: 403–408.

Bashaw GJ, Baker BS. 1997. The regulation of the *Drosophila msl-2* gene reveals a function for Sex-lethal in translational control. *Cell* **89**: 789–798.

Beach DL, Keene JD. 2008. Ribotrap: Targeted purification of RNA-specific RNPs from cell lysates through immunoaffinity precipitation to identify regulatory proteins and RNAs. *Methods Mol Biol* **419**: 69–91.

Beckmann K, Grskovic M, Gebauer F, Hentze MW. 2005. A dual inhibitory mechanism restricts *msl-2* mRNA translation for dosage compensation in *Drosophila*. *Cell* **122**: 529–540.

Brimacombe R, Stiege W, Kyriatsoulis A, Maly P. 1988. Intra-RNA and RNA–protein cross-linking techniques in *Escherichia coli* ribosomes. *Methods Enzymol* **164**: 287–309.

Castello A, Fischer B, Schuschke K, Horos R, Beckmann BM, Strein C, Davey NE, Humphreys DT, Preiss T, Steinmetz LM, et al. 2012. Insights into RNA biology from an atlas of mammalian mRNA-binding proteins. *Cell* (in press).

Clark IE, Wyckoff D, Gavis ER. 2000. Synthesis of the posterior determinant Nanos is spatially restricted by a novel cotranslational regulatory mechanism. *Curr Biol* **10**: 1311–1314.

Crucs S, Chatterjee S, Gavis ER. 2000. Overlapping but distinct RNA elements control repression and activation of *nanos* translation. *Mol Cell* **5**: 457–467.

Czaplinski K, Köcher T, Schelder M, Segref A, Wilm M, Mattaj IW. 2005. Identification of 40LoVe, a *Xenopus* hnRNP D family protein involved in localizing a TGF-β-related mRNA during oogenesis. *Dev Cell* **8**: 505–515.

Dahanukar A, Wharton RP. 1996. The Nanos gradient in *Drosophila* embryos is generated by translational regulation. *Genes Dev* **10**: 2610–2620.

Dahanukar A, Walker JA, Wharton RP. 1999. Smaug, a novel RNA-binding protein that operates a translational switch in *Drosophila*. *Mol Cell* **4**: 209–218.

De Gregorio E, Preiss T, Hentze MW. 1999. Translation driven by an eIF4G core domain in vivo. *EMBO J* **18**: 4865–4874.

Dreyfuss G, Choi YD, Adam SA. 1984. Characterization of heterogeneous nuclear RNA–protein complexes in vivo with monoclonal antibodies. *Mol Cell Biol* **4**: 1104–1114.

Duncan K. 2006. Sex-lethal imparts a sex-specific function to UNR by recruiting it to the *msl-2* mRNA 3′ UTR: Translational repression for dosage compensation. *Genes Dev* **20**: 368–379.

Duncan KE, Strein C, Hentze MW. 2009. The SXL–UNR corepressor complex uses a PABP-mediated mechanism to inhibit ribosome recruitment to *msl-2* mRNA. *Mol Cell* **36**: 571–582.

Filipovska A, Razif MFM, Nygård KKA, Rackham O. 2011. A universal code for RNA recognition by PUF proteins. *Nat Chem Biol* **7**: 425–427.

Forrest KM, Clark IE, Jain RA, Gavis ER. 2004. Temporal complexity within a translational control element in the *nanos* mRNA. *Development* **131**: 5849–5857.

Gavis ER, Lehmann R. 1994. Translational regulation of *nanos* by RNA localization. *Nature* **369**: 315–318.

Gavis ER, Lunsford L, Bergsten SE, Lehmann R. 1996. A conserved 90 nucleotide element mediates translational repression of *nanos* RNA. *Development* **122**: 2791–2800.

Gebauer F, Hentze M. 2004. Molecular mechanisms of translational control. *Nat Rev Mol Cell Biol* **5**: 827–835.

Gebauer F, Corona DF, Preiss T, Becker PB, Hentze MW. 1999. Translational control of dosage compensation in *Drosophila* by Sex-lethal: Cooperative silencing via the 5′ and 3′ UTRs of *msl-2* mRNA is independent of the poly(A) tail. *EMBO J* **18**: 6146–6154.

Gebauer F, Grskovic M, Hentze MW. 2003. *Drosophila* Sex-lethal inhibits the stable association of the 40S ribosomal subunit with *msl-2* mRNA. *Mol Cell* **11**: 1397–1404.

Gelbart ME, Kuroda MI. 2009. *Drosophila* dosage compensation: A complex voyage to the X chromosome. *Development* **136**: 1399–1410.

Graindorge A, Militti C, Gebauer F. 2011. Posttranscriptional control of X-chromosome dosage compensation. *Wiley Interdiscip Rev RNA* **2**: 534–545.

Green JB, Gardner CD, Wharton RP, Aggarwal AK. 2003. RNA recognition via the SAM domain of Smaug. *Mol Cell* **11**: 1537–1548.

Greenberg JR. 1979. Ultraviolet light-induced crosslinking of mRNA to proteins. *Nucleic Acids Res* **6:** 715–732.

Grskovic M, Hentze MW, Gebauer F. 2003. A co-repressor assembly nucleated by Sex-lethal in the 3′UTR mediates translational control of *Drosophila msl-2* mRNA. *EMBO J* **22:** 5571–5581.

Hafner M, Landthaler M, Burger L, Khorshid M, Hausser J, Berninger P, Rothballer A, Ascano M, Jungkamp A-C, Munschauer M, et al. 2010. Transcriptome-wide identification of RNA-binding protein and microRNA target sites by PAR-CLIP. *Cell* **141:** 129–141.

Hentze MW, Preiss T. 2010. The REM phase of gene regulation. *Trends Biochem Sci* **35:** 423–426.

Hentze MW, Caughman SW, Rouault TA, Barriocanal JG, Dancis A, Harford JB, Klausner RD. 1987. Identification of the iron-responsive element for the translational regulation of human ferritin mRNA. *Science* **238:** 1570–1573.

Hentze MW, Muckenthaler MU, Galy B, Camaschella C. 2010. Two to tango: Regulation of mammalian iron metabolism. *Cell* **142:** 24–38.

Hinnebusch AG. 2005. Translational regulation of GCN4 and the general amino acid control of yeast. *Ann Rev Microbiol* **59:** 407–450.

Hockensmith JW, Kubasek WL, Vorachek WR, von Hippel PH. 1986. Laser cross-linking of nucleic acids to proteins. Methodology and first applications to the phage T4 DNA replication system. *J Biol Chem* **261:** 3512–3518.

Hogg JR, Collins K. 2007. RNA-based affinity purification reveals 7SK RNPs with distinct composition and regulation. *RNA* **13:** 868–880.

Holcik M, Sonenberg N, Korneluk RG. 2000. Internal ribosome initiation of translation and the control of cell death. *Trends Genet* **16:** 469–473.

Igreja C, Izaurralde E. 2011. CUP promotes deadenylation and inhibits decapping of mRNA targets. *Genes Dev* **25:** 1955–1967.

Ingolia NT, Lareau LF, Weissman JS. 2011. Ribosome profiling of mouse embryonic stem cells reveals the complexity and dynamics of mammalian proteomes. *Cell* **147:** 789–802.

Jankowsky E, Bowers H. 2006. Remodeling of ribonucleoprotein complexes with DExH/D RNA helicases. *Nucleic Acids Res* **34:** 4181–4188.

Jeske M, Meyer S, Temme C, Freudenreich D, Wahle E. 2006. Rapid ATP-dependent deadenylation of *nanos* mRNA in a cell-free system from *Drosophila* embryos. *J Biol Chem* **281:** 25124–25133.

Jeske M, Moritz B, Anders A, Wahle E. 2011. Smaug assembles an ATP-dependent stable complex repressing *nanos* mRNA translation at multiple levels. *EMBO J* **30:** 90–103.

Jia J, Arif A, Ray PS, Fox PL. 2008. WHEP domains direct noncanonical function of glutamyl-Prolyl tRNA synthetase in translational control of gene expression. *Mol Cell* **29:** 679–690.

Johnson PE, Donaldson LW. 2006. RNA recognition by the Vts1p SAM domain. *Nat Struct Mol Biol* **13:** 177–178.

Kalifa Y, Huang T, Rosen LN, Chatterjee S, Gavis ER. 2006. Glorund, a *Drosophila* hnRNP F/H homolog, is an ovarian repressor of *nanos* translation. *Dev Cell* **10:** 291–301.

Kapasi P, Chaudhuri S, Vyas K, Baus D, Komar AA, Fox PL, Merrick WC, Mazumder B. 2007. L13a blocks 48S assembly: Role of a general initiation factor in mRNA-specific translational control. *Mol Cell* **25:** 113–126.

Keene JD. 2007. RNA regulons: Coordination of post-transcriptional events. *Nat Rev Genet* **8:** 533–543.

Kelley RL, Wang J, Bell L, Kuroda MI. 1997. Sex lethal controls dosage compensation in *Drosophila* by a non-splicing mechanism. *Nature* **387:** 195–199.

Kozak M. 1986. Influences of mRNA secondary structure on initiation by eukaryotic ribosomes. *Proc Natl Acad Sci* **83:** 2850–2854.

Leibold EA, Munro HN. 1987. Characterization and evolution of the expressed rat ferritin light subunit gene and its pseudogene family. Conservation of sequences within noncoding regions of ferritin genes. *J Biol Chem* **262:** 7335–7341.

Li JB, Levanon EY, Yoon JK, Aach J, Xie B, Leproust E, Zhang K, Gao Y, Church GM. 2009. Genome-wide identification of human RNA editing sites by parallel DNA capturing and sequencing. *Science* **324:** 1210–1213.

Mackay JP, Font J, Segal DJ. 2011. The prospects for designer single-stranded RNA-binding proteins. *Nat Struct Mol Biol* **18:** 256–261.

Martin KC, Ephrussi A. 2009. mRNA localization: Gene expression in the spatial dimension. *Cell* **136:** 719–730.

Mazumder B, Seshadri V, Imataka H, Sonenberg N, Fox PL. 2001. Translational silencing of ceruloplasmin requires the essential elements of mRNA circularization: Poly(A) tail, poly(A)-binding protein, and eukaryotic translation initiation factor 4G. *Mol Cell Biol* **21:** 6440–6449.

Mazumder B, Sampath P, Seshadri V, Maitra RK, DiCorleto PE, Fox PL. 2003. Regulated release of L13a from the 60S ribosomal subunit as a mechanism of transcript-specific translational control. *Cell* **115:** 187–198.

Medenbach J, Seiler M, Hentze MW. 2011. Translational control via protein-regulated upstream open reading frames. *Cell* **145:** 902–913.

Mihailovich M, Militti C, Gabaldon T, Gebauer F. 2010. Eukaryotic cold shock domain proteins: Highly versatile regulators of gene expression. *BioEssays* **32:** 109–118.

Miller MA, Olivas WM. 2011. Roles of Puf proteins in mRNA degradation and translation. *Wiley Interdiscip Rev RNA* **2:** 471–492.

Moore MJ, Proudfoot NJ. 2009. Pre-mRNA processing reaches back to transcription and ahead to translation. *Cell* **136:** 688–700.

Mueller PP, Hinnebusch AG. 1986. Multiple upstream AUG codons mediate translational control of GCN4. *Cell* **45:** 201–207.

Mukhopadhyay R, Ray PS, Arif A, Brady AK, Kinter M, Fox PL. 2008. DAPK–ZIPK–L13a axis constitutes a negative-feedback module regulating inflammatory gene expression. *Mol Cell* **32:** 371–382.

Mukhopadhyay R, Jia J, Arif A, Ray PS, Fox PL. 2009. The GAIT system: A gatekeeper of inflammatory gene expression. *Trends Biochem Sci* **34:** 324–331.

Namy O, Rousset JP, Napthine S, Brierley I. 2004. Reprogrammed genetic decoding in cellular gene expression. *Mol Cell* **13:** 157–168.

Nelson MR, Leidal AM, Smibert CA. 2004. *Drosophila* Cup is an eIF4E-binding protein that functions in Smaug-mediated translational repression. *EMBO J* **23**: 150–159.

Oberstrass FC, Lee A, Stefl R, Janis M, Chanfreau G, Allain FH. 2006. Shape-specific recognition in the structure of the Vts1p SAM domain with RNA. *Nat Struct Mol Biol* **13**: 160–167.

Pantopoulos K. 2004. Iron metabolism and the IRE/IRP regulatory system: An update. *Ann NY Acad Sci* **1012**: 1–13.

Parker BJ, Moltke I, Roth A, Washietl S, Wen J, Kellis M, Breaker R, Pedersen JS. 2011. New families of human regulatory RNA structures identified by comparative analysis of vertebrate genomes. *Genome Res* **21**: 1929–1943.

Patalano S, Mihailovich M, Belacortu Y, Paricio N, Gebauer F. 2009. Dual sex-specific functions of *Drosophila* upstream of N-ras in the control of X chromosome dosage compensation. *Development* **136**: 689–698.

Pelletier J, Sonenberg N. 1985. Insertion mutagenesis to increase secondary structure within the 5′ noncoding region of a eukaryotic mRNA reduces translational efficiency. *Cell* **40**: 515–526.

Ray PS, Fox PL. 2007. A post-transcriptional pathway represses monocyte VEGF-A expression and angiogenic activity. *EMBO J* **26**: 3360–3372.

Ray PS, Arif A, Fox PL. 2007. Macromolecular complexes as depots for releasable regulatory proteins. *Trends Biochem Sci* **32**: 158–164.

Richter JD, Sonenberg N. 2005. Regulation of cap-dependent translation by eIF4E inhibitory proteins. *Nature* **433**: 477–480.

Rouget C, Papin C, Boureux A, Meunier AC, Franco B, Robine N, Lai EC, Pelisson A, Simonelig M. 2010. Maternal mRNA deadenylation and decay by the piRNA pathway in the early *Drosophila* embryo. *Nature* **467**: 1128–1132.

Sachs AB, Buratowski S. 1997. Common themes in translational and transcriptional regulation. *Trends Biochem Sci* **22**: 189–192.

Saito T, Saetrom P. 2010. MicroRNAs—Targeting and target prediction. *N Biotechnol* **27**: 243–249.

Sampath P, Mazumder B, Seshadri V, Fox PL. 2003. Transcript-selective translational silencing by γ interferon is directed by a novel structural element in the ceruloplasmin mRNA 3′ untranslated region. *Mol Cell Biol* **23**: 1509–1519.

Sampath P, Mazumder B, Seshadri V, Gerber CA, Chavatte L, Kinter M, Ting SM, Dignam JD, Kim S, Driscoll DM, et al. 2004. Noncanonical function of glutamyl-prolyl-tRNA synthetase: Gene-specific silencing of translation. *Cell* **119**: 195–208.

Scherrer T, Mittal N, Janga SC, Gerber AP. 2010. A screen for RNA-binding proteins in yeast indicates dual functions for many enzymes. *PLoS ONE* **5**: e15499.

Schwanhausser B, Busse D, Li N, Dittmar G, Schuchhardt J, Wolf J, Chen W, Selbach M. 2011. Global quantification of mammalian gene expression control. *Nature* **473**: 337–342.

Smibert CA, Wilson JE, Kerr K, Macdonald PM. 1996. Smaug protein represses translation of unlocalized *nanos*

mRNA in the *Drosophila* embryo. *Genes Dev* **10**: 2600–2609.

Smibert CA, Lie YS, Shillinglaw W, Henzel WJ, Macdonald PM. 1999. Smaug, a novel and conserved protein, contributes to repression of *nanos* mRNA translation in vitro. *RNA* **5**: 1535–1547.

Sonenberg N, Hinnebusch AG. 2009. Regulation of translation initiation in eukaryotes: Mechanisms and biological targets. *Cell* **136**: 731–745.

Squires JE, Preiss T. 2010. Function and detection of 5-methylcytosine in eukaryotic RNA. *Epigenomics* **2**: 709–715.

Squires JE, Patel HR, Nousch M, Sibbritt T, Humphreys DT, Parker BJ, Suter CM, Preiss T. 2012. Widespread occurrence of 5-methylcytosine in human coding and noncoding RNA. *Nucleic Acids Res* doi: 10.1093/nar/gks144.

Tadros W, Goldman AL, Babak T, Menzies F, Vardy L, Orr-Weaver T, Hughes TR, Westwood JT, Smibert CA, Lipshitz HD. 2007. SMAUG is a major regulator of maternal mRNA destabilization in *Drosophila* and its translation is activated by the PAN GU kinase. *Dev Cell* **12**: 143–155.

Tarun SZ Jr, Sachs AB. 1996. Association of the yeast poly(A) tail binding protein with translation initiation factor eIF-4G. *EMBO J* **15**: 7168–7177.

Tenenbaum S, Carson C, Lager P, Keene J. 2000. Identifying mRNA subsets in messenger ribonucleoprotein complexes by using cDNA arrays. *Proc Natl Acad Sci* **97**: 14085–14090.

Tsai BP, Wang X, Huang L, Waterman ML. 2011. Quantitative profiling of in vivo–assembled RNA–protein complexes using a novel integrated proteomic approach. *Mol Cell Proteomics* **10**: M110.007385.

Tsvetanova NG, Klass DM, Salzman J, Brown PO. 2010. Proteome-wide search reveals unexpected RNA-binding proteins in *Saccharomyces cerevisiae*. *PLoS ONE* **5**: e12671.

Ule J. 2003. CLIP identifies Nova-regulated RNA networks in the brain. *Science* **302**: 1212–1215.

Valasek L, Szamecz B, Hinnebusch AG, Nielsen KH. 2007. In vivo stabilization of preinitiation complexes by formaldehyde cross-linking. *Methods Enzymol* **429**: 163–183.

Vyas K, Chaudhuri S, Leaman DW, Komar AA, Musiyenko A, Barik S, Mazumder B. 2009. Genome-wide polysome profiling reveals an inflammation-responsive posttranscriptional operon in γ interferon-activated monocytes. *Mol Cell Biol* **29**: 458–470.

Wells SE, Hillner PE, Vale RD, Sachs AB. 1998. Circularization of mRNA by eukaryotic translation initiation factors. *Mol Cell Biol* **2**: 135–140.

Zaessinger S, Busseau I, Simonelig M. 2006. Oskar allows *nanos* mRNA translation in *Drosophila* embryos by preventing its deadenylation by Smaug/CCR4. *Development* **133**: 4573–4583.

Zhang X, Liu Z, Yi J, Tang H, Xing J, Yu M, Tong T, Shang Y, Gorospe M, Wang W. 2012. The tRNA methyltransferase NSun2 stabilizes p16(INK4) mRNA by methylating the 3-untranslated region of p16. *Nat Commun* **3**: 712.

Zhao H, Yang Y, Zhou Y. 2011. Structure-based prediction of RNA-binding domains and RNA-binding sites and application to structural genomics targets. *Nucleic Acids Res* **39**: 3017–3025.

Regulation of mRNA Translation by Signaling Pathways

Philippe P. Roux[1,2,3] and Ivan Topisirovic[4,5]

[1]Institute for Research in Immunology and Cancer, Université de Montréal, Montréal, Québec H3C 3J7, Canada

[2]Molecular Biology Program, Université de Montréal, Montréal, Québec H3C 3J7, Canada

[3]Department of Pathology and Cell Biology, Faculty of Medicine, Université de Montréal, Québec H3C 3J7, Canada

[4]Lady Davis Institute for Medical Research, Jewish General Hospital, McGill University, Montréal, Québec H3T 1E2, Canada

[5]Department of Oncology, McGill University, Montréal, Québec H3T 1E2, Canada

Correspondence: philippe.roux@umontreal.ca; ivan.topisirovic@mcgill.ca

mRNA translation is the most energy consuming process in the cell. In addition, it plays a pivotal role in the control of gene expression and is therefore tightly regulated. In response to various extracellular stimuli and intracellular cues, signaling pathways induce quantitative and qualitative changes in mRNA translation by modulating the phosphorylation status and thus the activity of components of the translational machinery. In this work we focus on the phosphoinositide 3-kinase (PI3K)/AKT and the mitogen-activated protein kinase (MAPK) pathways, as they are strongly implicated in the regulation of translation in homeostasis, whereas their malfunction has been linked to aberrant translation in human diseases, including cancer.

Translation plays an evolutionarily conserved role in the regulation of gene expression (Mathews et al. 2007). In fact, recent findings indicate that translation plays a major part in determining protein expression levels in mammalian cells (Schwanhausser et al. 2011). Amperometric oxygen consumption measurements in rat thymocytes revealed that translation consumes ~20% of cellular ATP, and is thus thought to be the most energy demanding cellular process (Buttgereit and Brand 1995). Therefore, it is not surprising that translation represents a tightly regulated cellular process, dysregulation of which contributes to diverse human diseases, including cancer (see Silvera et al. 2010; Braun et al. 2012).

Translation occurs in three distinct steps: initiation, elongation, and termination (Sonenberg et al. 2012). Although all of the steps are highly regulated, most of the translational control occurs at the rate-limiting initiation step, during which mRNAs and $tRNA_i^{Met}$ are recruited to the ribosome through the highly orchestrated action of translation initiation factors (IFs) (reviewed by Sonenberg and Hinnebusch 2009). Various stimuli including environmental stress (e.g., heat-shock, UV irradiation), extracellular stimuli (e.g., nutrients, hormones, growth factors), and intracellular cues (e.g., energy status of the cell, intracellular availability of amino acids) (Wek et al. 2006; Ma and Blenis 2009; Sonenberg and Hinnebusch 2009) induce dramatic qualitative and quantitative changes in the translatome, i.e., the pools of cellular mRNAs that are being actively translated (Greenbaum et al. 2001). This is achieved in part via the phosphorylation of eukaryotic translation initiation factors (eIFs) and proteins that regulate their activity (e.g., 4E-BPs, PDCD4) by signaling pathways (reviewed by Sonenberg and Hinnebusch 2009; Jackson et al. 2010).

Here we summarize current knowledge on the role of cellular signaling pathways in translational control. In particular, we portray the role of the target of rapamycin (TOR) and the mitogen-activated protein kinase (MAPK) pathways in the regulation of translation, because these pathways regulate the phosphorylation and function of a multitude of eIFs and associated factors. Another important signaling node in translation involves the eIF2α kinases, discussed by Benham (2012).

TOR

TOR is an evolutionarily conserved Ser/Thr kinase, which regulates proliferation (increase in cell number) and growth (increase in volume/mass) in response to cellular energy status, growth factors, hormones, and nutrient availability (Wullschleger et al. 2006). TOR exists in two functionally and structurally distinct protein complexes referred to as TOR complex 1 and 2 (TORC1 and 2) (Wullschleger et al. 2006). In mammalian cells, mechanistic/mammalian TORC1 (mTORC1) consists of the catalytic component mTOR, the scaffolding protein Raptor (regulatory-associated protein of TOR, which is orthologous to KOG1 in yeast), the GTPase β-subunit like protein gβL (also known as mLST8), proline-rich AKT substrate of 40 kDa (PRAS40) and Deptor (disheveled, Egl-10, pleckstrin [DEP] domain containing mTOR interacting protein) (Fig. 1) (Guertin and Sabatini 2007; Peterson et al. 2009).

mTOR, gβL, and Deptor are also found in mTORC2 (Guertin and Sabatini 2007; Peterson et al. 2009). In turn, Rictor (rapamycin-insensitive companion of TOR, which is orthologous to AVO3 in yeast), mSIN1 (mammalian stress-activated protein kinase (SAPK)-interacting protein), and PRR5 (Proline-rich protein 5, also known as Protor) (Frias et al. 2006; Jacinto et al. 2006; Pearce et al. 2007; Thedieck et al. 2007; Woo et al. 2007) are found exclusively in mTORC2 (Fig. 2). mTORC1 and mTORC2 regulate disparate cellular functions by phosphorylating distinct sets of substrates. Several substrates of mTORC1 have been identified in the past two decades including the eIF4E-binding proteins (4E-BPs), 70 kDa ribosomal S6 kinases 1 and 2 (S6Ks), PRAS40, Ser/Thr kinase Ulk1 (also known as hATG1), and growth factor receptor-bound protein 10 (Grb10) (reviewed by Caron et al. 2010; Yea and Fruman 2011; Zoncu et al. 2011b). By modulating their activity, mTORC1 regulates a variety of cellular processes including growth, proliferation, translation, autophagy, as well as its own activation (Zoncu et al. 2011b). The function, upstream regulators, and associated substrates of mTORC2 are less well understood (Oh and Jacinto 2011). mTORC2 phosphorylates AGC kinase family members, e.g., AKT, protein kinase C (PKC) and serum/glucocorticoid regulated kinase 1 (SGK1) and thereby controls cytoskeletal organization and cell survival (Sarbassov et al. 2004; Sarbassov et al. 2005; Guertin and Sabatini 2007; Garcia-Martinez and Alessi 2008). mTORC2 also associates with the ribosome (Zinzalla et al. 2011) where it phosphorylates residues in nascent polypeptide chains that are important for optimal protein folding (Fig. 2) (Oh et al. 2010).

mTORC1 and mTORC2 exhibit differential sensitivity to rapamycin. Rapamycin is a naturally occurring allosteric inhibitor of mTORC1 (Hay and Sonenberg 2004; Petroulakis et al. 2006; Guertin and Sabatini 2009). It forms a complex with its intracellular receptor, FKBP12 (FK506-binding protein of 12 kDa), which binds to the FRB (FKBP12-rapamycin binding) domain of mTOR and inhibits mTORC1 function (Chen et al. 1995). Although the precise

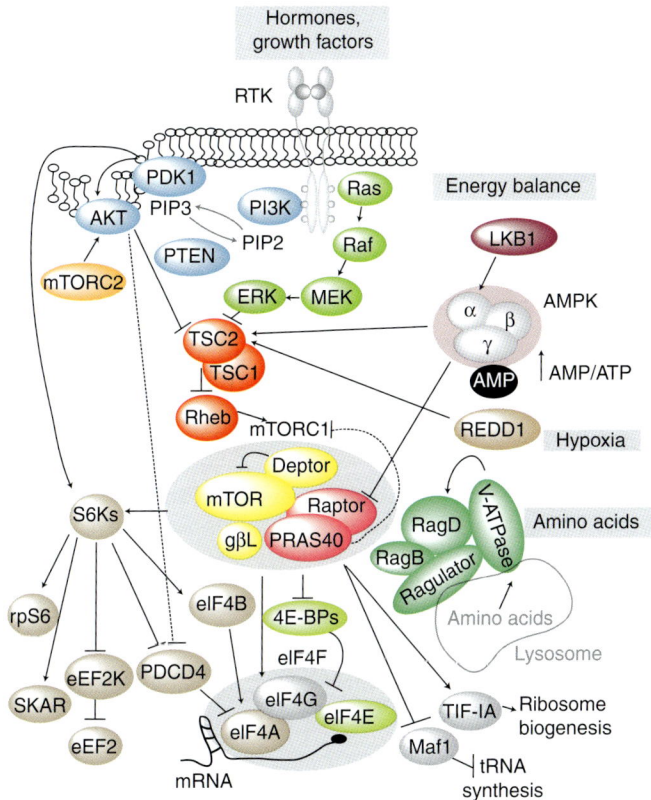

Figure 1. Schematic representation of mTORC1 signaling to the translational machinery. Growth factors and hormones stimulate mTORC1 by activating receptor tyrosine kinase (RTK) signaling via PI3K/AKT (light blue) and Ras/ERK (green) pathways. mTORC1 is also activated by amino acids via small Rag GTPases (aquamarine). Insufficient energy resources and hypoxia inactivate mTORC1 via the LKB1/AMPK pathway (purple) and REDD1 (brown), respectively. TSC1/2 (red) suppresses mTORC1 signaling by inhibiting GTPase activity of Rheb (red). mTORC1 modulates translation via the phosphorylation of downstream targets including 4E-BPs (olive), S6Ks (champagne), and their downstream effectors. In addition, mTORC1 stimulates ribosome biogenesis and tRNA synthesis by activating TIF-IA (gray) and inhibiting Maf1 (gray), respectively. mTOR, Deptor, and gβL (yellow) are found in both mTORC1 and mTORC2, whereas PRAS40 and Raptor are specific components of mTORC1 (pink). T bars represent inhibitory signals, whereas arrows indicate stimulatory signals. Abbreviations and detailed explanations are provided in the text.

mechanism of inhibition remains incompletely defined, rapamycin:FKBP12 binding to the FRB domain was shown to weaken the mTOR:-Raptor interaction and partially reduce intrinsic mTORC1 catalytic activity, as measured by mTOR autophosphorylation at Ser2481 (Soliman et al. 2010; Yip et al. 2010). Recent studies, however, revealed that the inhibition of mTORC1 by rapamycin is incomplete, inasmuch as it efficiently suppresses the phosphorylation of S6Ks, but not of 4E-BPs (Choo et al.

2008). In contrast to mTORC1, mTORC2 appears to be largely insensitive to the effects of rapamycin in acute treatment. However, it was proposed that prolonged rapamycin treatment results in inhibition of mTORC2 signaling in a cell type-specific manner, likely by blunting its de novo assembly (Sarbassov et al. 2006).

mTORC1 plays a central role in the regulation of cell growth and proliferation (Fingar and Blenis 2004; Ma and Blenis 2009; Zoncu et al. 2011b), cellular processes that are directly

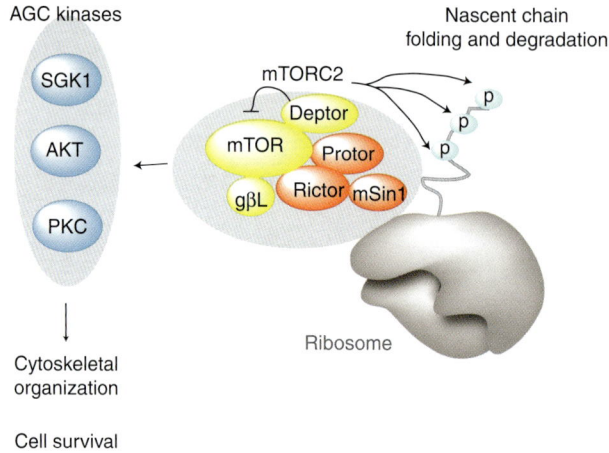

Figure 2. Schematic representation of mTORC2 signaling pathway. Cellular functions of mTORC2 are described less well than those of mTORC1. Protor, Rictor, and mSin1 (orange) are specific components of mTORC2. mTORC2 stimulates the activity of AGC kinase family members (blue) by phosphorylating residues localized in their hydrophobic motifs. Phosphorylation of PKC, AKT, and SGK1 by mTORC2 is thought to influence cytoskeletal organization and cell survival, respectively. In addition, it was recently reported that mTORC2 associates with the ribosome. It appears that ribosome-localized mTORC2 cotranslationally phosphorylates nascent polypeptide chains as they emerge from the ribosome thus facilitating their optimal folding. Upstream regulators of mTORC2 are still largely elusive. T bars represent inhibitory signals, whereas arrows indicate stimulatory signals. Abbreviations and detailed explanations are provided in the text.

proportional to translational activity (Brooks 1977; Zetterberg et al. 1995). Growth factors, hormones, branched-chain amino acids, and glucose stimulate mTORC1, up-regulate translation, and stimulate cellular growth and proliferation (reviewed by Zoncu et al. 2011b). Conversely, under conditions in which energy production, oxygen supply, and nutrients are inadequate, mTORC1 signaling is down-regulated, resulting in inhibition of translation, reduction in cellular growth proliferation, and induction of autophagy (reviewed by Zoncu et al. 2011b). Autophagy is the major catabolic process in the cell, during which cytoplasmic organelles and macromolecular complexes are degraded to replenish intracellular pools of amino acids (Stipanuk 2009).

Upstream of mTORC1

mTORC1 signaling to the translational apparatus is modulated via a multitude of cellular pathways including phosphoinositide 3-kinase (PI3K), Ras/MAPK, Rag GTPases, and adenosine monophosphate (AMP)-activated protein kinase (AMPK) (Fig. 1) (Hay and Sonenberg 2004; Laplante and Sabatini 2009).

Growth Factors and Hormones

Hormones and growth factors [e.g., insulin and insulin-like growth factors (IGFs)] activate PI3K via receptor tyrosine kinases (RTKs; such as insulin- or IGF-receptor) and associated adaptor molecules (e.g., IRS-1 and -2). PI3K converts phosphatidylinositol 4,5-bisphosphate (PIP_2) into phosphatidylinositol-3,4,5-triphosphate (PIP_3) (reviewed by Cantley 2002; Engelman et al. 2006), which is reversed by the tumor suppressor phosphatase and tensin homolog [PTEN; (Maehama and Dixon 1998)]. PIP_3 binds the pleckstrin homology domains of PDK1 and AKT resulting in their recruitment to the plasma membrane (Cantley 2002; Engelman et al. 2006). The Ser/Thr kinase PDK1 phosphorylates AKT and other AGC-kinases (e.g., PKC, RSK, and S6K) on residues localized in their activation loops (e.g., Thr308 in human AKT1) (reviewed by Pearce et al. 2010). In addition, mTORC2 modulates the activity of AKT by phosphorylating a serine residue in its hydrophobic motif (Ser473 in human AKT1) (Sarbassov et al. 2005). TSC1/2 is comprised of the scaffolding protein TSC1 (hamartin) and the GTPase activating protein (GAP) TSC2 (tuberin) (Kwiatkowski and Manning 2005). AKT phosphorylates TSC2 at multiple sites, which are thought to inhibit its GAP activity, thus reducing Ras homolog enriched in brain (Rheb)-GTP hydrolysis to its inactive GDP-bound form (Manning et al. 2002; Garami et al. 2003; Inoki et al. 2003). Rheb is a small GTPase (Yamagata et al. 1994) that activates mTORC1 through a poorly understood mechanism (Inoki et al. 2003). A single gene coding for Rheb is found in yeast and *Drosophila*, whereas mammalian cells express two variants of this protein, Rheb1 and Rheb2 (reviewed by Avruch et al. 2006).

PRAS40 is a recently identified suppressor of mTORC1 (Fonseca et al. 2007; Oshiro et al. 2007; Sancak et al. 2007; Vander Haar et al. 2007; Wang et al. 2007). PRAS40 is phosphorylated on Thr246 via PI3K/AKT, which suggests that it acts as an upstream inhibitor of mTORC1 (Sancak et al. 2007; Vander Haar et al. 2007). However, PRAS40 binds Raptor via a TOR signaling (TOS) motif, which is present in other mTORC1 substrates such as 4E-BPs and S6Ks (Schalm and Blenis 2002; Schalm et al. 2003) and is phosphorylated by mTORC1 at multiple sites, including rapamycin-sensitive (Ser183 and Ser221) and -insensitive (Ser212) residues (Oshiro et al. 2007; Wang et al. 2007). These findings suggest a model whereby PRAS40 acts as a downstream target of mTORC1, which competes with 4E-BPs and S6Ks for binding to Raptor (Fig. 1). Accordingly, phosphorylation and release of PRAS40 from mTORC1 increases substrate accessibility but does not increase its kinase activity (Rapley et al. 2011).

Growth factors also stimulate mTORC1 activity through a pathway involving the small Ras GTPases (reviewed by Shaw and Cantley 2006). Loss of the tumor suppressor gene *NF1*, which encodes a Ras GAP, results in increased mTORC1 signaling and tumorigenesis (Johannessen et al. 2005, 2008). Ras is an oncogene that triggers several signaling cascades, including the MAPK pathway, which consists of the sequential activation of the Raf, MEK, and ERK protein kinases (Rajalingam et al. 2007). At present, it is not clear how ERK1/2 stimulate mTORC1 activity, but TSC2 was shown to be a direct target of ERK1/2 and its downstream substrate, the 90 kDa ribosomal S6 kinase (RSK) (Roux et al. 2004; Ballif et al. 2005; Ma et al. 2005). Both protein kinases also phosphorylate Raptor at regulatory sites, correlating with increased mTORC1 activity and signaling to downstream substrates (Carriere et al. 2008a, 2011). Recent evidence suggests that RSK also phosphorylates Deptor within a degron sequence that is recognized by the E3-ubiquitin ligase $SCF^{\beta TrCP}$, thereby promoting its ubiquitination and proteosomal degradation (Zhao et al. 2011).

Nutrients, Oxygen, and Energy Status in the Cell

mTORC1 controls cellular proliferation and growth in response to nutrient availability and cellular energy balance (Fig. 1). Amino acids, in particular those with a branched side chain, are indispensable for mTORC1 signaling in cell culture (Hara et al. 1998; Wang et al. 1998). In yeast, amino acids stimulate Vam6/VPS39, which loads GTP onto Gtr1/Gtr2 GTPases (Binda et al. 2009). Gtr1/Gtr2 are components of the vacuolar-membrane-associated EGO complex, which binds to and activates TORC1 (Kim and Guan 2011). A similar mechanism has recently been described in mammals, whereby in response to amino acids Rag GTPases recruit mTORC1 to lysosomal membranes via interaction with Raptor (Sancak et al. 2008). This brings mTORC1 into close proximity of Rheb resulting in its activation (Kim et al. 2008; Sancak et al. 2008). Analogous to Gtr1/Gtr2, Rags form heterodimers comprising RagA or RagB bound to RagC or RagD, whose activity is governed by the scaffolding Ragulator complex (Sengupta et al. 2010), which consists of the p18, p14, and MP1 proteins and is required for the anchoring of Rags to the lysosome (Sancak et al. 2010). Vacuolar H^+-ATPase (v-ATPase) associates with the Ragulator complex and plays an essential role in conducting amino acid signals from the lysosomal lumen to Rags (Zoncu et al. 2011a). In addition, it has been reported that signaling adaptor p62 interacts with Raptor, mediates the interaction of mTOR with Rag GTPases, and stimulates translocation of mTORC1 to the lysosome (Duran et al. 2011).

Glucose deprivation leads to a decrease in glucose flux and cellular ATP levels, thus inhibiting mTORC1 signaling (Fig. 1). Changes in cellular energy balance impact on mTORC1 signaling via AMPK, which is a Ser/Thr kinase consisting of a catalytic α-subunit and two regulatory subunits, β and γ (Kahn et al. 2005; Shaw 2009). AMPK is activated under conditions in which the intracellular AMP/ATP ratio is increased (e.g., lack of nutrients, mitochondrial dysfunction) (Kahn et al. 2005; Shaw 2009). AMP directly associates with the γ-subunit of AMPK, thus facilitating the phosphorylation of the α-subunit on Thr172 by upstream kinases such as Serine/Threonine kinase 11 (STK11/LKB1) (Shaw 2009). On activation, AMPK suppresses anabolic processes including protein synthesis and restricts proliferation and growth (Kahn et al. 2005; Shaw 2009). This is partly achieved via inhibition of mTORC1 (Shaw et al. 2004). AMPK inhibits mTORC1 by phosphorylating and activating TSC2 (Corradetti et al. 2004) and by phosphorylating Raptor resulting in its sequestration by 14-3-3 proteins (Gwinn et al. 2008). Inhibition of mTORC1 signaling appears to be required for the block in proliferation exerted by AMPK (Gwinn et al. 2008), which is paralleled by a decrease in global translation (Dowling et al. 2007). These results suggest that the AMPK/mTOR pathway connects the regulation of intracellular energy balance with translational control and cell proliferation.

Under conditions where the supply of oxygen is limited, mTORC1 signaling is suppressed via multiple mechanisms (Fig. 1). In addition to inhibiting oxidative phosphorylation and, as a result, activating AMPK, hypoxia represses mTORC1 signaling through regulated in development and DNA damage response 1 (REDD1) (Brugarolas et al. 2004). REDD1 expression is stimulated by several cellular insults (Ellisen et al. 2002) and was shown to inhibit mTORC1 by stabilizing the TSC1/2 complex (DeYoung et al. 2008). Additional inhibitory mechanisms were described to occur during hypoxia, including the direct interaction and inhibition of Rheb by the hypoxia-inducible proapoptotic protein BNIP3 (BCl2/adenovirus E1B 19 kDa protein-interacting protein 3) (Li et al. 2007).

It is worthwhile to note that because of non-physiological conditions used in the aforementioned in vitro studies (i.e., glucose and amino acid deprivation followed by acute refeeding), the understanding of how mTORC1 is regulated by nutrients and alterations in the energy balance in vivo is still incomplete. Indeed, recent studies suggest that the regulation of mTORC1 signaling by nutrients at the organismal level is more complex than initially thought (reviewed by Howell and Manning 2011).

Emerging Mechanisms of mTORC1 Regulation

mTOR activity is regulated by the phosphorylation of residues located within its kinase domain (Ser2159 and Thr2164 in human mTOR) (Ekim et al. 2011). Phosphorylation of these residues stimulates mTOR autophosphorylation (on Ser2481 in human protein) (Soliman et al. 2010) and is required for its effects on cell growth and proliferation (Ekim et al. 2011). The identity of the mTOR Ser2159 and Thr2164 kinase(s) remains unknown. Phosphorylation of mTOR at Ser1261, which lies in a centrally located HEAT (Huntington, Elongation Factor 3, PR65/A, TOR)-repeat, also promotes mTORC1 signaling, mTOR autophosphorylation at Ser2461 and cell growth (Acosta-Jaquez et al. 2009). mTOR also phosphorylates Raptor on several sites (including Ser863 in human Raptor) that, in turn, increases mTOR activity toward downstream substrates (Foster et al. 2010). Interestingly, ERK1/2 were also shown to phosphorylate some of these sites (Carriere et al. 2011), but the physiological relevance of this additional layer of regulation remains unknown.

Cite this article as *Cold Spring Harb Perspect Biol* doi: 10.1101/cshperspect.a012252

mTORC1 Signaling to the Translational Machinery

mTORC1 stimulates global protein synthesis, as well as translation of a specific subset of mRNAs (Fig. 3). 4E-BPs and S6Ks are the most extensively studied and best-understood downstream effectors of mTORC1, which have been implicated in the regulation of translation (Hay and Sonenberg 2004).

4E-BPs

The first step of cap-dependent translation initiation is the assembly of the eIF4F complex on the 5′-mRNA cap structure (Mathews et al. 2007; Sonenberg and Hinnebusch 2009; Jackson et al. 2010; Topisirovic et al. 2011). The eIF4F complex comprises the cap-binding subunit eIF4E, the large scaffolding protein eIF4G, and the DEAD-box RNA helicase eIF4A, which unwinds secondary structure within the 5′-untranslated region (5′UTR) of the mRNA (Fig. 3A) (Gingras et al. 1999b; Sonenberg and Hinnebusch 2009; Jackson et al. 2010). 4E-BPs are small-molecular weight translational repressors (4E-BP1, 2, and 3 in mammals), which interfere with the assembly of the eIF4F complex by competing with eIF4G for binding to eIF4E (Pause et al. 1994). On activation, mTORC1 phosphorylates residues corresponding to Thr37 and Thr46 on human 4E-BP1, which act as priming sites for the phosphorylation of Ser65 and Thr70 (Fig. 3A) (Gingras et al. 1999a, 2001). Phosphorylation of 4E-BPs on these four residues, leads to their dissociation from eIF4E, thus allowing the assembly of the eIF4F complex (Pause et al. 1994; Gingras et al. 1999a, 2001).

eIF4E is a general translation initiation factor required for cap-dependent translation of all cellular mRNAs (reviewed by Sonenberg and Hinnebusch 2009). Nonetheless, it is well established that alterations in eIF4E levels and/or activity affect translation of a specific pool of "eIF4E-sensitive" mRNAs, but do not have a major impact on global protein synthesis (Graff and Zimmer 2003; De Benedetti and Graff 2004; Graff et al. 2008; Sonenberg and Hinnebusch

2009). It is thought that "eIF4E-sensitivity" of mRNAs is determined by the complexity of their 5′UTRs. "eIF4E-sensitive" mRNAs, which frequently encode proliferation and survival promoting proteins (e.g., Bcl-xL, cyclins, ornithine decarboxylase, c-myc, vascular, and endothelial growth factor), possess long and highly structured 5′UTRs and thus are strongly dependent on the unwinding activity of the eIF4A subunit of eIF4F, whereas "eIF4E-insensitive" mRNAs such as those encoding housekeeping proteins (e.g., actins and tubulins) bear short 5′UTRs and are only minimally sensitive to alterations in eIF4F levels (Fig. 3B) (Koromilas et al. 1992; Svitkin et al. 2001; De Benedetti and Graff 2004; Sonenberg 2008). Because eIF4E is the most limiting factor among eukaryotic translation initiation factors, it controls the levels of eIF4F (reviewed by Sonenberg and Hinnebusch 2009). 4E-BPs impede eIF4F complex assembly (Pause et al. 1994; Gingras et al. 1999a, 2001). Correspondingly, alterations in the expression and/or phosphorylation status of 4E-BPs only marginally affect global protein synthesis, while strongly influencing translation of a subset of mRNAs (e.g., IRF-7, Gas2, cyclin D3, ornithine decarboxylase, and vascular and endothelial growth factor) (Lynch et al. 2004; Colina et al. 2008; Petroulakis et al. 2009; Dowling et al. 2010a). Thus, in addition to its effects on global protein synthesis, mTORC1 selectively stimulates translation of "eIF4E-sensitive" mRNAs by phosphorylating and inactivating 4E-BPs. The vast majority of "eIF4E-sensitive" mRNAs encode proliferation, survival, and tumor-promoting proteins, and accordingly, 4E-BPs act as major mediators of the effects of mTORC1 on proliferation (Dowling et al. 2010a). In Drosophila, dS6K and d4E-BP play overlapping roles in regulating cell size and proliferation downstream from dTOR (Montagne et al. 1999; Miron et al. 2001). In contrast, mammalian 4E-BPs do not appear to significantly influence cell size, which is thought to be largely achieved via S6Ks (Pende et al. 2004; Ohanna et al. 2005; Dowling et al. 2010a). The evolutionary advantage of this "division of labor" between 4E-BPs and S6Ks downstream from mTORC1 is still unclear.

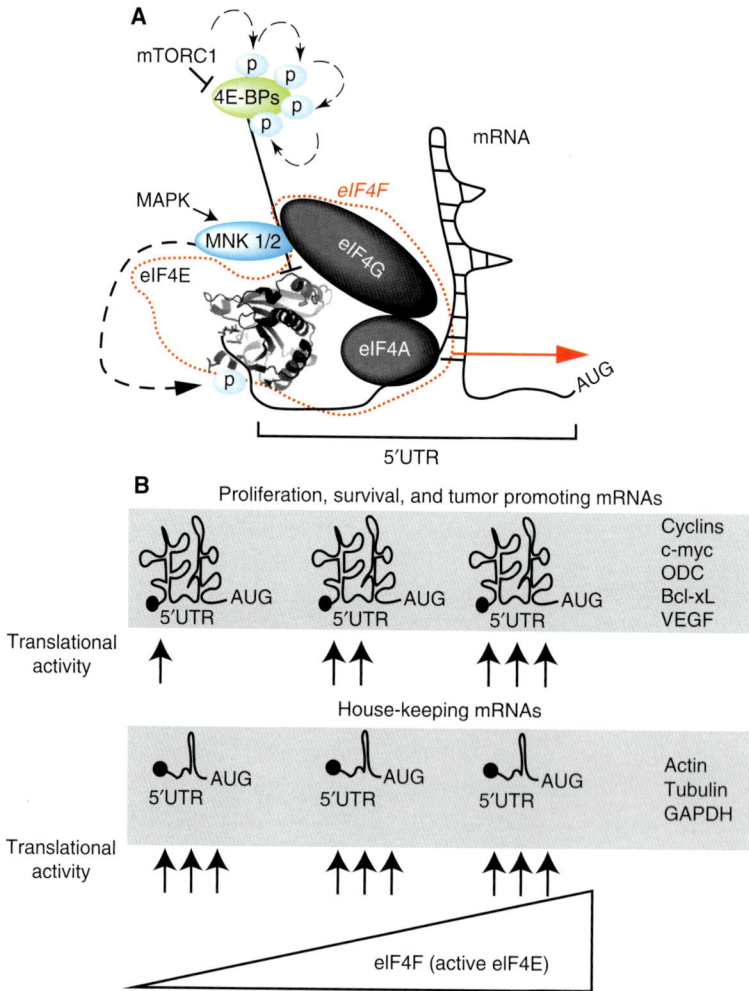

Figure 3. mTORC1 and MAPK pathways modulate translation of a specific subset of mRNAs by stimulating eIF4E activity. (*A*) Hierarchical phosphorylation of 4E-BPs (green) by mTORC1 (broken arrows) leads to their dissociation from eIF4E, thereby stimulating the interaction of eIF4E with eIF4G and the assembly of the eIF4F complex (red). eIF4E is the most limiting subunit of the eIF4F complex, which is critical for the recruitment of eIF4A to the 5'UTR of mRNA and unwinding of the secondary structure during scanning of the ribosome toward the initiation codon (AUG; red arrow). MAPK-interacting kinases 1 and 2 (MNK1/2) (blue) are recruited to eIF4E via eIF4G and phosphorylate eIF4E at a single Ser residue (Ser209 in human eIF4E). (*B*) mRNAs that contain long and highly structured 5'UTRs frequently encode proliferation, survival, and tumor promoting proteins (*top* panel). Translation of these mRNAs strongly depends on the unwinding activity of eIF4A, and is thus robustly stimulated by the increase in the amount of eIF4E, which is available for eIF4F complex assembly (further explanation provided in the text). In addition, phospho-eIF4E selectively stimulates translation of Mcl-1, MMP, and proinflammatory mRNAs by a hitherto unknown mechanism. In contrast, housekeeping proteins are typically encoded by mRNAs that bear short, unstructured 5' UTRs (*lower* panel) and their translational activity is only marginally influenced by the changes in the availability of eIF4E or its phosphorylation status. The eIF4E diagram was generated using PyMOL software (http://www.pymol.org). eIF4E-PDB accession number 1L8B.

S6Ks

In addition to 4E-BPs, TOR regulates translation by activating the S6Ks (Hay and Sonenberg 2004; Ma and Blenis 2009; Dowling et al. 2010b; Zoncu et al. 2011b). Although *Drosophila* expresses a single S6K protein (dS6K), mammals express two variants of S6K (S6K1 and S6K2; or S6Kα and S6Kβ, respectively), which are encoded by two distinct genes (*RPS6KB1* and *RPS6KB2*) and share a high degree of homology (reviewed by Fenton and Gout 2011). S6K1 and S6K2 exist in two different isoforms (p70 and p85, and p54 and p56, respectively), which are generated via alternative translational initiation sites from a common mRNA (Grove et al. 1991; Gout et al. 1998). p70S6K1 is the more abundant isoform of S6K1 and is predominantly cytoplasmic, whereas p85S6K1, p54S6K2, and p56S6K2 are localized in the nucleus (reviewed by Fenton and Gout 2011).

S6Ks belong to the AGC kinase family and are activated by PDK1 and mTORC1 via the phosphorylation of Thr residues localized in their activation loop (Thr229 in human p70S6K1) and hydrophobic motif (Thr389 in human p70S6K1), respectively (reviewed by Fenton and Gout 2011). Recent findings indicate that GSK3 also contributes to the activation of S6Ks through the phosphorylation of their turn motif (Ser371 in human p70S6K1) (Shin et al. 2011a). Several S6Ks substrates have been implicated in the regulation of translation including ribosomal protein S6 (rpS6) (Banerjee et al. 1990; Kozma et al. 1990), eukaryotic initiation factor 4B (eIF4B) (Raught et al. 2004; Shahbazian et al. 2006), and programmed cell death 4 protein (PDCD4) (Dorrello et al. 2006).

rpS6 was the first identified S6K substrate. Five phosphorylation sites (Ser235, Ser236, Ser240, Ser244, and Ser247 in humans and rodents) are clustered in the carboxyl terminus of rpS6 (Meyuhas 2008). It has been proposed that S6Ks phosphorylate rpS6 in a sequential fashion, whereby the phosphorylation of Ser236 is followed by the phosphorylation of Ser235, Ser240, Ser244, and Ser247 (Krieg et al. 1988; Ferrari et al. 1991; Bandi et al. 1993). Although S6K1 contributes to rpS6 phosphorylation, S6K2 appears to be the predominant kinase that phosphorylates rpS6 on these residues (Meyuhas 2008). In contrast, the RSKs phosphorylate rpS6 only on Ser235 and Ser236 (Pende et al. 2004; Roux et al. 2007), but the physiological relevance for this specificity remains unknown. Experiments obtained using mice in which wild-type rpS6 is replaced by a nonphosphorylatable mutant revealed that the loss of rpS6 phosphorylation mirrors defects in cell growth observed in S6K1/2 knockout mice (Ruvinsky et al. 2005). Notwithstanding the overlap in their physiological roles, the understanding of how S6Ks and rpS6 impact translation remains obscure. Loss of S6Ks only modestly affects global translation rates, whereas the expression of the nonphosphorylatable mutant of the rpS6 results in a moderate up-regulation of overall protein synthesis rates (Pende et al. 2004; Ruvinsky et al. 2005). 5′-terminal oligopyrimidine tract (5′-TOP) containing mRNAs encode components of the translational machinery and their translation is repressed under conditions where S6K activity and rpS6 phosphorylation are minimal, such as when cells are deprived of amino acids (Levy et al. 1991). Thus, it was proposed that S6Ks promote translation of the 5′-TOP-mRNAs via stimulation of rpS6 phosphorylation (Kawasome et al. 1998; Shima et al. 1998; Loreni et al. 2000). However, it was subsequently shown that neither the loss of S6Ks nor the phosphorylation status of rpS6 influences translation of 5′-TOP mRNAs (Tang et al. 2001; Pende et al. 2004; Ruvinsky et al. 2005).

PDCD4 plays an established role in apoptosis and has been suggested to possess tumor suppressor properties (reviewed by Lankat-Buttgereit and Goke 2009). PDCD4 binds to eIF4A via two conserved MA-3 domains (also found in eIF4G) and competes with eIF4G for eIF4A binding (Goke et al. 2002; Yang et al. 2003). This leads to the inhibition of eIF4A and the consequent repression of cap-dependent translation (Yang et al. 2003). S6Ks and AKT phosphorylate PDCD4 on Ser67 and Ser457 leading to its degradation by the E3-ubiquitin ligase SCF$^{\beta TrCP}$ (Dorrello et al. 2006).

eIF4B and eIF4H are two auxiliary factors that stimulate the RNA unwinding activity of eIF4A (Grifo et al. 1984; Rozen et al. 1990; Pause et al. 1994; Richter-Cook et al. 1998; Rogers et al. 2001). eIF4B stimulates cellular proliferation and survival by selectively up-regulating translation of mRNAs such as those encoding Cdc25, ODC, XIAP, and Bcl-2 (Shahbazian et al. 2010). eIF4B is phosphorylated by several AGC kinases on Ser406 (likely by RSK and S6K) and Ser422 (by S6K, AKT, and RSK) in a stimulus- and cell type-dependent manner (Raught et al. 2004; Shahbazian et al. 2006; van Gorp et al. 2009). It was suggested that eIF4B phosphorylation promotes its association with eIF3 and correlates with increased translation initiation (Holz et al. 2005; Shahbazian et al. 2006). eIF3 is the multisubunit complex that was also identified as a dynamic scaffold for mTORC1 and S6K1 binding (Holz et al. 2005). On activation, mTORC1 is recruited to the eIF3 complex where it phosphorylates S6K1, leading to its dissociation from eIF3 and subsequent phosphorylation by PDK1 (Holz et al. 2005).

S6Ks also phosphorylate eukaryotic elongation factor 2 (eEF2) kinase at Ser366 (in human eEF2K). eEF2K impedes translation elongation by phosphorylating and inhibiting eEF2 (Wang et al. 2001a), a GTPase that promotes translocation of the nascent polypeptide chain from the A-site to the P-site of the ribosome (Skogerson and Moldave 1968; Bermek and Matthaei 1971). Finally, S6K1 has been shown to promote translational efficiency of newly spliced mRNAs on its recruitment to the exon-junction complex (EJC) by its substrate and binding partner SKAR (Richardson et al. 2004). Recruitment of S6K1 and SKAR to the EJC leads to the phosphorylation of numerous mRNA binding proteins and correlates with increased translational efficiency of spliced mRNAs (Ma et al. 2008).

Additional mTOR Targets Implicated in Translational Control

Besides 4E-BPs and S6Ks, mTORC1 has been suggested to modulate translational initiation via phosphorylation of eIF4G at multiple residues (Raught et al. 2000). mTORC1 also stimulates ribosome biogenesis and tRNA synthesis, via stimulation of TIF-IA (Mayer et al. 2004)

and inhibition of the TF-IIIC repressor Maf1, respectively (Wei et al. 2009; Kantidakis et al. 2010; Michels et al. 2010; Shor et al. 2010).

MAPK SIGNALING TO THE TRANSLATIONAL MACHINERY

The MAPKs are Ser/Thr kinases that are among the most ancient signal transduction pathways and are widely used throughout evolution in many physiological processes (Widmann et al. 1999). All eukaryotic cells possess multiple MAPK pathways, which coordinate gene expression, mitosis, metabolism, motility, survival, apoptosis, and differentiation. In mammals, 14 MAPKs have been characterized, but the most extensively studied groups are the ERK1/2, JNKs, and p38 isoforms (reviewed by Chen et al. 2001; Kyriakis and Avruch 2001; Pearson et al. 2001). The wide range of functions regulated by these MAPKs is mediated through phosphorylation of several substrates, including members of a family of Ser/Thr kinases termed MAPK-activated protein kinase (MAPKAPK) (Gaestel 2006; Gaestel 2008; Cargnello and Roux 2011). Two MAPKAPKs have been directly implicated in the regulation of translation, namely the RSKs (Carriere et al. 2008b) and the MAPK-interacting kinases (MNKs) (Buxade et al. 2008). Although the RSK isoforms are strictly regulated by ERK1/2 downstream from growth factors and mitogens, the MNKs can be activated by either ERK1/2 or p38 isoforms, making them responsive to both mitogenic and stress stimuli (Fig. 4). The RSKs and MNKs become active on phosphorylation of a threonine residue followed by a proline, located in the activation loop of their kinase domain (Roux and Blenis 2004).

MNKs

In addition to its regulation by the 4E-BPs through mTORC1 signaling, the activity of eIF4E is regulated by MNK1 and MNK2, which phosphorylate a single serine residue in its carboxyl terminus (Fig. 4) (Ser209 in human eIF4E) (Fukunaga and Hunter 1997; Waskiewicz et al. 1997). MNK1 and 2 exist in two isoforms

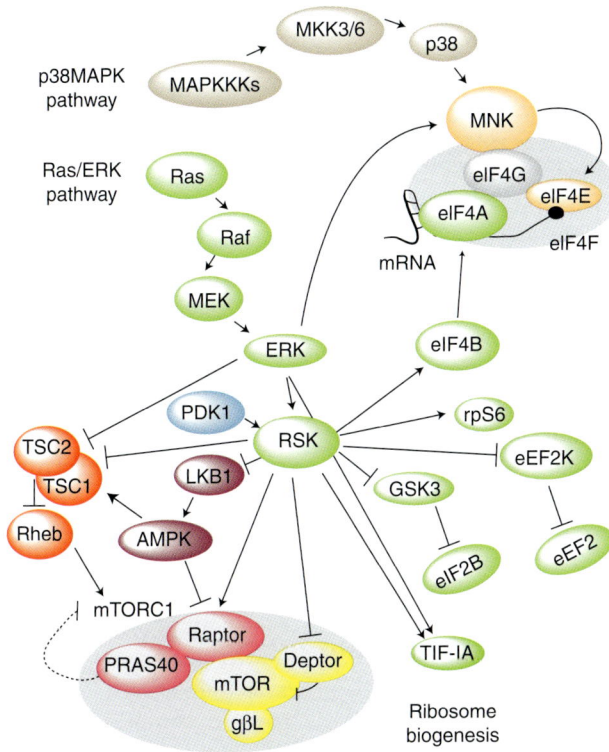

Figure 4. Schematic representation of MAPK signaling to the translational machinery. The Ras/ERK (green) and p38MAPK (champagne) pathways impinge at different levels on the translational machinery. Although Ras/ERK signaling stimulates the activity of both RSK (green) and MNK (orange), the latter is also responsive to agonists of the p38MAPK pathway. MNK interacts with eIF4G and on activation phosphorylates eIF4E (orange) at Ser209, a site that increases its oncogenic potential and facilitates the translation of specific mRNAs. Following stimulation of the Ras/ERK pathway, activated RSK phosphorylates rpS6, eIF4B, and eEF2K (green), which are important translational regulators. RSK also participates in the regulation of mTORC1 by inhibiting TSC2 (red) and Deptor (yellow), which are negative regulators of mTORC1. ERK and RSK regulate LKB1-dependent (purple) and -independent phosphorylation of Raptor (pink), resulting in increased mTORC1 signaling. ERK and RSK also collaborate in the regulation of ribosome biogenesis by promoting TIF-1A phosphorylation (green). T bars represent inhibitory signals, whereas arrows indicate stimulatory signals. Abbreviations and detailed explanations are provided in the text.

(MNK1a and b, and MNK2a and b), which are generated by alternative splicing. MNK1a and MNK2a have long C termini harboring MAPK-binding motifs and show predominantly cytoplasmic localization, whereas MNK1b and MNK2b, which contain short carboxyl termini, are equally distributed between the nucleus and cytoplasm (reviewed by Buxade et al. 2008). In most cell lines, the basal activity of MNK2a is high, which is thought to be a consequence of its ability to sustainably associate with activated ERK (reviewed by Buxade et al. 2008). Conversely, MNK1a has a low basal activity and it is activated by ERK and p38MAPK in response to stimuli such as growth factors and phorbol esters or cytokines and environmental stress, respectively (Scheper et al. 2001; Wang et al. 2001a). MNK1b exhibits high basal activity, whereas MNK2b, which is the only MNK variant that lacks the MAPK-binding domain, has a very low basal activity (reviewed by Buxade et al. 2008).

MNKs are recruited to eIF4E through association with the carboxy-terminal part of eIF4G, via a polybasic region located in their amino termini (Pyronnet et al. 1999). Phosphorylation of eIF4E appears to be restricted to metazoans as yeast lack MNK orthologs and a MNK interaction domain in eIF4G. Although initial genetic studies in *Drosophila* revealed that the phosphorylation of eIF4E is required for normal development, growth, and viability (Lachance et al. 2002), MNK1/2 double knockout (DKO) mice, and mice in which wild-type eIF4E was replaced with a nonphosphorylatable mutant, do not exhibit any conspicuous phenotype (Ueda et al. 2004; Furic et al. 2010). Nonetheless, the phosphorylation of eIF4E appears to be critical for its tumorigenic activity, inasmuch as the nonphosphorylatable mutant of eIF4E exhibits dramatically lower oncogenic potential in vivo and in vitro as compared to wild-type (Topisirovic et al. 2004; Wendel et al. 2007).

The effects of eIF4E phosphorylation on cap-dependent translation are still poorly understood. Because MNKs are recruited to eIF4E via eIF4G (Pyronnet et al. 1999), it is most likely that the phosphorylation of eIF4E occurs during or shortly after the assembly of the eIF4F complex on the 5′-mRNA cap (Fig. 3B). Because Ser209 is located in the proximity of the entrance of the cap-binding pocket of eIF4E, it is plausible that the phosphorylation status of eIF4E alters its cap-binding activity (Topisirovic et al. 2011). It was initially predicted that a salt bridge between the phosphate group of Ser209 and Lys159, forms a "clamp" that stabilizes the eIF4E:5′mRNA cap complex (Marcotrigiano et al. 1997; Matsuo et al. 1997). However, subsequent studies revealed that the phosphorylation of eIF4E reduces its affinity for the cap (Scheper et al. 2002; Slepenkov et al. 2006). Depending on the experimental conditions, eIF4E phosphorylation was shown to correlate with increased (Kaspar et al. 1990; Manzella et al. 1991; Walsh and Mohr 2004; Worch et al. 2004) or decreased global translation rates (Knauf et al. 2001; Morley and Naegele 2002; Naegele and Morley 2004). In addition to the variable effects on global translation, it was shown that the phosphorylation of eIF4E selectively affects translation of a subset of mRNAs, such as Mcl-1 (Wendel et al. 2007). Indeed, a genome-wide comparison of translational activity observed in wild-type mouse embryonic fibroblasts and their counterparts in which wild-type eIF4E was replaced by the nonphosphorylatable mutant, revealed that the phosphorylation status of eIF4E selectively affects translation of a subset of mRNAs including those encoding proteins that play roles in inflammation (e.g., Ccl2 and Ccl7) and tumor progression (e.g., MMP3 and MMP9) (Furic et al. 2010). These findings suggest an intriguing possibility that phospho-eIF4E modulates the inflammatory response and stimulates tumorigenesis by selectively up-regulating translation of mRNAs that encode proteins critical for these processes.

RSKs

In addition to being involved in the upstream regulation of mTORC1, several studies have shown a significant role for RSKs in various stages of translational control (Fig. 4). The RSK family is comprised of four Ser/Thr kinases (RSK1-4) that are directly activated by ERK1/2-mediated phosphorylation (reviewed by Carriere et al. 2008b). RSK family members exist in all vertebrate species and not so distant RSK orthologs have been identified in *Drosophila* and *C. elegans*. All RSK isoforms are expressed at relatively high levels during development and in adult tissues, with the exception of RSK4, which is more abundant during embryogenesis (Zeniou et al. 2002). A notable structural feature of the RSK family came about during evolution, when the genes for two distinct protein kinases fused, generating a single kinase capable of receiving an upstream activating signal from ERK1/2 to the RSK carboxy-terminal kinase domain (CTKD), and transmitting, with high efficiency and fidelity, an activating input to the RSK N-terminal kinase domain (NTKD). Both the NTKD and CTKD are distinct and functional (Jones et al. 1988; Fisher and Blenis 1996), but only the NTKD has thus far been implicated in the phosphorylation of exogenous substrates. This domain belongs to the AGC family of protein kinases, which also includes AKT, PKC, and S6Ks. These kinases phosphorylate very similar consensus sequences and, consistent with this, several bona fide RSK substrates were also shown to be targeted by AKT, PKC, and/or S6Ks.

The first clue suggesting that RSK may be involved in translational control came 25 years ago when it was identified as an rpS6 kinase in maturing *Xenopus laevis* oocytes (Erikson and Maller 1985, 1986). A subsequent study showed that activated RSK translocates to polysomes and stimulates the phosphorylation of several ribosome-associated proteins (Angenstein et al. 1998). With the use of rapamycin, the S6Ks were later found to be the predominant rpS6 kinases operating in somatic cells (Chung et al. 1992). Studies from S6K1/2-knockout mice confirmed these findings but also showed that there was residual MEK1/2-dependent phosphorylation at Ser235/236 (Pende et al. 2004). In accordance with this observation, RSK was shown to specifically phosphorylate rpS6 on Ser235/236 in vitro and in cells in response to agonists of the MAPK pathway (Roux et al. 2007). RSK-mediated phosphorylation of rpS6 correlates with eIF4F complex assembly and cap-dependent translation, suggesting that RSK provides an mTOR/S6K-independent input linking MAPK signaling to the regulation of translation initiation.

It is now understood that the Ras/MAPK pathway impinges on the PI3K/mTOR pathway at various steps to regulate translation. Aside from its role upstream of mTORC1, RSK phosphorylates additional components of the translational machinery, such as eIF4B in vitro and in vivo (Shahbazian et al. 2006). eIF4B stimulates the RNA-helicase activity of eIF4A (Rozen et al. 1990) and eIF4B phosphorylation was found to promote its interaction with eIF3 (Holz et al. 2005; Shahbazian et al. 2006). This interaction correlates with increased translation rates (Shahbazian et al. 2006) and also agrees with the finding that phosphorylated eIF4B stimulates cap-dependent translation in vivo (Holz et al. 2005). As indicated above, S6Ks also regulate eIF4B phosphorylation, which may explain the biphasic pattern of eIF4B phosphorylation observed in response to certain mitogenic cues.

RSK might also regulate translation through the phosphorylation of GSK3β (Sutherland et al. 1993). As with AKT and S6Ks, RSK-mediated phosphorylation of GSK3β on Ser9 inhibits its kinase activity and thereby releases the inhibition of eIF2B (Cohen and Frame 2001). In collaboration with ERK1/2, RSK2 was shown to participate in rRNA synthesis and cell growth. On serum stimulation, RSK2 phosphorylates a residue important for the function of TIF-1A, a transcription initiation factor required for RNA Pol I activity and rRNA transcription (Zhao et al. 2003). Finally, RSK was also shown to phosphorylate eEF2 kinase (Wang et al. 2001a), thereby underscoring the involvement of RSK at multiple levels of the pathway that leads to protein synthesis. Thus, RSK activation seems to coordinate crucial processes that are associated with the proper regulation of gene expression and protein synthesis.

FUTURE PERSPECTIVES

Over the past two decades, significant progress has been made in understanding how translation is modulated by signaling pathways in response to various extracellular stimuli and intracellular cues. Notwithstanding the large body of recently garnered data, in particular those uncovering the molecular mechanisms that underlie the effects of MAPK, TOR (this article), and eIF2α-kinases (Benham 2012) on translation, several outstanding issues await to be addressed. For instance, it has been known for more than a decade that rapamycin suppresses translation of 5′-TOP mRNAs (Jefferies et al. 1994). How mTORC1 stimulates translation of these transcripts is still unknown, however. Moreover, emerging data suggest that mTOR and MAPK act as pivotal regulators of energy metabolism (Shaw and Cantley 2006; Howell and Manning 2011). Protein synthesis is the most energy consuming process in the cell (Buttgereit and Brand 1995), which begs the question of whether mTOR and MAPK act as central nodes of the cellular networks that coordinate changes in the translatome and metabolome by orchestrating the activity of cellular factors involved in their respective processes. Finally, several signaling pathways other than MAPK, TOR, and eIF2α-kinases (e.g., Pak2,

GSK3, Cdk11, and CK2) have been implicated in the phosphorylation of the components of the translational apparatus and auxiliary factors (Table 1), but their physiological role in translational control is still largely unknown. In spite of these issues, there are good reasons to be optimistic as we enter a new era of research in signaling to translational machinery. Recent advances in pharmacological tools (e.g., compounds that specifically and potently target signaling to translational machinery such as active-site mTOR inhibitors [Malina et al. 2012]), and advancements in the technologies that enable monitoring of changes in the phosphoproteome (e.g., quantitative mass spectrometry), translatome (e.g., ribosome profiling/RNA-seq [Lasko 2012]) on a genome-wide scale will undoubtedly facilitate efforts to establish the role of signaling pathways in translational control in homeostasis and disease.

Table 1. Phosphorylation sites in mammalian translation factors and associated proteins, regulatory kinases, and functional consequences of the phosphorylation

Protein	Phosphorylation sites[a] (major kinase)	Function and references[b]
4E-BP1	Thr37,[1] Thr46[1] (mTORC1)[2]	Priming sites[3-5]
	Ser65[1] (mTORC1),[2] Thr70[1] (mTORC1?/CDK1?)	Dissociation from eIF4E[3-5]
	Ser84[1] (?), Ser101[6] (?), Ser112[7] (CDK?)	Unknown function
4E-BP2	Thr37,[8] Thr46[8] (mTORC1)	Priming sites (by analogy with 4E-BP1)
	Ser65 (mTORC1), Thr70 (?mTORC1/?CDK1)	Dissociation from eIF4E (by analogy with 4E-BP1)
eIF4E	Ser209[9] (MNK1/2)[10]	Unknown function[10,11]
		Increases oncogenic activity and promotes translation of a subset of mRNAs (e.g., Mcl-1, MMPs, CCLs)[12]
eIF4GI	Ser1186 (PKCα)[13]	Modulates MNK binding[13]
	Ser1108, Ser1148, Ser1192 (mTORC1)[14]	Stimulation of translation of mRNAs containing upstream open reading frame (uORF)[15] (?)
	Ser896 (Pak2)[16]	Inhibition of cap-dependent translation[16]
eIF2α	Ser51[17] (HRI; PKR; GCN2, PERK) reviewed in Ref. 18 and see Pavitt and Ron (2012)	Stabilizes the eIF2/GDP/eIF2B complex, thus preventing recycling of eIF2 (reviewed in Ref. 18 and see Pavitt and Ron 2012)
rpS6	Ser235[19] and Ser236[8] (S6K1/2,[8] RSK[20]), Ser240,[8] Ser244,[8] and Ser247[8] (S6K1/2)[8]	Unknown function[20-23]
		Global translation rates increased in MEFs expressing a nonphosphorylatable form of rpS6[24]
PDCD4	Ser67, Ser457 (S6K1/2; AKT)[25,26]	Degradation by the ubiquitin-proteasome system and subsequent activation of eIF4A[25,26]
eIF4B	Ser406(?),[27] Ser422 (S6K1/2,[28] AKT,[27] RSK[29])	Increases binding to eIF3[29,30]
eIF4H	Tyr12,[31] Tyr45,[31] Tyr101,[31] Ser193[31] (?)	?

Continued

Table 1. *Continued*

Protein	Phosphorylation sites[a] (major kinase)	Function and references[b]
eIF2Bε	Ser540 (GSK3)[32]	Inhibits recycling of eIF2[32]
	Ser544 (DYRK)[33]	Priming site for GSK3[33]
	Ser717/718 (CK2)[34]	Facilitates eIF2 binding[34]
eIF3 (29 phosphosites)[35]	eIF3b: Ser83,[36] Ser85,[36] Ser125[36](?)	?
	eIF3c: Ser39,[37] Ser166,[36] Thr524,[36] Ser909[38] (?)	?
	eIF3f: Ser46, Thr119 (CDK11)[39,40]	Regulation of protein synthesis and apoptosis[39,40]
	eIF3g: Thr41,[41] Ser42[41] (?)	?
	eIF3h: Ser183[42] (?)	Increased oncogenic activity[42]
	eIF3i: Tyr445[31] (?)	?
eIF1	Tyr30[31] (?)	?
eIF5	Ser389, Ser390 (CK2)[43]	Promotes cell cycle progression[43]
eIF5B	Ser107,[36] Ser113,[36] S135,[36] S137,[38] S164,[36] S182,[36] S183,[38] S186,[36] S190,[36] S214,[41] S1168[36] (?)	?
eIF6	Ser235 (PKCβII)[44]	Dissociation of eIF6 from the 60S; 80S assembly[44]
eEF1A	Thr431 (PKCδ)[45]	Activation (?)[45]
	Ser51 (PKCβI)[46]	?
	Ser300 (TβR-I)[47]	Inhibition of mRNA translation[47]
eEF2	Thr56 (eEF2K)[48]	Inhibits binding to the ribosome[49]
eEF2K	Ser78 (mTOR?)[50]	Inhibits CaM binding[50]
	Ser359 (SAPK/p38δ?)[51]	Inhibition (?)[51]
	Ser366 (S6K1; RSK)[52]	Inhibition[52]
	Ser398 (AMPK)[53]	Activation[53]
	Ser500 (PKA)[54]	Induces Ca^{2+}-independent activity[54]

This table includes selected phospho-acceptor sites identified in large-scale mass-spectrometry-based experiments that await functional characterization (e.g., eIF5B; unknown kinase/function indicated with ?), as well as phosphorylation sites with established role in translational control (e.g., 4E-BPs and eIF2α). Further information on the as-yet functionally noncharacterized phosphorylated residues of the components of the translational apparatus can be found on Phosphosite (www.phosphosite.org) or Uniprot (www.uniprot.org) Web sites.

[a]Amino acid numbering is based on human proteins.

[b]References: (1) Fadden et al. (1997); (2) Brunn et al. (1997); (3) Pause et al. (1994); (4) Gingras et al. (1999); (5) Gingras et al. (2001); (6) Wang et al. (2003); (7) Heesom et al. (1998); (8) Wang et al. (2005); (9) Joshi et al. (1995); (10) Waskiewicz et al. (1997); (11) Fukunaga et al. (1997); (12) Furic et al. (2010); (13) Dobrikov et al. (2011); (14) Raught et al. (2000); (15) Ramirez-Valle et al. (2008); (16) Ling et al. (2005); (17) Kudlicki et al. (1987); (18) Sonenberg and Hinnebusch (2009); (19) Krieg et al. (1988); (20) Roux et al. (2007); (21) Banerjee et al. (1990); (22) Kozma et al. (1990); (23) Pende et al. (2004); (24) Ruvinsky et al. (2005); (25) Dorrello et al. (2006); (26) Palamarchuk et al. (2005); (27) van Gorp et al. (2009); (28) Raught et al. (2004); (29) Shahbazian et al. (2006); (30) Holz et al. (2005); (31) Rush et al. (2005); (32) Welsh et al. (1998); (33) Woods et al. (2001); (34) Wang et al. (2001b); (35) Damoc et al. (2007); (36) Beausoleil et al. (2004); (37) Gevaert et al. (2005); (38) Kim et al. (2005); (39) Shi et al. (2003); (40) Shi et al. (2009); (41) Ballif et al. (2005); (42) Zhang et al. (2008); (43) Homma et al. (2005); (44) Ceci et al. (2003); (45) Kielbassa et al. (1995); (46) Piazzi et al. (2010); (47) Lin et al. (2010); (48) Nairn et al. (1987); (49) Price et al. (1991); (50) Browne et al. (2004); (51) Knebel et al. (2001); (52) Wang et al. (2001a); (53) Horman et al. (2002); (54) Diggle et al. (1998).

Abbreviations: CDK, cyclin-dependent kinase; PKC, protein kinase C; Pak2, p21-activated kinase 2; HRI, heme-regulated eIF2α kinase; PKR, double-stranded RNA-activated eIF2α kinase; GCN2, general control nonrepressed eIF2α kinase; PERK, double-stranded RNA-activated protein kinase-like ER kinase; DYRK, dual-specificity tyrosine phosphorylation-regulated kinase; CK2, protein kinase CK2 (formerly known as casein kinase II); TβR-I, TGF-β1 receptor; eEF2K, eukaryotic translation elongation factor 2-kinase; PKA, protein kinase A; SAPK, stress-activated protein kinase. Additional abbreviations are provided in the text.

ACKNOWLEDGMENTS

We apologize to those authors whose work was not cited because of space constraints. We thank Diane Fingar for invaluable suggestions and comments, Michael Witcher and members of Roux and Topisirovic's laboratories for critical reading of the manuscript, and Valerie Henderson for editing. Topisirovic is supported by grants from the Canadian Institutes of Health Research and Terry Fox Research Institute, whereas Roux is supported by grants from the Canadian Cancer Society Research Institute, the Cancer Research Society, and the Human Frontier Science Program.

REFERENCES

*Reference is also in this collection.

Acosta-Jaquez HA, Keller JA, Foster KG, Ekim B, Soliman GA, Feener EP, Ballif BA, Fingar DC. 2009. Site-specific mTOR phosphorylation promotes mTORC1-mediated signaling and cell growth. *Mol Cell Biol* **29:** 4308–4324.

Angenstein F, Greenough WT, Weiler IJ. 1998. Metabotropic glutamate receptor-initiated translocation of protein kinase p90rsk to polyribosomes: A possible factor regulating synaptic protein synthesis. *Proc Natl Acad Sci* **95:** 15078–15083.

Avruch J, Hara K, Lin Y, Liu M, Long X, Ortiz-Vega S, Yonezawa K. 2006. Insulin and amino-acid regulation of mTOR signaling and kinase activity through the Rheb GTPase. *Oncogene* **25:** 6361–6372.

Ballif BA, Roux PP, Gerber SA, MacKeigan JP, Blenis J, Gygi SP. 2005. Quantitative phosphorylation profiling of the ERK/p90 ribosomal S6 kinase-signaling cassette and its targets, the tuberous sclerosis tumor suppressors. *Proc Natl Acad Sci* **102:** 667–672.

Bandi HR, Ferrari S, Krieg J, Meyer HE, Thomas G. 1993. Identification of 40 S ribosomal protein S6 phosphorylation sites in Swiss mouse 3T3 fibroblasts stimulated with serum. *J Biol Chem* **268:** 4530–4533.

Banerjee P, Ahmad MF, Grove JR, Kozlosky C, Price DJ, Avruch J. 1990. Molecular structure of a major insulin/mitogen-activated 70-kDa S6 protein kinase. *Proc Natl Acad Sci* **87:** 8550–8554.

Beausoleil SA, Jedrychowski M, Schwartz D, Elias JE, Villen J, Li J, Cohn MA, Cantley LC, Gygi SP. 2004. Large-scale characterization of HeLa cell nuclear phosphoproteins. *Proc Natl Acad Sci* **101:** 12130–12135.

* Benham AM. 2012. Protein secretion and the endoplasmic reticulum. *Cold Spring Harb Perspect Biol* doi: 10.1101/cshperspect.a012872.

Bermek E, Matthaei H. 1971. Interactions between human translocation factor, guanosine triphosphate, and ribosomes. *Biochemistry* **10:** 4906–4912.

Binda M, Peli-Gulli MP, Bonfils G, Panchaud N, Urban J, Sturgill TW, Loewith R, De Virgilio C. 2009. The Vam6

GEF controls TORC1 by activating the EGO complex. *Mol Cell* **35:** 563–573.

* Braun JE, Huntzinger E, Izaurralde E. 2012. A molecular link between miRISCs and deadenylases provides new insight into the mechanism of gene silencing by microRNAs. *Cold Spring Harb Perspect Biol* doi: 10.1101/cshperspect.a012328.

Brooks RF. 1977. Continuous protein synthesis is required to maintain the probability of entry into S phase. *Cell* **12:** 311–317.

Browne GJ, Proud CG. 2004. A novel mTOR-regulated phosphorylation site in elongation factor 2 kinase modulates the activity of the kinase and its binding to calmodulin. *Mol Cell Biol* **24:** 2986–2997.

Brugarolas J, Lei K, Hurley RL, Manning BD, Reiling JH, Hafen E, Witters LA, Ellisen LW, Kaelin WG Jr. 2004. Regulation of mTOR function in response to hypoxia by REDD1 and the TSC1/TSC2 tumor suppressor complex. *Genes Dev* **18:** 2893–2904.

Brunn GJ, Hudson CC, Sekulic A, Williams JM, Hosoi H, Houghton PJ, Lawrence JC Jr, Abraham RT. 1997. Phosphorylation of the translational repressor PHAS-I by the mammalian target of rapamycin. *Science* **277:** 99–101.

Buttgereit F, Brand MD. 1995. A hierarchy of ATP-consuming processes in mammalian cells. *Biochem J* **312 (Pt 1):** 163–167.

Buxade M, Parra-Palau JL, Proud CG. 2008. The Mnks: MAP kinase-interacting kinases (MAP kinase signal-integrating kinases). *Front Biosci* **13:** 5359–5373.

Cantley LC. 2002. The phosphoinositide 3-kinase pathway. *Science* **296:** 1655–1657.

Cargnello M, Roux PP. 2011. Activation and Function of the MAPKs and Their Substrates, the MAPK-Activated Protein Kinases. *Microbiol Mol Biol Rev* **75:** 50–83.

Caron E, Ghosh S, Matsuoka Y, Ashton-Beaucage D, Therrien M, Lemieux S, Perreault C, Roux PP, Kitano H. 2010. A comprehensive map of the mTOR signaling network. *Mol Syst Biol* **6:** 453.

Carriere A, Cargnello M, Julien LA, Gao H, Bonneil E, Thibault P, Roux PP. 2008a. Oncogenic MAPK signaling stimulates mTORC1 activity by promoting RSK-mediated raptor phosphorylation. *Curr Biol* **18:** 1269–1277.

Carriere A, Ray H, Blenis J, Roux PP. 2008b. The RSK factors of activating the Ras/MAPK signaling cascade. *Front Biosci* **13:** 4258–4275.

Carriere A, Romeo Y, Acosta-Jaquez HA, Moreau J, Bonneil E, Thibault P, Fingar DC, Roux PP. 2011. ERK1/2 phosphorylate Raptor to promote Ras-dependent activation of mTOR complex 1 (mTORC1). *J Biol Chem* **286:** 567–577.

Ceci M, Gaviraghi C, Gorrini C, Sala LA, Offenhauser N, Marchisio PC, Biffo S. 2003. Release of eIF6 (p27BBP) from the 60S subunit allows 80S ribosome assembly. *Nature* **426:** 579–584.

Chen J, Zheng XF, Brown EJ, Schreiber SL. 1995. Identification of an 11-kDa FKBP12-rapamycin-binding domain within the 289-kDa FKBP12-rapamycin-associated protein and characterization of a critical serine residue. *Proc Natl Acad Sci* **92:** 4947–4951.

Chen Z, Gibson TB, Robinson F, Silvestro L, Pearson G, Xu B, Wright A, Vanderbilt C, Cobb MH. 2001. MAP kinases. *Chem Rev* **101:** 2449–2476.

Choo AY, Yoon SO, Kim SG, Roux PP, Blenis J. 2008. Rapamycin differentially inhibits S6Ks and 4E-BP1 to mediate cell-type-specific repression of mRNA translation. *Proc Natl Acad Sci* **105:** 17414–17419.

Chung J, Kuo CJ, Crabtree GR, Blenis J. 1992. Rapamycin-FKBP specifically blocks growth-dependent activation of and signaling by the 70 kd S6 protein kinases. *Cell* **69:** 1227–1236.

Cohen P, Frame S. 2001. The renaissance of GSK3. *Nat Rev Mol Cell Biol* **2:** 769–776.

Colina R, Costa-Mattioli M, Dowling RJ, Jaramillo M, Tai LH, Breitbach CJ, Martineau Y, Larsson O, Rong L, Svitkin YV, et al. 2008. Translational control of the innate immune response through IRF-7. *Nature* **452:** 323–328.

Corradetti MN, Inoki K, Bardeesy N, DePinho RA, Guan KL. 2004. Regulation of the TSC pathway by LKB1: Evidence of a molecular link between tuberous sclerosis complex and Peutz-Jeghers syndrome. *Genes Dev* **18:** 1533–1538.

De Benedetti A, Graff JR. 2004. eIF-4E expression and its role in malignancies and metastases. *Oncogene* **23:** 3189–3199.

DeYoung MP, Horak P, Sofer A, Sgroi D, Ellisen LW. 2008. Hypoxia regulates TSC1/2-mTOR signaling and tumor suppression through REDD1-mediated 14–3–3 shuttling. *Genes Dev* **22:** 239–251.

Diggle TA, Redpath NT, Heesom KJ, Denton RM. 1998. Regulation of protein-synthesis elongation-factor-2 kinase by cAMP in adipocytes. *Biochem J* **336:** 525–529.

Dobrikov M, Dobrikova E, Shveygert M, Gromeier M. 2011. Phosphorylation of eukaryotic translation initiation factor 4G1 (eIF4G1) by protein kinase Cα regulates eIF4G1 binding to Mnk1. *Mol Cell Biol* **31:** 2947–2959.

Dorrello NV, Peschiaroli A, Guardavaccaro D, Colburn NH, Sherman NE, Pagano M. 2006. S6K1- and βTRCP-mediated degradation of PDCD4 promotes protein translation and cell growth. *Science* **314:** 467–471.

Dowling RJ, Zakikhani M, Fantus IG, Pollak M, Sonenberg N. 2007. Metformin inhibits mammalian target of rapamycin-dependent translation initiation in breast cancer cells. *Cancer Res* **67:** 10804–10812.

Dowling RJ, Topisirovic I, Alain T, Bidinosti M, Fonseca BD, Petroulakis E, Wang X, Larsson O, Selvaraj A, Liu Y, et al. 2010a. mTORC1-mediated cell proliferation, but not cell growth, controlled by the 4E-BPs. *Science* **328:** 1172–1176.

Dowling RJ, Topisirovic I, Fonseca BD, Sonenberg N. 2010b. Dissecting the role of mTOR: Lessons from mTOR inhibitors. *Biochim Biophys Acta* **1804:** 433–439.

Duran A, Amanchy R, Linares JF, Joshi J, Abu-Baker S, Porollo A, Hansen M, Moscat J, Diaz-Meco MT. 2011. p62 is a key regulator of nutrient sensing in the mTORC1 pathway. *Mol Cell* **44:** 134–146.

Ekim B, Magnuson B, Acosta-Jaquez HA, Keller JA, Feener EP, Fingar DC. 2011. mTOR kinase domain phosphorylation promotes mTORC1 signaling, cell growth, and cell cycle progression. *Mol Cell Biol* **31:** 2787–2801.

Ellisen LW, Ramsayer KD, Johannessen CM, Yang A, Beppu H, Minda K, Oliner JD, McKeon F, Haber DA. 2002. REDD1, a developmentally regulated transcriptional target of p63 and p53, links p63 to regulation of reactive oxygen species. *Mol Cell* **10:** 995–1005.

Engelman JA, Luo J, Cantley LC. 2006. The evolution of phosphatidylinositol 3-kinases as regulators of growth and metabolism. *Nat Rev Genet* **7:** 606–619.

Erikson E, Maller JL. 1985. A protein kinase from Xenopus eggs specific for ribosomal protein S6. *Proc Natl Acad Sci* **82:** 742–746.

Erikson E, Maller JL. 1986. Purification and characterization of a protein kinase from Xenopus eggs highly specific for ribosomal protein S6. *J Biol Chem* **261:** 350–355.

Fadden P, Haystead TA, Lawrence JC Jr. 1997. Identification of phosphorylation sites in the translational regulator PHAS-I that are controlled by insulin and rapamycin in rat adipocytes. *J Biol Chem* **272:** 10240–10247.

Fenton TR, Gout IT. 2011. Functions and regulation of the 70kDa ribosomal S6 kinases. *Int J Biochem Cell Biol* **43:** 47–59.

Ferrari S, Bandi HR, Hofsteenge J, Bussian BM, Thomas G. 1991. Mitogen-activated 70K S6 kinase. Identification of in vitro 40 S ribosomal S6 phosphorylation sites. *J Biol Chem* **266:** 22770–22775.

Fingar DC, Blenis J. 2004. Target of rapamycin (TOR): An integrator of nutrient and growth factor signals and coordinator of cell growth and cell cycle progression. *Oncogene* **23:** 3151–3171.

Fisher TL, Blenis J. 1996. Evidence for two catalytically active kinase domains in pp90rsk. *Mol Cell Biol* **16:** 1212–1219.

Fonseca BD, Smith EM, Lee VH, MacKintosh C, Proud CG. 2007. PRAS40 is a target for mammalian target of rapamycin complex 1 and is required for signaling downstream of this complex. *J Biol Chem* **282:** 24514–24524.

Foster KG, Acosta-Jaquez HA, Romeo Y, Ekim B, Soliman GA, Carriere A, Roux PP, Ballif BA, Fingar DC. 2010. Regulation of mTOR complex 1 (mTORC1) by raptor Ser863 and multisite phosphorylation. *J Biol Chem* **285:** 80–94.

Frias MA, Thoreen CC, Jaffe JD, Schroder W, Sculley T, Carr SA, Sabatini DM. 2006. mSin1 is necessary for Akt/PKB phosphorylation, and its isoforms define three distinct mTORC2s. *Curr Biol* **16:** 1865–1870.

Fukunaga R, Hunter T. 1997. MNK1, a new MAP kinase-activate protein kinase, isolated by a novel expression screening method for identifying protein kinase substrates. *EMBO J* **16:** 1921–1997.

Furic L, Rong L, Larsson O, Koumakpayi IH, Yoshida K, Brueschke A, Petroulakis E, Robichaud N, Pollak M, Gaboury LA, et al. 2010. eIF4E phosphorylation promotes tumorigenesis and is associated with prostate cancer progression. *Proc Natl Acad Sci* **107:** 14134–14139.

Gaestel M. 2006. MAPKAP kinases—MKs—two's company, three's a crowd. *Nat Rev Mol Cell Biol* **7:** 120–130.

Gaestel M. 2008. Specificity of signaling from MAPKs to MAPKAPKs: Kinases' tango nuevo. *Front Biosci* **13:** 6050–6059.

Garami A, Zwartkruis FJ, Nobukuni T, Joaquin M, Roccio M, Stocker H, Kozma SC, Hafen E, Bos JL, Thomas G. 2003. Insulin activation of Rheb, a mediator of mTOR/

S6K/4E-BP signaling, is inhibited by TSC1 and 2. *Mol Cell* **11:** 1457–1466.

Garcia-Martinez JM, Alessi DR. 2008. mTOR complex 2 (mTORC2) controls hydrophobic motif phosphorylation and activation of serum- and glucocorticoid-induced protein kinase 1 (SGK1). *Biochem J* **416:** 375–385.

Gevaert K, Staes A, Van Damme J, De Groot S, Hugelier K, Demol H, Martens L, Goethals M, Vandekerckhove J. 2005. Global phosphoproteome analysis on human HepG2 hepatocytes using reversed-phase diagonal LC. *Proteomics* **5:** 3589–3599.

Gingras AC, Gygi SP, Raught B, Polakiewicz RD, Abraham RT, Hoekstra MF, Aebersold R, Sonenberg N. 1999a. Regulation of 4E-BP1 phosphorylation: A novel two-step mechanism. *Genes Dev* **13:** 1422–1437.

Gingras AC, Raught B, Sonenberg N. 1999b. eIF4 initiation factors: Effectors of mRNA recruitment to ribosomes and regulators of translation. *Annu Rev Biochem* **68:** 913–963.

Gingras AC, Raught B, Gygi SP, Niedzwiecka A, Miron M, Burley SK, Polakiewicz RD, Wyslouch-Cieszynska A, Aebersold R, Sonenberg N. 2001. Hierarchical phosphorylation of the translation inhibitor 4E-BP1. *Genes Dev* **15:** 2852–2864.

Goke A, Goke R, Knolle A, Trusheim H, Schmidt H, Wilmen A, Carmody R, Goke B, Chen YH. 2002. DUG is a novel homologue of translation initiation factor 4G that binds eIF4A. *Biochem Biophys Res Commun* **297:** 78–82.

Gout I, Minami T, Hara K, Tsujishita Y, Filonenko V, Waterfield MD, Yonezawa K. 1998. Molecular cloning and characterization of a novel p70 S6 kinase, p70 S6 kinase beta containing a proline-rich region. *J Biol Chem* **273:** 30061–30064.

Graff JR, Zimmer SG. 2003. Translational control and metastatic progression: Enhanced activity of the mRNA cap-binding protein eIF-4E selectively enhances translation of metastasis-related mRNAs. *Clin Exp Metastasis* **20:** 265–273.

Graff JR, Konicek BW, Carter JH, Marcusson EG. 2008. Targeting the eukaryotic translation initiation factor 4E for cancer therapy. *Cancer Res* **68:** 631–634.

Greenbaum D, Luscombe NM, Jansen R, Qian J, Gerstein M. 2001. Interrelating different types of genomic data, from proteome to secretome: 'oming in on function. *Genome Res* **11:** 1463–1468.

Grifo JA, Abramson RD, Satler CA, Merrick WC. 1984. RNA-stimulated ATPase activity of eukaryotic initiation factors. *J Biol Chem* **259:** 8648–8654.

Grove JR, Banerjee P, Balasubramanyam A, Coffer PJ, Price DJ, Avruch J, Woodgett JR. 1991. Cloning and expression of two human p70 S6 kinase polypeptides differing only at their amino termini. *Mol Cell Biol* **11:** 5541–5550.

Guertin DA, Sabatini DM. 2007. Defining the role of mTOR in cancer. *Cancer Cell* **12:** 9–22.

Guertin DA, Sabatini DM. 2009. The pharmacology of mTOR inhibition. *Sci Signal* **2:** e24.

Gwinn DM, Shackelford DB, Egan DF, Mihaylova MM, Mery A, Vasquez DS, Turk BE, Shaw RJ. 2008. AMPK phosphorylation of raptor mediates a metabolic checkpoint. *Mol Cell* **30:** 214–226.

Hara K, Yonezawa K, Weng QP, Kozlowski MT, Belham C, Avruch J. 1998. Amino acid sufficiency and mTOR regulate p70 S6 kinase and eIF-4E BP1 through a common effector mechanism. *J Biol Chem* **273:** 14484–14494.

Hay N, Sonenberg N. 2004. Upstream and downstream of mTOR. *Genes Dev* **18:** 1926–1945.

Heesom KJ, Avison MB, Diggle TA, Denton RM. 1998. Insulin-stimulated kinase from rat fat cells that phosphorylates initiation factor 4E-binding protein 1 on the rapamycin-insensitive site (serine-111). *Biochem J* **336** (Pt 1): 39–48.

Holz MK, Ballif BA, Gygi SP, Blenis J. 2005. mTOR and S6K1 mediate assembly of the translation preinitiation complex through dynamic protein interchange and ordered phosphorylation events. *Cell* **123:** 569–580.

Homma MK, Wada I, Suzuki T, Yamaki J, Krebs EG, Homma Y. 2005. CK2 phosphorylation of eukaryotic translation initiation factor 5 potentiates cell cycle progression. *Proc Natl Acad Sci* **102:** 15688–15693.

Horman S, Browne G, Krause U, Patel J, Vertommen D, Bertrand L, Lavoinne A, Hue L, Proud C, Rider M. 2002. Activation of AMP-activated protein kinase leads to the phosphorylation of elongation factor 2 and an inhibition of protein synthesis. *Curr Biol* **12:** 1419–1423.

Howell JJ, Manning BD. 2011. mTOR couples cellular nutrient sensing to organismal metabolic homeostasis. *Trends Endocrinol Metab* **22:** 94–102.

Inoki K, Li Y, Xu T, Guan KL. 2003. Rheb GTPase is a direct target of TSC2 GAP activity and regulates mTOR signaling. *Genes Dev* **17:** 1829–1834.

Jacinto E, Facchinetti V, Liu D, Soto N, Wei S, Jung SY, Huang Q, Qin J, Su B. 2006. SIN1/MIP1 maintains rictor-mTOR complex integrity and regulates Akt phosphorylation and substrate specificity. *Cell* **127:** 125–137.

Jackson RJ, Hellen CU, Pestova TV. 2010. The mechanism of eukaryotic translation initiation and principles of its regulation. *Nat Rev Mol Cell Biol* **11:** 113–127.

Jefferies HB, Reinhard C, Kozma SC, Thomas G. 1994. Rapamycin selectively represses translation of the "polypyrimidine tract" mRNA family. *Proc Natl Acad Sci* **91:** 4441–4445.

Johannessen CM, Reczek EE, James MF, Brems H, Legius E, Cichowski K. 2005. The NF1 tumor suppressor critically regulates TSC2 and mTOR. *Proc Natl Acad Sci* **102:** 8573–8578.

Johannessen CM, Johnson BW, Williams SM, Chan AW, Reczek EE, Lynch RC, Rioth MJ, McClatchey A, Ryeom S, Cichowski K. 2008. TORC1 is essential for NF1-associated malignancies. *Curr Biol* **18:** 56–62.

Jones SW, Erikson E, Blenis J, Maller JL, Erikson RL. 1988. A Xenopus ribosomal protein S6 kinase has two apparent kinase domains that are each similar to distinct protein kinases. *Proc Natl Acad Sci* **85:** 3377–3381.

Joshi B, Cai AL, Keiper BD, Minich WB, Mendez R, Beach CM, Stepinski J, Stolarski R, Darzynkiewicz E, Rhoads RE. 1995. Phosphorylation of eukaryotic protein synthesis initiation factor 4E at Ser-209. *J Biol Chem* **270:** 14597–14603.

Kahn BB, Alquier T, Carling D, Hardie DG. 2005. AMP-activated protein kinase: Ancient energy gauge provides clues to modern understanding of metabolism. *Cell Metab* **1**: 15–25.

Kantidakis T, Ramsbottom BA, Birch JL, Dowding SN, White RJ. 2010. mTOR associates with TFIIIC, is found at tRNA and 5S rRNA genes, and targets their repressor Maf1. *Proc Natl Acad Sci* **107**: 11823–11828.

Kaspar RL, Rychlik W, White MW, Rhoads RE, Morris DR. 1990. Simultaneous cytoplasmic redistribution of ribosomal protein L32 mRNA and phosphorylation of eukaryotic initiation factor 4E after mitogenic stimulation of Swiss 3T3 cells. *J Biol Chem* **265**: 3619–3622.

Kawasome H, Papst P, Webb S, Keller GM, Johnson GL, Gelfand EW, Terada N. 1998. Targeted disruption of p70(s6k) defines its role in protein synthesis and rapamycin sensitivity. *Proc Natl Acad Sci* **95**: 5033–5038.

Kielbassa K, Muller HJ, Meyer HE, Marks F, Gschwendt M. 1995. Protein kinase C delta-specific phosphorylation of the elongation factor eEF-alpha and an eEF-1 alpha peptide at threonine 431. *J Biol Chem* **270**: 6156–6162.

Kim J, Guan KL. 2011. Amino acid signaling in TOR activation. *Annu Rev Biochem* **80**: 1001–1032.

Kim JE, Tannenbaum SR, White FM. 2005. Global phosphoproteome of HT-29 human colon adenocarcinoma cells. *J Proteome Res* **4**: 1339–1346.

Kim E, Goraksha-Hicks P, Li L, Neufeld TP, Guan KL. 2008. Regulation of TORC1 by Rag GTPases in nutrient response. *Nat Cell Biol* **10**: 935–945.

Knauf U, Tschopp C, Gram H. 2001. Negative regulation of protein translation by mitogen-activated protein kinase-interacting kinases 1 and 2. *Mol Cell Biol* **21**: 5500–5511.

Knebel A, Morrice N, Cohen P. 2001. A novel method to identify protein kinase substrates: eEF2 kinase is phosphorylated and inhibited by SAPK4/p38delta. *EMBO J* **20**: 4360–4369.

Koromilas AE, Lazaris-Karatzas A, Sonenberg N. 1992. mRNAs containing extensive secondary structure in their 5′ noncoding region translate efficiently in cells overexpressing initiation factor eIF-4E. *EMBO J* **11**: 4153–4158.

Kozma SC, Ferrari S, Bassand P, Siegmann M, Totty N, Thomas G. 1990. Cloning of the mitogen-activated S6 kinase from rat liver reveals an enzyme of the second messenger subfamily. *Proc Natl Acad Sci* **87**: 7365–7369.

Krieg J, Hofsteenge J, Thomas G. 1988. Identification of the 40 S ribosomal protein S6 phosphorylation sites induced by cycloheximide. *J Biol Chem* **263**: 11473–11477.

Kudlicki W, Wettenhall RE, Kemp BE, Szyszka R, Kramer G, Hardesty B. 1987. Evidence for a second phosphorylation site on eIF-2 alpha from rabbit reticulocytes. *FEBS Lett* **215**: 16–20.

Kwiatkowski DJ, Manning BD. 2005. Tuberous sclerosis: a GAP at the crossroads of multiple signaling pathways. *Hum Mol Genet* **14** (Spec No 2): R251–R258.

Kyriakis JM, Avruch J. 2001. Mammalian mitogen-activated protein kinase signal transduction pathways activated by stress and inflammation. *Physiol Rev* **81**: 807–869.

Lachance PE, Miron M, Raught B, Sonenberg N, Lasko P. 2002. Phosphorylation of eukaryotic translation initiation factor 4E is critical for growth. *Mol Cell Biol* **22**: 1656–1663.

Lankat-Buttgereit B, Goke R. 2009. The tumour suppressor Pdcd4: Recent advances in the elucidation of function and regulation. *Biol Cell* **101**: 309–317.

Laplante M, Sabatini DM. 2009. mTOR signaling at a glance. *J Cell Sci* **122** (Pt 20): 3589–3594.

* Lasko P. 2012. mRNA localization and translational control in Drosophila oogenesis. *Cold Spring Harb Perspect Biol* doi: 10.1101/cshperspect.a012294.

Levy S, Avni D, Hariharan N, Perry RP, Meyuhas O. 1991. Oligopyrimidine tract at the 5′ end of mammalian ribosomal protein mRNAs is required for their translational control. *Proc Natl Acad Sci* **88**: 3319–3323.

Li Y, Wang Y, Kim E, Beemiller P, Wang CY, Swanson J, You M, Guan KL. 2007. Bnip3 mediates the hypoxia-induced inhibition on mammalian target of rapamycin by interacting with Rheb. *J Biol Chem* **282**: 35803–35813.

Lin KW, Yakymovych I, Jia M, Yakymovych M, Souchelnytskyi S. 2010. Phosphorylation of eEF1A1 at Ser300 by TbetaR-I results in inhibition of mRNA translation. *Curr Biol* **20**: 1615–1625.

Ling J, Morley SJ, Traugh JA. 2005. Inhibition of cap-dependent translation via phosphorylation of eIF4G by protein kinase Pak2. *EMBO J* **24**: 4094–4105.

Loreni F, Thomas G, Amaldi F. 2000. Transcription inhibitors stimulate translation of 5′ TOP mRNAs through activation of S6 kinase and the mTOR/FRAP signalling pathway. *Eur J Biochem* **267**: 6594–6601.

Lynch M, Fitzgerald C, Johnston KA, Wang S, Schmidt EV. 2004. Activated eIF4E-binding protein slows G1 progression and blocks transformation by c-myc without inhibiting cell growth. *J Biol Chem* **279**: 3327–3339.

Ma L, Chen Z, Erdjument-Bromage H, Tempst P, Pandolfi PP. 2005. Phosphorylation and functional inactivation of TSC2 by Erk implications for tuberous sclerosis and cancer pathogenesis. *Cell* **121**: 179–193.

Ma XM, Blenis J. 2009. Molecular mechanisms of mTOR-mediated translational control. *Nat Rev Mol Cell Biol* **10**: 307–318.

Ma XM, Yoon SO, Richardson CJ, Julich K, Blenis J. 2008. SKAR links pre-mRNA splicing to mTOR/S6K1-mediated enhanced translation efficiency of spliced mRNAs. *Cell* **133**: 303–313.

Maehama T, Dixon JE. 1998. The tumor suppressor, PTEN/MMAC1, dephosphorylates the lipid second messenger, phosphatidylinositol 3,4,5-trisphosphate. *J Biol Chem* **273**: 13375–13378.

* Malina A, Mills JR, Pelletier J. 2012. Emerging therapeutics targeting mRNA translation. *Cold Spring Harb Perspect Biol* doi: 10.1101/cshperspect.a012377.

Manning BD, Tee AR, Logsdon MN, Blenis J, Cantley LC. 2002. Identification of the tuberous sclerosis complex-2 tumor suppressor gene product tuberin as a target of the phosphoinositide 3-kinase/akt pathway. *Mol Cell* **10**: 151–162.

Manzella JM, Rychlik W, Rhoads RE, Hershey JW, Blackshear PJ. 1991. Insulin induction of ornithine decarboxylase. Importance of mRNA secondary structure and phosphorylation of eucaryotic initiation factors eIF-4B and eIF-4E. *J Biol Chem* **266**: 2383–2389.

Marcotrigiano J, Gingras AC, Sonenberg N, Burley SK. 1997. Cocrystal structure of the messenger RNA 5′ cap-binding protein (eIF4E) bound to 7-methyl-GDP. *Cell* **89**: 951–961.

Mathews MB, Sonenberg N, Hershey JWB. 2007. Translational control in biology and medicine. In *Translational control in biology and medicine* (ed. NS Michael, B Mathews, W John, B Hershey). Cold Spring Harbor Laboratory Press, Cold Spring, NY.

Matsuo H, Li H, McGuire AM, Fletcher CM, Gingras AC, Sonenberg N, Wagner G. 1997. Structure of translation factor eIF4E bound to m7GDP and interaction with 4E-binding protein. *Nat Struct Biol* **4**: 717–724.

Mayer C, Zhao J, Yuan X, Grummt I. 2004. mTOR-dependent activation of the transcription factor TIF-IA links rRNA synthesis to nutrient availability. *Genes Dev* **18**: 423–434.

Meyuhas O. 2008. Physiological roles of ribosomal protein S6: One of its kind. *Int Rev Cell Mol Biol* **268**: 1–37.

Michels AA, Robitaille AM, Buczynski-Ruchonnet D, Hodroj W, Reina JH, Hall MN, Hernandez N. 2010. mTORC1 directly phosphorylates and regulates human MAF1. *Mol Cell Biol* **30**: 3749–3757.

Miron M, Verdu J, Lachance PE, Birnbaum MJ, Lasko PF, Sonenberg N. 2001. The translational inhibitor 4E-BP is an effector of PI(3)K/Akt signalling and cell growth in Drosophila. *Nat Cell Biol* **3**: 596–601.

Montagne J, Stewart MJ, Stocker H, Hafen E, Kozma SC, Thomas G. 1999. Drosophila S6 kinase: A regulator of cell size. *Science* **285**: 2126–2129.

Morley SJ, Naegele S. 2002. Phosphorylation of eukaryotic initiation factor (eIF) 4E is not required for de novo protein synthesis following recovery from hypertonic stress in human kidney cells. *J Biol Chem* **277**: 32855–32859.

Naegele S, Morley SJ. 2004. Molecular cross-talk between MEK1/2 and mTOR signaling during recovery of 293 cells from hypertonic stress. *J Biol Chem* **279**: 46023–46034.

Nairn AC, Palfrey HC. 1987. Identification of the major Mr 100000 substrate for calmodulin-dependent protein kinase III in mammalian cells as elongation factor-2. *J Biol Chem* **262**: 17299–17303.

Oh WJ, Jacinto E. 2011. mTOR complex 2 signaling and functions. *Cell Cycle* **10**: 2305–2316.

Oh WJ, Wu CC, Kim SJ, Facchinetti V, Julien LA, Finlan M, Roux PP, Su B, Jacinto E. 2010. mTORC2 can associate with ribosomes to promote cotranslational phosphorylation and stability of nascent Akt polypeptide. *EMBO J* **29**: 3939–3951.

Ohanna M, Sobering AK, Lapointe T, Lorenzo L, Praud C, Petroulakis E, Sonenberg N, Kelly PA, Sotiropoulos A, Pende M. 2005. Atrophy of S6K1(-/-) skeletal muscle cells reveals distinct mTOR effectors for cell cycle and size control. *Nat Cell Biol* **7**: 286–294.

Oshiro N, Takahashi R, Yoshino K, Tanimura K, Nakashima A, Eguchi S, Miyamoto T, Hara K, Takehana K, Avruch J, et al. 2007. The proline-rich Akt substrate of 40 kDa (PRAS40) is a physiological substrate of mammalian target of rapamycin complex 1. *J Biol Chem* **282**: 20329–20339.

Palamarchuk A, Efanov A, Maximov V, Aqeilan RI, Croce CM, Pekarsky Y. 2005. Akt phosphorylates and regulates

Pdcd4 tumor suppressor protein. *Cancer Res* **65**: 11282–11286.

Pause A, Belsham GJ, Gingras AC, Donze O, Lin TA, Lawrence JC Jr, Sonenberg N. 1994. Insulin-dependent stimulation of protein synthesis by phosphorylation of a regulator of 5′-cap function. *Nature* **371**: 762–767.

* Pavitt GD, Ron D. 2012. New insights into translational regulation in the endoplasmic reticulum unfolded protein response. *Cold Spring Harb Perspect Biol* doi: 10.1101/cshperspect.a012278

Pearce LR, Huang X, Boudeau J, Pawlowski R, Wullschleger S, Deak M, Ibrahim AF, Gourlay R, Magnuson MA, Alessi DR. 2007. Identification of Protor as a novel Rictor-binding component of mTOR complex-2. *Biochem J* **405**: 513–522.

Pearce LR, Komander D, Alessi DR. 2010. The nuts and bolts of AGC protein kinases. *Nat Rev Mol Cell Biol* **11**: 9–22.

Pearson G, Robinson F, Beers Gibson T, Xu BE, Karandikar M, Berman K, Cobb MH. 2001. Mitogen-activated protein (MAP) kinase pathways: Regulation and physiological functions. *Endocr Rev* **22**: 153–183.

Pende M, Um SH, Mieulet V, Sticker M, Goss VL, Mestan J, Mueller M, Fumagalli S, Kozma SC, Thomas G. 2004. S6K1(-/-)/S6K2(-/-) mice exhibit perinatal lethality and rapamycin-sensitive 5′-terminal oligopyrimidine mRNA translation and reveal a mitogen-activated protein kinase-dependent S6 kinase pathway. *Mol Cell Biol* **24**: 3112–3124.

Peterson TR, Laplante M, Thoreen CC, Sancak Y, Kang SA, Kuehl WM, Gray NS, Sabatini DM. 2009. DEPTOR is an mTOR inhibitor frequently overexpressed in multiple myeloma cells and required for their survival. *Cell* **137**: 873–886.

Petroulakis E, Mamane Y, Le Bacquer O, Shahbazian D, Sonenberg N. 2006. mTOR signaling: Implications for cancer and anticancer therapy. *Br J Cancer* **94**: 195–199.

Petroulakis E, Parsyan A, Dowling RJ, LeBacquer O, Martineau Y, Bidinosti M, Larsson O, Alain T, Rong L, Mamane Y, et al. 2009. p53–dependent translational control of senescence and transformation via 4E-BPs. *Cancer Cell* **16**: 439–446.

Piazzi M, Bavelloni A, Faenza I, Blalock W, Urbani A, D'Aguanno S, Fiume R, Ramazzotti G, Maraldi NM, Cocco L. 2010. eEF1A phosphorylation in the nucleus of insulin-stimulated C2C12 myoblasts: Ser(3) is a novel substrate for protein kinase C betaI. *Mol Cell Proteomics* **9**: 2719–2728.

Price NT, Redpath NT, Severinov KV, Campbell DG, Russell JM, Proud CG. 1991. Identification of the phosphorylation sites in elongation factor-2 from rabbit reticulocytes. *FEBS Lett* **282**: 253–258.

Pyronnet S, Imataka H, Gingras AC, Fukunaga R, Hunter T, Sonenberg N. 1999. Human eukaryotic translation initiation factor 4G (eIF4G) recruits mnk1 to phosphorylate eIF4E. *EMBO J* **18**: 270–279.

Rajalingam K, Schreck R, Rapp UR, Albert S. 2007. Ras oncogenes and their downstream targets. *Biochim Biophys Acta* **1773**: 1177–1195.

Ramirez-Valle F, Braunstein S, Zavadil J, Formenti SC, Schneider RJ. 2008. eIF4GI links nutrient sensing by mTOR to cell proliferation and inhibition of autophagy. *J Cell Biol* **181**: 293–307.

Rapley J, Oshiro N, Ortiz-Vega S, Avruch J. 2011. The mechanism of insulin-stimulated 4E-BP protein binding to mammalian target of rapamycin (mTOR) complex 1 and its contribution to mTOR complex 1 signaling. *J Biol Chem* **286:** 38043–38053.

Raught B, Gingras AC, Gygi SP, Imataka H, Morino S, Gradi A, Aebersold R, Sonenberg N. 2000. Serum-stimulated, rapamycin-sensitive phosphorylation sites in the eukaryotic translation initiation factor 4GI. *EMBO J* **19:** 434–444.

Raught B, Peiretti F, Gingras AC, Livingstone M, Shahbazian D, Mayeur GL, Polakiewicz RD, Sonenberg N, Hershey JW. 2004. Phosphorylation of eucaryotic translation initiation factor 4B Ser422 is modulated by S6 kinases. *EMBO J* **23:** 1761–1769.

Richardson CJ, Broenstrup M, Fingar DC, Julich K, Ballif BA, Gygi S, Blenis J. 2004. SKAR is a specific target of S6 kinase 1 in cell growth control. *Curr Biol* **14:** 1540–1549.

Richter-Cook NJ, Dever TE, Hensold JO, Merrick WC. 1998. Purification and characterization of a new eukaryotic protein translation factor. Eukaryotic initiation factor 4H. *J Biol Chem* **273:** 7579–7587.

Rogers GW Jr, Richter NJ, Lima WF, Merrick WC. 2001. Modulation of the helicase activity of eIF4A by eIF4B, eIF4H, and eIF4F. *J Biol Chem* **276:** 30914–30922.

Roux PP, Blenis J. 2004. ERK and p38 MAPK-activated protein kinases: A family of protein kinases with diverse biological functions. *Microbiol Mol Biol Rev* **68:** 320–344.

Roux PP, Ballif BA, Anjum R, Gygi SP, Blenis J. 2004. Tumor-promoting phorbol esters and activated Ras inactivate the tuberous sclerosis tumor suppressor complex via p90 ribosomal S6 kinase. *Proc Natl Acad Sci* **101:** 13489–13494.

Roux PP, Shahbazian D, Vu H, Holz MK, Cohen MS, Taunton J, Sonenberg N, Blenis J. 2007. RAS/ERK signaling promotes site-specific ribosomal protein S6 phosphorylation via RSK and stimulates cap-dependent translation. *J Biol Chem* **282:** 14056–14064.

Rozen F, Edery I, Meerovitch K, Dever TE, Merrick WC, Sonenberg N. 1990. Bidirectional RNA helicase activity of eucaryotic translation initiation factors 4A and 4F. *Mol Cell Biol* **10:** 1134–1144.

Rush J, Moritz A, Lee KA, Guo A, Goss VL, Spek EJ, Zhang H, Zha XM, Polakiewicz RD, Comb MJ. 2005. Immunoaffinity profiling of tyrosine phosphorylation in cancer cells. *Nat Biotechnol* **23:** 94–101.

Ruvinsky I, Sharon N, Lerer T, Cohen H, Stolovich-Rain M, Nir T, Dor Y, Zisman P, Meyuhas O. 2005. Ribosomal protein S6 phosphorylation is a determinant of cell size and glucose homeostasis. *Genes Dev* **19:** 2199–2211.

Sancak Y, Thoreen CC, Peterson TR, Lindquist RA, Kang SA, Spooner E, Carr SA, Sabatini DM. 2007. PRAS40 is an insulin-regulated inhibitor of the mTORC1 protein kinase. *Mol Cell* **25:** 903–915.

Sancak Y, Peterson TR, Shaul YD, Lindquist RA, Thoreen CC, Bar-Peled L, Sabatini DM. 2008. The Rag GTPases bind raptor and mediate amino acid signaling to mTORC1. *Science* **320:** 1496–1501.

Sancak Y, Bar-Peled L, Zoncu R, Markhard AL, Nada S, Sabatini DM. 2010. Ragulator-Rag complex targets mTORC1 to the lysosomal surface and is necessary for its activation by amino acids. *Cell* **141:** 290–303.

Sarbassov DD, Ali SM, Kim DH, Guertin DA, Latek RR, Erdjument-Bromage H, Tempst P, Sabatini DM. 2004. Rictor, a novel binding partner of mTOR, defines a rapamycin-insensitive and raptor-independent pathway that regulates the cytoskeleton. *Curr Biol* **14:** 1296–1302.

Sarbassov DD, Guertin DA, Ali SM, Sabatini DM. 2005. Phosphorylation and regulation of Akt/PKB by the rictor-mTOR complex. *Science* **307:** 1098–1101.

Sarbassov DD, Ali SM, Sengupta S, Sheen JH, Hsu PP, Bagley AF, Markhard AL, Sabatini DM. 2006. Prolonged rapamycin treatment inhibits mTORC2 assembly and Akt/PKB. *Mol Cell* **22:** 159–168.

Schalm SS, Blenis J. 2002. Identification of a conserved motif required for mTOR signaling. *Curr Biol* **12:** 632–639.

Schalm SS, Fingar DC, Sabatini DM, Blenis J. 2003. TOS motif-mediated raptor binding regulates 4E-BP1 multisite phosphorylation and function. *Curr Biol* **13:** 797–806.

Scheper GC, Morrice NA, Kleijn M, Proud CG. 2001. The mitogen-activated protein kinase signal-integrating kinase Mnk2 is a eukaryotic initiation factor 4E kinase with high levels of basal activity in mammalian cells. *Mol Cell Biol* **21:** 743–754.

Scheper GC, van Kollenburg B, Hu J, Luo Y, Goss DJ, Proud CG. 2002. Phosphorylation of eukaryotic initiation factor 4E markedly reduces its affinity for capped mRNA. *J Biol Chem* **277:** 3303–3309.

Schwanhausser B, Busse D, Li N, Dittmar G, Schuchhardt J, Wolf J, Chen W, Selbach M. 2011. Global quantification of mammalian gene expression control. *Nature* **473:** 337–342.

Sengupta S, Peterson TR, Sabatini DM. 2010. Regulation of the mTOR complex 1 pathway by nutrients, growth factors, and stress. *Mol Cell* **40:** 310–322.

Shahbazian D, Roux PP, Mieulet V, Cohen MS, Raught B, Taunton J, Hershey JW, Blenis J, Pende M, Sonenberg N. 2006. The mTOR/PI3K and MAPK pathways converge on eIF4B to control its phosphorylation and activity. *EMBO J* **25:** 2781–2791.

Shahbazian D, Parsyan A, Petroulakis E, Topisirovic I, Martineau Y, Gibbs BF, Svitkin Y, Sonenberg N. 2010. Control of cell survival and proliferation by mammalian eukaryotic initiation factor 4B. *Mol Cell Biol* **30:** 1478–1485.

Shaw RJ. 2009. LKB1 and AMP-activated protein kinase control of mTOR signalling and growth. *Acta Physiol (Oxf)* **196:** 65–80.

Shaw RJ, Cantley LC. 2006. Ras, PI(3)K and mTOR signalling controls tumour cell growth. *Nature* **441:** 424–430.

Shaw RJ, Bardeesy N, Manning BD, Lopez L, Kosmatka M, DePinho RA, Cantley LC. 2004. The LKB1 tumor suppressor negatively regulates mTOR signaling. *Cancer Cell* **6:** 91–99.

Shi J, Feng Y, Goulet AC, Vaillancourt RR, Sachs NA, Hershey JW, Nelson MA. 2003. The p34cdc2-related cyclin-dependent kinase 11 interacts with the p47 subunit of eukaryotic initiation factor 3 during apoptosis. *J Biol Chem* **278:** 5062–5071.

Shi J, Hershey JW, Nelson MA. 2009. Phosphorylation of the eukaryotic initiation factor 3f by cyclin-dependent kinase 11 during apoptosis. *FEBS Lett* **583:** 971–977.

Shima H, Pende M, Chen Y, Fumagalli S, Thomas G, Kozma SC. 1998. Disruption of the p70(s6k)/p85(s6k) gene reveals a small mouse phenotype and a new functional S6 kinase. *EMBO J* **17:** 6649–6659.

Shin S, Wolgamott L, Yu Y, Blenis J, Yoon SO. 2011. Glycogen synthase kinase (GSK)-3 promotes p70 ribosomal protein S6 kinase (p70S6K) activity and cell proliferation. *Proc Natl Acad Sci* **108:** 1204–1213.

Shor B, Wu J, Shakey Q, Toral-Barza L, Shi C, Follettie M, Yu K. 2010. Requirement of the mTOR kinase for the regulation of Maf1 phosphorylation and control of RNA polymerase III-dependent transcription in cancer cells. *J Biol Chem* **285:** 15380–15392.

Silvera D, Formenti SC, Schneider RJ. 2010. Translational control in cancer. *Nat Rev Cancer* **10:** 254–266.

Skogerson L, Moldave K. 1968. Characterization of the interaction of aminoacyltransferase II with ribosomes. Binding of transferase II and translocation of peptidyl transfer ribonucleic acid. *J Biol Chem* **243:** 5354–5360.

Slepenkov SV, Darzynkiewicz E, Rhoads RE. 2006. Stopped-flow kinetic analysis of eIF4E and phosphorylated eIF4E binding to cap analogs and capped oligoribonucleotides: Evidence for a one-step binding mechanism. *J Biol Chem* **281:** 14927–14938.

Soliman GA, Acosta-Jaquez HA, Dunlop EA, Ekim B, Maj NE, Tee AR, Fingar DC. 2010. mTOR Ser-2481 autophosphorylation monitors mTORC-specific catalytic activity and clarifies rapamycin mechanism of action. *J Biol Chem* **285:** 7866–7879.

Sonenberg N. 2008. eIF4E, the mRNA cap-binding protein: From basic discovery to translational research. *Biochem Cell Biol* **86:** 178–183.

Sonenberg N, Hinnebusch AG. 2009. Regulation of translation initiation in eukaryotes: Mechanisms and biological targets. *Cell* **136:** 731–745.

* Sonenberg N, Mathews MB, Hershey JWB. 2012. Principles of translational control. *Cold Spring Harb Perspect Biol* doi: 10.1101/cshperspect.a011528.

Stipanuk MH. 2009. Macroautophagy and its role in nutrient homeostasis. *Nutr Rev* **67:** 677–689.

Sutherland C, Leighton IA, Cohen P. 1993. Inactivation of glycogen synthase kinase-3 beta by phosphorylation: New kinase connections in insulin and growth-factor signalling. *Biochem J* **296:** 15–19.

Svitkin YV, Pause A, Haghighat A, Pyronnet S, Witherell G, Belsham GJ, Sonenberg N. 2001. The requirement for eukaryotic initiation factor 4A (eIF4A) in translation is in direct proportion to the degree of mRNA 5′ secondary structure. *RNA* **7:** 382–394.

Tang H, Hornstein E, Stolovich M, Levy G, Livingstone M, Templeton D, Avruch J, Meyuhas O. 2001. Amino acid-induced translation of TOP mRNAs is fully dependent on phosphatidylinositol 3-kinase-mediated signaling, is partially inhibited by rapamycin, and is independent of S6K1 and rpS6 phosphorylation. *Mol Cell Biol* **21:** 8671–8683.

Thedieck K, Polak P, Kim ML, Molle KD, Cohen A, Jeno P, Arrieumerlou C, Hall MN. 2007. PRAS40 and PRR5-like protein are new mTOR interactors that regulate apoptosis. *PLoS ONE* **2:** e1217.

Topisirovic I, Ruiz-Gutierrez M, Borden KL. 2004. Phosphorylation of the eukaryotic translation initiation factor eIF4E contributes to its transformation and mRNA transport activities. *Cancer Res* **64:** 8639–8642.

Topisirovic I, Svitkin YV, Sonenberg N, Shatkin AJ. 2011. Cap and cap-binding proteins in the control of gene expression. *RNA* **2:** 277–298.

Ueda T, Watanabe-Fukunaga R, Fukuyama H, Nagata S, Fukunaga R. 2004. Mnk2 and Mnk1 are essential for constitutive and inducible phosphorylation of eukaryotic initiation factor 4E but not for cell growth or development. *Mol Cell Biol* **24:** 6539–6549.

Vander Haar E, Lee SI, Bandhakavi S, Griffin TJ, Kim DH. 2007. Insulin signalling to mTOR mediated by the Akt/PKB substrate PRAS40. *Nat Cell Biol* **9:** 316–323.

van Gorp AG, van der Vos KE, Brenkman AB, Bremer A, van den Broek N, Zwartkruis F, Hershey JW, Burgering BM, Calkhoven CF, Coffer PJ. 2009. AGC kinases regulate phosphorylation and activation of eukaryotic translation initiation factor 4B. *Oncogene* **28:** 95–106.

Walsh D, Mohr I. 2004. Phosphorylation of eIF4E by Mnk-1 enhances HSV-1 translation and replication in quiescent cells. *Genes Dev* **18:** 660–672.

Wang X, Campbell LE, Miller CM, Proud CG. 1998. Amino acid availability regulates p70 S6 kinase and multiple translation factors. *Biochem J* **334 (Pt 1):** 261–267.

Wang X, Li W, Williams M, Terada N, Alessi DR, Proud CG. 2001a. Regulation of elongation factor 2 kinase by p90(RSK1) and p70 S6 kinase. *EMBO J* **20:** 4370–4379.

Wang X, Paulin FE, Campbell LE, Gomez E, O'Brien K, Morrice N, Proud CG. 2001b. Eukaryotic initiation factor 2B: Identification of multiple phosphorylation sites in the epsilon-subunit and their functions in vivo. *EMBO J* **20:** 4349–4359.

Wang X, Li W, Parra JL, Beugnet A, Proud CG. 2003. The C terminus of initiation factor 4E-binding protein 1 contains multiple regulatory features that influence its function and phosphorylation. *Mol Cell Biol* **23:** 1546–1557.

Wang X, Beugnet A, Murakami M, Yamanaka S, Proud CG. 2005. Distinct signaling events downstream of mTOR cooperate to mediate the effects of amino acids and insulin on initiation factor 4E-binding proteins. *Mol Cell Biol* **25:** 2558–2572.

Wang L, Harris TE, Roth RA, Lawrence JC Jr. 2007. PRAS40 regulates mTORC1 kinase activity by functioning as a direct inhibitor of substrate binding. *J Biol Chem* **282:** 20036–20044.

Waskiewicz AJ, Flynn A, Proud CG, Cooper JA. 1997. Mitogen-activated protein kinases activate the serine/threonine kinases Mnk1 and Mnk2. *Embo J* **16:** 1909–1920.

Wei Y, Tsang CK, Zheng XF. 2009. Mechanisms of regulation of RNA polymerase III-dependent transcription by TORC1. *EMBO J* **28:** 2220–2230.

Wek RC, Jiang HY, Anthony TG. 2006. Coping with stress: eIF2 kinases and translational control. *Biochem Soc Trans* **34**(Pt 1)**:** 7–11.

Welsh GI, Miller CM, Loughlin AJ, Price NT, Proud CG. 1998. Regulation of eukaryotic initiation factor eIF2B: Glycogen synthase kinase-3 phosphorylates a conserved serine which undergoes dephosphorylation in response to insulin. *FEBS Lett* **421:** 125–130.

Wendel HG, Silva RL, Malina A, Mills JR, Zhu H, Ueda T, Watanabe-Fukunaga R, Fukunaga R, Teruya-Feldstein J, Pelletier J, et al. 2007. Dissecting eIF4E action in tumorigenesis. *Genes Dev* **21:** 3232–3237.

Widmann C, Gibson S, Jarpe MB, Johnson GL. 1999. Mitogen-activated protein kinase: Conservation of a three-kinase module from yeast to human. *Physiol Rev* **79:** 143–180.

Woo SY, Kim DH, Jun CB, Kim YM, Haar EV, Lee SI, Hegg JW, Bandhakavi S, Griffin TJ. 2007. PRR5, a novel component of mTOR complex 2, regulates platelet-derived growth factor receptor beta expression and signaling. *J Biol Chem* **282:** 25604–25612.

Woods YL, Cohen P, Becker W, Jakes R, Goedert M, Wang X, Proud CG. 2001. The kinase DYRK phosphorylates protein-synthesis initiation factor eIF2Bepsilon at Ser539 and the microtubule-associated protein tau at Thr212: Potential role for DYRK as a glycogen synthase kinase 3-priming kinase. *Biochem J* **355** (Pt 3)**:** 609–615.

Worch J, Tickenbrock L, Schwable J, Steffen B, Cauvet T, Mlody B, Buerger H, Koeffler HP, Berdel WE, Serve H, et al. 2004. The serine-threonine kinase MNK1 is post-translationally stabilized by PML-RARalpha and regulates differentiation of hematopoietic cells. *Oncogene* **23:** 9162–9172.

Wullschleger S, Loewith R, Hall MN. 2006. TOR signaling in growth and metabolism. *Cell* **124:** 471–484.

Yamagata K, Sanders LK, Kaufmann WE, Yee W, Barnes CA, Nathans D, Worley PF. 1994. rheb, a growth factor- and synaptic activity-regulated gene, encodes a novel Ras-related protein. *J Biol Chem* **269:** 16333–16339.

Yang HS, Jansen AP, Komar AA, Zheng X, Merrick WC, Costes S, Lockett SJ, Sonenberg N, Colburn NH. 2003. The transformation suppressor Pdcd4 is a novel eukaryotic translation initiation factor 4A binding protein that inhibits translation. *Mol Cell Biol* **23:** 26–37.

Yea SS, Fruman DA. 2011. Cell signaling. New mTOR targets Grb attention. *Science* **332:** 1270–1271.

Yip CK, Murata K, Walz T, Sabatini DM, Kang SA. 2010. Structure of the human mTOR complex I and its implications for rapamycin inhibition. *Mol Cell* **38:** 768–774.

Zeniou M, Ding T, Trivier E, Hanauer A. 2002. Expression analysis of RSK gene family members: The RSK2 gene, mutated in Coffin-Lowry syndrome, is prominently expressed in brain structures essential for cognitive function and learning. *Hum Mol Genet* **11:** 2929–2940.

Zetterberg A, Larsson O, Wiman KG. 1995. What is the restriction point? *Curr Opin Cell Biol* **7:** 835–842.

Zhang L, Smit-McBride Z, Pan X, Rheinhardt J, Hershey JW. 2008. An oncogenic role for the phosphorylated h-subunit of human translation initiation factor eIF3. *J Biol Chem* **283:** 24047–24060.

Zhao J, Yuan X, Frodin M, Grummt I. 2003. ERK-dependent phosphorylation of the transcription initiation factor TIF-IA is required for RNA polymerase I transcription and cell growth. *Mol Cell* **11:** 405–413.

Zhao Y, Xiong X, Sun Y. 2011. DEPTOR, an mTOR inhibitor, is a physiological substrate of SCF(betaTrCP) E3 ubiquitin ligase and regulates survival and autophagy. *Mol Cell* **44:** 304–316.

Zhou M, Sandercock AM, Fraser CS, Ridlova G, Stephens E, Schenauer MR, Yokoi-Fong T, Barsky D, Leary JA, Hershey JW, et al. 2008. Mass spectrometry reveals modularity and a complete subunit interaction map of the eukaryotic translation factor eIF3. *Proc Natl Acad Sci* **105:** 18139–18144.

Zinzalla V, Stracka D, Oppliger W, Hall MN. 2011. Activation of mTORC2 by association with the ribosome. *Cell* **144:** 757–768.

Zoncu R, Bar-Peled L, Efeyan A, Wang S, Sancak Y, Sabatini DM. 2011a. mTORC1 senses lysosomal amino acids through an inside-out mechanism that requires the vacuolar H-ATPase. *Science* **334:** 678–683.

Zoncu R, Efeyan A, Sabatini DM. 2011b. mTOR: From growth signal integration to cancer, diabetes and ageing. *Nat Rev Mol Cell Biol* **12:** 21–35.

Protein Secretion and the Endoplasmic Reticulum

Adam M. Benham

School of Biological and Biomedical Sciences, Durham University, Durham DH1 3LE,
United Kingdom

Correspondence: adam.benham@durham.ac.uk

In a complex multicellular organism, different cell types engage in specialist functions, and as a result, the secretory output of cells and tissues varies widely. Whereas some quiescent cell types secrete minor amounts of proteins, tissues like the pancreas, producing insulin and other hormones, and mature B cells, producing antibodies, place a great demand on their endoplasmic reticulum (ER). Our understanding of how protein secretion in general is controlled in the ER is now quite sophisticated. However, there remain gaps in our knowledge, particularly when applying insight gained from model systems to the more complex situations found in vivo. This article describes recent advances in our understanding of the ER and its role in preparing proteins for secretion, with an emphasis on glycoprotein quality control and pathways of disulfide bond formation.

How a cell controls its complex output and ensures that only properly folded and functional proteins reach their correct destination is a question of considerable importance in biology. It has been calculated that the products of 11% of the approximately 25,000 predicted human full-length open reading frames (ORFs) are soluble secretory proteins, with a further 20% being single-pass or multi-pass transmembrane proteins that get targeted to the ER (Kanapin et al. 2003). Many of these proteins can also be alternatively spliced, adding a further layer of complexity (Carninci et al. 2005). To ensure orderly ER exit, protein secretion from the ER is determined by several factors. These include targeting of the nascent protein to the ER by the ribosome, the translocation of the protein into the ER, the provision of sugars for glycoproteins, the folding and quality control of the protein, the availability of cofactors (particularly calcium), and the appropriate formation of disulfide bonds (Braakman and Bulleid 2011). Some proteins require assembly into complexes, and provision of lipids for ER membranes and the regulation of cholesterol content are also important considerations. Here, the focus is on recent advances in our understanding of the glycosylation machinery, the provision of disulfide bonds, and the ER exit strategies used by proteins to ensure their efficient secretion from the ER (Fig. 1).

TARGETING OF PROTEINS TO THE ER

In eukaryotic cells, most secreted proteins are cotranslationally translocated from ribosomes into the rough ER. However, some proteins,

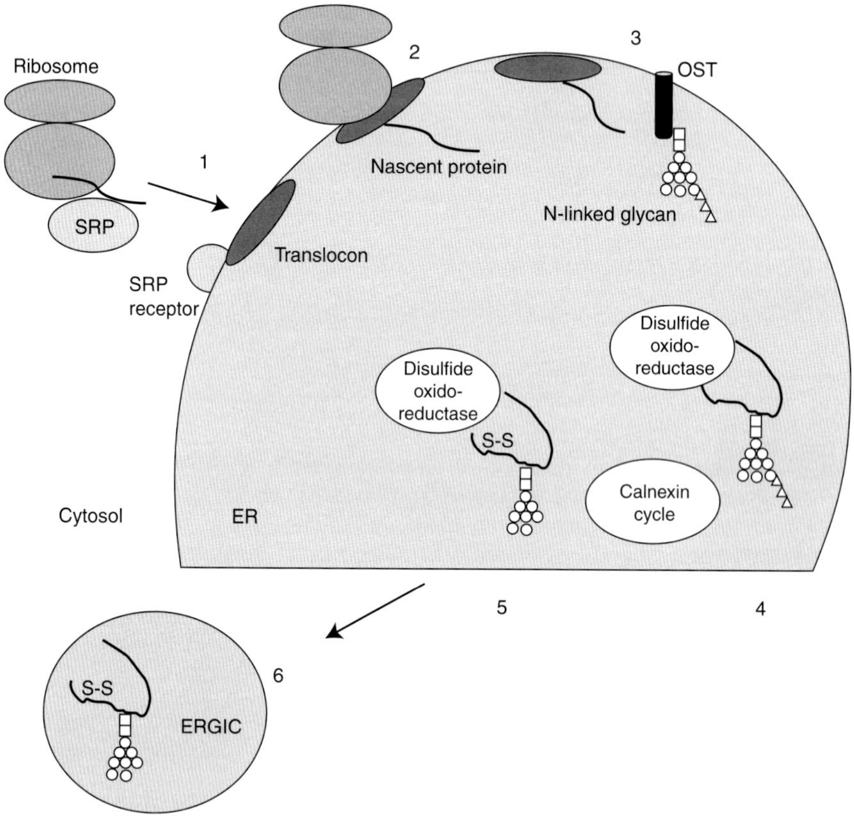

Figure 1. Overview of protein targeting to the ER and glycoprotein quality control. Six key points in the fidelity of protein secretion are illustrated. (1) Correct targeting of the glycoprotein to the ER as it emerges from the translocon. This is mediated by the signal recognition particle (SRP) and its receptor, which helps position the emerging protein at the translocon. (2) Translocation of the (glyco)protein into the ER by the translocon. (3) Addition of N-glycans to Asn residues of glycoproteins by oligosaccharyl transferase (OST). (4) Correct folding and quality control by the calnexin cycle. (5) Introduction/correction of disulfide bonds by protein disulfide isomerases/oxidoreductases. (6) Directing glycoproteins into ER exit sites followed by packaging into appropriate compartments, for example, the ER Golgi intermediate compartment (ERGIC). Note that protein folding and disulfide bond formation happen rapidly after the nascent protein emerges from the translocon. (S–S) Disulfide bond; (□) N-Acetyl glucosamine; (○) mannose; (△) glucose.

particularly small polypeptides or proteins with "weak" signal sequences, can access the ER post-translationally (Ng et al. 1996; Shao and Hegde 2011). The cotranslational targeting of secreted proteins to the ER is achieved by an amino-terminal signal sequence of 16–30 amino acids. Signal sequences are quite variable, but generally have 6 to 12 hydrophobic amino acids flanked by one or more positively charged residues. As protein synthesis begins, the signal sequence is recognized by a signal recognition

particle (SRP) (Saraogi and Shan 2011). SRP is composed of six proteins (P9/P14, P68/P72, P19, and P54) and a 300-nucleotide RNA scaffold. The P54 subunit of SRP is primarily responsible for binding to the signal sequence. Translation is temporarily halted until SRP54 interacts with the α-subunit of an SRP receptor that is embedded in the ER membrane through its β-subunit. Elongation arrest was first noted in one of Walter and Blobel's pioneering studies, in which they analyzed secretory protein

Cite this article as *Cold Spring Harb Perspect Biol* doi: 10.1101/cshperspect.a012872

synthesis using microsomal membranes in vitro (Walter and Blobel 1981). Further analysis and binding studies showed that the SRP9 and SRP14 proteins bind to the *Alu* domain (a repetitive element) of the SRP RNA as a heterodimer and are required for elongation arrest (Siegel and Walter 1986; Strub et al. 1991). Particles reconstituted with a carboxy-terminal truncation of SRP14 are unable to arrest properly, despite having normal targeting and ribosome-binding capabilities (Thomas et al. 1997; Mason et al. 2000). The observation that changes in SRP RNA tertiary structure occurred upon binding to truncated SRP9/14 led to the proposal that RNA structural features were involved in mediating translation arrest (Thomas et al. 1997). A positively charged patch of predominantly lysine residues was shown subsequently to be essential for mediating the interaction between SRP14 and the SRP RNA (Mary et al. 2010). An impressive cryo-electron microscopy study of a mammalian SRP bound to a ribosome with a signal sequence has revealed the arrangement of these components in an arrested state. In this structure, the *Alu* domain of the SRP RNA is positioned in the elongation-factor-binding site to bring about translational arrest by direct competition (Halic et al. 2004).

Why translational arrest should be needed during secretory protein synthesis has been food for thought for many years. In 2008, Strub's group showed that the amount of SRP receptor at the ER membrane was a major limiting factor in determining translation rates (Lakkaraju et al. 2008). Thus translational pausing is necessary because it helps to keep nascent proteins translocation competent, until an SRP receptor becomes available for docking, hence preventing secretory-pathway proteins from being synthesized in the wrong compartment. Further studies will be needed to determine exactly how nascent chain length influences targeting, to explain the molecular sequence of events that lead to resumption of translation upon SRP receptor engagement, and to explain how translationally paused transcripts avoid mRNA degradation by RNA quality-control mechanisms present in the cytosol (Becker et al. 2011).

After the ribosomal/nascent protein/SRP complex has docked to the SRP receptor, GTP hydrolysis releases SRP for another round of nascent polypeptide recruitment. Structural details of the interaction between the eukaryotic SRP, the SRP receptor, and the signal sequence are still awaited, but several structures of bacterial homologs involved in the translocation of polypeptides across their plasma membrane have been solved. For example, the heterodimeric structure of the GTPase domains of Ffh (a bacterial SRP54 homolog) and FtsY (a bacterial SRP receptor α-subunit homolog) is stabilized by the binding of two GTP molecules. This structure reveals how allosteric activation of the two proteins is coordinated (Focia et al. 2004). More recently, a full-length Ffh:FtsY complex in the pre-GTP hydrolysis state (cocrystallized with the GTP analog GMPPCP) has been solved to 3.9 Å (Ataide et al. 2011). This work suggests that a large-scale repositioning event occurs to hand over the signal sequence from the M (methionine-rich) domain of SRP54 (Ffh) to the translocon. It has proved difficult to obtain structures of SRP54 in combination with a signal sequence, but an elegant approach using an Ffh-signal sequence fusion protein has allowed a 3.5 Å crystal structure to be determined from dimers of the fusion protein (Janda et al. 2010). The structure is consistent with previous biochemical experiments, showing that the signal peptide interacts with the M domain of Ffh and that hydrophobic residues in the core are necessary for binding. Further studies using a range of different signal peptides should help explain how Ffh, and ultimately SRP54, can maintain selectivity while tolerating diversity in signal peptide length and sequence.

The docking of SRP to the SRP receptor positions the 60S subunit of the ribosome above the translocon. The translocon is a proteinaceous channel that translocates the nascent polypeptide chain into the ER lumen, and in mammals the translocon is composed of three principal subunits—Sec61α, β, and γ. The translocon is gated on the luminal side by the ER chaperone BiP (grp78), which helps to preserve the barrier function of the ER membrane (Sanders et al. 1992; Alder et al. 2005).

The translocon is also associated with signal peptidase, which removes signal peptides post-translationally from newly synthesized proteins, as the protein enters the ER lumen (Jackson and Blobel 1977). Several additional proteins assist the translocon, including translocating-chain-associated protein (TRAM), translocon-associating protein (TRAP), and RAMP4 (ribosome-associated membrane protein). You are referred to recent reviews for more detail regarding the function and regulation of the translocon (Osborne et al. 2005; Mandon et al. 2009; Park and Rapoport 2012).

THE PROVISION OF GLYCANS

A limiting step in the secretion of glycoproteins is the addition of sugars (Fig. 2). For N-linked glycosylation, this involves building a branching chain of glucose (Glc) and mannose (Man) sugars onto a "stem" of N-acetylglucosamine (NAc). The process of assembling the sugar donor starts in the cytosol rather than the ER. Dolichol (an isoprenylated hydrocarbon with a reactive alcohol group) acts as a carrier lipid that receives two GlcNAc moieties from UDP-GlcNAc and five mannose residues from GDP-mannose. The first GlcNAc residue is linked to dolichol by a high-energy pyrophosphate that can ultimately be transferred to the asparagine side chain of the accepting glycoprotein in the ER. Once the initial synthesis of the dolichol phosphate backbone has been achieved in the cytosol, the entire lipid intermediate, Man(5)-GlcNAc(2)-PP-dolichol, must be translocated from the cytosolic face to the luminal face of the ER membrane. The identity of the Man(5)-GlcNAc(2)-PP-dolichol "flippase" is still debated. Rft1, an essential ER membrane protein in yeast, was proposed to be the responsible Man(5)GlcNAc(2)-PP-dolichol translocator in 2002 (Helenius et al. 2002). The investigators showed that yeast with a defective RFT1 gene hypoglycosylated the model protein carboxypeptidase Y and accumulated Man(5)GlcNAc(2) precursors, whereas O-mannosylation of chitinase was unaffected. The Rft1 protein fits the bill as a candidate "flippase" because it resides in the ER, spanning the membrane multiple times,

and is conserved in eukaryotes that synthesize N-linked sugars. A key role for Rft1 in glycoprotein synthesis is supported by clinical studies, namely, that mutations in the human RFT1 gene lead to diseases of N-glycosylation. Patients present with severe developmental defects, have abnormal glycoprotein profiles on iso-electric focusing gels, and accumulate dolichol-linked Man(5)GlcNAc(2) precursors, with low levels of dolichol-linked Glc(3)Man(9)GlcNAc(2) (Haeuptle et al. 2008). However, the assertion that Rft1 is a stand-alone translocator has been questioned by Mennon and colleagues, who were unable to detect "flippase" activity of the protein when reconstituted vesicles were subjected to in vitro flipping experiments (Frank et al. 2008). Sanyal and Menon (2009) reconstituted "flippase" activity in proteoliposomes and established a tritiated substrate assay to show that the process was ATP dependent and required a protein that sediments at 4S, with high specificity for Man(5)GlcNAc(2)-PP-dolichol. However, it remains possible that the conditions used to prepare the membranes for these in vitro experiments lead to loss of activity of Rft1 or other important translocator components.

Once in the ER, the oligosaccharide donor Glc(3)Man(9)GlcNAc(2)-PP-dolichol is synthesized from Man(5)GlcNAc(2)-PP-dolichol by the sequential addition of four mannose and three glucose residues from dolichol-P-mannose and dolichol-P-glucose. How dolichol-P-glucose and dolichol-P-mannose are flipped from the cytosolic face to the ER side of the membrane is also not known. The protein responsible for dolichol-P-mannose transfer can be distinguished biochemically from the "flippase" that translocates Man(5)GlcNAc(2)-PP-dolichol (Sanyal and Menon 2010). Because phospholipid synthesis occurs in the cytoplasmic leaflet of the ER and "flip-flop" of phospholipids is required to equilibrate the bilayer, members of the phospholipid scramblase ("flippase") family might be potential candidates for this activity. A combination of in vitro studies, genetics, cell biology, and the development of robust reporters and substrates is likely to be needed to unambiguously identify the

Cite this article as *Cold Spring Harb Perspect Biol* doi: 10.1101/cshperspect.a012872

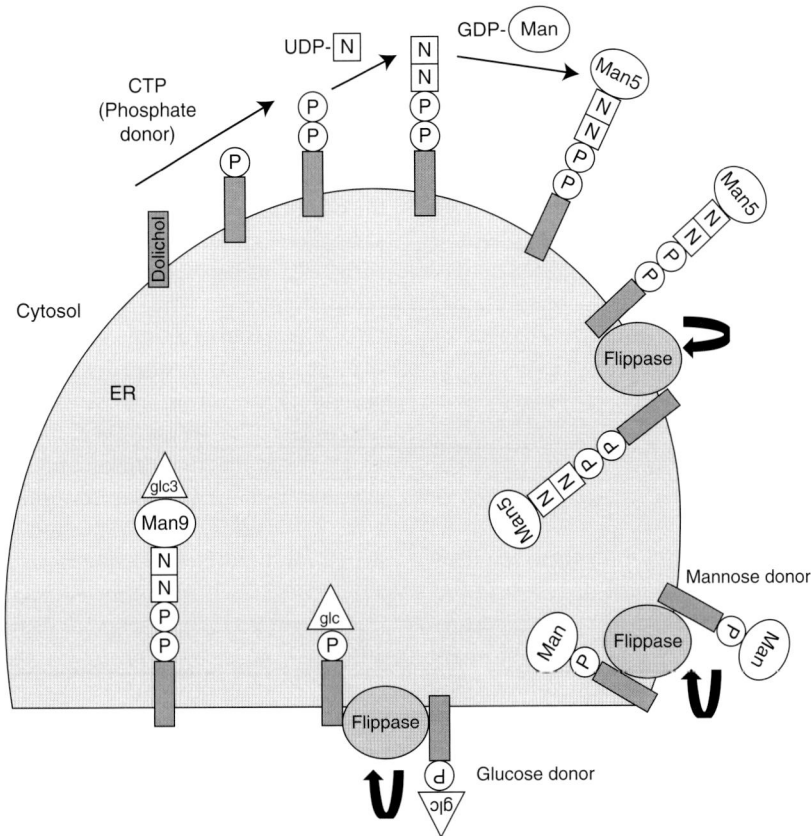

Figure 2. The *N*-linked glycosylation pathway. *N*-linked glycosylation begins in the cytosol. Phosphate (P) from CTP is used to charge dolichol in the lipid bilayer. Dolichol phosphate receives two *N*-acetyl glucosamine moieties (N) from UDP-*N*-acetyl glucosamine. Five GDP-mannose (Man) residues are added sequentially before the sugar donor is translocated across the ER membrane by the "flippase." In the ER lumen, four additional mannose and three glucose residues are added to yield a dolichol-linked Glc(3)Man(9)GlcNAc(2) structure that is transferred to an Asn residue of a nascent glycoprotein by OST. The identity of the "flippases" that translocate Man(5)GlcNAc(2)-P-P-dolichol, Man-P-dolichol, and Glc-P-dolichol are debated (see text for details). (□) *N*-Acetyl glucosamine; (○) mannose; (△) glucose.

sugar-precursor "flippases" involved in the *N*-glycosylation pathway.

Oligosaccharide transferase (OST) is a multi-subunit protein responsible for transferring the dolichol-linked oligosaccharide to an asparagine residue on an acceptor protein as it enters the ER through the translocon (Dempski and Imperiali 2002). In yeast, five OST subunits are essential: Nlt1p/Ost1p (Ribophorin I in mammals), Ost2p (DAD1), Stt3p, Swp1p (Ribophorin II), and Wbp1p (OST48). A major step forward in understanding OST function has

come with the arrival of an X-ray structure of a bacterial periplasmic oligosaccharyl transferase called PglB (Lizak et al. 2011). The active STT3 subunit of *Campylobacter lari* PglB has two large amino-terminal loops that reach into the periplasm and contain residues required for peptide binding and catalysis. The peptide-binding cleft is formed by both a membrane-binding and a periplasmic domain and sits opposite the catalytic site and the lipid donor-binding site. This arrangement enables the enzyme to accommodate both the lipid and the acceptor protein. The specificity of OST for serine or threonine is explained by the positioning of a Trp-Trp-Asp motif next to the Asn-X-Ser/Thr acceptor sequence. Higher-resolution structures, coupled with further examples of client and product cocrystals, will undoubtedly shed more light on the OST catalytic cycle (Gilmore 2011). It will also be important to decipher the molecular details of how and why the OST complex assembles with different accessory proteins, for example, in response to changes in secretory demand, nutrient availability (Kelleher et al. 2003), and client delivery (Wilson et al. 2008).

THE CALNEXIN/CALRETICULIN CYCLE

One of the landmark developments in the field of glycoprotein folding was the discovery of the calnexin cycle, in which two of the three terminal glucose residues of N-linked glycoproteins are sequentially removed by ER glucosidases (Parodi 2000). Glucose trimming allows monoglucosylated oligosaccharides to bind to the ER lectin-like chaperones, calnexin (membrane bound) and calreticulin (soluble). In collaboration with ERp57, a protein disulfide isomerase (PDI) (Zapun et al. 1998), the glycoprotein client is subjected to repeated cycles of folding until a native structure is reached. Glucose trimming by glucosidases competes with glucose addition by glucosyl transferase (GT; also known as UDP-glucose:glycoprotein glucosyltransferase 1) until the terminal glucose is removed and the protein is properly folded. In this state, the client no longer has affinity for calnexin/calreticulin or GT (Zapun et al. 1997; Ritter and Helenius 2000; Trombetta and Helenius 2000). Glucosidase II activity is a key regulatory step for entry into the calnexin cycle, because the location and number of glycans on a chain can influence its activity. Trimming of the second glucose by glucosidase II only occurs efficiently when a second glycan is present, at least for model glycopeptides and model substrates such as pancreatic RNase (Deprez et al. 2005).

Various studies have been performed to address the mechanism by which GT identifies misfolded proteins for exchange with the calnexin cycle, using RNase (Ritter et al. 2005), exo-(1,3)-β-glucanase (Taylor et al. 2004), chymotrypsin inhibitor-2 (Caramelo et al. 2003), and MHC class I molecules (Wearsch et al. 2011; Zhang et al. 2011b) as substrates. Overall, the closer a glycan is to an unstructured region of its protein, the better it is glucosylated by GT, with GT recognizing hydrophobic solvent-exposed patches. When RNase was tested, GT only reglucosylated the glycans when they were at or close to disordered regions of the protein (Ritter et al. 2005). However, in the case of exo-(1,3)-β-glucanase, an F280S point mutant that retained enzymatic activity and was only locally misfolded was glucosylated at a distant residue, 40 Å away from the site of structural disruption (Taylor et al. 2004). The differences observed between RNase and exo-(1,3)-β-glucanase may reflect substrate-specific differences in recognition by GT. This is supported by the observation that deletion of GT has no effect on the interaction of VSV G-protein with calnexin, but does compromise the folding of other proteins such as BACE501 and influenza HA by calnexin (Solda et al. 2007). More structural and biochemical studies using an expanded range of clients should help to clarify exactly how GT recognizes misfolded proteins. It should be noted that many ER proteins do not just require the calnexin cycle to fold properly and are assisted by various other chaperones, including hsp70s (e.g., BiP/grp78) and hsp90s (e.g., grp94/gp96) resident in the ER (Marzec et al. 2012).

Misfolded proteins that cannot be rescued by the calnexin cycle are removed from the ER and degraded in the cytosol by the proteasome in a pathway known as ER-associated degradation (ERAD). EDEM1, a mannosidase conserved in yeast (Clerc et al. 2009), is a key player in this process. EDEM1 singles out misfolded glycoproteins and works together with BiP and ERdJ5, a PDI family protein with a BiP-binding J-domain. The way in which EDEM1 recognizes misfolded glycoproteins has been subject to debate, but recent structural and biochemical evidence suggests that calnexin can directly hand over at least some substrates (e.g., misfolded α1 antitrypsin) to the ERdJ5/EDEM1 platform for reductive unfolding before removal of the spent client from the ER (Hagiwara et al. 2011). There are three different EDEM variants in mammals, and their exact contribution to glycoprotein quality control remains to be defined. Interestingly, the residues necessary for glycolytic activity are conserved, and it remains possible that one or more EDEM variants retain some mannosidase activity (Olivari and Molinari 2007).

BEYOND THE CALNEXIN CYCLE

Although the calnexin cycle is now well understood, some aspects of calnexin/calreticulin function still have to be clarified. The first crystal structure of calnexin was published in 2001 at 2.9 Å resolution and revealed a globular lectin domain and a P-domain that extended as a 140 Å arm (Schrag et al. 2001). However, further structural and biochemical work was required to reveal how binding of the lectin domain to mono-glucosylated glycans and binding of the arm domain to ERp57 could be reconciled with affinity for a range of client protein surfaces. A recent calreticulin structure has revealed a putative hydrophobic binding domain on the arm and shown that calreticulin can effectively prevent protein aggregation in the ER (Pocanschi et al. 2011). In addition, a high-resolution 1.95 Å structure of calreticulin in complex with a tetrasaccharide ligand has shown the molecular specificity of the lectin domain for glucose. The O_2 oxygen of glucose hydrogen bonds with the Lys111 side chain and the backbone of Gly-124 in a pocket that does not permit the entry of mannose (Kozlov et al. 2010b). Structural studies from the same group also support a role for the peptidyl-prolyl $cis-trans$ isomerase (PPIase) cyclophilin B in the calnexin cycle (Kozlov et al. 2010a). PPIases are required for proline isomerization, which is often considered to be a rate-limiting step in protein folding. PPIases have been implicated in folding collagen (Steinmann et al. 1991) and have been found associated with multimeric chaperone complexes in the ER (e.g., Meunier et al. 2002). In the clinic, mutations in cyclophilin B result in osteogenesis imperfecta (a cartilage

disease), by limiting type I procollagen chain association (Pyott et al. 2011). However, detailed analysis of the ER resident PPIase family and their function in ER quality control is lacking. The Kozlov study suggests that cyclophilin B occupies the same site in the P-domain as ERp57 (Kozlov et al. 2010a). Further work will be required to evaluate the specificity of cyclophilin B as a PPIase for folding mono-glucosylated glycoproteins, and to determine whether the P-domain of calreticulin itself, rich in prolines, requires cyclophilin B for its own quality control.

Both calnexin and calreticulin bind calcium and regulate its availability in the ER. Mice deficient in calnexin have particular problems with their nervous system, leading to dysmyelination (Kraus et al. 2010), whereas mice lacking calreticulin die during embryogenesis because of defective heart development (Mesaeli et al. 1999). The phenotypes of the knockout mice illustrate the physiological importance of these lectin-like ER chaperones (particularly calreticulin) in maintaining calcium homeostasis as well as being chaperones for glycoproteins (Michalak et al. 2009). The interaction between calnexin and its client is not always entirely sugar dependent, and calnexin may recognize different protein conformations, both in vitro and within the cell (Ihara et al. 1999; Danilczyk and Williams 2001; Brockmeier and Williams 2006). There are also two testis-specific homologs of calnexin and calreticulin, called calmegin and calsperin, respectively (Ikawa et al. 1997, 2011). Both proteins are required for male fertility in the mouse and are necessary for the quality control of specific ADAM proteins that mediate binding of sperm to the egg. The molecular basis for the restricted activity of calmegin and calsperin is not known, although calsperin has a divergent P-domain and lacks broad lectin activity (Ikawa et al. 2011).

The influence of O-linked sugars in determining a protein's capacity for secretion is often overlooked, because this modification occurs after the primary protein-folding events have taken place in the ER and the serine/threonine acceptor sites are hard to predict. Nevertheless, O-linked sugars may have a regulatory role to play for some proteins. Notch is one example. The Notch proteins are important signaling molecules that can control cell fate during development in several tissues. O-Fucosylation of the EGF repeats in the Notch extracellular domain by protein O-fucosyltransferase 1 (Pofut) occurs in the ER and is essential for Notch function. Subsequent modification of O-fucose in the Golgi by β1-3N-acetylglucosaminyltransferase (also known as Fringe, from mutational studies in *Drosophila*) can alter the activity of Notch; mutations in these transferase enzymes can cause skeletal malformations and other diseases of development (for a recent review, see Rana and Haltiwanger 2011). O-Glucosylation of Notch also occurs at EGF repeats, and the gene responsible has been traced to a protein O-glucosyltransferase (Poglut) called Rumi in *Drosophila* (Acar et al. 2008). Defects in Poglut/Rumi lead to loss of Notch function and Notch misfolding phenotypes in flies and mice. O-Glycosylation can influence protein quality control in more subtle ways. For example, proprotein convertases like furin can be O-glycosylated close to the Arg-X-X-Arg processing site, and this can inhibit their activity. It has been proposed that the large family of GalNAc transferases responsible for the initial steps of O-glycosylation may therefore indirectly regulate proprotein processing, and hence the release of some secreted proteins (Gram Schjoldager et al. 2011). Regulation of quality control may also extend beyond the ER. For example, the mannosidase ERMan1 is almost exclusively O-glycosylated in the Golgi apparatus. ERman1, like EDEM, can remove (hydrolyze) terminal mannose from glycoproteins and may therefore direct misfolded glycoproteins into the ERAD system. Further studies will be required to determine the full extent to which the Golgi apparatus contributes to "ER" quality control (Pan et al. 2011).

To fully understand how glycosylation limits protein secretion, better quantitative measurements of how N-linked, O-linked, and other types of glycans contribute to the folding process are required. Methods are now available to determine cellular dolichol phosphate levels, dolichol-linked oligosaccharides, and the gly-

cosylation of individual Asn-X-Ser/Thr sites (Hulsmeier et al. 2011). Using liquid chromatography–mass spectrometry (LC-MS/MS), it has been shown that congenital diseases of glycosylation have diminished N-glycosylation site occupancy that correlates with disease severity (Hulsmeier et al. 2007), and this approach could be used to determine the variability of site occupancy between proteins in different cell types.

MAKING DISULFIDE BONDS

Most secreted proteins require disulfide bonds for their correct structure and function (Fig. 3). Disulfide bond formation is supported in the oxidizing environment of the ER, which has a relatively low ratio of reduced to oxidized glutathione (GSH:GSSG between 1:1 and 3:1) (Hwang et al. 1992). The formation of a disulfide bond between two cysteine residues relies on catalysts of oxidative protein folding to enable the protein to reach its native state. In the late 1990s, it was established in *Saccharomyces cerevisiae* that a disulfide bond relay exists between PDI and endoplasmic reticulum oxidoreductase (Ero1p in yeast, Ero1α and Ero1β in mammals) (Frand and Kaiser 1998; Pollard et al. 1998; Sevier and Kaiser 2008). During the course of disulfide bond formation, electrons pass in the opposite direction, from substrate to PDI to Ero, and are delivered to oxygen via FAD. In *S. cerevisiae*, Ero1p may use alternative flavin electron acceptors, such as FMN, under anaerobic conditions (Gross et al. 2006). Crystal structures of Ero1p and human Ero1α have been solved (for review, see Araki and Inaba 2012) and reveal how the Cys-X-X-Cys active site within Ero1 communicates with FAD and transfers disulfides internally to an amino-terminal redox active site, for disulfide transfer to (and electron acceptance from) PDI. Ero–PDI cocrystals are awaited to see exactly how Ero1 transfers its disulfide to PDI; they should provide more detail of how electron exchange is achieved. Eros are under tight regulatory control to prevent hyperoxidation of the ER and to avoid the generation of potentially damaging reactive oxygen species. The mechanisms differ somewhat between Ero1p, Ero1α, and Ero1β,

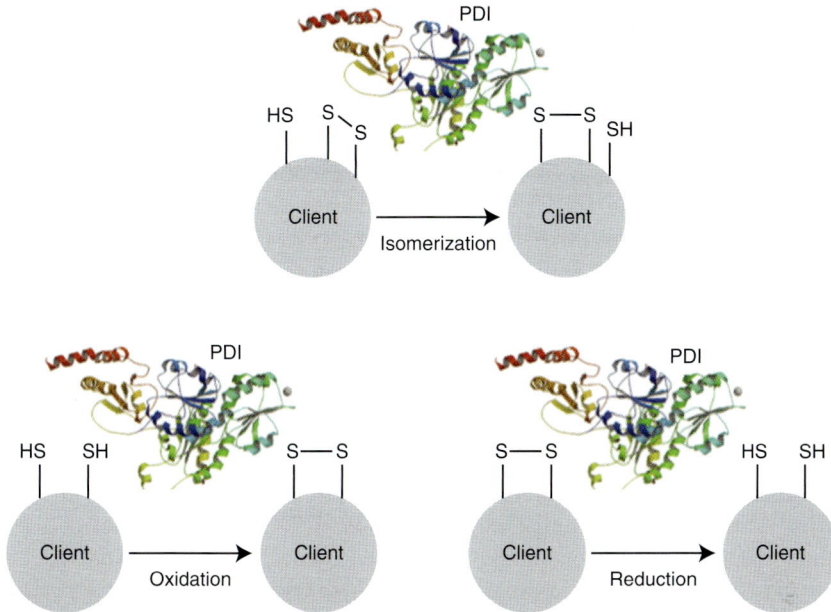

Figure 3. Disulfide bond formation in the ER. Disulfide bonds form between the −SH groups of two cysteine residues in a protein. The process is mostly confined to the ER in eukaryotes and is catalyzed by enzymes. The most abundant catalyst of disulfide bond formation is PDI, which can introduce (oxidize), remove (reduce), or swap (isomerize) disulfide bonds in a range of client proteins.

but each protein uses regulatory (noncatalytic) disulfides with a low reduction potential (for review, see Bulleid and Ellgaard 2011).

In yeast, functional Ero1p is essential for viability, but the situation is more complicated in higher organisms. A clue to potential differences between disulfide bond formation in the ER of yeast and higher eukaryotes came with the phenotype of flies that lacked the *Ero1L* gene. *Drosophila melanogaster Ero1L* mutants have a particular defect in the Notch pathway, with misfolded Notch accumulating in the ER (Tien et al. 2008). This phenotype was not expected but may relate to the requirement for complex disulfide bonds in the Lin12−Notch repeats in the extracellular domain of the protein. Mice deficient in Ero1α, Ero1β, or both Ero1α and Ero1β are also viable; the absence of Ero1β results in a mild diabetic phenotype (Zito et al. 2010a), and the absence of Ero1α results in heart abnormalities (Chin et al. 2011). The diabetic phenotype of the Ero1β-deficient mice is likely explained by inefficient production of in-

sulin, which is consistent with the high levels of Ero1β found in the pancreas (Dias-Gunasekara et al. 2005) and the regulation of Ero1β during insulin flux by the pancreatic transcription factor PDX1 (Khoo et al. 2011). How and why Ero1β preferentially controls the oxidative folding of insulin remain open questions. The Ero1α-deficient mice have defective cardiomyocyte calcium signaling, although they were protected to some extent against experimentally induced heart failure (Chin et al. 2011). This study, along with other reports (Wang et al. 2009; Anelli et al. 2012), suggests that Ero1α is important in linking calcium signaling to disulfide bond formation. Taken together with the phenotypes of mice lacking calnexin and calreticulin mentioned above, this highlights the close relationship between calcium homeostasis and protein folding in the ER. This area is likely to attract future interest, particularly with respect to disease, where misfolding and calcium have been linked with conditions as diverse as inflammation (Peters and Raghavan 2011), prion disorders (Torres et al. 2010), and familial hypercholesterolemia (Pena et al. 2010).

The relative well-being of Ero1-deficient mice can be partly explained by compensatory disulfide bond provision from peroxiredoxin IV (Zito et al. 2010b), which is ER localized but can also be secreted. The peroxiredoxin IV enzyme can salvage hydrogen peroxide to drive the oxidation of reduced PDI (Tavender et al. 2010). The source of peroxide for peroxiredoxin IV is likely to come from mitochondrial respiration, NADPH oxidase activity, and Ero1. Although it remains to be seen how peroxiredoxin IV is regulated, this pathway provides a neat solution to the problem of generating potentially harmful oxidants during the course of oxidative protein folding. Nevertheless, additional alternative routes for disulfide bond formation must exist, because peroxiredoxin IV–deficient mice are also viable. Male mice lacking peroxiredoxin IV are fertile, but they show elevated spermatogenic cell death (Iuchi et al. 2009). This suggests that peroxiredoxin IV may have a particular role in protection from oxidative stress during spermatogenesis. In this light, it may be informative to examine whether peroxiredoxin-deficient mice show any age-related phenotypes in other tissues as exposure to reactive oxygen species accumulates.

The difference between yeast and higher eukaryotes in their reliance on Ero proteins probably reflects the sheer range and diversity of proteins that need to be secreted by multicellular species. There are at least 21 human PDIs compared with five in *S. cerevisiae*, providing potential routes for alternative modes of disulfide bond formation and isomerization (Benham 2012). In addition, yeast Pdi1p is not functionally equivalent to mammalian PDI (Hatahet and Ruddock 2009). In this light, the role of glutathione in supporting disulfide bond formation is being revisited (Appenzeller-Herzog 2011) along with potential roles for other electron carriers (Bánhegyi et al. 2011). Oxidized glutathione (GSSG) was once considered to be the major source of disulfide bond equivalents in the ER, but a substantial fraction of glutathione in the ER is bound to protein as mixed disulfides (Bass et al. 2004). These glutathione–

protein mixed disulfides could be clients for PDI isomerase activity and therefore might influence the rate of oxidative protein folding. Although we remain unsure of the true concentration of GSSG in the ER, recent in vitro studies suggest that GSSG could still play a role in the catalytic cycle of PDI at physiologically relevant GSSG levels (Lappi and Ruddock 2011). Advances in our understanding of disulfide bond formation have led to the discovery of alternative pathways. The challenge now is to understand the physiological conditions under which these pathways service different clients.

HOW DOES ER EXIT REGULATE PROTEIN SECRETION?

Once a protein has folded and acquired native disulfide bonds, it is competent for ER exit (Fig. 4). The default pathway for ER exit to the Golgi is via COPII-coated vesicles (for review, see Dancourt and Barlowe 2010). The assembly of COPII-coated vesicles is best understood in yeast, where the guanine nucleotide exchange factor (GEF) Sec12p in the donor ER membrane binds inactive, soluble Sar1p-GDP, exchanging its GTP for GDP. The Sar1p-GTP exposes a fatty acid chain that inserts into the membrane and assembles Sec23–Sec24 subunits on the cytoplasmic face of the ER membrane. After recruiting ER cargo, the pre-budding complexes associate with an outer Sec13–Sec31 complex, inducing budding in the ER membrane and formation of the vesicle, which is usually 60–90 nm in diameter. COPII-mediated transport becomes problematic for very large substrates such as collagen molecules, which grow in the ER to diameters of 300–400 nm. This huge topological problem is dealt with by TANGO1 and cTAGE5, two proteins that are required for collagen VII secretion, dimerize at ER exit sites, and bind to Sec23–Sec24 (Malhotra and Erlmann 2011). TANGO1-null mice die at birth and are unable to secrete collagens, lending support to the notion that this protein is largely dedicated to the final steps of collagen quality control (Wilson et al. 2011).

COPII-coated vesicles subsequently fuse to form the ER Golgi intermediate compartment

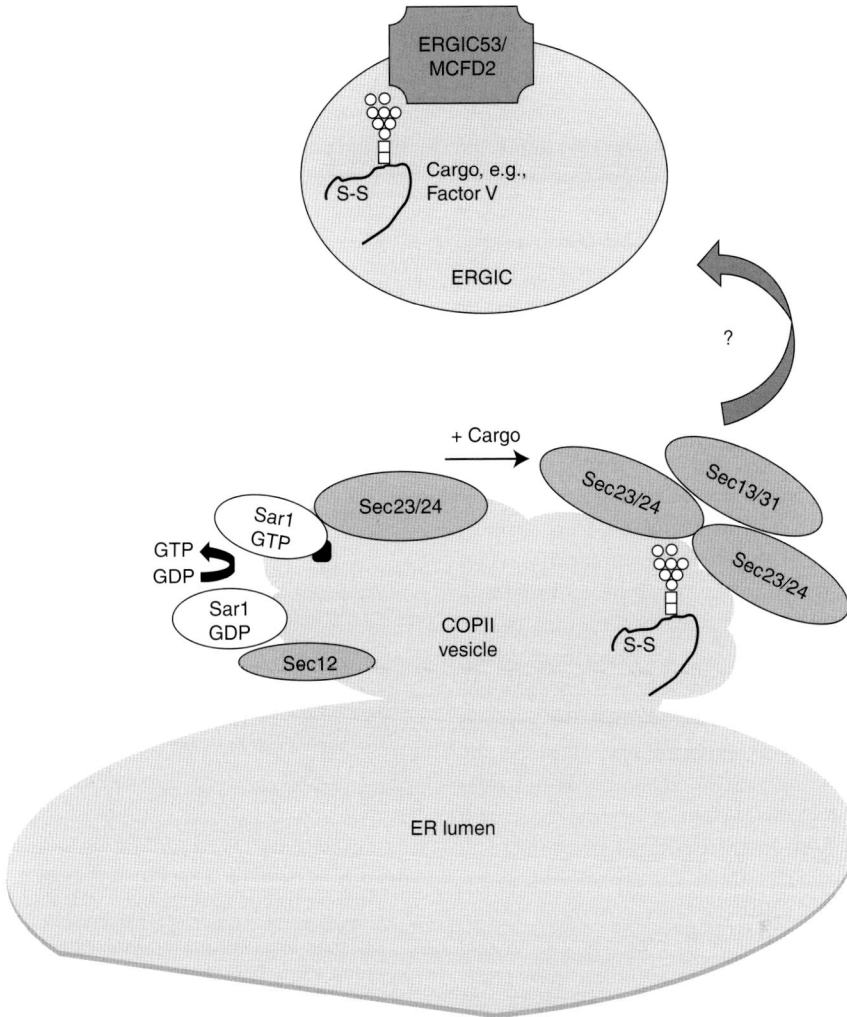

Figure 4. ER exit, COPII vesicles, and ERGIC. COPII exit vesicles form when Sec12p recruits Sar1p-GDP and exchanges GDP for GTP, enabling Sar1p to insert into the budding membrane. Sar1p facilitates the assembly of Sec23/24 and Sec13/31 at the membrane upon recruitment of cargo. The relationship between COPII vesicles and the ER Golgi intermediate compartment (ERGIC) is not comprehensively defined in higher eukaryotes, and some cargoes may escape the ER directly by bulk flow. ERGIC53 and MCFD2 are required to recruit at least some cargoes to the ERGIC (e.g., Factor V and VIII). (□) N-Acetyl glucosamine; (○) mannose.

(ERGIC). Although the mechanisms governing ERGIC function are not fully understood, the role of the type I transmembrane protein ER-GIC53 in mammals has become clear through studying bleeding disorders. ERGIC53 is a high mannose-binding lectin that cycles between the ER and the ERGIC. ERGIC53 works together with MCFD2, a soluble calcium-dependent protein, to capture its cargo dynamically (Nishio et al. 2010). Various inactivating mutations in the ERGIC53 gene (LMAN1) and in the MCDF2 gene lead to a multiple coagulation factor deficiency called MCFD1 (Nichols et al. 1998) because of a failure to package the heavily glycosylated Factor V and Factor VIII proteins (Zhang et al. 2003). $Lman1^{-/-}$ mice largely replicate human MCFD1 (although there is an unexplained strain-dependent partial lethal phenotype). Vesicles generated from $Lman1^{-/-}$ mice can support COPII-coated vesicle formation (Zhang et al. 2011a), emphasizing that COPII-mediated transport does not depend on ERGIC53's packaging function. On one hand, the expression of ERGIC53 is ubiquitous, suggesting that it may carry other cargoes. Indeed, it has been proposed that ERGIC53 is also a cargo receptor for α1-antitrypsin in the liver (Nyfeler et al. 2008) and IgM made by antibody-secreting B cells (Anelli et al. 2007). On the other hand, MCFD1 patients do not fail to secrete all of their glycoproteins, suggesting that other cargo receptors may exist. Alternatively, it is possible that ER-to-Golgi transport is only rate-limiting for a few select cargoes such as Factors V and VIII. This would lend support to "bulk flow" theories, which posit that secretion of abundant proteins can be achieved by a default pathway that does not require export signals (for review, see Klumperman 2000). The rate of protein secretion by bulk flow has been estimated in Chinese Hamster Ovary cells using pulse-chase experiments (Thor et al. 2009). The nonglycosylated Semliki Forest virus capsid protein was chosen as a model, because it was unlikely to interact with ER retention factors and cargo receptors. The capsid protein was secreted in a COPII-dependent manner within 15 min of synthesis, with a rate of 155 COPII vesicles per second. Taken together, these studies indicate that cargo receptor–mediated export and secretion by bulk flow are not mutually exclusive and may both coexist, depending on the cell type and its secretory load (Fig. 4).

CONCLUDING REMARKS

The last few years have seen major advances in our understanding of the details of quality control in the ER. Some major questions remain to be addressed, such as the identity of the "flippases" for sugar precursors in the ER membrane, the mechanisms for handing over folded glycoproteins to ER exit sites, and understanding what limits disulfide bond formation in different physiological contexts. A major challenge in the field has been quantifying the efficiency of protein secretion. On a sobering note, we are still some way from being able to engineer a cell line to optimally secrete a recombinant protein of choice for the production of a therapeutic. Improved computational strategies and modeling approaches are needed to better understand protein secretion networks, and their key "pinch points," in a variety of secretory cell types.

ACKNOWLEDGMENTS

I thank the BBSRC, Wellcome Trust, Arthritis Research UK, MRC, Leverhulme Trust, and Royal Society for funding projects in my laboratory.

REFERENCES

Acar M, Jafar-Nejad H, Takeuchi H, Rajan A, Ibrani D, Rana NA, Pan H, Haltiwanger RS, Bellen HJ. 2008. Rumi is a CAP10 domain glycosyltransferase that modifies Notch and is required for Notch signaling. *Cell* **132:** 247–258.

Alder NN, Shen Y, Brodsky JL, Hendershot LM, Johnson AE. 2005. The molecular mechanisms underlying BiP-mediated gating of the Sec61 translocon of the endoplasmic reticulum. *J Cell Biol* **168:** 389–399.

Anelli T, Ceppi S, Bergamelli L, Cortini M, Masciarelli S, Valetti C, Sitia R. 2007. Sequential steps and checkpoints in the early exocytic compartment during secretory IgM biogenesis. *EMBO J* **26:** 4177–4188.

Anelli T, Bergamelli L, Margittai E, Rimessi A, Fagioli C, Malgaroli A, Pinton P, Ripamonti M, Rizzuto R, Sitia R. 2012. Ero1α regulates Ca^{2+} fluxes at the endoplasmic reticulum-mitochondria interface (MAM). *Antioxid Redox Signal* **16:** 1077–1087.

Appenzeller-Herzog C. 2011. Glutathione- and non-glutathione-based oxidant control in the endoplasmic reticulum. *J Cell Sci* **124:** 847–855.

Araki K, Inaba K. 2012. Structure, mechanism and evolution of Ero1 family enzymes. *Antioxid Redox Signal* **16:** 790–799.

Ataide SF, Schmitz N, Shen K, Ke A, Shan SO, Doudna JA, Ban N. 2011. The crystal structure of the signal recognition particle in complex with its receptor. *Science* **331:** 881–886.

Bánhegyi G, Margittai E, Szarka A, Mandl J, Csala M. 2011. Crosstalk and barriers between the electron carriers of the endoplasmic reticulum. *Antioxid Redox Signal* **16:** 772–780.

Bass R, Ruddock LW, Klappa P, Freedman RB. 2004. A major fraction of endoplasmic reticulum-located glutathione is present as mixed disulfides with protein. *J Biol Chem* **279:** 5257–5262.

Becker T, Armache JP, Jarasch A, Anger AM, Villa E, Sieber H, Motaal BA, Mielke T, Berninghausen O, Beckmann R. 2011. Structure of the no-go mRNA decay complex Dom34–Hbs1 bound to a stalled 80S ribosome. *Nat Struct Mol Biol* **18:** 715–720.

Cite this article as *Cold Spring Harb Perspect Biol* doi: 10.1101/cshperspect.a012872

Benham AM. 2012. The protein disulfide isomerase family: Key players in health and disease. *Antioxid Redox Signal* **16:** 781–789.

Braakman I, Bulleid NJ. 2011. Protein folding and modification in the mammalian endoplasmic reticulum. *Annu Rev Biochem* **80:** 71–99.

Brockmeier A, Williams DB. 2006. Potent lectin-independent chaperone function of calnexin under conditions prevalent within the lumen of the endoplasmic reticulum. *Biochemistry* **45:** 12906–12916.

Bulleid NJ, Ellgaard L. 2011. Multiple ways to make disulfides. *Trends Biochem Sci* **36:** 485–492.

Caramelo JJ, Castro OA, Alonso LG, De Prat-Gay G, Parodi AJ. 2003. UDP-Glc:glycoprotein glucosyltransferase recognizes structured and solvent accessible hydrophobic patches in molten globule-like folding intermediates. *Proc Natl Acad Sci* **100:** 86–91.

Carninci P, Kasukawa T, Katayama S, Gough J, Frith MC, Maeda N, Oyama R, Ravasi T, Lenhard B, Wells C, et al. 2005. The transcriptional landscape of the mammalian genome. *Science* **309:** 1559–1563.

Chin KT, Kang G, Qu J, Gardner LB, Coetzee WA, Zito E, Fishman GI, Ron D. 2011. The sarcoplasmic reticulum luminal thiol oxidase ERO1 regulates cardiomyocyte excitation-coupled calcium release and response to hemodynamic load. *FASEB J* **25:** 2583–2591.

Clerc S, Hirsch C, Oggier DM, Deprez P, Jakob C, Sommer T, Aebi M. 2009. Htm1 protein generates the *N*-glycan signal for glycoprotein degradation in the endoplasmic reticulum. *J Cell Biol* **184:** 159–172.

Dancourt J, Barlowe C. 2010. Protein sorting receptors in the early secretory pathway. *Annu Rev Biochem* **79:** 777–802.

Danilczyk UG, Williams DB. 2001. The lectin chaperone calnexin utilizes polypeptide-based interactions to associate with many of its substrates in vivo. *J Biol Chem* **276:** 25532–25540.

Dempski RE Jr, Imperiali B. 2002. Oligosaccharyl transferase: Gatekeeper to the secretory pathway. *Curr Opin Chem Biol* **6:** 844–850.

Deprez P, Gautschi M, Helenius A. 2005. More than one glycan is needed for ER glucosidase II to allow entry of glycoproteins into the calnexin/calreticulin cycle. *Mol Cell* **19:** 183–195.

Dias-Gunasekara S, Gubbens J, van Lith M, Dunne C, Williams JA, Kataky R, Scoones D, Lapthorn A, Bulleid NJ, Benham AM. 2005. Tissue-specific expression and dimerization of the endoplasmic reticulum oxidoreductase Ero1β. *J Biol Chem* **280:** 33066–33075.

Focia PJ, Shepotinovskaya IV, Seidler JA, Freymann DM. 2004. Heterodimeric GTPase core of the SRP targeting complex. *Science* **303:** 373–377.

Frand AR, Kaiser CA. 1998. The ERO1 gene of yeast is required for oxidation of protein dithiols in the endoplasmic reticulum. *Mol Cell* **1:** 161–170.

Frank CG, Sanyal S, Rush JS, Waechter CJ, Menon AK. 2008. Does Rft1 flip an *N*-glycan lipid precursor? *Nature* **454:** E3–E5.

Gilmore R. 2011. Structural biology: Porthole to catalysis. *Nature* **474:** 292–293.

Gram Schjoldager KT, Vester-Christensen MB, Goth CK, Petersen TN, Brunak S, Bennett EP, Levery SB, Clausen H. 2011. A systematic study of site-specific GalNAc-type O-glycosylation modulating proprotein convertase processing. *J Biol Chem* **286:** 40122–40132.

Gross E, Sevier CS, Heldman N, Vitu E, Bentzur M, Kaiser CA, Thorpe C, Fass D. 2006. Generating disulfides enzymatically: Reaction products and electron acceptors of the endoplasmic reticulum thiol oxidase Ero1p. *Proc Natl Acad Sci* **103:** 299–304.

Haeuptle MA, Pujol FM, Neupert C, Winchester B, Kastaniotis AJ, Aebi M, Hennet T. 2008. Human RFT1 deficiency leads to a disorder of *N*-linked glycosylation. *Am J Hum Genet* **82:** 600–606.

Hagiwara M, Maegawa K, Suzuki M, Ushioda R, Araki K, Matsumoto Y, Hoseki J, Nagata K, Inaba K. 2011. Structural basis of an ERAD pathway mediated by the ER-resident protein disulfide reductase ERdj5. *Mol Cell* **41:** 432–444.

Halic M, Becker T, Pool MR, Spahn CM, Grassucci RA, Frank J, Beckmann R. 2004. Structure of the signal recognition particle interacting with the elongation-arrested ribosome. *Nature* **427:** 808–814.

Hatahet F, Ruddock LW. 2009. Protein disulfide isomerase: A critical evaluation of its function in disulfide bond formation. *Antioxid Redox Signal* **11:** 2807–2850.

Helenius J, Ng DT, Marolda CL, Walter P, Valvano MA, Aebi M. 2002. Translocation of lipid-linked oligosaccharides across the ER membrane requires Rft1 protein. *Nature* **415:** 447–450.

Hulsmeier AJ, Paesold-Burda P, Hennet T. 2007. *N*-Glycosylation site occupancy in serum glycoproteins using multiple reaction monitoring liquid chromatography-mass spectrometry. *Mol Cell Proteomics* **6:** 2132–2138.

Hulsmeier AJ, Welti M, Hennet T. 2011. Glycoprotein maturation and the UPR. *Methods Enzymol* **491:** 163–182.

Hwang C, Sinskey AJ, Lodish HF. 1992. Oxidized redox state of glutathione in the endoplasmic reticulum. *Science* **257:** 1496–1502.

Ihara Y, Cohen-Doyle MF, Saito Y, Williams DB. 1999. Calnexin discriminates between protein conformational states and functions as a molecular chaperone in vitro. *Mol Cell* **4:** 331–341.

Ikawa M, Wada I, Kominami K, Watanabe D, Toshimori K, Nishimune Y, Okabe M. 1997. The putative chaperone calmegin is required for sperm fertility. *Nature* **387:** 607–611.

Ikawa M, Tokuhiro K, Yamaguchi R, Benham AM, Tamura T, Wada I, Satouh Y, Inoue N, Okabe M. 2011. Calsperin is a testis-specific chaperone required for sperm fertility. *J Biol Chem* **286:** 5639–5646.

Iuchi Y, Okada F, Tsunoda S, Kibe N, Shirasawa N, Ikawa M, Okabe M, Ikeda Y, Fujii J. 2009. Peroxiredoxin 4 knockout results in elevated spermatogenic cell death via oxidative stress. *Biochem J* **419:** 149–158.

Jackson RC, Blobel G. 1977. Post-translational cleavage of presecretory proteins with an extract of rough microsomes from dog pancreas containing signal peptidase activity. *Proc Natl Acad Sci* **74:** 5598–5602.

Janda CY, Li J, Oubridge C, Hernandez H, Robinson CV, Nagai K. 2010. Recognition of a signal peptide by the signal recognition particle. *Nature* **465:** 507–510.

Kanapin A, Batalov S, Davis MJ, Gough J, Grimmond S, Kawaji H, Magrane M, Matsuda H, Schonbach C, Teasdale RD, et al. 2003. Mouse proteome analysis. *Genome Res* **13:** 1335–1344.

Kelleher DJ, Karaoglu D, Mandon EC, Gilmore R. 2003. Oligosaccharyltransferase isoforms that contain different catalytic STT3 subunits have distinct enzymatic properties. *Mol Cell* **12:** 101–111.

Khoo C, Yang J, Rajpal G, Wang Y, Liu J, Arvan P, Stoffers DA. 2011. Endoplasmic reticulum oxidoreductin-1-like β (ERO1lβ) regulates susceptibility to endoplasmic reticulum stress and is induced by insulin flux in β-cells. *Endocrinology* **152:** 2599–2608.

Klumperman J. 2000. Transport between ER and Golgi. *Curr Opin Cell Biol* **12:** 445–449.

Kozlov G, Bastos-Aristizabal S, Maattanen P, Rosenauer A, Zheng F, Killikelly A, Trempe JF, Thomas DY, Gehring K. 2010a. Structural basis of cyclophilin B binding by the calnexin/calreticulin P-domain. *J Biol Chem* **285:** 35551–35557.

Kozlov G, Pocanschi CL, Rosenauer A, Bastos-Aristizabal S, Gorelik A, Williams DB, Gehring K. 2010b. Structural basis of carbohydrate recognition by calreticulin. *J Biol Chem* **285:** 38612–38620.

Kraus A, Groenendyk J, Bedard K, Baldwin TA, Krause KH, Dubois-Dauphin M, Dyck J, Rosenbaum EE, Korngut L, Colley NJ, et al. 2010. Calnexin deficiency leads to dysmyelination. *J Biol Chem* **285:** 18928–18938.

Lakkaraju AK, Mary C, Scherrer A, Johnson AE, Strub K. 2008. SRP keeps polypeptides translocation-competent by slowing translation to match limiting ER-targeting sites. *Cell* **133:** 440–451.

Lappi AK, Ruddock LW. 2011. Reexamination of the role of interplay between glutathione and protein disulfide isomerase. *J Mol Biol* **409:** 238–249.

Lizak C, Gerber S, Numao S, Aebi M, Locher KP. 2011. X-ray structure of a bacterial oligosaccharyltransferase. *Nature* **474:** 350–355.

Malhotra V, Erlmann P. 2011. Protein export at the ER: Loading big collagens into COPII carriers. *EMBO J* **30:** 3475–3480.

Mandon EC, Trueman SF, Gilmore R. 2009. Translocation of proteins through the Sec61 and SecYEG channels. *Curr Opin Cell Biol* **21:** 501–507.

Mary C, Scherrer A, Huck L, Lakkaraju AK, Thomas Y, Johnson AE, Strub K. 2010. Residues in SRP9/14 essential for elongation arrest activity of the signal recognition particle define a positively charged functional domain on one side of the protein. *RNA* **16:** 969–979.

Marzec M, Eletto D, Argon Y. 2012. GRP94: An HSP90-like protein specialized for protein folding and quality control in the endoplasmic reticulum. *Biochim Biophys Acta* **1823:** 774–787.

Mason N, Ciufo LF, Brown JD. 2000. Elongation arrest is a physiologically important function of signal recognition particle. *EMBO J* **19:** 4164–4174.

Mesaeli N, Nakamura K, Zvaritch E, Dickie P, Dziak E, Krause KH, Opas M, MacLennan DH, Michalak M.

1999. Calreticulin is essential for cardiac development. *J Cell Biol* **144:** 857–868.

Meunier L, Usherwood YK, Chung KT, Hendershot LM. 2002. A subset of chaperones and folding enzymes form multiprotein complexes in endoplasmic reticulum to bind nascent proteins. *Mol Biol Cell* **13:** 4456–4469.

Michalak M, Groenendyk J, Szabo E, Gold LI, Opas M. 2009. Calreticulin, a multi-process calcium-buffering chaperone of the endoplasmic reticulum. *Biochem J* **417:** 651–666.

Ng DT, Brown JD, Walter P. 1996. Signal sequences specify the targeting route to the endoplasmic reticulum membrane. *J Cell Biol* **134:** 269–278.

Nichols WC, Seligsohn U, Zivelin A, Terry VH, Hertel CE, Wheatley MA, Moussalli MJ, Hauri HP, Ciavarella N, Kaufman RJ, et al. 1998. Mutations in the ER-Golgi intermediate compartment protein ERGIC-53 cause combined deficiency of coagulation factors V and VIII. *Cell* **93:** 61–70.

Nishio M, Kamiya Y, Mizushima T, Wakatsuki S, Sasakawa H, Yamamoto K, Uchiyama S, Noda M, McKay AR, Fukui K, et al. 2010. Structural basis for the cooperative interplay between the two causative gene products of combined factor V and factor VIII deficiency. *Proc Natl Acad Sci* **107:** 4034–4039.

Nyfeler B, Reiterer V, Wendeler MW, Stefan E, Zhang B, Michnick SW, Hauri HP. 2008. Identification of ERGIC-53 as an intracellular transport receptor of α1-antitrypsin. *J Cell Biol* **180:** 705–712.

Olivari S, Molinari M. 2007. Glycoprotein folding and the role of EDEM1, EDEM2 and EDEM3 in degradation of folding-defective glycoproteins. *FEBS Lett* **581:** 3658–3664.

Osborne AR, Rapoport TA, van den Berg B. 2005. Protein translocation by the Sec61/SecY channel. *Annu Rev Cell Dev Biol* **21:** 529–550.

Pan S, Wang S, Utama B, Huang L, Blok N, Estes MK, Moremen KW, Sifers RN. 2011. Golgi localization of ERManI defines spatial separation of the mammalian glycoprotein quality control system. *Mol Biol Cell* **22:** 2810–2822.

Park E, Rapoport TA. 2012. Mechanisms of Sec61/SecY-mediated protein translocation across membranes. *Annu Rev Biophys* doi: 10.1146/annurev-biophys-050511-102312.

Parodi AJ. 2000. Protein glucosylation and its role in protein folding. *Annu Rev Biochem* **69:** 69–93.

Pena F, Jansens A, van Zadelhoff G, Braakman I. 2010. Calcium as a crucial cofactor for low density lipoprotein receptor folding in the endoplasmic reticulum. *J Biol Chem* **285:** 8656–8664.

Peters LR, Raghavan M. 2011. Endoplasmic reticulum calcium depletion impacts chaperone secretion, innate immunity, and phagocytic uptake of cells. *J Immunol* **187:** 919–931.

Pocanschi CL, Kozlov G, Brockmeier U, Brockmeier A, Williams DB, Gehring K. 2011. Structural and functional relationships between the lectin and arm domains of calreticulin. *J Biol Chem* **286:** 27266–27277.

Pollard MG, Travers KJ, Weissman JS. 1998. Ero1p: A novel and ubiquitous protein with an essential role in oxidative protein folding in the endoplasmic reticulum. *Mol Cell* **1:** 171–182.

Pyott SM, Schwarze U, Christiansen HE, Pepin MG, Leistritz DF, Dineen R, Harris C, Burton BK, Angle B, Kim K, et al. 2011. Mutations in PPIB (cyclophilin B) delay type I procollagen chain association and result in perinatal lethal to moderate osteogenesis imperfecta phenotypes. *Hum Mol Genet* **20:** 1595–1609.

Rana NA, Haltiwanger RS. 2011. Fringe benefits: Functional and structural impacts of *O*-glycosylation on the extracellular domain of Notch receptors. *Curr Opin Struct Biol* **21:** 583–589.

Ritter C, Helenius A. 2000. Recognition of local glycoprotein misfolding by the ER folding sensor UDP-glucose:-glycoprotein glucosyltransferase. *Nat Struct Biol* **7:** 278–280.

Ritter C, Quirin K, Kowarik M, Helenius A. 2005. Minor folding defects trigger local modification of glycoproteins by the ER folding sensor GT. *EMBO J* **24:** 1730–1738.

Sanders SL, Whitfield KM, Vogel JP, Rose MD, Schekman RW. 1992. Sec61p and BiP directly facilitate polypeptide translocation into the ER. *Cell* **69:** 353–365.

Sanyal S, Menon AK. 2009. Specific transbilayer translocation of dolichol-linked oligosaccharides by an endoplasmic reticulum flippase. *Proc Natl Acad Sci* **106:** 767–772.

Sanyal S, Menon AK. 2010. Stereoselective transbilayer translocation of mannosyl phosphoryl dolichol by an endoplasmic reticulum flippase. *Proc Natl Acad Sci* **107:** 11289–11294.

Saraogi I, Shan SO. 2011. Molecular mechanism of co-translational protein targeting by the signal recognition particle. *Traffic* **12:** 535–542.

Schrag JD, Bergeron JJ, Li Y, Borisova S, Hahn M, Thomas DY, Cygler M. 2001. The structure of calnexin, an ER chaperone involved in quality control of protein folding. *Mol Cell* **8:** 633–644.

Sevier CS, Kaiser CA. 2008. Ero1 and redox homeostasis in the endoplasmic reticulum. *Biochim Biophys Acta* **1783:** 549–556.

Shao S, Hegde RS. 2011. A calmodulin-dependent translocation pathway for small secretory proteins. *Cell* **147:** 1576–1588.

Siegel V, Walter P. 1986. Removal of the *Alu* structural domain from signal recognition particle leaves its protein translocation activity intact. *Nature* **320:** 81–84.

Solda T, Galli C, Kaufman RJ, Molinari M. 2007. Substrate-specific requirements for UGT1-dependent release from calnexin. *Mol Cell* **27:** 238–249.

Steinmann B, Bruckner P, Superti-Furga A. 1991. Cyclosporin A slows collagen triple-helix formation in vivo: Indirect evidence for a physiologic role of peptidyl-prolyl *cis*–*trans*-isomerase. *J Biol Chem* **266:** 1299–1303.

Strub K, Moss J, Walter P. 1991. Binding sites of the 9- and 14-kilodalton heterodimeric protein subunit of the signal recognition particle (SRP) are contained exclusively in the *Alu* domain of SRP RNA and contain a sequence motif that is conserved in evolution. *Mol Cell Biol* **11:** 3949–3959.

Tavender TJ, Springate JJ, Bulleid NJ. 2010. Recycling of peroxiredoxin IV provides a novel pathway for disulphide formation in the endoplasmic reticulum. *EMBO J* **29:** 4185–4197.

Taylor SC, Ferguson AD, Bergeron JJ, Thomas DY. 2004. The ER protein folding sensor UDP-glucose glycoprotein-glucosyltransferase modifies substrates distant to local changes in glycoprotein conformation. *Nat Struct Mol Biol* **11:** 128–134.

Thomas Y, Bui N, Strub K. 1997. A truncation in the 14 kDa protein of the signal recognition particle leads to tertiary structure changes in the RNA and abolishes the elongation arrest activity of the particle. *Nucleic Acids Res* **25:** 1920–1929.

Thor F, Gautschi M, Geiger R, Helenius A. 2009. Bulk flow revisited: Transport of a soluble protein in the secretory pathway. *Traffic* **10:** 1819–1830.

Tien AC, Rajan A, Schulze KL, Ryoo HD, Acar M, Steller H, Bellen HJ. 2008. Ero1L, a thiol oxidase, is required for Notch signaling through cysteine bridge formation of the Lin12–Notch repeats in *Drosophila melanogaster*. *J Cell Biol* **182:** 1113–1125.

Torres M, Castillo K, Armisen R, Stutzin A, Soto C, Hetz C. 2010. Prion protein misfolding affects calcium homeostasis and sensitizes cells to endoplasmic reticulum stress. *PLoS ONE* **5:** e15658.

Trombetta ES, Helenius A. 2000. Conformational requirements for glycoprotein reglucosylation in the endoplasmic reticulum. *J Cell Biol* **148:** 1123–1129.

Walter P, Blobel G. 1981. Translocation of proteins across the endoplasmic reticulum III. Signal recognition protein (SRP) causes signal sequence-dependent and site-specific arrest of chain elongation that is released by microsomal membranes. *J Cell Biol* **91:** 557–561.

Wang L, Li SJ, Sidhu A, Zhu L, Liang Y, Freedman RB, Wang CC. 2009. Reconstitution of human Ero1-Lα/protein-disulfide isomerase oxidative folding pathway in vitro. Position-dependent differences in role between the α and α′ domains of protein-disulfide isomerase. *J Biol Chem* **284:** 199–206.

Wearsch PA, Peaper DR, Cresswell P. 2011. Essential glycan-dependent interactions optimize MHC class I peptide loading. *Proc Natl Acad Sci* **108:** 4950–4955.

Wilson CM, Roebuck Q, High S. 2008. Ribophorin I regulates substrate delivery to the oligosaccharyltransferase core. *Proc Natl Acad Sci* **105:** 9534–9539.

Wilson DG, Phamluong K, Li L, Sun M, Cao TC, Liu PS, Modrusan Z, Sandoval WN, Rangell L, Carano RA, et al. 2011. Global defects in collagen secretion in a Mia3/TANGO1 knockout mouse. *J Cell Biol* **193:** 935–951.

Zapun A, Petrescu SM, Rudd PM, Dwek RA, Thomas DY, Bergeron JJ. 1997. Conformation-independent binding of monoglucosylated ribonuclease B to calnexin. *Cell* **88:** 29–38.

Zapun A, Darby NJ, Tessier DC, Michalak M, Bergeron JJ, Thomas DY. 1998. Enhanced catalysis of ribonuclease B folding by the interaction of calnexin or calreticulin with ERp57. *J Biol Chem* **273:** 6009–6012.

Zhang B, Cunningham MA, Nichols WC, Bernat JA, Seligsohn U, Pipe SW, McVey JH, Schulte-Overberg U, de Bosch NB, Ruiz-Saez A, et al. 2003. Bleeding due to disruption of a cargo-specific ER-to-Golgi transport complex. *Nat Genet* **34:** 220–225.

Zhang B, Zheng C, Zhu M, Tao J, Vasievich MP, Baines A, Kim J, Schekman R, Kaufman RJ, Ginsburg D. 2011a.

Mice deficient in LMAN1 exhibit FV and FVIII deficiencies and liver accumulation of α1-antitrypsin. *Blood* **118:** 3384–3391.

Zhang W, Wearsch PA, Zhu Y, Leonhardt RM, Cresswell P. 2011b. A role for UDP-glucose glycoprotein glucosyltransferase in expression and quality control of MHC class I molecules. *Proc Natl Acad Sci* **108:** 4956–4961.

Zito E, Chin KT, Blais J, Harding HP, Ron D. 2010a. ERO1-β, a pancreas-specific disulfide oxidase, promotes insulin biogenesis and glucose homeostasis. *J Cell Biol* **188:** 821–832.

Zito E, Melo EP, Yang Y, Wahlander A, Neubert TA, Ron D. 2010b. Oxidative protein folding by an endoplasmic reticulum-localized peroxiredoxin. *Mol Cell* **40:** 787–797.

Cite this article as *Cold Spring Harb Perspect Biol* doi: 10.1101/cshperspect.a012872

New Insights into Translational Regulation in the Endoplasmic Reticulum Unfolded Protein Response

Graham D. Pavitt[1] and David Ron[2]

[1]Faculty of Life Sciences, University of Manchester, Oxford Road, Manchester M13 9PT, United Kingdom

[2]Metabolic Research Laboratories, University of Cambridge, Addenbrookes Hospital, Hills Road, Cambridge CB2 0QQ, United Kingdom

Correspondence: graham.pavitt@manchester.ac.uk and dr360@medschl.cam.ac.uk

Homeostasis of the protein-folding environment in the endoplasmic reticulum (ER) is maintained by signal transduction pathways that collectively constitute an unfolded protein response (UPR). These affect bulk protein synthesis and thereby the levels of ER stress, but also culminate in regulated expression of specific mRNAs, such as that encoding the transcription factor ATF4. Mechanisms linking eukaryotic initiation factor 2 (eIF2) phosphorylation to control of unfolded protein load in the ER were elucidated more than 10 years ago, but recent work has highlighted the diversity of processes that impinge on eIF2 activity and revealed that there are multiple mechanisms by which changes in eIF2 activity can modulate the translation of individual mRNAs. In addition, the potential for affecting this step of translation initiation pharmacologically is becoming clearer. Furthermore, it is now clear that another strand of the UPR, controlled by the endoribonuclease inositol-requiring enzyme 1 (IRE1), also affects rates of protein synthesis in stressed cells and that its effector function, mediated by the transcription factor X-box-binding protein 1 (XBP1), is subject to important mRNA-specific translational regulation. These new insights into the convergence of translational control and the UPR will be reviewed here.

In eukaryotes the early steps in biosynthesis of secreted proteins take place in the lumen of the endoplasmic reticulum (ER). Proteins are translocated cotranslationally in an unfolded state into the organelle, where a host of chaperones and enzymes assist in posttranslational modification and folding of the nascent chain of these ER client proteins. The flux of client proteins through the ER is subject to physiological regulation and can vary by up to an order of magnitude over a short period of time (Itoh and Okamoto 1980). These brisk fluctuations in the load of ER client proteins, which are all the more conspicuous in the specialized secretory cells of complex animals, has favored the evolution of homeostatic signaling pathways that couple the capacity of the ER to the load facing it. These pathways constitute an unfolded protein response (UPR) whose molecular details were first worked out in yeast (Ron and Walter 2007). The yeast UPR is a transcriptional response, by which cells up-regulate expression of ER chaperones in response to the threat of protein misfolding in the ER lumen. In higher eukaryotes the transcriptional response is complemented by feedback from the lumen of the ER to the translational apparatus to attenuate the flux of client proteins into the organelle (Fig. 1).

The eukaryotic initiation factor 2α (eIF2α) kinase PERK (RNA-dependent protein kinase–like ER kinase) is the key mediator of the translation arm of the UPR, and eIF2α phosphorylation has emerged as an evolutionarily

conserved node for regulating bulk protein synthesis and for affecting mRNA-specific changes in rates of translation. The molecular mechanisms affecting eIF2α phosphorylation's role in regulating the initiation step of mRNA translation have been well characterized and were subject to extensive review (Ron and Harding 2007). Here we will focus on new developments in this area. The discovery of a role for eIF5 as a antagonist of guanine nucleotide exchange by the eIF2B complex (Jennings and Pavitt 2010), working alongside eIF2α phosphorylation to restrain the re-cycling of eIF2 and thereby rates of translation initiation, will be highlighted. The development of genomic tools for profiling the disposition of ribosomes on mRNAs in vivo has provided new insight into how eIF2α phosphorylation regulates the rate of translation

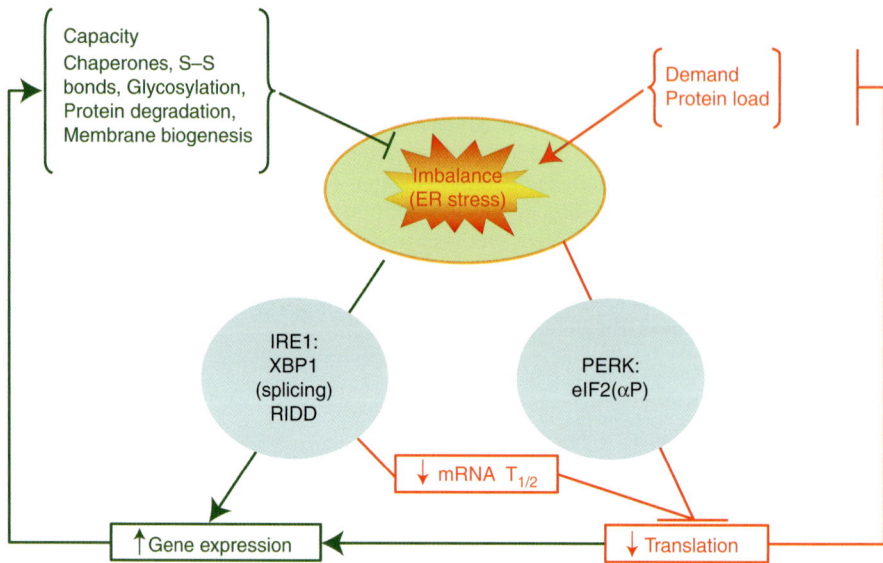

Figure 1. Overview of the unfolded protein response. Fluctuations in the unfolded protein load placed on the endoplasmic reticulum can give rise to transient imbalance between capacity and demand, which we refer to as ER stress. This elicits two major adaptations: attenuation of new protein synthesis (mediated by PERK and eIF2α phosphorylation), which rapidly decreases the load on the organelle; and induction of a gene expression program (effected by IRE1-mediated XBP1 splicing) that up-regulates the organelle's capacity to deal with incoming proteins. Regulated translation is also coupled to changes in gene expression, as indicated by the horizontal arrow connecting the two outputs. Destabilization of mRNAs encoding secreted proteins by regulated IRE1-dependent degradation (RIDD) contributes to the relief of ER load. Because these adaptations can be elicited experimentally by manipulations that perturb protein folding in the ER, they came to be known as the ER unfolded protein response (UPR). For simplicity, the third mediator of the UPR, ATF6, is omitted from this diagram.

initiation (Ingolia et al. 2009). The significance of these developments to our understanding of the physiological importance of eIF2α phosphorylation will be covered in detail.

Regulating the rate of global protein synthesis and with it the level of ER stress in cells has proven to be an important influence on cellular life/death decisions. The lessons provided from knockout of the eIF2(αP)-directed phosphatases will be reviewed. As pharmacological control of eIF2(αP) dephosphorylation has emerged as a potentially important way to bias life/death decisions; the implications of research in that area too will be highlighted.

The ancient and conserved transcriptional pathway of the UPR is mediated by the transmembrane ER protein inositol-requiring enzyme 1 (IRE1), which is activated in response to unfolded protein stress in the ER (Cox et al. 1993; Mori et al. 1993). Activated IRE1 is a sequence-specific endoribonuclease that participates in an unconventional cytoplasmic mRNA splicing event that culminates in the expression of a transcription factor, Hac1p in yeast and X-box-binding protein 1 (XBP1) in animals (Cox and Walter 1996; Yoshida et al. 2001; Calfon et al. 2002). Although the basic mechanisms underlying this unusual signal transduction pathway were worked out more than 10 years ago, recent studies have brought to light novel aspects of translational control that set up the XBP1 mRNA for IRE1-mediated splicing in mammalian cells. This remarkable example of mRNA-specific translational regulation in eukaryotes will be reviewed here in detail.

Recent work has also called attention to an alternative mode for IRE1 activity with relaxed specificity for substrates. This activity, which is apparently unique to animal cell IRE1, results in promiscuous degradation of membrane-bound mRNA and affects the repertoire of proteins synthesized in ER-stressed cells (Hollien and Weissman 2006; Hollien et al. 2009). This recently characterized activity of IRE1 has an interesting relationship to that of RNase L, a component of the innate immune response that evolved from the ancestral *IRE1* gene. Here we will review the significance of these newly discovered relationships.

eIF5 IS CRITICAL FOR THE eIF2(αP)-MEDIATED TRANSLATIONAL CONTROL RESPONSE

eIF2 is a G protein that cycles between an active form binding GTP and tRNA$_i^{Met}$ (called the ternary complex or TC) and an inactive GDP-bound state. It is this cycling that is interrupted by the PERK-mediated phosphorylation. The nucleotide status of eIF2 is controlled by two factors: eIF5 and eIF2B. It is well established that eIF2B is the guanine nucleotide exchange factor (GEF) that stimulates release of GDP (Rowlands et al. 1988; Nika et al. 2000), allowing GTP and tRNA$_i^{Met}$ to bind eIF2, and that eIF5 is the GTPase-accelerating factor that stimulates GTP hydrolysis (Das et al. 2001; Paulin et al. 2001) at or before AUG codon recognition. However, it was suggested recently that (at least in yeast) eIF2·GDP/eIF5 are released from initiating ribosomes together following GTP hydrolysis, because an abundant complex containing these factors was identified (Singh et al. 2006). It was then shown that eIF5 has a second function as a GDP-dissociation inhibitor (GDI) that prevents premature dissociation of GDP from eIF2 (Jennings and Pavitt 2010), thus providing an additional step in the translation initiation process, one that antagonizes eIF2B-mediated nucleotide exchange.

eIF2B GEF activity is inhibited by eIF2 phosphorylation (Pavitt et al. 1998). eIF2B forms a more stable complex with eIF2(αP), but it is suggested that binding is altered such that nucleotide exchange is not favored. A range of mutations in yeast eIF2B subunits can suppress the effects of eIF2(αP) (Pavitt et al. 1997), and where tested, these reduced affinity for phosphorylated eIF2α (Krishnamoorthy et al. 2001), preventing translational control. Similar effects were seen when equivalent mutations were introduced into mammalian eIF2B (Kimball et al. 1998) and were recently shown to enhance viral infection in a cell culture model in which the effects of eIF2B mutation were equivalent to knockdown of the relevant eIF2α kinase (Elsby et al. 2011). As eIF2B is less abundant than eIF2, phosphorylation of only a fraction of the total eIF2 can have a significant impact on eIF2B activity and consequently significantly reduce the amounts of active eIF2·GTP in affected cells.

It was found that eIF5 GDI is important to sensitize cells to eIF2 phosphorylation. In an experiment in which eIF2α phosphorylation was artificially elevated by a constitutively active kinase, total translation and cell growth were impaired. Remarkably, growth was restored simply by a single GDI mutant form of eIF5 (bearing a change from a conserved tryptophan to phenylalanine) that destabilizes the eIF2/eIF5 complex and thereby eliminates GDI activity. Thus, in normal cells, eIF5 GDI function was necessary to sensitize cells appropriately to eIF2 phosphorylation. Whether eIF5 GDI function is conserved in mammalian cells is not yet clear, but the critical elements identified in yeast are conserved in mammals including man. The experiment outlined above suggests that for mediating the response to eIF2(αP) it is the physical ability to form an eIF2·GDP/eIF5 complex that is the most critical step for involvement of eIF5 in eIF2(αP) regulation. The significance of this finding to the UPR and eIF2(αP) regulatory events other than amino acid starvation remains to be determined. However, the findings suggest that varying eIF2B and/or eIF5 levels/activity in different cell types could modulate their sensitivity to this regulatory mechanism. This could help explain tissue-specific defects observed in animal models and human disease pathology associated with alterations to these factors.

A relevant example of tissue-specific pathology is the link between eIF2B mutations and a fatal brain disorder, vanishing white matter disease (VWM) or childhood ataxia with CNS hypomyelination (CACH) (Pavitt and Proud 2009). VWM/CACH mutations cause cell-type-specific pathology, primarily affecting glial cells (astrocytes and oligodendrocytes), although some patients also present with premature ovarian failure (Fogli et al. 2003) and severe eIF2B mutations affect multiple organs (van der Knaap et al. 2003). Heightened signaling in the eIF2(αP)-dependent branch of the UPR has been observed in oligodendroglial cell lines (Kantor et al. 2008) and brain samples of VWM/CACH (van Kollenburg et al. 2006).

eIF2B GEF activity is also impaired in immortalized lymphocytes from patients, but a parallel branch of the UPR, XBP1 splicing, is unaffected (Horzinski et al. 2009, 2010). In contrast, a heightened UPR but apparently normal GEF activity was found with assays of primary fibroblasts (Kantor et al. 2005). Although the molecular basis underlying the cell-type specificity of this disorder remains a mystery, the generation of a disease model eIF2B mutant mouse should facilitate more rapid progress in distinguishing causes from consequences (Geva et al. 2010).

INDIVIDUAL mRNA TARGETS OF eIF2(αP) CONTROL REVEAL DIVERSITY IN uORF-DEPENDENT REGULATION OF TRANSLATION INITIATION

Phosphorylation of eIF2 reduces the level of eIF2·GTP·tRNA$_i^{Met}$ ternary complexes in cells, limiting its availability for translation initiation events. Although this reduces translation initiation rates for many mRNAs, it can also enhance translation of other mRNAs. One well-studied and important example is *ATF4* mRNA, which has been reviewed previously (see, e.g., Ron and Harding 2007; Jackson et al. 2010). ATF4 is a transcription factor with multiple target genes that are transcriptionally activated by ER and other stresses that induce eIF2 phosphorylation. *ATF4* translation increases with increasing eIF2 phosphorylation during stress and is dependent on eIF2α kinase activation. Mouse *PERK* kinase–knockout embryonic stem cells fail to induce *ATF4* translation in response to the ER stress agent thapsigargin, unlike wild-type cells similarly treated (Harding et al. 2000).

ATF4 is translationally controlled by a regulated reinitiation mechanism (Lu et al. 2004; Vattem and Wek 2004). Its mRNA bears two upstream open reading frames (uORFs) in the 5′ leader sequence before the main coding ORF. These uORFs restrict the flow of translating ribosomes scanning from the 5′ mRNA cap to the *ATF4* AUG initiation codon. The current model is based largely from reporter gene construct analyses and similarity to the well-documented model for regulated reinitiation on yeast *GCN4* mRNA (see Fig. 2). It proposes that ribosomes scan from the 5′ cap and translate uORF1, which is only three codons in length. The 40S subunit remains mRNA-associated following translation termination and resumes its migration downstream. Rapid acquisition of TC by the 40S subunit enables scanning and reinitiation downstream at uORF2. Translation of this uORF prevents translation of the *ATF4* ORF. During stress eIF2α kinase activation lowers TC levels. Reinitiation appears more sensitive to a modest reduction in TC levels than primary initiation, delaying the reacquisition of TC following uORF1 translation. Thus a portion of the ribosomes scan past uORF2 AUG before reacquiring TC and are unable to initiate translation of this repressive uORF. Some of these 40S ribosomes do acquire TC between the uORF2 AUG and *ATF4* AUG and translate the ATF4 protein. A related gene, *ATF5*, has an almost identical regulated reinitiation translational control mechanism (Zhou et al. 2008). Reinitiation generally requires that the first uORF be short (Poyry et al. 2004). For reinitiation to be regulated by eIF2(αP), uORF1 must be followed by a second uORF.

Alternative mechanisms for uORF-mediated regulation have been described for the UPR genes *CHOP* (Palam et al. 2011) and *GADD34* (Lee et al. 2009). Like *ATF4* and *ATF5*, both *CHOP* and *GADD34* mRNAs are transcriptionally induced following ER stress and require a translational control mechanism to ensure they are only translated for the duration of the stress response or period of elevated eIF2 phosphorylation. *CHOP* mRNA possesses a single 34-codon uORF in its 5′ leader sequence that contains two in-frame AUG codons at positions 1 and 4 in the uORF, both in poor recognition contexts. Reporter gene experiments showed that regulation is not dependent on the first AUG (uAUG1), but requires the second (uAUG2). The experimental findings support a regulated "leaky-scanning model" (Palam et al. 2011). Under conditions with low eIF2(αP), initiation at uAUG2 dominates, followed by termination and ribosome release so that CHOP translation is low, whereas when eIF2(αP) is high, a greater fraction of ribosomes bypass both uAUG1 and uAUG2 and instead initiate at the *CHOP* AUG, elevating CHOP expression. The sequence of the CHOP uORF (codons 24–34) was also found to be important under nonstress conditions: Translation of this element inhibits subsequent downstream initiation events. *GADD34* translation may be operationally similar to the *CHOP*-regulated

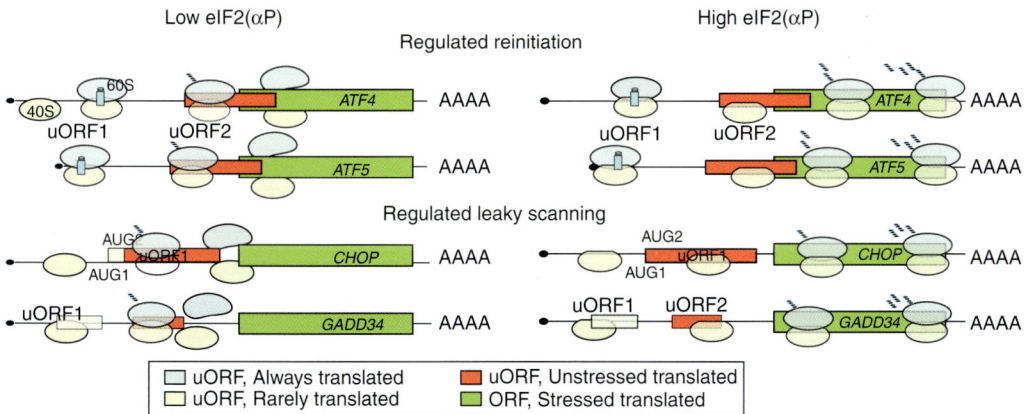

Figure 2. Translational control models for genes with uORFs translated during the UPR. Two states, normal conditions (*left*) and stress conditions with high eIF2 phosphorylation (*right*), are shown. Each mRNA is represented with the main coding ORF shown in green and uORFs with differing colors depending on whether they are always translated (cyan), translated in unstressed conditions only (red), or rarely translated (unfilled). Occupancy by scanning 40S ribosomes (yellow) and translating 80S (yellow and blue) is also indicated. See text for details.

leaky-scanning mechanism, although *GADD34* has two uORFs rather than a single uORF with two in-frame AUGs. Observations support a leaky-scanning model whereby in unstressed conditions most ribosomes scan through uORF1 and translate uORF2, whereas when eIF2(αP) is high, both uORFs are scanned through by a greater proportion of ribosomes and the *GADD34* ORF is now translated (see Fig. 2) (Lee et al. 2009).

It is not yet clear how leaky scanning is regulated by levels of eIF2(αP): Does phosphorylated eIF2 confer altered AUG recognition directly, or are the activities of other translation factors modulated by eIF2(αP)? In vitro experiments showed that eIF2α could be cross-linked to the -3 position of a model mRNA just upstream of the AUG codon (where A $=+1$) (Pisarev et al. 2006). It is not clear if this interaction is modulated by eIF2 phosphorylation. Other factors/elements that may be involved are eIF1, eIF5, and 18S rRNA. In yeast, mutations in eIFs 1, 1A, 2, and 5 alter the ability of ribosomes to recognize non-AUG initiator codons (giving rise to yeast Sui$^-$ mutations) and also poor-context AUG codons (Martin-Marcos et al. 2011), although it is not yet clear what role these factors play in eIF2(αP)-modulated leaky scanning, or why multiple mechanisms have evolved to regulate translation by eIF2(αP).

Thus, these four genes reveal at least two alternate mechanisms for translational control of their expression. All are also transcriptionally controlled, with stress increasing mRNA levels several-fold for each gene. What are the benefits of combining two levels of control? It was suggested that the special architecture of the 5′ untranslated region (UTR) ensures efficient translation during stress, when eIF2(αP) levels are high (Palam et al. 2011). However, constructs lacking uORFs are translated under stress conditions; thus transcriptional control alone should suffice to switch these genes on during stress. Alternatively, the combined transcription and translation control affords both activation during stress by a transcriptional mechanism and rapid inactivation of translation with the decline in eIF2(αP) levels once the stress has been resolved.

HOW WIDESPREAD IS uORF- AND eIF2(αP)-MEDIATED TRANSLATIONAL CONTROL?

It remains unclear how widespread eIF2(αP)-mediated uORF control of translation is. Recent genome-wide studies concluded that approximately half of human and mouse genes possess one or more uORFs (Calvo et al. 2009); this compares with $\sim$13% of yeast genes (Lawless et al. 2009). The yeast study revealed that mRNAs that are translationally up-regulated by stresses that cause increased eIF2(αP) have significantly longer 5′ UTRs with more uORFs than the reference pool of mRNAs. A separate study, using a new high-throughput sequencing method to determine ribosome-protected RNA fragments or "ribosome footprints" on total yeast mRNA, observed a significant increase in ribosome footprints in 5′ UTRs following induction of eIF2(αP) by amino acid deprivation (Ingolia et al. 2009). In many cases translation at the novel uORFs was found to initiate at non-AUG codons, suggesting that start codon recognition is altered by a stress-induced increase in eIF2 phosphorylation. Are non-AUG-initiating ribosomes important for translational control? This is not yet clear, but a recent study experimentally examined the two newly discovered non-AUG uORFs (termed nAuORFs) identified by Ingolia and colleagues in the *GCN4* leader sequence (Zhang and Hinnebusch 2011). They found that one of the *GCN4* nAuORFs was translated but not at a level sufficient to influence the well-established uORF control mechanism. Thus, nAuORFs are not important for *GCN4* translation; their significance across the transcriptome remains unclear and requires further experimentation. Similar genome-wide studies will be required to establish whether these findings in yeast apply to the mammalian UPR.

SEPARATING THE EFFECTS OF eIF2α PHOSPHORYLATION ON BULK PROTEIN SYNTHESIS FROM ITS EFFECTS ON GENE EXPRESSION

Gene knockout in mice and rare human mutations in *EIF2AK3*, the gene encoding PERK, point to the critical role of ER stress-induced

Cite this article as *Cold Spring Harb Perspect Biol* doi: 10.1101/cshperspect.a012278

eIF2α phosphorylation in the survival and well-being of secretory cells (Harding et al. 2001; Zhang et al. 2002; Julier and Nicolino 2010). These observations are most readily explained by the critical role of eIF2(αP) in attenuating the flux of proteins into the ER lumen. Secretory cells that are subject to powerful physiologically driven increase in rates of secretory protein synthesis are most heavily dependent on the moderating influence of eIF2α phosphorylation on unfolded protein stress (Trusina et al. 2008). This simple explanation, which emphasizes the control of bulk protein synthesis, fits nicely with the observed engorgement of the ER in secretory cells of PERK-knockout animals (Harding et al. 2001; Zhang et al. 2002).

Careful analysis of insulin-producing β cells suggests that PERK signaling affects aspects of ER function that go beyond protein folding. PERK-knockout cells exhibit defects in vesicular transport from the ER to Golgi and in the degradation of unfolded ER proteins (Gupta et al. 2010). These defects could well be explained by enhanced load of unfolded proteins in the ER of PERK-deficient cells. However, as eIF2α phosphorylation is also coupled to changes in gene transcription (mediated by ATF4 and ATF5, as noted above), it is difficult to exclude a role for altered developmental programs in the outcome of PERK deficiency. Resolution of this problem awaits the emergence of tools to inactivate PERK-dependent gene expression programs without affecting bulk protein synthesis.

THE BIPHASIC RELATIONSHIP BETWEEN eIF2α PHOSPHORYLATION AND FITNESS OF STRESSED CELLS

The benefits of linking ER stress to increased levels of eIF2α phosphorylation are made plain by the phenotype of PERK deficiency. However, it stands to reason that excess levels of phosphorylated eIF2α would interfere with the essential process of protein synthesis. This conjecture is strongly supported by the phenotype of loss-of-function mutations affecting the phosphatases that dephosphorylate eIF2α.

Two phosphatase complexes are known to dephosphorylate eIF2α. They consist of a single regulatory subunit, PPP1R15A (also known as GADD34 or MyD116; see above) or PPP1R15B (also known as CReP), and a common catalytic subunit, protein phosphatase 1 (PP1), which in mammals have four different isoforms encoded by three different genes. PPP1R15A/GADD34 is encoded by an inducible gene, whose expression is under the control of ATF4 and is therefore responsive to levels of phosphorylated eIF2α, and whose mRNA is subject to translational regulation, as noted above. Thus, PPP1R15A participates in a transcriptionally mediated negative-feedback loop that controls levels of phosphorylated eIF2α (Novoa et al. 2001, 2003; Brush et al. 2003). PPP1R15B (CReP) is encoded by a constitutively active gene and contributes to basal levels of eIF2(αP)-directed phosphatase activity (Jousse et al. 2003).

Gene knockout of these two PPP1R15 isoforms is informative in regard to the physiological role of eIF2(αP) dephosphorylation. PPP1-R15A knockout is well tolerated, with mice exhibiting an essentially normal phenotype. In fact, the knockout mice are resistant to some injuries induced by progressive ER stress (Marciniak et al. 2004). However, in response to injuries that result in rapid and sustained activation of PERK, such as the application of the toxin thapsigargin, the PPP1R15A-deficient cells show decreased fitness, which correlates with impaired recovery from PERK-mediated translation repression (Novoa et al. 2003). A role for unsustainably high levels of eIF2(αP) in death of severely ER-stressed cells was more recently suggested by the temporal profile of the activity of the known UPR transducers in response to severe ER stress; sustained and overexuberant PERK activity correlated with cell death (Lin et al. 2007; Calvo et al. 2009).

Inactivation of the constitutive PPP1R15B subunit was also informative. Early development was normal in knockout mice, but anemia developed late in gestation and the mice were stillborn. Compounding the PPP1R15B knockout with loss of PPP1R15A dramatically accelerated the embryonic lethality to the preimplantation stage. Importantly, the severely compromising compound mutant phenotype could be rescued completely by a mutation rendering eIF2α unphosphorylatable (eIF2α S51A) (Harding et al. 2009). These observations confirm the nonredundant role of the two PPP1R15 regulatory subunits in eIF2(αP) dephosphorylation and highlight the consequences of excessive levels of phosphorylated eIF2α.

Recent progress in chemical biology has provided further support for the biphasic relationship between levels of phosphorylated eIF2α and fitness and has suggested a way to manipulate eIF2(αP) dephosphorylation for the benefit of stressed cells. The small molecule guanabenz, originally developed as an antihypertensive drug, was found to selectively interfere with the formation of the complex between PPP1R15A and the catalytic subunit, thus selectively inhibiting the PPP1R15A-mediated stress-inducible dephosphorylation of eIF2(αP) while sparing the constitutive activity of PPP1R15B (Fig. 3) (Tsaytler et al. 2011). In keeping with this mechanism of action, guanabenz protected wild-type ER-stressed cells from death and had no effect on cells lacking PPP1R15A. But its application resulted in rapid loss of viability of cells lacking PPP1R15B, in which guanabenz led to unsustainable increases in levels of phosphorylated eIF2α.

The selectivity of guanabenz for PPP1R15A over PPP1R15B is in contrast to salubrinal, an inhibitor of eIF2(αP) dephosphorylation that targets both phosphatase complexes (Boyce et al. 2005). These differences may explain the ability of guanabenz to protect cells against ER stress over a broad range of concentrations, compared with the narrower window of utility of salubrinal. Unfortunately, guanabenz is far less active against the PPP1R15A complex than it is against α1-adrenergic receptors (the basis of its utility as an antihypertensive drug). Thus, in vivo toxicity limits the prospects of testing its ability to protect against physiologically relevant models of ER stress.

TRANSLATIONAL PAUSING ENABLES COLOCALIZATION OF THE *XBP1* mRNA AND IRE1'S EFFECTOR DOMAIN

IRE1-mediated cleavage of the mRNA encoding the transcription factor XBP1 and the subsequent splicing of the resulting 5' and 3' mRNA fragments (with elimination of the intron) is the key effector event in the most ancient arm of the UPR. The (unconventionally) spliced *XBP1* mRNA encodes a potent transcription factor that activates UPR target genes. ER stress is coupled to IRE1 activation by IRE1's self-directed protein kinase, thereby generating autophosphorylated IRE1 that acquires endoribonucleolytic activity (Bernales et al. 2006).

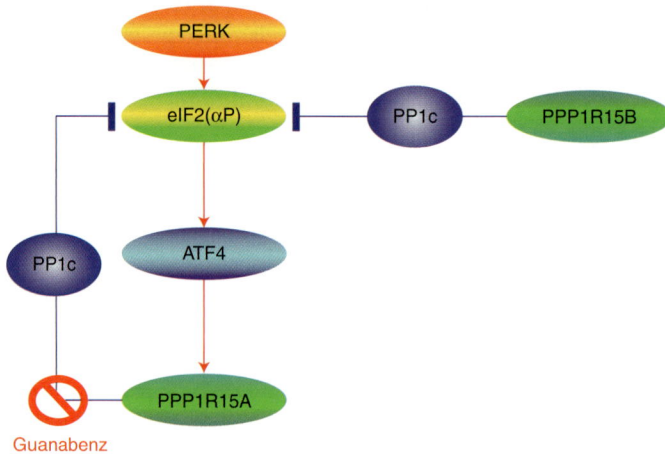

Figure 3. eIF2(αP) dephosphorylation. eIF2(αP) dephosphorylation is mediated by two phosphatase complexes: one with the constitutive regulatory subunit PPP1R15B and the other with the stress-inducible subunit PPP1R15A. The latter complex is subject to selective inhibition by guanabenz, whose application provides a measure of protection against the lethal consequences of ER stress.

 Cite this article as *Cold Spring Harb Perspect Biol* doi: 10.1101/cshperspect.a012278

In addition to the essential activation of IRE1 as an enzyme, recent work has pointed to an important role for targeted mRNA localization in IRE1-mediated *XBP1* mRNA cleavage. In mammals, localization of the *XBP1* mRNA has proven to depend on a novel mechanism of translational regulation.

IRE1 is an ER-localized transmembrane protein. Efficient cleavage of the *XBP1* mRNA by IRE1's effector endoribonuclease domain requires the colocalization of the two to the cytosolic face of the ER membrane. This, it turn, depends on the presence of a hydrophobic protein sequence at the carboxyl terminus encoded by the unspliced (inactive) *XBP1* mRNA. Interaction of the hydrophobic sequence (called hydrophobic region 2, or HR2) with the ER membrane is believed to recruit the XBP1 nascent chain and associated ribosome and unspliced mRNA to the cytosolic face of the ER membrane, in proximity to IRE1's effector domain (Fig. 4). The IRE1-mediated unconventional splicing event alters the *XBP1* coding sequence, eliminating the hydrophobic sequence. This untethers the ribosome–nascent chain complex encoding the spliced form of XBP1 from the ER (Yanagitani et al. 2009).

This proposed mechanism nicely explains the dependence of *XBP1* mRNA localization on the hydrophobic segment encoded by the unspliced mRNA. However, the HR2 segment is encoded at the extreme 3′ end of the coding sequence, 53 residues from the stop codon. Assuming normal elongation rates of 3–8 residues/sec, the dwell time of this hydrophobic sequence in the ribosome–nascent chain complex is expected to be very short, significantly limiting its utility as a tether. To counteract this limitation, the unspliced *XBP1* coding sequence evolved features that specify translational pausing. The pause is conveniently located very close to the stop codon, providing sufficient length for the preceding HR2 to make its way out of the ribosomal tunnel and interact with the ER membrane (Fig. 4) (Yanagitani et al. 2011). The molecular details of the pause have features in common with pausing in bacterial systems, in that it depends on the side chains of residues engaged in the ribosomal tunnel and may exploit the poor nucleophilicity of proline's secondary amine to position such a residue at the ribosome "A" site in the stalled complex (Nakatogawa and Ito 2002). Given the very high precision by which ribosome pauses can be mapped

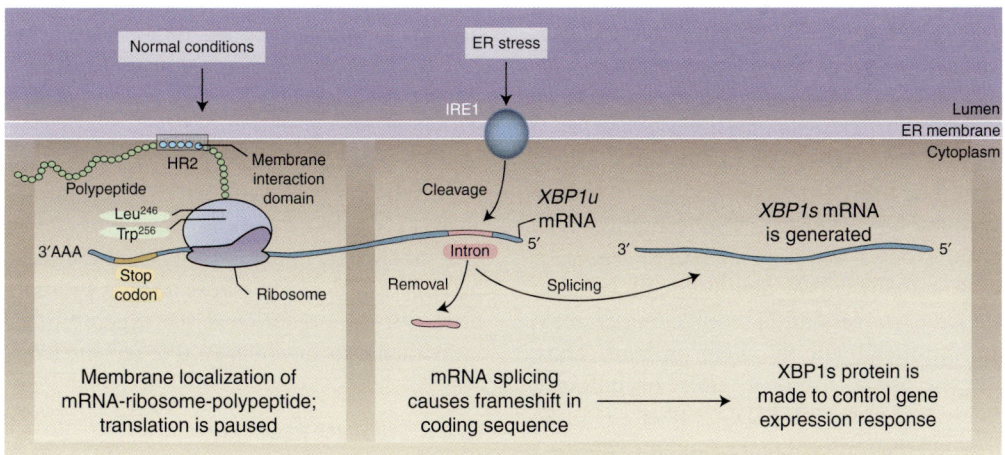

Figure 4. Translational pausing and IRE1-mediated splicing of *XBP1*. To be poised for rapid splicing in response to the accumulation of unfolded proteins in the ER, unspliced *XBP1u* mRNA is tethered to the ER membrane by the paused ribosome–nascent chain complex. Pausing requires interactions of Leu246 and Trp256 in the nascent chain with the ribosome. The pause stabilizes the complex, favoring cleavage of *XBP1u* by IRE1. This causes a frameshift in the coding sequence of the spliced *XBP1s* mRNA, which precludes translation of Leu246 and Trp256 and relieves the pause. (Reprinted, with permission, from Ron and Ito 2011.)

on mRNAs in vivo (see above), it will be interesting to see if this prediction is borne out.

IRE1'S PROMISCUOUS RNase ACTIVITY AND PROTEIN-FOLDING HOMEOSTASIS IN THE ER

In addition to its highly sequence-specific RNase activity, directed toward two sites in the XBP1 mRNA, observations first made in cultured insect cells point to an additional role for IRE1 in promiscuous degradation of membrane-bound mRNAs. It was noted that certain mRNAs, whose levels decreased in wild-type cells exposed to agents that promote ER stress, were spared this fate in IRE1-knockdown ER-stressed cells. Interestingly, this sparing effect was not observed in XBP1-knockdown cells, suggesting that decline in the levels of these mRNAs is not mediated by IRE1-dependent transcriptional programs. Furthermore, inhibiting the XRN1 and SKI2 exoribonucleases that processively degrade uncapped and deadenylated mRNAs led to appearance of fragments of the mRNAs targeted by this process. Importantly, such mRNA fragments were not observed in cells lacking IRE1. Therefore the process was termed regulated IRE1-dependent degradation (or RIDD) (Hollien and Weissman 2006).

Evidence has since mounted that RIDD is not an oddity confined to cultured insect cells but also occurs in other (mammalian) animal cells, but interestingly not in budding yeast (Hollien et al. 2009). The physiological role of RIDD is far from clear. It occurs relatively late in the course of the ER stress response and only in response to relatively high levels of ER stress. An important clue is provided by the observation that RIDD targets mRNAs encoding secreted and transmembrane ER client proteins. Thus, from the perspective of its effect on unfolded protein load in the ER, RIDD is allied to PERK-mediated translational repression, the latter as a rapid response to imbalance in the ER lumen and the former as a backup mechanism of longer latency. In addition to its predicted effects on the bulk flow of proteins into the ER, RIDD may also play a role in reprogramming the translational apparatus of the ER membrane.

By purging the cells of preexisting secreted protein, it favors the expression of a new repertoire of transcriptionally induced UPR target genes.

The two processes noted above are predicted to benefit the stressed cells, by lowering ER client protein load and promoting organellar adaptation to the stress. But there is also evidence that RIDD may play a role in the demise of ER-stressed cells. This is supported by the observation that hyperactivity of an IRE1 enzyme whose kinase domain has been modified to be less efficient in coupling ER stress to RIDD is better tolerated than hyperactivity of the wild-type, fully RIDD-proficient enzyme (Han et al. 2009).

It is unclear how RIDD would be coupled to cell death, but this link may not be entirely without precedent. The endoribonuclease most similar in sequence to IRE1 is RNase L, an interferon-inducible, promiscuous RNase whose fury is unleashed in virally infected cells as an antiviral strategy to compromise their function and viability (Bork and Sander 1993). The ancestral IRE1 may have been a sequence-specific RNase (and has remained so in budding yeast). In animal cells the enzyme evolved some promiscuous RNase activity. The intensity at which this RNase is activated contributes to a life/death decision in ER-stressed cells by controlling client protein load or effecting lethal levels of RNA degradation. A gene duplication yielded RNase L, whose function is directed toward the latter, destructive goal. This parallels the situation in regards to eIF2α phosphorylation. Simpler organisms possess eIF2α kinases that promote fitness of the individual cell (e.g., PERK to protect against ER stress). But more complex organisms evolved an eIF2α kinase, PKR, that is activated by viral infection and whose activity promotes fitness of the organism at the expense of the individual cells in which PKR is activated.

CONCLUDING REMARKS

The endoplasmic reticulum stress response exploits diverse mechanisms for translational regulation. These promote protein-folding homeostasis in the ER lumen of individual cells and tissue homeostasis at the level of the whole organism. Thus, translational regulation is an important facet of several nodes that influence important life-and-death decisions in ER-stressed cells and beyond.

ACKNOWLEDGMENTS

D.R. is a Principal Research Fellow of the Wellcome Trust. Work in G.D.P.'s laboratory is funded by grants from the Biotechnology and Biological Sciences Research Council and the European Union (Framework Program 7).

REFERENCES

Bernales S, Papa FR, Walter P. 2006. Intracellular signaling by the unfolded protein response. *Annu Rev Cell Dev Biol* **22:** 487–508.

Bork P, Sander C. 1993. A hybrid protein kinase-RNase in an interferon-induced pathway? *FEBS Lett* **334:** 149–152.

Boyce M, Bryant KF, Jousse C, Long K, Harding HP, Scheuner D, Kaufman RJ, Ma D, Coen D, Ron D, et al. 2005. A selective inhibitor of eIF2α dephosphorylation protects cells from ER stress. *Science* **307:** 935–939.

Brush MH, Weiser DC, Shenolikar S. 2003. Growth arrest and DNA damage-inducible protein GADD34 targets protein phosphatase 1α to the endoplasmic reticulum and promotes dephosphorylation of the α subunit of eukaryotic translation initiation factor 2. *Mol Cell Biol* **23:** 1292–1303.

Calfon M, Zeng H, Urano F, Till JH, Hubbard SR, Harding HP, Clark SG, Ron D. 2002. IRE1 couples endoplasmic reticulum load to secretory capacity by processing the XBP-1 mRNA. *Nature* **415:** 92–96.

Calvo SE, Pagliarini DJ, Mootha VK. 2009. Upstream open reading frames cause widespread reduction of protein expression and are polymorphic among humans. *Proc Natl Acad Sci* **106:** 7507–7512.

Cox JS, Walter P. 1996. A novel mechanism for regulating activity of a transcription factor that controls the unfolded protein response. *Cell* **87:** 391–404.

Cox JS, Shamu CE, Walter P. 1993. Transcriptional induction of genes encoding endoplasmic reticulum resident proteins requires a transmembrane protein kinase. *Cell* **73:** 1197–1206.

Das S, Ghosh R, Maitra U. 2001. Eukaryotic translation initiation factor 5 functions as a GTPase-activating protein. *J Biol Chem* **276:** 6720–6726.

Elsby R, Heiber JF, Reid P, Kimball SR, Pavitt GD, Barber GN. 2011. The α subunit of eukaryotic initiation factor 2B (eIF2B) is required for eIF2-mediated translational suppression of vesicular stomatitis virus. *J Virol* **85:** 9716–9725.

Fogli A, Rodriguez D, Eymard-Pierre E, Bouhour F, Labauge P, Meaney BF, Zeesman S, Kaneski CR, Schiffmann R, Boespflug-Tanguy O. 2003. Ovarian failure related to eukaryotic initiation factor 2B mutations. *Am J Hum Genet* **72:** 1544–1550.

Geva M, Cabilly Y, Assaf Y, Mindroul N, Marom L, Raini G, Pinchasi D, Elroy-Stein O. 2010. A mouse model for eukaryotic translation initiation factor 2B-leucodystrophy reveals abnormal development of brain white matter. *Brain* **133:** 2448–2461.

Gupta S, McGrath B, Cavener DR. 2010. PERK (EIF2AK3) regulates proinsulin trafficking and quality control in the secretory pathway. *Diabetes* **59:** 1937–1947.

Han D, Lerner AG, Vande Walle L, Upton JP, Xu W, Hagen A, Backes BJ, Oakes SA, Papa FR. 2009. IRE1α kinase activation modes control alternate endoribonuclease outputs to determine divergent cell fates. *Cell* **138:** 562–575.

Harding HP, Zhang Y, Bertolotti A, Zeng H, Ron D. 2000. *Perk* is essential for translational regulation and cell survival during the unfolded protein response. *Mol Cell* **5:** 897–904.

Harding H, Zeng H, Zhang Y, Jungreis R, Chung P, Plesken H, Sabatini D, Ron D. 2001. Diabetes mellitus and exocrine pancreatic dysfunction in $Perk^{-/-}$ mice reveals a role for translational control in survival of secretory cells. *Mol Cell* **7:** 1153–1163.

Harding H, Zhang Y, Zeng H, Novoa I, Lu P, Calfon M, Sadri N, Yun C, Popko B, Paules R, et al. 2003. An integrated stress response regulates amino acid metabolism and resistance to oxidative stress. *Mol Cell* **11:** 619–633.

Harding HP, Zhang Y, Scheuner D, Chen JJ, Kaufman RJ, Ron D. 2009. *Ppp1r15* gene knockout reveals an essential role for translation initiation factor 2 alpha (eIF2α) dephosphorylation in mammalian development. *Proc Natl Acad Sci* **106:** 1832–1837.

Hollien J, Weissman JS. 2006. Decay of endoplasmic reticulum-localized mRNAs during the unfolded protein response. *Science* **313:** 104–107.

Hollien J, Lin JH, Li H, Stevens N, Walter P, Weissman JS. 2009. Regulated Ire1-dependent decay of messenger RNAs in mammalian cells. *J Cell Biol* **186:** 323–331.

Horzinski L, Huyghe A, Cardoso MC, Gonthier C, Ouchchane L, Schiffmann R, Blanc P, Boespflug-Tanguy O, Fogli A. 2009. Eukaryotic initiation factor 2B (eIF2B) GEF activity as a diagnostic tool for EIF2B-related disorders. *PLoS ONE* **4:** e8318.

Horzinski L, Kantor L, Huyghe A, Schiffmann R, Elroy-Stein O, Boespflug-Tanguy O, Fogli A. 2010. Evaluation of the endoplasmic reticulum-stress response in eIF2B-mutated lymphocytes and lymphoblasts from CACH/VWM patients. *BMC Neurol* **10:** 94.

Ingolia NT, Ghaemmaghami S, Newman JR, Weissman JS. 2009. Genome-wide analysis in vivo of translation with nucleotide resolution using ribosome profiling. *Science* **324:** 218–223.

Itoh N, Okamoto H. 1980. Translational control of proinsulin synthesis by glucose. *Nature* **283:** 100–102.

Jackson RJ, Hellen CU, Pestova TV. 2010. The mechanism of eukaryotic translation initiation and principles of its regulation. *Nat Rev Mol Cell Biol* **11:** 113–127.

Jennings MD, Pavitt GD. 2010. eIF5 has GDI activity necessary for translational control by eIF2 phosphorylation. *Nature* **465:** 378–381.

Jousse C, Oyadomari S, Novoa I, Lu PD, Zhang Y, Harding HP, Ron D. 2003. Inhibition of a constitutive translation initiation factor 2α phosphatase, CReP, promotes survival of stressed cells. *J Cell Biol* **163:** 767–775.

Julier C, Nicolino M. 2010. Wolcott-Rallison syndrome. *Orphanet J Rare Dis* **5:** 29.

Kantor L, Harding HP, Ron D, Schiffmann R, Kaneski CR, Kimball SR, Elroy-Stein O. 2005. Heightened stress response in primary fibroblasts expressing mutant eIF2B genes from CACH/VWM leukodystrophy patients. *Hum Genet* **118:** 99–106.

Kantor L, Pinchasi D, Mintz M, Hathout Y, Vanderver A, Elroy-Stein O. 2008. A point mutation in translation initiation factor 2B leads to a continuous hyper stress state in oligodendroglial-derived cells. *PLoS ONE* **3:** e3783.

Kimball SR, Fabian JR, Pavitt GD, Hinnebusch AG, Jefferson LS. 1998. Regulation of guanine nucleotide exchange through phosphorylation of eukaryotic initiation factor eIF2α. Role of the α- and δ-subunits of eIF2b. *J Biol Chem* **273:** 12841–12845.

Krishnamoorthy T, Pavitt GD, Zhang F, Dever TE, Hinnebusch AG. 2001. Tight binding of the phosphorylated α subunit of initiation factor 2 (eIF2α) to the regulatory subunits of guanine nucleotide exchange factor eIF2B is required for inhibition of translation initiation. *Mol Cell Biol* **21:** 5018–5030.

Lawless C, Pearson RD, Selley JN, Smirnova JB, Grant CM, Ashe MP, Pavitt GD, Hubbard SJ. 2009. Upstream sequence elements direct post-transcriptional regulation of gene expression under stress conditions in yeast. *BMC Genomics* **10:** 7.

Lee YY, Cevallos RC, Jan E. 2009. An upstream open reading frame regulates translation of GADD34 during cellular stresses that induce eIF2α phosphorylation. *J Biol Chem* **284:** 6661–6673.

Lin JH, Li H, Yasumura D, Cohen HR, Zhang C, Panning B, Shokat KM, Lavail MM, Walter P. 2007. IRE1 signaling affects cell fate during the unfolded protein response. *Science* **318:** 944–949.

Lu PD, Harding HP, Ron D. 2004. Translation reinitiation at alternative open reading frames regulates gene expression in an integrated stress response. *J Cell Biol* **167:** 27–33.

Marciniak SJ, Yun CY, Oyadomari S, Novoa I, Zhang Y, Jungreis R, Nagata K, Harding HP, Ron D. 2004. CHOP induces death by promoting protein synthesis and oxidation in the stressed endoplasmic reticulum. *Genes Dev* **18:** 3066–3077.

Martin-Marcos P, Cheung YN, Hinnebusch AG. 2011. Functional elements in initiation factors 1, 1A and 2β discriminate against poor AUG context and non-AUG start codons. *Mol Cell Biol* **31:** 4814–4831.

Mori K, Ma W, Gething MJ, Sambrook J. 1993. A transmembrane protein with a cdc2$^+$/CDC28-related kinase activity is required for signaling from the ER to the nucleus. *Cell* **74:** 743–756.

Nakatogawa H, Ito K. 2002. The ribosomal exit tunnel functions as a discriminating gate. *Cell* **108:** 629–636.

Nika J, Yang W, Pavitt GD, Hinnebusch AG, Hannig EM. 2000. Purification and kinetic analysis of eIF2B from *Saccharomyces cerevisiae. J Biol Chem* **275:** 26011–26017.

Novoa I, Zeng H, Harding H, Ron D. 2001. Feedback inhibition of the unfolded protein response by GADD34-

mediated dephosphorylation of eIF2α. *J Cell Biol* **153:** 1011–1022.

Novoa I, Zhang Y, Zeng H, Jungreis R, Harding HP, Ron D. 2003. Stress-induced gene expression requires programmed recovery from translational repression. *EMBO J* **22:** 1180–1187.

Palam LR, Baird TD, Wek RC. 2011. Phosphorylation of eIF2 facilitates ribosomal bypass of an inhibitory upstream ORF to enhance CHOP translation. *J Biol Chem* **286:** 10939–10949.

Paulin FE, Campbell LE, O'Brien K, Loughlin J, Proud CG. 2001. Eukaryotic translation initiation factor 5 (eIF5) acts as a classical GTPase-activator protein. *Curr Biol* **11:** 55–59.

Pavitt GD, Proud CG. 2009. Protein synthesis and its control in neuronal cells with a focus on vanishing white matter disease. *Biochem Soc Trans* **37:** 1298–1310.

Pavitt GD, Yang W, Hinnebusch AG. 1997. Homologous segments in three subunits of the guanine nucleotide exchange factor eIF2B mediate translational regulation by phosphorylation of eIF2. *Mol Cell Biol* **17:** 1298–1313.

Pavitt GD, Ramaiah KV, Kimball SR, Hinnebusch AG. 1998. eIF2 independently binds two distinct eIF2B subcomplexes that catalyze and regulate guanine-nucleotide exchange. *Genes Dev* **12:** 514–526.

Pisarev AV, Kolupaeva VG, Pisareva VP, Merrick WC, Hellen CU, Pestova TV. 2006. Specific functional interactions of nucleotides at key $^-3$ and $^+4$ positions flanking the initiation codon with components of the mammalian 48S translation initiation complex. *Genes Dev* **20:** 624–636.

Poyry TA, Kaminski A, Jackson RJ. 2004. What determines whether mammalian ribosomes resume scanning after translation of a short upstream open reading frame? *Genes Dev* **18:** 62–75.

Ron D, Harding H. 2007. eIF2α phosphorylation in cellular stress responses and disease. In *Translational control* (ed. Sonenberg N, et al.), pp. 345–368. Cold Spring Harbor Laboratory Press, Cold Spring Harbor, NY.

Ron D, Ito K. 2011. Cell biology: A translational pause to localize. *Science* **331:** 543–544.

Ron D, Walter P. 2007. Signal integration in the endoplasmic reticulum unfolded protein response. *Nat Rev Mol Cell Biol* **8:** 519–529.

Rowlands AG, Panniers R, Henshaw EC. 1988. The catalytic mechanism of guanine nucleotide exchange factor action and competitive inhibition by phosphorylated eukaryotic initiation factor 2. *J Biol Chem* **263:** 5526–5533.

Singh CR, Lee B, Udagawa T, Mohammad-Qureshi SS, Yamamoto Y, Pavitt GD, Asano K. 2006. An eIF5/eIF2 complex antagonizes guanine nucleotide exchange by eIF2B during translation initiation. *EMBO J* **25:** 4537–4546.

Trusina A, Papa FR, Tang C. 2008. Rationalizing translation attenuation in the network architecture of the unfolded protein response. *Proc Natl Acad Sci* **105:** 20280–20285.

Tsaytler P, Harding HP, Ron D, Bertolotti A. 2011. Selective inhibition of a regulatory subunit of protein phosphatase 1 restores proteostasis. *Science* **332:** 91–94.

van der Knaap MS, van Berkel CG, Herms J, van Coster R, Baethmann M, Naidu S, Boltshauser E, Willemsen MA, Plecko B, Hoffmann GF, et al. 2003. eIF2B-related

Cite this article as *Cold Spring Harb Perspect Biol* doi: 10.1101/cshperspect.a012278

disorders: Antenatal onset and involvement of multiple organs. *Am J Hum Genet* **73:** 1199–1207.

van Kollenburg B, van Dijk J, Garbern J, Thomas AA, Scheper GC, Powers JM, van der Knaap MS. 2006. Glia-specific activation of all pathways of the unfolded protein response in vanishing white matter disease. *J Neuropathol Exp Neurol* **65:** 707–715.

Vattem KM, Wek RC. 2004. Reinitiation involving upstream ORFs regulates *ATF4* mRNA translation in mammalian cells. *Proc Natl Acad Sci* **101:** 11269–11274.

Yanagitani K, Imagawa Y, Iwawaki T, Hosoda A, Saito M, Kimata Y, Kohno K. 2009. Cotranslational targeting of XBP1 protein to the membrane promotes cytoplasmic splicing of its own mRNA. *Mol Cell* **34:** 191–200.

Yanagitani K, Kimata Y, Kadokura H, Kohno K. 2011. Translational pausing ensures membrane targeting and cytoplasmic splicing of *XBP1u* mRNA. *Science* **331:** 586–589.

Yoshida H, Matsui T, Yamamoto A, Okada T, Mori K. 2001. XBP1 mRNA is induced by ATF6 and spliced by IRE1 in response to ER stress to produce a highly active transcription factor. *Cell* **107:** 881–891.

Zhang F, Hinnebusch AG. 2011. An upstream ORF with non-AUG start codon is translated in vivo but dispensable for translational control of *GCN4* mRNA. *Nucleic Acids Res* **39:** 3128–3140.

Zhang P, McGrath B, Li S, Frank A, Zambito F, Reinert J, Gannon M, Ma K, McNaughton K, Cavener DR. 2002. The PERK eukaryotic initiation factor 2α kinase is required for the development of the skeletal system, postnatal growth, and the function and viability of the pancreas. *Mol Cell Biol* **22:** 3864–3874.

Zhou D, Palam LR, Jiang L, Narasimhan J, Staschke KA, Wek RC. 2008. Phosphorylation of eIF2 directs ATF5 translational control in response to diverse stress conditions. *J Biol Chem* **283:** 7064–7073.

P-Bodies and Stress Granules: Possible Roles in the Control of Translation and mRNA Degradation

Carolyn J. Decker and Roy Parker

Department of Molecular and Cellular Biology and Howard Hughes Medical Institute, University of Arizona, Tucson, Arizona 85721-0206

Correspondence: rrparker@u.arizona.edu

The control of translation and mRNA degradation is important in the regulation of eukaryotic gene expression. In general, translation and steps in the major pathway of mRNA decay are in competition with each other. mRNAs that are not engaged in translation can aggregate into cytoplasmic mRNP granules referred to as processing bodies (P-bodies) and stress granules, which are related to mRNP particles that control translation in early development and neurons. Analyses of P-bodies and stress granules suggest a dynamic process, referred to as the mRNA Cycle, wherein mRNPs can move between polysomes, P-bodies and stress granules although the functional roles of mRNP assembly into higher order structures remain poorly understood. In this article, we review what is known about the coupling of translation and mRNA degradation, the properties of P-bodies and stress granules, and how assembly of mRNPs into larger structures might influence cellular function.

The translation and decay of mRNAs play key roles in the control of eukaryotic gene expression. The determination of eukaryotic mRNA decay pathways has allowed insight into how translation and mRNA degradation are coupled. Degradation of eukaryotic mRNAs is generally initiated by shortening of the 3′ poly (A) tail (Fig. 1A) (reviewed in Parker and Song 2004; Garneau et al. 2007) by the major mRNA deadenylase, the Ccr4/Pop2/Not complex (Daugeron et al. 2001; Tucker et al. 2001; Thore et al. 2003). Following deadenylation, mRNAs can be degraded 3′ to 5′ by the exosome (Anderson and Parker 1998; Wang and Kiledjian 2001). However, more commonly, mRNAs are decapped by the Dcp1/Dcp2 decapping enzyme and then degraded 5′ to 3′ by the exonuclease, Xrn1 (Decker and Parker 1993; Hsu and Stevens 1993; Muhlrad et al. 1994, 1995; Dunckley and Parker 1999; van Dijk et al. 2002; Steiger et al. 2003). In metazoans, a second decapping enzyme, Nudt16, also contributes to mRNA turnover (Song et al. 2010).

The processes of mRNA decay and translation are interconnected in eukaryotic cells in many ways. For example, quality control mechanisms exist to detect aberrancies in translation, which then lead to mRNAs being degraded by specialized mRNA decay pathways (Fig. 1B). Nonsense-mediated decay (NMD) is one such mRNA quality control system that

A   General mRNA decay pathways

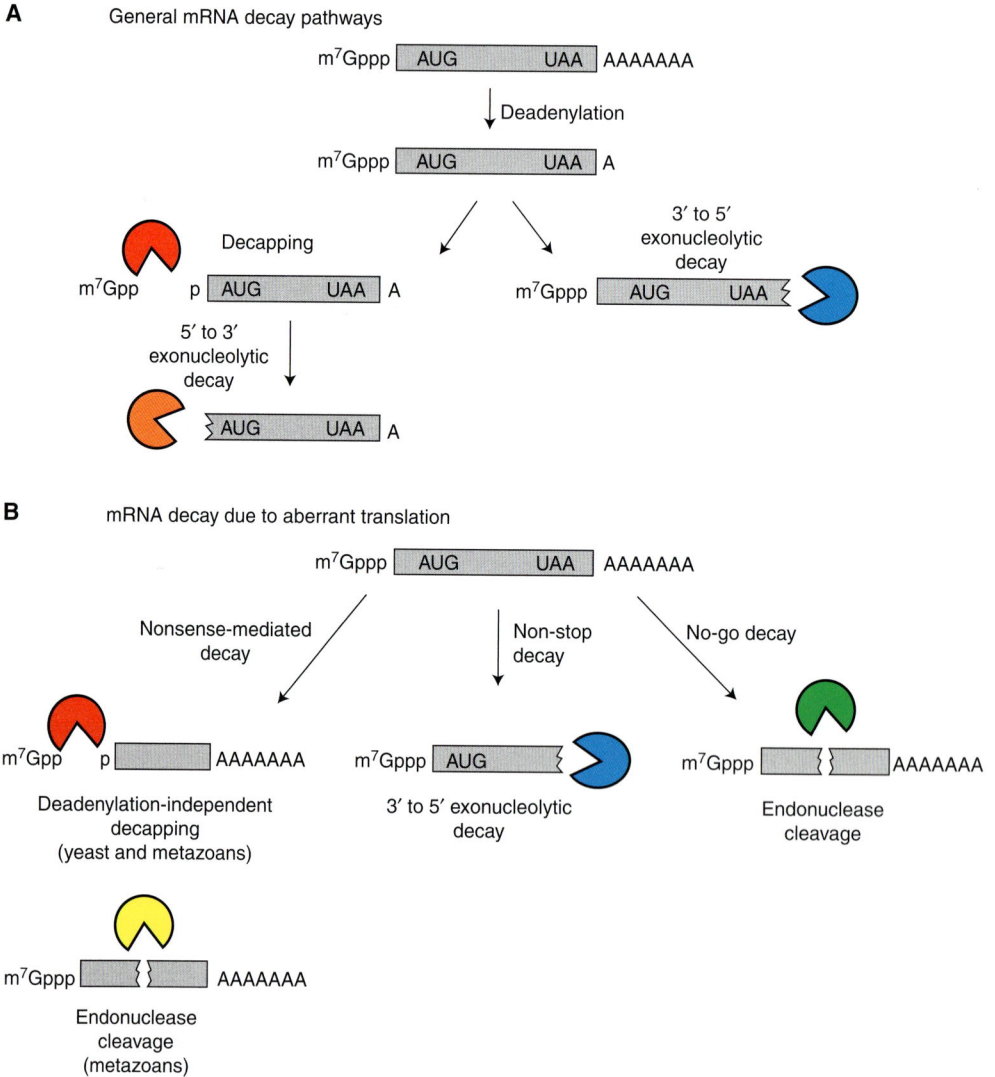

Figure 1. Eukaryotic mRNA decay pathways. (*A*) General mRNA decay pathways. (*B*) Specialized decay pathways that degrade translationally aberrant mRNAs.

degrades mRNAs that terminate translation aberrantly. In yeast, aberrant translation termination leads to deadenylation-independent decapping (Muhlrad and Parker 1994), whereas in metazoan cells NMD substrates can be both decapped and endonucleolytically cleaved and degraded (reviewed in Isken and Maquat 2007). A second quality control system for mRNA translation is referred to as no-go decay (NGD) and leads to endonucleolytic cleavage of mRNAs with strong stalls in translation elongation

(Doma and Parker 2006; reviewed in Harigaya and Parker 2010). Another mechanism of mRNA quality control is the rapid 3′ to 5′ degradation of mRNAs that do not contain translation termination codons, which is referred to as non-stop decay (NSD) (Frischmeyer et al. 2002; van Hoof et al. 2002). The available evidence suggests these specialized mechanisms function primarily on aberrant mRNAs that are produced by defects in splicing, 3′ end formation, or damage to RNAs.

The main pathway of mRNA degradation is also in competition with translation initiation. Competition between the two processes was first suggested by the observation that removal of the poly (A) tail and the cap structure, both of which stimulate translation initiation, were the key steps in mRNA degradation. In addition, inhibition of translation initiation by strong secondary structures in the 5′UTR, translation initiation inhibitors, a poor AUG context, or mutations in initiation factors increases the rates of deadenylation and decapping (Muhlrad et al. 1995; Muckenthaler et al. 1997; Lagrandeur and Parker 1999; Schwartz and Parker 1999). Moreover, the cap binding protein eIF4E, known to stimulate translation initiation, inhibits the decapping enzyme, Dcp1/Dcp2, both in vivo and in vitro (Schwartz and Parker 1999; Schwartz and Parker 2000). Finally, many mRNA specific regulatory factors, (e.g., miRNAs or PUF proteins), both repress translation and accelerate deadenylation and decapping (reviewed in Wickens et al. 2002; Behm-Ansmant et al. 2006; Franks and Lykke-Anderson 2008; Shyu et al. 2008).

In the simplest model, the competition between translation and mRNA degradation can be understood through changes in the proteins bound to the cap and poly (A) tail that then influence the accessibility of these structures to deadenylases and decapping enzymes. For example, given that the Ccr4/Pop2/Not deadenylase complex is inhibited by poly (A)-binding protein (Pab1) (Tucker et al. 2002), the effects of translation on deadenylation are most likely through dynamic changes in the association of Pab1 binding with the poly (A) tail. One possibility is that defects in translation initiation either directly or indirectly decrease Pab1 association with the poly (A) tail. Deadenylation is also affected by aspects of translation termination. For instance, premature translation termination in yeast accelerates poly (A) shortening as part of the process of NMD (Cao and Parker 2003; Mitchell and Tollervey 2003). The coupling of translation termination to deadenylation has been suggested to occur through direct interactions of the translation termination factor eRF3 with Pab1 (Cosson et al. 2002), which

may lead to Pab1 transiently dissociating from the poly (A) tail. Interestingly, in yeast, once the poly (A) tail reaches an oligo (A) length of 10–12 residues, a length that reduces the affinity of Pab1, the mRNA can become a substrate for decapping and for binding of the Pat1/Lsm1-7 complex (Tharun and Parker 2001; Chowdhury et al. 2007), which enhances the rate of decapping. This exchange of the Pab1 protein for the Pat1/Lsm1-7 complex is part of the mechanism that allows decapping to be promoted following deadenylation.

A similar mRNP dynamic is also likely to occur on the cap structure. Specifically, the competition between translation initiation and decapping suggests that prior to decapping, translation initiation factors are exchanged for decapping factors, thereby assembling a distinct "decapping" mRNP that is no longer capable of translation initiation (Tharun and Parker 2001). This idea is supported by the observation that some decapping activators also function as translational repressors (Coller and Parker 2005; Pilkington and Parker 2008; Nissan et al. 2010). Thus, mRNA decapping appears to occur in two steps, first inhibition of translation initiation and exchange of translation factors for the general repression/degradation machinery, and a second step whereby the mRNA is actually degraded. Thus, by understanding the changes in mRNP states between actively translating mRNAs and mRNAs that are translationally repressed and possibly stored or ultimately degraded we will better understand how the fate of mRNAs is controlled in the cytoplasm.

KEY PROTEINS PROMOTING DECAPPING AND REPRESSING TRANSLATION INITIATION

A conserved set of proteins function to modulate the transition between mRNAs engaged in translation initiation and an mRNP competent for mRNA decapping. Based on experiments in yeast, several of these conserved protein factors, referred to as either decapping enhancers or activators, function to stimulate the rate of decapping in vivo (Table 1). The core set of proteins affecting decapping includes Dhh1/Rck, a

Table 1. Components of P-bodies and stress granules

Name	Function	References
Components found predominantly in P-bodies		
Ccr4/Pop2/Not complex	Deadenylase	Sheth and Parker 2003; Cougot et al. 2004
Dcp1	Decapping enzyme subunit	Sheth and Parker 2003; Cougot et al. 2004
Dcp2	Decapping enzyme	Sheth and Parker 2003; Cougot et al. 2004
Edc1 and 2	Decapping activators	Neef and Thiele 2009
Edc3	Decapping activator	Kshirsagar and Parker 2004; Fenger-Gron et al. 2005
eIF4E-T	Translation repressor	Andrei et al. 2005; Ferraiuolo et al. 2005
eRF1 and eRF3	Translation termination	Buchan et al. 2008
GW182	miRNA function	Eystathioy et al. 2003
Hedls/Ge-1	Decapping activator	Fenger-Gron et al. 2005; Yu et al. 2005
Lsm1-7 complex	Decapping activator	Ingelfinger et al. 2002; Sheth and Parker 2003
Pat1/PatL1	Translation repressor/ decapping activator	Sheth and Parker 2003; Scheller et al. 2007
Upf1-3	Nonsense-mediated decay	Sheth and Parker 2006; Durand et al. 2007
Components found predominantly in stress granules		
40S ribosomal subunit	Translation	Kedersha et al. 2002; Grousl et al. 2009
Ataxin-2/Pbp1	Translation/mRNA processing	Nonhoff et al. 2007; Buchan et al. 2008
DDX3/Ded1	RNA helicase	Chalupnikova et al. 2008; Lai et al. 2008; Hilliker et al. 2011
eIF2α	Translation initiation	Kedersha et al. 2002; Kimball et al. 2003
eIF3	Translation initiation	Kedersha et al. 2002; Grousl et al. 2009
eIF4A	Translation initiation	Low et al. 2005; Buchan et al. 2011
eIF4B	Translation initiation	Low et al. 2005; Buchan et al. 2011
eIF4G	Translation initiation	Kedersha et al. 2005; Hoyle et al. 2007
FMRP	Translation, repression/ miRNA function	Mazroui et al. 2002; Kim et al. 2006
G3BP	Scaffolding protein, endoribonuclease	Tourriere et al. 2001
Pabp	polyA-binding protein	Kedersha et al. 1999; Hoyle et al. 2007
RACK1	Signaling scaffold protein	Arimoto et al. 2008
TIA-1/TIAR/Pub1/Ngr1	Translation repression/ mRNA stability	Kedersha et al. 1999; Buchan et al. 2008
Components found in both P-bodies and stress granules		
Agonaute proteins	miRNA function	Sen and Blau 2005; Leung et al. 2006
Dhh1/Rck/p54/	Translation repressor/ decapping activator	Sheth and Parker 2003; Wilczynska et al. 2005
eIF4E	Translation initiation	Andrei et al. 2005; Ferraiuolo et al. 2005; Hoyle et al. 2007
FAST	Fas activated serine/ threonine phosphoprotein	Kedersha et al. 2005
Rap55/Scd6	Translation repressor	Yang et al. 2006; Teixeira and Parker 2007
Xrn1	5′ to 3′ exonuclease	Sheth and Parker 2003; Kedersha et al. 2005

 Cite this article as *Cold Spring Harb Perspect Biol* doi: 10.1101/cshperspect.a012286

DEAD-box helicase, Pat1, Edc3, Scd6/Rap55, and the Lsm1-7 complex. Some of these decapping activators promote decapping by inhibiting translation initiation. For example, Dhh1, a member of the DEAD box family of ATPases, represses translation in vitro and its overexpression in cells inhibits translation and leads to the accumulation of cytoplasmic mRNP granules (Coller and Parker 2005; Swisher et al. 2010; Carroll et al. 2011). Similarly, Pat1, Scd6, and Stm1 (which affects the decapping of some mRNAs [Balagopal and Parker 2009]) repress translation both in vivo and in vitro (Pilkington and Parker 2008; Nissan et al. 2010; Balagopal and Parker 2011).

Decapping activators can inhibit translation at different steps. For example, the Pat1, Dhh1, and Scd6 proteins all appear to block translation before the formation of a 48S preinitiation complex (Coller and Parker 2005; Nissan et al. 2010). For Scd6, this translation repression appears to occur by direct binding to eIF4G and inhibition of the joining of the 43S complex (Rajyaguru et al. 2012). In contrast, the Stm1 protein, which promotes decapping of a subset of yeast mRNAs (Balagopal and Parker 2009), inhibits translation after formation of an 80S complex, likely through direct interactions with the ribosome (Balagopal and Parker 2011). An unresolved issue is how inhibition of translation initiation by these factors leads to decapping. One possibility is that by stalling initiation, it simply gives more time for dissociation of the translation initiation factors to allow for decapping complexes to associate with the mRNA. Alternatively, such a transition may involve an ordered exchange of factors on the mRNA, which is suggested by decapping activators, such as Pat1 and Scd6 that can directly interact with translation factors and the decapping enzyme (Nissan et al. 2010; Fromm et al. 2011). An important area for future research is determining how mRNPs are remodeled from a translationally active state to allow decapping complexes to form and degrade the mRNA. Moreover, a quantitative kinetic analysis of the binding interactions between mRNAs and translation initiation components and decapping factors would shed light on the apparent competition between translation initiation and decapping.

A second role of decapping activators is to promote the assembly of a larger decapping complex, which might also indirectly prevent translation initiation by limiting the interaction of translation initiation factors with the mRNA. The core set of decapping components shows an extensive network of direct interactions as determined by protein binding experiments with recombinant proteins and supported by coimmunoprecipitation and two hybrid analyses (Decker et al. 2007; Nissan et al. 2010). Based on coimmunoprecipitation results and the dependence of interactions on RNA, there appear to be two complexes that assemble on mRNAs targeted for decapping. One complex consists of the Pat1 protein, the Lsm1-7 complex, and Xrn1 (Bouveret et al. 2000; Tharun et al. 2000; Tharun and Parker 2001). This complex is thought to assemble on the 3′ end of deadenylated mRNAs based on its binding specificity in vitro (Chowdury et al. 2007) and the exonuclease trimming of deadenylated mRNAs in the absence of Pat1 or Lsm1 (Boeck et al. 1998; Tharun et al. 2000; He and Parker 2001). A second set of interacting proteins consists of Dcp1, Dcp2, Edc3 or Scd6 and Dhh1, although whether all these factors can associate at the same time remains to be determined. Within and between these complexes, Pat1 and Edc3 appear to play important scaffolding roles and interact with many components of the decapping machinery (Decker et al. 2007; Nissan et al. 2010).

A third role of decapping activators is to directly stimulate decapping by Dcp2. For example, the Edc3 and Pat1 proteins directly bind Dcp2 and enhance its activity in purified systems (Harigaya et al. 2010; Nissan et al. 2010). Similarly, the paralogs Edc1 and Edc2 in yeast, which are high copy suppressors of temperature-sensitive alleles in Dcp1 or Dcp2 (Dunckley et al. 2001), bind RNA and stimulate Dcp2 either in extracts or in reconstituted systems (Schwartz et al. 2003; Steiger et al. 2003). Edc1, and presumably Edc2 as well, directly bind Dcp1 to stimulate the decapping enzyme by enhancing both K_m and k_{cat} of Dcp2 (Borja et al. 2011).

Taken together, these proteins appear to modulate the mRNP composition on mRNAs to create two different classes of mRNPs, those associated with translation initiation factors and capable of recruiting ribosomes, and those associated with translational repressors and the mRNA degradation machinery. This dynamic between two different functional states also appears to be modulated by sequence specific RNA binding factors. For example, the Puf5 protein acts to repress translation and promote degradation at least in part by directly recruiting the Ccr4/Pop2/Not mRNA deadenylase complex through interactions with Pop2 (Goldstrum and Wickens 2006). Similarly, recent results show that miRNA-mediated translation repression and mRNA degradation are promoted by direct interactions between GW182 and the Not1 protein, which is a component of the Ccr4/Pop2/Not complex (Braun et al. 2011; Chekulaeva et al. 2011; Fabian et al. 2011).

NONTRANSLATING mRNAS CAN ASSEMBLE INTO RNA–PROTEIN GRANULES

In eukaryotic cells, nontranslating mRNAs can accumulate in two types of cytoplasmic mRNP granules: P-bodies, which contain the mRNA decay machinery (reviewed in Anderson and Kedersha 2006; Parker and Sheth 2007; Franks and Lykke-Andersen 2008), and stress granules, which contain many translation initiation components (Table 1) (reviewed in Buchan and Parker 2009). Moreover, stress granules and P-bodies are related to neuronal RNA granules and germ granules, which play important roles in the localization and control of mRNAs in neurons and embryos (reviewed in Kiebler and Bassell 2006; Seydoux and Braun 2006).

P-bodies are present in unstressed cells but are further induced in response to stresses or other conditions that lead to the inhibition of translation initiation (Kedersha et al. 2005; Teixeira et al. 2005). P-bodies are dynamic complexes whose assembly is dependent on, and proportional to, the pool of nontranslating mRNA (Liu et al. 2005a; Pillai et al. 2005; Teixeira et al. 2005). In addition to nontranslating mRNA, P-bodies contain the conserved core of proteins involved in mRNA decay and translation repression. These factors include the decapping enzyme complex Dcp1/Dcp2; the decapping activators Edc3 and the Lsm1-7 complex; factors that function to repress translation as well as to activate decapping including Dhh1/RCK/p54, Pat1 and Scd6/RAP55; the 5' to 3' exonuclease, Xrn1, and the Ccr4/Pop2/Not deadenylase complex (reviewed in Anderson and Kedersha 2006; Eulalio et al. 2007a; Parker and Sheth 2007). P-bodies can also contain mRNAs and proteins involved in NMD (Sheth and Parker 2006; reviewed in Franks and Lykke-Anderson 2008; Shyu et al. 2008). Metazoan P-bodies contain additional factors including proteins and miRNAs involved in the miRNA repression pathway (Eulalio et al. 2008; Lian et al. 2009). Translation initiation factors and ribosomal proteins are generally excluded from P-bodies with the exception of eIF4E in mammalian P-bodies. However, eIF4E is likely associated with repressed nontranslating mRNA in P-bodies given that P-bodies also contain eIF4E-T, which inhibits eIF4E function (Andrei et al. 2005; Ferraiuolo et al. 2005).

Stress granules are a second type of cytoplasmic mRNP granule that can be juxtaposed or overlap with P-bodies in both yeast and mammalian cells (Kedersha et al. 2005; Brengues et al. 2007; Hoyle et al. 2007; Buchan et al. 2008). Like P-bodies, stress granules are dynamic complexes whose assembly is dependent on the pool of nontranslating mRNAs. Although they share some components in common with P-bodies, stress granules typically contain translation initiation factors eIF4E, eIF4G, eIF4A, eIF4B, poly-A binding protein (Pabp), eIF3, eIF2, and the 40S ribosomal subunit (reviewed in Buchan and Parker 2009). Their composition suggests that stress granules are aggregates of mRNPs stalled in the process of translation initiation. Indeed, stress granules were first observed under stress conditions, in which translation initiation is often inhibited (Kedersha et al. 1999). However, it is now clear that stress granule formation is not limited to stress conditions, but can occur in response to a variety of blocks in translation initiation. For example, inhibition of translation initiation using drugs, knock down of translation initiation factors, or overexpression of translation repressors have all been shown to induce stress granules (reviewed in Buchan and Parker 2009). Interestingly, not all blocks to translation initiation induce stress granule assembly. For example, stress granules fail to assemble in response to depletion of eukaryotic initiation factor 3 (eIF3) subunits or to reduction in 60S subunit joining (Ohn et al. 2008; Mokas et al. 2009) suggesting there is a defined window within which translation needs to be stalled for an mRNP to be targeted to stress granules. Depending on conditions, stress granules can contain many other protein components including RNA helicases, regulators of translation and mRNA stability, and factors involved in cell signaling (reviewed in Buchan and Parker 2009; Kedersha and Anderson 2009).

HOW DO CYTOPLASMIC mRNP GRANULES ASSEMBLE?

The dependence on nontranslating mRNA on the formation of both P-bodies and stress granules suggests that they assemble first through the formation of translationally repressed mRNPs, which then aggregate into larger structures by specific protein–protein interaction domains. In yeast, P-body assembly may involve the recruitment of preexisting protein complexes to the mRNA given, as described above, that two sub-complexes of core P-body components copurify under a variety of conditions and appear to interact independent of RNA (Fig. 2). (Bouveret et al. 2000; Tharun et al. 2000; Tharun and Parker 2001; Fenger-Grøn et al. 2005; Gavin et al. 2006; Teixeira and Parker 2007). Together these observations suggest that the Dcp1/Dcp2/Dhh1/Edc3 complex, or an alternative Dcp1/Dcp2/Dhh1/Scd6 complex (Fromm et al. 2011), and the Pat1, Xrn1, and the Lsm1-7p complex are recruited onto mRNA as two groups, though the exact order of recruitment is unknown. Moreover, because Edc3, Dcp2, Scd6, and Dhh1 can interact with Pat1, these two complexes are proposed to interact to form a larger RNA–protein complex (Pilkington and Parker 2008; Nissan et al. 2010). Interestingly, because the Pat1-Lsm1-7p complex has

been proposed to bind the 3′ end (Chowdhury and Tharun 2009), and the decapping enzyme binds to the cap, this suggests a possible "closed-loop" model for mRNPs that assemble to form P-bodies. A similar rich interaction network between P-body components is likely to be the basis for formation of "P-body mRNPs" in metazoans although some of the specific interactions have been replaced or swapped between partners and additional factors also contribute.

P-bodies are then formed from these individual mRNPs aggregating into larger structures. In yeast, aggregation of mRNPs into P-bodies has been shown to be primarily dependent on a self-interaction domain (referred to as a Yjef-N domain) in the Edc3 protein and a glutamine/asparagine (Q/N) rich prion-like domain in the Lsm4 carboxyl terminus (Decker et al. 2007; Mazzoni et al. 2007; Reijns et al. 2008). Because the YjeF domain of Edc3 is conserved (Ling et al. 2008), it is likely that Edc3 will contribute to assembly of metazoan P-bodies. However, because depletion of Edc3 does not block P-body assembly in *Drosophila* S2 cells (Eulalio et al. 2007b), one anticipates that Q/N domains, and possibly other mechanisms, contribute to metazoan P-body assembly. Interestingly, multiple proteins in metazoan P-bodies contain Q/N rich domains including GW182, which functions in miRNA-mediated repression, and Ge-1/Hedls, a component of the metazoan decapping enzyme. Moreover, depletion of either of these proteins leads to decreased P-bodies in human and *Drosophila* cells (Liu et al. 2005b; Yu et al. 2005; Eulalio et al. 2007b). Finally, the Pat1 protein contributes to P-body aggregation (Buchan et al. 2008), possibly because of its role as a scaffold interacting with multiple P-body components including the Lsm1-7 complex, which is dependent on Pat1 for its localization to P-bodies (Texeira and Parker 2007). An unresolved and intriguing issue is why these protein components do not aggregate all the time. One possibility is that interaction with mRNA might in some manner promote the aggregation interactions that lead to P-bodies.

Multiple mechanisms involving protein–protein interactions between RNA-binding

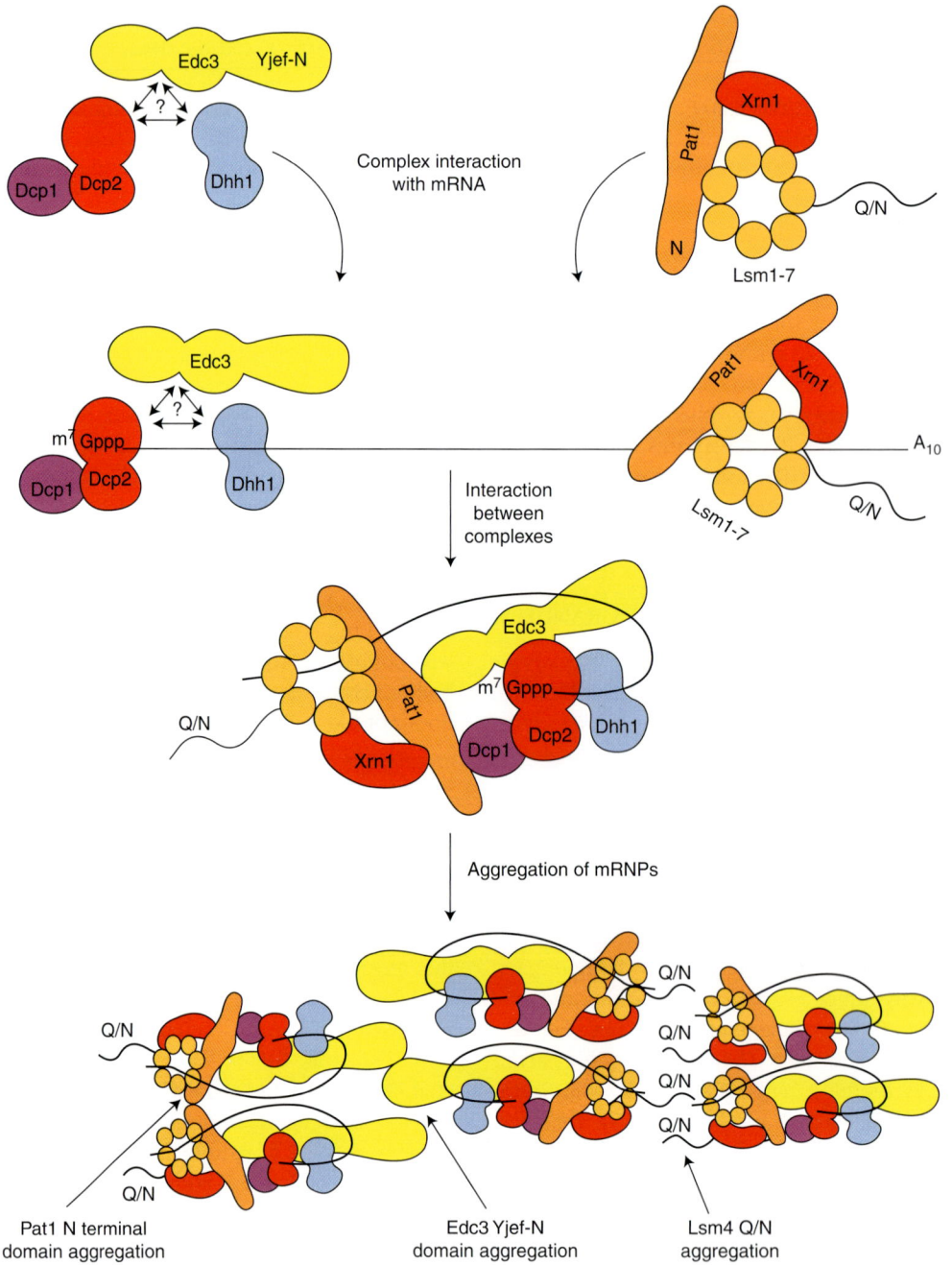

Figure 2. Model for P-body assembly in yeast. First, P-body factors are recruited to the mRNA as complexes. Second, interactions between P-body proteins in the complexes lead to the formation of a "closed-loop" structure. Finally, mRNPs aggregate via the Q/N domain of Lsm4, the Yjef-N domain of Edc3, or the amino-terminal domain of Pat1 to form microscopically visible cytoplasmic granules.

proteins have been implicated in the assembly of initiation-stalled mRNPs into stress granules. One mechanism of stress granule assembly is through the self-aggregation of QN-rich prion-like domains in the RNA binding proteins TIA-1 and TIA-R, and their orthologs (Kedersha et al. 1999, 2000; Gilks et al. 2004). Because aggregation of QN-rich prion domains is reversed by specific heat shock protein function (Rikhvanov et al. 2007), one possibility is that stress granule and P-body assembly may be promoted during stress because of accumulation of unfolded proteins, which may titrate heat shock proteins, thus driving the equilibrium of QN-rich domains toward an aggregated state. A second mechanism that contributes to stress granule assembly is the dimerization of G3BP protein (Tourriere et al. 2003). Assembly factors important under one stress condition can be unimportant during other stresses. For example, TIA-1, and its yeast homolog Pub1, facilitate stress granule assembly in response to arsenite and glucose deprivation, respectively (Gilks et al. 2004; Buchan et al. 2008), but not in response to other stresses such as heat shock (Lopez de Silanes 2005; Grousl et al. 2009). Therefore, the nature of the stress, which shapes the nontranslating mRNP pool, likely defines the assembly rules for stress granule formation.

Posttranslational modifications likely play a role in controlling the assembly of both P-bodies and stress granules (reviewed in Hilliker and Parker 2008; Buchan and Parker 2009). For example, phosphorylation of G3PB inhibits stress granule assembly (Tourriere et al. 2003). Recently, Pat1 was identified as a target for PKA phosphorylation and its phosphorylation interferes with the assembly of P-body aggregates (Ramachandran et al. 2011). Modification of proteins with O-Glc-NAc also enhances stress granule formation (Ohn et al. 2008). Finally, methylation, or the ability to bind methyl groups via Tudor domains, is necessary for localization of specific stress granule components (de Leeuw et al. 2007; Goulet et al. 2008), or their ability to drive stress granule formation when overexpressed (Hua and Zhou 2004). Methylation and Tudor domains have also been implicated in the assembly of other RNA granules (Thomson and Lasko 2004; Arkov et al. 2006; Chuma et al. 2006).

Posttranslational modification of mRNP components is an ideal mechanism to modulate mRNA function during a stress, in which rapid and reversible protein modifications allow adaptation to stress without new protein synthesis. Elucidating the key physiological targets of various modifications, and the mechanisms underlying their effects, will therefore be an important future goal.

DYNAMICS OF mRNPs IN THE CYTOPLASM

Several observations argue that cytoplasmic mRNAs can cycle between polysomes, P-bodies, and stress granules. First, inhibition of translation initiation by drugs, stresses, or mutations leads to loss of mRNAs from polysomes and a corresponding increase of mRNAs in P-bodies and stress granules (Kedersha et al. 2005; Teixeira et al. 2005; Anderson and Kedersha 2006). Second, trapping mRNAs in polysomes by blocking translation elongation decreases P-bodies and stress granules even during continued stress, which suggests that mRNAs in these compartments are in dynamic equilibrium with polysomes (Kedersha et al. 2000; Cougot et al. 2004; Teixeira et al. 2005; Mollet et al. 2008). This is consistent with the dynamic nature of P-bodies and stress granules based on fluorescence recovery after photobleaching (FRAP) studies (Andrei et al. 2005; Kedersha et al. 2005). Third, P-bodies and stress granules physically interact, often docking together in mammalian cells during stress (Kedersha et al. 2005; Wilczynska et al. 2005) or partially overlapping in yeast (Brengues and Parker 2007; Hoyle et al. 2007; Buchan et al. 2008). Finally, mRNAs within P-bodies and stress granules in yeast and mammalian cells can return to translation (Brengues et al. 2005; Anderson and Kedersha 2006; Bhattacharyya et al. 2006).

The mechanisms and directionality of mRNA movement between P-bodies and stress granules remain unresolved. During glucose deprivation in yeast, stress granules form after P-bodies, they primarily assemble on preexisting P-bodies, and are dependent on existing

P-bodies for their efficient assembly (Buchan et al. 2008). This suggests that yeast mRNAs exiting translation first form a P-body mRNP, and then mRNAs, which are targeted for reentry into translation, undergo mRNP remodeling to load translation initiation factors, thereby forming the type of mRNP that accumulates in stress granules. Some evidence however suggests that stress granules also form independently of P-bodies. First, in mammalian cells and with sodium azide stress in yeast, stress granules often form independently of visible P-bodies (Kedersha et al. 2005; Mollet et al. 2008, Buchan et al. 2011). Second, depletion of some factors in mammalian cells prevents P-body formation without affecting stress granule formation, suggesting the two processes can be uncoupled (Ohn et al. 2008). One possibility is that mRNAs may exchange in a bidirectional manner between stress granules and P-bodies and the specific mRNA, cell type, or condition may affect the predominant flow of bulk mRNA.

The movement of mRNAs between polysomes, stress granules, and P-bodies implies transitions between different mRNP states through specific rearrangements and exchanges of proteins on individual mRNAs. This may be facilitated by RNA helicases as exemplified by the Ded1 protein. Ded1 acts to assemble an mRNP intermediate that is stalled in translation initiation and accumulates in stress granules, which it then resolves in an ATP-dependent manner to allow the mRNA to reenter translation (Hilliker et al. 2011). Although P-bodies and stress granules represent microscopically visible aggregates of different mRNPs, the simplest model is that these mRNP transitions can occur independently of the larger aggregates. An important area of future work will be to determine how mRNPs in either stress granules or P-bodies are remodeled to affect their fate and how that impinges on the control of gene expression and response to stress.

A WORKING MODEL: THE mRNA CYCLE

The analyses of P-bodies and stress granules suggest a working model for the metabolism of cytoplasmic mRNA termed the mRNA Cycle

(Fig. 3). In this model, mRNAs present in polysomes undergo repeated rounds of translation initiation, elongation, and termination to produce polypeptides. In response to defects in translation initiation and/or termination, or through specific recruitment, mRNAs found in polysomes interact with proteins that repress translation initiation such as Dhh1/Rck and Pat1. At this stage, we envision run off of the elongating ribosomes, recruitment of the remainder of the decapping machinery including the decapping enzyme, and decapping followed by transcript degradation, although the relative timing of these events is unclear. After assembly of the decapping machinery individual mRNPs may aggregate into a P-body. mRNAs that are stalled in translation initiation but fail to recruit P-body components might accumulate in stress granules after elongating ribosomes run off the mRNA.

In this model, we suggest that mRNAs complexed with the decapping machinery can be degraded, aggregate into a P-body, or undergo an mRNP rearrangement wherein the degradation machinery is exchanged for translation initiation factors. Such mRNAs could then go on to initiate translation and enter polysomes. When translation initiation is inhibited, however, these mRNPs could accumulate in the stress granule state before eventually entering polysomes. Further, the state at which translation initiation is limiting might define the composition of the stress granule and therefore stress granule composition might vary in different organisms or in response to different stresses. One anticipates that specific mRNAs may preferentially accumulate in stress granules, P-bodies, or polysomes depending on their relative rates of transitions between these different biochemical states.

WHY DO mRNPs AGGREGATE?

An unresolved issue is the significance of mRNP aggregation into P-bodies and stress granules. The presence of the mRNA decay machinery and mRNA decay intermediates in P-bodies is consistent with the possibility that mRNA

 Cite this article as *Cold Spring Harb Perspect Biol* doi: 10.1101/cshperspect.a012286

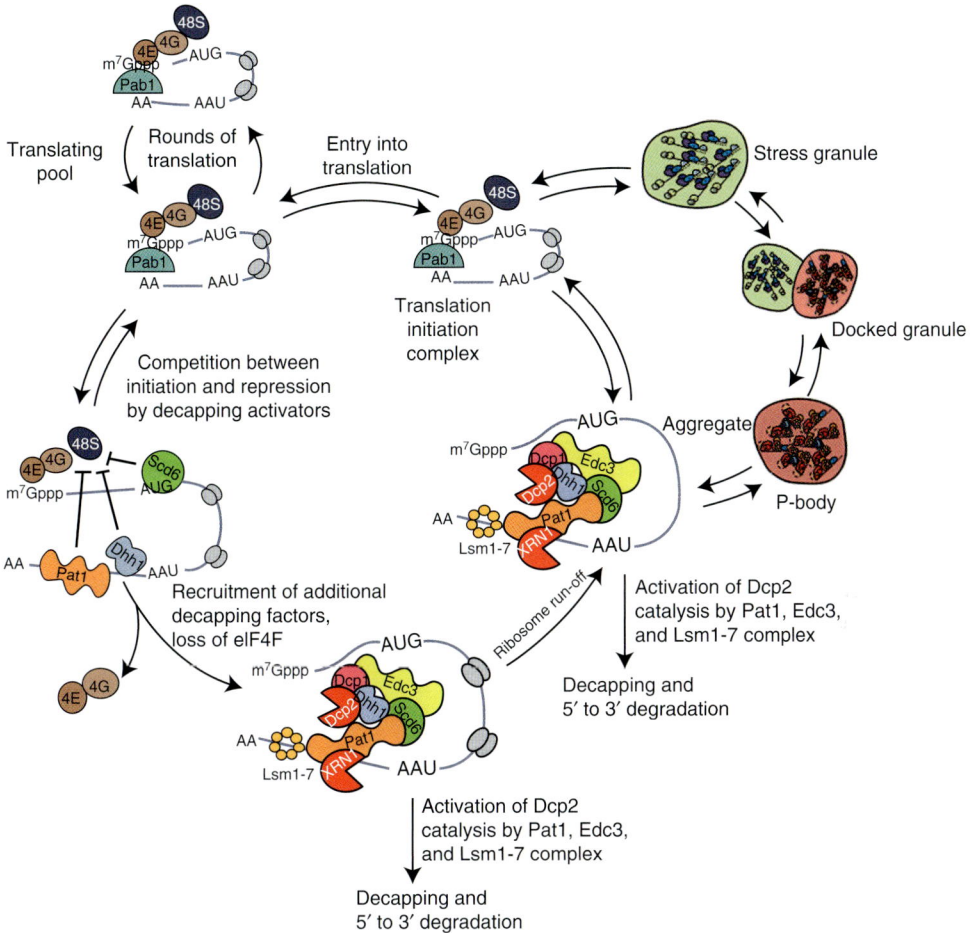

Figure 3. Model of the "mRNA cycle." Showing the dynamic movement of mRNA between polysomes, P-bodies, and stress granules, and the possible mRNP transitions between the different states of the mRNA.

degradation can occur in these structures (Sheth and Parker 2003). This possibility is supported by the observation that P-bodies increase in number and size when mRNA decay is inhibited at the stage of decapping by deletion of Dcp1 or after decapping by the deletion of Xrn1 (Sheth and Parker 2003; Cougot et al. 2004). However, to date, aggregation of mRNPs into microscopically visible P-bodies has been shown not to be required for mRNA decapping in yeast (Decker et al. 2007), for translation repression during stress in both yeast and mammals (Decker et al. 2007; Kwon et al. 2007; Buchan et al. 2008; Ohn et al. 2008), or for

mRNA stability during stress, at least in yeast (Buchan 2008). In addition, in metazoans, depletion of microscopically visible P-bodies does not seem to affect miRNA-mediated repression, decay of messages containing ARE elements, or decay of transcripts subject to NMD (Chu and Rana 2006; Stoecklin et al. 2006; Eulalio et al. 2007b). In contrast, deletion of the Lsm4 carboxy-terminal domain, which promotes P-body assembly (Decker et al. 2007; Mazzoni et al. 2007; Reijns et al. 2008), can affect mRNA degradation in at least some strains and/or conditions (Reijns et al. 2008), which suggests that aggregation of individual mRNPs into larger structures may have some role in mRNA degradation. In addition, there is some evidence that P-body aggregation may play a role in the long-term survival of yeast cells during stationary phase (Ramachandran et al. 2011). Moreover, because aggregation into RNP granules is a conserved feature of eukaryotic cells it is anticipated to have some role. One possibility is that mRNP aggregation has consequences for the control of translation and/or degradation but these functions are either limited to a subset of mRNAs or conditions, or are performed by granules below the detection limit of the light microscope.

More generally, the formation of RNP granules such as stress granules and P-bodies is expected to have specific consequences both by increasing the local concentration of factors within granules, and by depleting them from the bulk cytosol. For comparison, Cajal bodies improve the assembly of spliceosomal small nuclear ribonucleoprotein particles (snRNPs) by increasing the local concentrations of U4/U6 (Klingauf et al. 2006). By analogy, the concentration of Dcp2 in P-bodies might facilitate its interaction with mRNAs when Dcp2 is limiting, or the concentration of translation initiation factors in stress granules might drive the formation of productive translation complexes. In addition, an important role of RNP granules may be to remove factors from the cytosol. For example, formation of stress granules may sequester RACK1 away from MAP kinases, thereby limiting signal transduction and apoptosis (Arimoto et al. 2008). Moreover, the aggregation of mRNPs into stress granules and P-bodies may provide a buffering system for maintaining a proper ratio of translation capacity to the pool of mRNAs that are translating (discussed in Coller and Parker 2004). An excessive amount of mRNAs within the translating pool may compete for limiting translation factors and thereby prevent effective translation of many mRNAs.

CONCLUDING REMARKS

Although considerable advances have been made in the understanding of the mechanisms of mRNA decapping and the subcellular distri-

bution of different mRNPs, there are several outstanding questions that need to be addressed. One key issue will be to understand the molecular functions of decapping activators and how they affect translation mechanisms as well as the recruitment and stimulation of the decapping enzyme. A second important challenge is to understand the significance of the aggregation of mRNPs into P-bodies and stress granules, which is likely to contribute to our growing understanding of the importance of sub-cellular organization. Finally, it will be critical to understand the mechanisms and rates of the transitions of mRNPs between polysomes, P-bodies, and stress granules. Here it will be critical to understand the frequency and directionality of these exchanges, the molecular mechanisms that move mRNAs from one state to another, and how these states differ on individual mRNAs, thereby impacting the control of gene expression.

ACKNOWLEDGMENTS

We thank R. Buchan, S. Jain, and V. Balagopal for helpful discussions and A. Webb for technical support in the preparation of the manuscript. This work is supported by funds from the National Institutes of Health (Grant R37 GM45443) and the Howard Hughes Medical Institute.

REFERENCES

Anderson P, Kedersha N. 2006. RNA Granules. *J Cell Biol* **172:** 803–808.

Anderson JS, Parker RP. 1998. The 3′ to 5′ degradation of yeast mRNAs is a general mechanism for mRNA turnover that requires the SKI2 DEVH box protein and 3′ to 5′ exonucleases of the exosome complex. *EMBO J* **17:** 1497–1506.

Andrei MA, Ingelfinger D, Heintzmann R, Achsel T, Rivera-Pomar R, Luhrmann R. 2005. A role for eIF4E and eIF4E-transporter in targeting mRNPs to mammalian processing bodies. *RNA* **11:** 717–727.

Arimoto K, Fukuda H, Imajoh-Ohmi S, Saito H, Takekawa M. 2008. Formation of stress granules inhibits apoptosis by suppressing stress-responsive MAPK pathways. *Nat Cell Biol* **10:** 1324–1332.

Arkov A, Wang JY, Ramos A, Lehmann R. 2006. The role of Tudor domains in germline development and polar granule architecture. *Development* **133:** 4053–4062.

Balagopal V, Parker R. 2009. Stm1 modulates mRNA decay and Dhh1 function in *Saccharomyces cerevisae. Genetics* **181:** 93–103.

Balagopal V, Parker R. 2011. Stm1 modulates translation after 80S formation in *Saccharomyces cerevisiae. RNA* **17:** 835–842.

Behm-Ansmant I, Rehwinkel J, Doerks T, Stark A, Bork P, Izaurralde E. 2006. mRNA degradation by miRNAs and GW182 requires both CCR4:NOT deadenylase and DCP1:DCP2 decapping complexes. *Genes Dev* **20:** 1885–1898.

Bhattacharyya SN, Habermacher R, Martine U, Closs EI, Filipowicz W. 2006. Relief of microRNA-mediated translational repression in human cells subjected to stress. *Cell* **125:** 1111–1124.

Boeck R, Lapeyre B, Brown CE, Sachs AB. 1998. Capped mRNA degradation intermediates accumulate in the yeast spb8-2 mutant. *Mol Cell Biol* **18:** 5062–5072.

Borja MS, Piotukh K, Freund C, Gross JD. 2011. Dcp1 links coactivators of mRNA decapping to Dcp2 by proline recognition. *RNA* **17:** 278–290.

Bouveret E, Rigaut G, Shevchenko A, Wilm M. 2000. A Sm-like protein complex that participates in mRNA degradation. *EMBO J* **19:** 1661–1671.

Braun JE, Huntzinger E, Fauser M, Izzaurralde E. 2011. GW182 proteins directly recruit cytoplasmic deadenylase complexes to miRNA targets. *Mol Cell* **44:** 120–133.

Brengues M, Parker R. 2007. Accumulation of polyadenylated mRNA, Pab1p, eIF4E, and eIF4G with P-bodies in *Saccharomyces cerevisiae. Mol Biol Cell* **18:** 2592–2602.

Brengues M, Teixeira D, Parker R. 2005. Movement of eukaryotic mRNAs between polysomes and cytoplasmic processing bodies. *Science* **310:** 486–489.

Buchan JR, Parker R. 2009. Eukaryotic stress granules: The ins and outs of translation. *Mol Cell* **36:** 932–941.

Buchan JR, Muhlrad D, Parker R. 2008. P-bodies promote stress granule assembly in *Saccharomyces cerevisiae. J Cell Biol* **183:** 441–455.

Buchan JR, Yoon J-H, Parker R. 2011. Stress-specific composition, assembly and kinetics of stress granules in *Saccharomyces cerevisiae. J Cell Sci* **124:** 228–239.

Cao D, Parker R. 2003. Computational modeling and experimental analysis of nonsense-mediated decay in yeast. *Cell* **113:** 533–545.

Carroll JS, Munchel SE Weis K. 2011. The DExD/H box ATPase Dhh1 functions in translational repression, mRNA decay, and processing body dynamics. *J Cell Biol* **194:** 527–537.

Chalupnikova K, Lattmann S, Selak N, Iwamoto F, Fujiki Y, Nagamine Y. 2008. Recruitment of the RNA helicase RHAU to stress granules via a unique RNA-binding domain. *J Biol Chem* **283:** 35186–35198.

Chekulaeva M, Mathys H, Zipprich JT, Attig J, Colic M, Parker R, Filipowicz W. 2011. miRNA repression involves GW182-mediated recruitment of CCR4-NOT through conserved W-containing motifs. *Nat Struct Mol Biol* **18:** 1218–1226.

Chowdhury A, Tharun S. 2009. Activation of decapping involves binding of the mRNA and facilitation of the post-binding steps by the Lsm1-7-Pat1 complex. *RNA* **15:** 1837–1848.

Chowdhury A, Mukhopadhyay J, Tharun S. 2007. The decapping activator Lsm1p 7p-Pat1p complex has the intrinsic ability to distinguish between oligoadenylated and polyadenylated RNAs. *RNA* **13:** 998–1016.

Chu CY, Rana TM. 2006. Translation repression in human cells by microRNA-induced gene silencing requires RCK/p54. *PLoS Biol* **4:** e210.

Chuma S, Hosokawa M, Kitamura K, Kasai S, Fujioka M, Hiyoshi M, Takamune K, Noce T, Nakatsuji N. 2006. Tdrd1/Mtr-1, a tudor-related gene, is essential for male germ-cell differentiation and nuage/germinal granule formation in mice. *Proc Natl Acad Sci* **103:** 15894–15899.

Coller J, Parker R. 2004. Eukaryotic mRNA decapping. *Annu Rev Biochem* **73:** 861–890.

Coller J, Parker R. 2005. General translational repression by activators of mRNA decapping. *Cell* **122:** 875–886.

Cosson B, Couturier A, Chabelskaya S, Kiktev D, Inge-Vechtomov S, Philippe M, Zhouravleva G. 2002. Poly(A)-binding protein acts in translation termination via eukaryotic release factor 3 interaction and does not influence [PSI(+)] propagation. *Mol Cell Biol* **22:** 3301–3315.

Cougot N, Babajko S, Seraphin B. 2004. Cytoplasmic foci are sites of mRNA decay in human cells. *J Cell Biol* **165:** 31–40.

Daugeron MC, Mauxion F, Seraphin B. 2001. The yeast POP2 gene encodes a nuclease involved in mRNA deadenylation. *Nucleic Acids Res* **29:** 2448–2455.

Decker CJ, Parker R. 1993. A turnover pathway for both stable and unstable mRNAs in yeast: Evidence for a requirement for deadenylation. *Genes Dev* **7:** 1632–1643.

Decker CJ, Teixeira D, Parker R. 2007. Edc3p and a glutamine/asparagine-rich domain of Lsm4p function in processing body assembly in *Saccharomyces cerevisiae. J Cell Biol* **179:** 437–449.

De Leeuw F, Zhang T, Wauquier C, Huez G, Kruys V, Gueydan C. 2007. The cold-inducible RNA-binding protein migrates from the nucleus to cytoplasmic stress granules by a methylation-dependent mechanism and acts as a translational repressor. *Exp Cell Res* **313:** 4130–4144.

Doma MK, Parker R. 2006. Endonucleolytic cleavage of eukaryotic mRNAs with stalls in translation elongation. *Nature* **440:** 561–564.

Dunckley T, Parker R. 1999. The DCP2 protein is required for mRNA decapping in *Saccharomyces cerevisiae* and contains a functional MutT motif. *EMBO J* **18:** 5411–5422.

Dunckley T, Parker R. 2001. Two related proteins, Edc1p and Edc2p, stimulate mRNA decapping in *Saccharomyces cerevisiae. Genetics* **157:** 27–37.

Durand S, Cougot N, Mahuteau-Betzer F, Nguyen CH, Grierson DS, Bertrand E, Tazi J, Lejeune F. 2007. Inhibition of nonsense-mediated mRNA decay (NMD) by a new chemical molecule reveals the dynamic of NMD factors in P-bodies. *J Cell Biol* **178:** 1145–1160.

Eulalio A, Behm-Ansmant I, Izaurralde E. 2007a. P-bodies: At the crossroads of post-transcriptional pathways. *Nat Rev Mol Cell Biol* **8:** 9–22.

Eulalio A, Behm-Ansmant I, Schweizer D, Izaurralde E. 2007b. P-body formation is a consequence, not the cause, of RNA-mediated gene silencing. *Mol Cell Biol* **27:** 3970–3981.

Eulalio A, Huntzinger E, Izaurralde E. 2008. GW182 interaction with Argonaute is essential for miRNA-mediated translational repression and mRNA decay. *Nat Struct Mol Biol* **15:** 346–353.

Eystathioy T, Jakymiw A, Chan EK, Séraphin B, Cougot N, Fritzler MJ. 2003. The GW182 protein colocalizes with mRNA degradation associated proteins hDcp1 and hLsm4 in cytoplasmic GW bodies. *RNA* **9:** 1171–1173.

Fabian MR, Cieplak MK, Frank F, Morita M, Green J, Srikumar T, Nagar B, Yamamoto T, Raught B, Duchaine TF, et al. 2011. miRNA-mediated deadenylation is orchestrated by GW182 through two conserved motifs that interact with CCR4-NOT. *Nat Struct Mol Biol* **18:** 1211–1217.

Fenger-Grøn M, Fillman C, Norrild B, Lykke-Andersen J. 2005. Multiple processing body factors and the ARE binding protein TTP activate mRNA decapping. *Molecular Cell* **20:** 905–915.

Ferraiuolo MA, Basak S, Dostie J, Murray EL, Schoenberg DR, Sonenberg N. 2005. A role for the eIF4E-binding protein 4E-T in P-body formation and mRNA decay. *J Cell Biol* **170:** 913–24.

Franks TM, Lykke-Andersen J. 2008. The control of mRNA decapping and P body formation. *Mol Cell* **32:** 605–615.

Frischmeyer PA, van Hoof A, O'Donnell K, Guerrerio AL, Parker R, Dietz HC. 2002. An mRNA surveillance mechanism that eliminates transcripts lacking termination codons. *Science* **295:** 2258–2261.

Fromm SA, Truffault V, Kamenz J, Braun JE, Hoffmann NA, Izaurralde E, Sprangers R. 2011. The structural basis of Edc3- and Scd6-mediated activation of the Dcp1:Dcp2 mRNA decapping complex. *EMBO J* **31:** 279–290.

Garneau NL, Wilusz J, Wilusz CJ. 2007. The highways and byways of mRNA decay. *Nat Rev Mol Cell Biol* **8:** 113–126.

Gavin AC, Aloy P, Grandi P, Krause R, Boesche M, Marzioch M, Rau C, Jensen LJ, Bastuck S, Dümpelfeld B, et al. 2006. Proteome survey reveals modularity of the yeast cell machinery. *Nature* **440:** 631–636.

Gilks N, Kedersha N, Ayodele M, Shen L, Stoecklin G, Dember LM, Anderson P. 2004. Stress granule assembly is mediated by prion-like aggregation of TIA-1. *Mol Biol Cell* **15:** 5383–5398.

Goldstrohm AC, Hook BA, Seay DJ, Wickens M. 2006. PUF proteins bind Pop2p to regulate messenger RNAs. *Nat Struct Mol Biol* **13:** 533–539.

Goulet I, Boisvenue S, Mokas S, Mazroui R, Côté J. 2008. TDRD3, a novel Tudor domain-containing protein, localizes to cytoplasmic stress granules. *Hum Mol Genet* **17:** 3055–3074.

Grousl T, Ivanov P, Frydlova I, Vasicova P, Janda F, Vojtova J, Malinska K, Malcova I, Novakova L, Janoskova D, Hasek J, et al. 2009. Robust heat shock induces eIF2α-phosphorylation-independent assembly of stress granules containing eIF3 and 40S ribosomal subunits in budding yeast, *Saccharomyces cerevisiae*. *J Cell Sci* **122:** 2078–2088.

Harigaya Y, Parker R. 2010. No-go decay: A quality control mechanism for RNA in translation. *RNA* **1:** 132–141.

Harigaya Y, Jones BN, Muhlrad D, Gross JD, Parker R. 2010. Identification and analysis of the interaction between Edc3 and Dcp2 in *Saccharomyces cerevisiae*. *Mol Cell Biol* **30:** 1446–1456.

He W, Parker R. 2001. The yeast cytoplasmic Lsm1/Pat1p complex protects mRNA 3′ termini from partial degradation. *Genetics* **158:** 1445–1455.

Hilliker A, Parker R. 2008. Stressed out? Make some modifications! *Nat Cell Biol* **10:** 1129–1130.

Hilliker A, Gao Z, Jankowsky E, Parker R. 2011. The DEAD-box protein Ded1 modulates translation by the formation and resolution of an eIF4F-mRNA complex. *Mol Cell* **43:** 962–972.

Hoyle NP, Castelli LM, Campbell SG, Holmes LEA, Ashe MP. 2007. Stress-dependent relocalization of translationally primed mRNPs to cytoplasmic granules that are kinetically and spatially distinct from P-bodies. *J Cell Biol* **179:** 65–74.

Hsu CL, Stevens A. 1993. Yeast cells lacking 5′→3′ exoribonuclease 1 contain mRNA species that are poly(A) deficient and partially lack the 5′ cap structure. *Mol Cell Biol* **13:** 4826–4835.

Hua Y, Zhou J. 2004. Survival motor neuron protein facilitates assembly of stress granules. *FEBS Lett* **572:** 69–74.

Ingelfinger D, Arndt-Jovin DJ, Luhrmann R, Achsel T. 2002. The human LSm1-7 proteins colocalize with the mRNA-degrading enzymes Dcp1/2 and Xrnl in distinct cytoplasmic foci. *RNA* **8:** 1489–1501.

Isken O, Maquat LE. 2007. Quality control of eukaryotic mRNA: Safeguarding cells from abnormal mRNA function. *Genes Dev* **21:** 1833–1856.

Kedersha N, Anderson P. 2009. Regulation of translation by stress granules and processing bodies. *Prog Mol Biol Transl Sci* **90:** 155–185.

Kedersha NL, Gupta M, Li W, Miller I, Anderson P. 1999. RNA-binding proteins TIA-1 and TIAR link the phosphorylation of eIF-2 α to the assembly of mammalian Stress Granules. *J Cell Biol* **147:** 1431–1442.

Kedersha N, Cho MR, Li W, Yacono PW, Chen S, Gilks N, Golan DE, Anderson P. 2000. Dynamic shuttling of TIA-1 accompanies the recruitment of mRNA to mammalian stress granules. *J Cell Biol* **151:** 1257–1268.

Kedersha N, Chen S, Gilks N, Li W, Miller IJ, Stahl J, Anderson P. 2002. Evidence that ternary complex (eIF2-GTP-tRNA(i)(Met))-deficient preinitiation complexes are core constituents of mammalian stress granules. *Mol Biol Cell* **13:** 195–210.

Kedersha N, Stoecklin G, Ayodele M, Yacono P, Lykke-Andersen J, Fritzler MJ, Scheuner D, Kaufman RJ, Golan DE, Anderson P. 2005. Stress granules and processing bodies are dynamically linked sites of mRNP remodeling. *J Cell Biol* **169:** 871–884.

Kiebler MA, Bassell GJ. 2006. Neuronal RNA granules: Movers and makers. *Neuron* **51:** 685–90.

Kim SH, Dong WK, Weiler IJ, Greenough WT. 2006. Fragile X mental retardation protein shifts between polyribosomes and stress granules after neuronal injury by arsenite stress or in vivo hippocampal electrode insertion. *J Neurosci* **26:** 2413–2418.

Kimball SR, Horetsky RL, Ron D, Jefferson LS, Harding HP. 2003. Mammalian stress granules represent sites of accumulation of stalled translation initiation complexes. *Am J Physiol Cell Physiol* **284:** C273–84.

Klingauf M, Stanek D, Neugebauer KM. 2006. Enhancement of U4/U6 small nuclear ribonucleoprotein particle association in Cajal bodies predicted by mathematical modeling. *Mol Biol Cell* **17:** 4972–4981.

Kshirsagar M, Parker R. 2004. Identification of Edc3p as an enhancer of mRNA decapping in *Saccharomyces cerevisiae*. *Genetics* **166:** 729–739.

Kwon S, Zhang Y, Matthias P. 2007. The deacetylase HDAC6 is a novel critical component of stress granules involved in the stress response. *Genes Dev* **21:** 3381–3394.

Lagrandeur T, Parker R. 1999. The *cis* acting sequences responsible for the differential decay of the unstable MFA2 and stable PGK1 transcripts in yeast include the context of the translational start codon. *RNA* **5:** 420–433.

Lai MC, Lee YH, Tarn WY. 2008. The DEAD-box RNA helicase DDX3 associates with export messenger ribonucleoproteins as well as tip-associated protein and participates in translational control. *Mol Biol Cell* **19:** 3847–3858.

Leung AK, Calabrese JM, Sharp PA. 2006. Quantitative analysis of Argonaute protein reveals microRNA-dependent localization to stress granules. *Proc Natl Acad Sci* **103:** 18125–18130.

Lian S, Li S, Abadal G, Pauley B, Fritzler M, Chan EK. 2009. The C-terminal half of human Ago2 binds to multiple GW-rich regions of GW182 and requires GW182 to mediate silencing. *RNA* **15:** 804–813.

Ling SH, Decker CJ, Walsh MA, She M, Parker R, Song H. 2008. Crystal structure of human Edc3 and its functional implications. *Mol Cell Biol* **28:** 5965–5976.

Liu J, Valencia-Sanchez MA, Hannon GJ, Parker R. 2005a. MicroRNA-dependent localization of targeted mRNAs to mammalian P-bodies. *Nat Cell Biol* **7:** 719–723.

Liu J, Rivas FV, Wohlschlegel J, Yates JR 3rd, Parker R, Hannon GJ. 2005b. A role for the P-body component GW182 in microRNA function. *Nat Cell Biol* **7:** 1261–1266.

López de Silanes I, Galbán S, Martindale JL, Yang X, Mazan-Mamczarz K, Indig FE, Falco G, Zhan M, Gorospe M. 2005. Identification and functional outcome of mRNAs associated with RNA-binding protein TIA-1. *Mol Cell Biol* **25:** 9520–9531.

Low WK, Dang Y, Schneider-Poetsch T, Shi Z, Choi NS, Merrick WC, Romo D, Liu JO. 2005. Inhibition of eukaryotic translation initiation by the marine natural product pateamine A. *Mol Cell* **20:** 709–722.

Mazroui R, Huot ME, Tremblay S, Filion C, Labelle Y, Khandjian EW. 2002. Trapping of messenger RNA by Fragile X Mental Retardation protein into cytoplasmic granules induces translation repression. *Hum Mol Genet* **11:** 3007–3017.

Mazzoni C, D'Addario I, Falcone C. 2007. The C-terminus of the yeast Lsm4p is required for the association to P-bodies. *FEBS Lett* **581:** 4836–4840.

Mitchell P, Tollervey D. 2003. An NMD pathway in yeast involving accelerated deadenylation and exosome-mediated $3' \to 5'$ degradation. *Mol Cell* **11:** 1405–1413.

Mokas S, Mills JR, Garreau C, Fournier MJ, Robert F, Arya P, Kaufman RJ, Pelletier J, Mazroui R. 2009. Uncoupling stress granule assembly and translation initiation inhibition. *Mol Biol Cell* **20:** 2673–2683.

Mollet S, Cougot N, Wilczynska A, Dautry F, Kress M, Bertrand E, Weil D. 2008. Translationally repressed mRNA transiently cycles through stress granules during stress. *Mol Biol Cell* **19:** 4469–4479.

Muckenthaler M, Gunkel N, Stripecke R, Hentze MW. 1997. Regulated poly(A) tail shortening in somatic cells mediated by cap-proximal translation repressor proteins and ribosome association. *RNA* **3:** 983–995.

Muhlrad D, Parker R. 1992. Mutations affecting stability and deadenylation of the yeast MFA2 transcript. *Genes Dev* **6:** 2100–2111.

Muhlrad D, Parker R. 1994. Premature translational termination triggers mRNA decapping. *Nature* **370:** 578–581.

Muhlrad D, Decker CJ, Parker R. 1994. Deadenylation of the unstable mRNA encoded by the yeast MFA2 gene leads to decapping followed by $5' \to 3'$ digestion of the transcript. *Genes Dev* **8:** 855–866.

Muhlrad D, Decker CJ, Parker R. 1995. Turnover mechanisms of the stable yeast PGK1 mRNA. *Mol Cell Biol* **15:** 2145–2156.

Neef DW, Thiele DJ. 2009. Enhancer of decapping proteins 1 and 2 are important for translation during heat stress in *Saccharomyces cerevisiae*. *Mol Microbiol* **73:** 1032–1042.

Nissan T, Rajyaguru P, She M, Song H, Parker R. 2010. Decapping activators in *Saccharomyces cerevisiae* act by multiple mechanisms. *Mol Cell* **39:** 773–783.

Nonhoff U, Ralser M, Welzel F, Piccini I, Balzereit D, Yaspo ML, Lehrach H, Krobitsch S. 2007. Ataxin-2 interacts with the DEAD/H-box RNA helicase DDX6 and interferes with P-bodies and stress granules. *Mol Biol Cell* **18:** 1385–1396.

Ohn T, Kedersha N, Hickman T, Tisdale S, Anderson P. 2008. A functional RNAi screen links O-GlcNAc modification of ribosomal proteins to stress granule and processing body assembly. *Nat Cell Biol* **10:** 1224–1231.

Parker R, Sheth U. 2007. P-bodies and the control of mRNA translation and degradation. *Mol Cell* **25:** 635–646.

Parker R, Song H. 2004. The enzymes and control of eukaryotic mRNA turnover. *Nat Struct Mol Biol* **11:** 121–127.

Pilkington GR, Parker R. 2008. Pat1 contains distinct functional domains that promote P-body assembly and activation of decapping. *Mol Cell Biol* **28:** 1298–1312.

Pillai RS, Bhattacharyya SN, Artus CG, Zoller T, Cougot N, Basyuk E, Bertrand E, Filipowicz W. 2005. Inhibition of translational initiation by Let-7 MicroRNA in human cells. *Science* **309:** 1573–1576.

Rajyaguru P, She M, Parker R. 2012. Scd6 targets eIF4G to repress translation: RGG motif proteins as a class of eIF4G-binding proteins. *Mol Cell* **45:** 244–254.

Ramachandran V, Shah KH, Herman PK. 2011. The cAMP-dependent protein kinase signaling pathway is a key regulator of P body foci formation. *Mol Cell* **43:** 973–981.

Reijns MA, Alexander RD, Spiller MP, Beggs JD. 2008. A role for Q/N-rich aggregation-prone regions in P-body localization. *J Cell Sci* **121:** 2463–2472.

Rikhvanov EG, Romanova NV, Chernoff YO. 2007. Chaperone effects on prion and nonprion aggregates. *Prion* **1:** 217–222.

Scheller N, Resa-Infante P, de la Luna S, Galao RP, Albrecht M, Kaestner L, Lipp P, Lengauer T, Meyerhans A, Díez J. 2007. Identification of PatL1, a human homolog to yeast

P-body component Pat1. *Biochim Biophys Acta* **1773:** 1786–92.

Schwartz DC, Parker R. 1999. Mutations in translation initiation factors lead to increased rates of deadenylation and decapping of mRNAs in *Saccharomyces cerevisiae*. *Mol Cell Biol* **19:** 5247–5256.

Schwartz DC, Parker R. 2000. mRNA decapping in yeast requires dissociation of the cap binding protein, eukaryotic translation initiation factor 4E. *Mol Cell Biol* **20:** 7933–7942.

Schwartz D, Decker CJ, Parker R. 2003. The enhancer of decapping proteins, Edc1p and Edc2p, bind RNA and stimulate the activity of the decapping enzyme. *RNA* **9:** 239–251.

Sen GL, Blau HM. 2005. Argonaute 2/RISC resides in sites of mammalian mRNA decay known as cytoplasmic bodies. *Nat Cell Biol* **7:** 633–636.

Seydoux G, Braun RE. 2006. Pathway to totipotency: Lessons from germ cells. *Cell* **127:** 891–904.

Sheth U, Parker R. 2003. Decapping and decay of messenger RNA occur in cytoplasmic processing bodies. *Science* **300:** 805–808.

Sheth U, Parker R. 2006. Targeting of aberrant mRNAs to cytoplasmic processing bodies. *Cell* **125:** 1095–1109.

Shyu AB, Wilkinson MF, van Hoof A. 2008. Messenger RNA regulation: To translate or to degrade. *EMBO J* **27:** 471–481.

Song MG, Li Y, Kiledjian M. 2010. Multiple mRNA decapping enzymes in mammalian cells. *Mol Cell* **40:** 423–432.

Steiger M, Carr-Schmid A, Schwartz DC, Kiledjian M, Parker R. 2003. Analysis of recombinant yeast decapping enzyme. *RNA* **9:** 231–238.

Stoecklin G, Mayo T, Anderson P. 2006. ARE-mRNA degradation requires the $5'$–$3'$ decay pathway. *EMBO Rep* **7:** 72–77.

Swisher KD, Parker R. 2010. Localization to, and effects of Pbp1, Pbp4, Lsm12, Dhh1, and Pab1 on stress granules in *Saccharomyces cerevisiae*. *PLoS ONE* **5:** e10006.

Teixeira D, Parker R. 2007. Analysis of P-body assembly in *Saccharomyces cerevisiae*. *Mol Biol Cell* **18:** 2274–2287.

Teixeira D, Sheth U, Valencia-Sanchez MA, Brengues M, Parker R. 2005. Processing bodies require RNA for assembly and contain nontranslating mRNAs. *RNA* **11:** 371–382.

Tharun S, Parker R. 2001. Targeting an mRNA for decapping: Displacement of translation factors and association of the Lsm1p-7p complex on deadenylated yeast mRNAs. *Mol Cell* **8:** 1075–1083.

Tharun S, He W, Mayes AE, Lennertz P, Beggs JD, Parker R. 2000. Yeast Sm-like proteins function in mRNA decapping and decay. *Nature* **404:** 515–518.

Thomson T, Lasko P. 2004. *Drosophila tudor* is essential for polar granule assembly and pole cell specification, but not for posterior patterning. *Genesis* **40:** 164–170.

Thore S, Mauxion F, Seraphin B, Suck D. 2003. X-ray structure and activity of the yeast Pop2 protein: A nuclease subunit of the mRNA deadenylase complex. *EMBO Rep* **4:** 1150–1155.

Tourrière H, Gallouzi IE, Chebli K, Capony JP, Mouaikel J, van der Geer P, Tazi J. 2001. RasGAP-associated endoribonuclease G3BP: Selective RNA degradation and phosphorylation-dependent localization. *Mol Cell Biol* **21:** 7747–7760.

Tourrière H, Chebli K, Zekri L, Courslaud B, Blanchard JM, Bertrand E, Tazi J. 2003. The RasGAP-associated endoribonuclease G3BP assembles stress granules. *J Cell Biol* **160:** 823–831.

Tucker M, Valencia-Sanchez MA, Staples RR, Chen J, Denis CL, Parker R. 2001. The transcription factor associated Ccr4 and Caf1 proteins are components of the major cytoplasmic mRNA deadenylase in *Saccharomyces cerevisiae*. *Cell* **104:** 377–386.

Tucker M, Staples RR, Valencia-Sanchez MA, Muhlrad D, Parker R. 2002. Ccr4p is the catalytic subunit of a Ccr4p/Pop2p/Notp mRNA deadenylase complex in *Saccharomyces cerevisiae*. *EMBO J* **21:** 1427–1436.

van Dijk E, Cougot N, Meyer S, Babajko S, Wahle E, Séraphin B. 2002. Human Dcp2: A catalytically active mRNA decapping enzyme located in specific cytoplasmic structures. *EMBO J* **21:** 6915–6924.

Van Hoof A, Frischmeyer PA, Dietz HC, Parker R. 2002. Exosome-mediated recognition and degradation of mRNAs lacking a termination codon. *Science* **295:** 2262–2264.

Wang Z, Kiledjian M. 2001. Functional link between the mammalian exosome and mRNA decapping. *Cell* **107:** 751–762.

Wickens M, Bernstein DS, Kimble J, Parker R. 2002. A PUF family portrait: $3'$UTR regulation as a way of life. *Trends Genet* **18:** 150–157.

Wilczynska A, Aigueperse C, Kress M, Dautry F, Weil D. 2005. The translational regulator CPEB1 provides a link between dcp1 bodies and stress granules. *J Cell Sci* **118:** 981–992.

Yang WH, Yu JH, Gulick T, Bloch KD, Bloch DB. 2006. RNA-associated protein 55 (RAP55) localizes to mRNA processing bodies and stress granules. *RNA* **12:** 547–554.

Yu JH, Yang WH, Gulick T, Bloch KD, Bloch DB. 2005. Ge-1 is a central component of the mammalian cytoplasmic mRNA processing body. *RNA* **11:** 1795–1802.

mRNA Localization and Translational Control in *Drosophila* Oogenesis

Paul Lasko

Department of Biology, Bellini Life Sciences Building, McGill University, Montréal, Québec H3G 0B1, Canada

Correspondence: paul.lasko@mcgill.ca

Localization of an mRNA species to a particular subcellular region can complement translational control mechanisms to produce a restricted spatial distribution of the protein it encodes. mRNA localization has been studied most in asymmetric cells such as budding yeast, early embryos, and neurons, but the process is likely to be more widespread. This article reviews the current state of knowledge about the mechanisms of mRNA localization and its functions in early embryonic development, focusing on *Drosophila* where the relevant knowledge is most advanced. Links between mRNA localization and translational control mechanisms also are examined.

Cell polarization requires proteins to be asymmetrically localized, which can be achieved by localizing specific mRNAs to particular regions of the cytoplasm so that their translation occurs only there mRNA localization is often inefficient, thus it is usually coupled to translational control mechanisms that repress translation of unlocalized mRNA while allowing translation of the localized mRNA to proceed. Genome-wide analysis of mRNA localization in early *Drosophila* embryos showed that the majority of mRNAs are asymmetrically distributed (Lécuyer et al. 2007; Tomancak et al. 2007).

The *Drosophila* oocyte is a valuable model system to study mRNA localization and translational control. In organisms such as *Drosophila* in which zygotic transcription does not commence until many nuclear or cellular divisions have occurred, translational control of maternally encoded mRNAs necessarily has a widespread role in regulating gene expression so that the initial stages of development can proceed. *Drosophila* oocytes develop within multicellular entities called egg chambers (King 1970). Each egg chamber contains a syncytium of 16 germ line cells (called cystocytes), which are connected by cytoplasmic bridges (ring canals). Only one cystocyte adopts an oocyte fate and completes meiosis while its siblings develop into polyploid nurse cells. The nurse cells are highly active in transcription and translation, and mRNAs and proteins expressed in those cells are transferred to the oocyte through the ring canals to the oocyte, whereas the oocyte nucleus is largely quiescent. Toward the end of oogenesis, the nurse cells expel their cytoplasm into the oocyte and afterward undergo apoptosis. The germ line cyst is surrounded by a single layer of follicle cells (the follicular epithelium), which not only secrete the eggshell but also play pivotal roles in signaling pathways that help establish oocyte polarity.

FOUR mRNAs ESSENTIAL FOR EMBRYONIC PATTERN SPECIFICATION ARE LOCALIZED TO THREE CYTOPLASMIC REGIONS OF THE *DROSOPHILA* OOCYTE: ANTERIOR, POSTERIOR, AND ANTERODORSAL

The future embryonic body axes are specified during oogenesis, and mRNA localization and translational control are crucial for this (Bastock and St Johnston 2008; Kugler and Lasko 2009; Becalska and Gavis 2009). Four localized mRNAs, *oskar* (*osk*), *nanos* (*nos*), *bicoid* (*bcd*), and *gurken* (*grk*), are the key players in embryonic axis specification (Fig. 1), and for this reason their regulation has been especially well studied. The anterior–posterior axis is elaborated through localization of *bcd* mRNA to the anterior of the oocyte, and localization of *osk* and *nos* mRNAs to the posterior of the oocyte. In late-stage oocytes, *bcd* and *nos* are translationally repressed. This repression is relieved after fertilization, and the corresponding proteins are produced in opposing gradients that initiates a cascade of zygotic gene expression that directs anterior–posterior patterning. As will be discussed in more detail below, formation of the anterior-to-posterior Bcd gradient is primarily achieved through localization of its mRNA at the anterior pole, whereas formation of the posterior-to-anterior Nos gradient is achieved through translational repression of its mRNA by Bcd, and enrichment of its mRNA at the posterior where it is translationally active.

osk mRNA begins to be translated during mid-oogenesis to nucleate the formation of the pole plasm, a specialized cytoplasm at the

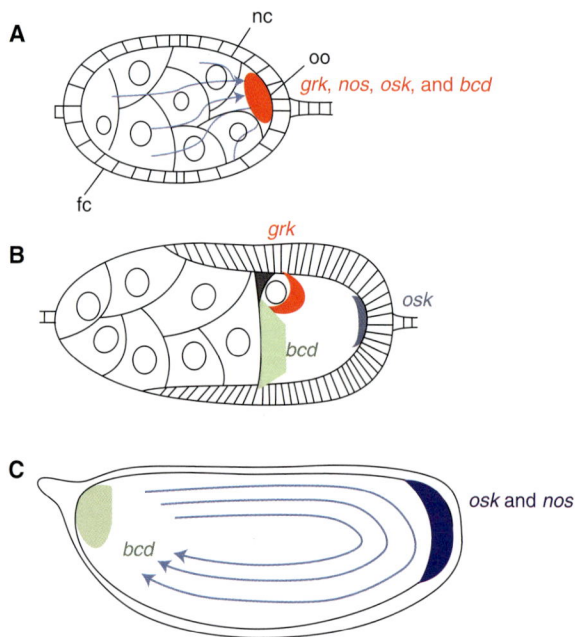

Figure 1. Localization of patterning mRNAs in *Drosophila* oogenesis. (*A*) In early oogenesis, several mRNAs, including *grk*, *nos*, *osk*, and *bcd*, are transported from the nurse cells through cytoplasmic bridges called ring canals into the oocyte. This involves minus-end directed transport along microtubules (blue arrows) mediated by the dynein motor complex. Abbreviations: nc, nurse cells, fc, follicle cells, oo, oocyte. (*B*) In mid-oogenesis, *osk* mRNA localizes to the posterior of the oocyte, *grk* mRNA localizes to the anterodorsal corner in close association with the oocyte nucleus, and *bcd* mRNA localizes to the anterior pole. (*C*) In late oogenesis, centrifugal cytoplasmic streaming (delineated by arrows) coupled with posterior anchoring brings about a further posterior enrichment of *osk* mRNA as well as posterior enrichment of *nos* mRNA. The distribution of *bcd* mRNA at the anterior pole is further refined.

     Cite this article as *Cold Spring Harb Perspect Biol* doi: 10.1101/cshperspect.a012294

posterior of the oocyte that contains large RNP complexes called polar granules that include posterior and germ cell determinants such as Nos, and which is therefore required in the embryo for posterior patterning and primordial germ cell specification. Like *nos*, *osk* mRNA localizes to the posterior pole where it is active, and is translationally silenced elsewhere. Grk, an epidermal growth factor receptor (EGFR) ligand, is crucial for the establishment of both the anterior–posterior and dorsal–ventral axes during oogenesis (González-Reyes et al. 1995; Roth et al. 1995). Grk is secreted from the oocyte to locally activate EGFR in adjacent follicle cells, and restricting its deployment enables it to specify spatial information. During early oogenesis, EGFR activation by Grk assigns posterior fate to a subpopulation of follicle cells that is essential for polarizing the oocyte and establishing anterior–posterior polarity. Later, Grk produced from localized mRNA at the antero-dorsal corner of the oocyte specifies the dorsal–ventral axis by inducing dorsal fate in the follicle cells immediately adjacent.

CIS-ACTING ELEMENTS THAT ARE ESSENTIAL FOR mRNA LOCALIZATION USUALLY INCLUDE STEM-LOOP STRUCTURES

Transport of many mRNAs, including *grk*, *bcd*, and *osk*, from the nurse cells to the oocyte occurs prior to overall cytoplasmic transfer, and proceeds via minus end-directed transport on the microtubule cytoskeleton that is driven by the dynein motor complex (Clark et al. 2007). Two proteins, Egalitarian (Egl) and Bicaudal-D (Bic-D), working in concert are directly responsible for linking mRNAs to dynein and to microtubules (Fig. 2) (Navarro et al. 2004; Dienstbier et al. 2009). Although associating with the localization element of the mRNA to be transported, the amino-terminal region of Egl directly binds to the carboxyl-terminal domain (CTD) of Bic-D (Dienstbier et al. 2009), which in turn interacts with dynein through dynactin (Hoogenraad et al. 2003). Egl also interacts with dynein light chain. A structural study using NMR spectroscopy indicates that a stem-loop with two double-stranded RNA

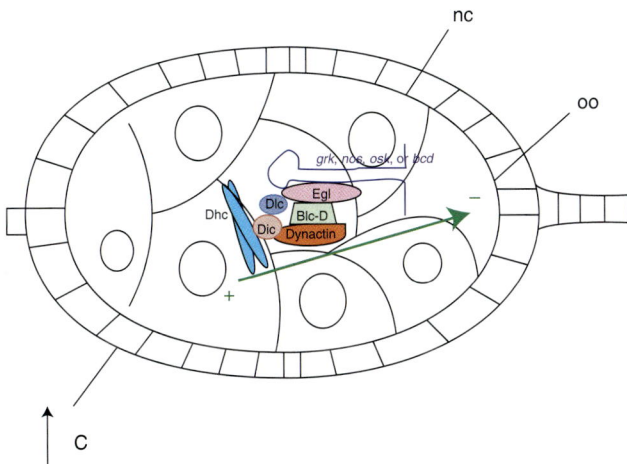

Figure 2. Model for linking mRNAs to the microtubule cytoskeleton for minus-end directed transport. Egalitarian (Egl) interacts directly with localization signals on mRNAs, with the carboxy-terminal end of Bicaudal-D (Bic-D), and with dynein light chain (Dlc). Bic-D interacts directly with dynactin, which in turn binds to dynein through its intermediate chain (Dic). Dynein heavy chain (Dhc) interacts with microtubules (green arrow) and catalyzes movement toward the minus-end. Although both in vivo and in vitro evidence exists to support this model for some instances of dynein-directed minus-end transport, and Egl and Bic-D are required for accumulation of *grk*, *nos*, *osk*, and *bcd* mRNAs into the oocyte, it has not yet been directly shown that this mechanism governs this particular localization event.

P. Lasko

helices in an unusual A′-form conformation (Arnott et al. 1972) is a crucial recognition site for this complex (Bullock et al. 2010).

ANTERIOR TARGETING, ANCHORING, AND TRANSLATIONAL REGULATION OF *bcd*

Localization of *bcd* mRNA proceeds through several steps (St Johnston et al. 1989). The initial phase of *bcd* localization to the anterior cortex of the oocyte requires Exuperantia (Exu) protein (Berleth et al. 1988; Cha et al. 2001; Mische et al. 2007). Exu-containing ribonucleoprotein particles (RNPs) display dynamic movements that are very similar to those displayed by injected fluorescent *bcd* mRNA, and GFP-Exu is recruited to injected *bcd* mRNA (Theurkauf and Hazelrigg 1998; Wilhelm et al. 2000). Exu is phosphorylated by the Par-1 kinase, and this post-translational modification of Exu is important for anterior Exu and *bcd* mRNA localization (Riechmann and Ephrussi 2004).

Genetic evidence has implicated several additional proteins as involved in anchoring of *bcd* at the anterior cortex, including Staufen (Stau), an RNA binding protein, Swallow (Swa), the γ-tubulin ring complex components γTub37C, dGrip75 and dGrip128, and the microtubule-associated protein Mini Spindles (Msps) (Ferrandon et al. 1994; Schnorrer et al. 2002; Moon and Hazelrigg 2004; Vogt et al. 2006; Weil et al. 2006). However, the roles of some of the molecules may be indirect. Super-resolution microscopy has shown that Swa, which was once thought to link *bcd* mRNA containing RNPs to the dynein motor complex, actually does not precisely colocalize in the same particles as *bcd* (Weil et al. 2010). Rather, Swa appears to regulate the actin cytoskeleton, which in turn could be essential for anchoring *bcd* mRNA. Unlike Swa, Stau is found in the same particles as *bcd* and appears to be directly involved in recruiting it to the dynein motor (Weil et al. 2010). *bcd* mRNA anchoring also requires the ESCRT-II complex. The three conserved ESCRT complexes (ESCRT-I, -II, and -III) collaborate to mediate endosomal sorting; only ESCRT-II is required for anterior *bcd* mRNA localization, however, suggesting that a different mechanism

is involved (Irion and St Johnston 2007). One subunit of ESCRT-II, VPS36, binds directly to sequences in the *bcd* 3′UTR and localizes to the anterior of the oocyte in a *bcd* mRNA-dependent but *stau*-independent manner, indicating that ESCRT-II acts upstream of Stau in the *bcd* localization pathway. A recent genetic screen has identified *short stop*, which encodes a spectroplakin protein that binds both actin and microtubules, and *Su(Mir)2*, whose identity is unknown, as encoding other potential factors involved in *bcd* anchoring (Chang et al. 2011).

FORMATION OF THE ANTERIOR–POSTERIOR BICOID PROTEIN GRADIENT FROM THE LOCALIZED *bcd* mRNA

As mentioned above, in the early *Drosophila* embryo, an anterior-to-posterior gradient of Bcd protein is established from its anteriorly localized mRNA. Bcd is a transcription factor, acting as a graded morphogen that influences developmental decisions in a concentration-dependent manner (Driever and Nüsslein-Volhard 1988). As nuclei migrate during syncytial divisions into different regions of the embryo, they activate expression of various sets of patterning genes based on the concentration of Bcd they encounter, and thus on their position along the anterior–posterior axis.

Despite indications that this classic model of Bcd function is insufficient to explain results observed when the Bcd gradient is physically perturbed, flattened, or abolished (Lucchetta et al. 2008; Löhr et al. 2009; Ochoa-Espinosa et al. 2009), it remains clear that Bcd is an important morphogen and the characteristics of its graded distribution need to be carefully controlled. Substantial attention has therefore been given in recent years as to how exactly the Bcd gradient is generated, taking into account the physical properties of the embryonic cytoplasm and the diffusion characteristics of *bcd* mRNA and Bcd protein. Initial attempts at modeling the Bcd gradient considered the mRNA as a point source and postulated that protein diffusion was the dominant means in which the gradient was produced (Houchmandzadeh et al. 2002; Gregor et al. 2007a,b), along with a constant

Cite this article as *Cold Spring Harb Perspect Biol* doi: 10.1101/cshperspect.a012294

amount of degradation. This model, however, does not fully agree with measured diffusion constants that predict a shorter length scale for the Bcd gradient than observed (Gregor et al. 2007b). Subsequently, it was realized that formation of the *bcd* mRNA gradient that presages the protein gradient is critical for establishing the latter (Spirov et al. 2009). Quantitative measurement of *bcd* mRNA and Bcd-GFP protein in real time indicates that the mRNA distribution is more tightly restricted to the anterior than the protein, implying that protein movement from the graded mRNA distribution makes an essential contribution to producing the protein gradient (Little et al. 2011). The recent discovery that Fates-shifted, a ubiquitin ligase substrate specificity receptor that targets Bcd for degradation, is required for formation of a normal Bcd gradient and for correct anterior–posterior patterning, makes it evident that regulation of Bcd protein stability is an important aspect of how the gradient is produced (Liu and Ma 2011).

Translational control appears not to be involved in establishing the Bcd gradient, but it is involved in temporal regulation because localized *bcd* mRNA is apparent from mid-oogenesis when Bcd protein is not detectable. Mutations in *pumilio* (*pum*), which encodes an RNA binding protein, or deletion of a consensus Pum binding site in the *bcd* 3′ UTR leads to increased Bcd expression during embryogenesis (Gamberi et al. 2002), but it is unknown whether this mechanism mediates translational repression during oogenesis.

TARGETING *osk* AND *nos* mRNAs TO THE POSTERIOR POLE PLASM

Both *osk* and *nos* are enriched at the posterior pole of the oocyte in a region termed the pole plasm, and their translation within the oocyte and syncytial embryo is restricted to that region.

osk Localization Is Microtubule-Dependent but Anchoring Requires F-Actin

As discussed earlier for *bcd*, initial loading of *osk* into the oocyte also proceeds through microtubule-dependent motor driven transport, and

like *bcd*, *osk* is initially transported via a minus-end directed dynein-mediated process. Beginning in mid-oogenesis, *osk* begins to accumulate in the posterior of the oocyte, and this localization is an essential first step for pole plasm assembly.

Localization of *osk* mRNA to the pole plasm requires *cis*-acting elements in its 3′ UTR and nuclear imprinting of unspliced *osk* with exon junction complex components (Mago Nashi, Y14, eIF4AIII) and Hrp48 (Hachet and Ephrussi 2004; Huynh et al. 2004; Palacios et al. 2004; Yano et al. 2004). Although splicing of the first intron of *osk* pre-mRNA is essential for its localization, reporter mRNAs lacking introns but containing *osk* 3′ UTR elements can localize via RNA:RNA dimerization with imprinted endogenous *osk*, even if that endogenous *osk* cannot be translated (Jambor et al. 2011). *osk* localization also requires a specific association with Stau, a RNA binding protein that interacts with certain stem-loop structures in the 3′ UTR (Micklem et al. 2000). Posterior localization of *osk* is microtubule dependent, but unlike the earlier phase, it is driven by the plus-end directed motor kinesin. Real-time analysis of the movements of individual *osk*-containing particles shows they are not highly directed, and that posterior enrichment is accomplished through a collection of random walks that is slightly biased toward the posterior, reflecting a similar weak enrichment of microtubule plus-ends at the oocyte posterior (Zimyanin et al. 2008).

A later stage of *osk* localization takes advantage of rapid movements of the oocyte cytoplasm that occur in later oogenesis and involves anchoring of the mRNA in the posterior pole plasm (Sinsimer et al. 2011). Anchoring *osk*-containing mRNPs at the posterior requires specifically the longer of two Osk protein isoforms, the endocytic pathway, and rearrangements of the F-actin cytoskeleton (Vanzo et al. 2007; Tanaka et al. 2011). Actomyosin-based transport is implicated in short-range movements that sharpen the polarization of *osk* mRNA distribution at the posterior pole (Krauss et al. 2009). Osk itself induces the formation of long F-actin projections from the posterior cortex into the pole plasm, corroborating the link between *osk* anchoring and the actin cytoskeleton (Babu et al. 2004).

Posterior Accumulation of *nos* Is Inefficient and Proceeds as a Consequence of Cytoplasmic Streaming

nos mRNA also specifically accumulates in the posterior pole plasm, but its localization is inefficient, with only an approximate 4% enrichment in the posterior half of early embryos (Bergsten and Gavis 1999). Translational repression is therefore the primary mechanism for excluding Nos outside the posterior. In fact, localization of *nos* mRNA is dispensable for somatic patterning, although it is required for germ cell development (Gavis et al. 2008). *nos* mRNA moves throughout the oocyte during a period of rapid cytoplasmic streaming that commences in mid-oogenesis, and gradually accumulates in the pole plasm through an anchoring mechanism (Forrest and Gavis 2003; Weil et al. 2006). Rumpelstiltskin (Rump), an hnRNP M homolog, binds to one of several 3′ UTR elements involved in *nos* mRNA localization and acts directly in its localization (Jain and Gavis 2008). Recently, mutations in *aubergine* (*aub*) were shown to affect *nos* localization, and Aub protein can be copurified with the *nos* 3′ UTR and with Rump (Becalska et al. 2011). Although Aub has been implicated in silencing of retrotransposons in the germline, its function in *nos* localization appears unrelated to this, as mutations in other genes involved in retrotransposon silencing do not have a similar effect on *nos*.

grk mRNA LOCALIZATION IS A MICROTUBULE-DEPENDENT PROCESS

grk mRNA, though mostly transcribed in the nurse cells, accumulates in the oocyte and co-localizes with the oocyte nucleus throughout much of oogenesis. In early oogenesis the oocyte nucleus is located at the posterior, and *grk* mRNA accumulates there. Later, when the oocyte nucleus moves to an anterodorsal position, *grk* mRNA forms a crescent between the apical surface of the nucleus and the neighboring region of the cortex. Transcription of *grk* from the

oocyte nucleus is not essential for this, because a similar distribution is observed in mosaic egg chambers in which the oocyte nucleus is homozygous for an RNA-null *grk* allele (Caceres and Nilson 2005). Further, *grk* transcription from the oocyte nucleus is not required for patterning, as dorsal follicle cell fates and the dorsal-ventral embryonic axis are specified in these mosaics.

Initial transport of *grk* mRNA from the nurse cells to the oocyte uses the dynein and Bic-D/Egl dependent pathway described above for *bcd* and *osk*. *grk* first accumulates along the anterior cortex, then it is transported laterally toward the oocyte nucleus (MacDougall et al. 2003; Jaramillo et al. 2008). This second phase of *grk* transport also depends on dynein and the microtubule cytoskeleton, and the oocyte nucleus appears to nucleate a distinct population of microtubules, which are thought to mediate lateral displacement (MacDougall et al. 2003; Januschke et al. 2006; Delanoue et al. 2007). There is some controversy about the nature of the *cis*-acting elements that are essential for *grk* localization. Studies of injected fluorescently-tagged *grk* mRNA implicated an element within the protein-coding region, termed the *grk* localization signal (GLS) as essential for both oocyte targeting and anterodorsal localization (Van De Bor et al. 2005). However, an analysis of localization of RNA produced from a series of modified *grk* transgenes indicates that the GLS is not sufficient for anterodorsal accumulation and that another element must be involved (Lan et al. 2010).

grk mRNA Anchoring also Requires Microtubules and Dynein

Microtubules and Dhc are required not only for *grk* mRNA transport but also for *grk* anchoring (Delanoue et al. 2007). How dynein switches from a dynamic to a static mode is not fully understood, but it clearly involves the activity of *squid* (*sqd*), which encodes an hnRNP, and perhaps *K10*, as mutations in either of those two genes abrogate stable *grk* accumulation at the anterodorsal corner (Jaramillo et al. 2008; Lan et al. 2010).

PROTECTION FROM MATERNAL DEGRADATION CAN RESULT IN GERM CELL ACCUMULATION OF SPECIFIC mRNAs

Another mechanism that can lead to asymmetric distribution of mRNAs in the early embryo involves protection from RNA degradation. This was first established as a mechanism for enrichment of *Hsp83* mRNA in the primordial germ cells (Bashirullah et al. 1999). Many maternally-expressed mRNAs are degraded at the maternal-to-zygotic transition, through the mediation of Smaug (Smg), a sequence-specific RNA binding protein that recruits the CCR4 deadenylase complex whose translation is drastically up-regulated on egg activation (Semotok et al. 2005; Tadros et al. 2007). Smg is also required for zygotic expression of the *miR-309* cluster microRNAs, that mediate destabilization of a large set of maternal mRNAs (Bushati et al. 2008; Benoit et al. 2009). Degradation of unlocalized *nos* mRNA by Smg also involves recruitment of two transposon-encoded piwi-associated RNAs (piRNAs) that are complementary to sequences in the *nos* 3′ UTR (Rouget et al. 2010). Pumilio, another RNA-binding protein that can recruit the CCR4 deadenylase complex, has also been implicated in maternal transcript destabilization (Gerber et al. 2006 and see below). As primordial germ cells remain transcriptionally silent throughout early embryogenesis, maternal mRNAs whose degradation involves the action of zygotically transcribed molecules such as the *miR-309* cluster may be preferentially stabilized in those cells (Walser and Lipshitz 2011).

TRANSLATIONAL CONTROL OF *osk* IS ELABORATE

osk mRNA is translated into two different isoforms, called Long Osk and Short Osk, that are expressed from different initiation codons in the *osk* mRNA (Markussen et al. 1995). Short Osk is sufficient to induce the accumulation of all other pole plasm components and to rescue the functions of *osk* in posterior patterning and germ cell specification, whereas Long Osk induces F-actin projections that are required for

anchoring its mRNA at the posterior pole (Vanzo et al. 2007; Tanaka and Nakamura 2008).

In early oogenesis *osk* translation is repressed by RNA interference (RNAi), as mutations in several genes (including *armitage, aubergine, cutoff, maelstrom, spindle-E, zucchini,* and *squash*) involved in piRNA processes cause precocious *osk* translation in early oocytes (Findley et al. 2003; Cook et al. 2004; Tomari et al. 2004; Chen et al. 2007; Lim and Kai 2007; Pane et al. 2007). However, the axis patterning defects also observed in these mutants appear not to result directly from *osk* overexpression, but rather from defects in microtubule organization resulting from inappropriate activation of DNA damage signaling (Klattenhoff et al. 2007). As *osk*-containing mRNPs begin to localize to the oocyte posterior, translation is blocked through a different mechanism, operating at the level of ribosome recruitment, by Cup, an eIF4E-binding protein that can interfere with the eIF4E-eIF4G interaction (Nakamura et al. 2004). Cup is recruited to *osk* by Bruno (Bru), an RNA binding protein with three RNA recognition motifs (RRMs). Through all three RRMs, Bru interacts directly with specific sequences (Bru-response elements, or BREs) in the *osk* 3′ UTR, and represses its translation (Snee et al. 2008). Surprisingly, however, recent evidence indicates that, although Cup indeed induces translational repression, this does not require its eIF4E-binding activity and thus does not involve competition for eIF4G (Igreja and Izaurralde 2011; Jeske et al. 2011). Rather, Cup recruits the CCR4 deadenylase complex to its target mRNAs and reduces *osk* poly(A) tail length. Cup-associated mRNAs are not subsequently degraded, however, as they are protected by an amino-terminal regulatory domain of Cup through a mechanism that prevents decapping and requires one of its two eIF4E binding motifs.

Bru also represses translation in another manner, by packaging *osk* mRNA into heavy particles that render it inaccessible to the translational machinery (Chekulaeva et al. 2006). Further insight into the nature of silencing complexes came from a study of polypyrimidine tract binding protein (PTB), which is required for translational repression of *osk* during early oogenesis (Besse et al. 2009). PTB binds with high affinity and cooperativity to the *osk* 3′ UTR, at several pyrimidine-rich sites, and catalyzes oligomerization of multiple *osk* mRNA molecules through bridging interactions.

Although mutation of BREs generally results in precocious *osk* translation, when endogenous *osk* mRNA is totally absent, translation from an *osk* transgene that lacks the distal pair of BREs but is otherwise complete (*osk* C⁻) is reduced. This was surprising as the opposite result would be expected from removing the BREs which were believed to be strictly repressor elements. In a genetic background in which *osk* mRNA with an intact 3′ UTR but an early stop codon is also expressed, *osk* C⁻ is translated at a higher level. This implies that the distal pair of BREs is bifunctional, operating in different contexts as a repressor or an activator element. Further, these results indicate that the presence of *osk* mRNA with an intact 3′ UTR in *osk* mRNPs can facilitate activation of *osk* C⁻ translation in trans, illustrating that mRNA molecules in the same RNP are coordinately regulated (Reveal et al. 2010).

Hrp48, an abundant RNA-binding protein that interacts with elements in both the 5′ and 3′ UTRs of *osk*, is essential for *osk* localization and also contributes to its translational regulation (Huynh et al. 2004; Yano et al. 2004; Norvell et al. 2005). Live imaging of *osk* in *hrp48* mutant ovaries implicate *hrp48* in assembling *osk* into cytoplasmic particles (Mhlanga et al. 2009) and for its subsequent association with Staufen, a translational activator of *osk* (Kim-Ha et al. 1995; Micklem et al. 2000; Braat et al. 2004; Mhlanga et al. 2009). Glorund, an hnRNP F/H family member, also associates with Hrp48 and may be another component of these particles (Kalifa et al. 2009). Another RNA binding protein, Bicaudal-C (Bic-C), has been implicated genetically as a negative regulator of *osk* translation (Saffman et al. 1998). Bic-C directly recruits the CCR4 deadenylase complex to target mRNAs through an association with its NOT3/5 subunit (Chicoine et al. 2007). These targets could potentially include *osk*.

Later in oogenesis, *osk* repression is alleviated, and translation activated, for the small portion of *osk* RNA that is localized to the pole plasm. A key activator of *osk* translation is Orb, the *Drosophila* homolog of *Xenopus* cytoplasmic polyadenylation element binding protein (CPEB). Orb directly associates with two poly(A) polymerases, PAP and Wispy (Wisp). PAP is required during mid-oogenesis to promote Osk expression, whereas Wisp functions only during late oogenesis and in the early embryo (Benoit et al. 2009). An RNA binding protein that promotes CCR4-mediated deadenylation, Bicaudal-C, interacts with Orb, PAP, and Wisp, and possibly inhibits their association with target mRNAs (Castagnetti and Ephrussi 2003; Chicoine et al. 2007; Cui et al. 2008; Benoit et al. 2009).

nos TRANSLATION IS ALSO HIGHLY REGULATED

Nos protein is restricted to the posterior germ plasm by RNA localization and by translational repression of *nos* mRNA outside that region. *nos* regulation is mediated by a 90 nt region of the 3′ UTR, termed the translational control element (TCE) (Crucs et al. 2000; Forrest et al. 2004). The TCE forms a complex secondary structure, and mutations that disrupt any portion of this structure prevent the binding of repressors of *nos* and render the entire element inactive (Forrest et al. 2004). Different parts of the TCE interact with different trans-acting factors at different developmental stages to ensure translational repression of unlocalized *nos* mRNA. During late oogenesis, repression is mediated by Glorund (Glo), an hnRNP F/H ortholog that binds to the stem of stem-loop III of the TCE (Kalifa et al. 2006). Another part of the TCE, the loop of stem-loop II, contains a Smaug Recognition Element (SRE), the binding site for Smg, which represses *nos* in early embryogenesis outside the pole plasm. Smg interacts with Cup, an eIF4E-binding protein that was discussed above in the context of *osk* regulation. The Cup-Smg interaction is required for Smg-mediated repression of SRE-containing mRNAs in embryo extracts (Nelson et al. 2004). Smg also interacts directly with the POP2 subunit of the CCR4 deadenylase complex, recruiting it to a large set of maternal mRNAs in the early embryo, including *nos*, and targeting them for decay (Semotok et al. 2005; Zaessinger et al. 2006; Tadros et al. 2007). Thus, *nos* mRNA is repressed in two distinct ways by Smg: by cap-dependent translational repression and by deadenylation of the silenced transcript (Fig. 3). Osk relieves Smg/CCR4-dependent deadenylation of *nos*, thus enabling its translation in the pole plasm (Zaessinger et al. 2006). Consistent with this, both the 5′ cap structure and the presence of a poly(A) tail are required for TCE-mediated repression of a reporter construct in cell-free extracts prepared from ovaries, although the poly(A) tail does not affect repression in similar extracts prepared from early embryos (Andrews et al. 2011). Mutational analysis sug-

gests that Glo is required for both the cap-dependent and poly(A)-dependent types of repression, although the mechanisms for its function remain unclear.

Translational regulation of ten other mRNAs that localize to the pole plasm at a similar developmental stage as *nos* was compared with the regulation of *nos* itself (Rangan et al. 2009). In all cases, the 3′ UTRs were sufficient to drive posterior localization and temporally restricted patterns of translation of the mRNAs. Often translational activation correlated with an increase in poly(A) tail length, but surprisingly for at least two of the mRNAs (*pgc* and *gcl*), reduction of *orb* activity had little effect on translation. Consistent with other results, this may indicate that it is more critical to regulate

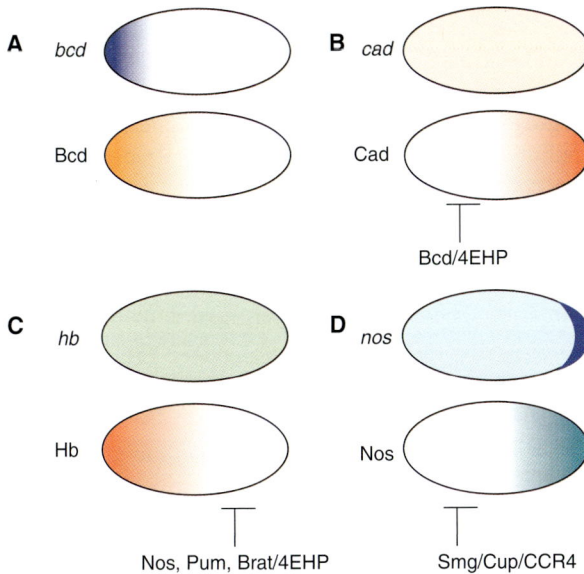

Figure 3. Mechanisms of establishing protein gradients in the early embryo prior to the onset of zygotic transcription. (*A*) Maternally expressed *bcd* mRNA (*top* panel) is localized in a steep gradient at the anterior pole. Bcd protein (*lower* panel) is translated from that localized mRNA and diffuses toward the posterior. (*B*) Maternally-expressed *cad* mRNA (*top* panel) is uniformly distributed. Translation of *cad* mRNA is, however, repressed by Bcd-mediated recruitment of 4EHP, resulting in a posterior-to-anterior gradient of Cad protein (*lower* panel) that is a mirror image of the Bcd gradient. (*C*) Maternally expressed *hb* mRNA (*top* panel) is uniformly distributed. Translation of *hb* mRNA is repressed by a complex of Nos, Pum, and Brat, that recruits 4EHP and probably other negative regulators to restrict Hb protein (*lower* panel) to the anterior half of the embryo. (*D*) Maternally expressed *nos* mRNA (*top* panel) is enriched at the posterior pole but present elsewhere. Translation of *nos* outside the posterior is repressed by Smg, which can recruit the repressor protein Cup and also the CCR4 deadenylase complex. Unlocalized *nos* is also targeted by piRNAs (not shown). Nos protein (lower panel) is translated from posteriorly-localized *nos* that is protected from degradation and repression.

deadenylation rather than polyadenylation in regulating translation.

4EHP, AN ALTERNATE CAP-BINDING PROTEIN, REPRESSES TRANSLATION OF *CAUDAL* AND *HUNCHBACK* mRNAs

Caudal (Cad), a transcription factor, is a master regulator of posterior patterning in many insects (Olesnicky et al. 2006). In *Drosophila* maternally expressed Cad forms a posterior-to-anterior gradient in the early embryo, and although its function is somewhat redundant with other regulatory factors it is nevertheless involved in activating posterior-specific zygotic genes (Schulz and Tautz 1995). In a role different from its function as a transcription factor, Bcd represses translation of *cad* mRNA, thus producing a gradient of Cad protein that is the mirror image of the Bcd gradient (Fig. 3) (Dubnau and Struhl 1996, Rivera-Pomar et al. 1996). Bcd binds to the 3′ UTR and recruits 4E homology protein (4EHP), an eIF4E-related cap binding protein that cannot bind eIF4G (Hernández et al. 2005), and thus cannot nucleate assembly of an active cap-binding complex (Cho et al. 2005). Females homozygous for a *4EHP* allele produce embryos with anterior defects, like those produced by *bcd* mutants, that fail to repress *cad* translation in the anterior. These phenotypes could be rescued by *4EHP* transgenic constructs, but not by constructs producing mutant forms of 4EHP that were abrogated for binding to the structure or to Bcd. Similarly, transgenically produced forms of Bcd that were abrogated for 4EHP binding could not repress *cad* translation. These results showed that a complex of 4EHP and Bcd, interacting with the cap structure and the 3′ UTR, respectively, circularizes *cad* mRNA and renders it translationally inactive.

4EHP was later shown to be involved in repression of *hb* mRNA in the posterior of the embryo (Fig. 3) (Cho et al. 2006), which had long been known to involve formation of a complex including Nos, Pum, Brain Tumor (Brat, an NHL-domain containing protein), and an element in the *hb* 3′ UTR called the Nanos-response element (NRE) (Sonoda and Wharton 2001). In the case of *hb*, 4EHP is recruited to the 5′ cap structure through an interaction with the NHL domain of Brat. As the binding sites for 4EHP on Bcd and Brat are not similar in sequence, and the latter does not resemble an eIF4E binding motif, it is possible that the interaction between 4EHP and Brat is indirect.

The translational repressor Pum is involved in many cellular and developmental processes in *Drosophila* other than embryonic patterning, including restriction of Cyclin B expression to the germline (Kadyrova et al. 2007), regulation of sodium current in motoneurons (Muraro et al. 2008), regulation of presynaptic morphology (Menon et al. 2004), regulation of dendrite morphogenesis in peripheral neurons (Ye et al. 2004), and maintenance of germline stem cell self-renewal (Gilboa and Lehmann 2004; Wang and Lin 2004; Szakmary et al. 2005; Li et al. 2009; Kim et al. 2010). Pum binds to a consensus sequence UGUANAUA (Gerber et al. 2006) and frequently operates in a complex with Nos. Often, Pum repression is independent of the cap structure and 4EHP, involving instead CCR4-mediated deadenylation of its target mRNAs (Wreden et al. 1997; Gamberi et al. 2002; Goldstrohm et al. 2006). A recent study provides evidence that Nos is not always required for Pum activity, and that regions outside the Pum C-terminal domain that binds Nos and Brat possess substantial translational repressor activity (Weidmann and Goldstrohm 2012).

VASA (VAS) IS A TRANSLATIONAL ACTIVATOR OF SPECIFIC GERM-LINE mRNAs

Activators of translation of specific mRNAs have not been identified as frequently as repressors, and less is known about their function. One translational activator that is involved in *Drosophila* embryonic development is Vas, a DEAD-box type RNA helicase. Complete loss of Vas blocks oogenesis and results in female sterility, whereas females homozygous for weaker *vas* alleles produce embryos that lack a germ line and posterior somatic segments. Vas binds to eIF5B, a translation factor that functions in ribosomal subunit joining, and mutations in *vas* and *eIF5B*

Cite this article as *Cold Spring Harb Perspect Biol* doi: 10.1101/cshperspect.a012294

genetically interact (Carrera et al. 2000). Severe *vas* mutations strongly reduce Grk accumulation in the oocyte (Styhler et al. 1998; Tomancak et al. 1998). A mutant form of Vas (VasΔ617) was generated whose ability to bind eIF5B was reduced 10-fold in yeast two-hybrid assays (Johnstone and Lasko 2004). Oocytes that express only VasΔ617 fail to accumulate normal levels of Grk, suggesting that Vas activates *grk* translation through an interaction with eIF5B. Vas has also been shown to bind specifically to a U-rich motif present in the 3′ UTR of another mRNA, *mei-P26*, and to positively regulate its translation in germ cells through that interaction (Liu et al. 2009). Again, the VasΔ617 mutation abrogates the effect.

Like many of the RNA binding proteins discussed in this review, Vas appears to have more than one function. It is a component of nuage, organelles that are involved in piRNA-mediated transposon silencing, and it has recently been implicated in regulating mitotic chromosome condensation in the *Drosophila* germline (Pek and Kai 2011). Dependent on the activities of *aub* and *spn-E*, two piRNA pathway genes, Vas accumulates in perichromosomal foci during mitosis, and facilitates the recruitment of Barren, which in turn is required for correct chromosome condensation and segregation. This function of Vas appears to be independent of translation, as the VasΔ617 mutant form operates normally in this regard.

FUTURE DIRECTIONS

In this field, some simple models have given way to more complicated ones over the past several years, and this trend will likely continue, because more extensive analysis has in many cases revealed novel unexpected activities for proteins and regulatory elements that were thought to be fully understood. Cup, the BREs, and perhaps Pum are just some examples of this. On first glance it seems perhaps illogical that many molecules involved in mRNA localization and translational control cannot be assigned a unitary function. What must be remembered, however, is that mRNAs and the proteins that regulate them are contained within a heterogeneous

and highly dynamic set of RNPs, and that a given mRNA or protein might be a component of many different species of RNP at different developmental times or in different cellular or spatial positions. If we consider that the RNPs, not individual molecules, are the real functional units in translational regulation, manipulation of a single gene that encodes a single RNP component might actually disrupt many different species of RNP, and thus lead to multiple effects. Advances in quantitative imaging and proteomics (see, for instance, Slobodin and Gerst 2011) will likely make it possible to more fully characterize the panoply of RNPs that are involved in mRNA localization and translational control in *Drosophila* oocytes in upcoming years, which will be a critical step forward in developing a deeper understanding of these processes.

REFERENCES

Andrews S, Snowflack DR, Clark IE, Gavis ER. 2011. Multiple mechanisms collaborate to repress nanos translation in the *Drosophila* ovary and embryo. *RNA* **17:** 967–977.

Arnott S, Hukins DW, Dover SD. 1972. Optimised parameters for RNA double-helices. *Biochem Biophys Res Commun* **48:** 1392–1399.

Babu K, Cai Y, Bahri S, Yang X, Chia W. 2004. Roles of Bifocal, Homer, and F-actin in anchoring Oskar to the posterior cortex of *Drosophila* oocytes. *Genes Dev* **18:** 138–143.

Bashirullah A, Halsell SR, Cooperstock RL, Kloc M, Karaiskakis A, Fisher WW, Fu W, Hamilton JK, Etkin LD, Lipshitz HD. 1999. Joint action of two RNA degradation pathways controls the timing of maternal transcript elimination at the midblastula transition in *Drosophila melanogaster*. *EMBO J* **18:** 2610–2620.

Bastock R, St Johnston D. 2008. *Drosophila* oogenesis. *Curr Biol* **18:** R1082–R1087.

Becalska AN, Gavis ER. 2009. Lighting up mRNA localization in *Drosophila* oogenesis. *Development* **136:** 2493–2503.

Becalska AN, Kim YR, Belletier NG, Lerit DA, Sinsimer KS, Gavis ER. 2011. Aubergine is a component of a *nanos* RNA localization complex. *Dev Biol* **349:** 46–52.

Benoit B, He CH, Zhang F, Votruba SM, Tadros W, Westwood JT, Smibert CA, Lipshitz HD, Theurkauf WE. 2009a. An essential role for the RNA-binding protein Smaug during the *Drosophila* maternal-to-zygotic transition. *Development* **136:** 923–932.

Benoit P, Papin C, Kwak JE, Wickens M, Simonelig M. 2009b. PAP- and GLD-2 type poly(A) polymerases are required sequentially in cytoplasmic polyadenylation and oogenesis in *Drosophila*. *Development* **135:** 1969–1979.

Bergsten SE, Gavis ER. 1999. Role for mRNA localization in translational activation but not spatial restriction of *nanos* mRNA. *Development* **126:** 659–669.

Berleth T, Burri M, Thoma G, Bopp D, Richstein S, Frigerio G, Noll M, Nüsslein-Volhard C. 1988. The role of localization of bicoid RNA in organizing the anterior pattern of the *Drosophila* embryo. *EMBO J* **7:** 1749–1756.

Besse F, Lopez de Quinto S, Marchand V, Trucco A, Ephrussi A. 2009. *Drosophila* PTB promotes formation of high-order RNP particles and represses *oskar* translation. *Genes Dev* **23:** 195–207.

Braat AK, Yan N, Arn E, Harrison D, Macdonald PM. 2004. Localization-dependent Oskar protein accumulation; control after the initiation of translation. *Dev Cell* **7:** 125–131.

Bullock SL, Ringel I, Ish-Horowicz D, Lukavsky PJ. 2010. A′-form RNA helices are required for cytoplasmic mRNA transport in *Drosophila*. *Nat Struct Mol Biol* **17:** 703–709.

Bushati N, Stark A, Brennecke J, Cohen SM. 2008. Temporal specificity of miRNAs and their targets during the maternal-to-zygotic transition in *Drosophila*. *Curr Biol* **18:** 501–506.

Caceres L, Nilson LA. 2005. Production of *gurken* in the nurse cells is sufficient for axis determination in the *Drosophila* oocyte. *Development* **132:** 2345–2353.

Carrera P, Johnstone O, Nakamura A, Casanova J, Jäckle H, Lasko P. 2000. VASA mediates translation through interaction with a *Drosophila* yIF2 homolog. *Mol Cell* **5:** 181–187.

Castagnetti S, Ephrussi A. 2003. Orb and a long poly(A) tail are required for efficient *oskar* translation at the posterior pole of the *Drosophila* oocyte. *Development* **130:** 835–843.

Cha BJ, Koppetsch BS, Theurkauf WE. 2001. In vivo analysis of *Drosophila* bicoid mRNA localization reveals a novel microtubule-dependent axis specification pathway. *Cell* **106:** 35–46.

Chang C-W, Nashchekin D, Wheatley L, Irion U, Dahlgaard K, Montague TG, Hall J, St Johnston D. 2011. Anterior-posterior axis specification in *Drosophila* oocytes: Identification of novel bicoid and oskar localization factors. *Genetics* **188:** 883–896.

Chekulaeva M, Hentze MW, Ephrussi A. 2006. Bruno acts as a dual repressor of *oskar* translation, promoting mRNA oligomerization and formation of silencing particles. *Cell* **124:** 521–533.

Chen Y, Pane A, Schüpbach T. 2007. *cutoff* and *aubergine* mutations result in retrotransposon upregulation and checkpoint activation in *Drosophila*. *Curr Biol* **17:** 637–642.

Chicoine J, Benoit P, Gamberi C, Paliouras M, Simonelig M, Lasko P. 2007. Bicaudal-C recruits CCR4-NOT deadenylase to target mRNAs and regulates oogenesis, cytoskeletal organization, and its own expression. *Dev Cell* **13:** 691–704.

Cho PF, Poulin F, Cho-Park YA, Cho-Park IB, Chicoine JD, Lasko P, Sonenberg N. 2005. A new paradigm for translational control: Inhibition via 5′-3′ mRNA tethering by Bicoid and the eIF4E cognate 4EHP. *Cell* **121:** 411–423.

Cho PF, Gamberi C, Cho-Park YA, Cho-Park IB, Lasko P, Sonenberg N. 2006. Cap-dependent translational inhibi-

tion establishes two opposing morphogen gradients in *Drosophila* embryos. *Curr Biol* **16:** 2035–2041.

Clark A, Meignin C, Davis I. 2007. A Dynein-dependent shortcut rapidly delivers axis determination transcripts into the *Drosophila* oocyte. *Development* **134:** 1955–1965.

Coll O, Villalba A, Bussotti G, Notredame C, Gebauer F. 2010. A novel, noncanonical mechanism of cytoplasmic polyadenylation operates in *Drosophila* embryogenesis. *Genes Dev* **24:** 129–134.

Cook HA, Koppetsch BS, Wu J, Theurkauf WE. 2004. The *Drosophila* SDE3 homolog *armitage* is required for *oskar* mRNA silencing and embryonic axis specification. *Cell* **116:** 817–829.

Crucs S, Chatterjee S, Gavis ER. 2000. Overlapping but distinct RNA elements control repression and activation of *nanos* translation. *Mol Cell* **5:** 457–467.

Cui J, Sackton KL, Homer VL, Kumar KE, Wolfner MF. 2008. Wispy, the *Drosophila* homolog of GLD-2, is required during oogenesis and egg activation. *Genetics* **178:** 2017–2029.

Delanoue R, Herpers B, Soetaert J, Davis I, Rabouille C. 2007. *Drosophila* Squid/hnRNP helps Dynein switch from a *gurken* mRNA transport motor to an ultrastructural static anchor in sponge bodies. *Dev Cell* **13:** 523–538.

Dienstbier M, Boehl F, Li X, Bullock SL. 2009. Egalitarian is a selective RNA-binding protein linking mRNA localization signals to the dynein motor. *Genes Dev* **23:** 1546–1558.

Driever W, Nüsslein-Volhard C. 1988. The Bicoid protein determines position in the *Drosophila* embryo in a concentration-dependent manner. *Cell* **54:** 95–104.

Dubnau J, Struhl G. 1996. RNA recognition and translational regulation by a homeodomain protein. *Nature* **379:** 694–699.

Ferrandon D, Elphick L, Nüsslein-Volhard C, St. Johnston D. 1994. Staufen protein associates with the 3′UTR of *bicoid* mRNA to form particles that move in a microtubule-dependent manner. *Cell* **79:** 1221–1232.

Findley SD, Tamanaha M, Clegg NJ, Ruohola-Baker H. 2003. *maelstrom*, a *Drosophila* spindle-class gene, encodes a protein that colocalizes with Vasa and RDE1/AGO1 homolog, Aubergine, in nuage. *Development* **130:** 859–871.

Forrest KM, Gavis ER. 2003. Live imaging of endogenous mRNA reveals a diffusion and entrapment mechanism for *nanos* mRNA localization in *Drosophila*. *Curr Biol* **13:** 1159–1168.

Forrest KM, Clark IE, Jain RA, Gavis ER. 2004. Temporal complexity within a translational control element in the *nanos* mRNA. *Development* **131:** 5849–5857.

Gamberi C, Peterson DS, He L, Gottlieb E. 2002. An anterior function for the *Drosophila* posterior determinant Pumilio. *Development* **129:** 2699–2710.

Gavis ER, Chatterjee S, Ford NR, Wolff LJ. 2008. Dispensability of *nanos* mRNA localization for abdominal patterning but not for germ cell development. *Mech Dev* **125:** 81–90.

Gerber AP, Luschnig S, Krasnow MA, Brown PO, Herschlag D. 2006. Genome-wide identification of mRNAs associated with the translational regulator PUMILIO in *Drosophila melanogaster*. *Proc Natl Acad Sci* **103**: 4487–4492.

Gilboa L, Lehmann R. 2004. Repression of primordial germ cell differentiation parallels germ line stem cell maintenance. *Curr Biol* **18**: 981–986.

Goldstrohm AC, Hook BA, Seay DJ, Wickens M. 2006. PUF proteins bind Pop2p to regulate messenger RNAs. *Nat Struct Mol Biol* **13**: 533–539.

González-Reyes A, Elliott H, St Johnston D. 1995. Polarization of both major body axes in *Drosophila* by *gurken-torpedo* signaling. *Nature* **375**: 654–658.

Gregor T, Wieschaus EF, McGregor AP, Bialek W, Tank DW. 2007a. Stability and nuclear dynamics of the Bicoid morphogen gradient. *Cell* **130**: 141–152.

Gregor T, Tank DW, Wieschaus EF, Bialek W. 2007b. Probing the limits to positional information. *Cell* **130**: 153–164.

Hachet O, Ephrussi A. 2004. Splicing of *oskar* RNA in the nucleus is coupled to its cytoplasmic localization. *Nature* **428**: 959–963.

Hernández G, Altmann M, Sierra JM, Urlaub H, Diez del Corral R, Schwartz P, Rivera-Pomar R. 2005. Functional analysis of seven genes encoding eight translation initiation factor 4E (eIF4E) isoforms in *Drosophila*. *Mech Dev* **122**: 529–543.

Hoogenraad CC, Wulf P, Schiefermeier N, Stepanova T, Galjart N, Small JV, Grosveld F, de Zeeuw CI, Akhmanova A. 2003. Bicaudal D induces selective dynein-mediated microtubule minus end-directed transport. *EMBO J* **22**: 6004–6015.

Houchmandzadeh B, Wieschaus E, Leibler S. 2002. Establishment of developmental precision and proportions in the early *Drosophila* embryo. *Nature* **415**: 798–802.

Huynh JR, Munro TP, Smith-Litière K, Lepesant JA, St. Johnston D. 2004. The *Drosophila* hnRNP A/B homolog, Hrp48, is specifically required for a distinct step in *osk* mRNA localization. *Dev Cell* **6**: 625–635.

Igreja C, Izaurralde E. 2011. CUP promotes deadenylation and inhibits decapping of mRNA targets. *Genes Dev* **25**: 1955–1967.

Irion U, St. Johnston D. 2007. *bicoid* RNA localization requires specific binding of an endosomal sorting complex. *Nature* **445**: 554–558.

Jain R, Gavis ER. 2008. The *Drosophila* hnRNP M homolog, Rumpelstiltskin, regulates *nanos* mRNA localization. *Development* **135**: 973–982.

Jambor H, Brunel C, Ephrussi A. 2011. Dimerization of *oskar* 3′ UTRs promotes hitchhiking for RNA localization in the *Drosophila* oocyte. *RNA* **17**: 2049–2057.

Januschke J, Gervais L, Gillet L, Keryer G, Bornens M, Guichet A. 2006. The centrosome–nucleus complex and microtubule organization in the *Drosophila* oocyte. *Development* **133**: 129–139.

Jaramillo AM, Weil TT, Goodhouse J, Gavis ER, Schüpbach T. 2008. The dynamics of fluorescently labelled endogenous *gurken* mRNA in *Drosophila*. *J Cell Sci* **121**: 887–894.

Jeske M, Moritz B, Anders A, Wahle E. 2011. Smaug assembles an ATP-dependent stable complex repressing *nanos* mRNA translation at multiple levels. *EMBO J* **30**: 90–103.

Johnstone O, Lasko P. 2004. Interaction with eIF5B is essential for Vasa function during development. *Development* **131**: 4167–4178.

Kadyrova LY, Habara Y, Lee TH, Wharton RP. 2007. Translational control of maternal *Cyclin B* mRNA by Nanos in the *Drosophila* germline. *Development* **134**: 1519–1527.

Kalifa Y, Huang T, Rosen LN, Chatterjee S, Gavis ER. 2006. Glorund, a *Drosophila* hnRNP F/H homolog, is an ovarian repressor of *nanos* translation. *Dev Cell* **10**: 291–301.

Kalifa Y, Armenti ST, Gavis ER. 2009. Glorund interactions in the regulation of *gurken* and *oskar* mRNAs. *Dev Biol* **326**: 68–74.

Kim JY, Lee YC, Kim C. 2010. Direct inhibition of Pumilio activity by Bam and Bgcn in *Drosophila* germ line stem cell differentiation. *J Biol Chem* **285**: 4741–4746.

Kim-Ha J, Kerr K, Macdonald PM. 1995. Translational regulation of *oskar* mRNA by Bruno, an ovarian RNA-binding protein, is essential. *Cell* **81**: 403–412.

King RC. 1970. *Ovarian development in Drosophila melanogaster*. Academic Press, New York.

Klattenhoff C, Bratu DP, McGinnis-Schultz N, Koppetsch BS, Cook HA, Theurkauf WE. 2007. *Drosophila* rasiRNA pathway mutations disrupt embryonic axis specification through activation of an ATR/Chk2 DNA damage response. *Dev Cell* **12**: 45–55.

Krauss J, López de Quinto S, Nüsslein-Volhard C, Ephrussi A. 2009. Myosin-V regulates *oskar* mRNA localization in the *Drosophila* oocyte. *Curr Biol* **19**: 1058–1063.

Kugler JM, Lasko P. 2009. Localization, anchoring, and translational control of *oskar*, *gurken*, *bicoid*, and *nanos* mRNA during *Drosophila* oogenesis. *Fly* **3**: 15–28.

Lan L, Lin S, Zhang S, Cohen RS. 2010. Evidence for a transport-trap mode of *Drosophila melanogaster gurken* mRNA localization. *PLoS ONE* **5**: e15448.

Lécuyer E, Yoshida H, Parthasarathy N, Alm C, Babak T, Cerovina T, Hughes TR, Tomancak P, Krause HM. 2007. Global analysis of mRNA localization reveals a prominent role in organizing cellular architecture and function. *Cell* **131**: 174–187.

Li Y, Minor NT, Park JK, McKearin DM, Maines JZ. 2009. Bam and Bgcn antagonize Nanos-dependent germ-line stem cell maintenance. *Proc Natl Acad Sci* **106**: 9304–9309.

Lim AK, Kai T. 2007. Unique germ-line organelle, nuage, functions to repress selfish genetic elements in *Drosophila melanogaster*. *Proc Natl Acad Sci* **104**: 6714–6719.

Little SC, Tkacik G, Kneeland TB, Wieschaus EF, Gregor T. 2011. The formation of the Bicoid gradient requires protein movement from anteriorly localized mRNA. *PLoS Biol* **9**: e1000596.

Liu J, Ma J. 2011. Fates-shifted is an F-box protein that targets Bicoid for degradation and regulates developmental fate determination in *Drosophila* embryos. *Nat Cell Biol* **13**: 22–29.

Liu N, Han H, Lasko P. 2009. Vasa promotes *Drosophila* germline stem cell differentiation by activating *mei-P26* translation by directly interacting with a (U)-rich motif in its 3′ UTR. *Genes Dev* **23**: 2742–2752.

Löhr U, Chung HR, Beller M, Jäckle H. 2009. Antagonistic action of Bicoid and the repressor Capicua determines the spatial limits of *Drosophila* head gene expression domains. *Proc Natl Acad Sci* **106**: 21695–21700.

Lucchetta EM, Vincent ME, Ismagilov RF. 2008. A precise Bicoid gradient is nonessential during cycles 11–13 for precise patterning in the *Drosophila* blastoderm. *PLoS ONE* **3**: e3651.

MacDougall N, Clark A, MacDougall E, Davis I. 2003. *Drosophila gurken* (TGFα) mRNA localizes as particles that move within the oocyte in two dynein-dependent steps. *Dev Cell* **4**: 307–319.

Markussen F-H, Michon AM, Breitwieser W, Ephrussi A. 1995. Translational control of *oskar* generates short OSK, the isoform that induces pole plasma assembly. *Development* **121**: 3723–3732.

Menon KP, Sanyal S, Habara Y, Sanchez R, Wharton RP, Ramaswami M, Zinn K. 2004. The translational repressor Pumilio regulates presynaptic morphology and controls postsynaptic accumulation of translation factor eIF-4E. *Neuron* **44**: 663–676.

Mhlanga MM, Bratu DP, Genovesio A, Rybarska A, Chenouard N, Nehrbass U, Olivo-Marin J-C. 2009. In vivo colocalisation of oskar mRNA and trans-acting proteins revealed by quantitative imaging of the *Drosophila* oocyte. *PLoS ONE* **4**: e6241.

Micklem DR, Adams J, Grünert S, St Johnston D. 2000. Distinct roles of two conserved Staufen domains in *oskar* mRNA localization and translation. *EMBO J* **19**: 1366–1377.

Mische S, Li M, Serr M, Hays TS. 2007. Direct observation of regulated ribonucleoprotein transport across the nurse cell/oocyte boundary. *Mol Biol Cell* **18**: 2254–2263.

Moon W, Hazelrigg T. 2004. The *Drosophila* microtubule-associated protein Mini spindles is required for cytoplasmic microtubules in oogenesis. *Curr Biol* **14**: 1957–1961.

Muraro NI, Weston AJ, Gerber AP, Luschnig S, Moffat KG, Baines RA. 2008. Pumilio binds *para* mRNA and requires Nanos and Brat to regulate sodium current in *Drosophila* motoneurons. *J Neurosci* **28**: 2099–2109.

Nakamura A, Sato K, Hanyu-Nakamura K. 2004. *Drosophila* Cup is an eIF4E binding protein that associates with Bruno and regulates *oskar* mRNA translation in oogenesis. *Dev Cell* **6**: 69–78.

Navarro C, Puthalakath H, Adams JM, Strasser A, Lehmann R. 2004. Egalitarian binds dynein light chain to establish oocyte polarity and maintain oocyte fate. *Nat Cell Biol* **6**: 427–435.

Nelson MR, Leidal AM, Smibert CA. 2004. *Drosophila* Cup is an eIF4E-binding protein that functions in Smaug-mediated translational repression. *EMBO J* **23**: 150–159.

Norvell A, Debec A, Finch D, Gibson L, Thoma B. 2005. Squid is required for efficient posterior localization of oskar mRNA during *Drosophila* oogenesis. *Dev Genes Evol* **215**: 340–349.

Ochoa-Espinosa A, Yu D, Tsirigos A, Struffi P, Small S. 2009. Anterior-posterior positional information in the absence of a strong Bicoid gradient. *Proc Natl Acad Sci* **106**: 3823–3828.

Olesnicky EC, Brent AE, Tonnes L, Walker M, Pultz MA, Leaf D, Desplan C. 2006. A *caudal* mRNA gradient controls posterior development in the wasp Nasonia. *Development* **133**: 3973–3982.

Palacios IM, Gatfield D, St. Johnston D, Izaurralde E. 2004. An eIF4AIII-containing complex required for mRNA localization and nonsense-mediated mRNA decay. *Nature* **427**: 753–757.

Pane A, Wehr K, Schüpbach T. 2007. *zucchini* and *squash* encode two putative nucleases required for rasiRNA production in the *Drosophila* germline. *Dev Cell* **12**: 851–862.

Pek JW, Kai T. 2011. A role for Vasa in regulating mitotic chromosome condensation in *Drosophila*. *Curr Biol* **21**: 39–44.

Rangan P, DeGennaro M, Jaime-Bustamante K, Coux RX, Martinho RG, Lehmann R. 2009. Temporal and spatial control of germ-plasm RNAs. *Curr Biol* **19**: 72–77.

Reveal B, Yan N, Snee MJ, Pai CI, Gim Y, Macdonald PM. 2010. BREs mediate both repression and activation of *oskar* mRNA translation and act in *trans*. *Dev Cell* **18**: 496–502.

Riechmann V, Ephrussi A. 2004. Par-1 regulates *bicoid* mRNA localisation by phosphorylating Exuperantia. *Development* **131**: 5897–5907.

Rivera-Pomar R, Niessing D, Schmidt-Ott U, Gehring WJ, Jäckle H. 1996. RNA binding and translational suppression by Bicoid. *Nature* **379**: 746–749.

Roth S, Neuman-Silberberg FS, Barcelo G, Schüpbach T. 1995. *cornichon* and the EGF receptor signalling process are necessary for both anterior-posterior and dorsal-ventral pattern formation in *Drosophila*. *Cell* **81**: 967–978.

Rouget C, Papin C, Boureux A, Meunier A-C, Franco B, Robine N, Lai EC, Pelisson A, Simonelig M. 2010. Maternal mRNA deadenylation and decay by the piRNA pathway in the early *Drosophila* embryo. *Nature* **467**: 1128–1132.

Saffman EE, Styhler S, Rother K, Li W, Richard S, Lasko P. 1998. Premature translation of *oskar* in oocytes lacking the RNA-binding protein Bicaudal-C. *Mol Cell Biol* **18**: 4855–4862.

Schnorrer F, Bohmann K, Nüsslein-Volhard C. 2000. The molecular motor Dynein is involved in targeting Swallow and *bicoid* RNA to the anterior pole of *Drosophila* oocytes. *Nat Cell Biol* **2**: 185–190.

Schnorrer F, Luschnig S, Koch I, Nüsslein-Volhard C. 2002. *Gamma-tubulin37C* and *gamma-tubulin ring complex protein 75* are essential for *bicoid* RNA localization during *Drosophila* oogenesis. *Dev Cell* **3**: 685–696.

Schulz C, Tautz D. 1995. Zygotic Caudal regulation by Hunchback and its role in abdominal segment formation of the *Drosophila* embryo. *Development* **121**: 1023–1028.

Semotok JL, Cooperstock RL, Pinder BD, Vari HK, Lipshitz HD, Smibert CA. 2005. Smaug recruits the CCR4/POP2/NOT deadenylase complex to trigger maternal transcript localization in the early *Drosophila* embryo. *Curr Biol* **15**: 284–294.

Sinsimer KS, Jain RA, Chatterjee S, Gavis ER. 2011. A late phase of germ plasm accumulation during *Drosophila* oogenesis requires Lost and Rumpelstiltskin. *Development* **138**: 3431–3440.

Slobodin B, Gerst JE. 2011. RaPID: An aptamer-based mRNA affinity purification technique for the identification of RNA and protein factors in ribonucleoprotein complexes. *Methods Mol Biol* **714**: 387–406.

Snee M, Benz D, Jen J, Macdonald PM. 2008. Two distinct domains of Bruno bind specifically to the *oskar* mRNA. *RNA Biol* **5:** 1–9.

Sonoda J, Wharton RP. 2001. *Drosophila* brain tumor is a translational repressor. *Genes Dev* **15:** 762–773.

Spirov A, Fahmy K, Schneider M, Frei E, Nöll M, Baumgartner S. 2009. Formation of the bicoid morphogen gradient: An mRNA gradient dictates the protein gradient. *Development* **136:** 605–614.

St Johnston D, Driever W, Berleth T, Richstein S, Nüsslein-Volhard C. 1989. Multiple steps in the localization of *bicoid* RNA to the anterior pole of the *Drosophila* oocyte. *Development* **107** (Suppl): 13–19.

Styhler S, Nakamura A, Swan A, Suter B, Lasko P. 1998. *vasa* is required for GURKEN accumulation in the oocyte, and is involved in oocyte differentiation and germline cyst development. *Development* **125:** 1569–1578.

Szakmary A, Cox DN, Wang Z, Lin H. 2005. Regulatory relationship among piwi, pumilio, and bag-of-marbles in *Drosophila* germline stem cell self-renewal and differentiation. *Curr Biol* **15:** 171–178.

Tadros W, Goldman AL, Babak T, Menzies T, Vardy L, Orr-Weaver T, Hughes TR, Westwood JT, Smibert CA, Lipshitz HD. 2007. SMAUG is a major regulator of maternal mRNA destabilization in *Drosophila* and its translation is activated by the PAN GU kinase. *Dev Cell* **12:** 143–155.

Tanaka T, Nakamura A. 2008. The endocytic pathway acts downstream of Oskar in *Drosophila* germ plasm assembly. *Development* **135:** 1107–1117.

Tanaka T, Kato Y, Matsuda K, Hanyu-Nakamura K, Nakamura A. 2011. *Drosophila* Mon2 couples Oskar-induced endocytosis with actin remodeling for cortical anchorage of the germ plasm. *Development* **138:** 2523–2532.

Theurkauf WE, Hazelrigg TI. 1998. In vivo analyses of cytoplasmic transport and cytoskeletal organization during *Drosophila* oogenesis: Characterization of a multi-step anterior localization pathway. *Development* **125:** 3655–3666.

Tomancak P, Guichet A, Zavorszky P, Ephrussi A. 1998. Oocyte polarity depends on regulation of *gurken* by Vasa. *Development* **125:** 1723–1732.

Tomancak P, Berman BP, Beaton A, Weiszmann R, Kwan E, Hartenstein V, Celniker SE, Rubin GM. 2007. Global analysis of patterns of gene expression during *Drosophila* embryogenesis. *Genome Biol* **8:** R145.

Tomari Y, Du T, Haley B, Schwarz DS, Bennett R, Cook HA, Koppetsch BS, Theurkauf WE, Zamore PD. 2004. RISC assembly defects in the *Drosophila* RNAi mutant *armitage*. *Cell* **119:** 831–841.

Van De Bor V, Hartswood E, Jones C, Finnegan D, Davis I. 2005. *gurken* and the *I* factor retrotransposon RNAs share common localization signals and machinery. *Dev Cell* **9:** 51–62.

Vanzo N, Oprins A, Xanthakis D, Ephrussi A, Rabouille C. 2007. Stimulation of endocytosis and actin dynamics by Oskar polarizes the *Drosophila* oocyte. *Dev Cell* **12:** 543–555.

Vogt N, Koch I, Schwarz H, Schnörrer F, Nüsslein-Volhard C. 2006. The gammaTuRC components Grip75 and Grip128 have an essential microtubule-anchoring function in the *Drosophila* germline. *Development* **133:** 3963–3972.

Walser CB, Lipshitz HD. 2011. Transcript clearance during the maternal-to-zygotic transition. *Curr Opin Genet Dev* **21:** 431–443.

Wang Z, Lin H. 2004. Nanos maintains germline stem cell self-renewal by preventing differentiation. *Science* **303:** 2016–2019.

Weidmann CA, Goldstrohm AC. 2012. *Drosophila* Pumilio protein contains multiple autonomous repression domains that regulate mRNAs independently of Nanos and Brain Tumor. *Mol Cell Biol* **32:** 527–540.

Weil TT, Forrest KM, Gavis ER. 2006. Localization of *bicoid* mRNA in late oocytes is maintained by continual active transport. *Dev Cell* **11:** 251–262.

Weil TT, Xanthakis D, Parton R, Dobbie I, Rabouille C, Gavis ER, Davis I. 2010. Distinguishing direct from indirect roles for bicoid mRNA localization factors. *Development* **137:** 169–176.

Wilhelm JE, Mansfield J, Hom-Booher N, Wang S, Turck CW, Hazelrigg T, Vale RD. 2000. Isolation of a ribonucleoprotein complex involved in mRNA localization in *Drosophila* oocytes. *J Cell Biol* **148:** 427–440.

Wreden C, Verrotti AC, Schisa JA, Lieberfarb ME, Strickland S. 1997. Nanos and Pumilio establish embryonic polarity in *Drosophila* by promoting posterior deadenylation of *hunchback* mRNA. *Development* **124:** 3015–3023.

Yano T, López de Quinto S, Matsui Y, Shevchenko A, Shevchenko A, Ephrussi A. 2004. Hrp48, a *Drosophila* hnRNP A/B homolog, binds and regulates translation of *oskar* mRNA. *Dev Cell* **6:** 637–648.

Ye B, Petritsch C, Clark IE, Gavis ER, Jan LY, Jan YN. 2004. Nanos and Pumilio are essential for dendrite morphogenesis in *Drosophila* peripheral neurons. *Curr Biol* **14:** 314–321.

Zaessinger S, Busseau I, Simonelig M. 2006. Oskar allows *nanos* mRNA translation in *Drosophila* embryos by preventing its deadenylation by Smaug/CCR4. *Development* **133:** 4573–4583.

Zimyanin VL, Belaya K, Pecreaux J, Gilchrist MJ, Clark A, Davis I, St. Johnston D. 2008. In vivo imaging of *oskar* mRNA transport reveals the mechanism of posterior localization. *Cell* **134:** 843–853.

Toward a Genome-Wide Landscape of Translational Control

Ola Larsson[1], Bin Tian[2], and Nahum Sonenberg[3]

[1]Department of Oncology–Pathology, Karolinska Institute, Stockholm SE-171 76, Sweden

[2]Department of Biochemistry and Molecular Biology, UMDNJ–New Jersey Medical School, Newark, New Jersey 07103

[3]Department of Biochemistry and Goodman Cancer Research Centre, McGill University, Montreal, Quebec H3A 1A3, Canada

Correspondence: ola.larsson@ki.se

Genome-wide analysis of translational control has taken strides in recent years owing to the advent of high-throughput technologies, including DNA microarrays and deep sequencing. Global studies have unraveled a principal role, among posttranscriptional mechanisms, for mRNA translation in determining protein levels in the cell. The impact of translational control in dynamic regulation of the proteome under different conditions is increasingly appreciated. Here we review genome-wide studies that use high-throughput techniques and bioinformatics to assess the role of mRNA translation in the regulation of protein levels; we also discuss how genome-wide data on mRNA translation can be obtained, analyzed, and used to identify mechanisms of translational control.

The gene expression pathway leading to protein production consists of many mechanistic layers that are subject to regulation. They are commonly grouped into transcriptional or posttranscriptional types. Some posttranscriptional mechanisms, including RNA splicing, mRNA editing, and posttranslational modification, determine the identity and activity of the protein products, whereas others control protein levels by regulating transport of mRNA from the nucleus to the cytoplasm, mRNA stability, translation, and protein stability. Determining how posttranscriptional regulation contributes to protein levels and, more precisely, how regulation of translation impacts gene expression have attracted substantial attention during the last decade.

POSTTRANSCRIPTIONAL MECHANISMS SUBSTANTIALLY AFFECT GENE EXPRESSION LEVELS AT A GENOME-WIDE SCALE

Several studies have examined the extent to which posttranscriptional mechanisms affect protein expression by comparing mRNA and protein levels, either in one cell state or across different conditions. This is typically based on Pearson or Spearman correlation coefficients, denoted as r_p or r_s, respectively (for examples, see Fig. 1). Both range from -1 to 1, where 0

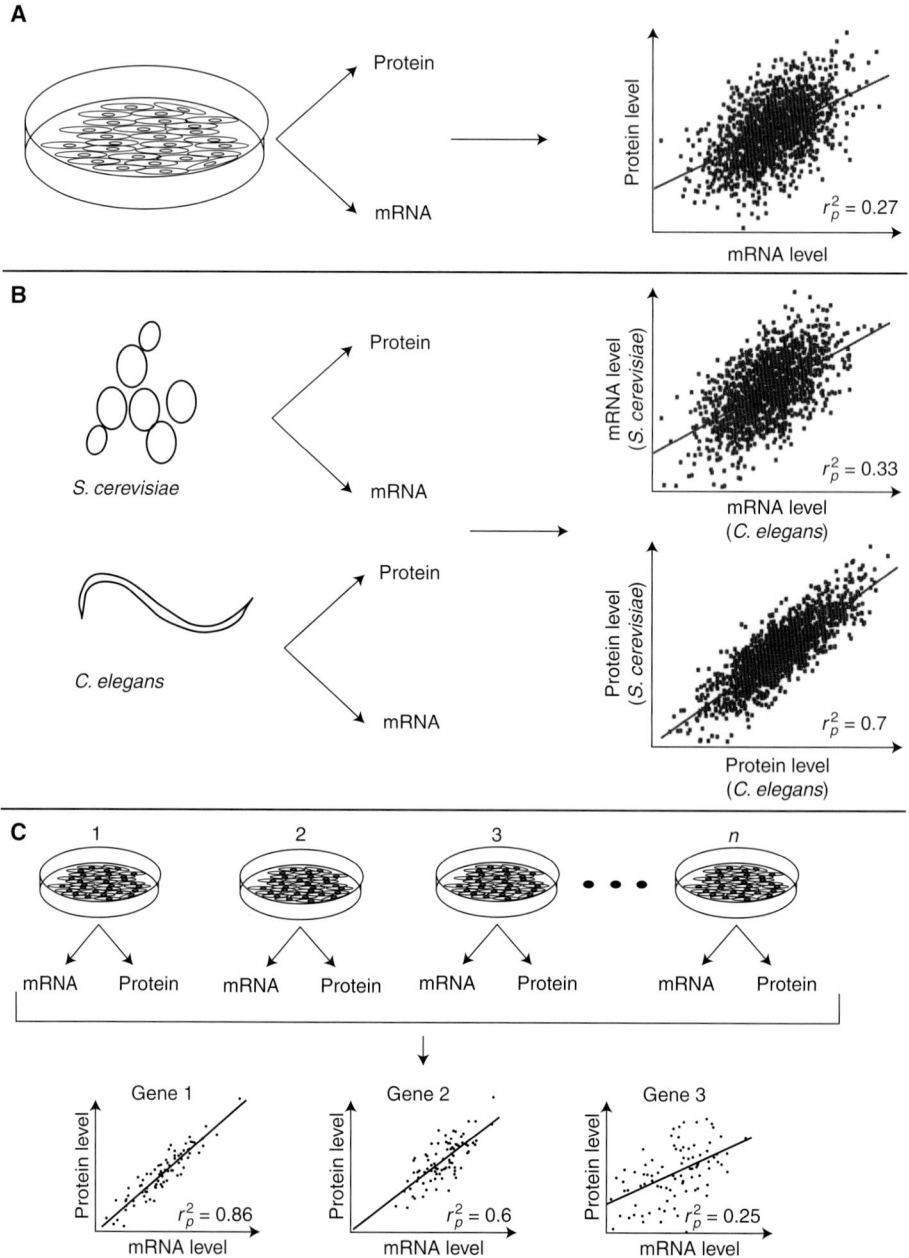

Figure 1. Approaches to examine the contribution of posttranscriptional regulation to protein expression. (*A*) Comparison of mRNA and protein levels from a single condition. Measured mRNA and protein levels are compared to indicate posttranscriptional regulation across genes using r_p^2, which describes the fraction of the observed protein levels that are explained by mRNA levels. (*B*) Cross-species analysis of mRNA and protein levels. mRNA and protein levels are obtained in parallel from two species, and cross-species levels are compared. Higher correlation of protein levels across species as compared with mRNA levels indicates posttranscriptional regulation that maintains conserved protein levels. (*C*) Parallel measurements of mRNA and protein levels across many conditions. Per gene mRNA and protein levels are compared across conditions to estimate dynamic regulation of gene expression by posttranscriptional mechanisms. As indicated by r_p^2, gene 3 shows more posttranscriptional regulation as compared with gene 1. Simulated numbers are shown.

Cite this article as *Cold Spring Harb Perspect Biol* doi: 10.1101/cshperspect.a012302

indicates no correlation and -1 and 1 indicate perfect negative and positive correlations, respectively. Whereas Pearson correlation is based on actual values, Spearman correlation uses ranks (ordered by values) and is therefore less influenced by extreme values (outliers). A squared Pearson correlation value (r_p^2; the coefficient of determination) describes how much of the variance of one factor is explained by that of the other factor. For example, a Pearson correlation between protein and mRNA levels of 0.6 indicates that 36% (0.6^2) of the protein levels can be explained by mRNA levels. Thus, a lower r_s or r_p^2 between mRNA and protein levels indicates more posttranscriptional regulation.

Comparison of mRNA and protein levels across genes under one condition (e.g., steady-state growth) has been conducted in species ranging from bacteria to humans. These studies attempted to assess intrinsic posttranscriptional regulation of gene expression (Fig. 1A). Pioneering studies in yeast established genome-wide differences in expression levels between mRNAs and proteins, although they assessed only a fraction of all yeast genes (Futcher et al. 1999; Gygi et al. 1999). Subsequent studies based on measurements of more genes estimated r_p^2 from 0.14 to 0.73 (Lu et al. 2007; Schmidt et al. 2007; Ingolia et al. 2009) or r_s of 0.57 and 0.58 (Ghaemmaghami et al. 2003; Beyer et al. 2004). In bacteria, r_p^2 between 0.20 and 0.47 has been reported (Nie et al. 2006; Lu et al. 2007; Jayapal et al. 2008). Moreover, in a recent single-cell study in *Escherichia coli*, mean mRNA copies and protein copies showed r_p^2 of 0.29 and 0.59 using deep sequencing of RNA (RNA-seq) and fluorescence in situ hybridization (FISH), respectively (Taniguchi et al. 2010). These studies indicate that posttranscriptional regulation is used in unicellular organisms, although estimates of its extent vary substantially across studies.

Similar studies performed in multicellular organisms, including *Arabidopsis thaliana* (Baerenfaller et al. 2008), *Drosophila melanogaster*, and *Caenorhabditis elegans* (Schrimpf et al. 2009), all indicated substantial posttranscriptional regulation (r_p^2 between 0.27 and 0.46 [Baerenfaller et al. 2008] and r_s of ~0.6 [Schrimpf et al. 2009]). Moreover, a recent

study of a human cancer cell line reported a modest r_p^2 of 0.29 from more than 1000 genes (Vogel et al. 2010). Interestingly, these investigators built a model that predicts protein levels from measured mRNA levels and a variety of mRNA properties that affect posttranscriptional regulation, such as the length of the 3' untranslated region (UTR). The r_p^2 between the predicted and the measured protein levels increased dramatically as compared with the initial r_p^2 between mRNA and protein levels (from 0.29–0.67). Because the data used to derive the model were also used to generate the predictions, creating the possibility of data overfitting (meaning that the prediction outcome is heavily influenced by the data used to construct the model), future studies will be needed to assess the generality of the model. Despite this limitation, the study suggests that mRNA sequence features systematically impact protein production through posttranscriptional control.

Cross-species comparisons provide an alternative approach for assessing the importance of posttranscriptional regulation (Fig. 1B). In such studies, mRNA and protein levels from one species are compared with their orthologs in a different species. Assuming that the measurement error is similar for mRNA and protein levels, a higher cross-species correlation between protein levels than between mRNA levels would suggest a role of posttranscriptional control in maintaining conserved protein levels. Based on this reasoning, Schrimpf et al. (2009) reported r_s of 0.79 and 0.47 for protein and mRNA levels, respectively, when comparing *C. elegans* with *D. melanogaster*. A follow-up study examined all paired comparisons between seven species and found a higher cross-species correlation for protein levels in 17 out of 21 comparisons (Laurent et al. 2010). These comparisons included mRNA measurements obtained by RNA-seq, which is thought to better reflect relative mRNA levels across genes as compared with DNA microarrays (Laurent et al. 2010). Cross-species comparisons thus provide further support for the idea that posttranscriptional mechanisms substantially contribute to determination of protein levels.

A common critique of correlation-based mRNA/protein comparative studies is that the magnitude of systematic and random variations inherent in mRNA and protein analysis tools, such as DNA microarrays, RNA-seq, and mass spectrometry, is often unknown. This is of significance because more variation will lead to lower correlation, giving the appearance of a greater degree of posttranscriptional regulation. This ambiguity in the interpretation of r_s or r_p^2 has been addressed in some studies by assessing how much variation affects the r_s or r_p^2 values. Another caveat is that the half-lives of proteins and their cognate mRNAs often differ, which can reduce r_s or r_p^2 if the mRNA or protein levels are obtained under non-steady-state conditions. Indeed, in mouse NIH/3T3 cells, mRNAs show a median half-life of 9 h, whereas proteins have a median half-life of 46 h (Schwanhausser et al. 2011), making protein level regulation linger over a longer time period relative to the corresponding mRNA. Caution is therefore needed when interpreting r_s or r_p^2 values.

DYNAMIC REGULATION OF GENE EXPRESSION AT THE POSTTRANSCRIPTIONAL LEVEL

In addition to the studies discussed above, which evaluate intrinsic protein levels in the cell, progress has been made in understanding the role of posttranscriptional mechanisms in dynamic regulation of gene expression. Three approaches have been applied to determine whether the protein product levels of individual genes can change independently of their mRNA levels.

In the first approach, mRNA and protein levels are measured under two conditions, and differences between the conditions are calculated for mRNA and protein levels separately. These per gene differences in mRNA and protein levels are then compared across all genes. This approach using yeast produced r_s of 0.21 or 0.45 (Griffin et al. 2002; Washburn et al. 2003). A similar study of two human cell lines reported an r_p^2 of 0.41 (Tian et al. 2004). Importantly, the latter study estimated the maximum obtainable r_p^2 to 0.81 (given the variation in measurements of mRNA and protein levels) using a simulation approach, suggesting a substantial contribution

from posttranscriptional mechanisms in the dynamic regulation of gene expression (r_p^2 of 0.41 vs. 0.81).

The second approach involves parallel measurement of mRNA and protein levels at several time points following a treatment. This approach also allows for assessment of the extent to which differences in half-lives between mRNAs and proteins can affect the result. A study using yeast monitored the mRNA and protein levels in untreated cells and at six time points following treatment with rapamycin (Fournier et al. 2010). The investigators found that for proteins whose expression had changed, their mRNA levels at 1 h after treatment showed the maximum correlation with protein levels 6 h after treatment—thus, a delayed adjustment of protein levels to mRNA levels. Yet, the r_p^2 only reached a maximum of 0.36 throughout the experiment. In a similarly designed experiment of mouse embryonic stem cell differentiation with four time points, Lu et al. (2009) concluded that only about half of the proteins that changed their levels also displayed concordant mRNA level changes. A proportion of the proteins that initially did not show concordant protein and mRNA levels did, however, show concordant levels at a later time point.

The third approach minimizes the potential bias arising from differences in mRNA and protein half-lives by studying mRNA and protein levels under steady-state conditions (Fig. 1C). In a recent study, 1066 mRNA and protein levels were measured in 23 human cell lines (Gry et al. 2009). The average r_s between mRNA and protein levels was 0.20 and 0.25 using cDNA microarrays or Affymetrix GeneChips, respectively. As a comparison, the average r_s between mRNA levels obtained from cDNA microarrays and Affymetrix GeneChips was 0.52. This is substantially higher than that observed between protein and mRNA levels (i.e., 0.2 or 0.25 as compared with 0.52), indicating that mRNA measurement error is not likely to explain the low correlations between protein and mRNA levels. In an extensive study using the approach shown in Figure 1C, mRNA and protein levels in livers from 97 inbred mice were measured (Ghazalpour et al. 2011). Out of 396 genes, only 21%

showed significant correlations between mRNA and protein levels. By replicating the experiment, the researchers stratified the genes based on their signal-to-noise ratio, thereby also assessing the impact of random variation on the reported correlations. As expected, the mean r_s increased as the signal-to-noise ratio increased and reached a maximum of ∼0.4. Thus, in this extensive study in which differences in half-lives between mRNAs and proteins are likely to have a minimal impact and only genes that could be measured with high confidence were analyzed, the results still support a substantial role for posttranscriptional mechanisms in dynamic regulation of protein levels.

GENOME-WIDE ANALYSIS OF TRANSLATIONAL ACTIVITY

The studies described above all indicate substantial posttranscriptional controls in different systems. A detailed, in-depth examination of posttranscriptional regulation was recently conducted by Schwanhausser et al. (2011), using a multi-omics approach in NIH/3T3 cells (Fig. 2A). They assumed a model in which mRNA levels are determined by transcription and mRNA stability, whereas protein levels are determined by mRNA levels, translational activity, and protein degradation (Fig. 2B). Accordingly, per gene translational activity and transcription could be inferred from measurements of mRNA levels, mRNA stability, protein levels, and protein degradation. Notably, the investigators used independently replicated data to assess the extent to which protein levels predicted by the model compared with the measured levels from the replicates. This allowed the researchers to determine the relative contribution of different gene expression mechanisms while avoiding overfitting. Strikingly, a principal role for mRNA translation among posttranscriptional mechanisms, was identified in determining intrinsic protein levels, strongly suggesting that most of the discrepancies between mRNA and protein levels result from translational control.

More direct evidence supporting the widespread role of translational control comes from studies of the global association between mRNAs and ribosomes. Because mRNAs that have a higher translational activity are associated with more ribosomes, the polysome microarray technique has been used to study genome-wide mRNA translation. For polysome preparation, translation elongation is inhibited by cycloheximide, the cytoplasmic lysate is isolated and applied to a sucrose gradient, and mRNAs associated with varying numbers of ribosomes are separated using ultracentrifugation according to their sedimentation velocity (Fig. 3A). Fractionated mRNAs are then extracted and subjected to DNA microarray analysis for

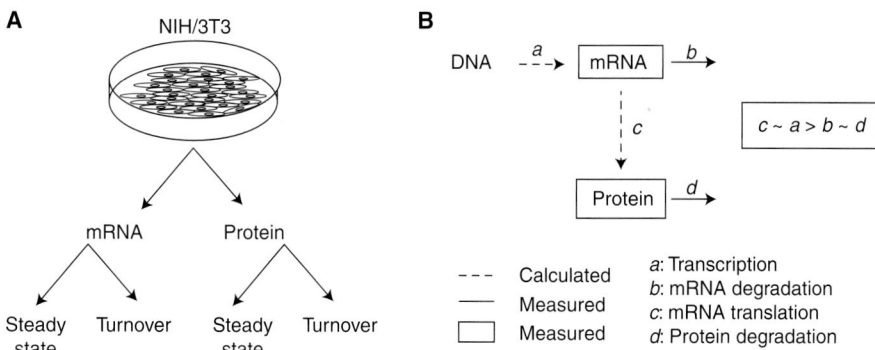

Figure 2. A multi-omics approach to examine relative contributions of posttranscriptional mechanisms to protein expression. (A) Levels and turnover rates for both mRNA and proteins are obtained in parallel. (B) A simple model is used to calculate transcription and translational efficiencies using measured levels and turnover rates for both mRNA and proteins.

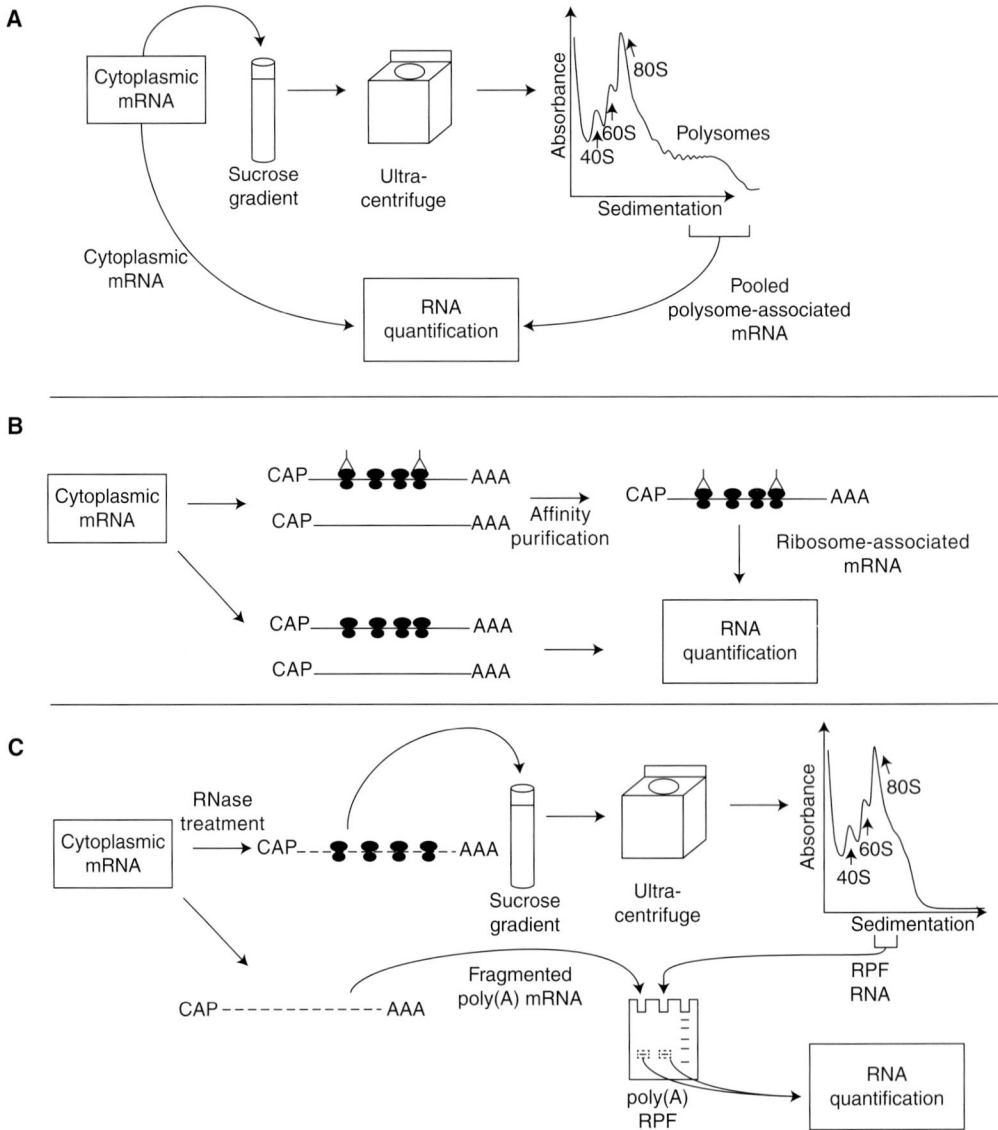

Figure 3. Techniques to obtain genome-wide data on mRNA translation. (*A*) The polysome technique. Polysome-associated mRNA is prepared in parallel with cytoplasmic mRNA and quantified using DNA microarrays or RNA-seq. (*B*) The affinity purification technique. mRNAs that are associated with tagged ribosomes are purified using an antibody-based affinity purification approach. Cytoplasmic mRNA is prepared in parallel, and both samples are quantified using DNA microarrays or RNA-seq. (*C*) The ribosome profiling technique. RPFs generated by RNase treatment are isolated from monosomes and subsequently purified by gel based on size. A randomly fragmented sample from poly(A) mRNA is prepared in parallel, and both samples are sequenced.

identification and quantification. Although a large number of fractions can be obtained to examine mRNA profiles across the entire polysome range (Arava et al. 2003), most studies pool fractions because of the high cost of

DNA microarrays (Johannes et al. 1999; Zong et al. 1999; Chen et al. 2002, 2011; Jechlinger et al. 2003; Preiss et al. 2003; Kitamura et al. 2004, 2008; Provenzani et al. 2006; Spence et al. 2006; Genolet et al. 2008; Parent and

Beretta 2008; Ramirez-Valle et al. 2008; Ceppi et al. 2009; Dhamija et al. 2010; Rivera-Ruiz et al. 2010; Di Florio et al. 2011). Commonly, fractions with two or more associated ribosomes are pooled, limiting the analysis of regulation to a shift from less than two to two or more associated ribosomes (i.e., "on–off" regulation), whereas other shifts (e.g., from three to nine associated ribosomes, i.e., "relative" regulation) are missed. An alternative approach involves pooling mRNAs that are associated with $>n$ ribosomes, where n is often 3 (Larsson et al. 2006, 2007; Mamane et al. 2007; Colina et al. 2008; Kim et al. 2009). This approach enables identification of differential translation involving both "on–off" and some "relative" regulation. Short mRNAs or mRNAs constantly associated with more than four ribosomes are, however, not studied. The polysome microarray technique has been applied to address a wide range of questions, including regulation of translation by the mTOR pathway (Rajasekhar et al. 2003; Tominaga et al. 2005; Larsson et al. 2006, 2007; Bilanges et al. 2007; Mamane et al. 2007; Kim et al. 2009; Furic et al. 2010), dynamic regulation of protein synthesis during cellular stress (Blais et al. 2004, 2006; Lu et al. 2006; Kumaraswamy et al. 2008; Dang Do et al. 2009; Matsuura et al. 2010), and the role of mRNA translation in development or differentiation (Iguchi et al. 2006; Grech et al. 2008; Parent and Beretta 2008; Sampath et al. 2008; Otulakowski et al. 2009) and disease (Larsson et al. 2008; Davidson et al. 2009; Treton et al. 2011).

A second set of techniques relies on affinity purification of ribosomes, analogous to the RNA-binding protein immunoprecipitation (RIP)–based techniques (Keene et al. 2006). These methods allow for identification and quantification of translating mRNAs (Fig. 3B). In one application, the ribosomal protein Rpl16 in yeast was modified by addition of a protein A tag to allow purification of mRNAs associated with ribosomes (Halbeisen et al. 2009). Similarly, a hemagglutinin (HA) tag was used to mark ribosomal protein Rpl22 in mouse (Sanz et al. 2009). Importantly, the HA tagging of Rpl22 was dependent on the activity of Cre recombinase, allowing analysis of mRNA translation

in a selected cell type in vivo when combined with cell-type-specific Cre expression (Sanz et al. 2009). Another approach is based on the capture of Hsp70 chaperones associated with polysomes (Kudo et al. 2010). Theoretically, the efficiency of immunoprecipitation depends on the number of associated ribosomes, thereby reflecting translational activity. However, the precise number of ribosomes per mRNA is unknown.

A new technique named ribosome profiling has recently been developed (Ingolia et al. 2009), which is designed to identify open reading frames (ORFs) and quantitatively examine ribosome association with mRNAs. The technique involves two steps: isolation of mRNA fragments that are protected by ribosomes (ribosome-protected fragments [RPF]) from RNase treatment, and identification and quantification of the fragments by RNA-seq (Ingolia et al. 2009). For simplicity, this method is called RPF-seq here. Because RNA fragments are size-selected to obtain those corresponding to the expected footprint of the ribosome (Fig. 3C), RPF data reveal the locations of ribosomes on mRNA. As such, detailed quantitative examination of all steps of mRNA translation, including initiation, elongation, and termination, becomes possible. RPF-seq was first used to examine translational control in budding yeast under rich and starvation conditions (Ingolia et al. 2009). More recent studies using RPF-seq have assessed translation in several systems, including mouse embryonic stem cells (Ingolia et al. 2011), meiosis in yeast (Brar et al. 2012), and microRNA-mediated suppression of gene expression (Bazzini et al. 2012).

ANALYSIS OF mRNA TRANSLATION DATA

A major advantage in studying translating mRNA (the "translatome") over steady-state mRNA (the "transcriptome") is the ability to obtain measurements that more closely correspond to protein levels (Ingolia et al. 2009), owing to fewer intermediate regulatory steps. Accurate assessment of translational control, however, requires adjustment for the influence of other steps in the gene expression pathway, including transcription, mRNA stability, and mRNA transport (Larsson et al. 2010). Because individual mRNAs can be regulated substantially at the level of mRNA transport (Rousseau et al. 1996), only comparison to cytoplasmic mRNA levels will allow for the precise analysis of differential translation, whereas comparison to whole-cell mRNA will lead to joint analysis of mRNA transport and translation. A common approach to examine translational control is to calculate translational efficiency scores—$\log_2$ [(translating mRNA)/(cytoplasmic mRNA)]— and compare these between conditions to identify differential translation. Because of a mathematical necessity, translational efficiency scores may correlate with the cytoplasmic mRNA abundance instead of solely describing mRNA translation (Larsson et al. 2010). This phenomenon is called spurious correlation (Pearson 1896), which leads to increased false-positive and false-negative rates when examining differential translation (Larsson et al. 2010). Indeed, an assessment of 20 studies using the polysome microarray technique or RPF-seq showed that spurious correlations are common (Larsson et al. 2010). This shortcoming of using translational efficiency scores prompted development of a method based on analysis of partial variance (APV, implemented in the program Anota) (Larsson et al. 2010), which eliminates spurious correlations. In APV, a linear regression model (between translating and cytoplasmic mRNA data) is applied. Fold-change for mRNA translation and associated *P*-values are calculated based on differences in intercepts and residual errors (Fig. 4). In addition, the method includes a range of quality criteria to judge whether the input data set violates model assumptions (Larsson et al. 2011). Notably, this method has been successfully applied to identify differential mRNA translation using both polysome microarray and RPF-seq data (Larsson et al. 2010).

TECHNIQUES TO REVEAL *CIS* AND *TRANS* REGULATORS OF POSTTRANSCRIPTIONAL CONTROL

Posttranscriptional regulation of gene expression, including translation, is believed to in-

volve sets of targeted mRNAs resembling the polycistronic operons present in bacteria (Spirin 1969; Keene and Tenenbaum 2002; Keene 2007). The theory posits that subsets of mRNAs can be regulated at the posttranscriptional level in a combinatorial fashion. Such selective regulation commonly involves RNA-binding proteins (RBPs) or microRNAs that associate with RNA elements within the target mRNA (Bartel 2004; Richter and Sonenberg 2005). RBP-associated mRNA elements are often defined by combinations of structure and sequence properties and usually reside in the mRNA UTR. Once associated with their target mRNA, the RBPs interact with translation initiation factors and sometimes other RBPs and/or microRNAs to positively or negatively regulate translation. Thus, active RNA elements and RBPs need to be identified to mechanistically reveal how differential translation takes place (Fig. 5).

Known RNA elements are often examined as the first step to explain differential translation. These are collected in general databases such as the UTRdb (Grillo et al. 2010), RBPDB (Cook et al. 2011), and CLIPZ (Khorshid et al. 2011) or element-specific databases such as the ARED (Halees et al. 2008) and IRESite (Mokrejs et al. 2010). Information regarding miRNA target sites can be obtained from many databases such as TargetScan (Lewis et al. 2005). Sometimes searching such databases leads to identification of active RNA elements as exemplified by identification of 5′-terminal oligopyrimidine tract (TOP) elements as targets of mTOR signaling (Bilanges et al. 2007; Mamane et al. 2007). More often, however, known RNA elements are not sufficient to explain observed mRNA translation patterns. This makes de novo discovery of regulatory RNA elements necessary.

Bioinformatic methods to uncover novel *cis* elements are based on the assumption that differentially translated mRNAs share common RNA elements (Larsson and Bitterman 2010). mRNA sequences, often in UTRs, are thus used as input to identify sequences or structures overrepresented in the regulated mRNAs, as compared with background ones (Larsson et al. 2006; Foat and Stormo 2009; Chen et al. 2011). A set of three algorithms detected ~50%

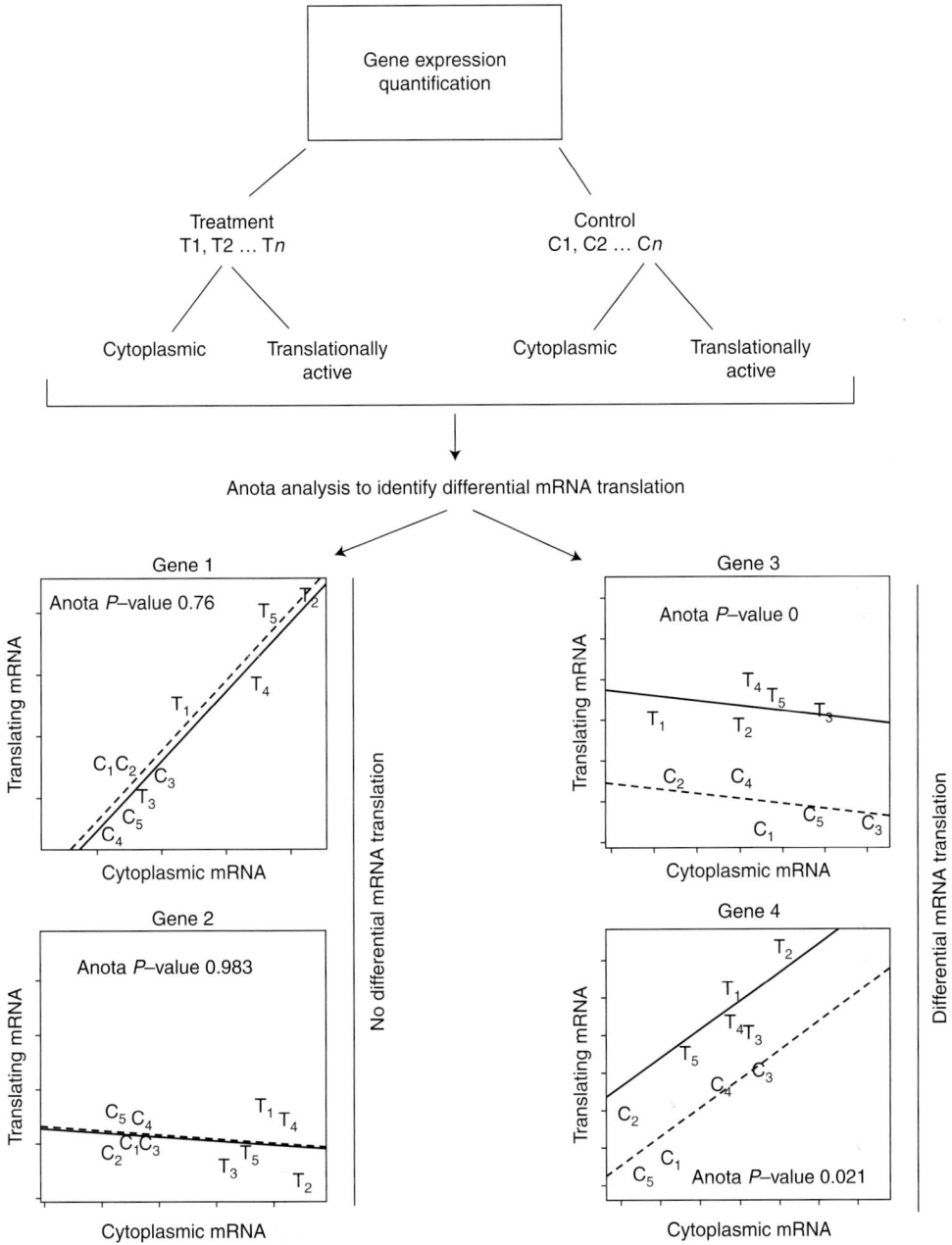

Figure 4. Genome-wide identification of differential mRNA translation by Anota. Replicated data from translating and cytoplasmic mRNAs is used as input in Anota. Anota performs linear regression using translating and cytoplasmic mRNAs for all conditions. Treatment condition (solid line); control condition (dotted line). The intercepts of the lines on the y-axis are compared to derive an mRNA translation fold-change and are related to the residual error of the regression to identify differential translation. Genes 3 and 4 are differentially translated, but genes 1 and 2 are not. Simulated numbers are shown.

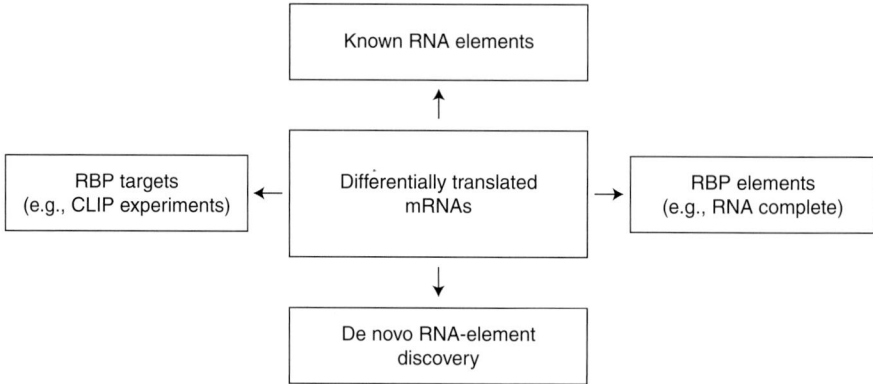

Figure 5. Approaches to examine mechanisms for differential mRNA translation.

of known RNA elements when UTRs with specific RNA elements were mixed with randomly selected, unrelated UTR sequences (Fan et al. 2009), highlighting the potential of this approach. There are, nonetheless, several limitations. First, the precise length of each UTR is often unknown, and the active RNA elements may therefore be located outside of the studied sequences. In addition, alternative cleavage and polyadenylation, which regulates 3′-UTR length, is widespread and dynamically regulated under different conditions and across tissue types (Tian et al. 2005; Sandberg et al. 2008). Examining UTRs that are not expressed in the studied cell type can lead to false identification of RNA elements. Moreover, there is a possibility that some regulation involves several players, including RBPs, microRNAs, and RNA elements, with linear or nonlinear interactions. Such a complex situation likely makes identification of any single mechanism more difficult (Fan et al. 2009).

Several high-throughput techniques have been developed to study RNA–RBP interactions. In vitro methods include RNAcompete, whereby a library of RNAs is synthesized and used in pull-down experiments with RBPs followed by detection using DNA microarrays (Ray et al. 2009), and SELEX-seq, whereby random RNA aptamers are selected based on interaction with a specific RBP and are deep-sequenced (Dittmar et al. 2012). Although these methods do not directly address which mRNAs

are targeted by the RBP under investigation, they elucidate binding specificities.

In vivo methods generally involve immuno-precipitation of RBPs from cells and identification of coimmunoprecipitated RNAs (Darnell and Richter 2012). Indeed, identification of RNA elements by ribonucleoprotein immunoprecipitation (RIP) followed by microarray (RIP-chip) has been successful for a number of RBPs (Gerber et al. 2004; Hogan et al. 2008). RBPs can also be UV-cross-linked to their binding RNAs in vivo, allowing purification of the RBP:RNA complex under denaturing conditions, for example, SDS-PAGE (polyacrylamide gel electrophoresis). Deep sequencing of RNA isolated by cross-linking and immunoprecipitation (HITS-CLIP) has been used to study RBPs (Licatalosi et al. 2008; Darnell et al. 2011) and microRNAs (Chi et al. 2009). The photoactivatable-ribonucleoside-enhanced cross-linking and immunoprecipitation (PAR-CLIP) technique, which also has been used to study microRNAs and RBPs (Hafner et al. 2010; Hoell et al. 2011; Lebedeva et al. 2011; Mukherjee et al. 2011), is similar to HITS-CLIP but allows cross-linking at a longer wavelength. Both HITS-CLIP and PAR-CLIP allow precise detection of the binding site because cross-linking sites lead to mutations and deletions in sequencing reads (Hafner et al. 2010; Zhang and Darnell 2011). A comparison between HITS-CLIP and PAR-CLIP indicated that experimental conditions, such as the extent of RNase digestion, need to be optimized to minimize the sequence bias of the identified RNA fragments (Kishore et al. 2011). Another CLIP-based approach, iCLIP, identifies sequencing reads that terminate at the cross-linked sites (Wang et al. 2010; Konig et al. 2011; Tollervey et al. 2011). One potential caveat of all of these approaches is that binding of an RBP to mRNA may depend on other RBPs. Moreover, separating transient interactions with limited effect on regulation from stable interactions that substantially affect regulation is a challenge (Mukherjee et al. 2011). Nevertheless, systematic studies of many RBPs will likely soon be available and could be very useful to mechanistically dissect genome-wide patterns of differential mRNA translation.

CONCLUDING REMARKS

Decades of research based on single genes have established that mRNA translation can profoundly control protein levels and thereby directly determine biological outcomes (Costa-Mattioli et al. 2009; Sonenberg and Hinnebusch 2009; Jackson et al. 2010; Silvera et al. 2010; Spriggs et al. 2010; Blagden and Willis 2011). Over the last few years, genome-wide analyses have shown that posttranscriptional regulation, translational control in particular, plays significant roles in determining protein levels in the cell. With the rapid development of current methodologies, especially deep-sequencing-based methods, we can expect many major discoveries to come from genome-wide studies. Integrating data on mRNA translation with information regarding RBP–mRNA interactions is an emerging area of research. In addition, given the widespread occurrence of mRNA isoforms in higher species, resulting from alternative initiation, splicing, and polyadenylation (Wang et al. 2008), sorting out the translational efficiency for each mRNA isoform will shed important light on the compendium of protein isoforms and reveal the connection between translational control and mRNA processing. On the other hand, to harness the power of genomics fully, we will need to better adopt high-throughput techniques and use rigorous data analysis approaches. It is exciting that a genome-wide landscape of translational control is now in sight.

ACKNOWLEDGMENTS

O.L. is supported by the Swedish Research Council, the Swedish Cancer Foundation, the Cancer Society in Stockholm, the Jeansson Foundation, and the Åke Wiberg Foundation. B.T. is supported by grants from the NIH. We thank Dr. Robert Nadon (McGill University) for comments on the manuscript.

REFERENCES

*Reference is also in this collection.

Arava Y, Wang Y, Storey JD, Liu CL, Brown PO, Herschlag D. 2003. Genome-wide analysis of mRNA translation profiles in *Saccharomyces cerevisiae*. *Proc Natl Acad Sci* **100:** 3889–3894.

Baerenfaller K, Grossmann J, Grobei MA, Hull R, Hirsch-Hoffmann M, Yalovsky S, Zimmermann P, Grossniklaus U, Gruissem W, Baginsky S. 2008. Genome-scale proteomics reveals *Arabidopsis thaliana* gene models and proteome dynamics. *Science* **320:** 938–941.

Bartel DP. 2004. MicroRNAs: Genomics, biogenesis, mechanism, and function. *Cell* **116:** 281–297.

Bazzini AA, Lee MT, Giraldez AJ. 2012. Ribosome profiling shows that miR-430 reduces translation before causing mRNA decay in zebrafish. *Science* **336:** 233–237.

Beyer A, Hollunder J, Nasheuer HP, Wilhelm T. 2004. Post-transcriptional expression regulation in the yeast *Saccharomyces cerevisiae* on a genomic scale. *Mol Cell Proteomics* **3:** 1083–1092.

Bilanges B, Argonza-Barrett R, Kolesnichenko M, Skinner C, Nair M, Chen M, Stokoe D. 2007. Tuberous sclerosis complex proteins 1 and 2 control serum-dependent translation in a TOP-dependent and -independent manner. *Mol Cell Biol* **27:** 5746–5764.

Blagden SP, Willis AE. 2011. The biological and therapeutic relevance of mRNA translation in cancer. *Nat Rev Clin Oncol* **8:** 280–291.

Blais JD, Filipenko V, Bi M, Harding HP, Ron D, Koumenis C, Wouters BG, Bell JC. 2004. Activating transcription factor 4 is translationally regulated by hypoxic stress. *Mol Cell Biol* **24:** 7469–7482.

Blais JD, Addison CL, Edge R, Falls T, Zhao H, Wary K, Koumenis C, Harding HP, Ron D, Holcik M, et al. 2006. Perk-dependent translational regulation promotes tumor cell adaptation and angiogenesis in response to hypoxic stress. *Mol Cell Biol* **26:** 9517–9532.

Brar GA, Yassour M, Friedman N, Regev A, Ingolia NT, Weissman JS. 2012. High-resolution view of the yeast meiotic program revealed by ribosome profiling. *Science* **335:** 552–557.

Ceppi M, Clavarino G, Gatti E, Schmidt EK, de Gassart A, Blankenship D, Ogola G, Banchereau J, Chaussabel D,

Pierre P. 2009. Ribosomal protein mRNAs are translationally-regulated during human dendritic cells activation by LPS. *Immunome Res* 5: 5.

Chen G, Gharib TG, Huang CC, Taylor JM, Misek DE, Kardia SL, Giordano TJ, Iannettoni MD, Orringer MB, Hanash SM, et al. 2002. Discordant protein and mRNA expression in lung adenocarcinomas. *Mol Cell Proteomics* 1: 304–313.

Chen J, Melton C, Suh N, Oh JS, Horner K, Xie F, Sette C, Blelloch R, Conti M. 2011. Genome-wide analysis of translation reveals a critical role for deleted in azoospermia-like (Dazl) at the oocyte-to-zygote transition. *Genes Dev* 25: 755–766.

Chi SW, Zang JB, Mele A, Darnell RB. 2009. Argonaute HITS-CLIP decodes microRNA–mRNA interaction maps. *Nature* 460: 479–486.

Colina R, Costa-Mattioli M, Dowling RJ, Jaramillo M, Tai LH, Breitbach CJ, Martineau Y, Larsson O, Rong L, Svitkin YV, et al. 2008. Translational control of the innate immune response through IRF-7. *Nature* 452: 323–328.

Cook KB, Kazan H, Zuberi K, Morris Q, Hughes TR. 2011. RBPDB: A database of RNA-binding specificities. *Nucleic Acids Res* 39: D301–308.

Costa-Mattioli M, Sossin WS, Klann E, Sonenberg N. 2009. Translational control of long-lasting synaptic plasticity and memory. *Neuron* 61: 10–26.

Dang Do AN, Kimball SR, Cavener DR, Jefferson LS. 2009. eIF2α kinases GCN2 and PERK modulate transcription and translation of distinct sets of mRNAs in mouse liver. *Physiol Genomics* 38: 328–341.

* Darnell JC, Richter JD. 2012. Cytoplasmic RNA binding proteins and the control of complex brain function. *Cold Spring Harb Perspect Biol* doi: 10.1101/cshperspect.a012344.

Darnell JC, Van Driesche SJ, Zhang C, Hung KY, Mele A, Fraser CE, Stone EF, Chen C, Fak JJ, Chi SW, et al. 2011. FMRP stalls ribosomal translocation on mRNAs linked to synaptic function and autism. *Cell* 146: 247–261.

Davidson LA, Wang N, Ivanov I, Goldsby J, Lupton JR, Chapkin RS. 2009. Identification of actively translated mRNA transcripts in a rat model of early-stage colon carcinogenesis. *Cancer Prev Res (Phila)* 2: 984–994.

Dhamija S, Doerrie A, Winzen R, Dittrich-Breiholz O, Taghipour A, Kuehne N, Kracht M, Holtmann H. 2010. IL-1-induced post-transcriptional mechanisms target overlapping post-transcriptional silencing and destabilizing elements in IκBζ mRNA. *J Biol Chem* 285: 29165–29178.

Di Florio A, Adesso L, Pedrotti S, Capurso G, Pilozzi E, Corbo V, Scarpa A, Geremia R, Delle Fave G, Sette C. 2011. Src kinase activity coordinates cell adhesion and spreading with activation of mammalian target of rapamycin in pancreatic endocrine tumour cells. *Endocr Relat Cancer* 18: 541–554.

Dittmar KA, Jiang P, Park JW, Amirikian K, Wan J, Shen S, Xing Y, Carstens RP. 2012. Genome-wide determination of a broad ESRP-regulated posttranscriptional network by high-throughput sequencing. *Mol Cell Biol* 32: 1468–1482.

Fan D, Bitterman PB, Larsson O. 2009. Regulatory element identification in subsets of transcripts: Comparison and integration of current computational methods. *RNA* 15: 1469–1482.

Foat BC, Stormo GD. 2009. Discovering structural *cis*-regulatory elements by modeling the behaviors of mRNAs. *Mol Syst Biol* 5: 268.

Fournier ML, Paulson A, Pavelka N, Mosley AL, Gaudenz K, Bradford WD, Glynn E, Li H, Sardiu ME, Fleharty B, et al. 2010. Delayed correlation of mRNA and protein expression in rapamycin-treated cells and a role for Ggc1 in cellular sensitivity to rapamycin. *Mol Cell Proteomics* 9: 271–284.

Furic L, Rong L, Larsson O, Koumakpayi IH, Yoshida K, Brueschke A, Petroulakis E, Robichaud N, Pollak M, Gaboury LA, et al. 2010. eIF4E phosphorylation promotes tumorigenesis and is associated with prostate cancer progression. *Proc Natl Acad Sci* 107: 14134–14139.

Futcher B, Latter GI, Monardo P, McLaughlin CS, Garrels JI. 1999. A sampling of the yeast proteome. *Mol Cell Biol* 19: 7357–7368.

Genolet R, Araud T, Maillard L, Jaquier-Gubler P, Curran J. 2008. An approach to analyse the specific impact of rapamycin on mRNA–ribosome association. *BMC Med Genomics* 1: 33.

Gerber AP, Herschlag D, Brown PO. 2004. Extensive association of functionally and cytotopically related mRNAs with Puf family RNA-binding proteins in yeast. *PLoS Biol* 2: e79.

Ghaemmaghami S, Huh WK, Bower K, Howson RW, Belle A, Dephoure N, O'Shea EK, Weissman JS. 2003. Global analysis of protein expression in yeast. *Nature* 425: 737–741.

Ghazalpour A, Bennett B, Petyuk VA, Orozco L, Hagopian R, Mungrue IN, Farber CR, Sinsheimer J, Kang HM, Furlotte N, et al. 2011. Comparative analysis of proteome and transcriptome variation in mouse. *PLoS Genet* 7: e1001393.

Grech G, Blazquez-Domingo M, Kolbus A, Bakker WJ, Mullner EW, Beug H, von Lindern M. 2008. Igbp1 is part of a positive feedback loop in stem cell factor–dependent, selective mRNA translation initiation inhibiting erythroid differentiation. *Blood* 112: 2750–2760.

Griffin TJ, Gygi SP, Ideker T, Rist B, Eng J, Hood L, Aebersold R. 2002. Complementary profiling of gene expression at the transcriptome and proteome levels in *Saccharomyces cerevisiae*. *Mol Cell Proteomics* 1: 323–333.

Grillo G, Turi A, Licciulli F, Mignone F, Liuni S, Banfi S, Gennarino VA, Horner DS, Pavesi G, Picardi E, et al. 2010. UTRdb and UTRsite (RELEASE 2010): A collection of sequences and regulatory motifs of the untranslated regions of eukaryotic mRNAs. *Nucleic Acids Res* 38: D75–D80.

Gry M, Rimini R, Stromberg S, Asplund A, Ponten F, Uhlen M, Nilsson P. 2009. Correlations between RNA and protein expression profiles in 23 human cell lines. *BMC Genomics* 10: 365.

Gygi SP, Rochon Y, Franza BR, Aebersold R. 1999. Correlation between protein and mRNA abundance in yeast. *Mol Cell Biol* 19: 1720–1730.

Hafner M, Landthaler M, Burger L, Khorshid M, Hausser J, Berninger P, Rothballer A, Ascano M Jr, Jungkamp AC, Munschauer M, et al. 2010. Transcriptome-wide identification of RNA-binding protein and microRNA target sites by PAR-CLIP. *Cell* 141: 129–141.

Halbeisen RE, Scherrer T, Gerber AP. 2009. Affinity purification of ribosomes to access the translatome. *Methods* **48:** 306–310.

Halees AS, El-Badrawi R, Khabar KS. 2008. ARED organism: Expansion of ARED reveals AU-rich element cluster variations between human and mouse. *Nucleic Acids Res* **36:** D137–D140.

Hoell JI, Larsson E, Runge S, Nusbaum JD, Duggimpudi S, Farazi TA, Hafner M, Borkhardt A, Sander C, Tuschl T. 2011. RNA targets of wild-type and mutant FET family proteins. *Nat Struct Mol Biol* **18:** 1428–1431.

Hogan DJ, Riordan DP, Gerber AP, Herschlag D, Brown PO. 2008. Diverse RNA-binding proteins interact with functionally related sets of RNAs, suggesting an extensive regulatory system. *PLoS Biol* **6:** e255.

Iguchi N, Tobias JW, Hecht NB. 2006. Expression profiling reveals meiotic male germ cell mRNAs that are translationally up- and down-regulated. *Proc Natl Acad Sci* **103:** 7712–7717.

Ingolia NT, Ghaemmaghami S, Newman JR, Weissman JS. 2009. Genome-wide analysis in vivo of translation with nucleotide resolution using ribosome profiling. *Science* **324:** 218–223.

Ingolia NT, Lareau LF, Weissman JS. 2011. Ribosome profiling of mouse embryonic stem cells reveals the complexity and dynamics of mammalian proteomes. *Cell* **147:** 789–802.

Jackson RJ, Hellen CU, Pestova TV. 2010. The mechanism of eukaryotic translation initiation and principles of its regulation. *Nat Rev Mol Cell Biol* **11:** 113–127.

Jayapal KP, Philp RJ, Kok YJ, Yap MG, Sherman DH, Griffin TJ, Hu WS. 2008. Uncovering genes with divergent mRNA–protein dynamics in *Streptomyces coelicolor*. *PLoS ONE* **3:** e2097.

Jechlinger M, Grunert S, Tamir IH, Janda E, Ludemann S, Waerner T, Seither P, Weith A, Beug H, Kraut N. 2003. Expression profiling of epithelial plasticity in tumor progression. *Oncogene* **22:** 7155–7169.

Johannes G, Carter MS, Eisen MB, Brown PO, Sarnow P. 1999. Identification of eukaryotic mRNAs that are translated at reduced cap binding complex eIF4F concentrations using a cDNA microarray. *Proc Natl Acad Sci* **96:** 13118–13123.

Keene JD. 2007. RNA regulons: Coordination of post-transcriptional events. *Nat Rev Genet* **8:** 533–543.

Keene JD, Tenenbaum SA. 2002. Eukaryotic mRNPs may represent posttranscriptional operons. *Mol Cell* **9:** 1161–1167.

Keene JD, Komisarow JM, Friedersdorf MB. 2006. RIP-Chip: The isolation and identification of mRNAs, microRNAs and protein components of ribonucleoprotein complexes from cell extracts. *Nat Protoc* **1:** 302–307.

Khorshid M, Rodak C, Zavolan M. 2011. CLIPZ: A database and analysis environment for experimentally determined binding sites of RNA-binding proteins. *Nucleic Acids Res* **39:** D245–D252.

Kim YY, Von Weymarn L, Larsson O, Fan D, Underwood JM, Peterson MS, Hecht SS, Polunovsky VA, Bitterman PB. 2009. Eukaryotic initiation factor 4E binding protein family of proteins: Sentinels at a translational control

checkpoint in lung tumor defense. *Cancer Res* **69:** 8455–8462.

Kishore S, Jaskiewicz L, Burger L, Hausser J, Khorshid M, Zavolan M. 2011. A quantitative analysis of CLIP methods for identifying binding sites of RNA-binding proteins. *Nat Methods* **8:** 559–564.

Kitamura H, Nakagawa T, Takayama M, Kimura Y, Hijikata A, Ohara O. 2004. Post-transcriptional effects of phorbol 12-myristate 13-acetate on transcriptome of U937 cells. *FEBS Lett* **578:** 180–184.

Kitamura H, Ito M, Yuasa T, Kikuguchi C, Hijikata A, Takayama M, Kimura Y, Yokoyama R, Kaji T, Ohara O. 2008. Genome-wide identification and characterization of transcripts translationally regulated by bacterial lipopolysaccharide in macrophage-like J774.1 cells. *Physiol Genomics* **33:** 121–132.

Konig J, Zarnack K, Rot G, Curk T, Kayikci M, Zupan B, Turner DJ, Luscombe NM, Ule J. 2011. iCLIP reveals the function of hnRNP particles in splicing at individual nucleotide resolution. *Nat Struct Mol Biol* **17:** 909–915.

Kudo K, Xi Y, Wang Y, Song B, Chu E, Ju J, Russo JJ. 2010. Translational control analysis by translationally active RNA capture/microarray analysis (TrIP–Chip). *Nucleic Acids Res* **38:** e104.

Kumaraswamy S, Chinnaiyan P, Shankavaram UT, Lu X, Camphausen K, Tofilon PJ. 2008. Radiation-induced gene translation profiles reveal tumor type and cancer-specific components. *Cancer Res* **68:** 3819–3826.

Larsson O, Bitterman P. 2010. Genome-wide analysis of translational control. In *mTOR pathway and mTOR inhibitors in cancer therapy (cancer drug discovery and development)* (ed. Polunovsky V, Houghton P), pp. 217–236. Humana Press, Totowa, NJ.

Larsson O, Perlman DM, Fan D, Reilly CS, Peterson M, Dahlgren C, Liang Z, Li S, Polunovsky VA, Wahlestedt C, et al. 2006. Apoptosis resistance downstream of eIF4E: Posttranscriptional activation of an anti-apoptotic transcript carrying a consensus hairpin structure. *Nucleic Acids Res* **34:** 4375–4386.

Larsson O, Li S, Issaenko OA, Avdulov S, Peterson M, Smith K, Bitterman PB, Polunovsky VA. 2007. Eukaryotic translation initiation factor 4E induced progression of primary human mammary epithelial cells along the cancer pathway is associated with targeted translational deregulation of oncogenic drivers and inhibitors. *Cancer Res* **67:** 6814–6824.

Larsson O, Diebold D, Fan D, Peterson M, Nho RS, Bitterman PB, Henke CA. 2008. Fibrotic myofibroblasts manifest genome-wide derangements of translational control. *PLoS ONE* **3:** e3220.

Larsson O, Sonenberg N, Nadon R. 2010. Identification of differential translation in genome wide studies. *Proc Natl Acad Sci* **107:** 21487–21492.

Larsson O, Sonenberg N, Nadon R. 2011. Anota: Analysis of differential translation in genome-wide studies. *Bioinformatics* **27:** 1440–1441.

Laurent JM, Vogel C, Kwon T, Craig SA, Boutz DR, Huse HK, Nozue K, Walia H, Whiteley M, Ronald PC, et al. 2010. Protein abundances are more conserved than mRNA abundances across diverse taxa. *Proteomics* **10:** 4209–4212.

Lebedeva S, Jens M, Theil K, Schwanhausser B, Selbach M, Landthaler M, Rajewsky N. 2011. Transcriptome-wide analysis of regulatory interactions of the RNA-binding protein HuR. *Mol Cell* **43:** 340–352.

Lewis BP, Burge CB, Bartel DP. 2005. Conserved seed pairing, often flanked by adenosines, indicates that thousands of human genes are microRNA targets. *Cell* **120:** 15–20.

Licatalosi DD, Mele A, Fak JJ, Ule J, Kayikci M, Chi SW, Clark TA, Schweitzer AC, Blume JE, Wang X, et al. 2008. HITS-CLIP yields genome-wide insights into brain alternative RNA processing. *Nature* **456:** 464–469.

Lu X, de la Pena L, Barker C, Camphausen K, Tofilon PJ. 2006. Radiation-induced changes in gene expression involve recruitment of existing messenger RNAs to and away from polysomes. *Cancer Res* **66:** 1052–1061.

Lu P, Vogel C, Wang R, Yao X, Marcotte EM. 2007. Absolute protein expression profiling estimates the relative contributions of transcriptional and translational regulation. *Nat Biotechnol* **25:** 117–124.

Lu R, Markowetz F, Unwin RD, Leek JT, Airoldi EM, MacArthur BD, Lachmann A, Rozov R, Ma'ayan A, Boyer LA, et al. 2009. Systems-level dynamic analyses of fate change in murine embryonic stem cells. *Nature* **462:** 358–362.

Mamane Y, Petroulakis E, Martineau Y, Sato TA, Larsson O, Rajasekhar VK, Sonenberg N. 2007. Epigenetic activation of a subset of mRNAs by eIF4E explains its effects on cell proliferation. *PLoS ONE* **2:** e242.

Matsuura H, Ishibashi Y, Shinmyo A, Kanaya S, Kato K. 2010. Genome-wide analyses of early translational responses to elevated temperature and high salinity in *Arabidopsis thaliana*. *Plant Cell Physiol* **51:** 448–462.

Mokrejs M, Masek T, Vopalensky V, Hlubucek P, Delbos P, Pospisek M. 2010. IRESite—A tool for the examination of viral and cellular internal ribosome entry sites. *Nucleic Acids Res* **38:** D131–D136.

Mukherjee N, Corcoran DL, Nusbaum JD, Reid DW, Georgiev S, Hafner M, Ascano M Jr, Tuschl T, Ohler U, Keene JD. 2011. Integrative regulatory mapping indicates that the RNA-binding protein HuR couples pre-mRNA processing and mRNA stability. *Mol Cell* **43:** 327–339.

Nie L, Wu G, Zhang W. 2006. Correlation of mRNA expression and protein abundance affected by multiple sequence features related to translational efficiency in *Desulfovibrio vulgaris*: A quantitative analysis. *Genetics* **174:** 2229–2243.

Otulakowski G, Duan W, O'Brodovich H. 2009. Global and gene-specific translational regulation in rat lung development. *Am J Respir Cell Mol Biol* **40:** 555–567.

Parent R, Beretta L. 2008. Translational control plays a prominent role in the hepatocytic differentiation of HepaRG liver progenitor cells. *Genome Biol* **9:** R19.

Pearson K. 1896. On a form of spurious correlation which may arise when indices are used in the measurement of organs. *Proc R Soc Lond* **60:** 489–498.

Preiss T, Baron-Benhamou J, Ansorge W, Hentze MW. 2003. Homodirectional changes in transcriptome composition and mRNA translation induced by rapamycin and heat shock. *Nat Struct Biol* **10:** 1039–1047.

Provenzani A, Fronza R, Loreni F, Pascale A, Amadio M, Quattrone A. 2006. Global alterations in mRNA polyso-mal recruitment in a cell model of colorectal cancer progression to metastasis. *Carcinogenesis* **27:** 1323–1333.

Rajasekhar VK, Viale A, Socci ND, Wiedmann M, Hu X, Holland EC. 2003. Oncogenic Ras and Akt signaling contribute to glioblastoma formation by differential recruitment of existing mRNAs to polysomes. *Mol Cell* **12:** 889–901.

Ramirez-Valle F, Braunstein S, Zavadil J, Formenti SC, Schneider RJ. 2008. eIF4GI links nutrient sensing by mTOR to cell proliferation and inhibition of autophagy. *J Cell Biol* **181:** 293–307.

Ray D, Kazan H, Chan ET, Castillo LP, Chaudhry S, Talukder S, Blencowe BJ, Morris Q, Hughes TR. 2009. Rapid and systematic analysis of the RNA recognition specificities of RNA-binding proteins. *Nat Biotechnol* **27:** 667–670.

Richter JD, Sonenberg N. 2005. Regulation of cap-dependent translation by eIF4E inhibitory proteins. *Nature* **433:** 477–480.

Rivera-Ruiz ME, Rodriguez-Quinones JF, Akamine P, Rodriguez-Medina JR. 2010. Post-transcriptional regulation in the myo1Δ mutant of *Saccharomyces cerevisiae*. *BMC Genomics* **11:** 690.

Rousseau D, Kaspar R, Rosenwald I, Gehrke L, Sonenberg N. 1996. Translation initiation of ornithine decarboxylase and nucleocytoplasmic transport of cyclin D1 mRNA are increased in cells overexpressing eukaryotic initiation factor 4E. *Proc Natl Acad Sci* **93:** 1065–1070.

Sampath P, Pritchard DK, Pabon L, Reinecke H, Schwartz SM, Morris DR, Murry CE. 2008. A hierarchical network controls protein translation during murine embryonic stem cell self-renewal and differentiation. *Cell Stem Cell* **2:** 448–460.

Sandberg R, Neilson JR, Sarma A, Sharp PA, Burge CB. 2008. Proliferating cells express mRNAs with shortened 3′ untranslated regions and fewer microRNA target sites. *Science* **320:** 1643–1647.

Sanz E, Yang L, Su T, Morris DR, McKnight GS, Amieux PS. 2009. Cell-type-specific isolation of ribosome-associated mRNA from complex tissues. *Proc Natl Acad Sci* **106:** 13939–13944.

Schmidt MW, Houseman A, Ivanov AR, Wolf DA. 2007. Comparative proteomic and transcriptomic profiling of the fission yeast *Schizosaccharomyces pombe*. *Mol Syst Biol* **3:** 79.

Schrimpf SP, Weiss M, Reiter L, Ahrens CH, Jovanovic M, Malmstrom J, Brunner E, Mohanty S, Lercher MJ, Hunziker PE, et al. 2009. Comparative functional analysis of the *Caenorhabditis elegans* and *Drosophila melanogaster* proteomes. *PLoS Biol* **7:** e1000048.

Schwanhausser B, Busse D, Li N, Dittmar G, Schuchhardt J, Wolf J, Chen W, Selbach M. 2011. Global quantification of mammalian gene expression control. *Nature* **473:** 337–342.

Silvera D, Formenti SC, Schneider RJ. 2010. Translational control in cancer. *Nat Rev Cancer* **10:** 254–266.

Sonenberg N, Hinnebusch AG. 2009. Regulation of translation initiation in eukaryotes: Mechanisms and biological targets. *Cell* **136:** 731–745.

Spence J, Duggan BM, Eckhardt C, McClelland M, Mercola D. 2006. Messenger RNAs under differential translational

Cite this article as *Cold Spring Harb Perspect Biol* doi: 10.1101/cshperspect.a012302

control in Ki-ras-transformed cells. *Mol Cancer Res* **4:** 47–60.

Spirin AS. 1969. The second Sir Hans Krebs Lecture. Informosomes. *Eur J Biochem* **10:** 20–35.

Spriggs KA, Bushell M, Willis AE. 2010. Translational regulation of gene expression during conditions of cell stress. *Mol Cell* **40:** 228–237.

Taniguchi Y, Choi PJ, Li GW, Chen H, Babu M, Hearn J, Emili A, Xie XS. 2010. Quantifying *E. coli* proteome and transcriptome with single-molecule sensitivity in single cells. *Science* **329:** 533–538.

Tian Q, Stepaniants SB, Mao M, Weng L, Feetham MC, Doyle MJ, Yi EC, Dai H, Thorsson V, Eng J, et al. 2004. Integrated genomic and proteomic analyses of gene expression in mammalian cells. *Mol Cell Proteomics* **3:** 960–969.

Tian B, Hu J, Zhang H, Lutz CS. 2005. A large-scale analysis of mRNA polyadenylation of human and mouse genes. *Nucleic Acids Res* **33:** 201–212.

Tollervey JR, Curk T, Rogelj B, Briese M, Cereda M, Kayikci M, Konig J, Hortobagyi T, Nishimura AL, Zupunski V, et al. 2011. Characterizing the RNA targets and position-dependent splicing regulation by TDP-43. *Nat Neurosci* **14:** 452–458.

Tominaga Y, Tamguney T, Kolesnichenko M, Bilanges B, Stokoe D. 2005. Translational deregulation in PDK-1$^{-/-}$ embryonic stem cells. *Mol Cell Biol* **25:** 8465–8475.

Treton X, Pedruzzi E, Cazals-Hatem D, Grodet A, Panis Y, Groyer A, Moreau R, Bouhnik Y, Daniel F, Ogier-Denis E. 2011. Altered endoplasmic reticulum stress affects translation in inactive colon tissue from patients with ulcerative colitis. *Gastroenterology* **141:** 1024–1035.

Vogel C, Abreu Rde S, Ko D, Le SY, Shapiro BA, Burns SC, Sandhu D, Boutz DR, Marcotte EM, Penalva LO. 2010. Sequence signatures and mRNA concentration can explain two-thirds of protein abundance variation in a human cell line. *Mol Syst Biol* **6:** 400.

Wang ET, Sandberg R, Luo S, Khrebtukova I, Zhang L, Mayr C, Kingsmore SF, Schroth GP, Burge CB. 2008. Alternative isoform regulation in human tissue transcriptomes. *Nature* **456:** 470–476.

Wang Z, Kayikci M, Briese M, Zarnack K, Luscombe NM, Rot G, Zupan B, Curk T, Ule J. 2010. iCLIP predicts the dual splicing effects of TIA–RNA interactions. *PLoS Biol* **8:** e1000530.

Washburn MP, Koller A, Oshiro G, Ulaszek RR, Plouffe D, Deciu C, Winzeler E, Yates JR III. 2003. Protein pathway and complex clustering of correlated mRNA and protein expression analyses in *Saccharomyces cerevisiae*. *Proc Natl Acad Sci* **100:** 3107–3112.

Zhang C, Darnell RB. 2011. Mapping in vivo protein–RNA interactions at single-nucleotide resolution from HITS-CLIP data. *Nat Biotechnol* **29:** 607–614.

Zong Q, Schummer M, Hood L, Morris DR. 1999. Messenger RNA translation state: The second dimension of high-throughput expression screening. *Proc Natl Acad Sci* **96:** 10632–10636.

Imaging Translation in Single Cells Using Fluorescent Microscopy

Jeffrey A. Chao, Young J. Yoon, and Robert H. Singer

Department of Anatomy and Structural Biology, Albert Einstein College of Medicine, Bronx, New York 10461

Correspondence: robert.singer@einstein.yu.edu

The regulation of translation provides a mechanism to control not only the abundance of proteins, but also the precise time and subcellular location that they are synthesized. Much of what is known concerning the molecular basis for translational control has been gleaned from experiments (e.g., luciferase assays and polysome analysis) that measure average changes in the protein synthesis of a population of cells, however, mechanistic insights can be obscured in ensemble measurements. The development of fluorescent microscopy techniques and reagents has allowed translation to be studied within its cellular context. Here we highlight recent methodologies that can be used to study global changes in protein synthesis or regulation of specific mRNAs in single cells. Imaging of translation has provided direct evidence for local translation of mRNAs at synapses in neurons and will become an important tool for studying translational control.

A recent genome-wide study of mRNA and protein abundance in mammalian cells determined that cellular levels of proteins are predominantly controlled at the level of translation (Schwanhausser et al. 2011). Although methods have been developed that allow transcription to be quantified at the single molecule level in living cells, similar measurements of translation in eukaryotic cells have not yet been possible. Only in bacteria has it been possible to measure the mRNA and protein levels of a gene with single molecule sensitivity in single cells (Taniguchi et al. 2010). Here we highlight what has been learned about the behavior of single mRNA molecules in gene expression and the current methodologies for imaging translation in single cells. We conclude by discussing what could

be performed to allow the translation of single mRNA molecules to be monitored in living cells.

SINGLE MOLECULE IMAGING OF mRNA IN GENE EXPRESSION

The development of imaging technologies capable of quantifying and tracking single RNA molecules has allowed gene expression to be characterized in living cells. The most common method for imaging single molecules of mRNA in living cells uses a unique RNA stem-loop sequence that is specifically recognized by the MS2 bacteriophage coat protein (MCP) (Valegard et al. 1994). By constructing a reporter mRNA that contains multiple copies of the MS2 operator RNA stem-loop and coexpressing a chimeric MS2 coat protein fusion with green fluorescent protein (GFP), single RNA molecules can be detected above the fluorescent background of the unbound MCP-GFP molecules (Fig. 1A) (Bertrand et al. 1998; Beach et al. 1999). Because the MCP binds to its cognate RNA stem-loop with high affinity, this interaction allows stable labeling of the reporter RNA in cells (Lowary and Uhlenbeck 1987; Lim and Peabody 1994). Although labeling of RNAs with the MS2 systems requires the addition of multiple RNA stem-loops and MCP-GFP molecules, this increase in size and mass does not significantly perturb RNA metabolism as evidenced by the creation of a transgenic mouse

that has an insertion of the MS2 stem-loops in the endogenous β-actin gene, which has no effect on viability (Lionnet et al. 2011). This system has been used to study many aspects of gene expression ranging from transcription to RNA localization in the cytoplasm in a variety of experimental systems.

Quantifying transcription in real time in living cells has revealed that the expression of individual genes is stochastic and has allowed direct measurement of the kinetics of the process. In bacteria, the counting of individual mRNAs in single cells showed that transcriptional bursting, infrequent transcription events that produce many transcripts within a short period of time,

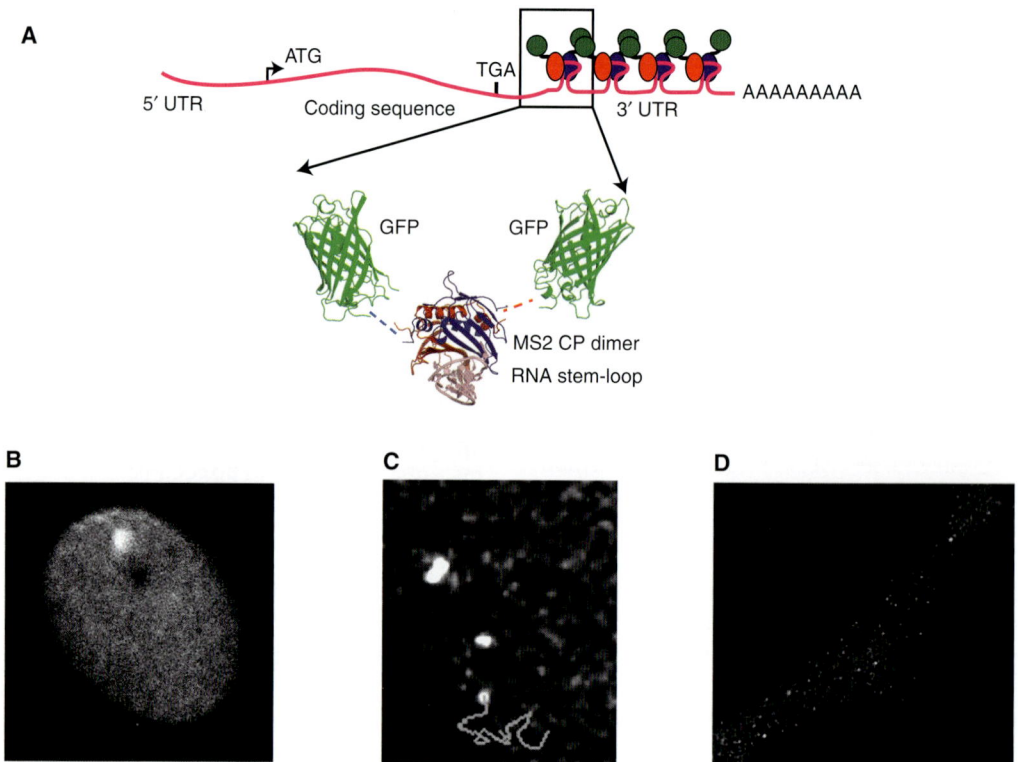

Figure 1. MS2 system for fluorescent labeling of mRNAs in living cells. (A) Schematic of a reporter mRNA that contains multiple copies of the MS2 stem-loop (boxed) that bind to the MS2 coat protein fused to GFP (arrows). The binding of many MCP-GFP molecules to a single transcript allows the RNA to be observed in living cells. (B) Transcription in the nucleus detected by binding of MCP-GFP to nascent mRNAs at the transcription site. (From Darzacq et al. 2007; reprinted, with permission, from the author.) (C) Tracking of single mRNAs (trajectory shown as line) in the nucleoplasm. (From Shav-tal et al. 2004; reprinted, with permission, from The American Association for the Advancement of Science © 2004.) (D) Single mRNAs detected in the cytoplasm. (From Fusco et al. 2003; reprinted, with permission, from Elsevier © 2003.)

Cite this article as *Cold Spring Harb Perspect Biol* doi: 10.1101/cshperspect.a012310

was responsible for the variability in gene expression (Golding and Cox 2004; Golding et al. 2005). Similarly, in the eukaryote *Dictyostelium*, transcription of an endogenous developmental gene, *dscA*, was observed to be expressed in discrete bursts of gene activity (Chubb et al. 2006). It has also been possible to measure the speed of the RNA polymerase II in mammalian cells and it was found to elongate at 4.3 kb min^{-1}. Surprisingly, gene expression was found to be inefficient, with only 1% of polymerases that contact the gene resulting in transcription of an mRNA (Fig. 1B) (Darzacq et al. 2007).

Once a transcript leaves the site of its transcription, its movements can be tracked within the nucleoplasm. Rapid time-lapse imaging and single particle tracking revealed that the diffusion of individual mRNAs is governed by simple diffusion and that transcripts can be partially corralled by the chromatin domains within the nucleoplasm (Fig. 1C) (Shav-Tal et al. 2004). Transcripts diffuse within the nucleoplasm for times ranging from 5 to 40 minutes; however, once a transcript engages a nuclear pore, export into the cytoplasm is fast (less than 0.2 sec) (Grunwald and Singer 2010; Mor et al. 2010). Interestingly, transport through the nuclear pore takes only 5–20 min and docking on the nuclear side and release into the cytoplasm both take approximately 80 min, indicating that translocation through the nuclear pore is not the rate-limiting step (Grunwald and Singer 2010).

In contrast to what was observed in the nucleus, once a transcript is in the cytoplasm, it can undergo rapid directional movements indicating active transport by molecular motors along the cytoskeleton. The MS2 system was first employed to characterize the localization of ASH1 mRNA to the bud tip in *Saccharomyces cerevisiae* (Bertrand et al. 1998). Time-lapse imaging of ASH1 mRNA tagged with MCP-GFP showed that it was transported at speeds of 0.20–0.44 µm sec^{-1} by Myo4p, a type V myosin motor, along the actin cytoskeleton. Characterization of mRNA movements in mammalian cells indicated that transcripts can undergo static, corralled, diffusive and directed movements in the cytoplasm, and individual transcripts frequently switch between these different modes of movement (Fig. 1D) (Fusco et al. 2003). The β-actin mRNA zipcode, a *cis*-acting element that is responsible for the localization of this transcript, was found to increase both the frequency and length of directed movements. In the *Drosophila melanogaster* oocyte, the posterior localization of *oskar* mRNA was found to be achieved by a biased random walk with a 14% excess of posterior directed movements by kinesin along weakly polarized microtubules (Zimyanin et al. 2008).

Characterization of mRNAs in the cytoplasm has also provided indirect evidence for how translation may affect these movements. In rat pheochromocytoma cells (PC12), a reporter mRNA that labeled peripherin mRNA with MCP-YFP (yellow fluorescent protein) indicated that RNA particles were motile with an average speed of 0.42 µm sec^{-1} and that often, when particles became stationary, the peripherin protein could subsequently be detected (Chang et al. 2006). This led to the hypothesis that motile mRNAs are translationally repressed. Further evidence for this model was obtained by tracking ARC mRNA with MCP-GFP in cultured rat neurons. ARC mRNA was observed to undergo rapid bidirectional movements while in transit, but would often reverse direction and decrease in velocity before becoming stationary suggesting that local signals in the dendrite regulated ARCs mRNA movement and determined the location for its translation (Dynes and Steward 2007). Although these experiments have provided hints as to how protein synthesis is regulated, more direct methods are required to study the translation of mRNAs in the cytoplasm.

GLOBAL MEASUREMENTS OF PROTEIN SYNTHESIS IN SINGLE CELLS

To distinguish newly synthesized proteins from ones that have been previously translated, cells can be incubated with radioisotopically labeled amino acids (^{35}S-methionine) that are then incorporated into nascent polypeptides. On cell lysis, the fraction of the proteome that has been synthesized after the addition of the radiolabeled amino acid can be measured by autoradiography. Although these types of experiments have been tremendously informative for deducing the mechanisms by which factors exert their translational control, they cannot be performed on intact cells and, consequently, information concerning the spatial regulation of protein synthesis within the cell is lost. Several fluorescent microscopy approaches have been developed that allow global changes in translation to be measured in single cells.

Fluorescent Noncanonical Amino Acid Tagging (FUNCAT)

Noncanonical amino acids that can be recognized by the appropriate aminoacyl tRNA synthetase and incorporated into elongating polypeptides by the ribosome provide a tool for fluorescent labeling of newly synthesized proteins. Schuman and colleagues devised a labeling strategy based on two methionine analogs, azidohomoalanine (AHA) and homopropargylglycine (HPG) (Dieterich et al. 2010). These noncanonical amino acids can be added at methionine codons. Because they contain chemical functionality that standard amino acids to do not have, they can be specifically conjugated to fluorescent dyes (Texas Red or carboxyfluoroscein) via "click chemistry" that uses a copper-catalyzed azide-alkyne $[3 + 2]$ cycloaddition (Fig. 2A) (Rostovtsev et al. 2002). This metabolic labeling strategy is conceptually very similar to radiolabeling with ^{35}S-methionine; however, it allows detection of newly synthesized proteins by fluorescent microscopy. Because AHA and HPG are not as efficient substrates for methionyl-tRNA-synthetase (metRS) as methionine, it is necessary to reduce the cellular concentration of methionine by starving the cells of this particular amino acid, so that the analogs can effectively compete for charging to tRNA and incorporation by the ribosome. The AHA and HPG amino acids are added to proteins in positions normally occupied by methionines, which can result in nonuniform labeling because the accessibility of the noncanonical side chains to participate in "click chemistry" may not be equivalent for every protein.

FUNCAT has been used to investigate the dynamics of protein synthesis in rat hippocampal neurons. Although this metabolic labeling strategy labels all newly synthesized cellular proteins after addition of AHA or HPG, the noncanonical amino acid can be locally microperfused using a small delivery pipet combined with a suction pipet to discrete subcellular regions. Using this approach FUNCAT was used to observe protein synthesis in individual dendrites that was diminished on addition of anisomycin, a translation inhibitor, and stimulated on addition of brain-derived neurotropic factor (BDNF), which induces translation-dependent enhancement of synaptic strength (Kang and Schuman 1996). The engineering of a mutant MetRS that is capable of charging the noncanonical amino acid azidonorleucine, which is not recognized by the cell's endogenous MetRS, can be used to restrict the incorporation of azidonorleucine to a defined population of cells (Ngo et al. 2009).

Fluorescent Puromycin Derivatives

Puromycin is a natural product derived from *Streptomyces alboniger* bacteria that causes premature termination of the elongating protein during translation. The small molecule contains

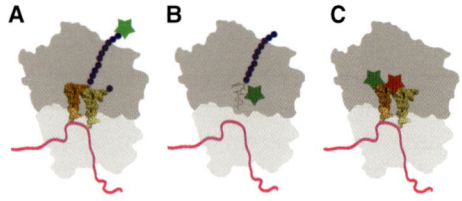

Figure 2. Global measurements of translation in single cells. (*A*) The incorporation of noncanonical amino acids using FUNCAT allows newly synthesized proteins to be fluorescently labeled (green star). Methionine analogs are incorporated into the elongating polypeptide chain that can be conjugated to fluorescent dyes. (*B*) Fluorescent analogs of puromycin (green star) prematurely terminate translation and allow nascent polypeptides to be fluorescently labeled. (*C*) FRET between adjacent fluorescent tRNAs (green and red stars are FRET pairs) allows the cellular location of active translation to be observed. Only when fluorescently labeled tRNAs are bound to the ribosome in adjacent sites, is the FRET signal observable.

 Cite this article as *Cold Spring Harb Perspect Biol* doi: 10.1101/cshperspect.a012310

a region that is structurally analogous to the $3'$-terminus of aminoacylated-tRNA and can enter the ribosome at the acceptor site. When puromycin is bound in this position, its primary amine forms a peptide bond with the carboxyl terminus of the nascent polypeptide, thereby stalling the ribosome and prematurely terminating protein synthesis (Yarmolinsky and Haba 1959; Nathans 1964). By conjugating a fluorescent dye to puromycin, newly synthesized polypeptide fragments can be detected in living cells. (Fig. 2B) It has been shown that puromycin substituted at the $5'$ OH has decreased efficacy as a translation inhibitor, so a deoxycytidine (dC) was used to link fluorescent dyes (Cy5 and fluorescein) to puromycin (Starck and Roberts 2002; Starck et al. 2004).

Fluorescein-dC-puromycin (F2P) has been used to measure global protein synthesis in the dendrites of rat hippocampal neurons. Addition of F2P to neuronal cultures resulted in an increase in fluorescence in both the soma and dendrites and this signal was attenuated by addition of anisomycin or unlabeled puromycin (Smith et al. 2005). When neurons were treated with a dopamine D1/D5 agonist, SKF-38393, a significant increase in fluorescence was measured in both dendrites and spines consistent with the ability of dopamine agonist to stimulate protein synthesis.

A second puromycin analog has been developed that contains a terminal alkyne group, O-propargyl-puromycin (OP-puro) (Liu et al. 2011). This substitution enables "click chemistry" to be performed on the polypeptide-OP-puro fusions allowing conjugation with fluorescent dyes in a similar manner as described for FUNCAT. Importantly, this labeling strategy can be used to fluorescently detect nascent proteins in whole animals. When OP-puro is injected intraperitoneally into mice, protein synthesis can be detected in tissue that has been fixed and stained with tetramethylrhodamine-azide. Protein synthesis was found to be highest in the crypts and at the base of intestinal villi in the small intestine, which is consistent with the function of these cells and shows the sensitivity of this method to detect differences in translation rates amongst and within different organs.

Fluorescent tRNA-Derivatives

An alternative strategy for measuring protein synthesis in living cells that does not label the protein product of translation has been shown using fluorescently labeled tRNA (fl-tRNA). In this experiment, bulk uncharged tRNA from yeast was labeled with either Cy3 or Rhodamine 110 (Rho110) in the D-loop (Betteridge et al. 2007; Pan et al. 2009). Most tRNAs contain dihydroU at multiple positions within the D-loop that can be selectively reduced by treatment with $NaBH_4$ and then conjugated to fluorescent dyes (Cerutti and Miller 1967). Modification of tRNAs in these positions does not alter their ability to bind to the ribosome and can be used to monitor translation (Pan et al. 2007).

When fl-tRNAs were transfected into Chinese hamster ovary (CHO) cells, they partially colocalized with several cellular proteins involved in tRNA charging or translation (Arg-RS, eEF1A, and rpS6) indicating that the fl-tRNAs were functional (Barhoom et al. 2011). When cells are transfected with both Cy3-tRNA and Rho110-tRNA, fluorescence resonance energy transfer (FRET) can be observed when the differentially labeled tRNAs occupy adjacent positions within the ribosome (Fig. 2C). Using this system, primary brain astrocytes that were activated with either bacterial lipopolysaccharides or interferon-γ were found to have increases in both the number of cells that had FRET signals as well as the intensity of the FRET signal per cell. These increases in FRET correlated with standard ^{35}S metabolic labeling, indicating that this approach could provide information on sites of active translation within cells.

TRANSCRIPT-SPECIFIC MEASUREMENTS OF TRANSLATION IN LIVING CELLS

To measure changes in protein synthesis for a particular protein of interest, radioisotope labeling with ^{35}S-methionine of all cellular proteins can be followed by immunoprecipitation with specific antibodies and standard electrophoresis. Although this approach allows the translation properties of endogenous protein to be measured, it is often of interest to know where and when specific proteins are synthesized within a cell. A number of approaches have been developed that allow the fluorescent detection of reporter constructs that have been exogenously expressed.

Fluorescent Protein Reporters

The discovery and development of fluorescent proteins has been a boon for scientists interested in visualizing intracellular gene expression. When a protein of interest is expressed as a chimeric fusion protein with a fluorescent protein, the intracellular fluorescence can report on the subcellular localization of the endogenous protein. A variety of fluorescent proteins with distinct spectral properties have been engineered (Giepmans et al. 2006; Wu et al. 2011). Recent advances have allowed fluorescent proteins to be used as translation reporters to dynamically visualize synthesis of new proteins in living cells. In neurons, the use of translation reporters has widely been used, due in large part to the cells' pronounced polarized morphology that clearly distinguishes the cell body from the processes. The neurites provide discrete subcellular compartments away from the cell body where protein synthesis can be spatially and temporally regulated. Furthermore, the observation that synaptic plasticity requires local synthesis of new proteins at specific synapses provides a rationale for devising and advancing effective translation reporters.

Our understanding of RNA localization and local protein synthesis has been central to the design of translation reporters. Schuman and colleagues created a dendritic translation reporter by flanking the coding sequence for GFP with the 5′ and 3′ untranslated regions (UTR) of the α subunit of Ca^{2+}/calmodulin-dependent kinase II (αCaMKII) (Fig. 3A.1) (Aakalu et al. 2001; Sutton et al. 2004). The 3′ UTR of αCaMKII has been shown to be sufficient for dendritic localization of the mRNA in neurons. Furthermore, to address the possibility that GFP synthesized in the cell body could diffuse into the dendrite, a myristoylation tag was included before the GFP coding sequence to anchor the newly synthesized reporter proteins

to the plasma membrane. The PEST sequence from ornithine decarboxylase was fused to the carboxyl terminus of GFP to reduce the half-life of the protein and enhance the temporal resolution of the reporter (Rogers et al. 1986; Corish and Tyler-Smith 1999). On treatment of hippocampal neurons with brain derived neurotrophic factor (BDNF), it was possible to visualize newly synthesized reporter fluorescence from distal regions of transected or optically isolated dendrites. Success of the dendritic translation reporter paved the way for efforts to design translation reporters targeted to different regions of the neuron. Flanagan and colleagues fused GFP with a 77-nucleotide conserved region from the 3′ UTR of the Eph2A receptor (Brittis et al. 2002). When expressed in commissural neurons of chick spinal cords, GFP fluorescence could be detected in distal axon segments. To address concerns that fast axonal transport may confound interpretation of the results, the GFP in the reporter was replaced with a fluorescent protein (Fluorescent Timer) that changes color from green to red over time (Terskikh et al. 2000). Comparison of the relative green and red fluorescence intensities from the soma and the axon growth cone gave results consistent with de novo protein synthesis within axons.

Continued development of fluorescent proteins that are capable of undergoing photoconversion and photoswitching has provided a means to distinguish newly synthesized proteins from the preexisting population of protein. A brief exposure of these fluorescent proteins to ultraviolet light (UV) within a confined focal volume results in the photoconversion of the fluorophore to different excitation and emission characteristics and allows detection of newly synthesized protein by visualizing the recovery of the unconverted fluorescent protein (Stepanenko et al. 2008; Wu et al. 2011). Experimental strategies that use these fluorescent proteins reduce concerns of somatic contamination of the reporter protein and provide unprecedented spatial and temporal resolution.

Localization of β-actin mRNA and its translation have been shown to be important for cell migration in fibroblasts and extending axonal growth cones in neurons (Kislauskis et al.

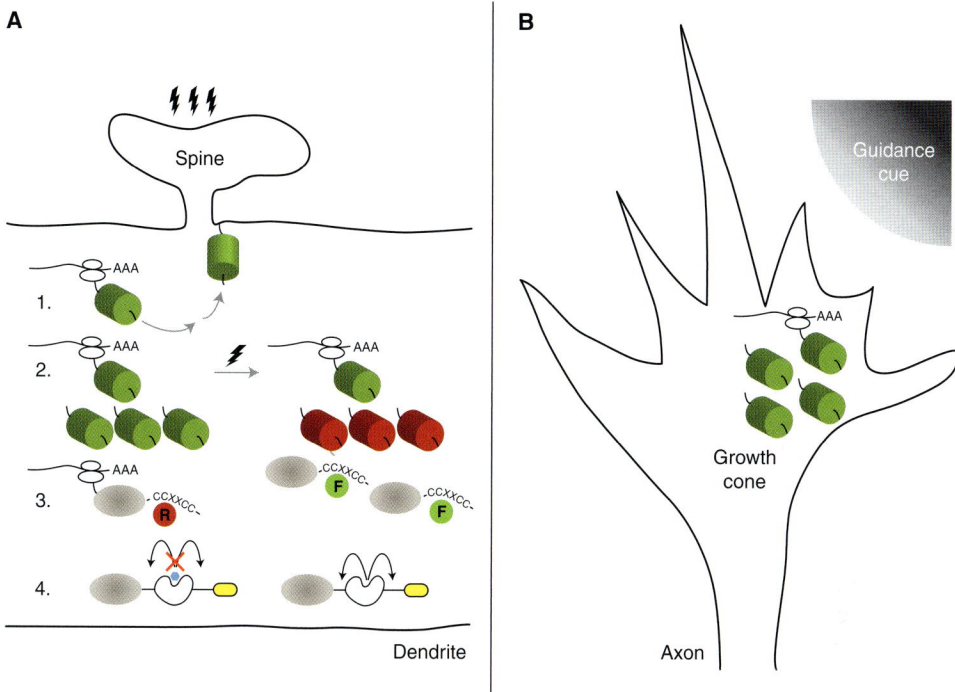

Figure 3. Strategies for visualization of newly synthesized proteins in neurons. Polarized cellular architecture of neurons provides an ideal system to study localized translation in discrete subcellular domains. In dendrites (*A*), a small structure called the spine receives signals and transmits the information to the cell body. During this process, localized protein synthesis can be visualized using a variety of strategies depicted in the illustration. 1. Anchored GFP: Postsynaptic stimulation leads to synthesis of myristoylated GFP (green cylinder) that becomes anchored to the membrane thereby demarcating the site of active translation. 2. Photoconvertible GFP: Conversion of the preexisting population of fluorescent proteins (cylinders) from green to red by a brief exposure to ultraviolet light allows visualization of the newly synthesized population of fluorescent proteins. 3. Biarsenical dyes: Association of cell-permeable biarsenical dyes, FlAsH (F) and ReAsH (R) to the tetracysteine motif (CCXXCC) permits detection of translation using a pulse-chase approach. 4. TimeSTAMP: A self-cleaving protease (white) cassette that carries an epitope tag (yellow) becomes detectable on addition of the protease inhibitor (blue) by immunofluorescence methods. In axon growth cones (*B*), the directionality of guidance cues (*upper right* quadrant) can lead to asymmetric translation thereby leading to rapid growth cone turning and extension. Most, if not all, of the strategies employed in dendrites can be used in growth cones and other subcellular compartments to further characterize localized GFP translation (green cylinder) in neurons.

1994; Bassell et al. 1998). To investigate the role of local β-actin mRNA translation in growth cone turning, Holt and colleagues constructed an axonal translation reporter by fusing the photoconvertible protein, Kaede, to the β-actin 3′ UTR, which contains *cis*-acting elements responsible for RNA localization and translational control (Leung et al. 2006). Kaede is a fluorescent protein that was isolated from the stony coral, *Trachyphyllia geoffroyi*, and undergoes an irreversible photoconversion (green to red) on irradiation with UV light that results in a 2000-fold increase in red-to-green fluorescent signal (Ando et al. 2002). In growth cones expressing the Kaede reporter, new translation was observed on the side of the growth cone that was closest to an external gradient of the netrin-1 guidance cue (Fig. 3B). These data indicate that local synthesis of β-actin protein is important for the attractive turning response of growth cones.

To investigate the somatodendritic presence of potassium channel Kv1.1, which has been shown to be predominantly localized to axons, Jan and colleagues designed a reporter with Kaede fused to the amino terminus of the potassium channel protein Kv1.1 and whose mRNA also contained the 3′ UTR of Kv1.1 (Raab-Graham et al. 2006). On photoconversion and rapamycin treatment, the investigators observed new Kaede-Kv1.1 expression in distal regions of the dendrites indicating that the 3′ UTR contributes to the dendritic localization of the mRNA and subsequent stimulation-dependent translation of Kv1.1 (Fig. 3A.2). Interestingly, they observed that the fluorescently labeled Kv1.1 was concentrated in stationary "hotspots" in dendrites similar to what was reported with other locally translated protein (Aakalu et al. 2001).

To directly measure the requirement for local translation in neuronal plasticity, Martin and colleagues constructed a translation reporter using the photoconvertible Dendra2 flanked by the 5′ and 3′ UTRs of sensorin (Wang et al. 2009). Dendra2, is a single point mutant (A224V) of Dendra which results in faster maturation and brighter fluorescence, that undergoes an irreversible photoconversion from green to red fluorescence on excitation with either UV or blue light (488 nm) (Gurskaya et al. 2006). Photoconversion with blue light instead of UV light reduces phototoxicity that can be induced by exposure to UV irradiation. Sensorin mRNA localizes to synapses between sensory neuron and motor neuron involved in the gill withdrawal reflex of the sea slug, *Aplysia*. The monosynaptic connection between a sensory neuron and a motor neuron in culture provides an elegant and yet simple system to directly test stimulus-specific responses and visualize plasticity in the synapses that express the reporter. Following photoconversion of the preexisting reporter population and serotonin stimulation, they observed local translation of the reporter in a synapse-specific and stimulus-dependent manner.

Biarsenical Dyes

Although fluorescent proteins have proven to be extremely useful for measuring translation in living cells, they are limited in their ability to measure rapid changes in protein synthesis because of the time it takes for the fluorophore to mature once the protein has been translated. For many fluorescent proteins, the maturation time has been measured to be greater than 10 minutes and can take up to several hours for fluorescence to become detectable (Evdokimov et al. 2006; Kremers et al. 2006; Shaner et al. 2008). This lag in time between when the fluorescent protein has been synthesized and when it becomes observable also prevents identification of the precise location of translation because the fluorescent protein can diffuse away from the site of its synthesis. The engineering of short peptides sequences that can significantly enhance the fluorescence of dyes or small molecules provides a means to detect proteins as they are being translated.

The Tsien laboratory developed two biarsenical fluorescent dyes based on fluorescein (FlAsH) and resorufin (ReAsH) that become fluorescent when bound by a peptide that contains four cysteine residues (Griffin et al. 1998; Adams et al. 2002). The peptide sequence (FLNCCPGCCMEP) adopts a specific β-hairpin turn structure that is able to bind the biarsenical dyes with picomolar binding affinities (Madani et al. 2009). The FlAsH and ReAsH dyes are also membrane permeable, which facilitates their use in imaging translation in living cells. This labeling strategy was used to show that the AMPA receptors, GluR1 and GluR2, are locally synthesized in dendrites and localized to synapses in response to activity-dependent changes in synaptic strength (Ju et al. 2004). To observe newly synthesized proteins, the FlAsH and ReAsH dyes were used in a pulse-chase experiment, so that all preexisting GluR1 proteins were labeled with ReAsH and only proteins synthesized during the chase would be labeled by FlAsH (Fig. 3A.3). To determine the location of protein synthesis, the tetracysteine peptide sequence was added to the amino terminus of a GFP-β-actin fusion protein (Rodriguez et al. 2006). By placing the tetracysteine peptide sequence at the beginning of the reporter protein, the location of polysomes carrying multiple nascent reporter proteins that are actively being translated could be identified. Recently, a peptide that contains three copies of the tetra-cysteine sequence was engineered that has increased brightness compared to the original sequence, which improves the signal-to-noise in these experiments and may facilitate detection of the cellular sites of translation (Van Engelenburg et al. 2010).

TimeSTAMP

In efforts to develop novel approaches to discern newly synthesized proteins from the preexisting population, the Tsien lab devised a drug-dependent epitope tagging approach that does not use fluorescent proteins. Time-specific tagging for the age measurement of proteins, or Time-STAMP, uses a viral protease whose activity can be controlled by a cell-permeant drug (Fig. 3A.4) (Lin et al. 2008). The crux of the idea was to design a self-cleaving cassette that can be regulated by a specific small molecule inhibitor. The cassette is composed of the NS3 protease from hepatitis C virus flanked by its cognate target sites along with an epitope tag on the opposite side of the protein of interest. In the absence of the inhibitor, BILN-2061, the protease will cleave itself, rendering the protein undetectable to antibodies against the tag. On addition of the inhibitor, the protease activity is blocked and the epitope tag then becomes available for detection. By capturing the immunofluorescence of the protein of interest and the epitope tag, it is possible to distinguish the population of newly synthesized protein during the time window of the drug administration.

The goal of TimeSTAMP was not only to visualize newly synthesized protein, but also to develop a method that circumvents some of the concerns with traditional fluorescent protein-mediated detection of new proteins. For example, phototoxicity during photoconversion or photobleaching of fluorescent proteins is a significant concern to those intending to study intracellular translation over a long period of time. Second, the time delay between synthesis and subsequent maturation of the chromophore reduces the temporal resolution of the onset of translation. In addition, limitations in detection of low copy number proteins and the inability to amplify fluorescence signal from fluorescent proteins allow only high abundance proteins as potential targets for analyses. In summary, although TimeSTAMP is not amenable to dynamic visualization of newly synthesized proteins, it provides an alternative strategy and has the potential to complement fluorescent protein-based reporter assays that are currently available. TimeSTAMP has been used to detect new protein synthesis of postsynaptic density protein (PSD)-95 and the trafficking of neuroligin (NLGN) 3 in cultured hippocampal neurons (Lin et al. 2008; De Jaco et al. 2010).

FUTURE DIRECTIONS

Thus far, the methodologies used to measure translation in single cells have focused on either the process of protein synthesis or its protein output, which has made detection of single molecules challenging. Because of the critical role of translation and its regulation in gene expression, new techniques and labeling strategies will need to be developed to more precisely interrogate protein synthesis. An alternate strategy that measures translation-dependent changes to reporter mRNAs themselves may allow translation to be monitored with the single molecule resolution in living cells. Besides the well-established fluorescent labeling method based on the MS2 bacteriophage coat protein, fluorescent RNA labeling has also been shown using the PP7 bacteriophage coat protein, the λN peptide and the U1A spliceosomal protein (Brodsky and Silver 2000; Takizawa and Vale 2000; Daigle and Ellenberg 2007; Larson et al. 2011). Because all four RNA-protein complexes are orthogonal, it should be possible, in principle, to specifically label a reporter mRNA in both the coding region as well as the 3′ UTR with spectrally distinct fluorescent proteins. In such a system, an untranslated RNA would labeled with two fluorescent proteins; on translation, however, the ribosome would displace the fluorescent protein that was bound to the coding region resulting in a singly fluorescently labeled RNA. Using this type of reporter, untranslated RNAs can be distinguished from ones that have encountered the ribosome in real time in living cells. Although this experimental methodology is only designed to allow observation of the first round of translation, much of the regulation of translation takes place at the initiation step. Such a system could be used to address a variety of outstanding questions including where the pioneer round of translation takes place for mRNAs that contain premature stop codons and are targets of nonsense mediated decay, and whether localized mRNAs are translationally repressed during transport (Maquat 2004; Chang et al. 2007; Sonenberg and Hinnebusch 2009; Donnelly et al. 2010). Continued development of reagents and techniques for quantifying translation of single molecules in living cells will lead to a mechanistic understanding of the regulation of translation and, undoubtedly, a few surprises along the way.

REFERENCES

Aakalu G, Smith WB, Nguyen N, Jiang C, Schuman EM. 2001. Dynamic visualization of local protein synthesis in hippocampal neurons. *Neuron* **30:** 489–502.

Adams SR, Campbell RE, Gross LA, Martin BR, Walkup GK, Yao Y, Llopis J, Tsien RY. 2002. New biarsenical ligands and tetracysteine motifs for protein labeling in vitro and in vivo: Synthesis and biological applications. *J Am Chem Soc* **124:** 6063–6076.

Ando R, Hama H, Yamamoto-Hino M, Mizuno H, Miyawaki A. 2002. An optical marker based on the UV-induced green-to-red photoconversion of a fluorescent protein. *Proc Natl Acad Sci* **99:** 12651–12656.

Barhoom S, Kaur J, Cooperman BS, Smorodinsky NI, Smilansky Z, Ehrlich M, Elroy-Stein O. 2011. Quantitative single cell monitoring of protein synthesis at subcellular resolution using fluorescently labeled tRNA. *Nucleic Acids Res* **39:** e129.

Bassell GJ, Zhang H, Byrd AL, Femino AM, Singer RH, Taneja KL, Lifshitz LM, Herman IM, Kosik KS. 1998. Sorting of β-actin mRNA and protein to neurites and growth cones in culture. *J Neurosci* **18:** 251–265.

Beach DL, Salmon ED, Bloom K. 1999. Localization and anchoring of mRNA in budding yeast. *Curr Biol* **9:** 569–578.

Bertrand E, Chartrand P, Schaefer M, Shenoy SM, Singer RH, Long RM. 1998. Localization of ASH1 mRNA particles in living yeast. *Mol Cell* **2:** 437–445.

Betteridge T, Liu H, Gamper H, Kirillov S, Cooperman BS, Hou YM. 2007. Fluorescent labeling of tRNAs for dynamics experiments. *RNA* **13:** 1594–1601.

Brittis PA, Lu Q, Flanagan JG. 2002. Axonal protein synthesis provides a mechanism for localized regulation at an intermediate target. *Cell* **110:** 223–235.

Brodsky AS, Silver PA. 2000. Pre-mRNA processing factors are required for nuclear export. *RNA* **6:** 1737–1749.

Cerutti P, Miller N. 1967. Selective reduction of yeast transfer ribonucleic acid with sodium borohydride. *J Mol Biol* **26:** 55–66.

Chang L, Shav-Tal Y, Trcek T, Singer RH, Goldman RD. 2006. Assembling an intermediate filament network by dynamic cotranslation. *J Cell Biol* **172:** 747–758.

Chang YF, Imam JS, Wilkinson MF. 2007. The nonsense-mediated decay RNA surveillance pathway. *Annu Rev Biochem* **76:** 51–74.

Chubb JR, Trcek T, Shenoy SM, Singer RH. 2006. Transcriptional pulsing of a developmental gene. *Curr Biol* **16:** 1018–1025.

Corish P, Tyler-Smith C. 1999. Attenuation of green fluorescent protein half-life in mammalian cells. *Protein Eng* **12:** 1035–1040.

Daigle N, Ellenberg J. 2007. λN-GFP: An RNA reporter system for live-cell imaging. *Nat Methods* **4:** 633–636.

Darzacq X, Shav-Tal Y, de Turris V, Brody Y, Shenoy SM, Phair RD, Singer RH. 2007. In vivo dynamics of RNA polymerase II transcription. *Nat Struct Mol Biol* **14:** 796–806.

De Jaco A, Lin MZ, Dubi N, Comoletti D, Miller MT, Camp S, Ellisman M, Butko MT, Tsien RY, Taylor P. 2010. Neuroligin trafficking deficiencies arising from mutations in the α/β-hydrolase fold protein family. *J Biol Chem* **285:** 28674–28682.

Dieterich DC, Hodas JJ, Gouzer G, Shadrin IY, Ngo JT, Triller A, Tirrell DA, Schuman EM. 2010. In situ visualization and dynamics of newly synthesized proteins in rat hippocampal neurons. *Nat Neurosci* **13:** 897–905.

Donnelly CJ, Fainzilber M, Twiss JL. 2010. Subcellular communication through RNA transport and localized protein synthesis. *Traffic* **11:** 1498–1505.

Dynes JL, Steward O. 2007. Dynamics of bidirectional transport of Arc mRNA in neuronal dendrites. *J Comp Neurol* **500:** 433–447.

Evdokimov AG, Pokross ME, Egorov NS, Zaraisky AG, Yampolsky IV, Merzlyak EM, Shkoporov AN, Sander I, Lukyanov KA, Chudakov DM. 2006. Structural basis for the fast maturation of Arthropoda green fluorescent protein. *EMBO Rep* **7:** 1006–1012.

Fusco D, Accornero N, Lavoie B, Shenoy SM, Blanchard JM, Singer RH, Bertrand E. 2003. Single mRNA molecules demonstrate probabilistic movement in living mammalian cells. *Curr Biol* **13:** 161–167.

Giepmans BN, Adams SR, Ellisman MH, Tsien RY. 2006. The fluorescent toolbox for assessing protein location and function. *Science* **312:** 217–224.

Golding I, Cox EC. 2004. RNA dynamics in live *Escherichia coli* cells. *Proc Natl Acad Sci* **101:** 11310–11315.

Golding I, Paulsson J, Zawilski SM, Cox EC. 2005. Real-time kinetics of gene activity in individual bacteria. *Cell* **123:** 1025–1036.

Griffin BA, Adams SR, Tsien RY. 1998. Specific covalent labeling of recombinant protein molecules inside live cells. *Science* **281:** 269–272.

Grunwald D, Singer RH. 2010. In vivo imaging of labelled endogenous β-actin mRNA during nucleocytoplasmic transport. *Nature* **467:** 604–607.

Gurskaya NG, Verkhusha VV, Shcheglov AS, Staroverov DB, Chepurnykh TV, Fradkov AF, Lukyanov S, Lukyanov KA. 2006. Engineering of a monomeric green-to-red photoactivatable fluorescent protein induced by blue light. *Nat Biotechnol* **24:** 461–465.

Ju W, Morishita W, Tsui J, Gaietta G, Deerinck TJ, Adams SR, Garner CC, Tsien RY, Ellisman MH, Malenka RC. 2004. Activity-dependent regulation of dendritic synthesis and trafficking of AMPA receptors. *Nat Neurosci* **7:** 244–253.

Kang H, Schuman EM. 1996. A requirement for local protein synthesis in neurotrophin-induced hippocampal synaptic plasticity. *Science* **273:** 1402–1406.

Kislauskis EH, Zhu X, Singer RH. 1994. Sequences responsible for intracellular localization of β-actin messenger RNA also affect cell phenotype. *J Cell Biol* **127:** 441–451.

Kremers GJ, Goedhart J, van Munster EB, Gadella TW Jr. 2006. Cyan and yellow super fluorescent proteins with improved brightness, protein folding, and FRET Forster radius. *Biochemistry* **45:** 6570–6580.

Larson DR, Zenklusen D, Wu B, Chao JA, Singer RH. 2011. Real-time observation of transcription initiation and elongation on an endogenous yeast gene. *Science* **332:** 475–478.

Leung KM, van Horck FP, Lin AC, Allison R, Standart N, Holt CE. 2006. Asymmetrical β-actin mRNA translation in growth cones mediates attractive turning to netrin-1. *Nat Neurosci* **9:** 1247–1256.

Lim F, Peabody DS. 1994. Mutations that increase the affinity of a translational repressor for RNA. *Nucleic Acids Res* **22:** 3748–3752.

Lin MZ, Glenn JS, Tsien RY. 2008. A drug-controllable tag for visualizing newly synthesized proteins in cells and whole animals. *Proc Natl Acad Sci* **105:** 7744–7749.

Lionnet T, Czaplinski K, Darzacq X, Shav-Tal Y, Wells AL, Chao JA, Park HY, de Turris V, Lopez-Jones M, Singer RH. 2011. A transgenic mouse for in vivo detection of endogenous labeled mRNA. *Nat Methods* **8:** 165–170.

Liu J, Xu Y, Stoleru D, Salic A. 2011. Imaging protein synthesis in cells and tissues with an alkyne analog of puromycin. *Proc Natl Acad Sci* doi: 10.1073/pnas.1111561108.

Lowary PT, Uhlenbeck OC. 1987. An RNA mutation that increases the affinity of an RNA-protein interaction. *Nucleic Acids Res* **15:** 10483–10493.

Madani F, Lind J, Damberg P, Adams SR, Tsien RY, Graslund AO. 2009. Hairpin structure of a biarsenical-tetracysteine motif determined by NMR spectroscopy. *J Am Chem Soc* **131:** 4613–4615.

Maquat LE. 2004. Nonsense-mediated mRNA decay: Splicing, translation and mRNP dynamics. *Nat Rev Mol Cell Biol* **5:** 89–99.

Mor A, Suliman S, Ben-Yishay R, Yunger S, Brody Y, Shav-Tal Y. 2010. Dynamics of single mRNP nucleocytoplasmic transport and export through the nuclear pore in living cells. *Nat Cell Biol* **12:** 543–552.

Nathans D. 1964. Puromycin inhibition of protein synthesis: Incorporation of puromycin into peptide chains. *Proc Natl Acad Sci* **51:** 585–592.

Ngo JT, Champion JA, Mahdavi A, Tanrikulu IC, Beatty KE, Connor RE, Yoo TH, Dieterich DC, Schuman EM, Tirrell DA. 2009. Cell-selective metabolic labeling of proteins. *Nat Chem Biol* **5:** 715–717.

Pan D, Kirillov SV, Cooperman BS. 2007. Kinetically competent intermediates in the translocation step of protein synthesis. *Mol Cell* **25:** 519–529.

Pan D, Qin H, Cooperman BS. 2009. Synthesis and functional activity of tRNAs labeled with fluorescent hydrazides in the D-loop. *RNA* **15:** 346–354.

Raab-Graham KF, Haddick PC, Jan YN, Jan LY. 2006. Activity- and mTOR-dependent suppression of Kv1.1 channel mRNA translation in dendrites. *Science* **314:** 144–148.

Rodriguez AJ, Shenoy SM, Singer RH, Condeelis J. 2006. Visualization of mRNA translation in living cells. *J Cell Biol* **175:** 67–76.

Rogers S, Wells R, Rechsteiner M. 1986. Amino acid sequences common to rapidly degraded proteins: The PEST hypothesis. *Science* **234:** 364–368.

Rostovtsev VV, Green LG, Fokin VV, Sharpless KB. 2002. A stepwise huisgen cycloaddition process: Copper(I)-catalyzed regioselective "ligation" of azides and terminal alkynes. *Angew Chem Int Ed Engl* **41:** 2596–2599.

Schwanhausser B, Busse D, Li N, Dittmar G, Schuchhardt J, Wolf J, Chen W, Selbach M. 2011. Global quantification of mammalian gene expression control. *Nature* **473:** 337–342.

Shaner NC, Lin MZ, McKeown MR, Steinbach PA, Hazelwood KL, Davidson MW, Tsien RY. 2008. Improving the photostability of bright monomeric orange and red fluorescent proteins. *Nat Methods* **5:** 545–551.

Shav-Tal Y, Darzacq X, Shenoy SM, Fusco D, Janicki SM, Spector DL, Singer RH. 2004. Dynamics of single mRNPs in nuclei of living cells. *Science* **304:** 1797–1800.

Smith WB, Starck SR, Roberts RW, Schuman EM. 2005. Dopaminergic stimulation of local protein synthesis enhances surface expression of GluR1 and synaptic transmission in hippocampal neurons. *Neuron* **45:** 765–779.

Sonenberg N, Hinnebusch AG. 2009. Regulation of translation initiation in eukaryotes: Mechanisms and biological targets. *Cell* **136:** 731–745.

Starck SR, Roberts RW. 2002. Puromycin oligonucleotides reveal steric restrictions for ribosome entry and multiple modes of translation inhibition. *RNA* **8:** 890–903.

Starck SR, Green HM, Alberola-Ila J, Roberts RW. 2004. A general approach to detect protein expression in vivo using fluorescent puromycin conjugates. *Chem Biol* **11:** 999–1008.

Stepanenko OV, Verkhusha VV, Kuznetsova IM, Uversky VN, Turoverov KK. 2008. Fluorescent proteins as biomarkers and biosensors: Throwing color lights on molecular and cellular processes. *Curr Protein Pept Sci* **9:** 338–369.

Sutton MA, Wall NR, Aakalu GN, Schuman EM. 2004. Regulation of dendritic protein synthesis by miniature synaptic events. *Science* **304:** 1979–1983.

Takizawa PA, Vale RD. 2000. The myosin motor, Myo4p, binds Ash1 mRNA via the adapter protein, She3p. *Proc Natl Acad Sci* **97:** 5273–5278.

Taniguchi Y, Choi PJ, Li GW, Chen H, Babu M, Hearn J, Emili A, Xie XS. 2010. Quantifying *E. coli* proteome and transcriptome with single-molecule sensitivity in single cells. *Science* **329:** 533–538.

Terskikh A, Fradkov A, Ermakova G, Zaraisky A, Tan P, Kajava AV, Zhao X, Lukyanov S, Matz M, Kim S, et al. 2000. "Fluorescent timer": Protein that changes color with time. *Science* **290:** 1585–1588.

Valegard K, Murray JB, Stockley PG, Stonehouse NJ, Liljas L. 1994. Crystal structure of an RNA bacteriophage coat protein-operator complex. *Nature* **371:** 623–626.

Van Engelenburg SB, Nahreini T, Palmer AE. 2010. FACS-based selection of tandem tetracysteine peptides with improved ReAsH brightness in live cells. *Chembiochem* **11:** 489–493.

Wang DO, Kim SM, Zhao Y, Hwang H, Miura SK, Sossin WS, Martin KC. 2009. Synapse- and stimulus-specific local translation during long-term neuronal plasticity. *Science* **324:** 1536–1540.

Wu B, Piatkevich KD, Lionnet T, Singer RH, Verkhusha VV. 2011. Modern fluorescent proteins and imaging technologies to study gene expression, nuclear localization, and dynamics. *Curr Opin Cell Biol* **23:** 310–317.

Yarmolinsky MB, Haba GL. 1959. Inhibition by puromycin of mino acid incorporation into protein. *Proc Natl Acad Sci* **45:** 1721–1729.

Zimyanin VL, Belaya K, Pecreaux J, Gilchrist MJ, Clark A, Davis I, St Johnston D. 2008. In vivo imaging of oskar mRNA transport reveals the mechanism of posterior localization. *Cell* **134:** 843–853.

A Molecular Link between miRISCs and Deadenylases Provides New Insight into the Mechanism of Gene Silencing by MicroRNAs

Joerg E. Braun, Eric Huntzinger, and Elisa Izaurralde

Department of Biochemistry, Max Planck Institute for Developmental Biology, Spemannstrasse 35, 72076 Tübingen, Germany

Correspondence: elisa.izaurralde@tuebingen.mpg.de

MicroRNAs (miRNAs) are a large family of endogenous noncoding RNAs that, together with the Argonaute family of proteins (AGOs), silence the expression of complementary mRNA targets posttranscriptionally. Perfectly complementary targets are cleaved within the base-paired region by catalytically active AGOs. In the case of partially complementary targets, however, AGOs are insufficient for silencing and need to recruit a protein of the GW182 family. GW182 proteins induce translational repression, mRNA deadenylation and exonucleolytic target degradation. Recent work has revealed a direct molecular link between GW182 proteins and cellular deadenylase complexes. These findings shed light on how miRNAs bring about target mRNA degradation and promise to further our understanding of the mechanism of miRNA-mediated repression.

MicroRNAs (miRNAs) are genome-encoded, ~21–23-nucleotide noncoding RNAs that posttranscriptionally silence mRNA targets containing complementary sequences (Bartel 2009). To exert their regulatory functions, miRNAs associate with Argonaute (AGO) proteins in effector complexes known as miRNA-induced silencing complexes (miRISCs). These complexes promote endonucleolytic cleavage of fully complementary targets or translational repression, mRNA deadenylation, and exonucleolytic decay of targets with partial complementarity (see Figs. 1 and 2 and Bartel 2009 for a detailed description of miRNA-target recognition; Djuranovic et al. 2011; Huntzinger and Izaurralde 2011). Invertebrate genomes contain at least 100 miRNA genes, whereas vertebrate and plant genomes possess 500 to 1000 miRNA genes (Bartel 2009; Voinnet 2009). Computational predictions and functional studies indicate that the highly expressed miRNAs can potentially regulate hundreds of different mRNAs, suggesting that a significant proportion of eukaryotic transcriptomes (~50% in humans) is subject to miRNA regulation (Bartel 2009; Voinnet 2009).

Given the large number of potential targets, it is not surprising that miRNAs play roles in nearly all developmental and cellular processes investigated thus far (Sayed and Abdellatif 2011). It is clear that changes in miRNA expression levels are associated with many human

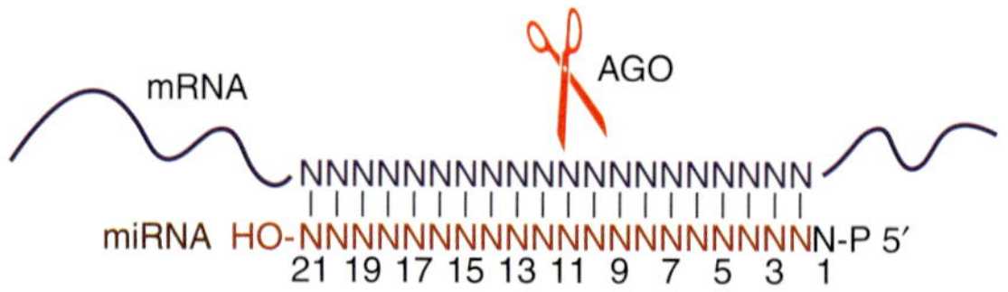

Figure 1. Mechanism of miRNA-mediated gene silencing (fully or nearly complementary targets). A miRNA bound to an AGO protein recognizes mRNA targets containing fully or nearly complementary binding sites. In plants, these binding sites are predominantly located within the ORF. The AGO protein cleaves the mRNA within the base-paired region (between nucleotides opposite to nucleotides 10 and 11 of the miRNA strand, which should be base-paired for AGO to cleave). Following cleavage, the mRNA decay intermediates are degraded from the newly generated 3′ and 5′ ends by the exosome and XRN1 or its plant ortholog XRN4 (not shown; Voinnet 2009). The miRNA 5′ terminal nucleotide (shown in black) is buried in the 5′-phosphate binding pocket of AGOs and is not available for pairing with the target.

diseases, such as cancer and metabolic disorders (Esteller 2011; Sayed and Abdellatif 2011).

During the past decade, remarkable progress has been made in understanding miRNA biogenesis and function (Krol et al. 2010; Esteller 2011; Sayed and Abdellatif 2011); however, the mechanisms by which miRNAs regulate gene expression remain unclear and are still a source of scientific debate (Djuranovic et al. 2011; Huntzinger and Izaurralde 2011). In this review, we focus on the effector step of silencing, that is, on what happens after a mRNA target is recognized by miRISCs. First, the emerging model of the molecular mechanism used by miRNAs to silence mRNA targets is described.

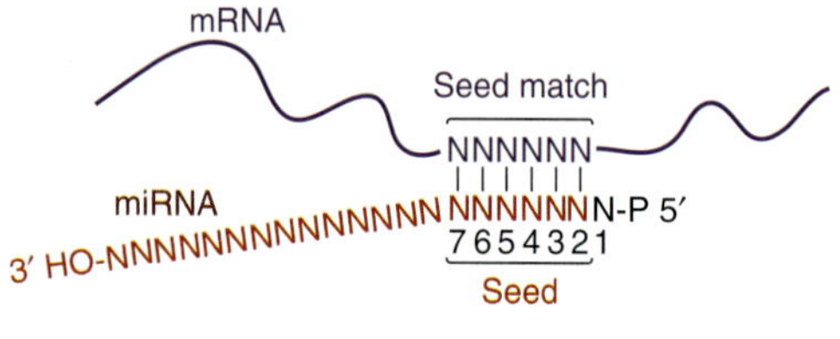

Figure 2. miRNA target recognition in animals. In animals, miRNAs typically recognize partially complementary binding sites, which are generally located in 3′ UTRs. Complementarity to the miRNA "seed" sequence, containing nucleotides 2–7 or 2–8, is a major determinant in target recognition and is sufficient to trigger silencing. In some cases, complementarity to the 3′ region of the miRNA might contribute to target binding (not shown, see Bartel 2009). However, even for these sites, miRNA nucleotides 9–12 are generally not complementary to the target, preventing endonucleolytic cleavage by AGOs. In these cases, AGOs recruit a protein of the GW182 family (see Fig. 3).

We then evaluate evidence supporting this model and discuss some key questions that remain to be answered, particularly regarding the mechanistic connections between miRNA-mediated translation repression and mRNA degradation. Studies on the GW182 family proteins, which are key components of miRISCs in animals, have led to many insights into the biochemical mechanisms of silencing. Therefore, in this review, we focus on recent data that have deepened our understanding of how this protein family cooperates with AGOs and cellular factors to bring about silencing.

EMERGING MODEL OF miRNA-MEDIATED GENE SILENCING

Accumulated data in the miRNA field suggest a model of silencing that begins with the recognition of the target by a miRNA in complex with an AGO protein. In instances when the complementarity between the target and the miRNA is extensive and the AGO protein is catalytically active, the target is cleaved by AGO within the base-paired region (between nucleotides 10 and 11, opposite the miRNA strand) (Fig. 1) (Bartel 2009; Jínek and Doudna 2009; Voinnet 2009). This mechanism appears to be most prominent in plants where miRNAs recognize fully or nearly complementary binding sites, which are generally located in the mRNA open reading frames (ORFs) (Voinnet 2009).

In animals, miRNAs recognize partially complementary binding sites, which are generally

Cite this article as *Cold Spring Harb Perspect Biol* doi: 10.1101/cshperspect.a012328

located in the mRNA 3′ untranslated region (UTR) (Bartel 2009). Complementarity to the 5′ end of the miRNA — the so-called "seed" sequence — is a major determinant in target recognition and is sufficient to trigger silencing (Fig. 2) (Bartel 2009). Even when target complementarity is not limited to the seed sequence, miRNA nucleotides 9–12 are generally not complementary to the target in animals, preventing endonucleolytic cleavage by AGOs (Bartel 2009; Jínek and Doudna 2009). This is important because in these cases, the AGO proteins are insufficient to mediate silencing and require interaction with additional proteins, including members of the GW182 family (Huntzinger and Izaurralde 2011).

Recruitment of a GW182 protein to a miRNA target triggers translational repression and mRNA deadenylation (Fig. 3). The mechanism of translational repression has yet to be elucidated, although increasing evidence points to an inhibition of translation initiation (Djuranovic et al. 2011; Huntzinger and Izaurralde 2011). In contrast, much is known about the mechanism of miRNA-mediated mRNA deadenylation. It is known that deadenylation is performed by the sequential action of two cytoplasmic deadenylase complexes (the PAN2–PAN3 and the CCR4–NOT complex) (Fig. 3) (Behm-Ansmant et al. 2006; Chen et al. 2009; Eulalio et al. 2009a; Piao et al. 2010). Both complexes directly interact with GW182 proteins (Braun et al. 2011; Chekulaeva et al. 2011; Fabian et al. 2011).

Depending on the cell type and/or specific target, the deadenylated mRNA target can be stored in a translationally repressed state, as observed, for example, in *Caenorhabditis elegans* embryos and in cell free extracts that recapitulate silencing (Mathonnet et al. 2007; Thermann and Hentze 2007; Wakiyama et al. 2007; Iwasaki

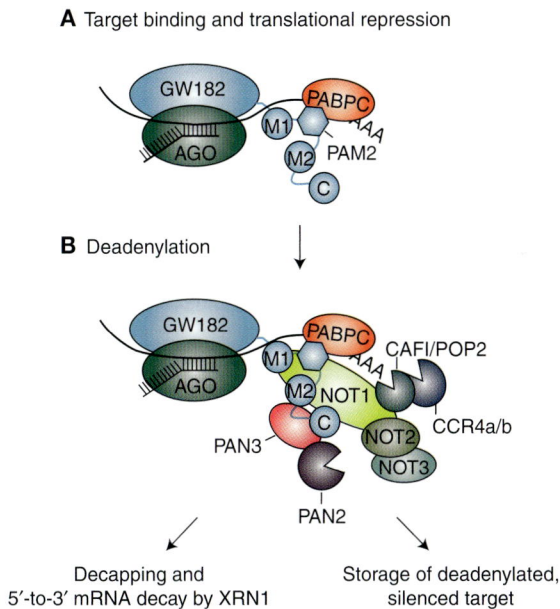

Figure 3. Mechanism of miRNA-mediated gene silencing in animals (partially complementary targets). (*A,B*) The AGO–GW182 complex represses translation through an unknown mechanism and directs the mRNA to deadenylation. Human GW182 proteins interact with PABPC through the PAM2 motif, with PAN3 through the M2 and carboxy-terminal (C-term) regions and with NOT1 through the M1, M2, and C-term regions. Although translational repression and deadenylation are shown as consecutive steps, the order of events remains controversial, and it is unclear whether the two processes are linked or independent. Depending on the cell type and/or specific target, deadenylated mRNAs can be stored in a deadenylated, translationally repressed state. In animal cell cultures, deadenylated mRNAs are generally decapped and rapidly degraded by the major 5′-to-3′ exonuclease XRN1.

et al. 2009; Zdanowicz et al. 2009; Wu et al. 2010). However, in diverse organisms and cell types, a reduction in mRNA levels is observed (Bagga et al. 2005; Lim et al. 2005; Wu and Belasco 2005; Behm-Ansmant et al. 2006; Giraldez et al. 2006; Mishima et al. 2006; Rehwinkel et al. 2006; Schmitter et al. 2006; Wu et al. 2006). This can be explained by the observation that deadenylated mRNAs are in generally unstable and are rapidly decapped and degraded by the major 5′-to-3′ exonuclease XRN1. Thus, miRNAs accelerate target destruction by recruiting the enzymes involved in one of the two major cellular mRNA decay pathways, namely the 5′-to-3′ mRNA decay pathway (Rehwinkel et al. 2005; Behm-Ansmant et al. 2006; Eulalio et al. 2007a, 2009a; Chen et al. 2009; Piao et al. 2010). In this pathway, mRNAs are first deadenylated, then decapped and, finally, exonucleolytically degraded from the 5′-end by XRN1.

Although it is now clear how miRNAs cause target degradation, there is more to the story that needs to be understood. For example, GW182 proteins interact with the cytoplasmic poly(A)-binding protein (PABPC, where C stands for cytoplasmic), but it is not known how this interaction contributes to silencing (Fabian et al. 2009; Zekri et al. 2009; Huntzinger et al. 2010; Jínek et al. 2010; Kozlov et al. 2010). Another key question that remains is whether translational repression and deadenylation are interconnected or represent two independent mechanisms used by miRNAs to silence mRNA targets. Finally, understanding how translational repression is achieved remains a major challenge for future studies.

THE GW182 PROTEIN FAMILY: DOMAIN ORGANIZATION

GW182 proteins were first identified as antigens recognized by serum from a patient with motor and sensory neuropathy (Eystathioy et al. 2002). Their role in the miRNA pathway was revealed in subsequent studies showing that GW182 proteins copurify with AGO-containing complexes and their depletion inhibits silencing of miRNA targets (Jakymiw et al. 2005; Liu et al. 2005; Meister et al. 2005; Rehwinkel et al. 2005). In

independent studies, the proteins were identified based on mutant phenotypes that reflect the inactivation of the miRNA pathway in *C. elegans* or in RNAi screens for suppressors of silencing in *Drosophila melanogaster* (Ding et al. 2005; Behm-Ansmant et al. 2006). Subsequently, it was shown that GW182 proteins act downstream of AGOs, in the effector step of silencing (Eulalio et al. 2008). In vertebrates and several invertebrate species, there are up to three GW182 paralogs with partially redundant functions (also known as TNRC6A, B and C), whereas there is only one orthologous protein in *D. melanogaster* (*Dm* GW182).

GW182 proteins typically contain an amino-terminal region (N-term) with multiple glycine-tryptophan (GW) repeats (i.e., GW, WG, or GWG), a central ubiquitin-associated (UBA) domain, and an RNA recognition motif (RRM) (Fig. 4) (Eulalio et al. 2009b). Additional regions of the protein include a glutamine-rich (Q-rich) region, which is located between the UBA domain and the RRM, and middle (Mid) and carboxy-terminal (C-term) regions containing fewer or no GW repeats (Fig. 4) (Eulalio et al. 2009b). Interestingly, this domain organization is not conserved in the *C. elegans* proteins AIN-1 and AIN-2, which are highly divergent members of the protein family (Ding et al. 2005; Zhang et al. 2007; Eulalio et al. 2009b).

A fascinating aspect of the GW182 family is that the protein regions that are functionally relevant for silencing are predicted to be disordered. For example, the N-term GW repeat region of the proteins mediates binding to the AGO proteins and thus is essential for miRNA-mediated gene silencing (Behm-Ansmant et al. 2006; Till et al. 2007; Eulalio et al. 2008, 2009c; Lazzaretti et al. 2009; Lian et al. 2009; Takimoto et al. 2009). This region is predicted to be unstructured and interacts with AGOs through multiple GW repeats, which contribute to the interaction in an additive manner. The exact location and number of repeats are, however, not conserved.

The Mid and C-term regions (collectively termed the silencing domain), both predicted to be disordered, are also essential for silencing (Chekulaeva et al. 2009; Eulalio et al. 2009c; Lazzaretti et al. 2009; Zipprich et al. 2009;

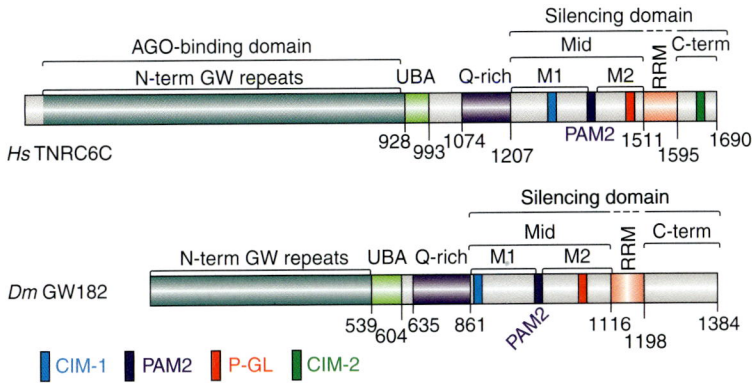

Figure 4. GW182 protein family. Domain organization of GW182 proteins. The amino-terminal AGO-binding domains contain multiple GW repeats (not shown). The silencing domain includes the Mid and carboxy-terminal regions but not the RRM, which is dispensable for silencing. The Mid and C-term regions contain a variable number of GW repeats (not shown) and additional motifs termed CIM-1, CIM-2, and P-GL. Human TNRC6C and *Dm* GW182 are shown as representative family members of the GW182 proteins. Abbreviations: UBA, ubiquitin-associated domain; Q-rich, region rich in glutamine; Mid, middle domain with PAM2 motif (dark blue), which divides the Mid region into the M1 and M2 regions; RRM, RNA recognition motif; C-term, carboxy-terminal region; N-term, amino-terminal.

Huntzinger et al. 2010). For instance, GW182 protein mutants lacking the silencing domain generally do not rescue silencing in cells depleted of endogenous GW182, even though these proteins interact with AGOs and are active in tethering assays (Chekulaeva et al. 2009; Eulalio et al. 2009c; Zekri et al. 2009; Huntzinger et al. 2010; Braun et al. 2011). Recent work has shown that the silencing domains of human GW182 proteins provide a binding platform for PABPC and subunits of the two major cytoplasmic deadenylase complexes (the PAN2–PAN3 and CCR4–NOT complexes) (Fabian et al. 2009, 2011; Zekri et al. 2009; Huntzinger et al. 2010; Jínek et al. 2010; Kozlov et al. 2010; Braun et al. 2011; Chekulaeva et al. 2011). We discuss the contribution of these interactions to miRNA-mediated repression below.

THE GW182 PROTEIN FAMILY: INTERACTION NETWORK

GW182 Proteins Are PABP-Interacting Proteins (Paips)

The interaction of GW182 proteins with PABPC is mediated by a conserved PAM2 motif (PABP-binding motif 2) located in the Mid region

(Fig. 4) (Fabian et al. 2009; Huntzinger et al. 2010; Jínek et al. 2010; Kozlov et al. 2010). The PAM2 motif was first identified in the translational regulators Paip1 and Paip2 (PABP-interacting proteins 1 and 2) (Derry et al. 2006). This motif confers direct binding to the PABPC carboxy-terminal domain termed the MLLE domain (Fabian et al. 2009; Huntzinger et al. 2010; Jínek et al. 2010; Kozlov et al. 2010). Moreover, GW182 protein sequences downstream of the PAM2 motif (termed M2) together with the C-term region mediate indirect binding to PABPC in vivo (Huntzinger et al. 2010). Although indirect, this interaction is dominant over that mediated by the PAM2 motif in *D. melanogaster* cells (Zekri et al. 2009; Huntzinger et al. 2010).

Interestingly, in addition to PABPC, the PAM2 motif of mammalian GW182 proteins interacts with the EDD protein (also known as hyperplastic discs [HYD]), which like PABPC, contains an MLLE domain (Su et al. 2011). Because EDD also interacts indirectly with decapping factors and the CCR4–NOT complex (Su et al. 2011), it could help to recruit these factors to miRNA targets. The extent of the contribution of EDD to silencing and the conservation of its role in other species remains to be established. Furthermore, because PABPC and EDD interact with GW182 proteins in a mutually exclusive manner, it will be important to determine whether EDD and PABPC play redundant or distinct roles in silencing, for example, leading to different outcomes.

GW182 Proteins Interact Directly with Deadenylase Complexes

In addition to PABPC, the silencing domains of human GW182 proteins confer direct binding to PAN3 and NOT1, which are subunits of the PAN2–PAN3 and CCR4–NOT deadenylase complexes, respectively (Braun et al. 2011; Chekulaeva et al. 2011; Fabian et al. 2011). Binding to PAN3 requires the M2 and C-term regions of the silencing domain (Braun et al. 2011), whereas NOT1 binding is mediated by the M1, M2, and C-term regions (Fig. 4), which contribute to the interaction in an additive manner (Braun et al. 2011; Chekulaeva et al. 2011; Fabian et al. 2011).

A precise mapping of the interaction with the CCR4–NOT complex identified two distinct short linear motifs within the M1 and C-term regions of human GW182 silencing domains. These motifs were termed CCR4 interaction motifs (CIM)-1 and 2, respectively (Fig. 4) (Fabian et al. 2011). CIM-2 is characterized by a LWG repeat, suggesting that GW-repeats can mediate interactions with protein partners other than AGOs, depending on the context. Accordingly, alanine substitutions of all tryptophan residues in the M2 and C-term regions of the human TNRC6C silencing domain abolished both CCR4–NOT and PAN2–PAN3 binding (Chekulaeva et al. 2011).

Interestingly, although CIM-1 and CIM-2 recruit the CCR4–NOT complex to mRNA targets, only CIM-2 supports full deadenylation, suggesting that these motifs are functionally distinct (Fabian et al. 2011). These findings also indicate that recruitment of the CCR4–NOT complex is not sufficient to promote processive deadenylation and that an "activation step" is involved. The CIM-2 motif, but not CIM-1, can mediate this activation. This opens the perspec-

tive that GW182 proteins do not merely recruit the CCR4–NOT complex but enhance deadenylation rates (Fabian et al. 2011).

The interactions with the deadenylase complexes are conserved in *D. melanogaster* (Braun et al. 2011; Chekulaeva et al. 2011); however, the mode of interaction is not. For instance, only the CIM-1 motif is present in *Dm* GW182, whereas CIM-2 is absent (Fig. 4) (Fabian et al. 2011). Furthermore, in contrast to human GW182s, deletion of the silencing domain in *Dm* GW182 does not abolish binding to NOT1 (Braun et al. 2011; Chekulaeva et al. 2011), suggesting that additional binding motifs, upstream of the silencing domain, are present in *Dm* GW182.

GW182 Proteins Contain a Conserved Proline-Rich Motif

A recent study using zebrafish led to the identification of a proline rich motif (P-GL motif) (Fig. 4) in the M2 region of the silencing domains that, together with the PAM2 motif, contributes to translational repression and the deadenylation of polyadenylated targets, as well as to translational repression in the absence of deadenylation (i.e., targets lacking a poly(A) tail) (Mishima et al. 2012). Interestingly, this motif is not required for PABPC or deadenylase binding. This finding adds another layer of complexity to GW182 protein function, as it suggests that the proteins interact with another, yet unidentified partner. The P-GL motif is conserved in vertebrate and insect GW182 proteins; thus, it would be of interest to identify its potential binding partners.

Plasticity of the GW182 Protein Interaction Network

The study of the GW182 protein interaction network in different species has revealed that these proteins interact with their partners (e.g., AGOs, PABPC, PAN3, and NOT1) through short linear motifs (SLiMs) (Davey et al. 2012) embedded in unstructured regions. This provides one explanation for why the binding of GW182 proteins to their partners requires multiple and partially redundant motifs (e.g., multiple GW repeats), as SLiMs provide low-affinity binding and multiple motifs are required for high-affinity (avidity) interactions (Davey et al. 2012). This also explains why the partners are conserved but the molecular details of the interaction change throughout evolution (e.g., *Dm* GW182 lacks CIM-2). Indeed, linear motifs are evolutionarily plastic, as only a small number of point mutations in a disordered region of a protein sequence can result in a gain, loss, or relocation of SLiMs. In this context, it is interesting to note that the *C. elegans* GW182 protein AIN-1 interacts with PABPC, PAN3, NOT1, and NOT2, despite that it lacks all the motifs described above (Kuzuoglu-Öztürk et al. 2012). Therefore, it is possible that AIN-1 has evolved different ways to interact with the same partners.

ROLE OF PABPC

The PAM2 motif is highly conserved among vertebrate and insect GW182 proteins, and the interaction with PABPC is well documented by biochemical and structural studies (Fabian et al. 2009; Zekri et al. 2009; Huntzinger et al. 2010; Jínek et al. 2010; Kozlov et al. 2010). However, little is known regarding how the GW182–PABPC interaction contributes to silencing. Several mechanisms have been proposed, which are not mutually exclusive.

The first model is that the interaction of GW182 proteins with PABPC interferes with mRNA circularization (Fabian et al. 2009; Zekri et al. 2009). It is known that PABPC (bound to the mRNA poly(A) tail) interacts with eIF4G (associated with the cap structure through interaction with the cap-binding protein eIF4E), giving rise to circular mRNAs that are efficiently translated and protected from degradation (Derry et al. 2006). By analogy with Paip2, GW182 proteins may compete with eIF4G for binding to PABPC, thereby preventing mRNA circularization and consequently inhibiting translation.

The second possibility is that the PABPC–GW182 interaction reduces the affinity of PABPC for the poly(A) tail, as described for Paip2 (Derry et al. 2006). The third model is that the PABPC–GW182 interaction accelerates miRNA-mediated deadenylation. This last model is supported by the observation that in cell-free extracts from mouse Krebs-2 ascites cells, PABPC depletion or mutations in the PAM2 motif of TNRC6C reduce the rate, but not the extent, of deadenylation (Fabian et al. 2009, 2011; Jínek et al. 2010).

It is also unclear whether PABPC is required for silencing, as conflicting lines of evidence have been reported (Fabian et al. 2009; Zekri et al. 2009; Huntzinger et al. 2010; Jínek et al. 2010; Braun et al. 2011; Fukaya and Tomari, 2011; Mishima et al. 2012). A role for PABPC in silencing is supported by three observations: (1) overexpressing PABPC or an excess of Paip2–PAM2 peptide reduces silencing under conditions in which general translation is not affected (Fabian et al. 2009; Zekri et al. 2009; Walters et al. 2010); (2) depletion of PABPC from Krebs-2 ascites cell extracts inhibits miRNA-mediated deadenylation (Fabian et al. 2009, 2011; Jínek et al. 2010); and (3) in complementation assays, human GW182 mutants that do not interact with PABPC (i.e., carrying mutations in the PAM2 motif) are impaired in rescuing silencing in cells depleted of the endogenous GW182 proteins (Huntzinger et al. 2010; Braun et al. 2011). These results indicate that the PABPC–GW182 interaction is important, although not essential, for silencing in these cells. It has also been shown that PABPC-binding is not sufficient for silencing in vivo (Huntzinger et al. 2010; Braun et al. 2011) or for GW182-mediated deadenylation in vitro (Fabian et al. 2011).

Arguing against a role for PABPC in silencing is the observation that mRNAs lacking a poly(A) tail (that is, they cannot circularize and are not deadenylated) are nevertheless silenced (Humphreys et al. 2005; Pillai et al. 2005; Wu et al. 2006; Eulalio et al. 2008, 2009a; Chekulaeva et al. 2011). This suggests that PABPC is either not required for silencing or is only required for repression of polyadenylated targets. Alternatively, PABPC could be recruited to unadenylated targets through interactions with translation factors and bind to GW182 in the absence of a poly(A) tail, but how this interaction might contribute to silencing is currently unknown.

Remarkably, in tethering assays, multiple and nonoverlapping fragments of *Dm* GW182, including amino-terminal fragments that do not interact with PABPC, trigger translational repression and mRNA degradation (Chekulaeva et al. 2009; Fukuya and Tomari 2011; Yao et al. 2011). These results were interpreted as evidence that interactions of GW182 proteins with PABPC are not required for silencing (Fukaya and Tomari 2011). There are alternative explanations for this observation, including that tethering assays may not recapitulate all steps of silencing. Indeed, in complementation assays, amino-terminal fragments of GW182 (lacking the silencing domain) fail to complement the silencing of a large number of targets (Eulalio et al. 2009c). Thus, although tethering assays are an invaluable tool in the dissection of the role of GW182 protein domains in silencing, conclusions from these assays should be validated in complementation assays.

Fukaya and Tomari (2011) also observed silencing in extracts in which PABPC was inhibited from functioning in translation by the addition of an excess of Paip2, which displaces PABPC from the mRNA poly(A) tail and competes with GW182 for binding to PABPC. Based on these observations, they concluded that PABPC is dispensable for silencing. Yet, if the role of the GW182–PABPC interaction were to facilitate PABPC dissociation from the poly(A) tail, then this interaction would become dispensable in extracts in which PABPC has been removed from the poly(A) tail by Paip2.

The most compelling evidence supporting a nonessential role for PABPC is that its depletion does not affect translational repression or target degradation in zebrafish embryos (Mishima et al. 2012). Paradoxically, in this system, simultaneous mutations in the PAM2 and P-GL motifs of zebrafish GW182 (TNRC6A) strongly impaired the silencing activity of the protein in tethering assays.

A potential explanation for the disparate observations outlined above is that PABPC might be required for silencing of specific targets and/or under specific cellular conditions (see below). This possibility, together with alternative roles for PABPC in silencing, need to be explored in future studies. Until the molecular mechanism of silencing is fully understood, PABPC involvement cannot be ruled out.

ROLE OF CYTOPLASMIC DEADENYLASE COMPLEXES

In eukaryotes, the mRNA poly(A) tail is removed by the sequential action of two cytoplasmic deadenylase complexes (Chen et al. 2011). The dimeric PAN2–PAN3 complex deadenylates mRNAs in a distributive manner and is responsible for shortening long poly(A) tails to ~50–110 nucleotides, depending on the specific mRNA and organism of study (Chen et al. 2011). The PAN2–PAN3 complex is recruited to mRNA targets via interaction PAN3 with PABPC. PAN3 also recruits PAN2, the catalytic subunit of the complex to mRNA targets (Siddiqui et al. 2007). The second, processive phase of deadenylation is catalyzed by the CCR4–NOT complex, which can rescue cytoplasmic mRNA deadenylation in the absence of the PAN2–PAN3 complex (Chen et al. 2011). The conserved core of the metazoan CCR4–NOT complex consists of five subunits: NOT1, NOT2, NOT3 (also known as NOT3/5), and two catalytically active subunits, CCR4a or its paralog CCR4b and CAF1 or its paralog POP2 (Fig. 3) (Chen et al. 2011).

The relative roles of the two deadenylase complexes in the miRNA pathway are well defined. The CCR4–NOT complex provides a major contribution to miRNA target decay, whereas the PAN2–PAN3 complex plays only a minor role. For example, PAN2 depletion does not prevent miRNA target degradation, whereas the depletion of components of the CCR4–NOT complex inhibits miRNA target deadenylation and subsequent degradation (Behm-Ansmant et al. 2006; Eulalio et al. 2009a; Piao et al. 2010; Braun et al. 2011). Accordingly, overexpression of catalytically inactive CCR4a, CAF1, or POP2 suppresses silencing in a dominant-negative manner in human cells (Chen et al. 2009; Piao et al. 2010). By contrast, although overexpression of a catalytically inactive PAN2 mutant slows the initial phase of miRNA target deadenylation (Chen et al. 2009), it does

Cite this article as *Cold Spring Harb Perspect Biol* doi: 10.1101/cshperspect.a012328

not suppress silencing, supporting the conclusion that the PAN2–PAN3 complex is involved in but is not essential for deadenylation of miRNA targets.

Despite the important role of the deadenylase complexes in the miRNA pathway, until recently, the molecular mechanism underlying their recruitment remained controversial. A translational-repression-only model for miRNA silencing suggested that deadenylases are recruited to miRNA targets by default, as an indirect consequence of a primary inhibitory effect of miRISCs on translation (reviewed by Djuranovic et al. 2011). The new finding that human GW182 proteins interact with PAN3 and NOT1 (Braun et al. 2011; Chekulaeva et al. 2011; Fabian et al. 2011) definitively shows that the silencing machinery recruits these complexes directly and thus that deadenylation is a direct effect of miRNA regulation and not a mere consequence of translational repression.

ROLE OF THE DECAPPING COMPLEX AND DECAPPING ACTIVATORS

Generally, deadenylated mRNAs are committed to decapping and 5′-to-3′ exonucleolytic degradation in somatic cells. The decapping enzyme DCP2 requires cofactors for full activity. In metazoa, these include DCP1, EDC4 (also known as Ge-1), Pat, and the DEAD-box protein RCK/p54. These decapping activators are also involved in miRNA-mediated mRNA destabilization (Huntzinger and Izaurralde 2011). For instance, mRNA levels of predicted and validated miRNA targets increase when decapping activators are depleted or when dominant-negative forms are overexpressed (Rehwinkel et al. 2005; Behm-Ansmant et al. 2006; Eulalio et al. 2007a, 2009a; Chen et al. 2009; Piao et al. 2010). However, target protein levels are not fully restored (Eulalio et al. 2007a). This is consistent with the observation that decapping inhibition causes accumulation of deadenylated decay intermediates (because deadenylation precedes decapping), and these deadenylated mRNAs are not translated efficiently, providing one explanation for why protein levels are not rescued.

One question that remains is whether decapping of miRNA targets occurs as a consequence of deadenylation or whether miRISCs can also recruit components of the decapping machinery independent of deadenylation. For example, RCK/p54 coimmunoprecipitates with AGOs in human cells (Chu and Rana 2006), although it is not known whether this interaction is direct. The next question is whether decapping activators are required for silencing. In cell-free extracts, decapping does not occur even when miRNA targets are deadenylated, indicating that decapping per se is not required for the establishment of silencing in vitro. Nonetheless, decapping activators act as general repressors of translation even in the absence of decapping (Nissan et al. 2010); thus, it would be of interest to determine whether they play a more direct role in translational repression of miRNA targets.

GW182 INTERACTION NETWORK: THE ROLE OF REDUNDANT AND COMBINATORIAL INTERACTIONS

The information available on the GW182 interaction network reveals some important features, including redundant and combinatorial interactions, which must be taken into account in considering how GW182 proteins exert their repressive functions.

Redundancy is manifested at three different levels. The first level is because of gene duplication events that generated two to three GW182 paralogs in almost all the animal species investigated thus far. These paralogs are, in addition, present in multiple isoforms. Likewise, many species contain multiple AGO and PABPC paralogs. Moreover, two paralogs of each of the catalytic subunits of the CCR4–NOT complex are expressed in vertebrates (CCR4a and CCR4b and CAF1 and POP2), leading to the assembly of at least four CCR4–NOT complexes. Redundancy at the level of the individual silencing factors enables the assembly of multiple alternate miRISCs, which might be functionally distinct, resulting in different outputs.

Another layer of redundancy is observed in the connections between network components. GW182 proteins interact with deadenylases,

PABPC, EDD, and probably other unknown partners. At the same time, PABPC and deadenylases interact with each other. For instance, PAN3 also contains a PAM2 motif and directly interacts with PABPC (Siddiqui et al. 2007), thereby providing an indirect link between GW182 and PABPC (Braun et al. 2011). Furthermore, it has been suggested that the two deadenylase complexes interact and are part of a larger multiprotein complex in vivo (Chen et al. 2011). Finally, EDD interacts directly with the PAM2 motif of GW182 and indirectly with the decapping activator RCK/p54 and the CCR4–NOT complex (Su et al. 2011). These examples imply that GW182 proteins can recruit deadenylases either directly or indirectly through PABPC or EDD and can interact with PABPC directly or indirectly through PAN3. Again, although these examples indicate high redundancy, the functional outcomes of direct or indirect recruitment of partners might differ.

A third layer of redundancy is observed at the level of GW182 protein sequences, which is consistent with the existence of alternative ways to recruit partners mentioned above. For instance, individual deletion of *Dm* GW182 M1, M2, or C-term regions or of the PAM2 motif does not abolish silencing, but a combination of two or more deletions reduces or abrogates silencing activity in complementation assays (Eulalio et al. 2009c; Huntzinger et al. 2010; Braun et al. 2011; Chekulaeva et al. 2011).

Additionally, other proteins may influence the composition of GW182 complexes in a target-specific manner. For example, the Puf-9 protein facilitates silencing of miRNA targets by different mechanisms (reviewed by Pasquinelli 2012). Puf proteins interact with the CCR4–NOT complex, raising the possibility that for these targets, the interaction of GW182 with the CCR4–NOT complex might become dispensable.

In sum, the assembly of functional miRISC complexes is achieved through redundant and combinatorial interactions among silencing factors. This opens the possibility that depending on the exact combination of these factors and the cellular context (which specifies the relative expression levels of different factors), the func-

tional output might differ, providing a potential explanation for the different modes of miRNA-mediated regulation reported in the literature. Additionally, depending on the relative levels of all factors in a particular cell or tissue, a specific GW182 partner might become essential or dispensable.

GW182 PROTEINS IN PLANTS

In general, plant miRNAs guide AGOs to nearly perfectly complementary targets, which are then cleaved by the endonucleolytic activity of the AGO proteins (Voinnet 2009). However, evidence exists that plant miRNAs can also repress translation (Aukerman et al. 2003; Chen et al. 2004; Gandikota et al. 2007; Brodersen et al. 2008; Dugas and Bartel 2008). Until recently, it has remained unclear how much translational control is exerted by miRNAs over the plant proteome and how this repression is achieved, as plants lack GW182 orthologs. A recent report presents evidence that plants contain a functional GW182 analog, termed SUO (Yang et al. 2012).

SUO is a large protein with an amino-terminal bromo-adjacent homology (BAH) and transcription elongation factor S-II (TFS2N) domains (Yang et al. 2012). The only feature SUO shares with GW182 proteins is the presence of two carboxy-terminal GW repeats. In contrast to animal GW182 proteins, SUO is predominantly nuclear, although a small fraction of the protein localizes to cytoplasmic mRNA-processing bodies (P-bodies), as shown for animal GW182s (Yang et al. 2012). Indeed, in animal cells, GW182 proteins are mainly cytoplasmic but they are also detected in P-bodies, which are cytoplasmic foci where proteins involved in translational repression, mRNA decapping and decay accumulate (Eulalio et al. 2007b). Notably, the functional significance of this localization remains unclear, as GW182 proteins that fail to localize to P-bodies are active in complementation assays (Eulalio et al. 2007c, 2009b).

Evidence that SUO plays a role in miRNA-mediated translational repression is based on the phenotype of *suo* mutants, which is reminiscent of the phenotype observed in plants in which the miRNA pathway is impaired (Yang et al. 2012). Additionally, in *suo*-mutant plants, the expression of miRNA targets that are known to be regulated at the translational level increases without changes in their mRNA levels (Yang et al. 2012). It remains to be seen whether SUO interacts with AGOs or with any of the known GW182 interaction partners (e.g., PABPC and deadenylases) and whether it uses a similar mechanism to repress mRNA targets.

Evidence for similarities in the mechanism of silencing between plants and animals includes the identification of enhancer of decapping-4 (EDC4) in screens for suppressors of miRNA-mediated gene silencing both in *D. melanogaster* cells and *Arabidopsis thaliana* (Eulalio et al. 2007a; Brodersen et al. 2008). However, the molecular details may differ between these two kingdoms; in *D. melanogaster*, EDC4 plays a role in miRNA target degradation through deadenylation and decapping, whereas in plants, EDC4 may play a role in translational repression (Brodersen et al. 2008).

The study by Yang et al. (2012) also provides important information regarding the contribution of miRNA-mediated translational repression to silencing in plants, by showing that putative null alleles of *suo* have a weak phenotype compared to the phenotypes of null alleles of DCL1 (the dicer protein that produces miRNAs in *A. thaliana*). Thus, translational repression may not be essential for miRNA target silencing in plants. However, at present, we cannot exclude the possibility that plant genomes encode additional GW182 analogs (Yang et al. 2012).

MECHANISMS OF TRANSLATIONAL REPRESSION

Earlier studies indicated that animal miRNAs predominantly repress translation, and this repression was proposed to occur in four distinct ways: (1) cotranslational protein degradation (Nottrott et al. 2006); (2) inhibition of translation elongation (Olsen and Ambros 1999; Seggerson et al. 2002; Maroney et al. 2006); (3) premature ribosome dissociation (Petersen et al. 2006); and (4) inhibition of translation initiation (Humphreys et al. 2005; Pillai et al. 2005).

Subsequent studies in different cell types and cell-free extracts of diverse origins supported a role for miRNAs in inhibiting translation initiation (Wang et al. 2006; Mathonnet et al. 2007; Thermann and Hentze 2007; Ding and Grosshans 2009; Iwasaki et al. 2009). More recently, ribosome profiling data and translational rate measurements indicate that translational repression occurs predominantly at initiation (Hendrickson et al. 2009; Guo et al. 2010). However, to date, the precise step that is affected remains elusive.

INTERPLAY BETWEEN TRANSLATIONAL REPRESSION AND DEADENYLATION

An important question that remains unresolved is whether translational repression and deadenylation are linked. Much evidence exists to suggest that deadenylation and subsequent degradation do not require active translation. For example, miRNA-mediated deadenylation and decay can be observed in the presence of translation inhibitors (e.g., cycloheximide or hippuristanol) (Eulalio et al. 2007a; Wakiyama et al. 2007; Fabian et al. 2009). Furthermore, miRNA-target reporters that are poorly translated because of a defective cap structure (ApppG-cap) or the presence of a stable stem-loop structure in the 5′ UTR are nevertheless deadenylated and degraded in an miRNA-dependent manner (Mishima et al. 2006; Wu et al. 2006; Wakiyama et al. 2007; Eulalio et al. 2009a). In fact, an open reading frame is dispensable for miRNA-directed deadenylation as short, ApppG-capped RNAs containing miRNA-binding sites and a poly(A) tail can be efficiently deadenylated in cell-free extracts from mouse Krebs-2 ascites cells (Fabian et al. 2009). Finally, miRISCs directly interact with deadenylases via the GW182 proteins (Braun et al. 2011; Chekulaeva et al. 2011; Fabian et al. 2011). Collectively, these results indicate that miRNAs trigger deadenylation and decay directly and independently of the translation status of the mRNA target. Additionally, the observation that the zebrafish protein DAZL relieves miRNA-mediated repression by counteracting deadenylation, indicates that for some targets deadenylation provides a major contribution to silencing (Takeda et al. 2009).

Conversely, there is compelling evidence that translational repression can occur in the absence of deadenylation. For example, in some cell-free extracts, zebrafish embryos and *Drosophila* cells, translational repression precedes deadenylation (Mathonnet et al. 2007; Thermann and Hentze 2007; Fabian et al. 2009; Zdanowicz et al. 2009; Bazzini et al. 2012; Djuranovic et al. 2012). Another finding suggesting that translational repression occurs in the absence of deadenylation is that miRNA reporters in which the poly(A) tail is replaced by a histone mRNA stem-loop structure or a self-cleavable ribozyme are still repressed by miRNAs (Wu et al. 2006; Eulalio et al. 2008, 2009a). Similarly, mRNA targets containing a blocked poly(A) tail (a poly(A) tail followed by an unrelated sequence that prevents deadenylation) are repressed by miRNAs in the absence of deadenylation (Fukaya and Tomari 2011; Bazzini et al. 2012; Mishima et al. 2012). Finally, in NOT1-depleted cells, some miRNA targets remain translationally repressed even though mRNA degradation is inhibited (Behm-Ansmant et al. 2006; Braun et al. 2011). These examples show the existence of deadenylation-independent mechanisms of translational repression.

In sum, translational repression and deadenylation can occur independently. Consequently, mRNA targets can be silenced at the level of translation or mRNA deadenylation and degradation or by a combination of both. Nevertheless, translational repression and deadenylation can still be connected and might represent two distinct outcomes of a single molecular mechanism that simultaneously interferes with translation and triggers deadenylation. The relative contributions of these two processes to silencing may then be determined by the repertoire of proteins bound to the mRNA target and/or the cellular context. Furthermore, the order of events may depend on the relative rates of inhibition of translation and mRNA deadenylation, which in turn may also be target dependent, resulting in targets silenced predominantly at the translational or mRNA levels.

What might be this initial triggering mechanism? Although completely speculative at this time, an interesting possibility is that the recruitment of deadenylase complexes to the 3′ UTR of miRNA targets triggers both translational repression and deadenylation. This possibility is suggested by the intriguing observation that depletion of PAN3 and NOT1 suppresses the silencing of unadenylated reporters, suggesting that deadenylase complexes could also contribute to translational repression in addition to promoting deadenylation (Braun et al. 2011; Chekulaeva et al. 2011). Furthermore, a catalytically inactive CAF1 mutant induces translational repression of a reporter mRNA to which it is tethered in the absence of deadenylation (Cooke et al. 2010; Chekulaeva et al. 2011). These data suggest that deadenylase complexes could contribute to translational repression independent of their role in deadenylation. Future work should determine whether deadenylase complexes repress translation and by which mechanism(s).

CONCLUDING REMARKS

miRNAs have evolved two divergent ways of promoting target silencing: AGO-mediated endonucleolytic mRNA cleavage (fully or nearly complementary targets) and GW182-mediated translational repression and deadenylation (which may or may not be followed by mRNA destabilization). Despite the significant progress made in dissecting the functions of GW182 proteins in recent years, there is still much to learn about this protein family, including identifying their full repertoire of interacting partners. This information will help clarify whether translational repression and deadenylation represent independent outcomes triggered by GW182 proteins through interactions with distinct protein partners or are induced by a single molecular mechanism. In this context, the study of the more divergent members of the GW182 protein family (e.g., *C. elegans* AIN-1 and AIN-2 or *A. thaliana* SUO) is likely to lead to a much deeper understanding of these issues in the coming years. It is also not known what types of regulation GW182 protein activity is subjected to and whether RNA-binding proteins bound to specific targets modulate their functions.

Another important task for future research in the miRNA field will be to understand how translational repression is achieved. We expect that answers to this question will emerge as more studies examine the molecular structures and functions of silencing factors and how they interact to assemble into active effector complexes.

In summary, in recent years significant advances in our understanding of the mechanism of silencing have been made, both in animals and plants. Although the molecular details remain to be elucidated, new findings revealed an unanticipated direct connection to the CCR4–NOT complex, which is a master posttranscriptional regulator in eukaryotic cells. Investigating how this complex interacts with the silencing machinery promises to move the field forward and will be an exciting area for future studies.

ACKNOWLEDGMENTS

The research from this laboratory is supported by the Max Planck Society, by grants from the Deutsche Forschungsgemeinschaft (DFG, FOR855, and the Gottfried Wilhelm Leibniz Program awarded to E.I.).

REFERENCES

Aukerman MJ, Sakai H. 2003. Regulation of flowering time and floral organ identity by a MicroRNA and its APE-TALA2-like target genes. *Plant Cell* **15**: 2730–2741.

Bagga S, Bracht J, Hunter S, Massirer K, Holtz J, Eachus R, Pasquinelli AE. 2005. Regulation by let-7 and lin-4 miRNAs results in target mRNA degradation. *Cell* **122**: 553–563.

Bartel DP. 2009. MicroRNAs: Target recognition and regulatory functions. *Cell* **136**: 215–233.

Bazzini AA, Lee MT, Giraldez AJ. 2012. Ribosome profiling shows that miR-430 reduces translation before causing mRNA decay in zebrafish. *Science* **336**: 233–237.

Behm-Ansmant I, Rehwinkel J, Doerks T, Stark A, Bork P, Izaurralde E. 2006. mRNA degradation by miRNAs and GW182 requires both CCR4:NOT deadenylase and DCP1:DCP2 decapping complexes. *Genes Dev* **20**: 1885–1898.

Braun JE, Huntzinger E, Fauser M, Izaurralde E. 2011. GW182 proteins recruit cytoplasmic deadenylase complexes to miRNA targets. *Mol Cell* **44**: 120–133.

Brodersen P, Sakvarelidze-Achard L, Bruun-Rasmussen M, Dunoyer P, Yamamoto YY, Sieburth L, Voinnet O. 2008. Widespread translational inhibition by plant miRNAs and siRNAs. *Science* **320**: 1185–1190.

Chekulaeva M, Filipowicz W, Parker R. 2009. Multiple independent domains of dGW182 function in miRNA-mediated repression in *Drosophila*. *RNA* **15**: 794–803.

Chekulaeva M, Mathys H, Zipprich JT, Attig J, Colic M, Parker R, Filipowicz W. 2011. miRNA repression involves GW182-mediated recruitment of CCR4-NOT through conserved W-containing motifs. *Nat Struct Mol Biol* **18**: 1218–1226.

Chen X. 2004. A microRNA as a translational repressor of APETALA2 in *Arabidopsis* flower development. *Science* **303**: 2022–2025.

Chen CY, Shyu AB. 2011. Mechanisms of deadenylation-dependent decay. *RNA* **2**: 167–183.

Chen CY, Zheng D, Xia Z, Shyu AB. 2009. Ago-TNRC6 triggers microRNA-mediated decay by promoting two deadenylation steps. *Nat Struct Mol Biol* **16**: 1160–1166.

Chu CY, Rana TM. 2006. Translation repression in human cells by microRNA-induced gene silencing requires RCK/p54. *PLoS Biol* **4**: e210.

Cooke A, Prigge A, Wickens M. 2010. Translational repression by deadenylases. *J Biol Chem* **285**: 28506–28513.

Davey NE, Van Roey K, Weatheritt RJ, Toedt G, Uyar B, Altenberg B, Budd A, Diella F, Dinkel H, Gibson TJ. 2012. Attributes of shirt linear motifs. *Mol Biosyst* **8**: 268–281.

Derry MC, Yanagiya A, Martineau Y, Sonenberg N. 2006. Regulation of poly(A)-binding protein through PABP-interacting proteins. *Cold Spring Harb Symp Quant Biol* **71**: 537–543.

Ding XC, Großhans H. 2009. Repression of *C. elegans* microRNA targets at the initiation level of translation requires GW182 proteins. *EMBO J* **28**: 213–222.

Ding L, Spencer A, Morita K, Han M. 2005. The developmental timing regulator AIN-1 interacts with miRISCs and may target the argonaute protein ALG-1 to cytoplasmic P bodies in *C. elegans*. *Mol Cell* **19**: 437–447.

Djuranovic S, Nahvi A, Green R. 2011. A parsimonious model for gene regulation by miRNAs. *Science* **331**: 550–553.

Djuranovic S, Nahvi A, Green A. 2012. miRNA-mediated gene silencing by translational repression followed by mRNA deadenylation and decay. *Science* 336: 237–240.

Dugas DV, Bartel B. 2008. Sucrose induction of *Arabidopsis* miR398 represses two Cu/Zn superoxide dismutases. *Plant Mol Biol* **67**: 403–417.

Esteller M. 2011. Non-coding RNAs in human disease. *Nat Rev Genet* **12**: 861–874.

Eulalio A, Rehwinkel J, Stricker M, Huntzinger E, Yang SF, Doerks T, Dorner S, Bork P, Boutros M, Izaurralde E. 2007a. Target-specific requirements for enhancers of decapping in miRNA-mediated gene silencing. *Genes Dev* **21**: 2558–2570.

Eulalio A, Behm-Ansmant I, Izaurralde E. 2007b. P bodies: at the crossroads of post-transcriptional pathways. *Nat Rev Mol Cell Biol* **8**: 9–22.

Eulalio A, Behm-Ansmant I, Schweizer D, Izaurralde E. 2007c. P-body formation is a consequence, not the cause of RNA-mediated gene silencing. *Mol Cell Biol* **27:** 3970–3981.

Eulalio A, Huntzinger E, Izaurralde E. 2008. GW182 interaction with Argonaute is essential for miRNA-mediated translational repression and mRNA decay. *Nat Struct Mol Biol* **15:** 346–353.

Eulalio A, Huntzinger E, Nishihara T, Rehwinkel J, Fauser M, Izaurralde E. 2009a. Deadenylation is a widespread effect of miRNA regulation. *RNA* **15:** 21–32.

Eulalio A, Tritschler F, Izaurralde E. 2009b. The GW182 protein family in animal cells: New insights into domains required for miRNA mediated gene silencing. *RNA* **15:**1433–1442.

Eulalio A, Helms S, Fritzsch C, Fauser M, Izaurralde E. 2009c. A C-terminal silencing domain in GW182 is essential for miRNA function. *RNA* **15:** 1067–1077.

Eystathioy T, Chan EK, Tenenbaum SA, Keene JD, Griffith K, Fritzler MJ. 2002. A phosphorylated cytoplasmic autoantigen, GW182, associates with a unique population of human mRNAs within novel cytoplasmic speckles. *Mol Biol Cell* **13:** 1338–1351.

Fabian MR, Mathonnet G, Sundermeier T, Mathys H, Zipprich JT, Svitkin YV, Rivas F, Jinek M, Wohlschlegel J, Doudna JA, et al. 2009. Mammalian miRNA RISC recruits CAF1 and PABP to affect PABP-dependent deadenylation. *Mol Cell* **35:** 868–880.

Fabian MR, Cieplak MK, Frank F, Morita M, Green J, Srikumar T, Nagar B, Yamamoto T, Raught B, Duchaine TF, et al. 2011. miRNA-mediated deadenylation is orchestrated by GW182 through two conserved motifs that interact with CCR4–NOT. *Nat Struct Mol Biol* **18:** 1211–1217.

Fukaya T, Tomari Y. 2011. PABP is not essential for microRNA-mediated translational repression and deadenylation in vitro. *EMBO J* **30:** 4998–5009.

Gandikota M, Birkenbihl RP, Höhmann S, Cardon GH, Saedler H, Huijser P. 2007. The miRNA156/157 recognition element in the 3′ UTR of the *Arabidopsis* SBP box gene SPL3 prevents early flowering by translational inhibition in seedlings. *Plant J* **49:** 683–693.

Giraldez AJ, Mishima Y, Rihel J, Grocock RJ, Van Dongen S, Inoue K, Enright AJ, Schier AF. 2006. Zebrafish MiR-430 promotes deadenylation and clearance of maternal mRNAs. *Science* **312:** 75–79.

Guo H, Ingolia NT, Weissman JS, Bartel DP. 2010. Mammalian microRNAs predominantly act to decrease target mRNA levels. *Nature* **466:** 835–840.

Hendrickson DG, Hogan DJ, McCullough HL, Myers JW, Herschlag D, Ferrell JE, Brown PO. 2009. Concordant regulation of translation and mRNA abundance for hundreds of targets of a human microRNA. *PLoS Biol* **7:** e1000238.

Humphreys DT, Westman BJ, Martin DI, Preiss T. 2005. MicroRNAs control translation initiation by inhibiting eukaryotic initiation factor 4E/cap and poly(A) tail function. *Proc Natl Acad Sci* **102:** 16961–16966.

Huntzinger E, Izaurralde E. 2011. Gene silencing by microRNAs: Contributions of translational repression and mRNA decay. *Nat Rev Genet* **12:** 99–110.

Huntzinger E, Braun JE, Heimstädt S, Zekri L, Izaurralde E. 2010. Two PABPC-binding sites in GW182 proteins promote miRNA-mediated gene silencing. *EMBO J* **29:** 4146–4160.

Iwasaki S, Kawamata T, Tomari Y. 2009. *Drosophila* argonaute1 and argonaute2 employ distinct mechanisms for translational repression. *Mol Cell* **34:** 58–67.

Jakymiw A, Lian S, Eystathioy T, Li S, Satoh M, Hamel JC, Fritzler MJ, Chan EK. 2005. Disruption of GW bodies impairs mammalian RNA interference. *Nat Cell Biol* **7:** 1267–1274.

Jínek M, Doudna JA. 2009. A three-dimensional view of the molecular machinery of RNA interference. *Nature* **457:** 405–412.

Jínek M, Fabian MR, Coyle SM, Sonenberg N, Doudna JA. 2010. Structural insights into the human GW182-PABC interaction in microRNA-mediated deadenylation. *Nat Struct Mol Biol* **17:** 238–240.

Kozlov G, Safaee N, Rosenauer A, Gehring K. 2010. Structural basis of binding of P-body associated protein GW182 and Ataxin-2 by the MLLE domain of poly(A)-binding protein. *J Biol Chem* **285:** 13599–13606.

Krol J, Loedige I, Filipowicz W. 2010. The widespread regulation of microRNA biogenesis, function and decay. *Nature Rev Genet* **11:** 597–610.

Kuzuoglu-Öztürk D, Huntzinger E, Schmidt S, Izaurralde E. 2012. The *Caenorhabditis elegans* GW182 protein AIN-1 interacts with PAB-1 and subunits of the PAN2-PAN3 and CCR4-NOT deadenylase complexes. *Nucleic Acids Res* doi: 10.1093/nar/gks218.

Lazzaretti D, Tournier I, Izaurralde E. 2009. The C-terminal domains of human TNRC6A, B and C silence bound transcripts independently of the Argonaute proteins. *RNA* **15:** 1059–1066.

Lian SL, Abadal GX, Pauley BA, Fritzler MJ, Chan EKL. 2009. The C-terminal half of human Ago2 binds to multiple GW-rich regions of GW182 and requires GW182 to mediate silencing. *RNA* **15:** 804–813.

Lim LP, Lau NC, Garrett-Engele P, Grimson A, Schelter JM, Castle J, Bartel DP, Linsley PS, Johnson JM. 2005. Microarray analysis shows that some microRNAs downregulate large numbers of target mRNAs. *Nature* **433:** 769–773.

Liu J, Rivas FV, Wohlschlegel J, Yates JR III, Parker R, Hannon GJ. 2005. A role for the P-body component GW182 in microRNA function. *Nat Cell Biol* **7:** 1261–1266.

Maroney PA, Yu Y, Fisher J, Nilsen TW. 2006. Evidence that microRNAs are associated with translating messenger RNAs in human cells. *Nat Struct Mol Biol* **13:** 1102–1107.

Mathonnet G, Fabian MR, Svitkin YV, Parsyan A, Huck L, Murata T, Biffo S, Merrick WC, Darzynkiewicz E, Pillai RS, et al. 2007. MicroRNA inhibition of translation initiation in vitro by targeting the cap-binding complex eIF4F. *Science* **317:** 1764–1767.

Meister G, Landthaler M, Peters L, Chen PY, Urlaub H, Luhrmann R, Tuschl T. 2005. Identification of novel argonaute-associated proteins. *Curr Biol* **15:** 2149–2155.

Mishima Y, Giraldez AJ, Takeda Y, Fujiwara T, Sakamoto H, Schier AF, Inoue K. 2006. Differential regulation of germline mRNAs in soma and germ cells by zebrafish miR-430. *Curr Biol* **16:** 2135–2142.

Mishima Y, Fukao A, Kishimoto T, Sakamoto H, Fujiwara T, Inoue K. 2012. Translational inhibition by deadenylation-independent mechanisms is central to microRNA-mediated silencing in zebrafish. *Proc Natl Acad Sci* doi: 10.1073/pnas.1113350109.

Nissan T, Rajyaguru P, She M, Song H, Parker R. 2010. Decapping activators in *Saccharomyces cerevisiae* act by multiple mechanisms. *Mol Cell* **39**: 773–783.

Nottrott S, Simard MJ, Richter JD. 2006. Human let-7a miRNA blocks protein production on actively translating polyribosomes. *Nature Struct Mol Biol* **13**: 1108–1114.

Olsen PH, Ambros V. 1999. The lin-4 regulatory RNA controls developmental timing in *Caenorhabditis elegans* by blocking LIN-14 protein synthesis after the initiation of translation. *Dev Biol* **216**: 671–680.

Pasquinelli AE. 2012. A team effort blocks the ribosome in its tracks. *Nature Struct Mol Biol* **19**: 133–134.

Petersen CP, Bordeleau ME, Pelletier J, Sharp PA. 2006. Short RNAs supress translation after initiation in mammalian cells. *Mol. Cell* **21**: 533–542.

Piao X, Zhang X, Wu L, Belasco JG. 2010. CCR4-NOT deadenylates mRNA associated with RNA-induced silencing complexes in human cells. *Mol Cell Biol* **30**: 1486–1494.

Pillai RS, Bhattacharyya SN, Artus CG, Zoller T, Cougot N, Basyuk E, Bertrand E, Filipowicz W. 2005. Inhibition of translational initiation by Let-7 MicroRNA in human cells. *Science* **309**: 1573–1576.

Rehwinkel J, Behm-Ansmant I, Gatfield D, Izaurralde E. 2005. A crucial role for GW182 and the DCP1:DCP2 decapping complex in miRNA-mediated gene silencing. *RNA* **11**: 1640–1647.

Rehwinkel J, Natalin P, Stark A, Brennecke J, Cohen SM, Izaurralde E. 2006. Genome-wide analysis of mRNAs regulated by Drosha and Argonaute proteins in *Drosophila melanogaster*. *Mol Cell Biol* **26**: 2965–2975.

Sayed D, Abdellatif M. 2011. MicroRNAs in development and disease. *Physiol Rev* **91**: 827–887.

Schmitter D, Filkowski J, Sewer A, Pillai RS, Oakeley EJ, Zavolan M, Svoboda P, Filipowicz W. 2006. Effects of Dicer and Argonaute down-regulation on mRNA levels in human HEK293 cells. *Nucleic Acids Res* **34**: 4801–4815.

Seggerson K, Tang L, Moss EG. 2002. Two genetic circuits repress the *Caenorhabditis elegans* heterochronic gene lin-28 after translation initiation. *Dev Biol* **243**: 215–225.

Siddiqui N, Mangus DA, Chang TC, Palermino JM, Shyu AB, Gehring K. 2007. Poly(A) nuclease interacts with the C-terminal domain of polyadenylate-binding protein domain from poly(A)-binding protein. *J Biol Chem* **282**: 25067–25075.

Su H, Meng S, Lu Y, Trombly MI, Chen J, Lin C, Turk A, Wang X. 2011. Mammalian hyperplastic discs homolog EDD regulates miRNA-mediated gene silencing. *Mol Cell* **43**: 97–109.

Takeda Y, Mishima Y, Fujiwara T, Sakamoto H, Inoue K. 2009. DAZL relieves miRNA-mediated repression of germline mRNAs by controlling poly(A) tail length in zebrafish. *PLoS ONE* **4**: e7513.

Takimoto K, Wakiyama M, Yokoyama S. 2009. Mammalian GW182 contains multiple Argonaute binding sites and functions in microRNA-mediated translational repression. *RNA* **15**: 1078–1089.

Thermann R, Hentze MW. 2007. Drosophila miR2 induces pseudo-polysomes and inhibits translation initiation. *Nature* **447**: 875–858.

Till S, Lejeune E, Thermann R, Bortfeld M, Hothorn M, Enderle D, Heinrich C, Hentze MW, Ladurner AG. 2007. A conserved motif in Argonaute-interacting proteins mediates functional interactions through the Argonaute PIWI domain. *Nat Struct Mol Biol* **14**: 897–903.

Voinnet O. 2009. Origin, biogenesis, and activity of plant microRNA. *Cell* **136**: 669–687.

Walters RW, Bradrick SS, Gromeier M. 2010. Poly(A)-binding protein modulates mRNA susceptibility to cap-dependent miRNA-mediated repression. *RNA* **16**: 239–250.

Wang B, Love TM, Call ME, Doench JG, Novina CD. 2006. Recapitulation of short RNA-directed translational gene silencing in vitro. *Mol Cell* **22**: 553–560.

Wakiyama M, Takimoto K, Ohara O, Yokoyama S. 2007. Let-7 microRNA-mediated mRNA deadenylation and translational repression in a mammalian cell-free system. *Genes Dev* **21**: 1857–1862.

Wu L, Belasco JG. 2005. Micro-RNA regulation of the mammalian lin-28 gene during neuronal differentiation of embryonal carcinoma cells. *Mol Cell Biol* **25**: 9198–9208.

Wu L, Fan J, Belasco JG. 2006. MicroRNAs direct rapid deadenylation of mRNA. *Proc Natl Acad Sci* **103**: 4034–4039.

Wu E, Thivierge C, Flamand M, Mathonnet G, Vashisht AA, Wohlschlegel J, Fabian MR, Sonenberg N, Duchaine TF. 2010. Pervasive and cooperative deadenylation of 3′UTRs by embryonic microRNA families. *Mol Cell* **40**: 558–570.

Yang L, Wu G, Poethig RS. 2012. Mutations in the GW-repeat protein SUO reveal a developmental function for microRNA-mediated translational repression in *Arabidopsis*. *Proc Natl Acad Sci* **109**: 315–320.

Yao B, Li S, Jung HM, Lian SL, Abadal GX, Han F, Fritzler MJ, Chan EK. 2011. Divergent GW182 functional domains in the regulation of translational silencing. *Nucleic Acids Res* **39**: 2534–2547.

Zdanowicz A, Thermann R, Kowalska J, Jemielity J, Duncan K, Preiss T, Darzynkiewicz E, Hentze MW. 2009. *Drosophila* miR2 primarily targets the m7GpppN cap structure for translational repression. *Mol Cell* **35**: 881–888.

Zekri L, Huntzinger E, Heimstädt S, Izaurralde E. 2009. The silencing domain of GW182 interacts with PABPC1 to promote translational repression and degradation of miRNA targets and is required for target release. *Mol Cell Biol* **29**: 6220–6231.

Zhang L, Ding L, Cheung TH, Dong M-Q, Che J, Sewell AK, Liu X, Yates JR, Han M. 2007. Systematic identification of *C. elegans* miRISC proteins, miRNAs, and mRNA targets by their interactions with GW182 proteins AIN-1 and AIN-2. *Mol Cell* **28**: 598–613.

Zipprich JT, Bhattacharyya S, Mathys H, Filipowicz W. 2009. Importance of the C-terminal domain of the human GW182 protein TNRC6C for translational repression. *RNA* **15**: 781–793.

Translational Control in Cancer Etiology

Davide Ruggero

Helen Diller Cancer Center, School of Medicine, University of California, San Francisco, California 94158

Correspondence: davide.ruggero@ucsf.edu

The link between perturbations in translational control and cancer etiology is becoming a primary focus in cancer research. It has now been established that genetic alterations in several components of the translational apparatus underlie spontaneous cancers as well as an entire class of inherited syndromes known as "ribosomopathies" associated with increased cancer susceptibility. These discoveries have illuminated the importance of deregulations in translational control to very specific cellular processes that contribute to cancer etiology. In addition, a growing body of evidence supports the view that deregulation of translational control is a common mechanism by which diverse oncogenic pathways promote cellular transformation and tumor development. Indeed, activation of these key oncogenic pathways induces rapid and dramatic translational reprogramming both by increasing overall protein synthesis and by modulating specific mRNA networks. These translational changes promote cellular transformation, impacting almost every phase of tumor development. This paradigm represents a new frontier in the multihit model of cancer formation and offers significant promise for innovative cancer therapies. Current research, in conjunction with cutting edge technologies, will further enable us to explore novel mechanisms of translational control, functionally identify translationally controlled mRNA groups, and unravel their impact on cellular transformation and tumorigenesis.

Given the fact that translation is the ultimate step for producing a functional protein, it is surprising how much the cancer biology community has historically overlooked the importance of its deregulation toward cancer development. Indeed, decades of research into the molecular programs that govern cellular transformation have mainly focused on the cancer transcriptome (van 't Veer et al. 2002; Hawkins and Ren 2006). The microarray era enabled the research community to catalog genome-wide variations in the repertoire of transcriptional outputs downstream of specific oncogenic signaling pathways. Perhaps the reluctance in hypothesizing that changes in translational control may coordinate major events underlying cancer formation is because of the misconception that specificity in gene expression is primarily due to transcription regulation. An exciting body of research now shows that changes in mRNA translation control distinct cellular processes including metabolism, cell migration, cell adhesion, cell growth, cell-cycle control, and tumorigenesis (Silvera et al. 2010). These studies have been important in revealing not only novel general mechanisms of translational control, but also new paradigms for cellular transformation and cancer development (Fig. 1). For

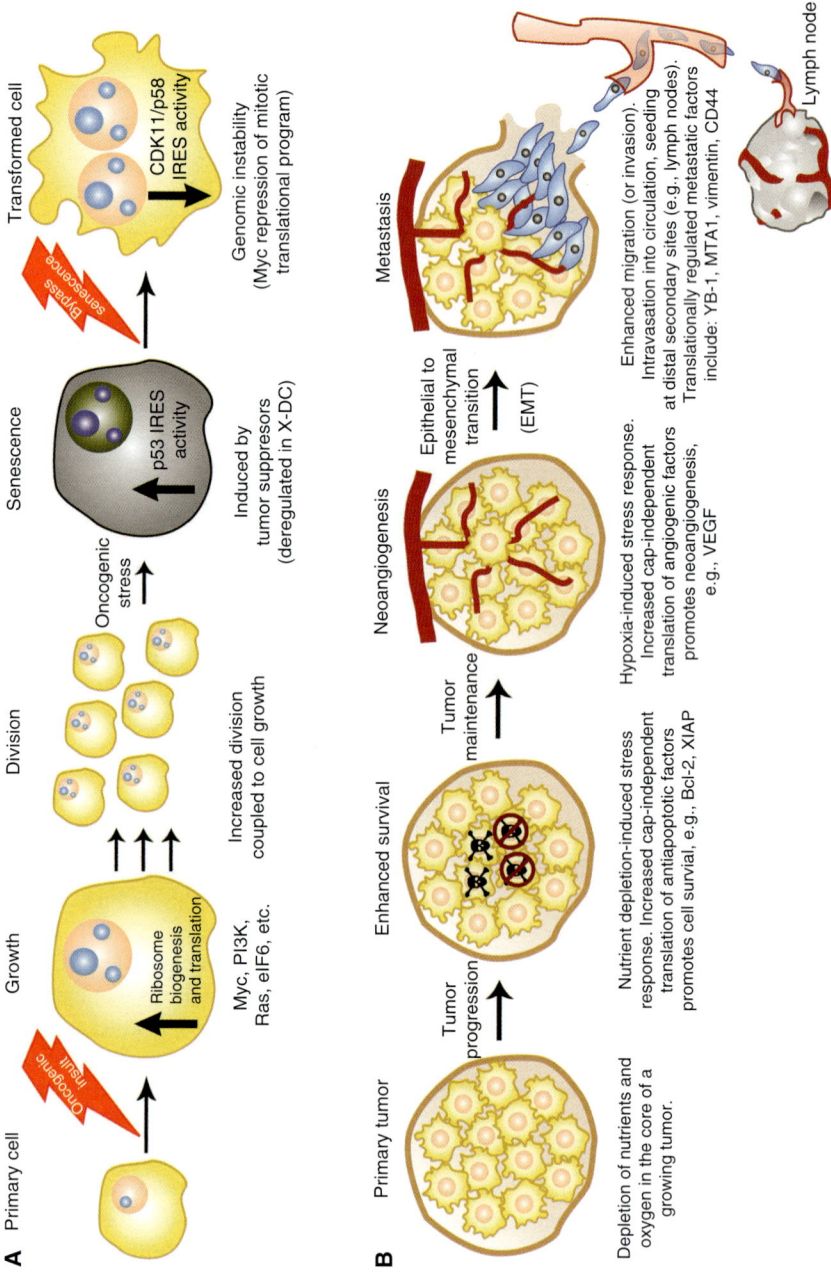

Figure 1. Deregulations in translational control contribute to each step of cellular transformation and tumor progression. (A) Upon receiving an oncogenic insult (lightning bolt) such as Myc overexpression or PI3K hyperactivation, ribosome biogenesis and global protein synthesis are augmented resulting in increased cell size coupled to cell division. Upon this oncogenic stress, cells initiate a tumor-suppressive response associated with increased internal ribosome entry site (IRES)-mediated translation, leading to cell-cycle arrest and senescence. To overcome the barrier of oncogene-induced senescence (OIS), cells acquire additional mutations known as secondary hits (lightning bolt). (*Legend continues on facing page.*)

example, key oncogenic pathways such as Myc and PI3K monopolize the levels and activity of specific components of the translational machinery (e.g., initiation factors, ribosomal proteins) to direct specific posttranscriptional changes in gene expression directly at the level of protein production (Hannan et al. 2011). In addition, an entire class of inherited syndromes collectively referred to as "ribosomopathies" harbor mutations in distinct ribosome components and is characterized by increased cancer susceptibility (Ganapathi and Shimamura 2008). The realization that there are posttranscriptional control mechanisms in cancer development has provided a strong rationale in the design of new cancer therapies that may ultimately eradicate the abnormal translational program of cancer cells. I provide here an analysis and perspective on how the regulation of protein synthesis plays a critical and pivotal role in the etiology of cancer.

THE CANCEROUS TRANSLATION MACHINERY

Initiation Factors (eIFs)

Genetic studies of human cancers have made a compelling case for the role of deregulations in translational control within the multihit model of cancer etiology. Genes encoding initiation factors are aberrantly expressed during cancer progression (Table 1). Most importantly, clinical research reveals that these genes are gained or lost in human tumors, consistent with a possible causal role in cancer etiology (Table 1). However, it will be critical to determine to what extent these genetic changes in eIFs directly contribute to cancer development.

Initiation is one of the most regulated steps in translation (Sonenberg and Hinnebusch 2009). During initiation, many checkpoints are coordinated by several initiation factors to control not only whether a specific mRNA is translated but also the rate at which mRNA translation occurs. This process ultimately contributes to the overall abundance of the specific protein in the cell. Genes encoding initiation factors (e.g., eIF3, eIF4G, eIF4E, eIF5A2) are amplified in a variety of human cancers and regulate specific steps of translation initiation including the ribosome–mRNA interaction and initiator Met-tRNAi binding (Table 1). For example, the eIF3 subunit eIF3h is amplified along with the Myc oncogene in breast cancer (Nupponen et al. 1999). Importantly, high-level amplification of eIF3h is found in advanced stage, androgen-independent, and poor-prognosis prostate cancer, but no copy number changes are found in nonmalignant (benign prostatic hyperplasia) or premalignant (prostatic intraepithelial neoplasia) prostates (Saramaki et al. 2001). Intriguingly, loss of heterozygosity and gene copy number analyses showed that loss of the gene encoding eIF3f, another subunit of the eIF3 initiation complex, occurs in pancreatic cancer as well as

Figure 1. (*Continued*) One such mechanism to overcome tumor-suppressive checkpoints such as OIS is the promotion of genome instability by decreasing IRES-dependent translation of the CDK11/p58 tumor suppressor. (*B*) Once established, a primary tumor will undergo unrestrained growth, which triggers a stress response including depletion of oxygen and essential nutrients from the core of the tumor. Lack of key nutrients such as growth factors often leads to increased apoptosis (skulls). Tumor cells up-regulate cap-independent translation of antiapoptotic factors (e.g., Bcl-2 and XIAP) as a mechanism to promote survival (blocked skulls). To bypass stress caused by low levels of oxygen in the tumor, cells induce cap-independent translation of neoangiogenesis promoting factors such as vascular endothelial growth factor (VEGF). Primary tumors metastasize to secondary organs. Epithelial-to-mesenchymal transition (EMT) facilitates metastatic dissemination; specifically, cancer cells lose their epithelial characteristics (yellow cells) and acquire mesenchymal traits (blue cells) including enhanced mobility and invasiveness. Multiple stages of metastatic cell dissemination include degradation of the basement membrane, intravasation into the circulatory system, and extravasation at the distal site. eIF4E regulates the translation of an invasive messenger RNA (mRNA) signature (i.e., YB-1, MTA1, vimentin, and CD44) that promotes metastasis formation. Increased translation of prometastatic factors localized at the leading edge of the tumor's invasive front may facilitate metastasis colonization.

Table 1. Genetic alterations in translational components are associated with increased cancer susceptibility

	Gene	Genetic lesion	Expression/ activity	Cancer association	Reference(s)
Ribosomopathies					
X-linked dyskeratosis congenita	DKC1	Mutation/ deletion	Decrease	Increased incidence of hematological and solid tumors	Heiss et al. 1998
Diamond-Blackfan anemia	RPS 7, 10, 15, 17, 19, 24, 26, 27A RPL 5, 11, 35A, 36	Mutation/ deletion	Decrease	Increased incidence of hematological and solid tumors	Draptchinskaia et al. 1999; Gazda et al. 2006, 2008; Cmejla et al. 2007, 2009; Farrar et al. 2008; Doherty et al. 2010
5q deletion syndrome	RPS14	Deletion	Decrease	Increased incidence of hematological tumors	Ebert et al. 2008
Shwachman-Diamond syndrome	SDBS	Mutation	Decrease	Increased incidence of hematological tumors	Boocock et al. 2003; Austin et al. 2005
Cartilage-hair hypoplasia	RMRP	Mutation	Decrease	Increased incidence of hematological and solid tumors	Ridanpaa et al. 2001
Translation initiation factors					
	eIF2a	ND	Increase	In hematological and solid tumors	Wang et al. 1999; Rosenwald et al. 2001
	eIF3a	ND	Increase	In solid tumors	Bachmann et al. 1997; Dellas et al. 1998
	eIF3c	ND	Increase	In solid tumors	Rothe et al. 2000
	eIF3e	ND	Increase	In solid tumors	Marchetti et al. 2001
	eIF3f	LOH	Decrease	In solid tumors	Shi et al. 2006; Doldan et al. 2008a,b
	eIF3h	Amplification	Increase	In solid tumors	Saramaki et al. 2001
	eIF3i	ND	Increase	In solid tumors	Rauch et al. 2004; Ahlemann et al. 2006
	eIF4A	ND	Increase	In solid tumors	Eberle et al. 1997; Shuda et al. 2000
	eIF4E	Amplification	Increase	In hematological and solid tumors	De Benedetti and Graff 2004
	eIF4G	Amplification	Increase	In solid tumors	Brass et al. 1997; Bauer et al. 2002
	eIF5A	ND	Increase	In hematological tumors	Balabanov et al. 2007
	eIF5A2	Amplification	Increase	In solid tumors	Guan et al. 2001
	eIF6	ND	Increase	In hematological and solid tumors	Sanvito et al. 2000; Harris et al. 2004
Translation elongation factors					
	eEF2	ND	Increase	In solid tumors	Nakamura et al. 2009
	eEF1A2	Amplification	Increase	In solid tumors	Anand et al. 2002; Lee 2003
IRES *trans*-acting factors					
	hnRNP A1	ND	Increase	In solid tumors	Ma et al. 2009

Continued

Cite this article as *Cold Spring Harb Perspect Biol* doi: 10.1101/cshperspect.a012336

Table 1. *Continued*

Gene	Genetic lesion	Expression/ activity	Cancer association	Reference(s)
hnRNP E1	ND	Decrease	In solid tumors	Wang et al. 2010; Zhang et al. 2010
hnRNP E2	ND	Decrease	In solid tumors	Roychoudhury et al. 2007
La	ND	Increase	In solid tumors	Trotta et al. 2003
PTB	ND	Increase	In solid tumors	Jin et al. 2000
YB-1	ND	Increase	Correlates with higher grade tumors	Stratford et al. 2007
Ribosome biogenesis factors				
Ubf1	ND	Increase	In solid tumors	Huang et al. 2002
Bop1	Amplification	Increase	In solid tumors	Killian et al. 2006
Npm	Mutation	Decrease	In hematological tumors	Yun et al. 2007

Abbreviations: RPS, ribosomal protein small subunit; RPL, ribosomal protein large subunit; ND, not determined; PTB, polypyrimidine tract-binding protein; IRES, internal ribosome entry site; LOH, loss of heterozygosity.

in melanoma (Doldan et al. 2008a,b). These data suggest that discrete subunits of the same initiation complex (eIF3) could differentially modulate protein synthesis (see below), acting either as oncogenes or tumor suppressors. The genetic loci of eIF4E and eIF4G, two key initiation factors involved in cap-dependent translation, were originally found amplified in breast cancer and squamous cell lung carcinoma (Sorrells et al. 1998; Bauer et al. 2001) (see section below). Furthermore, gene amplification and overexpression of eIF5A2, which enhances the formation of the first peptide bond, has been reported in ovarian cancer as well as in other solid tumors including breast cancer, hepatocellular carcinoma, and nonsmall cell lung cancer (Guan et al. 2004; Clement et al. 2006; Tang et al. 2010; He et al. 2011). The genetic lesions of the initiation factors described above have been shown to affect the overall expression levels of their mRNA and/or protein. In addition, many other components of the translation machinery involved in the initiation step are also misexpressed in a variety of hematological malignancies and solid tumors (see Table 1), although the genetic bases for some of these abnormalities are unknown.

What is the evidence that these changes exert direct and causal effects in cancer? Proving causality is challenging. One approach is to define the mechanism by which alterations in

these initiation factors modulate gene expression at the translational level, ultimately contributing to cellular transformation and cancer development. The challenge in unraveling this mechanism resides in the fact that overexpression of some of these initiation factors modulates both global protein synthesis and translation efficiency of specific mRNAs. For example, in NIH 3T3 cells, overexpression of five of the 12 individual subunits that comprise the eIF3 initiation complex (eIF3a, eIF3b, eIF3c, eIF3h, and eIF3i) results in a modest increase in general protein synthesis rate (Zhang et al. 2007). In the same studies, the investigators reported that overexpression of eIF3c or eIF3h results in a shift of cyclin D1, Myc, ODC, and FGF2 mRNAs into larger polysomes, suggesting that overexpression of a transforming eIF3 subunit can also affect the translational efficiency of specific oncogenic mRNAs. Furthermore, in inflammatory breast cancer (IBC), a rare but lethal form of the disease, overexpression of eIF4G has also been shown to specifically control the internal ribosome entry site (IRES)-dependent translation (see below) of two oncogenic mRNAs (VEGF and p120 catenin) important for IBC tumor survival and dissemination (Silvera et al. 2009).

Overall, it remains uncertain whether misexpression of eIFs causes changes in overall protein synthesis (quantitative), or alters the expression of specific mRNAs (qualitative)—distinct outcomes that could both contribute to key steps in cancer formation. Based on the role attributed to these general initiation factors, it stands to reason that increases in their activity would in turn augment global protein synthesis. I will discuss below how this quantitative increase might impact cell growth during cellular transformation, representing the first "hit" for cancer formation. On the other hand, up-regulating the rate of translation initiation through overexpression of these eIFs may also qualitatively enhance the translation efficiency of the poorly translated mRNAs described above. Importantly, many of these mRNAs encode key proteins that critically impact tumorigenesis by modulating the cell cycle and survival. However, the causal contribution of quantitative and qualitative changes in translational control toward cancer development still needs to be shown in a more physiological context. To this end, future studies should use inducible tissue-specific mouse models that harbor the gain or loss of function mutation for these specific initiation factors. These in vivo approaches would be invaluable for discriminating whether the genomic alterations of these translation factors found in human cancers represent driver mutations in tumor initiation or rather contribute to progression and/or metastatic stages of the disease. It is also very important to determine whether the putative target mRNAs of eIF3, eIF4E, eIF4G, and/or eIF5A2 play a functional role during tumor development.

Although many questions still need to be addressed, it is nevertheless an exciting moment in the field of translation and cancer. It is clear that even initiation factors widely assumed to play only a housekeeping function likely show activity that would impinge on specific gene expression programs underlying distinct stages of cancer initiation and/or progression. There are now new technologies such as ribosome profiling (see below) that may shed light on important questions currently pursued in the field and facilitate the examination of the translational landscape of the cancer genome in a tissue- and cell-specific manner.

Do Mutations in the Ribosome Cause Cancer?

One of the most surprising recent discoveries in cancer biology has been the realization that mutations in ribosomal components are directly linked to tumorigenesis. Elucidating the mechanistic basis of this novel relationship has proven to be a conceptual challenge. Even a decade ago, it would have been inconceivable that impaired ribosomal activity would alter specific cellular processes, as it was a commonly held belief that perturbations in the ribosome inevitably would trigger cell death. Human genetic research has now identified ribosomopathies, which are characterized by increased cancer susceptibility. These syndromes harbor mutations in distinct ribosomal components (Table 1). One of the most interesting aspects of these ribosomopathies is the high degree of cell and tissue specificity in the presentation of their pathological features. These findings challenge the dogma that the activity and composition of ribosomes are identical in each cell and tissue in an organism—in other words, that the ribosome is just a static factory to produce proteins and not a regulatory element of the translational apparatus. The discovery of ribosomopathies as well as recent findings in the developmental biology field (see below) strongly suggests that we should reevaluate our perspective on the ribosome and its role in translational control.

One clear example of how defects in the ribosome may alter the translational landscape and contribute to specific disease pathologies and cancer susceptibility is the ribosomopathy X-linked dyskeratosis congenita (X-DC). X-DC is the most common and severe form of dyskeratosis congenita (DC), and is invariably associated with mutations of the DKC1 gene, encoding dyskerin (Table 1) (Heiss et al. 1998; Ruggero et al. 2003). Although prominent features of X-DC pathogenesis include bone marrow failure and skin abnormalities, a wide variety of tumor types including carcinomas and hematopoietic malignancies are also manifest. Dyskerin is an evolutionarily conserved enzyme responsible for the modification of approximately 100 specific uridine residues in ribosomal RNA (rRNA) into pseudouridine (Ni et al. 1997; Lafontaine et al. 1998). The role of these rRNA modifications in translational control has historically been poorly understood. The unexpected human genetic mutations associated with this key enzyme responsible for site-specific rRNA modifications were enigmatic, especially with respect to the tissue-specific phenotypes present in X-DC. The specific features of X-DC pathogenesis, including cancer susceptibility, can be explained, at least in part, by the finding that a distinct subset of mRNAs is not efficiently translated in an X-DC mouse model, X-DC patient cells, or dyskerin-knockdown cancer cells (Yoon et al. 2006; Bellodi et al. 2010a,b; Montanaro et al. 2010). On the other hand, general protein synthesis appears to be unperturbed (Yoon et al. 2006; Montanaro et al. 2010). Surprisingly, these studies have uncovered an unexpected requirement for rRNA modifications in translational control. Specifically, this subset of mRNAs shares a common regulatory motif in their 5′ untranslated region (5′ UTR): an IRES element that acts to initiate translation in a cap-independent manner. It has been proposed that IRES elements may exist in ∼10% of mRNAs, including many tumor suppressor genes and antiapoptotic factors (Fig. 2) (Graber and Holcik 2007; Komar and Hatzoglou 2011). Several studies have highlighted the importance of RNA-binding proteins known as IRES *trans*-acting factors (ITAFs) in modulating IRES-mediated translation of several cellular mRNAs (Spriggs et al. 2005; Lewis and Holcik 2008; Komar and Hatzoglou 2011). IRES-dependent translation acts to fine-tune gene expression at the post-transcriptional level (Holcik and Sonenberg 2005; Spriggs et al. 2010; Komar and Hatzoglou 2011). In this context, it has been shown that defects in the IRES-mediated translation of a subset of mRNAs, including important tumor suppressors such as p53 and p27, in DKC1 mutant cells contribute to cellular transformation and tumor development (Bellodi et al. 2010a,b; Montanaro et al. 2010). Recent findings also show an evolutionarily conserved role for rRNA pseudouridylation in faithfully maintaining the translational reading frame. Specifically, impaired pseudouridylation of rRNA promotes increased programmed ribosomal frameshifting (PRF) of both viral and cellular mRNAs containing a PRF signal (Jack et al. 2011). It would be interesting to assess whether rRNA pseudouridylation regulates, at the translational level, the expression of mRNAs containing a PRF signal, perhaps in a cell- and/or tissue-specific manner. An emerging role for rRNA modifications in regulating specific modes of translation has recently been highlighted by findings that methylation of rRNA is also implicated in IRES-dependent translation of specific cellular mRNAs (Chaudhuri et al. 2007; Belin et al. 2009; Basu et al. 2011). Furthermore, loss of rRNA methylation at specific sites in zebrafish leads to severe developmental defects (Higa-Nakamine et al. 2012).

Taken as a whole, these studies raise several important questions that remain to be addressed. In particular, what regulates ribosomal RNA (rRNA) modifications? Can rRNA modifications modulate the recruitment of specific factors to the ribosome that would enhance translation specificity? What specific mRNAs are sensitive to ribosome modifications? For example, in addition to IRES elements, do rRNA modifications affect the translation of mRNAs possessing distinct *cis*-acting regulatory elements?

Diamond-Blackfan anemia (DBA) is another ribosomopathy in which mutations in several different ribosomal proteins (RPs) lead to bone marrow failure and an increased risk of leukemia and solid tumors (Table 1) (Flygare and Karlsson 2007; Lipton and Ellis 2010; Narla et al. 2011). This is consistent with studies in zebrafish showing that mutations in RPs can cause cancer (Amsterdam et al. 2004). In DBA, alterations in rRNA processing have been described and proposed to trigger a stress response marked by p53 activation, which would lead to cell-cycle arrest or programmed cell death (Danilova et al. 2008; McGowan et al. 2008; Dutt et al. 2011). Although the induction of p53 may explain certain pathological features of DBA, it does not provide a molecular basis for the increased cancer susceptibility associated with the disease, the cause of which remains almost completely unknown. One possibility is

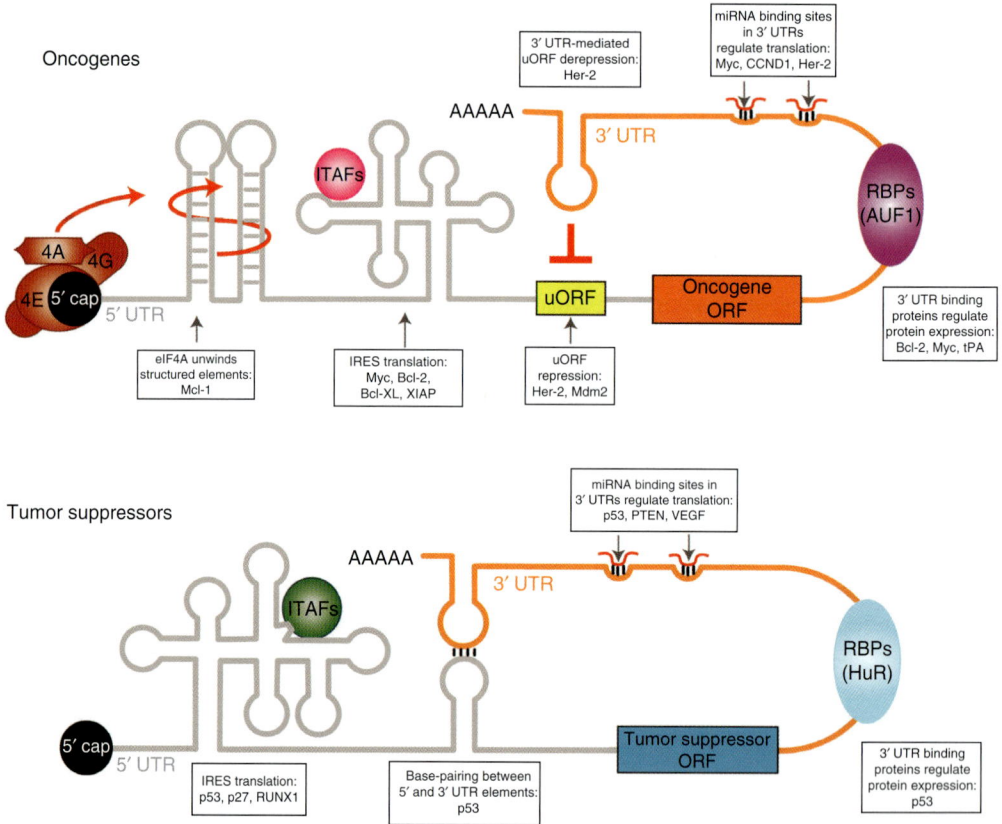

Figure 2. Oncogenes and tumor suppressors are exquisitely regulated at the translational level through specific regulatory elements in their mRNAs. Depictions of *cis*-regulatory elements present in oncogene (*top*) and tumor suppressor (*bottom*) mRNAs including key cell-cycle and survival factors. Examples of regulated mRNAs are given (boxes). The cap-binding protein eIF4E binds the 5′ cap of mRNAs. Increased eIF4E activity recruits eIF4G and the eIF4A helicase to unwind (red arrow) structured elements in the 5′ UTR of poorly translated mRNAs, increasing the expression of many growth promoting and prosurvival genes such as Mcl-1. IRES elements are structures that also direct translational regulation by promoting cap-independent protein synthesis of both tumor suppressors (p53) and prosurvival factors (XIAP) during different stages of tumor development (see Fig. 1). IRES *trans*-acting factors (ITAFs) modulate translational regulation directed by specific IRES elements. Upstream open reading frames (uORF) are present in select mRNAs (including growth promoting factors such as Mdm2 and Her-2) and inhibit translation initiation by preventing the ribosome from scanning to the start codon. Additionally, a region in the 3′ UTR of the Her-2 mRNA can inhibit uORF-mediated repression of translation. 3′ UTRs contain many distinct elements that interact with RNA-binding proteins (RBPs) and microRNAs (miRNAs) to promote or inhibit translation. One interesting example is the regulation of p53 translation mediated by base-pairing interactions between elements in the 5′ and 3′ UTRs of p53.

that RPs may exert more specialized functions in translational control, either on or off the ribosome, altering gene expression to contribute to the cancer susceptibility associated with DBA (Mazumder et al. 2003; Landry et al. 2009). Loss of function of RPS25 does not affect general protein synthesis but rather impairs IRES-dependent translation (Landry et al. 2009). Very recent research has shown that a single ribosomal protein, RPL38, regulates transcript-specific translational control of an important class of Hox mRNAs. In this study, the authors showed that RPL38$^{+/-}$ mice harbor homeotic transformations due to loss of function of Hox mRNA translational control (Kondrashov et al. 2011). Moreover, it appears that ribosomal protein expression may be dynamically regulated in a tissue-specific manner, suggesting that the protein composition of the ribosomal subunits may differ from cell to cell, even within the same organ (Kondrashov et al. 2011). As such, it is tempting to speculate that mutations in ribosomal proteins may have more specific roles in translational control that underlie the increased cancer susceptibility observed in DBA (Xue and Barna 2012).

It is becoming increasingly apparent that the role of ribosomal proteins in cancer development is more complex than originally envisioned, as overexpression of many RPs has been associated with cancer (Ruggero and Pandolfi 2003; Lai and Xu 2007). It is still unknown whether increases in ribosomal protein levels augment the translational capacity of a cell and whether RP overexpression is a cause or just a consequence of cellular transformation. In addition, research data suggest that extraribosomal functions exerted by a few RPs may also play a tumor-suppressive role in cancer development. For example, it has been shown that RPL11 negatively modulated c-Myc levels and activity in cultured cells (Dai et al. 2007; Challagundla et al. 2011). Other studies suggest that the L11-Mdm2-p53 pathway may act as a tumor-suppressive barrier against c-Myc-induced tumorigenesis (Macias et al. 2010).

Additional notable examples of "ribosomopathies" include cartilage–hair hypoplasia (CHH) syndrome, Shwachman-Diamond syndrome (SDS), and 5q deletion syndrome (−5q), which also show increased risk of developing certain cancers of the skin (basal cell carcinoma) as well as blood (leukemia and lymphoma) (Barlow et al. 2010; Shimamura and Alter 2010). In all of these syndromes, mutations in regulatory factors important for ribosome biogenesis and translational control have been identified (Table 1). In certain cases, these factors may also show extraribosomal functions that might contribute to disease pathogenesis (Austin et al. 2008; Dutt et al. 2011). These are intriguing lines of research that go beyond the scope of this article and are further reviewed in (Ganapathi and Shimamura 2008; Burwick

et al. 2011; Fumagalli and Thomas 2011). Recent findings have uncovered a role for SBDS (the gene mutated in SDS syndrome) in ribosomal subunit joining and translational control (Finch et al. 2011; Sen et al. 2011; Wong et al. 2011). In hematopoietic stem cells from −5q patients in which RPS14 is found mutated, alterations in expression of key components of the translational machinery have been reported (Pellagatti et al. 2008). Therefore, it is likely that defects in translational control can contribute to features of these ribosomopathies including increased incidences of cancer.

Collectively, inherited ribosomopathies reveal that mutations in regulatory factors important for ribosome activity may produce mutant ribosomes lacking important constituents such as ribosomal proteins or rRNA modifications (Table 1). Can such mutant forms of the translational machinery be referred to as "cancer ribosomes"? If so, an outstanding question is the mechanism by which the "cancer ribosome" could promote cancer development via aberrant translational control. This is an important question to resolve, as at first glance it may appear counterintuitive that alterations in ribosome function could cause cancer, especially given the important connection between increased protein synthesis and cell growth (see below and Fig. 1). However, recent findings show surprising specificity in the classes of mRNAs that are specifically deregulated and underlie cancer susceptibility as a consequence of perturbations in ribosome function, as illustrated by X-DC. These advances also reflect an emerging appreciation for and increased knowledge of more specialized and dynamic regulation of translational control in vivo at an organismal level (Xue and Barna 2012).

CANCER CELLS EXPLOIT TRANSLATIONAL CONTROL FOR THEIR ONCOGENIC PROGRAM

Oncogenic Signaling Pathways Perturb Specific Translational Components

There is a growing body of evidence supporting the idea that deregulation of translational control serves as a common mechanism by which diverse oncogenic pathways (e.g., PI3K, Myc, and Ras) promote cellular transformation and tumor development (Silvera et al. 2010; van Riggelen et al. 2010; Hannan et al. 2011). In a normal cell, these pathways act as sensors of energy, stress, and nutrient availability, as well as growth factor signals, and integrate these inputs to direct control of ribosome production and gene expression at the translational level. One of the primary reasons for this cross talk is to couple these external stimuli with the execution of cell growth, division, and survival, which are directly coupled to protein synthesis (see below). Importantly, all these signals become oncogenic when hyperactivated. How do these signaling pathways alter translational control when mutated in cancer? This question has been answered, at least in part, by the discovery that the oncogenic signals directly modulate the activity and expression of specific translational components (Fig. 3). Notable examples include the PI3K-AKT-mTOR and Ras-MAPK signal transduction pathways, as well as transcriptional programs regulated by oncogenic Myc (Frederickson et al. 1992; Waskiewicz et al. 1999; Schuhmacher et al. 2001; Zeller et al. 2006; Furic et al. 2010; Silvera et al. 2010; van Riggelen et al. 2010; Hannan et al. 2011). I discuss the role of Myc in protein synthesis and cell growth in the next section.

One of the best-studied examples of oncogenic signaling impinging on translational control is the PI3K-AKT-mTOR pathway, which directly modulates translation initiation largely through activation of the kinase mammalian target of rapamycin complex 1 (mTORC1) (Zoncu et al. 2011). The PI3K-AKT-mTOR pathway is one of the most commonly mutated pathways in cancer (Vogt et al. 2011). mTORC1 phosphorylates ribosomal protein S6 kinase 1/ 2 (S6K1/2) and the eIF4E-binding proteins (4E-BPs), which negatively regulate the major cap-binding protein eIF4E (Dennis et al. 1996; Hara et al. 1997; Gingras et al. 1999). Phosphorylation of 4E-BPs leads to a conformational change that frees eIF4E, which, with the help of other eIFs, ultimately recruits the 40S ribosomal subunit to the $5'$ cap of mRNAs during translation initiation (Gingras et al. 1999; Rug-

gero and Sonenberg 2005). Initiation is thought to be the rate-limiting step of cap-dependent translation and eIF4E is considered the key factor in controlling this step (Duncan et al. 1987). This deduction is in part based on the fact that eIF4E activity is highly regulated at both the mRNA and protein level. Indeed, eIF4E is up-regulated at the mRNA level by a number of transcription factors including the oncogene Myc (Jones et al. 1996; De Benedetti and Graff 2004; Ruggero and Sonenberg 2005). At the protein level, eIF4E activity is also controlled through phosphorylation at serine 209 by the MAP kinase targets MNK1/2, in addition to inhibitory interactions with 4E-BPs (see above and Fig. 3) (Lachance et al. 2002; Ueda et al. 2004, 2010; Wendel et al. 2007; Furic et al. 2010). This tight regulation of eIF4E activity provides a rapid mechanism for cells to modulate translation initiation in response to numerous stimuli, including growth factor and oncogenic signaling.

The 4E-BPs/eIF4E axis is among the most well-characterized nodes in translation control and cancer. For example, eIF4E gene amplification has been reported in human breast, head, and neck cancer specimens, and eIF4E is also found overexpressed in a variety of tumors (Sorrells et al. 1998, 1999; Crew et al. 2000; Berkel et al. 2001; Rosenwald et al. 2001; Salehi and Mashayekhi 2006; Wang et al. 2009a; Yoshizawa et al. 2010). The oncogenic potential of eIF4E hyperactivity has been faithfully recapitulated both in vitro and in vivo. Overexpression of eIF4E is sufficient to induce transformation of immortalized murine fibroblasts and human epithelial cells (Lazaris-Karatzas et al. 1990; Avdulov et al. 2004). Constitutive overexpression of eIF4E in a mouse model that mimics the human oncogenic lesion leads to increased cancer susceptibility; eIF4E transgenic mice develop lymphomas, angiosarcomas, lung carcinomas, and hepatomas (Ruggero et al. 2004). Furthermore, in vivo overexpression of eIF4E cooperates with c-Myc to drive lymphomagenesis in part via a mechanism by which eIF4E overcomes Myc-induced apoptosis, a cellular barrier to tumor formation (Ruggero et al. 2004; Wendel et al. 2004). Phosphorylation of eIF4E also

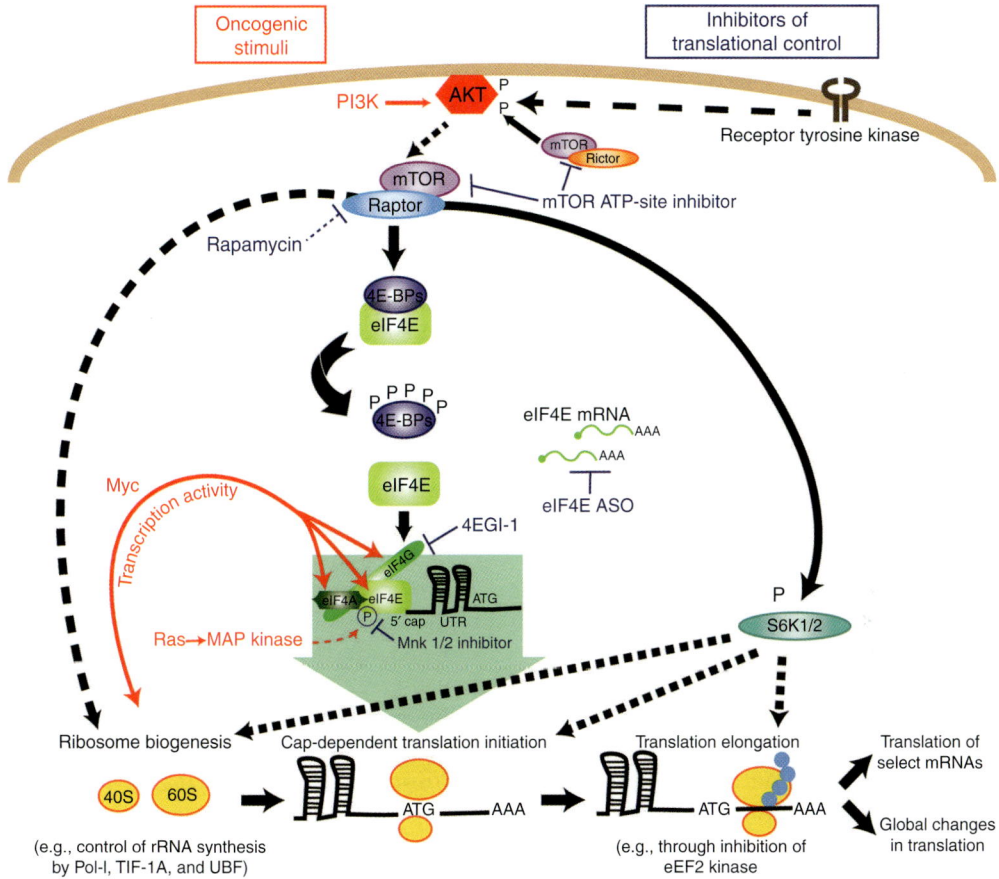

Figure 3. Oncogenic signals regulate each stage of translation. Oncogenic stimuli (red) such as PI3K-AKT-mTOR, Myc, and Ras promote protein synthesis by coordinating the regulation of ribosome biogenesis, translation initiation, and translation elongation. The PI3K-AKT-mTOR signaling pathway promotes ribosome biogenesis through both enhanced rRNA synthesis and enhanced ribosomal protein production (Hannan et al. 2003; Martin et al. 2004; Mayer et al. 2004). This signaling pathway stimulates translation initiation predominantly through mTORC1-dependent hyperactivation of eIF4E. In the absence of signaling, hypophosphorylated 4E-BPs bind to and inhibit eIF4E, blocking its ability to interact with eIF4G. PI3K-AKT signaling activates mTORC1, initiating a series of phosphorylations that release 4E-BPs from eIF4E. This allows for eIF4G binding to eIF4E and the subsequent recruitment of the 40S ribosomal subunit. Furthermore, S6K1/2 downstream mammalian target of rapamycin (mTOR) affects the efficiency of translation initiation and elongation (Ma and Blenis 2009). In addition, Ras-MAP kinase signaling up-regulates eIF4E activation via phosphorylation at serine 209. Myc promotes protein synthesis by increasing the transcription of multiple translational components including eIF4E mRNA. Together, these oncogenic stimuli regulate the multiple stages of translation to drive both global changes in protein synthesis as well as selective changes in the translation of specific mRNAs. Multiple approaches (blue) are used to therapeutically target the translational apparatus including rapamycin, ATP-active site inhibitors of mTOR, MNK1/2 kinase inhibitors, 4EGI-1, and eIF4E antisense oligonucleotides (ASO).

plays an important role in cancer formation. In fact, by using an adoptive transfer method in vivo, it has been shown that phosphorylated eIF4E contributes to Myc-induced tumorigenesis mainly by suppressing apoptosis (Wendel et al. 2007). In addition, whole body expression of a knockin mutant of eIF4E, which can no longer be phosphorylated at the serine 209 residue, was found to decrease the incidence and grade of prostatic intraepithelial neoplasia in a mouse prostate cancer model driven by PTEN loss (Furic et al. 2010).

Although these findings support the notion that eIF4E is a bona fide oncogene, they do not address the extent to which specific oncogenic signaling (i.e., PI3K-AKT-mTOR pathway) relies on eIF4E translational activity for tumor development. Some of the first evidence for such a connection came from a study showing that pharmacological inhibition of oncogenic Ras and AKT in glioblastoma cells caused a rapid and profound change in mRNA translation that far outweighed transcriptional changes. These translational changes were associated with loss of mTORC1-dependent phosphorylation of 4E-BPs (Rajasekhar et al. 2003). In addition, transfection of a 4E-BP1 phosphorylation site mutant into breast carcinoma cells suppressed their tumorigenicity by inducing apoptosis (Avdulov et al. 2004). It has recently been shown in vitro and in vivo that the 4E-BPs/eIF4E axis exerts significant control over cap-dependent translation, cell growth, cancer initiation, and progression downstream of mTOR hyperactivation (Dowling et al. 2010; Hsieh et al. 2010). Most importantly, restoring eIF4E oncogenic activity to normal levels downstream of AKT-mTOR hyperactivation results in blockage of tumor progression in a mouse model for lymphomagenesis associated with a drastic increase in overall survival (Hsieh et al. 2010).

Do these findings indicate that cells transformed by mTOR hyperactivation become "addicted" to oncogenic eIF4E-mediated translation control? Oncogenic addiction describes a tumor's dependence on specific oncogenic lesions for its survival and growth. This represents a potential "Achilles heel" in tumor formation by which inactivation of a specific oncogene can lead to apoptosis, senescence, and tumor regression (Weinstein 2000, 2002; Weinstein and Joe 2006). The importance of this concept resides in the fact that although cancer cells harbor many oncogenic lesions, targeting the mediators of oncogenic addiction would herald a landmark improvement in cancer therapy. Although the mechanisms underlying this "Achilles heel" are poorly understood, it is tempting to speculate

that the 4E-BPs/eIF4E axis may represent a novel node of oncogenic addiction to target for therapeutic intervention (see below). At the mechanistic level, targeting eIF4E activity in cancer cells can be a key to initiate a cell death program resulting in tumor regression. In support of this hypothesis, it has been shown that eIF4E hyperactivation is able to enhance the translation of select mRNAs (De Benedetti and Graff 2004), including some encoding antiapoptotic factors (Mamane et al. 2007). The 5' UTR of these mRNAs are believed to harbor regulatory elements, such as complex secondary structures, that impart this selectivity (Fig. 2). One important example is Mcl-1, an antiapoptotic factor containing a complex 5' UTR that is translationally up-regulated specifically upon eIF4E hyperactivation, leading to enhanced survival of cancer-initiating cells (Mills et al. 2008; Hsieh et al. 2010). Therefore, cancer cells may be addicted to eIF4E-promoted survival, which would serve as a promising target for therapeutic intervention.

Regulation of eIF4E is not the only node in which information from signaling pathways is received by the translational machinery. It is now clear that an entire repertoire of translational components may be co-opted to promote cancer initiation. For example, AKT hyperactivation also modulates translation elongation (Fig. 3; Table 1) (Wang et al. 2001). Additional regulated translational components include eIF2α, which is part of the ternary complex required to chaperone the initiator transfer RNA (tRNA) to the ribosome. Phosphorylation of eIF2α by several kinases including the double-stranded RNA-dependent protein kinase (PKR) and the endoplasmic reticulum kinase PERK leads to consequent inhibition of protein synthesis and this is a major cell growth checkpoint (Ron and Harding 2007). Disruption of the eIF2α checkpoint can lead to the transformation of immortalized rodent and human cells (Koromilas et al. 1992b; Meurs et al. 1993; Barber et al. 1995; Donze et al. 1995). However, prevention of eIF2α phosphorylation is not tumorigenic in vivo (Yang et al. 1995; Abraham et al. 1999; Scheuner et al. 2001). In addition, it has been shown that the PERK-eIF2α axis represents a survival pathway for cancer cells that need to adapt and overcome hypoxic stress during tumor progression (Fels and Koumenis 2006). Transformed cells with inactivating PERK or eIF2α mutations form smaller tumors that grow more slowly in nude mice that grow more slowly, and show higher levels of apoptosis in hypoxic areas compared to control cells with intact eIF2α phosphorylation (Bi et al. 2005). eIF2α is also commonly overexpressed in cancers and may thereby provide an uncontrolled stimulus leading to increased rates of protein synthesis (Wang et al. 1999). Interestingly, overexpression of the initiator tRNA itself is able to drive cellular transformation (Marshall et al. 2008). Another translation factor that promotes cellular transformation is eIF6, which regulates the joining of the 60S ribosomal subunit to the 48S preinitiation complex to begin translation (Gandin et al. 2008). Importantly, eIF6 has been shown to be rate limiting for translation, cell growth, and transformation (Gandin et al. 2008). This initiation factor also interacts with RACK1, a ribosome-associated scaffolding protein that coordinates signaling by PKC and src kinases (Ceci et al. 2003). Therefore, signaling through RACK1 to eIF6 may be another key node of oncogenic regulation.

Although a wealth of research (described above), including elegant genetic studies, has shown that eIF4E hyperactivation and other nodes of translational control play a key role in cancer development, the repertoire of mRNAs that are specifically sensitive to translational perturbations is not completely understood. There are now emerging technologies that may facilitate their identification. In particular, the ability to deep-sequence ribosome-protected mRNAs will enable codon-by-codon resolution of ribosome occupancy on specific mRNAs (Ingolia et al. 2009). Furthermore, through deep sequencing, it is now possible to determine the secondary structures of mRNAs by using a novel strategy termed parallel analysis of RNA structures (PARS) (Kertesz et al. 2010). The combination of these two technologies may provide a very accurate portrait of how mRNA secondary structures control the translation of the cancer genome. Furthermore, techniques

such as high-throughput sequencing of RNA isolated by crosslinking immunoprecipitation (HITS-CLIP) and affinity RNA purification followed by mass spectrometry have successfully identified cis-regulatory elements and specific regulatory factors associated with RNAs (Ji et al. 2004; Darnell et al. 2011; Tsai et al. 2011; Darnell and Richter 2012). Such technologies could be used to characterize the mRNAs and the translational complex directly bound to eIFs such as eIF3, eIF4G, eIF5B, and eIF4A. Undertaking these approaches in a time course manner during the different steps in tumorigenesis would be valuable in identifying novel translational regulatory networks and might help to reveal the interplay between these networks during the stages of tumor development. Fluorescence-based living cell approaches such as fluorescence resonance energy transfer (FRET) could also enable the analysis of molecular interactions of a specific translational complex in real time during cellular transformation (Huranova et al. 2009; Lorenz 2009). Simply analyzing a static complex paints an incomplete picture. By studying the dynamic assembly and disassembly of these regulatory nodes, we could gain crucial insight into their ability to promote distinct translation initiation modes, ultimately contributing to cancer development.

Why Has the Translational Control of Specific mRNAs Been Selected for Cancer Development?

To consider this question, it is perhaps important to consider mechanisms that drive cancer evolution, whereby altering specific cellular processes may lead to uncontrolled growth. In this respect, perturbing translational control could provide a rapid change in gene expression upon environmental cues that allows the cancer cell to adapt and survive even under unfavorable conditions.

But how does the cancer program control specific mRNAs to support the growth of cancer cells? A central principle emerging from detailed molecular studies is that important regulatory elements (e.g., IRES, complex 5′ and 3′ UTRs, RNA-binding proteins, and microRNA sites) may render mRNAs encoding key tumor suppressors and protooncogenes (including cell-cycle and survival factors) exquisitely vulnerable to any perturbations in translational control (Fig. 2). Many of these mRNAs possess long 5′ UTRs that contain upstream ORFs and/or stable secondary structures (Stoneley and Willis 2003; Pickering and Willis 2005; Sonenberg and Hinnebusch 2009). These specific *cis* features act as inhibitory elements that interfere with the activity of the eIF4F initiation complex and ribosome scanning, thereby maintaining the translational efficiency of these mRNAs at a level that does not exceed a threshold necessary for cellular homeostasis. Importantly, these mRNAs are aberrantly translated as a consequence of overexpressed eIF4E or the other components of the eIF4F complex, which overcome the translational inhibitory effects imposed by their 5′ UTR *cis* elements (Fig. 2) (Koromilas et al. 1992a; Stoneley and Willis 2003; Pickering and Willis 2005; Ruggero and Sonenberg 2005; Mills et al. 2008; Hsieh et al. 2010). In addition, overexpression of RNA-binding proteins such as CPEB4 has been shown to increase the translation of specific oncogenic mRNAs by binding to *cis* elements present in their 3′ UTRs (Ortiz-Zapater et al. 2011; for further review, please see van Kouwenhove et al. 2011). Furthermore, the sequence and structural integrity of the 5′ and 3′ UTRs in the p53 tumor suppressor mRNA are required for its translation (Fig. 2) (Chen and Kastan 2010).

Do genetic perturbations in mRNA regulatory elements directly lead to cancer formation? This is an intriguing line of research that remains to be fully explored. Alterations in the 5′ UTR of MDM2, δ-catenin, or Myc oncogenes increase the translational level of their mRNAs in multiple myeloma, Burkitt's lymphoma, and prostate cancer (Paulin et al. 1996; Landers et al. 1997; Capoulade et al. 1998; Stoneley et al. 1998; Brown et al. 1999; Jin et al. 2003; Wang et al. 2009b). A mutation in the 5′ UTR of the CDKN2a tumor suppressor (also known as p16^{INK4A}) is associated with increased susceptibility to melanoma. This mutation creates an aberrant upstream translation initiation codon impairing translation initiation efficiency from

the wild-type AUG (Liu et al. 1999). Additionally, cancer-derived mutations in the p53 gene may alter an IRES element present in the 5′ UTR of its mRNA decreasing p53 protein expression (Grover et al. 2011).

Altogether, the remarkable repertoire of deregulated translational components, whose activity is directly controlled downstream of specific oncogenic signals, strongly supports their critical and causal role in cancer initiation and progression. What also emerges from these studies is that perturbations in translational control provide highly specific outcomes for gene expression that impinge on distinct steps along the pathway toward cancer development.

EXAMINING THE MECHANISMS BY WHICH THE TRANSLATIONAL MACHINERY PROMOTES THE MULTISTEP PROCESSES OF CELLULAR TRANSFORMATION AND TUMOR DEVELOPMENT

Cellular Transformation and Tumor Initiation

Increases in cell mass (cell growth), which occur mainly in the G1 phase of the cell cycle, is a prerequisite for accurate cell division (Fig 1) (Hall et al. 2004). Ribosome biogenesis and global protein synthesis are tightly and dynamically regulated to accommodate the growth demands of a cell (Hall et al. 2004). Therefore, an important relationship exists between the cell cycle, ribosome production, and translational control. This balance is maintained in the cell through key checkpoints. For example, downregulation of ribosome formation and activity is required during M phase to ensure proper cytokinesis (Pyronnet et al. 2001; Boisvert et al. 2007; Barna et al. 2008; Marash et al. 2008; Sivan et al. 2011). As such, it appears that a translational program that interfaces with the cell-cycle machinery should ensure the translation of specific mRNAs at appropriate levels during each window of cell growth and division. However, we have yet to determine which portion of the genome is translationally regulated during each phase of the cell cycle.

In cancer cells the balance between growth and division is broken, leading to unrestrained increases in protein synthesis and cell size (Fig. 1). Studies on the oncogenic activity of Myc provide an excellent window for understanding this mechanism of cellular transformation and tumor initiation. The Myc transcription factor, which is commonly deregulated in human cancers, is an exquisite regulator of ribosome biogenesis, protein synthesis, and cell growth (Ruggero 2009; van Riggelen et al. 2010). Myc directly increases protein synthesis rates by controlling the expression of multiple components of the protein synthetic machinery, including ribosomal proteins, initiation factors, Pol III, and recombinant DNA (rDNA) (Fernandez et al. 2003; Gomez-Roman et al. 2003; Grandori et al. 2005; Zeller et al. 2006). Through its transcriptional regulation of the translational apparatus, Myc activity leads to increased cell size often associated with changes in nucleolar architecture (Fig. 1) (Kim et al. 2000; Arabi et al. 2003; Grandori et al. 2005; Grewal et al. 2005; Shiue et al. 2010). Interestingly, one of the initial markers for cancer cells, discovered more than 100 years ago, is an increase in the size and number of nucleoli (Pianese 1896). However, whether the increases in ribosome biogenesis and cell growth are a direct cause of cellular transformation or, as many researchers have long assumed, a consequence of the need to sustain elevated rates of proliferation in tumor development, remains a critical and challenging question. This query has been in part addressed with novel genetic strategies that restore aberrant protein synthesis to normal levels before tumor formation in Myc transgenic mice. These studies showed that the ability of Myc to increase protein synthesis directly augments cell growth (Ruggero 2009). Is it possible that, on its own, increased cell growth represents one of the oncogenic "hits" that leads to cellular transformation? For example, cell growth may provide a significant competitive advantage to a preneoplastic cell in a specific tissue type. In this situation, a larger cell may outcompete neighboring normal cells in uptake of nutrients and growth factors. Consequently, a somatic clone harboring such a cancerous lesion may possess an early competitive advantage. This advantage may result in cell death or starvation of normal cells and the de-

struction of the organ boundary concomitantly with the expansion of the neoplastic clone.

Aberrant cell growth due to oncogene-dependent increases in protein synthesis rates may directly accelerate cell-cycle progression. As mentioned above, the two processes are intimately linked. In this case, the oversized cells may be more susceptible to entering unscheduled cell division, as an aberrant increase in cell growth may mimic the normal G1 phase of the cell cycle (Fig. 1). This hypothesis may explain why restricting protein synthesis downstream of Myc results in restoration of normal cell growth as well as cell division (Barna et al. 2008). Importantly, Myc's oncogenic potential is dependent on its ability to up-regulate protein synthesis (Barna et al. 2008). Other recent genetic studies have provided compelling evidence that the manipulation of the translational machinery by Myc is integral to its oncogenic activity. Inactivation of one copy of eIF6 limits Myc-driven lymphomagenesis (Miluzio et al. 2011). Furthermore, Myc also synergizes with AKT to regulate Pol I activity, another important determinant of cell growth and cancer development (Chan et al. 2011). The significance of Pol I for cancer therapy is underscored by the fact that Cylene Pharmaceuticals has developed two potent and specific inhibitors of this RNA polymerase, which show antitumor activity in xenograft models (Drygin et al. 2009, 2011).

However, we have yet to understand the exact molecular mechanisms by which oncogenic signaling, affecting ribosome biogenesis and translational control, couples cell growth with cell division during cellular transformation. For example, we do not know whether an increase in ribosome number and/or nucleolar size, as a consequence of Myc overexpression, sends a "signal" to the cell division machinery to facilitate cell-cycle progression, or whether the translational program downstream of oncogenic activation directly changes the expression of key components of cell growth and division. In this latter scenario, it stands to reason that specific mRNAs (e.g., cyclins or replication enzymes) would be more sensitive to translational perturbation, representing key translational nodes for controlling cell growth and division.

Cite this article as *Cold Spring Harb Perspect Biol* doi: 10.1101/cshperspect.a012336

Increased cell division as a consequence of increased cell growth is not sufficient to induce cellular transformation. Indeed, even upon aberrant cell-cycle progression, a critical cellular response can counteract cellular transformation. This response is known as oncogene-induced senescence (OIS) (Fig. 1). OIS is characterized by cell-cycle arrest and induction of cell-cycle inhibitors such as p15, p16, and p53, which restrain the proliferative potential of preneoplastic clones (Serrano et al. 1997; Gil and Peters 2006). Importantly, during OIS, a switch between cap- and IRES-dependent translation occurs (Bellodi et al. 2010a). During this switch, an IRES element positioned in the 5' UTR of p53 is engaged to promote p53 translation (Bellodi et al. 2010a). Therefore specialized translational control of mRNAs such as p53 provides a molecular barrier for cellular transformation (Takagi et al. 2005; Ray et al. 2006; Bellodi et al. 2010a; Montanaro et al. 2010). Interestingly, defects in rRNA modifications specifically perturb p53 IRES-dependent translation, resulting in defective OIS and expansion of preneoplastic clones (Bellodi et al. 2010a). This may be a critical mechanism that underlies cancer susceptibility in X-DC, discussed above (Bellodi et al. 2010a; Montanaro et al. 2010). Most recently, another translational control mechanism that modulates p53 expression in the senescence phenotype has been discovered. In this case, p53 mRNA polyadenylation/translation is hierarchically regulated by the interplay between noncanonical poly(A) polymerases, cytoplasmic polyadenylation element-binding protein (CPEB), and miR-122 (Burns and Richter 2008; Burns et al. 2011). Ultimately, alterations in translational control of p53 mRNAs would appear to be an essential step leading to transformation.

A subsequent step to overcoming tumor-suppressive barriers such as OIS, which is important to achieve full cellular transformation, is the acquisition of additional genetic lesions. These lesions are commonly referred to as secondary hits (Fig. 1). However, the mechanisms by which alterations in translational control results in the accumulation of such hits remain poorly understood. We have recently learned that deregulations in mitotic translational control play an important role at this step by promoting genomic instability (Wilker et al. 2007; Barna et al. 2008). During mitosis, only a small fraction of mRNA is translated in a cap-independent manner (Pyronnet et al. 2000; Qin and Sarnow 2004; Komar and Hatzoglou 2011). CDK11/p58 is a well-characterized endogenous mRNA that is only translated during mitosis by an IRES element (Cornelis et al. 2000; Wilker et al. 2007). Distinct ITAFs, including PTB (polypyrimidine tract-binding protein) and UNR (upstream of N-RAS), have been shown to regulate CDK11/p58 expression (Tinton et al. 2005; Ohno et al. 2011). In the context of an oncogenic lesion such as Myc hyperactivation, an aberrant increase in cap-dependent translation specifically impairs the mitotic translational switch from cap- to IRES-dependent translation (Fig. 1). This results in reduced expression of CDK11/p58, which leads to cytokinesis defects associated with increased centrosome numbers and genome instability (Barna et al. 2008). Given that deregulations in mitotic translational control directly promote genome instability, further studies will be important to illuminate the precise relationship between specific genetic lesions such as Myc and the altered "oncogenic" translational machinery.

Progression and Metastasis

Once a tumor has been established it will progress through a number of stages and may ultimately become metastatic (Fig. 1). The functional role of translational control during these stages of tumorigenesis is only now beginning to be explored. Indeed, much of the data to date are derived primarily from expression analysis of translation factors in fixed human samples. For example, the incidences of eIF5A2 overexpression increase in human colorectal tumor specimens from benign tumors to primary carcinomas to metastatic lesions (Xie et al. 2008). The expression and activity of eIF4E has also been associated with malignant progression in many human cancers including prostate, head and neck, bladder, colon, breast, lung, and lymphoma (De Benedetti and Graff

Cite this article as *Cold Spring Harb Perspect Biol* doi: 10.1101/cshperspect.a012336

2004; Graff et al. 2009). In addition, hyperactivation of specific signal transduction pathways (see above) appears to directly increase translation factor activity during tumor progression. Nevertheless, we still have an incomplete understanding of the precise impact of translational control in the development of metastatic features in cancer cells. Unraveling the mechanistic basis of this process will shed light on new aspects of translational control and help identify novel targets for cancer therapy. In this respect, it appears that tumor cells have co-opted specific modes of translational regulation to express key survival factors for their continued development. For example, during cancer progression, tumor cells may survive stress conditions such as nutrient and oxygen deprivation through their ability to promote cap-independent translation of specific antiapoptotic factors such as Bcl-2 and XIAP (Fig. 1). In addition, IRES-dependent translation of the neoangiogenic factor VEGF promotes tumor size by enhancing blood flow to the tumor (Holcik and Korneluk 2000; Sherrill et al. 2004; Silvera et al. 2009; Silvera and Schneider 2009). Overexpression of eIF4G and 4E-BP1 is functionally linked to the promotion of IRES-dependent mRNA translation during cancer progression (Silvera et al. 2009).

Why is translational control of specific mRNAs important for the coordination of cell invasion and metastasis? To address this question, it is useful to consider the common features underlying these tumorigenic steps. For instance, cancer cells need to maintain a high capacity for migration and invasion to by-pass tissue barriers, intravasate into the blood stream, and extravasate at distal secondary sites (Valastyan and Weinberg 2011). This multistep process requires rapid and specific modulation of gene expression; in this context, it is conceivable that translational control plays an important role. In support of this hypothesis, recent studies have shown that eIF4E and eIF4GI were associated with the Golgi apparatus and membrane microdomains such as the perinuclear region and lamellipodia, which are essential for cell motility. Most importantly, a fraction of these initiation factors localizes at sites of

active translation near the leading edge of migrating cells (Willett et al. 2011). It is tempting to speculate that this phenomenon may represent a rapid mechanism to achieve a critical threshold of migratory proteins surrounding the leading edge of the cell. This localized translation may also facilitate the secretion of key proteins such as metalloproteases, which are necessary to degrade the basement membrane, an obligatory step before the cancer cells enter the bloodstream. Therefore, this spatial increase in translational control at the leading edge could aid the invasion of cancer cells from the primary tumor to secondary sites.

Another cellular process associated with increased metastatic potential is known as epithelial-to-mesenchymal transition (EMT), in which epithelial cancer cells decrease their epithelial properties and acquire mesenchymal features (Fig. 1B) (Floor et al. 2011; Valastyan and Weinberg 2011). This process is associated with increases in cell motility and changes in cell–cell or cell–matrix adhesion. Importantly, very recent studies using ribosome profiling have delineated at a codon-by-codon resolution that oncogenic mTOR signaling has a striking effect on the translational landscape of the cancer genome of metastatic cells (Hsieh et al. 2012). This work has functionally characterized a novel mRNA signature associated with cancer cell invasion and metastasis in vivo. This signature is comprised of vimentin, an intermediate filament protein highly up-regulated during EMT (Lahat et al. 2010); MTA1 (metastasis associated 1), a putative chromatin-remodeling protein that has been shown to drive cancer metastasis by promoting neoangiogenesis (Yoo et al. 2006); CD44, commonly overexpressed in tumor-initiating cells and is implicated in cancer metastasis (Liu et al. 2011); and YB-1 (Y-box binding protein), also called YBX1. Interestingly, it has been recently shown that the YB-1 protein promotes the cap-independent translation of specific subsets of mRNAs encoding key proteins underlying EMT (Evdokimova et al. 2009). Mechanistically, eIF4E regulates the translation of these mRNAs encoding key players in cancer metastasis at least in part through a novel regulatory element present in their 5′ UTR known as pyrimidine-rich translation element (PRTE) (Hsieh et al. 2012). Significantly, a novel clinical ATP-site inhibitor of mTOR (INK128, see below) exquisitely targets this translational signature with therapeutic benefit at all stages of prostate cancer progression, including metastasis (Hsieh et al. 2012). However, the identity and number of the key components of the translational apparatus that may regulate metastatic potential demand further investigation. Furthermore, both the repertoire of translationally regulated mRNAs that contributes to tumor invasion and metastatic dissemination, and the mechanisms that regulate their expression, remain an exciting frontier that merits further exploration.

FROM THE BENCH TO THE PATIENT: CLINICAL-"TRANSLATIONAL" ADVANCES

After key findings establishing the causal role of the deregulated translational machinery within the multihit model of cancer etiology, we are arriving at an exciting crossroad between discovery and clinical application. I will briefly summarize some of the main preclinical and clinical strategies that directly or indirectly target the protein synthesis apparatus (Hsieh and Ruggero 2010; see also Malina et al. 2012). The eIF4E oncogene, which is frequently hyperactivated in human cancer, represents an attractive target for rational drug design (Fig. 3). There are currently several approaches being pursued to therapeutically inhibit eIF4E, but perhaps the most direct of these approaches is the use of specific antisense oligonucleotides (ASOs) that bind to eIF4E mRNA and mediate its destruction by RNase H. The eIF4E-ASO is currently being tested in combination with chemotherapy in new phase I/II clinical trials of metastatic lung cancer and prostate cancer (Graff et al. 2007). Additional attempts to target eIF4E have focused on blocking its ability to interact with eIF4G. The interaction between eIF4E and eIF4G is dependent on an eIF4G Y(X)$_4$LΦ motif, in which X is variable and Φ is hydrophobic (Marcotrigiano et al. 1999). High-throughput screens for inhibitors that could prevent eIF4E from binding to the Y(X)$_4$LΦ motif identified 4EGI-1 as a candidate

compound. 4EGI-1 is both a cytotoxic and cytostatic agent acting across multiple cell lines. Importantly, 4EGI-1 inhibits proliferation and clonogenic growth of transformed cells more effectively than untransformed cells (Moerke et al. 2007; Tamburini et al. 2009). 4EGI-1 is licensed to Eugenix, which is working toward an investigational new drug application for this compound (G Wagner, pers. comm.). Concurrently, indirect approaches to target oncogenic eIF4E activity are proving particularly effective (Hsieh et al. 2011). For example, inhibiting the mTOR kinase holds tremendous promise for the clinic. In vitro and in vivo characterizations of recently developed mTOR ATP-site inhibitors have shown greater efficacy than allosteric mTOR inhibitors (rapalogs) (see also Malina et al. 2012; Roux and Topisirovic 2012). Unlike rapalogs, mTOR ATP-site inhibitors target 4E-BP phosphorylation as well as the mTORC2 complex (Benjamin et al. 2011). PP242, the first of a series of reported ATP-site mTOR inhibitors (Feldman et al. 2009), has been used in a preclinical trial for AKT-driven lymphomagenesis (Hsieh et al. 2010). In this study, PP242 (but not rapamycin) suppressed tumor growth by inhibiting 4E-BP1 phosphorylation. In line with these studies, the antiproliferative effect of both PP242 and Torin (another ATP-site inhibitor) was attenuated in cultured 4E-BP null cells (Thoreen et al. 2009; Dowling et al. 2010). Altogether, these findings strongly suggest that the therapeutic efficacy of ATP-site inhibitors may be in large part associated with their ability to block mTORC1-dependent 4E-BP phosphorylation and eIF4E oncogenic activity (Hsieh et al. 2010). Perhaps most notably, INK128, a more potent derivative of PP242, as well as several other ATP-site inhibitors, are currently in phase I/II clinical trials in patients with advanced solid tumors and hematological malignancies (Benjamin et al. 2011; Hsieh et al. 2012).

Equally significant for the development of future cancer treatments, recent studies have shown that amplification of the eIF4E gene leads to drug resistance in immortalized human mammary epithelial cells (HMECs) that were treated with the dual PI3K/mTOR inhibitor, BEZ235 (Ilic et al. 2011). Specifically, BEZ235 elicited a cytostatic effect only in HMECs that do not display eIF4E amplification and the associated increases in cap-dependent translation (Ilic et al. 2011). Therefore, deregulation in translational control may also represent a mechanism for drug resistance. For example, increases in expression and activity of distinct translational components might evade the inhibitory activity of compounds that target key upstream regulators of the translational machinery.

It remains to be seen how the deregulation of other translation components, in addition to eIF4E, may act either as therapeutic targets or mechanisms of resistance to mTOR inhibitors. Ultimately, to understand these contributions, we need to characterize the translationally deregulated mRNAs downstream of these events.

CONCLUDING REMARKS

It is clear that the study of deregulations in the translational machinery has moved into the center stage of cancer research. Indeed, even if many questions remain to be addressed, we are now closer than ever to understanding the causal relationship between alterations in translational control and the etiology of cancer development. The critical function of translational control in cancer is underscored by the discovery of mutations in translational components underlying genetic syndromes associated with cancer susceptibility. In addition, several studies have described gene amplification of distinct key regulators of protein synthesis in somatic cancers (Table 1). Furthermore, there is a growing appreciation for the exquisite specificity in translational regulation mediated by the core components of the translational apparatus. For example, the unexpected discovery that only specific mRNAs harboring IRES elements are translationally affected by impairments in rRNA modifications has changed our perception of how the translation machinery regulates gene expression. Additionally, mutant ribosomes lacking important constituents such as ribosomal proteins may be largely functionally active, but preferentially defective in specific aspects of translational control, contributing to cancer development.

Altering the translational landscape of the cancer genome appears to be a key driver in tumor formation associated with hyperactivation of signal transduction pathways. Indeed, distinct oncogenic signals appear to monopolize translational control at almost every stage of cancer initiation and development, resulting in specific and distinct cellular outcomes (Fig. 1). Important studies have also uncovered a surprising level of specificity for controlling gene expression at the posttranscriptional level downstream of the activation of these oncogenic signals. This translational specificity is also dictated by distinct regulatory elements within key mRNAs, such as IRES elements or structured 5' UTRs, which control the expression of these mRNAs during specific steps of cellular transformation and tumor development (Fig. 2). These translational control mechanisms would provide cancer cells ample opportunity to grow, survive, and expand indefinitely even under unfavorable conditions. Future research can build on these recent significant findings to elucidate the dynamic relationship between the regulatory components of the translational machinery and oncogenic signaling. Taken as a whole, alterations in translational control may represent an "oncogenic addiction" node. Such a node could serve as a potential "Achilles heel" for tumor suppression (Fig. 3). Ultimately, the significance of this research is further reflected by a number of novel and promising therapeutic approaches to target specific translational components in cancer that are presently in clinical trials.

ACKNOWLEDGMENTS

I am indebted to Kimhouy Tong for all her help with this article. I am grateful to Maria Barna and Craig Stumpf and the other members of the Ruggero Laboratory for many helpful discussions and for critically reviewing this manuscript. I apologize to those whose work I was unable to cite. This work is supported by National Institutes of Health grants R01 HL085572, R01 CA140456, and R01 CA154916, and Phi Beta Psi Sorority. Davide Ruggero is a Leukemia and Lymphoma Society Research Scholar.

D. Ruggero

REFERENCES

*Reference is also in this collection.

Abraham N, Stojdl DF, Duncan PI, Methot N, Ishii T, Dube M, Vanderhyden BC, Atkins HL, Gray DA, McBurney MW, et al. 1999. Characterization of transgenic mice with targeted disruption of the catalytic domain of the double-stranded RNA-dependent protein kinase, PKR. *J Biol Chem* **274:** 5953–5962.

Ahlemann M, Zeidler R, Lang S, Mack B, Munz M, Gires O. 2006. Carcinoma-associated eIF3i overexpression facilitates mTOR-dependent growth transformation. *Mol Carcinog* **45:** 957–967.

Amsterdam A, Sadler KC, Lai K, Farrington S, Bronson RT, Lees JA, Hopkins N. 2004. Many ribosomal protein genes are cancer genes in zebrafish. *PLoS Biol* **2:** E139.

Anand N, Murthy S, Amann G, Wernick M, Porter LA, Cukier IH, Collins C, Gray JW, Diebold J, Demetrick DJ, et al. 2002. Protein elongation factor EEF1A2 is a putative oncogene in ovarian cancer. *Nat Genet* **31:** 301–305.

Arabi A, Rustum C, Hallberg E, Wright AP. 2003. Accumulation of c-Myc and proteasomes at the nucleoli of cells containing elevated c-Myc protein levels. *J Cell Sci* **116:** 1707–1717.

Austin KM, Leary RJ, Shimamura A. 2005. The Shwachman-Diamond SBDS protein localizes to the nucleolus. *Blood* **106:** 1253–1258.

Austin KM, Gupta ML, Coats SA, Tulpule A, Mostoslavsky G, Balazs AB, Mulligan RC, Daley G, Pellman D, Shimamura A. 2008. Mitotic spindle destabilization and genomic instability in Shwachman-Diamond syndrome. *J Clin Invest* **118:** 1511–1518.

Avdulov S, Li S, Michalek V, Burrichter D, Peterson M, Perlman DM, Manivel JC, Sonenberg N, Yee D, Bitterman PB, et al. 2004. Activation of translation complex eIF4F is essential for the genesis and maintenance of the malignant phenotype in human mammary epithelial cells. *Cancer Cell* **5:** 553–563.

Bachmann F, Banziger R, Burger MM. 1997. Cloning of a novel protein overexpressed in human mammary carcinoma. *Cancer Res* **57:** 988–994.

Balabanov S, Gontarewicz A, Ziegler P, Hartmann U, Kammer W, Copland M, Brassat U, Priemer M, Hauber I, Wilhelm T, et al. 2007. Hypusination of eukaryotic initiation factor 5A (eIF5A): A novel therapeutic target in BCR-ABL-positive leukemias identified by a proteomics approach. *Blood* **109:** 1701–1711.

Barber GN, Wambach M, Thompson S, Jagus R, Katze MG. 1995. Mutants of the RNA-dependent protein kinase (PKR) lacking double-stranded RNA binding domain I can act as transdominant inhibitors and induce malignant transformation. *Mol Cell Biol* **15:** 3138–3146.

Barlow JL, Drynan LF, Trim NL, Erber WN, Warren AJ, McKenzie AN. 2010. New insights into 5q- syndrome as a ribosomopathy. *Cell Cycle* **9:** 4286–4293.

Barna M, Pusic A, Zollo O, Costa M, Kondrashov N, Rego E, Rao PH, Ruggero D. 2008. Suppression of Myc oncogenic activity by ribosomal protein haploinsufficiency. *Nature* **456:** 971–975.

Basu A, Das P, Chaudhuri S, Bevilacqua E, Andrews J, Barik S, Hatzoglou M, Komar AA, Mazumder B. 2011. Requirement of rRNA methylation for 80S ribosome assembly on a cohort of cellular internal ribosome entry sites. *Mol Cell Biol* **31:** 4482–4499.

Bauer C, Diesinger I, Brass N, Steinhart H, Iro H, Meese EU. 2001. Translation initiation factor eIF-4G is immunogenic, overexpressed, and amplified in patients with squamous cell lung carcinoma. *Cancer* **92:** 822–829.

Bauer C, Brass N, Diesinger I, Kayser K, Grasser FA, Meese E. 2002. Overexpression of the eukaryotic translation initiation factor 4G (eIF4G-1) in squamous cell lung carcinoma. *Int J Cancer* **98:** 181–185.

Belin S, Beghin A, Solano-Gonzalez E, Bezin L, Brunet-Manquat S, Textoris J, Prats AC, Mertani HC, Dumontet C, Diaz JJ. 2009. Dysregulation of ribosome biogenesis and translational capacity is associated with tumor progression of human breast cancer cells. *PLoS ONE* **4:** e7147.

Bellodi C, Kopmar N, Ruggero D. 2010a. Deregulation of oncogene-induced senescence and p53 translational control in X-linked dyskeratosis congenita. *EMBO J* **29:** 1865–1876.

Bellodi C, Krasnykh O, Haynes N, Theodoropoulou M, Peng G, Montanaro L, Ruggero D. 2010b. Loss of function of the tumor suppressor DKC1 perturbs p27 translation control and contributes to pituitary tumorigenesis. *Cancer Res* **70:** 6026–6035.

Benjamin D, Colombi M, Moroni C, Hall MN. 2011. Rapamycin passes the torch: A new generation of mTOR inhibitors. *Nat Rev Drug Discov* **10:** 868–880.

Berkel HJ, Turbat-Herrera EA, Shi R, de Benedetti A. 2001. Expression of the translation initiation factor eIF4E in the polyp-cancer sequence in the colon. *Cancer Epidemiol Biomarkers Prev* **10:** 663–666.

Bi M, Naczki C, Koritzinsky M, Fels D, Blais J, Hu N, Harding H, Novoa I, Varia M, Raleigh J, et al. 2005. ER stress-regulated translation increases tolerance to extreme hypoxia and promotes tumor growth. *EMBO J* **24:** 3470–3481.

Boisvert FM, van Koningsbruggen S, Navascues J, Lamond AI. 2007. The multifunctional nucleolus. *Nat Rev Mol Cell Biol* **8:** 574–585.

Boocock GR, Morrison JA, Popovic M, Richards N, Ellis L, Durie PR, Rommens JM. 2003. Mutations in SBDS are associated with Shwachman-Diamond syndrome. *Nat Genet* **33:** 97–101.

Brass N, Heckel D, Sahin U, Pfreundschuh M, Sybrecht GW, Meese E. 1997. Translation initiation factor eIF-4γ is encoded by an amplified gene and induces an immune response in squamous cell lung carcinoma. *Hum Mol Genet* **6:** 33–39.

Brown CY, Mize GJ, Pineda M, George DL, Morris DR. 1999. Role of two upstream open reading frames in the translational control of oncogene mdm2. *Oncogene* **18:** 5631–5637.

Burns DM, Richter JD. 2008. CPEB regulation of human cellular senescence, energy metabolism, and p53 mRNA translation. *Genes Dev* **22:** 3449–3460.

Burns DM, D'Ambrogio A, Nottrott S, Richter JD. 2011. CPEB and two poly(A) polymerases control miR-122

Cite this article as *Cold Spring Harb Perspect Biol* doi: 10.1101/cshperspect.a012336

stability and p53 mRNA translation. *Nature* **473**: 105–108.

Burwick N, Shimamura A, Liu JM. 2011. Non-Diamond Blackfan anemia disorders of ribosome function: Shwachman Diamond syndrome and 5q- syndrome. *Semin Hematol* **48**: 136–143.

Capoulade C, Bressac-de Paillerets B, Lefrere I, Ronsin M, Feunteun J, Tursz T, Wiels J. 1998. Overexpression of MDM2, due to enhanced translation, results in inactivation of wild-type p53 in Burkitt's lymphoma cells. *Oncogene* **16**: 1603–1610.

Ceci M, Gaviraghi C, Gorrini C, Sala LA, Offenhauser N, Marchisio PC, Biffo S. 2003. Release of eIF6 (p27BBP) from the 60S subunit allows 80S ribosome assembly. *Nature* **426**: 579–584.

Challagundla KB, Sun XX, Zhang X, DeVine T, Zhang Q, Sears RC, Dai MS. 2011. Ribosomal protein L11 recruits miR-24/miRISC to repress c-Myc expression in response to ribosomal stress. *Mol Cell Biol* **31**: 4007–4021.

Chan JC, Hannan KM, Riddell K, Ng PY, Peck A, Lee RS, Hung S, Astle MV, Bywater M, Wall M, et al. 2011. AKT promotes rRNA synthesis and cooperates with c-MYC to stimulate ribosome biogenesis in cancer. *Sci Signal* **4**: ra56.

Chaudhuri S, Vyas K, Kapasi P, Komar AA, Dinman JD, Barik S, Mazumder B. 2007. Human ribosomal protein L13a is dispensable for canonical ribosome function but indispensable for efficient rRNA methylation. *RNA* **13**: 2224–2237.

Chen J, Kastan MB. 2010. 5′-3′-UTR interactions regulate p53 mRNA translation and provide a target for modulating p53 induction after DNA damage. *Genes Dev* **24**: 2146–2156.

Clement PM, Johansson HE, Wolff EC, Park MH. 2006. Differential expression of eIF5A-1 and eIF5A-2 in human cancer cells. *FEBS J* **273**: 1102–1114.

Cmejla R, Cmejlova J, Handrkova H, Petrak J, Pospisilova D. 2007. Ribosomal protein S17 gene (RPS17) is mutated in Diamond-Blackfan anemia. *Hum Mutat* **28**: 1178–1182.

Cmejla R, Cmejlova J, Handrkova H, Petrak J, Petrtylova K, Mihal V, Stary J, Cerna Z, Jabali Y, Pospisilova D. 2009. Identification of mutations in the ribosomal protein L5 (RPL5) and ribosomal protein L11 (RPL11) genes in Czech patients with Diamond-Blackfan anemia. *Hum Mutat* **30**: 321–327.

Cornelis S, Bruynooghe Y, Denecker G, Van Huffel S, Tinton S, Beyaert R. 2000. Identification and characterization of a novel cell cycle-regulated internal ribosome entry site. *Mol Cell* **5**: 597–605.

Crew JP, Fuggle S, Bicknell R, Cranston DW, de Benedetti A, Harris AL. 2000. Eukaryotic initiation factor-4E in superficial and muscle invasive bladder cancer and its correlation with vascular endothelial growth factor expression and tumour progression. *Br J Cancer* **82**: 161–166.

Dai MS, Arnold H, Sun XX, Sears R, Lu H. 2007. Inhibition of c-Myc activity by ribosomal protein L11. *EMBO J* **26**: 3332–3345.

Danilova N, Sakamoto KM, Lin S. 2008. Ribosomal protein S19 deficiency in zebrafish leads to developmental abnormalities and defective erythropoiesis through activation of p53 protein family. *Blood* **112**: 5228–5237.

* Darnell JC, Richter JD. 2012. Cytoplasmic RNA binding proteins and the control of complex brain function. *Cold Spring Harb Perspect Biol* doi: 10.1101/cshperspect.a012344.

Darnell JC, Van Driesche SJ, Zhang C, Hung KY, Mele A, Fraser CE, Stone EF, Chen C, Fak JJ, Chi SW, et al. 2011. FMRP stalls ribosomal translocation on mRNAs linked to synaptic function and autism. *Cell* **146**: 247–261.

De Benedetti A, Graff JR. 2004. eIF-4E expression and its role in malignancies and metastases. *Oncogene* **23**: 3189–3199.

Dellas A, Torhorst J, Bachmann F, Banziger R, Schultheiss E, Burger MM. 1998. Expression of p150 in cervical neoplasia and its potential value in predicting survival. *Cancer* **83**: 1376–1383.

Dennis PB, Pullen N, Kozma SC, Thomas G. 1996. The principal rapamycin-sensitive p70(s6k) phosphorylation sites, T-229 and T-389, are differentially regulated by rapamycin-insensitive kinase kinases. *Mol Cell Biol* **16**: 6242–6251.

Doherty L, Sheen MR, Vlachos A, Choesmel V, O'Donohue MF, Clinton C, Schneider HE, Sieff CA, Newburger PE, Ball SE, et al. 2010. Ribosomal protein genes RPS10 and RPS26 are commonly mutated in Diamond-Blackfan anemia. *Am J Hum Genet* **86**: 222–228.

Doldan A, Chandramouli A, Shanas R, Bhattacharyya A, Cunningham JT, Nelson MA, Shi J. 2008a. Loss of the eukaryotic initiation factor 3f in pancreatic cancer. *Mol Carcinog* **47**: 235–244.

Doldan A, Chandramouli A, Shanas R, Bhattacharyya A, Leong SP, Nelson MA, Shi J. 2008b. Loss of the eukaryotic initiation factor 3f in melanoma. *Mol Carcinog* **47**: 806–813.

Donze O, Jagus R, Koromilas AE, Hershey JW, Sonenberg N. 1995. Abrogation of translation initiation factor eIF-2 phosphorylation causes malignant transformation of NIH 3T3 cells. *EMBO J* **14**: 3828–3834.

Dowling RJ, Topisirovic I, Alain T, Bidinosti M, Fonseca BD, Petroulakis E, Wang X, Larsson O, Selvaraj A, Liu Y, et al. 2010. mTORC1-mediated cell proliferation, but not cell growth, controlled by the 4E-BPs. *Science* **328**: 1172–1176.

Draptchinskaia N, Gustavsson P, Andersson B, Pettersson M, Willig TN, Dianzani I, Ball S, Tchernia G, Klar J, Matsson H, et al. 1999. The gene encoding ribosomal protein S19 is mutated in Diamond-Blackfan anaemia. *Nat Genet* **21**: 169–175.

Drygin D, Siddiqui-Jain A, O'Brien S, Schwaebe M, Lin A, Bliesath J, Ho CB, Proffitt C, Trent K, Whitten JP, et al. 2009. Anticancer activity of CX-3543: A direct inhibitor of rRNA biogenesis. *Cancer Res* **69**: 7653–7661.

Drygin D, Lin A, Bliesath J, Ho CB, O'Brien SE, Proffitt C, Omori M, Haddach M, Schwaebe MK, Siddiqui-Jain A, et al. 2011. Targeting RNA polymerase I with an oral small molecule CX-5461 inhibits ribosomal RNA synthesis and solid tumor growth. *Cancer Res* **71**: 1418–1430.

Duncan R, Milburn SC, Hershey JW. 1987. Regulated phosphorylation and low abundance of HeLa cell initiation factor eIF-4F suggest a role in translational control. Heat shock effects on eIF-4F. *J Biol Chem* **262**: 380–388.

Dutt S, Narla A, Lin K, Mullally A, Abayasekara N, Meger-dichian C, Wilson FH, Currie T, Khanna-Gupta A, Berliner N, et al. 2011. Haploinsufficiency for ribosomal protein genes causes selective activation of p53 in human erythroid progenitor cells. *Blood* **117**: 2567–2576.

Eberle J, Krasagakis K, Orfanos CE. 1997. Translation initiation factor eIF-4A1 mRNA is consistently overexpressed in human melanoma cells in vitro. *Int J Cancer* **71**: 396–401.

Ebert BL, Pretz J, Bosco J, Chang CY, Tamayo P, Galili N, Raza A, Root DE, Attar E, Ellis SR, et al. 2008. Identification of RPS14 as a 5q- syndrome gene by RNA interference screen. *Nature* **451**: 335–339.

Evdokimova V, Tognon C, Ng T, Ruzanov P, Melnyk N, Fink D, Sorokin A, Ovchinnikov LP, Davicioni E, Triche TJ, et al. 2009. Translational activation of snail1 and other developmentally regulated transcription factors by YB-1 promotes an epithelial-mesenchymal transition. *Cancer Cell* **15**: 402–415.

Farrar JE, Nater M, Caywood E, McDevitt MA, Kowalski J, Takemoto CM, Talbot CC Jr, Meltzer P, Esposito D, Beggs AH, et al. 2008. Abnormalities of the large ribosomal subunit protein, Rpl35a, in Diamond-Blackfan anemia. *Blood* **112**: 1582–1592.

Feldman ME, Apsel B, Uotila A, Loewith R, Knight ZA, Ruggero D, Shokat KM. 2009. Active-site inhibitors of mTOR target rapamycin-resistant outputs of mTORC1 and mTORC2. *PLoS Biol* **7**: e38.

Fels DR, Koumenis C. 2006. The PERK/eIF2α/ATF4 module of the UPR in hypoxia resistance and tumor growth. *Cancer Biol Ther* **5**: 723–728.

Fernandez PC, Frank SR, Wang L, Schroeder M, Liu S, Greene J, Cocito A, Amati B. 2003. Genomic targets of the human c-Myc protein. *Genes Dev* **17**: 1115–1129.

Finch AJ, Hilcenko C, Basse N, Drynan LF, Goyenechea B, Menne TF, Gonzalez Fernandez A, Simpson P, D'Santos CS, Arends MJ, et al. 2011. Uncoupling of GTP hydrolysis from eIF6 release on the ribosome causes Shwachman-Diamond syndrome. *Genes Dev* **25**: 917–929.

Floor S, van Staveren WC, Larsimont D, Dumont JE, Maenhaut C. 2011. Cancer cells in epithelial-to-mesenchymal transition and tumor-propagating-cancer stem cells: Distinct, overlapping or same populations. *Oncogene* **30**: 4609–4621.

Flygare J, Karlsson S. 2007. Diamond-Blackfan anemia: Erythropoiesis lost in translation. *Blood* **109**: 3152–3154.

Frederickson RM, Mushynski WE, Sonenberg N. 1992. Phosphorylation of translation initiation factor eIF-4E is induced in a ras-dependent manner during nerve growth factor-mediated PC12 cell differentiation. *Mol Cell Biol* **12**: 1239–1247.

Fumagalli S, Thomas G. 2011. The role of p53 in ribosomopathies. *Semin Hematol* **48**: 97–105.

Furic L, Rong L, Larsson O, Koumakpayi IH, Yoshida K, Brueschke A, Petroulakis E, Robichaud N, Pollak M, Gaboury LA, et al. 2010. eIF4E phosphorylation promotes tumorigenesis and is associated with prostate cancer progression. *Proc Natl Acad Sci* **107**: 14134–14139.

Ganapathi KA, Shimamura A. 2008. Ribosomal dysfunction and inherited marrow failure. *Br J Haematol* **141**: 376–387.

Gandin V, Miluzio A, Barbieri AM, Beugnet A, Kiyokawa H, Marchisio PC, Biffo S. 2008. Eukaryotic initiation factor 6 is rate-limiting in translation, growth and transformation. *Nature* **455**: 684–688.

Gazda HT, Grabowska A, Merida-Long LB, Latawiec E, Schneider HE, Lipton JM, Vlachos A, Atsidaftos E, Ball SE, Orfali KA, et al. 2006. Ribosomal protein S24 gene is mutated in Diamond-Blackfan anemia. *Am J Hum Genet* **79**: 1110–1118.

Gazda HT, Sheen MR, Vlachos A, Choesmel V, O'Donohue MF, Schneider H, Darras N, Hasman C, Sieff CA, Newburger PE, et al. 2008. Ribosomal protein L5 and L11 mutations are associated with cleft palate and abnormal thumbs in Diamond-Blackfan anemia patients. *Am J Hum Genet* **83**: 769–780.

Gil J, Peters G. 2006. Regulation of the INK4b-ARF-INK4a tumour suppressor locus: All for one or one for all. *Nat Rev Mol Cell Biol* **7**: 667–677.

Gingras AC, Gygi SP, Raught B, Polakiewicz RD, Abraham RT, Hoekstra MF, Aebersold R, Sonenberg N. 1999. Regulation of 4E-BP1 phosphorylation: A novel two-step mechanism. *Genes Dev* **13**: 1422–1437.

Gomez-Roman N, Grandori C, Eisenman RN, White RJ. 2003. Direct activation of RNA polymerase III transcription by c-Myc. *Nature* **421**: 290–294.

Graber TE, Holcik M. 2007. Cap-independent regulation of gene expression in apoptosis. *Mol Biosyst* **3**: 825–834.

Graff JR, Konicek BW, Vincent TM, Lynch RL, Monteith D, Weir SN, Schwier P, Capen A, Goode RL, Dowless MS, et al. 2007. Therapeutic suppression of translation initiation factor eIF4E expression reduces tumor growth without toxicity. *J Clin Invest* **117**: 2638–2648.

Graff JR, Konicek BW, Lynch RL, Dumstorf CA, Dowless MS, McNulty AM, Parsons SH, Brail LH, Colligan BM, Koop JW, et al. 2009. eIF4E activation is commonly elevated in advanced human prostate cancers and significantly related to reduced patient survival. *Cancer Res* **69**: 3866–3873.

Grandori C, Gomez-Roman N, Felton-Edkins ZA, Ngouenet C, Galloway DA, Eisenman RN, White RJ. 2005. c-Myc binds to human ribosomal DNA and stimulates transcription of rRNA genes by RNA polymerase I. *Nat Cell Biol* **7**: 311–318.

Grewal SS, Li L, Orian A, Eisenman RN, Edgar BA. 2005. Myc-dependent regulation of ribosomal RNA synthesis during *Drosophila* development. *Nat Cell Biol* **7**: 295–302.

Grover R, Sharathchandra A, Ponnuswamy A, Khan D, Das S. 2011. Effect of mutations on the p53 IRES RNA structure: Implications for de-regulation of the synthesis of p53 isoforms. *RNA Biol* **8**: 132–142.

Guan XY, Sham JS, Tang TC, Fang Y, Huo KK, Yang JM. 2001. Isolation of a novel candidate oncogene within a frequently amplified region at 3q26 in ovarian cancer. *Cancer Res* **61**: 3806–3809.

Guan XY, Fung JM, Ma NF, Lau SH, Tai LS, Xie D, Zhang Y, Hu L, Wu QL, Fang Y, et al. 2004. Oncogenic role of eIF-5A2 in the development of ovarian cancer. *Cancer Res* **64**: 4197–4200.

Hall MN, Raff M, Thomas G, eds. 2004. *Cell growth: Control of cell size.* Cold Spring Harbor Laboratory Press, Cold Spring Harbor, NY.

Hannan KM, Brandenburger Y, Jenkins A, Sharkey K, Cavanaugh A, Rothblum L, Moss T, Poortinga G, McArthur GA, Pearson RB, et al. 2003. mTOR-dependent regulation of ribosomal gene transcription requires S6K1 and is mediated by phosphorylation of the carboxy-terminal activation domain of the nucleolar transcription factor UBF. *Mol Cell Biol* **23:** 8862–8877.

Hannan KM, Sanij E, Hein N, Hannan RD, Pearson RB. 2011. Signaling to the ribosome in cancer—It is more than just mTORC1. *IUBMB Life* **63:** 79–85.

Hara K, Yonezawa K, Kozlowski MT, Sugimoto T, Andrabi K, Weng QP, Kasuga M, Nishimoto I, Avruch J. 1997. Regulation of eIF-4E BP1 phosphorylation by mTOR. *J Biol Chem* **272:** 26457–26463.

Harris MN, Ozpolat B, Abdi F, Gu S, Legler A, Mawuenyega KG, Tirado-Gomez M, Lopez-Berestein G, Chen X. 2004. Comparative proteomic analysis of all-trans-retinoic acid treatment reveals systematic posttranscriptional control mechanisms in acute promyelocytic leukemia. *Blood* **104:** 1314–1323.

Hawkins RD, Ren B. 2006. Genome-wide location analysis: Insights on transcriptional regulation. *Hum Mol Genet* **15** (Spec No 1): R1–R7.

He LR, Zhao HY, Li BK, Liu YH, Liu MZ, Guan XY, Bian XW, Zeng YX, Xie D. 2011. Overexpression of eIF5A-2 is an adverse prognostic marker of survival in stage I non-small cell lung cancer patients. *Int J Cancer* **129:** 143–150.

Heiss NS, Knight SW, Vulliamy TJ, Klauck SM, Wiemann S, Mason PJ, Poustka A, Dokal I. 1998. X-linked dyskeratosis congenita is caused by mutations in a highly conserved gene with putative nucleolar functions. *Nat Genet* **19:** 32–38.

Higa-Nakamine S, Suzuki T, Uechi T, Chakraborty A, Nakajima Y, Nakamura M, Hirano N, Kenmochi N. 2012. Loss of ribosomal RNA modification causes developmental defects in zebrafish. *Nucleic Acids Res* **40:** 391–398.

Holcik M, Korneluk RG. 2000. Functional characterization of the X-linked inhibitor of apoptosis (XIAP) internal ribosome entry site element: Role of La autoantigen in XIAP translation. *Mol Cell Biol* **20:** 4648–4657.

Holcik M, Sonenberg N. 2005. Translational control in stress and apoptosis. *Nat Rev Mol Cell Biol* **6:** 318–327.

Hsieh AC, Ruggero D. 2010. Targeting eukaryotic translation initiation factor 4E (eIF4E) in cancer. *Clin Cancer Res* **16:** 4914–4920.

Hsieh AC, Costa M, Zollo O, Davis C, Feldman ME, Testa JR, Meyuhas O, Shokat KM, Ruggero D. 2010. Genetic dissection of the oncogenic mTOR pathway reveals druggable addiction to translational control via 4EBP-eIF4E. *Cancer Cell* **17:** 249–261.

Hsieh AC, Truitt ML, Ruggero D. 2011. Oncogenic AKTivation of translation as a therapeutic target. *Br J Cancer* **105:** 329–336.

Hsieh AC, Liu Y, Edlind MP, Ingolia NT, Janes MR, Sher A, Shi EY, Stumpf CR, Christensen C, Bonham MJ, et al. 2012. The translational landscape of mTOR signalling steers cancer initiation and metastasis. *Nature* **485:** 55–61.

Huang R, Wu T, Xu L, Liu A, Ji Y, Hu G. 2002. Upstream binding factor up-regulated in hepatocellular carcinoma is related to the survival and cisplatin-sensitivity of cancer cells. *FASEB J* **16:** 293–301.

Huranova M, Jablonski JA, Benda A, Hof M, Stanek D, Caputi M. 2009. In vivo detection of RNA-binding protein interactions with cognate RNA sequences by fluorescence resonance energy transfer. *RNA* **15:** 2063–2071.

Ilic N, Utermark T, Widlund HR, Roberts TM. 2011. PI3K-targeted therapy can be evaded by gene amplification along the MYC-eukaryotic translation initiation factor 4E (eIF4E) axis. *Proc Natl Acad Sci* **108:** E699–E708.

Ingolia NT, Ghaemmaghami S, Newman JR, Weissman JS. 2009. Genome-wide analysis in vivo of translation with nucleotide resolution using ribosome profiling. *Science* **324:** 218–223.

Jack K, Bellodi C, Landry DM, Niederer RO, Meskauskas A, Musalgaonkar S, Kopmar N, Krasnykh O, Dean AM, Thompson SR, et al. 2011. rRNA pseudouridylation defects affect ribosomal ligand binding and translational fidelity from yeast to human cells. *Mol Cell* **44:** 660–666.

Ji H, Fraser CS, Yu Y, Leary J, Doudna JA. 2004. Coordinated assembly of human translation initiation complexes by the hepatitis C virus internal ribosome entry site RNA. *Proc Natl Acad Sci* **101:** 16990–16995.

Jin W, McCutcheon IE, Fuller GN, Huang ES, Cote GJ. 2000. Fibroblast growth factor receptor-1 α-exon exclusion and polypyrimidine tract-binding protein in glioblastoma multiforme tumors. *Cancer Res* **60:** 1221–1224.

Jin X, Turcott E, Englehardt S, Mize GJ, Morris DR. 2003. The two upstream open reading frames of oncogene mdm2 have different translational regulatory properties. *J Biol Chem* **278:** 25716–25721.

Jones RM, Branda J, Johnston KA, Polymenis M, Gadd M, Rustgi A, Callanan L, Schmidt EV. 1996. An essential E box in the promoter of the gene encoding the mRNA cap-binding protein (eukaryotic initiation factor 4E) is a target for activation by c-myc. *Mol Cell Biol* **16:** 4754–4764.

Kertesz M, Wan Y, Mazor E, Rinn JL, Nutter RC, Chang HY, Segal E. 2010. Genome-wide measurement of RNA secondary structure in yeast. *Nature* **467:** 103–107.

Killian A, Sarafan-Vasseur N, Sesboue R, Le Pessot F, Blanchard F, Lamy A, Laurent M, Flaman JM, Frebourg T. 2006. Contribution of the BOP1 gene, located on 8q24, to colorectal tumorigenesis. *Genes Chromosomes Cancer* **45:** 874–881.

Kim S, Li Q, Dang CV, Lee LA. 2000. Induction of ribosomal genes and hepatocyte hypertrophy by adenovirus-mediated expression of c-Myc in vivo. *Proc Natl Acad Sci* **97:** 11198–11202.

Komar AA, Hatzoglou M. 2011. Cellular IRES-mediated translation: The war of ITAFs in pathophysiological states. *Cell Cycle* **10:** 229–240.

Kondrashov N, Pusic A, Stumpf CR, Shimizu K, Hsieh AC, Xue S, Ishijima J, Shiroishi T, Barna M. 2011. Ribosome-mediated specificity in Hox mRNA translation and vertebrate tissue patterning. *Cell* **145:** 383–397.

Koromilas AE, Lazaris-Karatzas A, Sonenberg N. 1992a. mRNAs containing extensive secondary structure in their 5′ non-coding region translate efficiently in cells overexpressing initiation factor eIF-4E. *EMBO J* **11:** 4153–4158.

Koromilas AE, Roy S, Barber GN, Katze MG, Sonenberg N. 1992b. Malignant transformation by a mutant of the IFN-inducible dsRNA-dependent protein kinase. *Science* **257:** 1685–1689.

Lachance PE, Miron M, Raught B, Sonenberg N, Lasko P. 2002. Phosphorylation of eukaryotic translation initiation factor 4E is critical for growth. *Mol Cell Biol* **22:** 1656–1663.

Lafontaine DL, Bousquet-Antonelli C, Henry Y, Caizergues-Ferrer M, Tollervey D. 1998. The box H + ACA snoRNAs carry Cbf5p, the putative rRNA pseudouridine synthase. *Genes Dev* **12:** 527–537.

Lahat G, Zhu QS, Huang KL, Wang S, Bolshakov S, Liu J, Torres K, Langley RR, Lazar AJ, Hung MC, et al. 2010. Vimentin is a novel anti-cancer therapeutic target; insights from in vitro and in vivo mice xenograft studies. *PLoS ONE* **5:** e10105.

Lai MD, Xu J. 2007. Ribosomal proteins and colorectal cancer. *Curr Genomics* **8:** 43–49.

Landers JE, Cassel SL, George DL. 1997. Translational enhancement of mdm2 oncogene expression in human tumor cells containing a stabilized wild-type p53 protein. *Cancer Res* **57:** 3562–3568.

Landry DM, Hertz MI, Thompson SR. 2009. RPS25 is essential for translation initiation by the Dicistroviridae and hepatitis C viral IRESs. *Genes Dev* **23:** 2753–2764.

Lazaris-Karatzas A, Montine KS, Sonenberg N. 1990. Malignant transformation by a eukaryotic initiation factor subunit that binds to mRNA 5′ cap. *Nature* **345:** 544–547.

Lee JM. 2003. The role of protein elongation factor eEF1A2 in ovarian cancer. *Reprod Biol Endocrinol* **1:** 69.

Lewis SM, Holcik M. 2008. For IRES trans-acting factors, it is all about location. *Oncogene* **27:** 1033–1035.

Lipton JM, Ellis SR. 2010. Diamond Blackfan anemia 2008–2009: Broadening the scope of ribosome biogenesis disorders. *Curr Opin Pediatr* **22:** 12–19.

Liu L, Dilworth D, Gao L, Monzon J, Summers A, Lassam N, Hogg D. 1999. Mutation of the CDKN2A 5′ UTR creates an aberrant initiation codon and predisposes to melanoma. *Nat Genet* **21:** 128–132.

Liu C, Kelnar K, Liu B, Chen X, Calhoun-Davis T, Li H, Patrawala L, Yan H, Jeter C, Honorio S, et al. 2011. The microRNA miR-34a inhibits prostate cancer stem cells and metastasis by directly repressing CD44. *Nat Med* **17:** 211–215.

Lorenz M. 2009. Visualizing protein-RNA interactions inside cells by fluorescence resonance energy transfer. *RNA* **15:** 97–103.

Ma XM, Blenis J. 2009. Molecular mechanisms of mTOR-mediated translational control. *Nat Rev Mol Cell Biol* **10:** 307–318.

Ma YL, Peng JY, Zhang P, Huang L, Liu WJ, Shen TY, Chen HQ, Zhou YK, Zhang M, Chu ZX, et al. 2009. Heterogeneous nuclear ribonucleoprotein A1 is identified as a potential biomarker for colorectal cancer based on differential proteomics technology. *J Proteome Res* **8:** 4525–4535.

Macias E, Jin A, Deisenroth C, Bhat K, Mao H, Lindstrom MS, Zhang Y. 2010. An ARF-independent c-MYC-activated tumor suppression pathway mediated by ribosomal protein-Mdm2 interaction. *Cancer Cell* **18:** 231–243.

* Malina A, Mills JR, Pelletier J. 2012. Emerging therapeutics targeting mRNA translation. *Cold Spring Harb Perspect Biol* doi: 10.1101/cshperspect.a012377.

Mamane Y, Petroulakis E, Martineau Y, Sato TA, Larsson O, Rajasekhar VK, Sonenberg N. 2007. Epigenetic activation of a subset of mRNAs by eIF4E explains its effects on cell proliferation. *PLoS ONE* **2:** e242.

Marash L, Liberman N, Henis-Korenblit S, Sivan G, Reem E, Elroy-Stein O, Kimchi A. 2008. DAP5 promotes cap-independent translation of Bcl-2 and CDK1 to facilitate cell survival during mitosis. *Mol Cell* **30:** 447–459.

Marchetti A, Buttitta F, Pellegrini S, Bertacca G, Callahan R. 2001. Reduced expression of INT-6/eIF3-p48 in human tumors. *Int J Oncol* **18:** 175–179.

Marcotrigiano J, Gingras AC, Sonenberg N, Burley SK. 1999. Cap-dependent translation initiation in eukaryotes is regulated by a molecular mimic of eIF4G. *Mol Cell* **3:** 707–716.

Marshall L, Kenneth NS, White RJ. 2008. Elevated tRNA(i-Met) synthesis can drive cell proliferation and oncogenic transformation. *Cell* **133:** 78–89.

Martin DE, Soulard A, Hall MN. 2004. TOR regulates ribosomal protein gene expression via PKA and the Forkhead transcription factor FHL1. *Cell* **119:** 969–979.

Mayer C, Zhao J, Yuan X, Grummt I. 2004. mTOR-dependent activation of the transcription factor TIF-IA links rRNA synthesis to nutrient availability. *Genes Dev* **18:** 423–434.

Mazumder B, Sampath P, Seshadri V, Maitra RK, DiCorleto PE, Fox PL. 2003. Regulated release of L13a from the 60S ribosomal subunit as a mechanism of transcript-specific translational control. *Cell* **115:** 187–198.

McGowan KA, Li JZ, Park CY, Beaudry V, Tabor HK, Sabnis AJ, Zhang W, Fuchs H, de Angelis MH, Myers RM, et al. 2008. Ribosomal mutations cause p53-mediated dark skin and pleiotropic effects. *Nat Genet* **40:** 963–970.

Meurs EF, Galabru J, Barber GN, Katze MG, Hovanessian AG. 1993. Tumor suppressor function of the interferon-induced double-stranded RNA-activated protein kinase. *Proc Natl Acad Sci* **90:** 232–236.

Mills JR, Hippo Y, Robert F, Chen SM, Malina A, Lin CJ, Trojahn U, Wendel HG, Charest A, Bronson RT, et al. 2008. mTORC1 promotes survival through translational control of Mcl-1. *Proc Natl Acad Sci* **105:** 10853–10858.

Miluzio A, Beugnet A, Grosso S, Brina D, Mancino M, Campaner S, Amati B, de Marco A, Biffo S. 2011. Impairment of cytoplasmic eIF6 activity restricts lymphomagenesis and tumor progression without affecting normal growth. *Cancer Cell* **19:** 765–775.

Moerke NJ, Aktas H, Chen H, Cantel S, Reibarkh MY, Fahmy A, Gross JD, Degterev A, Yuan J, Chorev M, et al. 2007. Small-molecule inhibition of the interaction between the translation initiation factors eIF4E and eIF4G. *Cell* **128:** 257–267.

Montanaro L, Calienni M, Bertoni S, Rocchi L, Sansone P, Storci G, Santini D, Ceccarelli C, Taffurelli M, Carnicelli D, et al. 2010. Novel dyskerin-mediated mechanism of p53 inactivation through defective mRNA translation. *Cancer Res* **70:** 4767–4777.

Cite this article as *Cold Spring Harb Perspect Biol* doi: 10.1101/cshperspect.a012336

Nakamura J, Aoyagi S, Nanchi I, Nakatsuka S, Hirata E, Shibata S, Fukuda M, Yamamoto Y, Fukuda I, Tatsumi N, et al. 2009. Overexpression of eukaryotic elongation factor eEF2 in gastrointestinal cancers and its involvement in G2/M progression in the cell cycle. *Int J Oncol* **34:** 1181–1189.

Narla A, Hurst SN, Ebert BL. 2011. Ribosome defects in disorders of erythropoiesis. *Int J Hematol* **93:** 144–149.

Ni J, Tien AL, Fournier MJ. 1997. Small nucleolar RNAs direct site-specific synthesis of pseudouridine in ribosomal RNA. *Cell* **89:** 565–573.

Nupponen NN, Porkka K, Kakkola L, Tanner M, Persson K, Borg A, Isola J, Visakorpi T. 1999. Amplification and overexpression of p40 subunit of eukaryotic translation initiation factor 3 in breast and prostate cancer. *Am J Pathol* **154:** 1777–1783.

Ohno S, Shibayama M, Sato M, Tokunaga A, Yoshida N. 2011. Polypyrimidine tract-binding protein regulates the cell cycle through IRES-dependent translation of CDK11 (p58) in mouse embryonic stem cells. *Cell Cycle* **10:** 3706–3713.

Ortiz-Zapater E, Pineda D, Martinez-Bosch N, Fernandez-Miranda G, Iglesias M, Alameda F, Moreno M, Eliscovich C, Eyras E, Real FX, et al. 2011. Key contribution of CPEB4-mediated translational control to cancer progression. *Nat Med* **18:** 83–90.

Paulin FE, West MJ, Sullivan NF, Whitney RL, Lyne L, Willis AE. 1996. Aberrant translational control of the c-myc gene in multiple myeloma. *Oncogene* **13:** 505–513.

Pellagatti A, Hellstrom-Lindberg E, Giagounidis A, Perry J, Malcovati L, Della Porta MG, Jadersten M, Killick S, Fidler C, Cazzola M, et al. 2008. Haploinsufficiency of RPS14 in 5q- syndrome is associated with deregulation of ribosomal- and translation-related genes. *Br J Haematol* **142:** 57–64.

Pianese G. 1896. Beitrag zur histologie und aetiologie der carcinoma. Histologische und experimentelle untersuchungen. *Beitr Pathol Anat Allg Pathol*: 1–193.

Pickering BM, Willis AE. 2005. The implications of structured 5′ untranslated regions on translation and disease. *Semin Cell Dev Biol* **16:** 39–47.

Pyronnet S, Pradayrol L, Sonenberg N. 2000. A cell cycle-dependent internal ribosome entry site. *Mol Cell* **5:** 607–616.

Pyronnet S, Dostie J, Sonenberg N. 2001. Suppression of cap-dependent translation in mitosis. *Genes Dev* **15:** 2083–2093.

Qin X, Sarnow P. 2004. Preferential translation of internal ribosome entry site-containing mRNAs during the mitotic cycle in mammalian cells. *J Biol Chem* **279:** 13721–13728.

Rajasekhar VK, Viale A, Socci ND, Wiedmann M, Hu X, Holland EC. 2003. Oncogenic Ras and Akt signaling contribute to glioblastoma formation by differential recruitment of existing mRNAs to polysomes. *Mol Cell* **12:** 889–901.

Rauch J, Ahlemann M, Schaffrik M, Mack B, Ertongur S, Andratschke M, Zeidler R, Lang S, Gires O. 2004. Allogenic antibody-mediated identification of head and neck cancer antigens. *Biochem Biophys Res Commun* **323:** 156–162.

Ray PS, Grover R, Das S. 2006. Two internal ribosome entry sites mediate the translation of p53 isoforms. *EMBO Rep* **7:** 404–410.

Ridanpaa M, van Eenennaam H, Pelin K, Chadwick R, Johnson C, Yuan B, vanVenrooij W, Pruijn G, Salmela R, Rockas S, et al. 2001. Mutations in the RNA component of RNase MRP cause a pleiotropic human disease, cartilage-hair hypoplasia. *Cell* **104:** 195–203.

Ron D, Harding HP. 2007. eIF2α phosphorylation in cellular stress-responses and disease. In *Translational control in biology and medicine* (ed. Matthews MB, Sonenberg N, Hershey JW), pp. 345–368. Cold Spring Harbor Laboratory Press, Cold Spring Harbor, NY.

Rosenwald IB, Hutzler MJ, Wang S, Savas L, Fraire AE. 2001. Expression of eukaryotic translation initiation factors 4E and 2α is increased frequently in bronchioloalveolar but not in squamous cell carcinomas of the lung. *Cancer* **92:** 2164–2171.

Rothe M, Ko Y, Albers P, Wernert N. 2000. Eukaryotic initiation factor 3 p110 mRNA is overexpressed in testicular seminomas. *Am J Pathol* **157:** 1597–1604.

* Roux PP, Topisirovic I. 2012. Regulation of mRNA translation by signaling pathways. *Cold Spring Harb Perspect Biol* doi: 10.1101/cshperspect.a012252.

Roychoudhury P, Paul RR, Chowdhury R, Chaudhuri K. 2007. HnRNP E2 is downregulated in human oral cancer cells and the overexpression of hnRNP E2 induces apoptosis. *Mol Carcinog* **46:** 198–207.

Ruggero D. 2009. The role of Myc-induced protein synthesis in cancer. *Cancer Res* **69:** 8839–8843.

Ruggero D, Pandolfi PP. 2003. Does the ribosome translate cancer? *Nat Rev Cancer* **3:** 179–192.

Ruggero D, Sonenberg N. 2005. The Akt of translational control. *Oncogene* **24:** 7426–7434.

Ruggero D, Grisendi S, Piazza F, Rego E, Mari F, Rao PH, Cordon-Cardo C, Pandolfi PP. 2003. Dyskeratosis congenita and cancer in mice deficient in ribosomal RNA modification. *Science* **299:** 259–262.

Ruggero D, Montanaro L, Ma L, Xu W, Londei P, Cordon-Cardo C, Pandolfi PP. 2004. The translation factor eIF-4E promotes tumor formation and cooperates with c-Myc in lymphomagenesis. *Nat Med* **10:** 484–486.

Salehi Z, Mashayekhi F. 2006. Expression of the eukaryotic translation initiation factor 4E (eIF4E) and 4E-BP1 in esophageal cancer. *Clin Biochem* **39:** 404–409.

Sanvito F, Vivoli F, Gambini S, Santambrogio G, Catena M, Viale E, Veglia F, Donadini A, Biffo S, Marchisio PC. 2000. Expression of a highly conserved protein, p27BBP, during the progression of human colorectal cancer. *Cancer Res* **60:** 510–516.

Saramaki O, Willi N, Bratt O, Gasser TC, Koivisto P, Nupponen NN, Bubendorf L, Visakorpi T. 2001. Amplification of EIF3S3 gene is associated with advanced stage in prostate cancer. *Am J Pathol* **159:** 2089–2094.

Scheuner D, Song B, McEwen E, Liu C, Laybutt R, Gillespie P, Saunders T, Bonner-Weir S, Kaufman RJ. 2001. Translational control is required for the unfolded protein response and in vivo glucose homeostasis. *Mol Cell* **7:** 1165–1176.

Schuhmacher M, Kohlhuber F, Holzel M, Kaiser C, Burtscher H, Jarsch M, Bornkamm GW, Laux G, Polack

A, Weidle UH, et al. 2001. The transcriptional program of a human B cell line in response to Myc. *Nucleic Acids Res* **29:** 397–406.

Sen S, Wang H, Nghiem CL, Zhou K, Yau J, Tailor CS, Irwin MS, Dror Y. 2011. The ribosome-related protein, SBDS, is critical for normal erythropoiesis. *Blood* **118:** 6407–6417.

Serrano M, Lin AW, McCurrach ME, Beach D, Lowe SW. 1997. Oncogenic ras provokes premature cell senescence associated with accumulation of p53 and p16INK4a. *Cell* **88:** 593–602.

Sherrill KW, Byrd MP, Van Eden ME, Lloyd RE. 2004. BCL-2 translation is mediated via internal ribosome entry during cell stress. *J Biol Chem* **279:** 29066–29074.

Shi J, Kahle A, Hershey JW, Honchak BM, Warneke JA, Leong SP, Nelson MA. 2006. Decreased expression of eukaryotic initiation factor 3f deregulates translation and apoptosis in tumor cells. *Oncogene* **25:** 4923–4936.

Shimamura A, Alter BP. 2010. Pathophysiology and management of inherited bone marrow failure syndromes. *Blood Rev* **24:** 101–122.

Shiue CN, Arabi A, Wright AP. 2010. Nucleolar organization, growth control and cancer. *Epigenetics* **5:** 200–205.

Shuda M, Kondoh N, Tanaka K, Ryo A, Wakatsuki T, Hada A, Goseki N, Igari T, Hatsuse K, Aihara T, et al. 2000. Enhanced expression of translation factor mRNAs in hepatocellular carcinoma. *Anticancer Res* **20:** 2489–2494.

Silvera D, Schneider RJ. 2009. Inflammatory breast cancer cells are constitutively adapted to hypoxia. *Cell Cycle* **8:** 3091–3096.

Silvera D, Arju R, Darvishian F, Levine PH, Zolfaghari L, Goldberg J, Hochman T, Formenti SC, Schneider RJ. 2009. Essential role for eIF4GI overexpression in the pathogenesis of inflammatory breast cancer. *Nat Cell Biol* **11:** 903–908.

Silvera D, Formenti SC, Schneider RJ. 2010. Translational control in cancer. *Nat Rev Cancer* **10:** 254–266.

Sivan G, Aviner R, Elroy-Stein O. 2011. Mitotic modulation of translation elongation factor 1 leads to hindered tRNA delivery to ribosomes. *J Biol Chem* **286:** 27927–27935.

Sonenberg N, Hinnebusch AG. 2009. Regulation of translation initiation in eukaryotes: Mechanisms and biological targets. *Cell* **136:** 731–745.

Sorrells DL, Black DR, Meschonat C, Rhoads R, De Benedetti A, Gao M, Williams BJ, Li BD. 1998. Detection of eIF4E gene amplification in breast cancer by competitive PCR. *Ann Surg Oncol* **5:** 232–237.

Sorrells DL, Ghali GE, Meschonat C, DeFatta RJ, Black D, Liu L, De Benedetti A, Nathan CO, Li BD. 1999. Competitive PCR to detect eIF4E gene amplification in head and neck cancer. *Head Neck* **21:** 60–65.

Spriggs KA, Bushell M, Mitchell SA, Willis AE. 2005. Internal ribosome entry segment-mediated translation during apoptosis: The role of IRES-trans-acting factors. *Cell Death Differ* **12:** 585–591.

Spriggs KA, Bushell M, Willis AE. 2010. Translational regulation of gene expression during conditions of cell stress. *Mol Cell* **40:** 228–237.

Stoneley M, Willis AE. 2003. Aberrant regulation of translation initiation in tumorigenesis. *Curr Mol Med* **3:** 597–603.

Stoneley M, Paulin FE, Le Quesne JP, Chappell SA, Willis AE. 1998. C-Myc 5′ untranslated region contains an internal ribosome entry segment. *Oncogene* **16:** 423–428.

Stratford AL, Habibi G, Astanehe A, Jiang H, Hu K, Park E, Shadeo A, Buys TP, Lam W, Pugh T, et al. 2007. Epidermal growth factor receptor (EGFR) is transcriptionally induced by the Y-box binding protein-1 (YB-1) and can be inhibited with Iressa in basal-like breast cancer, providing a potential target for therapy. *Breast Cancer Res* **9:** R61.

Takagi M, Absalon MJ, McLure KG, Kastan MB. 2005. Regulation of p53 translation and induction after DNA damage by ribosomal protein L26 and nucleolin. *Cell* **123:** 49–63.

Tamburini J, Green AS, Bardet V, Chapuis N, Park S, Willems L, Uzunov M, Ifrah N, Dreyfus F, Lacombe C, et al. 2009. Protein synthesis is resistant to rapamycin and constitutes a promising therapeutic target in acute myeloid leukemia. *Blood* **114:** 1618–1627.

Tang DJ, Dong SS, Ma NF, Xie D, Chen L, Fu L, Lau SH, Li Y, Guan XY. 2010. Overexpression of eukaryotic initiation factor 5A2 enhances cell motility and promotes tumor metastasis in hepatocellular carcinoma. *Hepatology* **51:** 1255–1263.

Thoreen CC, Kang SA, Chang JW, Liu Q, Zhang J, Gao Y, Reichling LJ, Sim T, Sabatini DM, Gray NS. 2009. An ATP-competitive mammalian target of rapamycin inhibitor reveals rapamycin-resistant functions of mTORC1. *J Biol Chem* **284:** 8023–8032.

Tinton SA, Schepens B, Bruynooghe Y, Beyaert R, Cornelis S. 2005. Regulation of the cell-cycle-dependent internal ribosome entry site of the PITSLRE protein kinase: Roles of Unr (upstream of N-ras) protein and phosphorylated translation initiation factor eIF-2α. *Biochem J* **385:** 155–163.

Trotta R, Vignudelli T, Candini O, Intine RV, Pecorari L, Guerzoni C, Santilli G, Byrom MW, Goldoni S, Ford LP, et al. 2003. BCR/ABL activates mdm2 mRNA translation via the La antigen. *Cancer Cell* **3:** 145–160.

Tsai BP, Wang X, Huang L, Waterman ML. 2011. Quantitative profiling of in vivo-assembled RNA-protein complexes using a novel integrated proteomic approach. *Mol Cell Proteomics* **10:** M110 007385.

Ueda T, Watanabe-Fukunaga R, Fukuyama H, Nagata S, Fukunaga R. 2004. Mnk2 and Mnk1 are essential for constitutive and inducible phosphorylation of eukaryotic initiation factor 4E but not for cell growth or development. *Mol Cell Biol* **24:** 6539–6549.

Ueda T, Sasaki M, Elia AJ, Chio II, Hamada K, Fukunaga R, Mak TW. 2010. Combined deficiency for MAP kinase-interacting kinase 1 and 2 (Mnk1 and Mnk2) delays tumor development. *Proc Natl Acad Sci* **107:** 13984–13990.

Valastyan S, Weinberg RA. 2011. Tumor metastasis: Molecular insights and evolving paradigms. *Cell* **147:** 275–292.

van Kouwenhove M, Kedde M, Agami R. 2011. MicroRNA regulation by RNA-binding proteins and its implications for cancer. *Nat Rev Cancer* **11:** 644–656.

van Riggelen J, Yetil A, Felsher DW. 2010. MYC as a regulator of ribosome biogenesis and protein synthesis. *Nat Rev Cancer* **10:** 301–309.

van 't Veer LJ, Dai H, van de Vijver MJ, He YD, Hart AA, Mao M, Peterse HL, van der Kooy K, Marton MJ,

Cite this article as *Cold Spring Harb Perspect Biol* doi: 10.1101/cshperspect.a012336

Witteveen AT, et al. 2002. Gene expression profiling predicts clinical outcome of breast cancer. *Nature* **415:** 530–536.

Vogt PK, Rommel C, Vanhaesebroeck B, eds. 2011. *Phosphoinositide 3-kinase in health and disease.* Springer-Verlag, Berlin/Heidelberg.

Wang S, Rosenwald IB, Hutzler MJ, Pihan GA, Savas L, Chen JJ, Woda BA. 1999. Expression of the eukaryotic translation initiation factors 4E and 2α in non-Hodgkin's lymphomas. *Am J Pathol* **155:** 247–255.

Wang X, Li W, Williams M, Terada N, Alessi DR, Proud CG. 2001. Regulation of elongation factor 2 kinase by p90(RSK1) and p70 S6 kinase. *EMBO J* **20:** 4370–4379.

Wang R, Geng J, Wang JH, Chu XY, Geng HC, Chen LB. 2009a. Overexpression of eukaryotic initiation factor 4E (eIF4E) and its clinical significance in lung adenocarcinoma. *Lung Cancer* **66:** 237–244.

Wang T, Chen YH, Hong H, Zeng Y, Zhang J, Lu JP, Jeansonne B, Lu Q. 2009b. Increased nucleotide polymorphic changes in the 5′-untranslated region of δ-catenin (CTNND2) gene in prostate cancer. *Oncogene* **28:** 555–564.

Wang H, Vardy LA, Tan CP, Loo JM, Guo K, Li J, Lim SG, Zhou J, Chng WJ, Ng SB, et al. 2010. PCBP1 suppresses the translation of metastasis-associated PRL-3 phosphatase. *Cancer Cell* **18:** 52–62.

Waskiewicz AJ, Johnson JC, Penn B, Mahalingam M, Kimball SR, Cooper JA. 1999. Phosphorylation of the cap-binding protein eukaryotic translation initiation factor 4E by protein kinase Mnk1 in vivo. *Mol Cell Biol* **19:** 1871–1880.

Weinstein IB. 2000. Disorders in cell circuitry during multistage carcinogenesis: The role of homeostasis. *Carcinogenesis* **21:** 857–864.

Weinstein IB. 2002. Cancer. Addiction to oncogenes—The Achilles heal of cancer. *Science* **297:** 63–64.

Weinstein IB, Joe AK. 2006. Mechanisms of disease: Oncogene addiction—A rationale for molecular targeting in cancer therapy. *Nat Clin Pract Oncol* **3:** 448–457.

Wendel HG, De Stanchina E, Fridman JS, Malina A, Ray S, Kogan S, Cordon-Cardo C, Pelletier J, Lowe SW. 2004. Survival signalling by Akt and eIF4E in oncogenesis and cancer therapy. *Nature* **428:** 332–337.

Wendel HG, Silva RL, Malina A, Mills JR, Zhu H, Ueda T, Watanabe-Fukunaga R, Fukunaga R, Teruya-Feldstein J, Pelletier J, et al. 2007. Dissecting eIF4E action in tumorigenesis. *Genes Dev* **21:** 3232–3237.

Wilker EW, van Vugt MA, Artim SA, Huang PH, Petersen CP, Reinhardt HC, Feng Y, Sharp PA, Sonenberg N, White FM, et al. 2007. 14-3-3sigma controls mitotic translation to facilitate cytokinesis. *Nature* **446:** 329–332.

Willett M, Brocard M, Davide A, Morley SJ. 2011. Translation initiation factors and active sites of protein synthesis co-localize at the leading edge of migrating fibroblasts. *Biochem J* **438:** 217–227.

Wong CC, Traynor D, Basse N, Kay RR, Warren AJ. 2011. Defective ribosome assembly in Shwachman-Diamond syndrome. *Blood* **118:** 4305–4312.

Xie D, Ma NF, Pan ZZ, Wu HX, Liu YD, Wu GQ, Kung HF, Guan XY. 2008. Overexpression of EIF-5A2 is associated with metastasis of human colorectal carcinoma. *Hum Pathol* **39:** 80–86.

Xue S, Barna M. 2012. Specialized ribosomes: A new frontier in gene regulation and organismal biology. *Nat Rev Mol Cell Biol* **13:** 357–371.

Yang YL, Reis LF, Pavlovic J, Aguzzi A, Schafer R, Kumar A, Williams BR, Aguet M, Weissmann C. 1995. Deficient signaling in mice devoid of double-stranded RNA-dependent protein kinase. *EMBO J* **14:** 6095–6106.

Yoo YG, Kong G, Lee MO. 2006. Metastasis-associated protein 1 enhances stability of hypoxia-inducible factor-1α protein by recruiting histone deacetylase 1. *EMBO J* **25:** 1231–1241.

Yoon A, Peng G, Brandenburger Y, Zollo O, Xu W, Rego E, Ruggero D. 2006. Impaired control of IRES-mediated translation in X-linked dyskeratosis congenita. *Science* **312:** 902–906.

Yoshizawa A, Fukuoka J, Shimizu S, Shilo K, Franks TJ, Hewitt SM, Fujii T, Cordon-Cardo C, Jen J, Travis WD. 2010. Overexpression of phospho-eIF4E is associated with survival through AKT pathway in non-small cell lung cancer. *Clin Cancer Res* **16:** 240–248.

Yun JP, Miao J, Chen GG, Tian QH, Zhang CQ, Xiang J, Fu J, Lai PB. 2007. Increased expression of nucleophosmin/B23 in hepatocellular carcinoma and correlation with clinicopathological parameters. *Br J Cancer* **96:** 477–484.

Zeller KI, Zhao X, Lee CW, Chiu KP, Yao F, Yustein JT, Ooi HS, Orlov YL, Shahab A, Yong HC, et al. 2006. Global mapping of c-Myc binding sites and target gene networks in human B cells. *Proc Natl Acad Sci* **103:** 17834–17839.

Zhang L, Pan X, Hershey JW. 2007. Individual overexpression of five subunits of human translation initiation factor eIF3 promotes malignant transformation of immortal fibroblast cells. *J Biol Chem* **282:** 5790–5800.

Zhang T, Huang XH, Dong L, Hu D, Ge C, Zhan YQ, Xu WX, Yu M, Li W, Wang X, et al. 2010. PCBP-1 regulates alternative splicing of the CD44 gene and inhibits invasion in human hepatoma cell line HepG2 cells. *Mol Cancer* **9:** 72.

Zoncu R, Efeyan A, Sabatini DM. 2011. mTOR: From growth signal integration to cancer, diabetes and ageing. *Nat Rev Mol Cell Biol* **12:** 21–35.

Cytoplasmic RNA-Binding Proteins and the Control of Complex Brain Function

Jennifer C. Darnell[1] and Joel D. Richter[2]

[1]Department of Molecular Neuro-Oncology, Rockefeller University, New York, New York 10065

[2]Program in Molecular Medicine, University of Massachusetts Medical School, Worcester, Massachusetts 01605

Correspondence: darneje@mail.rockefeller.edu; joel.richter@umassmed.edu

The formation and maintenance of neural circuits in the mammal central nervous system (CNS) require the coordinated expression of genes not just at the transcriptional level, but at the translational level as well. Recent evidence shows that regulated messenger RNA (mRNA) translation is necessary for certain forms of synaptic plasticity, the cellular basis of learning and memory. In addition, regulated translation helps guide axonal growth cones to their targets on other neurons or at the neuromuscular junction. Several neurologic syndromes have been correlated with and indeed may be caused by aberrant translation; one important example is the fragile X mental retardation syndrome. Although translation in the CNS is regulated by multiple mechanisms and factors, we focus this review on regulatory mRNA-binding proteins with particular emphasis on fragile X mental retardation protein (FMRP) and cytoplasmic polyadenylation element binding (CPEB) because they have been shown to be at the nexus of translational control and brain function in health and disease.

The integrated circuit that is the mammalian CNS controls complex cognitive processes such as learning, memory, and behavior. Synapses, the points of communication between neurons, are essential for closing this circuit. They are morphologically dynamic structures that respond to neurotransmitter stimulation by increasing or decreasing the strength of their response, a phenomenon referred to as synaptic plasticity. In the most basic sense, synapses are composed of pre- and postsynaptic compartments; presynaptic boutons are axonal substructures that transmit signals, whereas postsynaptic domains primarily occur on the dendrites and cell bodies of the neurons that receive these signals. The adult presynaptic compartment generally lacks identifiable polysomes, suggesting that translation in this compartment is unlikely to play a substantial role in mature neuronal function. However, lack of detection of polysomes by electron microscopy is not proof of their absence, and indeed if only a few ribosomes are engaged with each messenger RNA (mRNA), then ultrastructural analysis may underestimate the translational capacity of this compartment. On the other hand, the postsynaptic compartment clearly contains polysomes, indicating that translation in this region could be important for higher-order brain activity (Steward and Levy 1982). Indeed, some forms of synaptic plasticity, which comprise the underlying cellular basis of learning and memory, almost certainly require mRNA translation in the postsynaptic region (reviewed by Kandel 2001; Sutton and Schuman 2006; Richter and Klann 2009). Moreover, this translation is "local" because at least with Schaffer collateral CA1 neurons of the hippocampus, it takes place in the dendritic layer even when it is severed from the nucleus-containing cell body (Kang and Schuman 1996; Huber et al. 2000). Translational control and RNA localization in neurons have been reviewed by a number of investigators (Holt and Bullock 2009; Wang et al. 2010; Doyle and Kiebler 2011); here, we focus on RNA-binding proteins as regulators of translation and complex brain function.

TRANSLATIONAL REGULATION IN NEURONS

It is now accepted wisdom that memory consolidation requires protein synthesis during a defined period relative to the training stimulus (Flexner et al. 1962; Flexner and Flexner 1968; Sutton and Schuman 2006). The foundation of this consolidation requires long-term changes in synaptic efficacy, measured electrophysiologically as long-term potentiation (LTP) or long-term depression (LTD), or ultrastructurally as shape changes in dendritic spines (Ostroff et al. 2002, 2010; Mishchenko et al. 2010; Dent et al. 2011). Before discussing particular mechanisms of translational control by RNA-binding proteins in neurons, however, synaptic plasticity should be placed into a molecular and cellular context that addresses not only how translation is regulated, but why. Consider that a typical neuron in the CNS may have thousands of synaptic inputs, yet when stimulated, the neuron can distinguish between experienced (stimulated) and naïve (unstimulated) synapses. How it does so has been a subject of intense investigation, but it is generally thought that the neuron "tags" the stimulated synapses, which results in changes in synaptic efficacy (Frey and Morris 1997). The nature of the tag may be complex, but at least in some circumstances, it involves de novo translation in dendrites. Thus, cellular

memory, which when placed within a complex circuitry forms the basis for organismal memory, requires regulated translation. Moreover, certain stimulation protocols elicit changes in the synthesis of specific proteins (Scheetz et al. 2000), suggesting that general translational control mechanisms may not be sufficient for changes in plasticity. For reviews of the signaling pathways leading to changes in translation, see recent articles by Costa-Mattioli et al. (2009) and Hoeffer and Klann (2010).

FMRP AND THE FRAGILE X SYNDROME

Although the brain contains a multitude of mRNA-binding proteins, only a few have been shown to be required for proper neurologic function. Perhaps the most well known of these proteins is fragile X mental retardation protein (FMRP), the product of *FMR1*, the fragile X mental retardation syndrome gene. Individuals with the fragile X syndrome display a range of afflictions including epileptic-like cognitive deficits, autistic behaviors, epileptic-like seizures in childhood, and morphological anomalies such as elongated faces and large ears (Hagerman and Hagerman 2002; Penagarikano et al. 2007; Hagerman et al. 2009; Santoro et al. 2011). The syndrome is caused by a CGG triplet repeat expansion in the 5' UTR of the *FMR1* gene, resulting in abnormal DNA methylation and transcriptional silencing (Pieretti et al. 1991; Verkerk et al. 1991). FMRP is a complex RNA-binding protein that contains two KH (hnRNP K homology) domains and an RGG (arginine-glycine-glycine) box (Siomi et al. 1993), of which KH2 is perhaps the most critical for function (De Boulle et al. 1993; Zang et al. 2009). FMRP likely represses translation in dendrites as well as the cell body, yet when brain lysates are centrifuged through sucrose gradients, FMRP sediments with polysomes (Feng et al. 1997b; Khandjian et al. 2004; Stefani et al. 2004; Darnell et al. 2011), an unexpected association as translation is frequently repressed at the level of initiation and therefore most translational repressors are associated with mRNAs that are not polysome associated. The polyribosomal association of FMRP is dependent on the KH2 domain because a mutation in that region (I304N) abrogates RNA binding and polysome association (Feng et al. 1997a; Darnell et al. 2005a; Zang et al. 2009); as a consequence, expression of this nonfunctional protein elicits symptoms of the fragile X syndrome even when there is no CGG repeat (De Boulle et al. 1993). In addition, an I304N knockin mouse model also shows characteristics of the syndrome nearly identical to that of an FMRP knockout mouse (Zang et al. 2009). Taken together, these and other data indicate that the loss of FMRP function in regulating polysome-associated translational inhibition forms the molecular basis of the fragile X syndrome. It therefore follows that two keys for understanding the etiology of the syndrome lie with the identification of target mRNAs and the biochemical mechanism for their translational regulation.

The identification of mRNA targets and binding sites for RNA-binding proteins can be technically challenging processes. Two often-used approaches are in vitro RNA selection (Ellington and Szostak 1990) (also known as SELEX, selective enrichment of ligands by exponential enrichment [Irvine et al. 1991]) or coimmunoprecipitation of protein-RNA complexes followed by microarray analysis or RT-PCR identification of specific transcripts (RIP-Chip) (Tenenbaum et al. 2000). Although often successful, these methods have their limitations; SELEX-identified sequences do not take into account the in vivo milieu that can help determine sequence specificity and RIP-Chip has been criticized for its low stringency and a high rate of both false positives and false negatives (Mukherjee et al. 2011). To circumvent these issues, an in vivo UV cross-linking procedure (CLIP; when combined with high throughout sequencing, it is referred to as HITS-CLIP or CLIP-seq) has been devised that is highly specific with respect to both mRNA that is bound by a particular protein and the precise *cis* element that it recognizes (Ule et al. 2003; Licatalosi et al. 2008; Chi et al. 2009; Darnell et al. 2011; reviewed in Darnell 2010). HITS-CLIP uses UV irradiation to covalently link RNA-binding proteins and RNA, followed by stringent immunoprecipitation of the protein of interest, limited RNase digestion to

reduce the size of the RNA that is "CLIPed," addition of linkers, and high-throughput sequencing (Fig. 1). The details of the procedure have been described elsewhere (Ule et al. 2005; Jensen and Darnell 2008), as has an outline of the advantages this method has over other techniques (Darnell 2010; Zhang and Darnell 2011). Suffice it to say that CLIP and subsequent modifications (Granneman et al. 2009; Hafner et al. 2010; Konig et al. 2010) have now been widely used to analyze several RNA-binding proteins from a number of tissues (Darnell 2010).

FMRP offers a particularly illustrative example of the power of CLIP and the results that can be obtained. Previously, in vitro SELEX analysis of FMRP had shown that the RGG box bound a G-quadruplex structure (Darnell et al. 2001, 2004) and KH2 bound a "kissing complex" (Darnell et al. 2005b). Moreover, RIP-Chip analysis from mouse brain found that the protein interacted with ∼430 mRNAs (Brown et al. 2001). Application of the HITS-CLIP technique to polysome-associated FMRP revealed that FMRP interacted with a large and specific set of mRNAs highly enriched for those encoding components of both the pre- and postsynaptic compartments (Darnell et al. 2011). Interestingly, about half of the RIP-Chip-identified FMRP target mRNAs could not be confirmed by in vivo UV cross-linking, possibly illustrating the issue of false positives in the RIP-Chip assay (as in any method, CLIP may also have false positives as well). Moreover, only about a quarter of the ∼840 FMRP UV cross-linked mRNAs were found by the RIP-Chip approach, indicating a high false negative rate for RIP-Chip, likely because of their loss during immunoprecipitation owing to the lack of a covalent bond between protein and RNA. In addition and quite surprisingly, the preponderance of FMRP-binding sites (CLIP sequence tags) resided in mRNA coding sequences, and not in 5′ or 3′ UTRs as might be expected based on analogy with other RNA-binding proteins such as the neuronal Hu family (Darnell et al. 2011). Moreover, FMRP interacted with its target mRNAs with an even distribution throughout the coding sequence, similar to the distribution of ribosomes, with no discernible preference for sequence or structural

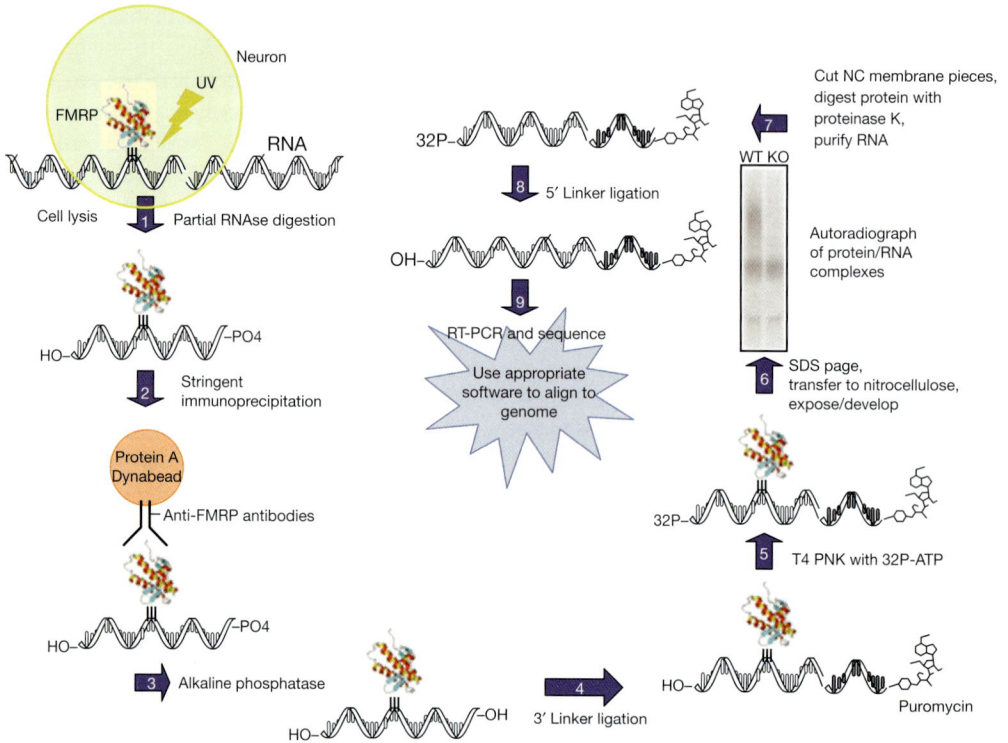

Figure 1. Basic scheme of the HITS-CLIP method. Beginning at the upper left and moving counterclockwise, tissue (e.g., brain) or cells is UV irradiated at 254 nm introducing covalent cross-links between RNA-protein complexes in living cells. 1. Cell lysis and partial RNase digestion reduces the modal size of cross-linked RNA "tags" to a size determined by the experimenter; 50–100 nucleotides is ideal. RNase A and T1 leave a 5'-OH and a 3' phosphate group on the digested RNA. 2. Cross-linking allows stringent immunoprecipitation of RNA-binding protein:RNA complexes. 3 and 4. The 3' phosphate is removed by alkaline phosphatase, to prevent intramolecular RNA circularization during the ligation of a linker to the 3' end of the RNA tags in step 4. This linker is blocked at the 3' end with a puromycin molecule to prevent competing linker-linker ligation reactions. 5. The RNA tags are then labeled at the 5' end with T4 polynucleotide kinase and ^{32}P-γ-ATP. 6. RNABP:RNA complexes are released from beads, run on denaturing SDS-PAGE, transferred to nitrocellulose, and imaged by autoradiography. SDS-PAGE and transfer to nitrocellulose are two important additional purification steps to separate the desired RNABP:RNA complexes away from other RNABPs or free RNA. 7. The radioactive RNABP:RNA complex is excised from the nitrocellulose filter, digested with proteinase K to remove the RNABP and to elute the RNA, which is then isolated by phenol-chloroform extraction and ethanol precipitation. 8. A second linker is added to the 5' end of the RNA. 9. The RNA is amplified by RT-PCR and sequenced by high-throughput sequencing methods. The sequences of the tags can then be aligned with the genome of interest to produce a genome-wide map of where the RNABP was bound to RNA in vivo.

motifs (Fig. 2, the placement of the ribosomes on the mRNA is for illustrative purposes only and is not meant to convey a particular congregation at the 3' end). These seemly paradoxical observations lead one to ask how does FMRP associate with polysomes yet still repress translation, and how does FMRP bind certain mRNAs in an apparently *cis* element-independent man-

ner? Finally, do the CLIP results give insight into the fragile X syndrome or autism spectrum disorders (ASDs)?

The observation that FMRP represses translation (Laggerbauer et al. 2001; Li et al. 2001) but does so while associating with polysomes (Khandjian et al. 1996, 2004; Corbin et al. 1997; Feng et al. 1997a,b; Stefani et al. 2004) has

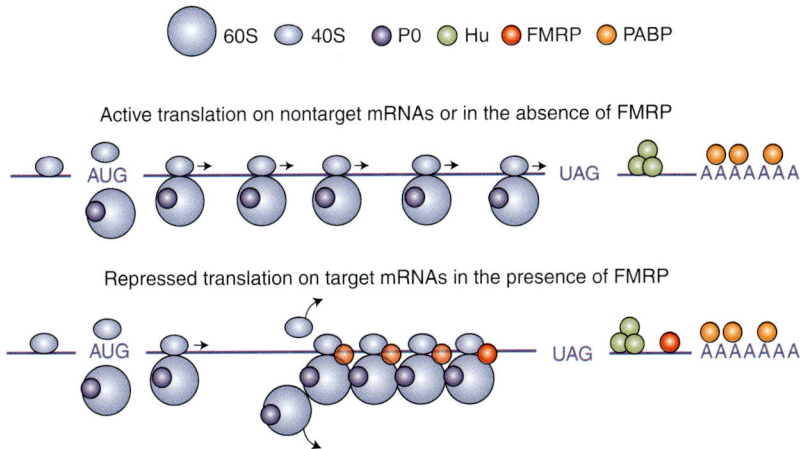

Figure 2. FMRP stalls ribosomes during elongation to repress protein synthesis. (*Upper* panel) Active translation: In the absence of FMRP, brain transcripts are shown being translated into protein by translocating ribosomes (made up of 40S and 60S subunits shown in light blue), which assemble at the start codon (i.e., AUG, initiation) and dissociate at the stop codon (i.e., UAG, termination). Ribosomal protein P0 is shown as a darker blue sphere on the 60S subunits. The poly(A)-binding protein (PABP) and the Hu family of RNABPs interacting with specific binding sites in 3′ UTRs are depicted with orange and green spheres, respectively. All four of these RNA-binding proteins are polyribosome associated, each by a different mechanism, and for this reason PABP, Hu, and P0 are shown as contrast for the properties and function of FMRP. (*Lower* panel) Repressed translation is associated with FMRP interaction with target mRNAs. FMRP preferentially interacts with specific mRNAs and in this context inhibits protein synthesis by stalling ribosomal translocation on those transcripts as part of a micrococcal nuclease-resistant multiribosome complex. This inhibition appears to be reversible, as it can be acutely relieved by competing FMRP off of polyribosomes with an RNA decoy; it is unknown whether this might occur in vivo owing to changes in the phosphorylation state of FMRP, its degradation by the proteasome, interactions with other proteins, noncoding RNAs, or other physiologic effectors. The stoichiometry of FMRP and stalled ribosomes remains to be determined. We have drawn a minimum of one (red) FMRP present in the stalled complex, recognizing the possibility that additional FMRP molecules (illustrated by transparent red figures) may be present. The presence of some FMRP in the UTRs (depicted on the 3′ UTR) is consistent with FMRP HITS-CLIP results. (From Darnell et al. 2011; reprinted with modifications, with permission, from Elsevier © 2011.)

given rise to the idea that it inhibits polypeptide elongation (ribosome transit) (Ceman et al. 2003) but much of the initial data supporting that idea was based on overexpression of FMRP, cultured cells, and the use of copy DNA (cDNA) reporter constructs. The identification and validation of a robust set of in vivo mRNA targets of FMRP, and an equally high-confidence set of nontarget mRNAs as controls, allowed evaluation of this hypothesis in a physiologically relevant setting (Darnell et al. 2011). Darnell et al. devised an in vitro assay of translation using the same starting material used to measure FMRP mRNA binding (a polyribosomal extract from mouse brain) so that mRNA binding could be

correlated with function with statistical significance. They found that FMRP target mRNAs were associated with stalled ribosomes that were resistant to runoff, whereas mRNAs that were not bound to FMRP (as assessed by HITS-CLIP) showed a typical run-off profile and shifted from the polysome region to lighter fractions of the sucrose gradient. Although the mechanism by which FMRP stalls ribosomes remains to be determined, it seems clear that this is the primary mechanism by which FMRP inhibits translation (illustrated in Fig. 2).

Presumably related to ribosome stalling is the fact that FMRP interacts with coding sequences. Does FMRP act as a "roadblock" for transiting ribosomes as has been suggested for ash1p mRNA in yeast (Chartrand et al. 2002)? Does it act in a manner analogous to SRP, which causes reversible ribosome stalling on mRNAs encoding secreted or membrane-associated proteins (Wolin and Walter 1988)? Perhaps FMRP blocks peptide elongation through interactions with eukaryotic elongation factor 2 (eEF2) (Sutton et al. 2007; Park et al. 2008). Much remains to be determined about FMRP's mechanism of action and whether its apparently reversible action to repress translation is mediated by its phosphorylation state (Ceman et al. 2003), turnover (Hou et al. 2006), or another mechanism.

Finally, the set of mRNA substrates to which FMRP cross-links begins to offer some insight into the etiology of the fragile X syndrome specifically, and ASDs more generally. For example, of the 842 FMRP target mRNAs, 28 have also been linked to ASDs including neuroligin 3, neurexin 1, shank3, PTEN, TSC2, and neurofibromatosis 1, suggesting that alterations in the functional levels of these proteins in neurons may underlie the common symptoms in fragile X syndrome and autism (Darnell et al. 2011). Other FMRP target mRNAs whose encoded proteins are implicated in ASDs include those involved in the ERK and mTOR pathways, NMDA receptor complexes, regulators of small GTPases, and cell–cell adhesion molecules (Kelleher and Bear 2008; Hoeffer and Klann 2010). The FMRP CLIP data have revealed other potential insights into FMRP function. Consider that about one-third of the FMRP CLIP targets encode presynaptic proteins (Darnell et al. 2011). Indeed, a presynaptic function for FMRP has been suggested by localization studies that have found FMRP in axons and growth cones (Feng et al. 1997b; Antar et al. 2006; Christie et al. 2009) as well as the characterization of axonal growth cone motility (Antar et al. 2006; Li et al. 2009), elongation (Tessier and Broadie 2008), and pathfinding (Michel et al. 2004) defects in both fly and mouse models of fragile X syndrome. In addition, there is experimental support for a presynaptic role in synapse formation and the establishment of circuitry (Zhang et al. 2001b; Hanson and Madison 2007; Bureau et al. 2008; Gibson et al. 2008). Recent studies on the role of

the Aplysia FMRP homolog (ApFMRP) in sensory to motor neuron synaptic plasticity supports both a pre- and postsynaptic role for FMRP in regulating protein synthesis in response to synaptic stimulation (Till et al. 2011). A fruitful area for further research is to connect these observations with the set of presynaptic mRNAs whose translation is regulated by FMRP (Darnell et al. 2011).

CPEB

The cytoplasmic polyadenylation element-binding protein CPEB is an mRNA-specific translational control factor that was initially identified for its role in regulating cytoplasmic polyadenylation during Xenopus oocyte maturation (reviewed in Mendez and Richter 2001). CPEB has two RNA-recognition domains (RRM motifs) and two zinc-finger motifs (Hake and Richter 1994; Hake et al. 1998), and by binding to the 3' UTR cytoplasmic polyadenylation element (CPE) found in specific mRNAs, it recruits a number of interacting proteins to modulate poly(A) tail length and as a result, translation (Richter 2007). CPEB activity begins in the nucleus, where it binds the CPE of pre-mRNAs, which like most nuclear pre-mRNAs, probably have long poly(A) tails (Lin et al. 2010). Following export to the cytoplasm, CPEB associates with a number of factors including Gld-2, a noncanonical poly(A) polymerase, and PARN, a poly(A) ribonuclease. PARN is the more active of the enzymes and thus shortens the poly(A) tail to usually ~20–40 nucleotides. Hormonal stimulation leads to CPEB phosphorylation, which expels PARN from the RNP complex; as a consequence, the poly(A) tail is elongated by Gld2 (Barnard et al. 2004; Kim and Richter 2006, 2007). The poly(A) tail is then bound by poly(A)-binding protein (PABP), which in turn, helps recruit the initiation complex to the 5' end of the mRNA. Indeed in a general sense, genome-wide analysis has shown that poly(A) tail length is positively correlated with both the ribosome density on a transcript and the degree of association of PABP, providing additional support for the model that poly(A) tail length regulates translation through PABP and the initiation complex, primarily eIF4G (Halbeisen et al. 2008).

CPEB also represses translation by recruiting specific eIF4E-binding proteins (4E-BPs, neuroguidin in neurons, maskin in oocytes) to the mRNA (Stebbins-Boaz et al. 1999; Richter and Sonenberg 2005; Jung et al. 2006). 4E-BPs repress translation by binding eIF4E, the cap-binding protein, which prevents eIF4E from associating with eIF4G and thus inhibits initiation. Neuroguidin localizes to puncta in dendrites and growth cones, and is frequently seen at the very tip of filopodia and at the leading edges of growth cones. Knockdown of the neuroguidin homolog in Xenopus causes a failure of neural tube closure and inhibition of neural crest cell migration, suggesting an important role for CPEB- and neuroguidin-mediated translational repression in neurons during development (Jung et al. 2006).

Vertebrates contain four CPEB paralogs; CPEB1 described above may be functionally distinct from CPEB proteins 2–4. Inhibition of CPEB1 activity, through expression of dominant–negative forms of the protein or in Cpeb1 knockout (KO) mice, leads to defects in learning, memory, synaptic plasticity, dendritic arborization, and neuronal circuit formation (Alarcon et al. 2004; Berger-Sweeney et al. 2006; McEvoy et al. 2007; Bestman and Cline 2008). CPEB1 inhibits translation of its target mRNAs until glutamatergic activation (Wells et al. 2001) stimulates its phosphorylation by either Aurora kinase A (Mendez et al. 2000a,b; Huang et al. 2002) or Camk2a (Atkins et al. 2004, 2005), resulting in increased mRNA polyadenylation and translation at synapses. The strongest evidence for participation of CPEB1 in *local* translation is its localization to postsynaptic sites in dendrites (Wu et al. 1998) and enrichment of phospho-CPEB1 in a postsynaptic density (PSD) fraction purified from rat neurons (Atkins et al. 2004), although it is not entirely clear that the antibodies used did not cross-react with either the CPEB2-4s or other phosphoproteins because a Cpeb1 KO was not available as a negative control at that time.

Recently it was suggested that CPEB1 may also regulate local translation in hippocampal growth cones in response to neurotrophin stimulation through regulation of CPE sites in β-catenin mRNA (Ctnnb1) (Kundel et al. 2009). This raises an interesting issue in the context of a related study by Holt and colleagues, also addressing whether CPEB regulates local translation in the growth cone (Lin et al. 2009). In the latter study both the CPE and cytoplasmic polyadenylation of mRNAs were required for translation-dependent chemotropic responses, but CPEB1 itself was not. Because other RNA-binding proteins can potentially bind the CPE and regulate cytoplasmic polyadenylation (Slevin et al. 2007), studies using overexpression of the CPEB RNA-binding domain (RBD) to inhibit CPEB activity may also block CPE binding by other RNABPs. The CPE, with a consensus of $U_{4-6}AUU$, is similar to other U-rich or UA-rich binding sites for the large family of ARE-binding proteins, including Hu (discussed below), and, in fact, the Xenopus homolog of Hu, ElrA, has been reported to bind the CPE (Slevin et al. 2007). The extent to which mammalian neuronal Hu isoforms may interact with CPEB1 to control translation is an interesting question for further study.

In sum, although it is likely that CPEB1 fulfills an important role in regulating neuronal translation it is not fully established that CPEB1 is responsible for local translational control underlying long-term synaptic plasticity changes. CPEB2–4 are also expressed in neurons and there is significant data from Aplysia and fly homologs to suggest that they may also regulate local translation. CPEB2–4 differ from CPEB1 in that they do not bind the CPE and do not regulate polyadenylation but instead recognize a different RNA consensus sequence (Huang et al. 2006). They also lack the Aurora kinase A phosphorylation site of CPEB1 (Theis et al. 2003). They have been shown to regulate translation in an N-methyl-D-aspartate receptor (NMDAR)-dependent manner in neurons, through a mechanism independent of CPE binding and polyadenylation (Huang et al. 2006).

The Aplysia CPEB homolog, ApCPEB, is neuron specific and most similar to CPEB3 (Liu and Schwartz 2003; Si et al. 2003a; Theis et al. 2003). Like CPEB3 and 4 in mouse, ApCPEB is induced in response to synaptic activation (Theis et al. 2003) and is required for long-term facilitation at the activated synapse, strengthening the model that "type 2" neuronal CPEBs regulate protein synthesis-dependent forms of long-term synaptic plasticity. Owing to an amino-terminal glutamine-rich domain, ApCPEB has the additional property of forming prionlike particles that may be related to synaptic changes underlying long-term memory storage or form a synaptic "tag" marking that synapse (Darnell 2003; Si et al. 2003b, 2010).

Notably, loss of function of the fly homolog of CPEB1 (Orb) has phenotypes in the fly oocyte (Lantz et al. 1994) but has not been linked to a neuronal defect. However, the CPEB2–4 homolog, Orb2, expressed predominantly in neurons and also containing a prionlike domain, is necessary for some forms of long-term memory (LTM). Intriguingly, its effects on LTM are dependent on the prionlike domain (Keleman et al. 2007). Taken together, results from Aplysia and flies suggest that elucidation of the mRNA targets and functions of CPEB2-4 in addition to CPEB1 will clarify mechanisms underlying some forms of LTM.

PUMILIO

The regulation of cytoplasmic polyadenylation is an important point of translational control for another family of RNABPs, the pumilio (pum) or PUF family (Wickens et al. 2002; Quenault et al. 2011). PUF proteins harbor a conserved Pumilio homology domain that binds RNA, and the proteins regulate cytoplasmic poly(A) length and may compete or cooperate with CPEB1. Like CPEB1, Pumilio regulates gene expression through two mechanisms, in this case by (1) affecting the stability of target mRNAs through recruitment of a deadenylase complex to shorten poly(A) tails and promote mRNA turnover (reviewed in Quenault et al. 2011) and (2) repressing translation directly. A carboxy-terminal RBD in Pumilio, first described by Zamore and colleagues (1997), recognizes the 16 nt "nanos response element" (NRE) in the 3′ UTR of hunchback mRNA in fly embryos (Wharton and Struhl 1991). Pumilio recruits two other proteins: Nanos, which is also involved in RNA localization, and Brat, which binds the eIF4E-like cap-binding protein d4EHP (Cho et al. 2006). Unlike eIF4E, d4EHP does not interact with deIF4G; consequently initiation does not occur because without eIF4G, the 40S ribosomal subunit is not recruited to the mRNA (Cho et al. 2006). Pumilio also has been shown to have cap-binding activity itself and so may compete with eIF4E to block initiation (Cao et al. 2010). Pumilio activates translation in some cases through cooperative binding with CPEB1 on transcripts containing both NRE and CPE elements, and may stabilize CPEB binding, but translational activation does not appear to involve altered polyadenylation of target mRNAs (Pique et al. 2008). Pumilio has been found to associate with a Nanos homolog and CPEB in Xenopus (Nakahata et al. 2001), supporting the conservation of this mechanism.

There are two human homologs, Pum1 and 2, present in the brain. To date, neither has been linked to a human cognitive disease, and there are no reports of neuronal defects in KO mice to date. Nonetheless, compelling evidence exists for an important role in mammalian neuronal function (reviewed in Baines 2005). Loss of Pumilio leads to defective LTM in flies (Dubnau et al. 2003), likely owing to its effects on dendrite morphogenesis as well as synapse growth and function (Mee et al. 2004; Menon et al. 2004; Ye et al. 2004). Pum2 is present in RNPs in the cell body and dendrites of rodent hippocampal neurons (Vessey et al. 2006). Studies using bidirectional manipulation of Pum2 levels show reciprocal changes in dendrite outgrowth, the density and morphology of dendritic spines, and the frequency of miniature excitatory postsynaptic currents (mEPSCs) (Vessey et al. 2010), revealing that Pum2 negatively regulates all three phenotypes. The observed increased mEPSC frequency appears to be owing to an increased number of excitatory synapses on dendritic shafts in the absence of Pum2 (Vessey et al. 2010).

Zip Code Binding Protein

Zip code binding protein 1 (ZBP1) is an interesting example of a neuronal RNA-binding protein recognized for its role in mRNA localization that also represses translation. ZBP1 harbors 4 KH-type RBDs. The homologs of chicken ZBP1 (mammalian IMP1-3 or IGF2BP1-3) are oncofetal proteins that are highly expressed during development, absent postnatally, and are reexpressed in a high percentage of tumors (Tessier et al. 2004). Nonetheless, ZBP1 appears to have an important role in developing neurons. It is present in granules in the somatodendritic compartment together with β-actin mRNA, actb, through interactions with a 54-nucleotide binding site (the "zip code") in the 3′ UTR (Zhang et al. 2001a). These granules contain components of the exon junction complex, indicating that they are translationally repressed on exiting the nucleus (Jonson et al. 2007). In primary neurons, the levels of ZBP1 in growth cones (Zhang et al. 2001a), dendrites, and spines (Tiruchinapalli et al. 2003) is positively correlated with neuronal activity. Significantly, knockdown of ZBP1 in cultured neurons eliminates a BDNF-stimulated increase in actin-rich dendritic spines (Eom et al. 2003). Knockdown or overexpression of ZBP1 in cultured neurons was found to decrease complexity of the dendritic arbor and this was dependent on the ZBP1 RBDs (Perycz et al. 2011). No effect was found on established arbors, suggesting an important role for ZBP1 during the critical period of dendritogenesis (Perycz et al. 2011).

Direct translational control by ZBP1 in neurons was shown by Huttelmaier and colleagues who found that ZBP1 repressed initiation on neuronal actb mRNA by binding to the 3′ UTR zip-code sequence and interfering with 80S complex formation (Huttelmaier et al. 2005). Mechanistic studies revealed that phosphorylation of ZBP1 by Src kinase on Y396 decreased binding to the zip code and allowed initiation, resulting in local synthesis of β-actin (Huttelmaier et al. 2005). This mechanism was also reported to regulate local translation of actb in growth cones to allow turning in response to brain-derivid neurotrophic factor (BDNF) (Sasaki et al. 2010). Furthermore, nonphosphorylatable Y396F-ZBP1 was unable to rescue the ZBP1 knockdown effect on dendritic arborization (Perycz et al. 2011). ZBP1 has also been reported to increase internal ribosome entry site

(IRES) -dependent initiation of the hepatitis C virus RNA by interacting with both 5′ and 3′ UTRs (Weinlich et al. 2009); however, a function to regulate IRES-dependent initiation in neurons has not been established.

CAPRIN

Caprins1 and 2 (also known as RNG105 and RNG140) are among the newest additions to the list of potential translational regulators in neurons. Caprins bind RNA through an amino-terminal coiled-coil motif and a carboxy-terminal RGG box. Caprin1 (Grill et al. 2004) associates with polyribosomes in rat cortex (Angenstein et al. 2005) and is a component of the same ribosome-containing granules as the fragile X mental retardation autosomal homolog proteins FXRPs (Shiina et al. 2005; Shiina and Tokunaga 2010), suggesting a role in translational regulation similar to that of the FMRP family. Knockdown of caprin1 or caprin2 in cultured mouse neurons causes a reduction in the number, length, and branching of dendrites (Shiina and Tokunaga 2010) and studies on primary neuronal cultures from caprin1 KO mice show defects in synapse development (Shiina et al. 2010). BDNF treatment caused the dissociation of caprin1 and its mRNA targets from these granules into polyribosomes, suggesting that neural activity relieves translational repression imposed by caprin1 (Shiina et al. 2005). Caprin1 and caprin2 localize to different granules in dendrites and so appear to have distinct functions in neurons, although both appear to be involved in dendrite and spine development.

The caprins nonspecifically repress translation in in vitro translation systems and globally repress translation when overexpressed in vivo. These findings might suggest that the caprins repress translation nonspecifically. However, loss of function of caprin does not increase global translation rates, raising a more general issue regarding overexpression of RNABPs in functional assays, which has been addressed in the context of caprin-induced stress granule (SG) formation (Solomon et al. 2007). SGs are the result of clustering of mRNAs, stalled preinitiation complexes, and RNABPs caused by phosphorylation of eIF2α or inhibition of the eIF4F complex (see Decker and Parker 2012). Caprins belong to a group of mRNA-binding proteins (including FMRP) that nucleate SGs when expressed at greater than physiological levels. Solomon et al. (2007) found that caprin overexpression caused eIF2α phosphorylation resulting in both global translational repression and the formation of SGs. Although it is possible that this might be related to the physiologic role of caprins, it raises a red flag concerning the assignment of the role of translational repressor based solely on overexpression studies.

HU

The mammalian neuronal homologs of the *Drosophila* ELAV protein (Hu B, C, and D) have been reported to regulate mRNA stability (Bolognani and Perrone-Bizzozero 2008), translation (Fukao et al. 2009), and pre-mRNA splicing (Zhu et al. 2006) to regulate nervous system development and function (Okano and Darnell 1997; Akamatsu et al. 1999; reviewed by Antic and Keene 1997; Hinman and Lou 2008). ELAV is known to be essential for nervous system development and function in flies (Campos et al. 1985; Robinow et al. 1988). The mammalian homologs were first cloned as antigens of paraneoplastic antibodies in the human neurologic Hu syndrome (Szabo et al. 1991; Posner and Dalmau 1997) and studies using mice in which these proteins have been knocked out or overexpressed show a number of neuronal, learning, and memory phenotypes (Akamatsu et al. 2005; Bolognani et al. 2006, 2007a,b; Tanner et al. 2008). Hu has been consistently shown to be up-regulated in hippocampal neurons after contextual (Bolognani et al. 2004) or spatial learning tasks (Quattrone et al. 2001; Pascale et al. 2004) and after glutamate receptor activation (Tiruchinapalli et al. 2008). The roles of the neuronal Hu proteins in development, plasticity, and memory have been recently reviewed (Deschenes-Furry et al. 2006; Pascale et al. 2008).

Hu proteins have three RRM domains and bind to AU- and GU-rich elements in the 3′ UTRs of specific mRNA transcripts. These ele-

ments, also known as AREs, are also binding sites for a number of RNABPs that promote turnover of these mRNAs, including KSRP, TTP, AUF1, and BRF1. Hu is thought to bind the same or similar elements to protect ARE-containing transcripts from degradation. As a group, the ARE-binding proteins regulate mRNA stability in response to cell stimulation including stress, proliferative stimulation, immune signaling, and developmental signals.

Hu can directly regulate the translation of target mRNAs as well. Neuronal Hu's are localized to somatic and dendritic granules containing ribosomes (Bolognani et al. 2004) and are polyribosome-associated in human medulloblastoma cells (Gao and Keene 1996), PC12 cells (Fukao et al. 2009), primary neurons (Tiruchinapalli et al. 2008), and brain (Bolognani et al. 2004; Darnell et al. 2009). Polyribosome-associated Hu levels increase after KCl stimulation (Tiruchinapalli et al. 2008) or learning (Bolognani et al. 2004), suggesting that Hu increases translation of its mRNA targets in response to neuronal activity by promoting initiation (Chen and Shyu 2009; Fukao et al. 2009) supported by the increased association of its mRNA targets with polyribosomes (Antic et al. 1999; Mazan-Mamczarz et al. 2003; Kawai et al. 2006; Galban et al. 2008). Stimulation of translation by HuD has been shown to be dependent on both the poly-A tail and 5′ cap and occurs through direct interaction of HuD with the helicase eIF4A (Fukao et al. 2009). This mechanism of facilitating initiation is critical for Hu-dependent neurite outgrowth in PC12 cells (Fukao et al. 2009). Hu has also been proposed to inhibit translation in some cases (reviewed in Hinman and Lou 2008).

CONCERTED FUNCTION OF TRANSLATIONAL REGULATORY PROTEINS IN NEURONS

In sum, intriguing evidence supports the idea that specific mRNA-binding proteins regulate translation that in turn mediates neuronal function including synaptic plasticity. Moreover, mounting evidence suggests that such regulation is likely to require the concerted action of the translational regulators discussed in this review. Although identification of important neuronal translational regulatory proteins through genetic and biochemical approaches has been quite successful, identification of the mRNAs whose expression is controlled by these binding proteins remains a major hurdle limiting our understanding of the mechanisms involved. Application of HITS-CLIP and related techniques to FMRP, CPEB, Pumilio, ZBP, caprin, and Hu in brain, and their use in quantifying changes in RNA binding in response to activity, or in subcellular fractions such as polyribosomes or purifiable granules, or during development will likely lead to dramatic advances in our understanding of how these proteins fine tune the synthesis of key neuronal proteins.

RELEVANCE TO THERAPY FOR HUMAN COGNITIVE AND BEHAVIORAL DISEASES

One approach to modulating translation in neurons to ameliorate symptoms of diseases such as fragile X syndrome or some cases of autism, both thought to be caused by "runaway translation," is to target the general signaling pathways that transduce engagement of cell-surface receptors to increased translation, including PI3K/Akt/mTOR and MEK/ERK (Kelleher and Bear 2008; Hoeffer and Klann 2010). These pathways are already the subject of anticancer therapy research because the cell cycle uses similar pathways to transduce growth factor signals into translational output as do neurons (Guertin and Sabatini 2005; Tee and Blenis 2005; Sabatini 2006; Frost et al. 2009; Dowling et al. 2010; Livingstone et al. 2010; Silvera et al. 2010). Hopefully, very specific therapies may result from understanding binding sites and mechanism of translational control at the molecular level, both of which can be addressed through HITS-CLIP "mapping" of physiologic RNA-binding protein:RNA interactions paired with appropriately designed mechanistic studies using endogenous mRNA targets. The majority of the mRNA translational regulatory proteins are thought to *repress* translation whether at the level of elongation, like FMRP, or initiation, like CPEB, pumilio, and ZBP. Replacement of the

repression lost in disease might be achieved by delivery of a mimic of FMRP activity to stall ribosomes, for example. Furthermore, it's possible that generally slowing translation with the off-target use of antibiotics might take the edge off of excessive translation and provide therapeutic benefit, as has been suggested for fragile X syndrome (Darnell et al. 2011).

REFERENCES

*Reference is also in this collection.

Akamatsu W, Okano HJ, Osumi N, Inoue T, Nakamura S, Sakakibara S, Miura M, Matsuo N, Darnell RB, Okano H, et al. 1999. Mammalian ELAV-like neuronal RNA-binding proteins HuB and HuC promote neuronal development in both the central and the peripheral nervous systems. *Proc Natl Acad Sci* **96**: 9885–9890.

Akamatsu W, Fujihara H, Mitsuhashi T, Yano M, Shibata S, Hayakawa Y, Okano HJ, Sakakibara S, Takano H, Takano T, et al. 2005. The RNA-binding protein HuD regulates neuronal cell identity and maturation. *Proc Natl Acad Sci* **102**: 4625–4630.

Alarcon JM, Hodgman R, Theis M, Huang YS, Kandel ER, Richter JD. 2004. Selective modulation of some forms of Schaffer collateral-CA1 synaptic plasticity in mice with a disruption of the CPEB-1 gene. *Learn Mem* **11**: 318–327.

Angenstein F, Evans AM, Ling SC, Settlage RE, Ficarro S, Carrero-Martinez FA, Shabanowitz J, Hunt DF, Greenough WT. 2005. Proteomic characterization of messenger ribonucleoprotein complexes bound to nontranslated or translated poly(A) mRNAs in the rat cerebral cortex. *J Biol Chem* **280**: 6496–6503.

Antar LN, Li C, Zhang H, Carroll RC, Bassell GJ. 2006. Local functions for FMRP in axon growth cone motility and activity-dependent regulation of filopodia and spine synapses. *Mol Cell Neurosci* **32**: 37–48.

Antic D, Keene JD. 1997. Embryonic lethal abnormal visual RNA-binding proteins involved in growth, differentiation, and posttranscriptional gene expression. *Am J Hum Genet* **61**: 273–278.

Antic D, Lu N, Keene JD. 1999. ELAV tumor antigen, HelN1, increases translation of neurofilament M mRNA and induces formation of neurites in human teratocarcinoma cells. *Genes Dev* **13**: 449–461.

Atkins CM, Nozaki N, Shigeri Y, Soderling TR. 2004. Cytoplasmic polyadenylation element binding protein-dependent protein synthesis is regulated by calcium/calmodulin-dependent protein kinase II. *J Neurosci* **24**: 5193–5201.

Atkins CM, Davare MA, Oh MC, Derkach V, Soderling TR. 2005. Bidirectional regulation of cytoplasmic polyadenylation element-binding protein phosphorylation by $Ca2^+$/calmodulin-dependent protein kinase II and protein phosphatase 1 during hippocampal long-term potentiation. *J Neurosci* **25**: 5604–5610.

Baines RA. 2005. Neuronal homeostasis through translational control. *Mol Neurobiol* **32**: 113–121.

Barnard DC, Ryan K, Manley JL, Richter JD. 2004. Symple-kin and xGLD-2 are required for CPEB-mediated cyto-plasmic polyadenylation. *Cell* 119: 641–651.

Berger-Sweeney J, Zearfoss NR, Richter JD. 2006. Reduced extinction of hippocampal-dependent memories in CPEB knockout mice. *Learn Mem* 13: 4–7.

Bestman JE, Cline HT. 2008. The RNA binding protein CPEB regulates dendrite morphogenesis and neuronal circuit assembly in vivo. *Proc Natl Acad Sci* 105: 20494–20499.

Bolognani F, Perrone-Bizzozero NI. 2008. RNA-protein in-teractions and control of mRNA stability in neurons. *J Neurosci Res* 86: 481–489.

Bolognani F, Merhege MA, Twiss J, Perrone-Bizzozero NI. 2004. Dendritic localization of the RNA-binding protein HuD in hippocampal neurons: Association with poly-somes and upregulation during contextual learning. *Neurosci Lett* 371: 152–157.

Bolognani F, Tanner DC, Merhege M, Deschenes-Furry J, Jasmin B, Perrone-Bizzozero NI, et al. 2006. In vivo post-transcriptional regulation of GAP-43 mRNA by overex-pression of the RNA-binding protein HuD. *J Neurochem* 96: 790–801.

Bolognani F, Qiu S, Tanner DC, Paik J, Perrone-Bizzozero NI, Weeber EJ, et al. 2007a. Associative and spatial learn-ing and memory deficits in transgenic mice overexpress-ing the RNA-binding protein HuD. *Neurobiol Learn Mem* 87: 635–643.

Bolognani F, Tanner DC, Nixon S, Okano HJ, Okano H, Perrone-Bizzozero NI, et al. 2007b. Coordinated expres-sion of HuD and GAP-43 in hippocampal dentate gran-ule cells during developmental and adult plasticity. *Neu-rochem Res* 32: 2142–2151.

Brown V, Jin P, Ceman S, Darnell JC, O'Donnell WT, Tenenbaum SA, Jin X, Feng Y, Wilkinson KD, Keene JD, et al. 2001. Microarray identification of FMRP-associated brain mRNAs and altered mRNA translational profiles in Fragile X Syndrome. *Cell* 107: 477–487.

Bureau I, Shepherd GM, Svoboda K. 2008. Circuit and plas-ticity defects in the developing somatosensory cortex of FMR1 knock-out mice. *J Neurosci* 28: 5178–5188.

Campos AR, Grossman D, White K. 1985. Mutant alleles at the locus *elav* in *Drosophila melonagaster* lead to nervous system defects. A developmental-genetic analysis. *J Neu-rogenet* 2: 197–218.

Cao Q, Padmanabhan K, Richter JD. 2010. Pumilio 2 con-trols translation by competing with eIF4E for 7-methyl guanosine cap recognition. *RNA* 16: 221–227.

Ceman S, O'Donnell WT, Reed M, Patton S, Pohl J, Warren ST. 2003. Phosphorylation influences the translation state of FMRP-associated polyribosomes. *Hum Mol Genet* 12: 3295–3305.

Chartrand P, Meng XH, Huttelmaier S, Donato D, Singer RH. 2002. Asymmetric sorting of ash1p in yeast results from inhibition of translation by localization elements in the mRNA. *Mol Cell* 10: 1319–1330.

Chen CY, Shyu AB. 2009. HuD stimulates translation via eIF4A. *Mol Cell* 36: 920–921.

Chi SW, Zang JB, Mele A, Darnell RB. 2009. Argonaute HITS-CLIP decodes microRNA-mRNA interaction maps. *Nature* 460: 479–486.

Cho PF, Gamberi C, Cho-Park YA, Cho-Park IB, Lasko P, Sonenberg N. 2006. Cap-dependent translational inhibi-tion establishes two opposing morphogen gradients in *Drosophila* embryos. *Curr Biol* 16: 2035–2041.

Christie SB, Akins MR, Schwob JE, Fallon JR. 2009. The FXG: A presynaptic fragile X granule expressed in a subset of developing brain circuits. *J Neurosci* 29: 1514–1524.

Corbin F, Bouillon M, Fortin A, Morin S, Rousseau F, Khandjian EW. 1997. The fragile X mental retarda-tion protein is associated with poly(A)$^+$ mRNA in active-ly translating polyribosomes. *Hum Mol Genet* 6: 1465–1472.

Costa-Mattioli M, Sossin WS, Klann E, Sonenberg N. 2009. Translational control of long-lasting synaptic plasticity and memory. *Neuron* 61: 10–26.

Darnell RB. 2003. Memory, synaptic translation, and pri-ons? *Cell* 115: 767–768.

Darnell RB. 2010. HITS-CLIP: Panoramic views of protein-RNA regulation in living cells. *RNA* 1: 266–286.

Darnell JC, Jensen KB, Jin P, Brown V, Warren ST, Darnell RB. 2001. Fragile X mental retardation protein targets G Quartet mRNAs important for neuronal function. *Cell* 107: 489–499.

Darnell JC, Warren ST, Darnell RB. 2004. The fragile X men-tal retardation protein, FMRP, recognizes G-quartets. *Ment Retard Dev Disabil Res Rev* 10: 49–52.

Darnell JC, Mostovetsky O, Darnell RB. 2005a. FMRP RNA targets: Identification and validation. *Genes Brain Behav* 4: 341–349.

Darnell JC, Fraser CE, Mostovetsky O, Stefani G, Jones TA, Eddy SR, Darnell RB. 2005b. Kissing complex RNAs me-diate interaction between the Fragile-X mental retarda-tion protein KH2 domain and brain polyribosomes. *Genes Dev* 19: 903–918.

Darnell JC, Fraser CE, Mostovetsky O, Darnell RB. 2009. Discrimination of common and unique RNA-binding activities among Fragile X mental retardation protein paralogs. *Hum Mol Genet* 18: 3164–3177.

Darnell JC, Van Driesche SJ, Zhang C, Hung KY, Mele A, Fraser CE, Stone EF, Chen C, Fak JJ, Chi SW, et al. 2011. FMRP stalls ribosomal translocation on mRNAs linked to synaptic function and autism. *Cell* 146: 247–261.

De Boulle K, Verkerk AJ, Reyniers E, Vits L, Hendrickx J, Van Roy B, Van den Bos F, de Graaff E, Oostra BA, Willems PJ, et al. 1993. A point mutation in the FMR-1 gene associ-ated with fragile X mental retardation. *Nat Genet* 3: 31–35.

* Decker CJ, Parker R. 2012. P bodies and stress granules: Possible roles in the control of translation and mRNA degradation. *Cold Spring Harb Perspect Biol* doi: 10.1101/cshperspect.a012286.

Dent EW, Merriam EB, Hu X. 2011. The dynamic cytoskel-eton: Backbone of dendritic spine plasticity. *Curr Opin Neurobiol* 21: 175–181.

Deschenes-Furry J, Perrone-Bizzozero N, Jasmin BJ. 2006. The RNA-binding protein HuD: A regulator of neuronal differentiation, maintenance and plasticity. *Bioessays* 28: 822–833.

Dowling RJ, Topisirovic I, Fonseca BD, Sonenberg N. 2010. Dissecting the role of mTOR: Lessons from mTOR in-hibitors. *Biochim Biophys Acta* 1804: 433–439.

Doyle M, Kiebler MA. 2011. Mechanisms of dendritic mRNA transport and its role in synaptic tagging. *EMBO J* **30:** 3540–3552.

Dubnau J, Chiang AS, Grady L, Barditch J, Gossweiler S, McNeil J, Smith P, Buldoc F, Scott R, Certa U, et al. 2003. The staufen/pumilio pathway is involved in *Drosophila* long-term memory. *Curr Biol* **13:** 286–296.

Ellington AD, Szostak JW. 1990. In vitro selection of RNA molecules that bind specific ligands. *Nature* **346:** 818–822.

Eom T, Antar LN, Singer RH, Bassell GJ. 2003. Localization of a β-actin messenger ribonucleoprotein complex with zipcode-binding protein modulates the density of dendritic filopodia and filopodial synapses. *J Neurosci* **23:** 10433–10444.

Feng Y, Absher D, Eberhart DE, Brown V, Malter HE, Warren ST. 1997a. FMRP associates with polyribosomes as an mRNP, and the I304N mutation of severe fragile X syndrome abolishes this association. *Mol Cell* **1:** 109–118.

Feng Y, Gutekunst CA, Eberhart DE, Yi H, Warren ST, Hersch SM. 1997b. Fragile X mental retardation protein: Nucleocytoplasmic shuttling and association with somatodendritic ribosomes. *J Neurosci* **17:** 1539–1547.

Flexner LB, Flexner JB. 1968. Intracerebral saline: Effect on memory of trained mice treated with puromycin. *Science* **159:** 330–331.

Flexner JB, Flexner LB, Stellar E, De Lahabag, Roberts RB. 1962. Inhibition of protein synthesis in brain and learning and memory following puromycin. *J Neurochem* **9:** 595–605.

Frey U, Morris RG. 1997. Synaptic tagging and long-term potentiation. *Nature* **385:** 533–536.

Frost P, Shi Y, Hoang B, Gera J, Lichtenstein A. 2009. Regulation of D-cyclin translation inhibition in myeloma cells treated with mammalian target of rapamycin inhibitors: Rationale for combined treatment with extracellular signal-regulated kinase inhibitors and rapamycin. *Mol Cancer Ther* **8:** 83–93.

Fukao A, Sasano Y, Imataka H, Inoue K, Sakamoto H, Sonenberg N, Thoma C, Fujiwara T. 2009. The ELAV protein HuD stimulates cap-dependent translation in a Poly(A)- and eIF4A-dependent manner. *Mol Cell* **36:** 1007–1017.

Galban S, Kuwano Y, Pullmann RJ, Martindale JL, Kim HH, Lal A, Abdelmohsen K, Yang X, Dang Y, Liu JO, et al. 2008. RNA-binding proteins HuR and PTB promote the translation of hypoxia-inducible factor 1α. *Mol Cell Biol* **28:** 93–107.

Gao FB, Keene JD. 1996. Hel-N1/Hel-N2 proteins are bound to poly(A)$^+$ mRNA in granular RNP structures and are implicated in neuronal differentiation. *J Cell Sci* **109:** 579–589.

Gibson JR, Bartley AF, Hays SA, Huber KM. 2008. Imbalance of neocortical excitation and inhibition and altered UP states reflect network hyperexcitability in the mouse model of fragile X syndrome. *J Neurophysiol* **100:** 2615–2626.

Granneman S, Kudla G, Petfalski E, Tollervey D. 2009. Identification of protein binding sites on U3 snoRNA and pre-rRNA by UV cross-linking and high-throughput analysis of cDNAs. *Proc Natl Acad Sci* **106:** 9613–9618.

Grill B, Wilson GM, Zhang KX, Wang B, Doyonnas R, Quadroni M, Schrader JW. 2004. Activation/division of lymphocytes results in increased levels of cytoplasmic activation/proliferation-associated protein-1: Prototype of a new family of proteins. *J Immunol* **172:** 2389–2400.

Guertin DA, Sabatini DM. 2005. An expanding role for mTOR in cancer. *Trends Mol Med* **11:** 353–361.

Hafner M, Landthaler M, Burger L, Khorshid M, Hausser J, Berninger P, Rothballer A, Ascano MJ Jr, Jungkamp AC, Munschauer M, et al. 2010. Transcriptome-wide identification of RNA-binding protein and microRNA target sites by PAR-CLIP. *Cell* **141:** 129–141.

Hagerman RJ, Hagerman PJ, eds. 2002. *Fragile X syndrome: Diagnosis, treatment, and research.* The Johns Hopkins University Press, Baltimore and London.

Hagerman RJ, Berry-Kravis E, Kaufmann WE, Ono MY, Tartaglia N, Lachiewicz A, Kronk R, Delahunty C, Hessl D, Visootsak J, et al. 2009. Advances in the treatment of fragile X syndrome. *Pediatrics* **123:** 378–390.

Hake LE, Richter JD. 1994. CPEB is a specificity factor that mediates cytoplasmic polyadenylation during Xenopus oocyte maturation. *Cell* **79:** 617–627.

Hake LE, Mendez R, Richter JD. 1998. Specificity of RNA binding by CPEB: Requirement for RNA recognition motifs and a novel zinc finger. *Mol Cell Biol* **18:** 685–693.

Halbeisen RE, Galgano A, Scherrer T, Gerber AP. 2008. Posttranscriptional gene regulation: From genome-wide studies to principles. *Cell Mol Life Sci* **65:** 798–813.

Hanson JE, Madison DV. 2007. Presynaptic FMR1 genotype influences the degree of synaptic connectivity in a mosaic mouse model of fragile X syndrome. *J Neurosci* **27:** 4014–4018.

Hinman MN, Lou H. 2008. Diverse molecular functions of Hu proteins. *Cell Mol Life Sci* **65:** 3168–3181.

Hoeffer CA, Klann E. 2010. mTOR signaling: At the crossroads of plasticity, memory and disease. *Trends Neurosci* **33:** 67–75.

Holt CE, Bullock SL. 2009. Subcellular mRNA localization in animal cells and why it matters. *Science* **326:** 1212–1216.

Hou L, Antion MD, Hu D, Spencer CM, Paylor R, Klann E. 2006. Dynamic translational and proteasomal regulation of fragile X mental retardation protein controls mGluR-dependent long-term depression. *Neuron* **51:** 441–454.

Huang YS, Jung MY, Sarkissian M, Richter JD. 2002. N-methyl-D-aspartate receptor signaling results in Aurora kinase-catalyzed CPEB phosphorylation and α CaMKII mRNA polyadenylation at synapses. *EMBO J* **21:** 2139–2148.

Huang YS, Kan MC, Lin CL, Richter JD. 2006. CPEB3 and CPEB4 in neurons: Analysis of RNA-binding specificity and translational control of AMPA receptor GluR2 mRNA. *EMBO J* **25:** 4865–4876.

Huber KM, Kayser MS, Bear MF. 2000. Role for rapid dendritic protein synthesis in hippocampal mGluR-dependent long-term depression. *Science* **288:** 1254–1257.

Huttelmaier S, Zenklusen D, Lederer M, Dictenberg J, Lorenz M, Meng X, Bassell GJ, Condeelis J, Singer RH. 2005. Spatial regulation of β-actin translation by Src-dependent phosphorylation of ZBP1. *Nature* **438:** 512–515.

Irvine D, Tuerk C, Gold L. 1991. SELEXION. Systematic evolution of ligands by exponential enrichment with integrated optimization by non-linear analysis. *J Mol Biol* **222:** 739–761.

Jensen KB, Darnell RB. 2008. CLIP: Crosslinking and immunoprecipitation of in vivo RNA targets of RNA-binding proteins. *Methods Mol Biol* **488:** 85–98.

Jonson L, Vikesaa J, Krogh A, Nielsen LK, Hansen T, Borup R, Johnsen AH, Christiansen J, Nielsen FC. 2007. Molecular composition of IMP1 ribonucleoprotein granules. *Mol Cell Proteomics* **6:** 798–811.

Jung MY, Lorenz L, Richter JD. 2006. Translational control by neuroguidin, a eukaryotic initiation factor 4E and CPEB binding protein. *Mol Cell Biol* **26:** 4277–4287.

Kandel ER. 2001. The molecular biology of memory storage: A dialogue between genes and synapses. *Science* **294:** 1030–1038.

Kang H, Schuman EM. 1996. A requirement for local protein synthesis in neurotrophin-induced hippocampal synaptic plasticity. *Science* **273:** 1402–1406.

Kawai T, Lal A, Yang X, Galban S, Mazan-Mamczarz K, Gorospe M. 2006. Translational control of cytochrome c by RNA-binding proteins TIA-1 and HuR. *Mol Cell Biol* **26:** 3295–3307.

Keleman K, Kruttner S, Alenius M, Dickson BJ. 2007. Function of the *Drosophila* CPEB protein Orb2 in long-term courtship memory. *Nat Neurosci* **10:** 1587–1593.

Kelleher R Jr, Bear MF. 2008. The autistic neuron: Troubled translation? *Cell* **135:** 401–406.

Khandjian EW, Corbin F, Woerly S, Rousseau F. 1996. The fragile X mental retardation protein is associated with ribosomes. *Nat Genet* **12:** 91–93.

Khandjian EW, Huot ME, Tremblay S, Davidovic L, Mazroui R, Bardoni B. 2004. Biochemical evidence for the association of fragile X mental retardation protein with brain polyribosomal ribonucleoparticles. *Proc Natl Acad Sci* **101:** 13357–13362.

Kim JH, Richter JD. 2006. Opposing polymerase-deadenylase activities regulate cytoplasmic polyadenylation. *Mol Cell* **24:** 173–183.

Kim JH, Richter JD. 2007. RINGO/cdk1 and CPEB mediate poly(A) tail stabilization and translational regulation by ePAB. *Genes Dev* **21:** 2571–2579.

Konig J, Zarnack K, Rot G, Curk T, Kayikci M, Zupan B, Turner DJ, Luscombe NM, Ule J. 2010. iCLIP reveals the function of hnRNP particles in splicing at individual nucleotide resolution. *Nat Struct Mol Biol* **17:** 909–915.

Kundel M, Jones KJ, Shin CY, Wells DG. 2009. Cytoplasmic polyadenylation element-binding protein regulates neurotrophin-3-dependent β-catenin mRNA translation in developing hippocampal neurons. *J Neurosci* **29:** 13630–13639.

Laggerbauer B, Ostareck D, Keidel EM, Ostareck-Lederer A, Fischer U. 2001. Evidence that fragile X mental retardation protein is a negative regulator of translation. *Hum Mol Genet* **10:** 329–338.

Lantz V, Chang JS, Horabin JI, Bopp D, Schedl P. 1994. The *Drosophila* orb RNA-binding protein is required for the formation of the egg chamber and establishment of polarity. *Genes Dev* **8:** 598–613.

Li Z, Zhang Y, Ku L, Wilkinson KD, Warren ST, Feng Y. 2001. The fragile X mental retardation protein inhibits translation via interacting with mRNA. *Nucleic Acids Res* **29:** 2276–2283.

Li C, Bassell GJ, Sasaki Y. 2009. Fragile X mental retardation protein is involved in protein synthesis-dependent collapse of growth cones induced by Semaphorin-3A. *Front Neural Circuits* **3:** 11.

Licatalosi DD, Mele A, Fak JJ, Ule J, Kayikci M, Chi SW, Clark TA, Schweitzer AC, Blume JE, Wang X, et al. 2008. HITS-CLIP yields genome-wide insights into brain alternative RNA processing. *Nature* **456:** 464–469.

Lin AC, Tan CL, Lin CL, Strochlic L, Huang YS, Richter JD, Holt CE. 2009. Cytoplasmic polyadenylation and cytoplasmic polyadenylation element-dependent mRNA regulation are involved in Xenopus retinal axon development. *Neural Dev* **4:** 8.

Lin CL, Evans V, Shen S, Xing Y, Richter JD. 2010. The nuclear experience of CPEB: Implications for RNA processing and translational control. *RNA* **16:** 338–348.

Liu J, Schwartz JH. 2003. The cytoplasmic polyadenylation element binding protein and polyadenylation of messenger RNA in Aplysia neurons. *Brain Res* **959:** 68–76.

Livingstone M, Atas E, Meller A, Sonenberg N. 2010. Mechanisms governing the control of mRNA translation. *Phys Biol* **7:** 021001.

Mazan-Mamczarz K, Galban S, Lopez de Silanes I, Martindale JL, Atasoy U, Keene JD, Gorospe M. 2003. RNA-binding protein HuR enhances p53 translation in response to ultraviolet light irradiation. *Proc Natl Acad Sci* **100:** 8354–8359.

McEvoy M, Cao G, Montero Llopis P, Kundel M, Jones K, Hofler C, Shin C, Wells DG. 2007. Cytoplasmic polyadenylation element binding protein 1-mediated mRNA translation in Purkinje neurons is required for cerebellar long-term depression and motor coordination. *J Neurosci* **27:** 6400–6411.

Mee CJ, Pym EC, Moffat KG, Baines RA. 2004. Regulation of neuronal excitability through pumilio-dependent control of a sodium channel gene. *J Neurosci* **24:** 8695–8703.

Mendez R, Richter JD. 2001. Translational control by CPEB: A means to the end. *Nat Rev Mol Cell Biol* **2:** 521–529.

Mendez R, Hake LE, Andresson T, Littlepage LE, Ruderman JV, Richter JD. 2000a. Phosphorylation of CPE binding factor by Eg2 regulates translation of c-mos mRNA. *Nature* **404:** 302–307.

Mendez R, Murthy KG, Ryan K, Manley JL, Richter JD. 2000b. Phosphorylation of CPEB by Eg2 mediates the recruitment of CPSF into an active cytoplasmic polyadenylation complex. *Mol Cell* **6:** 1253–1259.

Menon KP, Sanyal S, Habara Y, Sanchez R, Wharton RP, Ramaswami M, Zinn K. 2004. The translational repressor Pumilio regulates presynaptic morphology and controls postsynaptic accumulation of translation factor eIF-4E. *Neuron* **44:** 663–676.

Michel CI, Kraft R, Restifo LL. 2004. Defective neuronal development in the mushroom bodies of *Drosophila* fragile X mental retardation 1 mutants. *J Neurosci* **24:** 5798–5809.

Mishchenko Y, Hu T, Spacek J, Mendenhall J, Harris KM, Chklovskii DB. 2010. Ultrastructural analysis of hippocampal neuropil from the connectomics perspective. *Neuron* **67**: 1009–1020.

Mukherjee N, Corcoran DL, Nusbaum JD, Reid DW, Georgiev S, Hafner M, Ascano MJ, Tuschl T, Ohler U, Keene JD, et al. 2011. Integrative regulatory mapping indicates that the RNA-binding protein HuR couples pre-mRNA processing and mRNA stability. *Mol Cell* **43**: 327–339.

Nakahata S, Katsu Y, Mita K, Inoue K, Nagahama Y, Yamashita M. 2001. Biochemical identification of Xenopus Pumilio as a sequence-specific cyclin B1 mRNA-binding protein that physically interacts with a Nanos homolog, Xcat-2, and a cytoplasmic polyadenylation element-binding protein. *J Biol Chem* **276**: 20945–20953.

Okano HJ, Darnell RB. 1997. A hierarchy of Hu RNA binding proteins in developing and adult neurons. *J Neurosci* **17**: 3024–3037.

Ostroff LE, Fiala JC, Allwardt B, Harris KM. 2002. Polyribosomes redistribute from dendritic shafts into spines with enlarged synapses during LTP in developing rat hippocampal slices. *Neuron* **35**: 535–545.

Ostroff LE, Cain CK, Bedont J, Monfils MH, Ledoux JE. 2010. Fear and safety learning differentially affect synapse size and dendritic translation in the lateral amygdala. *Proc Natl Acad Sci* **107**: 9418–9423.

Park S, Park JM, Kim S, Kim JA, Shepherd JD, Smith-Hicks CL, Chowdhury S, Kaufmann W, Kuhl D, Ryazanov AG, et al. 2008. Elongation factor 2 and fragile X mental retardation protein control the dynamic translation of Arc/Arg3.1 essential for mGluR-LTD. *Neuron* **59**: 70–83.

Pascale A, Gusev PA, Amadio M, Dottorini T, Govoni S, Alkon DL, Quattrone A. 2004. Increase of the RNA-binding protein HuD and posttranscriptional up-regulation of the GAP-43 gene during spatial memory. *Proc Natl Acad Sci* **101**: 1217–1222.

Pascale A, Amadio M, Quattrone A. 2008. Defining a neuron: Neuronal ELAV proteins. *Cell Mol Life Sci* **65**: 128–140.

Penagarikano O, Mulle JG, Warren ST. 2007. The pathophysiology of fragile x syndrome. *Annu Rev Genomics Hum Genet* **8**: 109–129.

Perycz M, Urbanska AS, Krawczyk PS, Parobczak K, Jaworski J. 2011. Zipcode binding protein 1 regulates the development of dendritic arbors in hippocampal neurons. *J Neurosci* **31**: 5271–5285.

Pieretti M, Zhang F, Fu Y, Warren ST, Oostra BA, Caskey CT, Nelson DL. 1991. Absence of expression of the FMR-1 gene in fragile X syndrome. *Cell* **66**: 817–822.

Pique M, Lopez JM, Foissac S, Guigo R, Mendez R. 2008. A combinatorial code for CPE-mediated translational control. *Cell* **132**: 434–448.

Posner JB, Dalmau J. 1997. Paraneoplastic syndromes. *Curr Opin Immunol* **9**: 723–729.

Quattrone A, Pascale A, Nogues X, Zhao W, Gusev P, Pacini A, Alkon DL. 2001. Posttranscriptional regulation of gene expression in learning by the neuronal ELAV-like mRNA-stabilizing proteins. *Proc Natl Acad Sci* **98**: 11668–11673.

Quenault T, Lithgow T, Traven A. 2011. PUF proteins: Repression, activation and mRNA localization. *Trends Cell Biol* **21**: 104–112.

Richter JD. 2007. CPEB: A life in translation. *Trends Biochem Sci* **32**: 279–285.

Richter JD, Klann E. 2009. Making synaptic plasticity and memory last: Mechanisms of translational regulation. *Genes Dev* **23**: 1–11.

Richter JD, Sonenberg N. 2005. Regulation of cap-dependent translation by eIF4E inhibitory proteins. *Nature* **433**: 477–480.

Robinow S, Campos AR, Yao KM, White K. 1988. The elav gene product of *Drosophila*, required in neurons, has three RNP consensus motifs. *Science* **242**: 1570–1572.

Sabatini DM. 2006. mTOR and cancer: Insights into a complex relationship. *Nat Rev Cancer* **6**: 729–734.

Santoro MR, Bray SM, Warren ST. 2011. Molecular mechanisms of fragile X syndrome: A twenty-year perspective. *Annu Rev Pathol* **7**: 219–245.

Sasaki Y, Welshhans K, Wen Z, Yao J, Xu M, Goshima Y, Zheng JQ, Bassell GJ. 2010. Phosphorylation of zipcode binding protein 1 is required for brain-derived neurotrophic factor signaling of local β-actin synthesis and growth cone turning. *J Neurosci* **30**: 9349–9358.

Scheetz AJ, Nairn AC, Constantine-Paton M. 2000. NMDA receptor-mediated control of protein synthesis at developing synapses. *Nat Neurosci* **3**: 211–216.

Shiina N, Tokunaga M. 2010. RNA granule protein 140 (RNG140), a paralog of RNG105 localized to distinct RNA granules in neuronal dendrites in the adult vertebrate brain. *J Biol Chem* **285**: 24260–24269.

Shiina N, Shinkura K, Tokunaga M. 2005. A novel RNA-binding protein in neuronal RNA granules: Regulatory machinery for local translation. *J Neurosci* **25**: 4420–4434.

Shiina N, Yamaguchi K, Tokunaga M. 2010. RNG105 deficiency impairs the dendritic localization of mRNAs for Na^+/K^+ ATPase subunit isoforms and leads to the degeneration of neuronal networks. *J Neurosci* **30**: 12816–12830.

Si K, Giustetto M, Etkin A, Hsu R, Janisiewicz AM, Miniaci MC, Kim JH, Zhu H, Kandel ER. 2003a. A neuronal isoform of CPEB regulates local protein synthesis and stabilizes synapse-specific long-term facilitation in aplysia. *Cell* **115**: 893–904.

Si K, Lindquist S, Kandel ER. 2003b. A neuronal isoform of the aplysia CPEB has prion-like properties. *Cell* **115**: 879–891.

Si K, Choi YB, White-Grindley E, Majumdar A, Kandel ER. 2010. Aplysia CPEB can form prion-like multimers in sensory neurons that contribute to long-term facilitation. *Cell* **140**: 421–435.

Silvera D, Formenti SC, Schneider RJ. 2010. Translational control in cancer. *Nat Rev Cancer* **10**: 254–266.

Siomi H, Siomi MC, Nussbaum RL, Dreyfuss G. 1993. The protein product of the fragile X gene, FMR1, has characteristics of an RNA-binding protein. *Cell* **74**: 291–298.

Slevin MK, Gourronc F, Hartley RS. 2007. ElrA binding to the 3′UTR of cyclin E1 mRNA requires polyadenylation elements. *Nucleic Acids Res* **35**: 2167–2176.

Solomon S, Xu Y, Wang B, David MD, Schubert P, Kennedy D, Schrader JW. 2007. Distinct structural features of caprin-1 mediate its interaction with G3BP-1 and its induction of phosphorylation of eukaryotic translation initiation factor 2α, entry to cytoplasmic stress granules, and selective interaction with a subset of mRNAs. *Mol Cell Biol* **27:** 2324–2342.

Stebbins-Boaz B, Cao Q, de Moor CH, Mendez R, Richter JD. 1999. Maskin is a CPEB-associated factor that transiently interacts with eIF-4E. *Mol Cell* **4:** 1017–1027.

Stefani G, Fraser CE, Darnell JC, Darnell RB. 2004. Fragile X mental retardation protein is associated with translating polyribosomes in neuronal cells. *J Neurosci* **24:** 7272–7276.

Steward O, Levy WB. 1982. Preferential localization of polyribosomes under the base of dendritic spines in granule cells of the dentate gyrus. *J Neurosci* **2:** 284–291.

Sutton MA, Schuman EM. 2006. Dendritic protein synthesis, synaptic plasticity, and memory. *Cell* **127:** 49–58.

Sutton MA, Taylor AM, Ito HT, Pham A, Schuman EM. 2007. Postsynaptic decoding of neural activity: eEF2 as a biochemical sensor coupling miniature synaptic transmission to local protein synthesis. *Neuron* **55:** 648–661.

Szabo A, Dalmau J, Manley G, Rosenfeld M, Wong E, Henson J, Posner JB, Furneaux HM. 1991. HuD, a paraneoplastic encephalomyelitis antigen, contains RNA-binding domains and is homologous to Elav and Sex-lethal. *Cell* **67:** 325–333.

Tanner DC, Qiu S, Bolognani F, Partridge LD, Weeber EJ, Perrone-Bizzozero NI. 2008. Alterations in mossy fiber physiology and GAP-43 expression and function in transgenic mice overexpressing HuD. *Hippocampus* **18:** 814–823.

Tee AR, Blenis J. 2005. mTOR, translational control and human disease. *Semin Cell Dev Biol* **16:** 29–37.

Tenenbaum SA, Carson CC, Lager PJ, Keene JD. 2000. Identifying mRNA subsets in messenger ribonucleoprotein complexes by using cDNA arrays. *Proc Natl Acad Sci* **97:** 14085–14090.

Tessier CR, Broadie K. 2008. *Drosophila* fragile X mental retardation protein developmentally regulates activity-dependent axon pruning. *Development* **135:** 1547–1557.

Tessier CR, Doyle GA, Clark BA, Pitot HC, Ross J. 2004. Mammary tumor induction in transgenic mice expressing an RNA-binding protein. *Cancer Res* **64:** 209–214.

Theis M, Si K, Kandel ER. 2003. Two previously undescribed members of the mouse CPEB family of genes and their inducible expression in the principal cell layers of the hippocampus. *Proc Natl Acad Sci* **100:** 9602–9607.

Till SM, Li HL, Miniaci MC, Kandel ER, Choi YB. 2011. A presynaptic role for FMRP during protein synthesis-dependent long-term plasticity in Aplysia. *Learn Mem* **18:** 39–48.

Tiruchinapalli DM, Oleynikov Y, Kelic S, Shenoy SM, Hartley A, Stanton PK, Singer RH, Bassell GJ. 2003. Activity-dependent trafficking and dynamic localization of zipcode binding protein 1 and β-actin mRNA in dendrites and spines of hippocampal neurons. *J Neurosci* **23:** 3251–3261.

Tiruchinapalli DM, Ehlers MD, Keene JD. 2008. Activity-dependent expression of RNA binding protein HuD and its association with mRNAs in neurons. *RNA Biol* **5:** 157–168.

Ule J, Jensen KB, Ruggiu M, Mele A, Ule A, Darnell RB. 2003. CLIP identifies Nova-regulated RNA networks in the brain. *Science* **302:** 1212–1215.

Ule J, Jensen K, Mele A, Darnell RB. 2005. CLIP: A method for identifying protein-RNA interaction sites in living cells. *Methods* **37:** 376–386.

Verkerk AJ, Pieretti M, Sutcliffe JS, Fu YH, Kuhl DP, Pizzuti A, Reiner O, Richards S, Victoria MF, Zhang FP. 1991. Identification of a gene (FMR-1) containing a CGG repeat coincident with a breakpoint cluster region exhibiting length variation in fragile X syndrome. *Cell* **65:** 905–914.

Vessey JP, Vaccani A, Xie Y, Dahm R, Karra D, Kiebler MA, Macchi P. 2006. Dendritic localization of the translational repressor Pumilio 2 and its contribution to dendritic stress granules. *J Neurosci* **26:** 6496–6508.

Vessey JP, Schoderboeck L, Gingl E, Luzi E, Riefler J, Di Leva F, Karra D, Thomas S, Kiebler MA, Macchi P, et al. 2010. Mammalian Pumilio 2 regulates dendrite morphogenesis and synaptic function. *Proc Natl Acad Sci* **107:** 3222–3227.

Wang DO, Martin KC, Zukin RS. 2010. Spatially restricting gene expression by local translation at synapses. *Trends Neurosci* **33:** 173–182.

Weinlich S, Huttelmaier S, Schierhorn A, Behrens SE, Ostareck-Lederer A, Ostareck DH. 2009. IGF2BP1 enhances HCV IRES-mediated translation initiation via the 3′UTR. *RNA* **15:** 1528–1542.

Wells DG, Dong X, Quinlan EM, Huang YS, Bear MF, Richter JD, Fallon JR. 2001. A role for the cytoplasmic polyadenylation element in NMDA receptor-regulated mRNA translation in neurons. *J Neurosci* **21:** 9541–9548.

Wharton RP, Struhl G. 1991. RNA regulatory elements mediate control of *Drosophila* body pattern by the posterior morphogen nanos. *Cell* **67:** 955–967.

Wickens M, Bernstein DS, Kimble J, Parker R. 2002. A PUF family portrait: 3′UTR regulation as a way of life. *Trends Genet* **18:** 150–157.

Wolin SL, Walter P. 1988. Ribosome pausing and stacking during translation of a eukaryotic mRNA. *EMBO J* **7:** 3559–3569.

Wu L, Wells D, Tay J, Mendis D, Abbott MA, Barnitt A, Quinlan E, Heynen A, Fallon JR, Richter JD, et al. 1998. CPEB-mediated cytoplasmic polyadenylation and the regulation of experience-dependent translation of α-CaMKII mRNA at synapses. *Neuron* **21:** 1129–1139.

Ye B, Petritsch C, Clark IE, Gavis ER, Jan LY, Jan YN. 2004. Nanos and Pumilio are essential for dendrite morphogenesis in *Drosophila* peripheral neurons. *Curr Biol* **14:** 314–321.

Zamore PD, Williamson JR, Lehmann R. 1997. The Pumilio protein binds RNA through a conserved domain that defines a new class of RNA-binding proteins. *RNA* **3:** 1421–1433.

Zang JB, Nosyreva ED, Spencer CM, Volk LJ, Musunuru K, Zhong R, Stone EF, Yuva-Paylor LA, Huber KM, Paylor R, et al. 2009. A mouse model of the human Fragile X syndrome I304N mutation. *PLoS Genet* **5:** e1000758.

Zhang C, Darnell RB. 2011. Mapping in vivo protein-RNA interactions at single-nucleotide resolution from HITS-CLIP data. *Nat Biotechnol* **29:** 607–614.

Zhang HL, Eom T, Oleynikov Y, Shenoy SM, Liebelt DA, Dictenberg JB, Singer RH, Bassell GJ. 2001a. Neurotrophin-induced transport of a β-actin mRNP complex increases β-actin levels and stimulates growth cone motility. *Neuron* **31:** 261–275.

Zhang YQ, Bailey AM, Matthies HJ, Renden RB, Smith MA, Speese SD, Rubin GM, Broadie K. 2001b. *Drosophila* fragile X-related gene regulates the MAP1B homolog Futsch to control synaptic structure and function. *Cell* **107:** 591–603.

Zhu H, Hasman RA, Barron VA, Luo G, Lou H. 2006. A nuclear function of Hu proteins as neuron-specific alternative RNA processing regulators. *Mol Biol Cell* **17:** 5105–5114.

Tinkering with Translation: Protein Synthesis in Virus-Infected Cells

Derek Walsh[1], Michael B. Mathews[2], and Ian Mohr[1]

[1]Department of Microbiology, New York University School of Medicine, New York, New York 10016

[2]Department of Biochemistry and Molecular Biology, UMDNJ–New Jersey Medical School, University of Medicine and Dentistry New Jersey, Newark, New Jersey 07103-1709

Correspondence: derek.walsh@med.nyu.edu; mathews@umdnj.edu; ian.mohr@med.nyu.edu

Viruses are obligate intracellular parasites, and their replication requires host cell functions. Although the size, composition, complexity, and functions encoded by their genomes are remarkably diverse, all viruses rely absolutely on the protein synthesis machinery of their host cells. Lacking their own translational apparatus, they must recruit cellular ribosomes in order to translate viral mRNAs and produce the protein products required for their replication. In addition, there are other constraints on viral protein production. Crucially, host innate defenses and stress responses capable of inactivating the translation machinery must be effectively neutralized. Furthermore, the limited coding capacity of the viral genome needs to be used optimally. These demands have resulted in complex interactions between virus and host that exploit ostensibly virus-specific mechanisms and, at the same time, illuminate the functioning of the cellular protein synthesis apparatus.

The dependence of viruses on the host translation system imposes constraints that are central to virus biology and have led to specialized mechanisms and intricate regulatory interactions. Failure to translate viral mRNAs and to modulate host mRNA translation would have catastrophic effects on virus replication, spread, and evolution. Accordingly, a wide assortment of virus-encoded functions is dedicated to commandeering and controlling the cellular translation apparatus. Viral strategies to dominate the host translation machinery target the initiation, elongation, and termination steps and include mechanisms ranging from the manipulation of key eukaryotic translation factors to the evolution of specialized *cis*-acting elements that recruit ribosomes or modify genome-coding capacity. Because many of these strategies have likely been pirated from their hosts and because virus genetic systems can be manipulated with relative ease, the study of viruses has been a preeminent source of information on the mechanism and regulation of the protein synthesis machinery. In this article, we focus on select viruses that infect mammalian or plant cells and review the mechanisms they use to exploit and control the cellular protein synthesis machinery.

VIRAL REPLICATION AND TRANSLATION STRATEGIES

The prodigious diversity of viruses, their unparalleled rate of evolution, and the wide repertoire of their interactions with their hosts, all contribute to their utility as biological tools to study the translation system. Thus, viral genomes are composed of DNA or RNA, either of which may be single- or double-stranded; replication may take place in the nucleus or cytoplasm of the cell; infection may give rise to acute or persistent infection; viruses may be agents of disease or innocuous; and the consequences of infection may vary in different organisms or cell types. For a discussion of the underlying biological principles, you are referred to virology reviews (e.g., Pe'ery and Mathews 2007). Here, we cannot do more than briefly introduce the relationship between virus genome structure and mRNA biogenesis.

Viruses are classified into families according to their genome structure, which dictates the mode of synthesis of their mRNA, its structure, and often its translation. In some viral families, the viral RNA is single-stranded and of the same sense as the mRNA, termed (+)-stranded, and thus it can be directly translated. Viruses with RNA genomes that are of the opposite polarity, that is, (−)-stranded or double-stranded (ds), require virus-encoded RNA-dependent RNA polymerases to generate their mRNA. Exceptionally, retroviruses use viral reverse transcriptase to convert their (+)-stranded RNA genomes into a dsDNA form, which then serves as template for mRNA synthesis via cellular pathways. Most viruses with DNA genomes take advantage of the cellular machinery for mRNA production, but others (such as the poxviruses) use their own enzyme systems. As a consequence, virus families are characterized by mRNAs that have a range of structures at their 5′ ends (capped, not capped, cap-substituted) and 3′ ends (polyadenylated or not), and carry various cis-acting sequences and elements (internal ribosome entry sites [IRES], protein binding sites, etc.)—features that determine many aspects of viral mRNA translation (Fig. 1).

IMPAIRING HOST TRANSLATION

The vast majority of eukaryotic mRNAs are translated in a cap-dependent manner that involves regulated recruitment of a 40S ribosome-containing pre-initiation complex (43S complex) by eukaryotic translation initiation factors (eIFs) (Hershey et al. 2012; Hinnebusch and Lorsch 2012). By impairing cap-dependent ribosome recruitment to host mRNAs, many viruses globally interfere with host mRNA translation—a phenomenon termed "host shut-off"—thereby crippling host antiviral responses and favoring viral protein synthesis. Viral mRNA translation in these instances proceeds via an alternative initiation strategy that relies on cis-acting RNA elements. Other viruses target host mRNA metabolism, impairing host mRNA processing, stability, and/or export to the cytoplasm.

Direct Effects on Cellular Translation Factors

Some viruses directly target cellular translation factors to prevent ribosome recruitment by host mRNAs (Fig. 2A). Poliovirus (an enterovirus), feline calicivirus, and retroviruses encode proteases that cleave eIF4G, separating its (amino-terminal) eIF4E-interacting domain from its eIF4A- and eIF3-binding segment (Etchison et al. 1982; Ventoso et al. 2001; Alvarez et al. 2003; Willcocks et al. 2004). In poliovirus-infected cells, both eIF4GI and eIF4GII are cleaved, and host shut-off correlates more closely with the protracted cleavage of eIF4GII (Gradi et al. 1998). In addition to eIF4G proteolysis, enterovirus 71 infection induces host micro-RNA (miRNA) miR-141 expression, which reduces eIF4E abundance and suppresses host protein synthesis (Ho et al. 2010). Vesicular stomatitis virus (VSV), influenza virus, and adenovirus (Ad) decrease eIF4E phosphorylation (Huang and Schneider 1991; Feigenblum and Schneider 1993; Connor and Lyles 2002). By binding to eIF4G, the Ad 100K protein displaces the kinase Mnk1 (see below), resulting in the accumulation of unphosphorylated eIF4E late in infection (Cuesta et al. 2000b, 2004). This stimulates selective late viral mRNA translation (discussed below) and may contribute to host

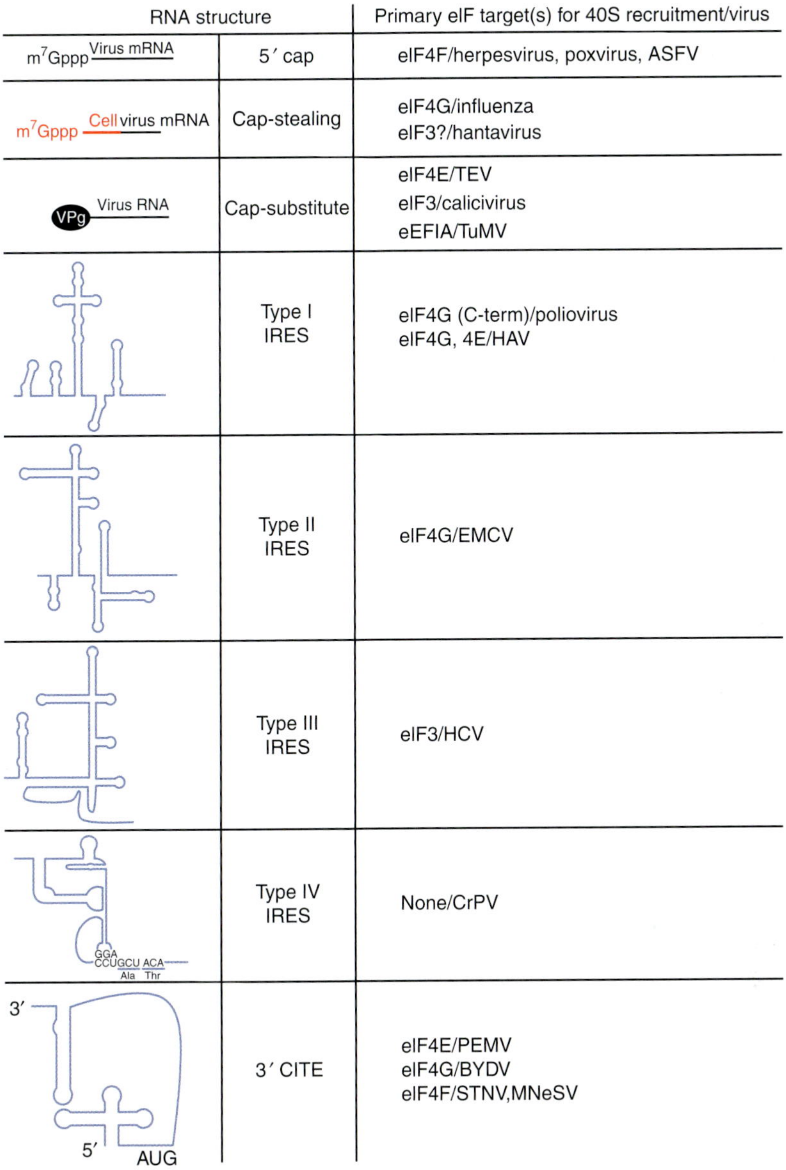

RNA structure		Primary eIF target(s) for 40S recruitment/virus
m⁷Gppp—Virus mRNA	5′ cap	eIF4F/herpesvirus, poxvirus, ASFV
m⁷Gppp—Cell virus mRNA	Cap-stealing	eIF4G/influenza eIF3?/hantavirus
VPg—Virus RNA	Cap-substitute	eIF4E/TEV eIF3/calicivirus eEFIA/TuMV
	Type I IRES	eIF4G (C-term)/poliovirus eIF4G, 4E/HAV
	Type II IRES	eIF4G/EMCV
	Type III IRES	eIF3/HCV
GGA CCUGCU ACA / Ala Thr	Type IV IRES	None/CrPV
3′ ... 5′ AUG	3′ CITE	eIF4E/PEMV eIF4G/BYDV eIF4F/STNV,MNeSV

Figure 1. Recruitment of 40S ribosome subunits to viral mRNAs: structural features and initiation factor targets. RNA structural elements involved in recruiting cellular 40S ribosome subunits to the 5′ end of viral mRNAs are shown on the *left*, with their corresponding name immediately to the *right*. The primary eukaryotic translation initiation factor (eIF) targets involved in recognizing each structural element and recruiting additional factors or mediating 40S subunit binding are indicated, with representative viruses that use them. (?) indicates that the precise factor targeted by hantavirus N protein remains unknown. Abbreviations: MNeSV, Maize necrotic streak virus; BYDV, Barley yellow dwarf virus; PEMV, Pea enation mosaic virus.

shut-off (Huang and Schneider 1991). During infection with some RNA viruses, however, accumulation of unphosphorylated eIF4E represents a host antiviral response, which induces interferon (IFN)-β production through translational activation of NF-κB (Herdy et al. 2012).

Enteroviral, apthoviral, caliciviral, and retroviral proteases also cleave the poly(A)-binding protein (PABP) (Kuyumcu-Martinez et al. 2004a,b; Alvarez et al. 2006; Rodriguez Pulido et al. 2007; Zhang et al. 2007; Bonderoff et al. 2008). PABP cleavage in enterovirus-infected cells does not always correlate with host shut-off, however (discussed below). Rather than cleaving PABP, the rotavirus NSP3 protein disrupts the PABP–eIF4G association (Piron et al. 1998; Groft and Burley 2002), allowing viral mRNA translation to proceed while host mRNA translation, which requires PABP, is suppressed (Montero et al. 2006). The large DNA-containing poxviruses produce small non-coding polyadenylated RNAs (POLADs) that have been reported to impair host PABP-dependent translation, but their precise function in infected cell biology remains unknown (Cacoullos and Bablanian 1991; Lu and Bablanian 1996).

Some RNA viruses also induce eIF2α phosphorylation to impair host responses (see below), or—in the case of rabies virus (M protein) (Komarova et al. 2007), measles virus (N protein) (Sato et al. 2007), and coronavirus (spike protein) (Xiao et al. 2008)—impair cap-dependent translation via virus-encoded eIF3-interacting proteins. How viral mRNAs are translated under the latter conditions remains unknown, as are the contributions of the eIF3-binding proteins to viral biology.

Indirect Effects on Cellular Translation Factors

Other viruses impact initiation factors indirectly (Fig. 2B). Many RNA viruses, including encephalomyocarditis virus (EMCV), poliovirus, cricket paralysis virus (CrPV), VSV, Sindbis virus (SINV), Dengue virus (DENV), and reovirus (Gingras et al. 1996; Connor and Lyles 2002; Villas-Boas et al. 2009; Garrey et al. 2010; Mohankumar et al. 2011), as well as small DNA viruses such as SV40 (Yu et al. 2005), induce the accumulation of hypophosphorylated 4E-BP1, which sequesters the cap-binding subunit eIF4E and prevents eIF4F assembly (Hinnebusch and Lorsch 2012; Roux and Topisirovic 2012). In the case of VSV, this requires the viral M protein, which suppresses Akt-signaling to prevent mTORC1-mediated inactivation of 4E-BP1 (Dunn and Connor 2011). SV40

small T-antigen, however, uses a protein phosphatase 2A (PP2A)–dependent mechanism to dephosphorylate 4E-BP1 (Yu et al. 2005). Inhibition of eIF4F via 4E-BP1 can also suppress host antiviral responses. Mouse embryo fibroblasts (MEFs) lacking 4E-BP1 and 4E-BP2 produce high levels of type I IFN because of increased *IRF7* mRNA translation, and 4E-BP1/2-deficient knockout mice are more resistant to EMCV, VSV, influenza, or SINV infection (Colina et al. 2008).

Changing the subcellular distribution of translation factors in virus-infected cells represents another way in which host cap-dependent mRNA translation can be altered by viruses. SINV redistributes eIF3 and eEF2 to cytoplasmic replication compartments while excluding eIF4G (Sanz et al. 2009). Because SINV mRNAs contain an IRES, this selective redistribution likely suppresses host translation and fosters selective viral protein synthesis. eIF4E is redistributed to the nucleus by poliovirus (Sukarieh et al. 2010). Although the 2A protein encoded by EMCV, another picornavirus, has a nuclear localization signal and binds to eIF4E (Groppo et al. 2010), whether eIF4E accumulates within the nuclei of infected cells remains unknown. PABP distribution can be similarly altered upon virus infection. Bunyavirus NSS (Blakqori et al. 2009) and rotavirus NSP3 proteins (Harb et al. 2008) cause nuclear PABP accumulation. Herpes simplex virus 1 (HSV-1) and Kaposi's sarcoma–associated herpesvirus (KSHV) do not stimulate PABP recruitment into eIF4F complexes but redistribute it to the nucleus, which may contribute to host shut-off by some herpesviruses. PABP redistribution in HSV-1-infected cells involves the viral proteins ICP27 and UL47 (Dobrikova et al. 2010; Salaun et al. 2010), and SOX and/or K8.1 in KSHV-infected cells (Arias et al. 2009; Covarrubias et al. 2009; Lee and Glaunsinger 2009; Kumar and Glaunsinger 2010; Kumar et al. 2011). Notably, both viruses induce host shut-off and, in the case of KSHV, SOX mutants that fail to redistribute PABP to the nucleus do not impair host translation (Covarrubias et al. 2009). In contrast, human cytomegalovirus (HCMV), a β-herpesvirus that does not impair host translation, does not redistribute PABP to the nucleus but does recruit it into eIF4F complexes (Walsh et al. 2005; Perez et al. 2011). Redistribution of host factors also occurs in cells infected with vaccinia virus (VacV), a poxvirus, or African swine fever virus (ASFV), an asfarvirus (Katsafanas and Moss 2007; Walsh et al. 2008; Castelló et al. 2009). Both of these large DNA viruses replicate in the cytoplasm within specialized compartments termed viral factories. Redistribution of a number of eIFs to these sites may contribute to the suppression of host protein synthesis while at the same time favoring viral mRNA translation.

Controlling Translation by Manipulating RNA

Beyond targeting translation factors, mRNA structures (Fig. 1), metabolism, and trafficking have all been targeted by viruses to interfere with cellular protein production. Capped mRNAs produced by influenza virus, hantavirus, and the yeast L-A virus contain m^7GTP caps derived from host mRNAs (Plotch et al. 1981; Mir et al. 2008; Reguera et al. 2010; Fujimura and Esteban 2011). Although L-A virus transfers only the host cap to the viral mRNA 5′ end, influenza virus and hantavirus use a viral endonuclease to cleave host mRNAs 10–18 nucleotides downstream from the m^7GTP cap (Plotch et al. 1981; Shih and Krug 1996; Guilligay et al. 2008; Mir et al. 2008; Dias et al. 2009). The resulting capped oligonucleotides prime viral RNA synthesis (Plotch et al. 1981; Garcin et al. 1995; Reguera et al. 2010). At the same time, this process destabilizes host mRNAs and inhibits cellular mRNA translation. Notably, the severe acute respiratory virus (SARS) coronavirus protein Nsp1 associates with 40S ribosomes and selectively cleaves host mRNAs to induce host shut-off (Kamitani et al. 2006, 2009; Huang et al. 2011). An alternate strategy used by VSV, which replicates in the cytoplasm, suppresses nuclear export of cellular mRNAs to preclude the synthesis of host defense-related proteins (Faria et al. 2005).

Among DNA viruses that produce capped, polyadenylated mRNAs, poxvirus and asfarvirus decapping enzymes destabilize mRNAs (Parrish and Moss 2007; Parrish et al. 2007).

In Ad-infected cells, cellular mRNA nuclear export is inhibited, and the E1B 55K and E4 ORF6 proteins selectively export viral mRNAs from the nucleus via an NXF1/TAP-dependent mechanism (Kratzer et al. 2000; Yatherajam et al. 2011). The endoribonuclease encoded by the HSV-1 virion host shut-off (vhs) gene associates with eIF4F by interacting with eIF4A and eIF4B/eIF4H to increase mRNA turnover of host and viral mRNAs (Feng et al. 2005). Although stimulating mRNA decay impairs translation of host mRNAs, including those encoding host defense functions, accelerating viral mRNA turnover helps demarcate different kinetic populations of virus-encoded mRNAs expressed at immediate-early, early, or late times post-infection. The KSHV SOX protein also cleaves mRNA but recruits XRN1 to mediate target mRNA degradation (Covarrubias et al. 2011; Richner et al. 2011). The HSV-1-encoded protein ICP27 inhibits host mRNA splicing and transport, while promoting intronless viral mRNA export (Sandri-Goldin 2011). ICP27 also reportedly interacts with PABP to mediate selective translation of a small subset of viral mRNAs (Ellison et al. 2005; Larralde et al. 2006; Fontaine-Rodriguez and Knipe 2008), but the precise mechanism is unclear. A related protein encoded by KSHV ORF57 binds a host factor, PYM, to load ribosomes onto viral mRNAs, potentially coupling viral mRNA export and translation (Boyne et al. 2010; Jackson et al. 2011). Herpesviruses also encode miRNAs that selectively suppress translation of cellular and viral target mRNAs (Pfeffer et al. 2004; Cai et al. 2005; Murphy et al. 2008; Tang et al. 2008; Umbach et al. 2008; Wang et al. 2008; Bellare and Ganem 2009; Nachmani et al. 2009; Santhakumar et al. 2010). For HSV-1, these miRNAs may prevent lytic transcript accumulation in latently infected cells.

A key component of the cellular antiviral response relies on RNase L, which inactivates both rRNA and mRNA. In response to dsRNA, a molecular signature associated with virus infection, oligoadenylate (OA) polynucleotide chains with a unique 2′–5′ linkage, are generated by the IFN-inducible OA synthetase (OAS). 2′–5′ OA is a potent RNase L activator. To preserve the integrity of viral mRNAs and cellular rRNAs,

A

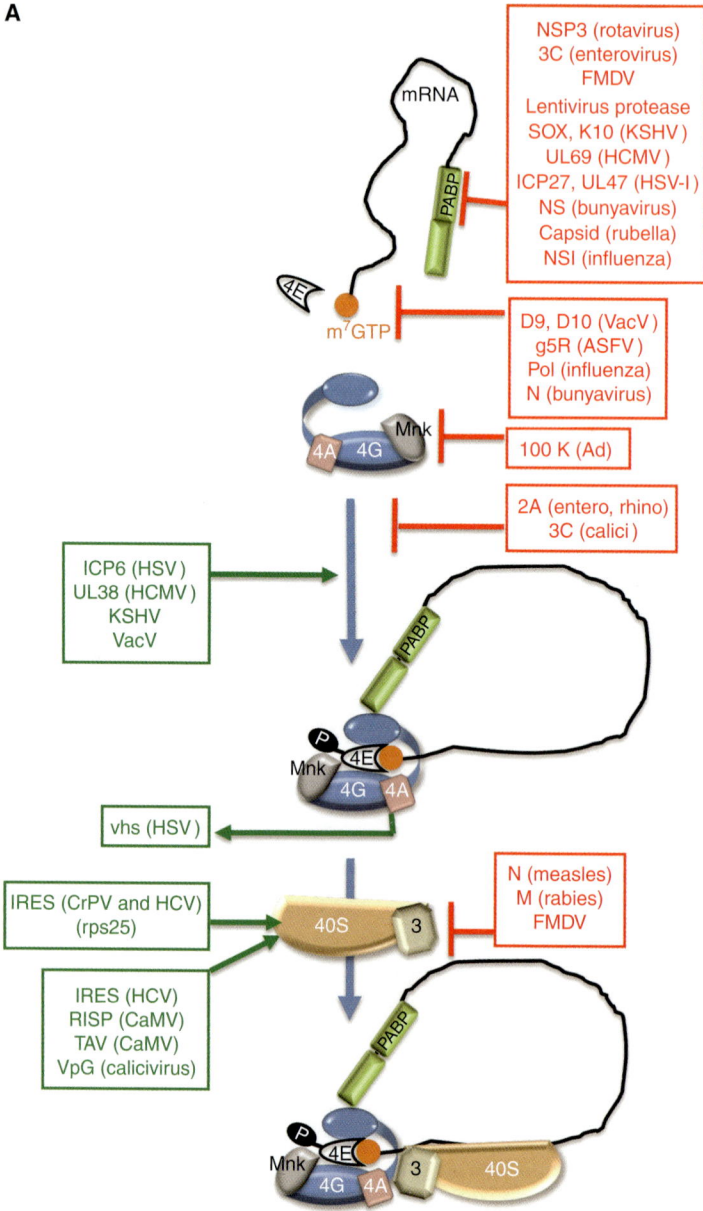

Figure 2. Control of translation by virus-encoded functions that regulate assembly of eIF4F. (*A*) eIF4E (4E) is able to interact with eIF4G and assemble the multi-subunit eIF4F complex (4E, 4A, 4G) on the m^7GTP-capped (orange ball) mRNA 5′ end. eIF4F assembly typically results in eIF4E phosphorylation by the eIF4G-associated kinase Mnk and recruits eIF3 bound to the 40S ribosome subunit together with associated factors in the 43S complex. PABP is depicted bound to the 3′-poly(A) tail and associates with eIF4G to stimulate translation. Virus-encoded factors that activate (green, on the *left*) and repress (red, to the *right*) cellular functions are shown. (*B*) Cell signaling pathways that control the activity of the translational repressor 4E-BP1 and regulate eIF4F assembly allow for rapid changes in gene expression programs in response to a variety of physiological cues, including viral infection. In response to growth factor–stimulated receptor tyrosine kinases (RTK), PI-3-kinase (PI3K) and mTORC2 activate Akt by phosphorylation on T308 and S473, which, in turn, represses the tuberous sclerosis complex (TSC) by phosphorylating the TSC2 subunit. (*Legend continues on facing page.*)

Cite this article as *Cold Spring Harb Perspect Biol* doi: 10.1101/cshperspect.a012351

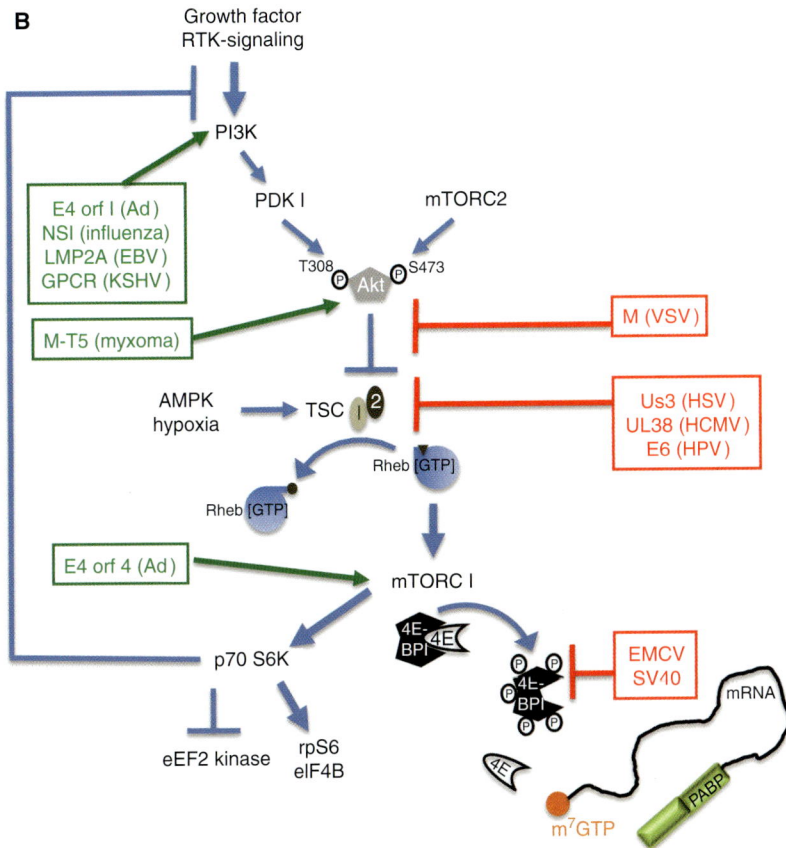

Figure 2. (*Continued*) TSC is a GTPase-activating protein (GAP) that limits the activity of Rheb by promoting Rheb·GDP accumulation, which represses mTORC1. Whereas AMPK and hypoxia stimulate TSC GAP activity to inhibit mTORC1, Receptor Tyrosine Kinase (RTK)–mediated Akt activation inhibits TSC, resulting in mTORC1 activation. Inhibiting TSC allows rheb·GTP accumulation and mTORC1 activation, and results in p70S6K and 4E-BP1 phosphorylation. By stimulating ribosomal protein S6 phosphorylation (rpS6), p70 S6K activation by mTORC1 stimulates the eIF4A-accessory factor eIF4B and inhibits eukaryotic elongation factor 2 (eEF2) kinase, stimulating elongation. Normally, p70 S6K activation represses PI3-kinase (PI3K) to feedback and limit mTORC1 activation. 4E-BP1 hyperphosphorylation (depicted as circled P) relieves translational repression and releases eIF4E.

viruses have evolved an array of countermeasures discussed below (Banerjee et al. 2000; Sanchez and Mohr 2007; Silverman 2007; Chakrabarti et al. 2011). Cytoplasmic structures associated with mRNA degradation may also play roles in infection, although they remain poorly understood. Poliovirus and CrPV suppress processing (P)–body formation (Dougherty et al. 2011; Khong and Jan 2011), which could represent another strategy to antagonize host responses because P-bodies restrict human immunodeficiency virus (HIV) infection (Nathans et al. 2009). It should be noted, however, that P-bodies facilitate the replication of brome mosaic virus (Beckham et al. 2007).

CAP-INDEPENDENT TRANSLATION

For viruses that interfere with eIF4F to suppress host protein synthesis and antiviral responses, an alternative, non-canonical mode of translation initiation is required, releasing viral mRNAs from regulatory constraints that normally control cellular mRNA translation.

Protein-Linked 5′ Ends

Instead of a m⁷GTP cap, some viral mRNAs contain a small virus-encoded protein, VPg (viral protein genome-linked), covalently attached to the 5′ end of their (+)-strand RNA genomes (Fig. 1). Calicivirus VPg interacts with eIF4E or eIF3 to mediate ribosome recruitment to viral mRNAs (Daughenbaugh et al. 2003, 2006; Goodfellow et al. 2005; Chaudhry et al. 2006). Interestingly, distinct plant virus-encoded VPg proteins selectively associate with different translation factors (Khan et al. 2006, 2008; Beauchemin et al. 2007). Tobacco mosaic virus (TMV) VPg binds eEF1A and concentrates it in membrane-associated viral replication sites (Thivierge et al. 2008), which may promote viral protein synthesis and/or impair host translation, whereas potyvirus potato virus A VPg promotes both viral mRNA stability and translation, effectively functioning as a proteinaceous cap (Eskelin et al. 2011). Free VPg can also suppress reporter mRNA translation, suggesting that it may sequester host factors to dampen host translation (Khan et al. 2008; Eskelin et al. 2011).

IRES-Dependent Mechanisms in Picornaviruses

Several RNA viruses contain an IRES within their 5′-untranslated regions (UTRs) that directs initiation through interactions with eIFs and/or ribosomal proteins (Fig. 1) (Jackson 2012). First discovered in poliovirus (Pelletier and Sonenberg 1988; Trono et al. 1988) and subsequently identified in other viruses including related picornaviruses (rhinovirus, hepatitis A [HAV], EMCV) (Jang et al. 1988) and hepatitis C virus (HCV), IRESs are divided into four structurally distinct classes, although a fifth related to class I and II IRESs has recently been proposed (Sweeney et al. 2011). Both class I and class II IRESs interact with the carboxy-terminal half of eIF4G (Kolupaeva et al. 1998; Bordeleau et al. 2006; de Breyne et al. 2009), which

recruits the 40S pre-initiation complex through its binding to eIF3. Class I and II IRESs also use specific host factors, termed IRES *trans*-activating factors (ITAFs), which alter IRES conformation and enhance eIF4G binding to facilitate ribosome recruitment (Kolupaeva et al. 1996; Kafasla et al. 2010; Yu et al. 2011a). During infection with poliovirus, the viral 2A protease cleaves eIF4G, inhibiting host cap-dependent translation, whereas its class I IRES uses the large eIF4G fragment that contains eIF4A and eIF3 binding sites to promote selective viral mRNA translation (Krausslich et al. 1987; Kempf and Barton 2008; Willcocks et al. 2011). The related picornaviruses EMCV and HAV contain a class II IRES whose function does not involve eIF4G cleavage. Instead, hypophosphorylated 4E-BP1 accumulates in EMCV-infected cells, which, in turn, sequester the cap-binding protein eIF4E and suppresses cap-dependent translation (Gingras et al. 1996). The HAV IRES is unusual because it uses intact eIF4F. However, the eIF4E cap-binding slot must be unoccupied, rendering HAV IRES initiation cap-independent and suggesting that it interacts with eIF4E or that eIF4E induces conformational changes in eIF4G that facilitate interaction with the IRES (Ali et al. 2001; Borman et al. 2001). The Aichivirus IRES is unusual in its absolute requirement for the helicase DHX59 to expose the start codon buried in a stable hairpin (Yu et al. 2011b).

Translation Mediated by Class III and IV IRES

The class III IRES in HCV and pestivirus RNAs uses a prokaryote-like mode of translation initiation by positioning ribosomes on the mRNA through interactions with both eIF3 and the ribosome itself, thereby circumventing any need for eIF4F (Pestova et al. 1998; Sizova et al. 1998; Ji et al. 2004; Otto and Puglisi 2004; Pisarev et al. 2004; Fraser and Doudna 2007; Locker et al. 2007; Babaylova et al. 2009; Berry et al. 2010). Translation of HCV RNA is also enhanced by the binding of host miR-122 to two target sites in the 5′ UTR, thereby promoting ribosome recruitment (Jopling et al. 2005, 2008; Henke et al. 2008; Jangra et al. 2010). In HIV-1, an IRES within the Gag open reading frame (ORF) also interacts with eIF3 and the 40S ribosome (Locker et al. 2010). The simian picornavirus (SPV9) IRES binds eIF3 and the 40S ribosome, yet its activity is enhanced by eIF4F (de Breyne et al. 2008b). In a striking departure, the class IV IRESs of the Dicistroviridae, including CrPV, recruit 40S subunits, assemble 80S ribosomes, and direct initiation in the absence of any eIFs and even of the initiator Met-tRNA$_i$, requiring only eEFs for polypeptide chain formation (Wilson et al. 2000; Jan and Sarnow 2002; Spahn et al. 2004; Cevallos and Sarnow 2005; Kamoshita et al. 2009). By positioning a CCU codon that is not decoded into the P-site, the CrPV IRES is able to initiate protein synthesis from an alanine codon positioned in the A-site after initial pseudotranslocation of the ribosome. The small ribosomal subunit protein RPS25 is essential for initiation from the CrPV and HCV IRES (Landry et al. 2009). RPS25 deletion in yeast or depletion in mammalian cells has minimal effects on cellular protein synthesis, implying that a ribosomal protein can be selectively required for IRES-mediated translation. Moreover, deficiencies in rRNA pseudouridylation reduce CrPV IRES binding to 40S ribosome subunits and inhibit IRES-dependent translation (Jack et al. 2011).

IRES-Like Mechanisms in Other Viruses

The 5′ UTRs of some plant potyviruses contain IRES-like structures that bind initiation factors (Fig. 1). For example, tobacco etch virus (TEV) interacts with eIF4G (Gallie 2001). Plant viruses also contain *cis*-acting cap-independent translational elements (CITEs) in their 3′ UTRs that interact with the 5′ UTR, bind translation factors, and place them proximal to the 5′ initiation site (Miller et al. 2007; Treder et al. 2008). The 3′ CITE in RNA2 of pea enation mosaic virus contains a pseudoknot that occupies the cap-binding slot in eIF4E, showing another mode by which the cap-binding protein can recognize mRNA (Wang et al. 2011). The turnip crinkle virus (TCV) CITE contains a 102-bp tRNA-like structural element that re-

cruits ribosomes (Zuo et al. 2010). A related strategy for recruiting factors to the 3′ UTR is used by DENV, where non-polyadenylated 3′-UTR sequences bind PABP to promote viral mRNA translation (Polacek et al. 2009).

Some RNA viruses that produce mRNAs with m^{7}GTP-capped 5′ ends—either by encoding methyltransferases and other capping enzymes or by acquiring caps from host mRNAs—can still use cap-independent modes of translation initiation. SINV provides an example of this latter strategy. In contrast to many RNA viruses, all known DNA viruses produce capped, polyadenylated, predominantly monocistronic mRNAs and generally use canonical cap-dependent modes of translation initiation. Rare instances of cap-independent strategies occur, however, typically involving polycistronic viral mRNAs. Thus, initiation on the SV40 late 19S mRNA is mediated by an IRES (Yu and Alwine 2006), and IRES-mediated translation of several herpesvirus mRNAs has also been reported (Bieleski and Talbot 2001; Grundhoff and Ganem 2001; Low et al. 2001; Griffiths and Coen 2005; Grainger et al. 2010).

MODULATION OF CAP-DEPENDENT INITIATION

As noted above, most dsDNA virus mRNAs are translated by conventional mechanisms used by the majority of host mRNAs, and many, but not all, of these viruses suppress host mRNA translation by altering mRNA metabolism (Clyde and Glaunsinger 2010). Unlike many RNA viruses, numerous DNA viruses stimulate factors, such as eIF4F to foster viral protein synthesis and replication irrespective of whether they inhibit host protein synthesis (Kudchodkar et al. 2004; Walsh and Mohr 2004; Walsh et al. 2005, 2008; Arias et al. 2009; Castelló et al. 2009). The mechanisms used in several viral families are discussed here.

Polyoma-, Papilloma-, and Adenoviruses

To activate mTORC1 and stimulate 4E-BP1 hyperphosphorylation, human papilloma virus (HPV) E6 inhibits the tuberous sclerosis complex (TSC) (Lu et al. 2004; Zheng et al. 2008; Spangle and Munger 2010). Merkel cell polyomavirus small T antigen also activates mTORC1, and its activity is required for transformation (Shuda et al. 2011). Ad E4-ORF1 stimulates phosphatidylinositol 3-kinase (PI3K) (Feigenblum and Schneider 1996; Gingras and Sonenberg 1997; O'Shea et al. 2005). In addition, Ad E4-ORF4 stimulates mTORC1 via PP2A independently of TSC (Fig. 2B) (O'Shea et al. 2005).

Late Ad mRNAs contain a common 200-nucleotide sequence at their 5′ ends. Termed the tripartite leader (TPL), it contains sequences complementary to 18S rRNA that are critical for ribosome shunting (Yueh and Schneider 1996, 2000). After loading onto the capped mRNA, the 40S ribosome is able to bypass large segments of the 5′ UTR via a nonlinear translocation mechanism before recognizing the AUG. Shunting is stimulated by the Ad 100 K protein, which binds TPL sequences and eIF4G (Xi et al. 2004, 2005). A shunting mechanism also controls production of the HPV E1 replication protein from a polycistronic mRNA (Remm et al. 1999).

Herpesviruses

Among the largest of the DNA viruses, herpesviruses replicate in the nucleus and exist in two discrete developmental states. Infection results in lifelong latency characterized by a restricted gene expression program, punctuated by episodic reactivation events in which productive, lytic viral replication ensues. Different herpesvirus subfamily members colonize specialized differentiated cell types. For example, α-herpesviruses such as HSV-1 are neurotrophic; HCMV, a β-herpesvirus, colonizes myeloid progenitors; and KHSV and Epstein-Barr virus (EBV), both γ-herpesviruses, colonize lymphoid cells and are associated with malignancies. During their productive or lytic replication cycle, these viruses induce 4E-BP1 hyperphosphorylation, eIF4F assembly, and eIF4E phosphorylation to promote viral protein production and virus replication (Kudchodkar et al. 2004; Walsh and Mohr 2004; Walsh et al. 2005; Arias et al. 2009). Re-

markably, different mechanistic strategies are enlisted to achieve this goal (Fig. 2A,B). For example, both HSV-1 and HCMV stimulate mTORC1 by targeting TSC. The HSV-1 Us3 ser/thr kinase directly phosphorylates substrates on sites targeted by Akt, including TSC2, functioning as an Akt mimic (Chuluunbaatar et al. 2010). The HCMV UL38 protein, however, physically associates with TSC2 and inactivates it via an unknown mechanism that does not involve direct phosphorylation (Moorman et al. 2008). HCMV also alters mTOR substrate specificity and subcellular localization (Kudchodkar et al. 2006; Clippinger et al. 2011). Two γ-herpesvirus proteins, KSHV vGPCR (Sodhi et al. 2006; Nichols et al. 2011) and EBV LMP2A (Moody et al. 2005), also activate signaling pathways upstream of mTORC1. mTOR active-site inhibitors PP242 and Torin1, which target both mTORC1 and 2 (Feldman et al. 2009; Thoreen et al. 2009), impair viral protein synthesis and replication (Moorman and Shenk 2010; McMahon et al. 2011), and their effect requires the translation repressor 4E-BP1 (Chuluunbaatar et al. 2010; Moorman and Shenk 2010; Perez et al. 2011). Besides stimulating cap-dependent initiation, mTORC1 activation during infection likely coordinately increases elongation rates by inhibiting eEF2 kinase (Browne and Proud 2002, 2004; Browne et al. 2004). Viruses can also directly modify elongation factors. A kinase conserved among different herpesviruses stimulates eEF1Bα phosphorylation (Kawaguchi et al. 1999); however, the significance to host or viral protein synthesis remains unknown.

In addition to its phosphorylation, 4E-BP1 is degraded by the proteasome in cells infected with HSV-1 or VacV (Walsh and Mohr 2004; Walsh et al. 2008). Ubiquitination and proteasomal degradation of 4E-BP1 are also observed in uninfected cells (Elia et al. 2008; Braunstein et al. 2009). Inactivation of 4E-BP1 is not sufficient to promote eIF4F assembly in HSV-1-infected cells, however. During virus-induced stress, binding of eIF4E to eIF4G requires the virus-encoded ICP6 protein (Walsh and Mohr 2006), which shares a small domain with homology to hsp27, a cellular chaperone that regulates eIF4F formation during stress and recovery (Carper et al. 1997; Cuesta et al. 2000a). Thus ICP6 may represent a strategy to foster eIF4F formation under the stress of infection.

HCMV uses different strategies to control eIF4F activity. Although unchanged in HSV1-infected cells, the steady-state levels of eIF4E, eIF4G, eIF4A, and PABP are increased by HCMV (Walsh et al. 2005). Increased eIF4E, eIF4G, and eIF4A mRNA levels accompany the rise in abundance of the corresponding proteins, whereas PABP abundance, along with other TOP mRNA-encoded host proteins, increases through an mTORC1-dependent translational control mechanism that requires UL38 (Perez et al. 2011; McKinney et al. 2012) and promotes viral protein accumulation, eIF4F assembly, and viral replication. This represents the first example of a virus-directed increase in translation initiation factor concentration that contributes to viral replication, and it contrasts with poliovirus biology, where full-length translation initiation factor abundance is selectively reduced by virus-encoded functions. Additionally, an HCMV RNA-binding protein (UL69) important for viral mRNA export from the nucleus (Lischka et al. 2006) reportedly interacts with eIF4A and PABP and stimulates eIF4E release from 4E-BP1 (Aoyagi et al. 2010), although the underlying mechanism remains unknown.

Poxviruses and Asfarviruses

Unlike herpesviruses, poxviruses and asfarviruses (ASFV) cause acute infections and replicate in the cytoplasm. Both viruses encode their own mRNA capping enzymes (Ensinger et al. 1975; Dixon et al. 1994). Notably, VacV and ASFV redistribute eIF4E and eIF4G to cytoplasmic viral factories, where replication occurs (Katsafanas and Moss 2007; Walsh et al. 2008; Castelló et al. 2009). As neither virus increases eIF4F subunit abundance (unlike HCMV), redistribution raises the effective local concentration of these factors to promote eIF4F assembly, perhaps by altering the eIF4E:eIF4G interaction equilibrium. Consistent with this, inhibiting viral factory formation and eIF4G redistribution suppressed eIF4F assembly in poxvirus-infected

cells (Zaborowska et al. 2012). In addition, 4EGi-1, an inhibitor of eIF4F and ribosome binding, has potent antiviral activity against VacV and HSV-1 (McMahon et al. 2011), and siRNA-mediated eIF4G depletion suppresses VacV infection (Welnowska et al. 2009). Concentrating eIF4G within replication factories may involve the VacV ssDNA-binding protein I3, which binds eIF4G and accumulates within replication factories (Zaborowska et al. 2012). VacV and ASFV infection also promotes 4E-BP1 hyperphosphorylation by mTORC1 (Walsh et al. 2008; Castello et al. 2009). Although the mechanism by which VacV and ASFV stimulate mTORC1 signaling remains unknown, the rabbit poxvirus myxoma M-T5 protein directly activates Akt signaling (Werden et al. 2007), yet its potential role in translational control has not been explored.

Mimi- and Megaviruses

The recently discovered giant viruses that infect *Acanthamoeba* (mimivirus, megavirus) and zooplankton (*Cafeteria roenbergensis* virus) encode orthologs of translation factors, including eIF4E (Raoult et al. 2004; Saini and Fischer 2007; Fischer et al. 2010; Arslan et al. 2011). Whether these viral orthologs function to control translation in infected cells is unknown.

eIF4E Phosphorylation Promotes DNA Virus Replication

By stimulating eIF4F assembly, the eIF4G-associated kinase Mnk1 is positioned near its substrate eIF4E in poxvirus-infected (Walsh et al. 2008; Zaborowska and Walsh 2009), asfarvirus-infected (Castelló et al. 2009), and herpesvirus-infected (Walsh and Mohr 2004, 2006; Walsh et al. 2005; Arias et al. 2009) cells, resulting in eIF4E phosphorylation. Mnk1 activation and subsequent eIF4E phosphorylation is dependent on ERK and/or p38 MAPK signaling. The small molecule Mnk1 inhibitor, CGP57380, impairs HSV1 (Walsh and Mohr 2004), HCMV (Walsh et al. 2005), ASFV (Castelló et al. 2009), and VacV (Walsh et al. 2008) protein synthesis, and VacV replication is reduced in Mnk1-deficient MEFs (Walsh et al.

2008), indicating that this kinase plays an important role in controlling viral protein synthesis and replication. HSV-1 and VacV replication is also reduced in MEFs containing an eIF4E-substituted allele in which eIF4E cannot be phosphorylated, directly establishing that eIF4E phosphorylation regulates virus replication (Herdy et al. 2012). For KSHV, Mnk1 activity is important in reactivation from latency because CGP57380 inhibits the transition from latency to lytic replication, suggesting a role for eIF4E phosphorylation in this developmental switch (Arias et al. 2009). Because Mnk1 is not essential for translation initiation but likely plays a regulatory role, it represents a potential target for therapeutic intervention against a wide range of large DNA viruses.

Cap-Dependent Translation in RNA Viruses

Some RNA viruses, including coronaviruses (SARS), orthomyxoviruses (influenza), rhabdoviruses (VSV, rabies virus), reoviruses, hantavirus, and alphaviruses, also produce capped mRNAs. Coronavirus mRNAs, which share a common capped 5′-leader and are polyadenylated, rely on eIF4F, because chemical inhibitors of the eIF4E·eIF4G interaction impair coronavirus replication in cultured cells (Cencic et al. 2011). Furthermore, coronavirus infection stimulates eIF4E phosphorylation (Banerjee et al. 2002; Mizutani et al. 2004). In influenza virus-infected cells, the viral cap-binding polymerase subunit proteins PB2 and NS1 reportedly interact with eIF4G and PABP (Aragon et al. 2000; Burgui et al. 2003, 2007; Yánguez et al. 2012), recruiting them to viral mRNAs. NS1 is not absolutely required for viral protein synthesis, however (Salvatore et al. 2002). Remarkably, the hantavirus N protein has been reported to have cap-binding, RNA-binding, RNA helicase, and ribosome-recruiting activities that functionally substitute for the entire eIF4F complex (Mir and Panganiban 2008, 2010). Finally, the dsRNA reoviruses and rotaviruses encode capping enzymes, and reovirus mRNAs are translated by a cap-dependent mechanism in vitro (Muthukrishnan et al. 1976; Sonenberg et al. 1979). Recently, modest reductions in eIF4E, eIF4G, eIF4B, and 4E-BP1 phosphorylation, along with enhanced eEF2 phosphorylation, have been reported in avian reovirus-infected cells or in cells expressing the viral p17 protein (Ji et al. 2009; Chulu et al. 2010). These modifications, which suppress translation, have been proposed to selectively impair translation of host mRNAs, although further investigation is required to test this hypothesis.

eIF2: A CENTRAL PLAYER IN THE INNATE IMMUNE RESPONSE TO INFECTION

For most host and viral mRNAs, the methionine-charged initiator tRNA is loaded into the ribosome by a ternary complex (TC) composed of eIF2, GTP, and an initiator tRNA (Met-tRNA$_i$) (Fig. 3). In addition to its critical role in initiation and AUG selection (Hinnebusch and Lorsch 2012), TC formation is targeted by innate host antiviral immune defenses. Like OAS (discussed above), the IFN-induced eIF2α kinase PKR acts as a pattern recognition sentinel that detects dsRNA or viral replication intermediates. Once activated, phosphorylated eIF2 accumulates and binds with high affinity to the eIF2B guanine nucleotide exchange factor (GEF), preventing exchange of GDP for GTP and inhibiting translation initiation. Small changes in eIF2α phosphorylation can result in a large suppression of ongoing protein synthesis and even incapacitate the cellular translation machinery because eIF2B is present in limiting quantities.

Eukaryotes contain three additional eIF2α kinases (Fig. 3), each of which responds to a discrete set of stresses. Both GCN2, which responds to UV irradiation and amino acid deprivation, and PERK, which can be triggered by ER overload in infected cells (Pavitt and Ron 2012), can have antiviral effects (Berlanga et al. 2006; Won et al. 2011). eIF2α kinases have also been implicated in autophagy induction (Talloczy et al. 2002). This powerful host response has forced viruses to evolve countermeasures such that they can sustain protein synthesis and complete their growth cycles in infected cells. Although many of these functions target PKR, others have broader capabilities and can

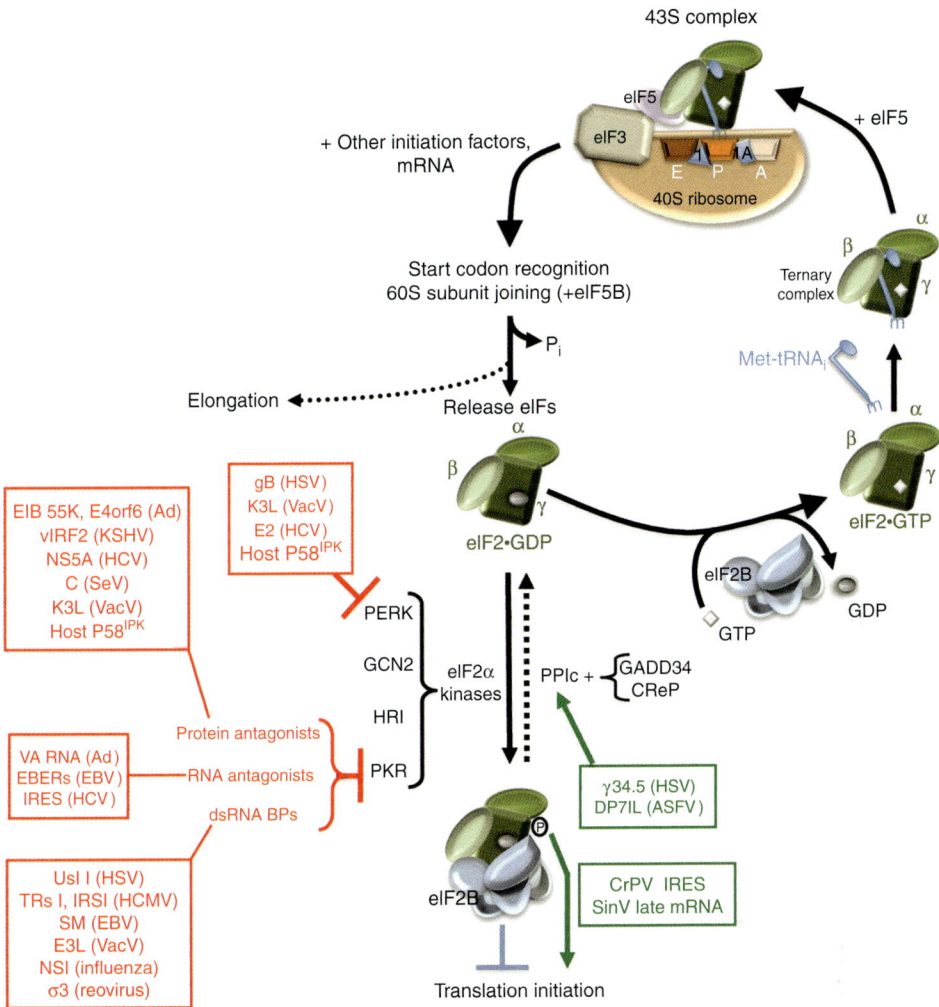

Figure 3. Antagonism of host antiviral defenses that inactivate eIF2 by viral functions. eIF2·GTP forms a ternary complex (TC) with Met-tRNA$_i$ and is loaded into the P-site in the 40S ribosome to form a 43S pre-initiation complex (*top right*). Upon engaging mRNA via other initiation factors (e.g., eIF4F) or via a specialized *cis*-acting RNA element (e.g., an IRES), the AUG start codon is identified (typically via scanning), the GTPase-activating protein eIF5 stimulates GTP hydrolysis, and 60S subunit joining triggers eIF2·GDP + phosphate (P$_i$) release. Translation elongation is executed by the resulting 80S ribosome. Inactive eIF2 (α, β, γ subunits depicted in *center*) bound to GDP (eIF2·GDP) is recycled to the active GTP-bound form (eIF2·GTP) by the five-subunit guanine nucleotide exchange factor eIF2B. Site-specific eIF2 phosphorylation on its α-subunit (S51) by either of four different cellular eIF2α-kinases (see text), each of which is activated by discrete stress, prevents eIF2 recycling. Phosphorylated eIF2 (*bottom*) binds tightly to and inhibits eIF2B, blocking translation initiation. The host protein phosphatase 1 catalytic (PP1c) subunit can dephosphorylate eIF2 when partnered with either an inducible (GADD34) or constitutively active (CReP) regulatory component. Virus-encoded factors that antagonize (red, to the *left*) specific host eIF2α kinases are indicated to the *left*. Viral PKR antagonists have been subdivided according to their mechanisms of action (see text; [dsRNA BPs] double-stranded RNA binding proteins). Virus-encoded factors that activate cellular phosphatases to dephosphorylate eIF2α, or allow translation to proceed without eIF2 or in the presence of phosphorylated eIF2α, are shown (green, to the *right*).

act directly on eIF2 or other eIF2α kinases. Viral functions that hinder eIF2α phosphorylation are key components of viral pathogenesis, and mutations that compromise these functions often result in attenuated variants. Adaptive evolutionary changes in PKR, under pressure to avoid viral substrate mimicry while maintaining natural substrate (eIF2α) recognition, emphasizes the ongoing nature of this struggle between host defenses and viral countermeasures for control over eIF2 (Elde et al. 2009; Rothenburg et al. 2009). eIF2 itself is regulated by IFN-stimulated genes (ISGs) 54 and 56, which antagonize ribosome binding to eIF4F and TC loading by binding to eIF3e and eIF3c (Guo et al. 2000; Hui et al. 2003, 2005; Terenzi et al. 2006). Viral ISG54 and 56 antagonists, however, have not been reported.

Viral Inhibitors of eIF2α Phosphorylation

Perhaps the most common strategy for preventing PKR activation and safeguarding eIF2 in virus-infected cells involves virus-encoded dsRNA-binding proteins that sequester dsRNA, preventing PKR from recognizing its activating ligand. In addition to dsRNA binding, many of these proteins also associate with and inhibit PKR, and prevent OAS activation (for review, see Walsh and Mohr 2011). A variation on this theme uses a structured, non-coding RNA lure to inhibit PKR activation, such as Ad VA RNA$_I$, HIV-1 TAR, and EBV EBER. VacV also encodes a pseudosubstrate, K3L, that effectively deflects PKR and PERK away from its substrate, eIF2α. The host molecular chaperone p58^{IPK}, a PKR/PERK inhibitor, is recruited to prevent eIF2α phosphorylation in cells infected with influenza virus, TMV, or TEV (Yan et al. 2002; Bilgin et al. 2003; Goodman et al. 2007, 2009). Finally, eIF2α itself can be targeted for dephosphorylation by viral functions. The ASFV DP17L-, HPV E6-, and HSV1 γ34.5-encoded phosphatase regulatory subunits target cellular catalytic PP1 subunits to phosphorylated eIF2α (He et al. 1997; Kazemi et al. 2004; Zhang et al. 2010; Li et al. 2011). Recruiting a phosphatase to dephosphorylate eIF2α is advantageous, conferring resistance to any cellular eIF2α kinase, not just PKR.

Combinatorial Strategies to Prevent eIF2α Phosphorylation

Some viruses use multiple, independent strategies to prevent eIF2α phosphorylation. VacV, for example, combines the dsRNA-binding protein/PKR inhibitor E3L with the PERK and PKR pseudosubstrate K3L (Chang et al. 1992; Davies et al. 1992, 1993; Kawagishi-Kobayashi et al. 1997; Sood et al. 2000; Ramelot et al. 2002; Pavio et al. 2003; Seo et al. 2008; Rothenburg et al. 2011). HSV-1, on the other hand, relies on the Us11 dsRNA-binding/PKR antagonist (Mulvey et al. 1999), the γ34.5 eIF2α phosphatase subunit, and a viral glycoprotein B (gB) PERK antagonist that prevents ER stress-induced eIF2α phosphorylation (Mulvey et al. 2003, 2007). Genomic elements can also play a role in inhibiting PKR. Although HCV E2 is a PKR and PERK pseudosubstrate (Taylor et al. 1999, 2001; Pavio et al. 2003), both the HCV NS5A protein (Gale et al. 1998; Gale and Foy 2005) and the HCV IRES itself (Vyas et al. 2003) directly inhibit PKR. Additionally, HIV-1 Tat and the cellular TAR RNA-binding protein (TRBP) bind the viral TAR element, and TRBP inhibits PKR (Gunnery et al. 1992).

Exploiting eIF2α Phosphorylation and Bypassing eIF2

Instead of preventing eIF2α phosphorylation, some RNA viruses induce eIF2α phosphorylation (O'Malley et al. 1989; Jordan et al. 2002; Gorchakov et al. 2004; McInerney et al. 2005; Smith et al. 2006; Montero et al. 2008; Garrey et al. 2010; White et al. 2010). Although stimulating the phosphorylation of eIF2α facilitates the host antiviral response by globally inhibiting protein synthesis in infected cells, some viruses use an alternative mode of translation initiation that does not absolutely require eIF2. These viruses derive benefit from eIF2α phosphorylation, which can effectively cause host shut-off and prevent IFN synthesis. Classical swine fever virus (CSFV) uses both eIF2-dependent and -independent modes of translation initiation (Pestova et al. 2008). In SINV-infected cells, a stem–loop structure in the late 26S mRNA stalls the ribosome at the initiation site and requires eIF2A to initiate translation in the presence of phosphorylated eIF2α (Ventoso et al. 2006). As noted above, the CrPV IRES loads ribosomes and initiates translation without eIF2, other eIFs, or initiator tRNA. With high magnesium concentrations in vitro, the HCV IRES also directs eIF2-independent initiation (Lancaster et al. 2006), whereas, it can use eIF2A-mediated Met-tRNA$_i$ delivery under stress conditions (Kim et al. 2011). However, although eIF2α phosphorylation is induced during infection (Garaigorta and Chisari 2009; Kang et al. 2009; Arnaud et al. 2010), the HCV IRES has been suggested to compete efficiently for eIF2 and Met-tRNA$_i$ under limiting conditions (Robert et al. 2006). Furthermore, as discussed above, HCV encodes three distinct functions to inhibit PKR and eIF2α phosphorylation. As with Ad (O'Malley et al. 1989), one possible explanation posits that HCV prevents eIF2 phosphorylation specifically at intracellular viral replication sites while allowing eIF2α phosphorylation to occur in the cytoplasm to impair host protein synthesis.

eIF2α is also phosphorylated as poliovirus infection progresses, and this correlates with declining viral protein synthesis (O'Neill and Racaniello 1989). Although poliovirus mRNA translation is sensitive to eIF2α phosphorylation in vitro, viral mRNA translation late in infection is partially resistant to stress-induced eIF2α phosphorylation (Redondo et al. 2011; Welnowska et al. 2011; White et al. 2011). The underlying mechanism is unknown but may involve cleavage of eIF5B (de Breyne et al. 2008a; White et al. 2011) and local control of active eIF2 pools or cellular factors. Indeed, host factors such as ligatin (eIF2D) and MCT1-DENR can recruit Met-tRNA$_i$ to ribosomes and function when the AUG start codon is directly positioned in the P-site (Dmitriev et al. 2010; Skabkin et al. 2010), potentially explaining eIF2-independent initiation mechanisms used by HCV and SINV.

eIF2α phosphorylation is also implicated in the formation of stress granules (SGs), stress-induced cytoplasmic sites of translationally inactive mRNA accumulation. Many viruses disrupt SG formation to evade translational inhibition (Emara and Brinton 2007; White et al. 2007; Montero et al. 2008; Khong and Jan 2011; Simpson-Holley et al. 2011), whereas others induce their formation to impair host translation or even facilitate virus assembly (McInerney et al. 2005; Raaben et al. 2007; Qin et al. 2009; Lindquist et al. 2010; Piotrowska et al. 2010). Notably, mammalian orthoreovirus escapes translational suppression by disrupting SGs, but does so independently of the status of eIF2α phosphorylation or PKR (Qin et al. 2011).

REGULATED TERMINATION AND REINITIATION: TRANSLATIONAL STRATEGIES FOR MAXIMIZING GENOME CODING CAPACITY

The small genome size of many RNA viruses constrains their coding capacity. Consequently, many viral mRNAs contain overlapping ORFs regulated by frameshifting or multiple ORFs regulated by termination and reinitiation events. Thus, translational control of ORF decoding plays an indispensable role in regulating viral protein production from polycistronic mRNAs. Influenza and feline calicivirus (FCV) use 45- to 87-nucleotide stretches termed TURBS (termination upstream ribosome binding site) that base-pair with terminating ribosomes to promote reinitiation at nearby downstream ORFs encoding viral BM2 and VP2 proteins, respectively (Horvath et al. 1990; Luttermann and Meyers 2009; Powell et al. 2011), as well as binding the ribosome and eIF3 for FCV VP1 reinitiation (Poyry et al. 2007). Murine norovirus VP2 is also translated by a coupled termination and reinitiation process (Napthine et al. 2009).

Viral proteins also regulate reinitiation; the retroviral Gag protein of Moloney murine leukemia virus (M-MuLV) binds eRF1 to enhance reinitiation or translational readthrough to synthesize the Gag-Pol precursor protein (Orlova et al. 2003). Reinitiation frequency is fine-tuned by pH-dependent conformational changes in the M-MuLV recoding signal sequence to ensure the balance of gag versus gag-pol fusion protein synthesis (Houck-Loomis et al. 2011). Reinitiation also protects HIV-1 retrovirus mRNA from nonsense-mediated decay (Hogg and Goff

2010). In an intricate regulatory mechanism, Borna virus X and P proteins are encoded by an mRNA that contains an additional uORF. X translation is regulated by a coupled termination and reinitiation event at the uORF, and the efficiency of reinitiation on the X ORF is enhanced by P through interactions with a host helicase, DDX21 (Watanabe et al. 2009).

Reinitiation on cauliflower mosaic virus (CaMV) polycistronic mRNA involves viral and host factors that manipulate eIF3. The viral transactivator viroplasmin (TAV) associates with the plant protein RISP, which supports reinitiation, and subsequently binds to eIF3g (Thiebeauld et al. 2009). Interestingly, eIF4B and TAV-bound RISP bind to the same eIF3g site. RISP associates with eIF3a, eIF3c, and ribosomal protein L24 to link TAV with 60S and eIF3-bound 40S subunits. TAV also recruits the plant cell TOR, which phosphorylates RISP, promoting reinitiation and viral replication (Schepetilnikov et al. 2011).

Most mRNAs encoded by large DNA viruses are not polycistronic, limiting the need for termination/reinitiation strategies. Nevertheless, termination within a small HCMV uORF plays a regulatory role restricting reinitiation at downstream cistrons. Sequence-dependent ribosomal stalling induced by the uORF2 peptide prevents scanning ribosomes from reaching the downstream UL4 ORF (Janzen et al. 2002). By binding eRF1, the uORF2 peptide inhibits translation at its own stop codon, and the stalled ribosome disengages the mRNA. Cryo-EM reconstruction showed uORF2 stabilized at the ribosomal exit tunnel constriction, where it directly interacts with ribosomal proteins L4 and L17. Peptide-induced conformational changes in the peptidyl transfer center were also observed, potentially contributing to translational stalling (Bhushan et al. 2010).

BALANCING TRANSLATION, GENOME REPLICATION, AND ENCAPSIDATION

Although (+)-strand RNA virus genomes such as those of poliovirus and HCV serve immediately as mRNAs to produce proteins required for viral replication, they are also genome rep-

lication templates. Production of (−)-strand RNA replication intermediates from (+)-strand templates is incompatible with ongoing (+)-strand translation (Andino et al. 1999). By cleaving PABP, which stimulates some IRESs (Michel et al. 2001; Bradrick et al. 2007), enteroviruses suppress mRNA translation to stimulate RNA replication. Rubella virus capsid protein similarly binds PABP to inhibit viral translation (Ilkow et al. 2008). Alternatively, the potyvirus 3′CITE can also suppress translation when genome replication begins (Miller et al. 2007).

Packaging translation-competent (+)-strand RNA genomes creates a different set of obstacles. Following reverse transcription and integration into the host cell genome, retroviruses such as HIV use the host RNA polymerase II transcription machinery to produce viral mRNAs. Reactivation and replication of HIV-1 are regulated by cyclin T1, which itself is translationally controlled by nuclear factor 90 (NF90) (Hoque et al. 2011). Although some of these genome-length viral mRNAs are translated, others are encapsidated into infectious particles. Importantly, binding of the HIV-1 Gag protein to eEF1A impairs viral mRNA translation to promote RNA packaging (Cimarelli and Luban 1999).

CONCLUDING REMARKS

Because of their reliance on the host for protein synthesis, the intimate relationship between viral genomes and their mRNAs, and the exigencies imposed by host defense mechanisms, viruses have evolved a vast array of strategies for exploiting and dominating the cellular translation machinery. Studies of translational regulation during virus infection have illuminated many features of the cellular translation system. They have also led to the discovery of translation mechanisms that figure prominently in virus-infected cells, and some that may even be virus specific. Notwithstanding the wealth of detail already amassed, there are many open questions, some of them noted above, and there is every reason to believe that we have only scratched the surface.

One prediction is that new principles will continue to emerge as the exploration of virus–host cell interactions expands in breadth and depth. For example, the control of translation in virus-infected cells shares notable similarities with stress-induced translational control mechanisms operative in uninfected cells. To prevent the accumulation of phosphorylated eIF2α, HSV1 encodes a phosphatase regulatory subunit similar to the ER stress-induced GADD34 (He et al. 1997). Cap-dependent translation is impaired in heat-shocked cells, necessitating a mechanism for heat shock protein synthesis much in the same way that viral translation proceeds in infected cells (Yueh and Schneider 1996, 2000). Indeed, both Ad late RNA and HSP70 mRNA contain a ribosome shunt to enable translation when eIF4F is limiting. Many of the key integrators controlling translation in response to nutrient availability, growth factors, and energy supplies are hijacked in infected cells to ensure high-level synthesis of viral proteins.

Another strong prediction is that the intensive analysis of virus–host interactions will have practical value, leading to therapeutics. Idiosyncratic mechanisms evolved by viruses, as well as the interplay between the virus and its host's translation system, represent potentially exploitable weaknesses. Remarkably, most, if not all, recessive virus resistance alleles in plants map to eIF4E or eIF4G, highlighting the importance of translation factors in the biology of plant RNA viruses (Nieto et al. 2007; Yeam et al. 2007; Truniger and Aranda 2009). As a potential host defense mechanism, pokeweed antiviral protein associates with eIF4G to access and depurinate viral RNA (Wang and Hudak 2006). Virus-specific mechanisms such as IRESs and frameshifting present direct targets for small-molecule drugs (Novac et al. 2004; Parsons et al. 2009; Gasparian et al. 2010; Paulsen et al. 2010). Cellular molecules such as ITAFs (Fontanes et al. 2009) and signaling pathways are also susceptible. Inhibitors of mTOR and eIF4F effectively reduce infection by many DNA as well as some RNA viruses, and may also serve to treat the tumors associated with some infections (Cen and Longnecker 2011), whereas chemical inducers of eIF2α phosphorylation reduce HSV-1 infection in vitro and in vivo (Boyce et al. 2005). Looking somewhat further afield, translational control is also being exploited in the development of new viral oncolytic therapies (Taneja et al. 2001; Mohr 2005; Stanford et al. 2007; Oliere et al. 2008; Thomas et al. 2009; Alain et al. 2010; Goetz et al. 2010).

REFERENCES

*Reference is also in this collection.

Alain T, Lun X, Martineau Y, Sean P, Pulendran B, Petroulakis E, Zemp FJ, Lemay CG, Roy D, Bell JC, et al. 2010. Vesicular stomatitis virus oncolysis is potentiated by impairing mTORC1-dependent type I IFN production. *Proc Natl Acad Sci* **107:** 1576–1581.

Ali IK, McKendrick L, Morley SJ, Jackson RJ. 2001. Activity of the hepatitis A virus IRES requires association between the cap-binding translation initiation factor (eIF4E) and eIF4G. *J Virol* **75:** 7854–7863.

Alvarez E, Menendez-Arias L, Carrasco L. 2003. The eukaryotic translation initiation factor 4GI is cleaved by different retroviral proteases. *J Virol* **77:** 12392–12400.

Alvarez E, Castello A, Menendez-Arias L, Carrasco L. 2006. HIV protease cleaves poly(A)-binding protein. *Biochem J* **396:** 219–226.

Andino R, Boddeker N, Silvera D, Gamarnik AV. 1999. Intracellular determinants of picornavirus replication. *Trends Microbiol* **7:** 76–82.

Aoyagi M, Gaspar M, Shenk TE. 2010. Human cytomegalovirus UL69 protein facilitates translation by associating with the mRNA cap-binding complex and excluding 4EBP1. *Proc Natl Acad Sci* **107:** 2640–2645.

Aragon T, de la Luna S, Novoa I, Carrasco L, Ortin J, Nieto A. 2000. Eukaryotic translation initiation factor 4GI is a cellular target for NS1 protein, a translational activator of influenza virus. *Mol Cell Biol* **20:** 6259–6268.

Arias C, Walsh D, Harbell J, Wilson AC, Mohr I. 2009. Activation of host translational control pathways by a viral developmental switch. *PLoS Pathog* **5:** e1000334.

Arnaud N, Dabo S, Maillard P, Budkowska A, Kalliampakou KI, Mavromara P, Garcin D, Hugon J, Gatignol A, Akazawa D, et al. 2010. Hepatitis C virus controls interferon production through PKR activation. *PLoS ONE* **5:** e10575.

Arslan D, Legendre M, Seltzer V, Abergel C, Claverie JM. 2011. Distant mimivirus relative with a larger genome highlights the fundamental features of Megaviridae. *Proc Natl Acad Sci* **108:** 17486–17491.

Babaylova E, Graifer D, Malygin A, Stahl J, Shatsky I, Karpova G. 2009. Positioning of subdomain IIId and apical loop of domain II of the hepatitis C IRES on the human 40S ribosome. *Nucleic Acids Res* **37:** 1141–1151.

Banerjee S, An S, Zhou A, Silverman RH, Makino S. 2000. RNase L-independent specific 28S rRNA cleavage in murine coronavirus-infected cells. *J Virol* **74:** 8793–8802.

Banerjee S, Narayanan K, Mizutani T, Makino S. 2002. Murine coronavirus replication-induced p38 mitogen-activated protein kinase activation promotes interleukin-6 production and virus replication in cultured cells. *J Virol* **76:** 5937–5948.

Beauchemin C, Boutet N, Laliberte JF. 2007. Visualization of the interaction between the precursors of VPg, the viral protein linked to the genome of turnip mosaic virus, and the translation eukaryotic initiation factor iso 4E in Planta. *J Virol* **81:** 775–782.

Beckham CJ, Light HR, Nissan TA, Ahlquist P, Parker R, Noueiry A. 2007. Interactions between brome mosaic virus RNAs and cytoplasmic processing bodies. *J Virol* **81:** 9759–9768.

Bellare P, Ganem D. 2009. Regulation of KSHV lytic switch protein expression by a virus-encoded microRNA: An evolutionary adaptation that fine-tunes lytic reactivation. *Cell Host Microbe* **6:** 570–575.

Berlanga JJ, Ventoso I, Harding HP, Deng J, Ron D, Sonenberg N, Carrasco L, de Haro C. 2006. Antiviral effect of the mammalian translation initiation factor 2α kinase GCN2 against RNA viruses. *EMBO J* **25:** 1730–1740.

Berry KE, Waghray S, Doudna JA. 2010. The HCV IRES pseudoknot positions the initiation codon on the 40S ribosomal subunit. *RNA* **16:** 1559–1569.

Bhushan S, Meyer H, Starosta AL, Becker T, Mielke T, Berninghausen O, Sattler M, Wilson DN, Beckmann R. 2010. Structural basis for translational stalling by human cytomegalovirus and fungal arginine attenuator peptide. *Mol Cell* **40:** 138–146.

Bieleski L, Talbot SJ. 2001. Kaposi's sarcoma-associated herpesvirus vCyclin open reading frame contains an internal ribosome entry site. *J Virol* **75:** 1864–1869.

Bilgin DD, Liu Y, Schiff M, Dinesh-Kumar SP. 2003. P58(IPK), a plant ortholog of double-stranded RNA-dependent protein kinase PKR inhibitor, functions in viral pathogenesis. *Dev Cell* **4:** 651–661.

Blakqori G, van Knippenberg I, Elliott RM. 2009. Bunyamwera orthobunyavirus S-segment untranslated regions mediate poly(A) tail-independent translation. *J Virol* **83:** 3637–3646.

Bonderoff JM, Larey JL, Lloyd RE. 2008. Cleavage of poly(A)-binding protein by poliovirus 3C proteinase inhibits viral internal ribosome entry site-mediated translation. *J Virol* **82:** 9389–9399.

Bordeleau ME, Mori A, Oberer M, Lindqvist L, Chard LS, Higa T, Belsham GJ, Wagner G, Tanaka J, Pelletier J. 2006. Functional characterization of IRESes by an inhibitor of the RNA helicase eIF4A. *Nat Chem Biol* **2:** 213–220.

Borman AM, Michel YM, Kean KM. 2001. Detailed analysis of the requirements of hepatitis A virus internal ribosome entry segment for the eukaryotic initiation factor complex eIF4F. *J Virol* **75:** 7864–7871.

Boyce M, Bryant KF, Jousse C, Long K, Harding HP, Scheuner D, Kaufman RJ, Ma D, Coen DM, Ron D, et al. 2005. A selective inhibitor of eIF2α dephosphorylation protects cells from ER stress. *Science* **307:** 935–939.

Boyne JR, Jackson BR, Taylor A, Macnab SA, Whitehouse A. 2010. Kaposi's sarcoma-associated herpesvirus ORF57 protein interacts with PYM to enhance translation of viral intronless mRNAs. *EMBO J* **29:** 1851–1864.

Bradrick SS, Dobrikova EY, Kaiser C, Shveygert M, Gromeier M. 2007. Poly(A)-binding protein is differentially re-

quired for translation mediated by viral internal ribosome entry sites. *RNA* **13:** 1582–1593.

Braunstein S, Badura ML, Xi Q, Formenti SC, Schneider RJ. 2009. Regulation of protein synthesis by ionizing radiation. *Mol Cell Biol* **29:** 5645–5656.

Browne GJ, Proud CG. 2002. Regulation of peptide-chain elongation in mammalian cells. *Eur J Biochem* **269:** 5360–5368.

Browne GJ, Proud CG. 2004. A novel mTOR-regulated phosphorylation site in elongation factor 2 kinase modulates the activity of the kinase and its binding to calmodulin. *Mol Cell Biol* **24:** 2986–2997.

Browne GJ, Finn SG, Proud CG. 2004. Stimulation of the AMP-activated protein kinase leads to activation of eukaryotic elongation factor 2 kinase and to its phosphorylation at a novel site, serine 398. *J Biol Chem* **279:** 12220–12231.

Burgui I, Aragon T, Ortin J, Nieto A. 2003. PABP1 and eIF4GI associate with influenza virus NS1 protein in viral mRNA translation initiation complexes. *J Gen Virol* **84:** 3263–3274.

Burgui I, Yanguez E, Sonenberg N, Nieto A. 2007. Influenza virus mRNA translation revisited: Is the eIF4E cap-binding factor required for viral mRNA translation? *J Virol* **81:** 12427–12438.

Cacoullos N, Bablanian R. 1991. Polyadenylated RNA sequences produced in vaccinia virus-infected cells under aberrant conditions inhibit protein synthesis in vitro. *Virology* **184:** 747–751.

Cai X, Lu S, Zhang Z, Gonzalez CM, Damania B, Cullen BR. 2005. Kaposi's sarcoma-associated herpesvirus expresses an array of viral microRNAs in latently infected cells. *Proc Natl Acad Sci* **102:** 5570–5575.

Carper SW, Rocheleau TA, Cimino D, Storm FK. 1997. Heat shock protein 27 stimulates recovery of RNA and protein synthesis following a heat shock. *J Cell Biochem* **66:** 153–164.

Castelló A, Quintas A, Sánchez EG, Sabina P, Nogal M, Carrasco L, Revilla Y. 2009. Regulation of host translational machinery by African Swine Fever Virus. *PLoS Pathog* **5:** e1000562.

Cen O, Longnecker R. 2011. Rapamycin reverses splenomegaly and inhibits tumor development in a transgenic model of Epstein-Barr virus-related Burkitt's lymphoma. *Mol Cancer Ther* **10:** 679–686.

Cencic R, Desforges M, Hall DR, Kozakov D, Du Y, Min J, Dingledine R, Fu H, Vajda S, Talbot PJ, et al. 2011. Blocking eIF4E–eIF4G interaction as a strategy to impair coronavirus replication. *J Virol* **85:** 6381–6389.

Cevallos RC, Sarnow P. 2005. Factor-independent assembly of elongation-competent ribosomes by an internal ribosome entry site located in an RNA virus that infects penaeid shrimp. *J Virol* **79:** 677–683.

Chakrabarti A, Jha BK, Silverman RH. 2011. New insights into the role of RNase L in innate immunity. *J Interferon Cytokine Res* **31:** 49–57.

Chang HW, Watson JC, Jacobs BL. 1992. The E3L gene of vaccinia virus encodes an inhibitor of the interferon-induced, double-stranded RNA-dependent protein kinase. *Proc Natl Acad Sci* **89:** 4825–4829.

Chaudhry Y, Nayak A, Bordeleau ME, Tanaka J, Pelletier J, Belsham GJ, Roberts LO, Goodfellow IG. 2006. Caliciviruses differ in their functional requirements for eIF4F components. *J Biol Chem* **281:** 25315–25325.

Chulu JL, Huang WR, Wang L, Shih WL, Liu HJ. 2010. Avian reovirus nonstructural protein p17–induced G₂/M cell cycle arrest and host cellular protein translation shutoff involve activation of p53-dependent pathways. *J Virol* **84:** 7683–7694.

Chuluunbaatar U, Roller R, Feldman ME, Brown S, Shokat KM, Mohr I. 2010. Constitutive mTORC1 activation by a herpesvirus Akt surrogate stimulates mRNA translation and viral replication. *Genes Dev* **24:** 2627–2639.

Cimarelli A, Luban J. 1999. Translation elongation factor 1-α interacts specifically with the human immunodeficiency virus type 1 Gag polyprotein. *J Virol* **73:** 5388–5401.

Clippinger AJ, Maguire TG, Alwine JC. 2011. The changing role of mTOR kinase in the maintenance of protein synthesis during human cytomegalovirus infection. *J Virol* **85:** 3930–3939.

Clyde K, Glaunsinger BA. 2010. Getting the message direct manipulation of host mRNA accumulation during gammaherpesvirus lytic infection. *Adv Virus Res* **78:** 1–42.

Colina R, Costa-Mattioli M, Dowling RJ, Jaramillo M, Tai LH, Breitbach CJ, Martineau Y, Larsson O, Rong L, Svitkin YV, et al. 2008. Translational control of the innate immune response through IRF-7. *Nature* **452:** 323–328.

Connor JH, Lyles DS. 2002. Vesicular stomatitis virus infection alters the eIF4F translation initiation complex and causes dephosphorylation of the eIF4E binding protein 4E-BP1. *J Virol* **76:** 10177–10187.

Covarrubias S, Richner JM, Clyde K, Lee YJ, Glaunsinger BA. 2009. Host shutoff is a conserved phenotype of gammaherpesvirus infection and is orchestrated exclusively from the cytoplasm. *J Virol* **83:** 9554–9566.

Covarrubias S, Gaglia MM, Kumar GR, Wong W, Jackson AO, Glaunsinger BA. 2011. Coordinated destruction of cellular messages in translation complexes by the gammaherpesvirus host shutoff factor and the mammalian exonuclease Xrn1. *PLoS Pathog* **7:** e1002339.

Cuesta R, Laroia G, Schneider RJ. 2000a. Chaperone hsp27 inhibits translation during heat shock by binding eIF4G and facilitating dissociation of cap-initiation complexes. *Genes Dev* **14:** 1460–1470.

Cuesta R, Xi Q, Schneider RJ. 2000b. Adenovirus-specific translation by displacement of kinase Mnk1 from cap-initiation complex eIF4F. *EMBO J* **19:** 3465–3474.

Cuesta R, Xi Q, Schneider RJ. 2004. Structural basis for competitive inhibition of eIF4G–Mnk1 interaction by the adenovirus 100-kilodalton protein. *J Virol* **78:** 7707–7716.

Daughenbaugh KF, Fraser CS, Hershey JW, Hardy ME. 2003. The genome-linked protein VPg of the Norwalk virus binds eIF3, suggesting its role in translation initiation complex recruitment. *EMBO J* **22:** 2852–2859.

Daughenbaugh KF, Wobus CE, Hardy ME. 2006. VPg of murine norovirus binds translation initiation factors in infected cells. *Virol J* **3:** 33.

Davies MV, Elroy-Stein O, Jagus R, Moss B, Kaufman RJ. 1992. The vaccinia virus K3L gene product potentiates translation by inhibiting double-stranded-RNA-activat-ed protein kinase and phosphorylation of the alpha subunit of eukaryotic initiation factor 2. *J Virol* **66:** 1943–1950.

Davies MV, Chang HW, Jacobs BL, Kaufman RJ. 1993. The E3L and K3L vaccinia virus gene products stimulate translation through inhibition of the double-stranded RNA-dependent protein kinase by different mechanisms. *J Virol* **67:** 1688–1692.

de Breyne S, Bonderoff JM, Chumakov KM, Lloyd RE, Hellen CU. 2008a. Cleavage of eukaryotic initiation factor eIF5B by enterovirus 3C proteases. *Virology* **378:** 118–122.

de Breyne S, Yu Y, Pestova TV, Hellen CU. 2008b. Factor requirements for translation initiation on the Simian picornavirus internal ribosomal entry site. *RNA* **14:** 367–380.

de Breyne S, Yu Y, Unbehaun A, Pestova TV, Hellen CU. 2009. Direct functional interaction of initiation factor eIF4G with type 1 internal ribosomal entry sites. *Proc Natl Acad Sci* **106:** 9197–9202.

Dias A, Bouvier D, Crepin T, McCarthy AA, Hart DJ, Baudin F, Cusack S, Ruigrok RW. 2009. The cap-snatching endonuclease of influenza virus polymerase resides in the PA subunit. *Nature* **458:** 914–918.

Dixon LK, Twigg SR, Baylis SA, Vydelingum S, Bristow C, Hammond JM, Smith GL. 1994. Nucleotide sequence of a 55 kbp region from the right end of the genome of a pathogenic African Swine Fever Virus isolate (Malawi LIL20/1). *J Gen Virol* **75:** 1655–1684.

Dmitriev SE, Terenin IM, Andreev DE, Ivanov PA, Dunaevsky JE, Merrick WC, Shatsky IN. 2010. GTP-independent tRNA delivery to the ribosomal P-site by a novel eukaryotic translation factor. *J Biol Chem* **285:** 26779–26787.

Dobrikova E, Shveygert M, Walters R, Gromeier M. 2010. Herpes simplex virus proteins ICP27 and UL47 associate with polyadenylate-binding protein and control its subcellular distribution. *J Virol* **84:** 270–279.

Dougherty JD, White JP, Lloyd RE. 2011. Poliovirus-mediated disruption of cytoplasmic processing bodies. *J Virol* **85:** 64–75.

Dunn EF, Connor JH. 2011. Dominant inhibition of Akt/protein kinase B signaling by the matrix protein of a negative-strand RNA virus. *J Virol* **85:** 422–431.

Elde NC, Child SJ, Geballe AP, Malik HS. 2009. Protein kinase R reveals an evolutionary model for defeating viral mimicry. *Nature* **457:** 485–489.

Elia A, Constantinou C, Clemens MJ. 2008. Effects of protein phosphorylation on ubiquitination and stability of the translational inhibitor protein 4E-BP1. *Oncogene* **27:** 811–822.

Ellison KS, Maranchuk RA, Mottet KL, Smiley JR. 2005. Control of VP16 translation by the herpes simplex virus type 1 immediate-early protein ICP27. *J Virol* **79:** 4120–4131.

Emara MM, Brinton MA. 2007. Interaction of TIA-1/TIAR with West Nile and dengue virus products in infected cells interferes with stress granule formation and processing body assembly. *Proc Natl Acad Sci* **104:** 9041–9046.

Ensinger MJ, Martin SA, Paoletti E, Moss B. 1975. Modification of the 5'-terminus of mRNA by soluble guanylyl and methyl transferases from vaccinia virus. *Proc Natl Acad Sci* **72:** 2525–2529.

Eskelin K, Hafren A, Rantalainen KI, Makinen K. 2011. Potyviral VPg enhances viral RNA translation and inhibits reporter mRNA translation in planta. *J Virol* **85:** 9210–9221.

Etchison D, Milburn SC, Edery I, Sonenberg N, Hershey JW. 1982. Inhibition of HeLa cell protein synthesis following poliovirus infection correlates with the proteolysis of a 220,000-dalton polypeptide associated with eucaryotic initiation factor 3 and a cap binding protein complex. *J Biol Chem* **257:** 14806–14810.

Faria PA, Chakraborty P, Levay A, Barber GN, Ezelle HJ, Enninga J, Arana C, van Deursen J, Fontoura BM. 2005. VSV disrupts the Rae1/mrnp41 mRNA nuclear export pathway. *Mol Cell* **17:** 93–102.

Feigenblum D, Schneider RJ. 1993. Modification of eukaryotic initiation factor 4F during infection by influenza virus. *J Virol* **67:** 3027–3035.

Feigenblum D, Schneider RJ. 1996. Cap-binding protein (eukaryotic initiation factor 4E) and 4E-inactivating protein BP-1 independently regulate cap-dependent translation. *Mol Cell Biol* **16:** 5450–5457.

Feldman ME, Apsel B, Uotila A, Loewith R, Knight ZA, Ruggero D, Shokat KM. 2009. Active-site inhibitors of mTOR target rapamycin-resistant outputs of mTORC1 and mTORC2. *PLoS Biol* **7:** e1000038.

Feng P, Everly DN Jr, Read GS. 2005. mRNA decay during herpes simplex virus (HSV) infections: Protein–protein interactions involving the HSV virion host shutoff protein and translation factors eIF4H and eIF4A. *J Virol* **79:** 9651–9664.

Fischer MG, Allen MJ, Wilson WH, Suttle CA. 2010. Giant virus with a remarkable complement of genes infects marine zooplankton. *Proc Natl Acad Sci* **107:** 19508–19513.

Fontaine-Rodriguez EC, Knipe DM. 2008. Herpes simplex virus ICP27 increases translation of a subset of viral late mRNAs. *J Virol* **82:** 3538–3545.

Fontanes V, Raychaudhuri S, Dasgupta A. 2009. A cell-permeable peptide inhibits hepatitis C virus replication by sequestering IRES transacting factors. *Virology* **394:** 82–90.

Fraser CS, Doudna JA. 2007. Structural and mechanistic insights into hepatitis C viral translation initiation. *Nat Rev* **5:** 29–38.

Fujimura T, Esteban R. 2011. Cap-snatching mechanism in yeast L-A double-stranded RNA virus. *Proc Natl Acad Sci* **108:** 17667–17671.

Gale M Jr, Foy EM. 2005. Evasion of intracellular host defence by hepatitis C virus. *Nature* **436:** 939–945.

Gale M Jr, Blakely CM, Kwieciszewski B, Tan SL, Dossett M, Tang NM, Korth MJ, Polyak SJ, Gretch DR, Katze MG. 1998. Control of PKR protein kinase by hepatitis C virus nonstructural 5A protein: Molecular mechanisms of kinase regulation. *Mol Cell Biol* **18:** 5208–5218.

Gallie DR. 2001. Cap-independent translation conferred by the 5′ leader of tobacco etch virus is eukaryotic initiation factor 4G dependent. *J Virol* **75:** 12141–12152.

Garaigorta U, Chisari FV. 2009. Hepatitis C virus blocks interferon effector function by inducing protein kinase R phosphorylation. *Cell Host Microbe* **6:** 513–522.

Garcin D, Lezzi M, Dobbs M, Elliott RM, Schmaljohn C, Kang CY, Kolakofsky D. 1995. The 5′ ends of Hantaan virus (Bunyaviridae) RNAs suggest a prime-and-realign mechanism for the initiation of RNA synthesis. *J Virol* **69:** 5754–5762.

Garrey JL, Lee YY, Au HH, Bushell M, Jan E. 2010. Host and viral translational mechanisms during cricket paralysis virus infection. *J Virol* **84:** 1124–1138.

Gasparian AV, Neznanov N, Jha S, Galkin O, Moran JJ, Gudkov AV, Gurova KV, Komar AA. 2010. Inhibition of encephalomyocarditis virus and poliovirus replication by quinacrine: Implications for the design and discovery of novel antiviral drugs. *J Virol* **84:** 9390–9397.

Gingras AC, Sonenberg N. 1997. Adenovirus infection inactivates the translational inhibitors 4E-BP1 and 4E-BP2. *Virology* **237:** 182–186.

Gingras AC, Svitkin Y, Belsham GJ, Pause A, Sonenberg N. 1996. Activation of the translational suppressor 4E-BP1 following infection with encephalomyocarditis virus and poliovirus. *Proc Natl Acad Sci* **93:** 5578–5583.

Goetz C, Everson RG, Zhang LC, Gromeier M. 2010. MAPK signal-integrating kinase controls cap-independent translation and cell type-specific cytotoxicity of an oncolytic poliovirus. *Mol Ther* **18:** 1937–1946.

Goodfellow I, Chaudhry Y, Gioldasi I, Gerondopoulos A, Natoni A, Labrie L, Laliberte JF, Roberts L. 2005. Calicivirus translation initiation requires an interaction between VPg and eIF 4 E. *EMBO Rep* **6:** 968–972.

Goodman AG, Smith JA, Balachandran S, Perwitasari O, Proll SC, Thomas MJ, Korth MJ, Barber GN, Schiff LA, Katze MG. 2007. The cellular protein P58IPK regulates influenza virus mRNA translation and replication through a PKR-mediated mechanism. *J Virol* **81:** 2221–2230.

Goodman AG, Fornek JL, Medigeshi GR, Perrone LA, Peng X, Dyer MD, Proll SC, Knoblaugh SE, Carter VS, Korth MJ, et al. 2009. P58[IPK]: A novel "CIHD" member of the host innate defense response against pathogenic virus infection. *PLoS Pathog* **5:** e1000438.

Gorchakov R, Frolova E, Williams BR, Rice CM, Frolov I. 2004. PKR-dependent and -independent mechanisms are involved in translational shutoff during Sindbis virus infection. *J Virol* **78:** 8455–8467.

Gradi A, Svitkin YV, Imataka H, Sonenberg N. 1998. Proteolysis of human eukaryotic translation initiation factor eIF4GII, but not eIF4GI, coincides with the shutoff of host protein synthesis after poliovirus infection. *Proc Natl Acad Sci* **95:** 11089–11094.

Grainger L, Cicchini L, Rak M, Petrucelli A, Fitzgerald KD, Semler BL, Goodrum F. 2010. Stress-inducible alternative translation initiation of human cytomegalovirus latency protein pUL138. *J Virol* **84:** 9472–9486.

Griffiths A, Coen DM. 2005. An unusual internal ribosome entry site in the herpes simplex virus thymidine kinase gene. *Proc Natl Acad Sci* **102:** 9667–9672.

Groft CM, Burley SK. 2002. Recognition of eIF4G by rotavirus NSP3 reveals a basis for mRNA circularization. *Mol Cell* **9:** 1273–1283.

Groppo R, Brown BA, Palmenberg AC. 2010. Mutational analysis of the EMCV 2A protein identifies a nuclear localization signal and an eIF4E binding site. *Virology* **410:** 257–267.

Grundhoff A, Ganem D. 2001. Mechanisms governing expression of the v-FLIP gene of Kaposi's sarcoma-associated herpesvirus. *J Virol* **75:** 1857–1863.

Guilligay D, Tarendeau F, Resa-Infante P, Coloma R, Crepin T, Sehr P, Lewis J, Ruigrok RW, Ortin J, Hart DJ, et al. 2008. The structural basis for cap binding by influenza virus polymerase subunit PB2. *Nat Struct Mol Biol* **15:** 500–506.

Gunnery S, Green SR, Mathews MB. 1992. Tat-responsive region RNA of human immunodeficiency virus type 1 stimulates protein synthesis in vivo and in vitro: Relationship between structure and function. *Proc Natl Acad Sci* **89:** 11557–11561.

Guo J, Hui DJ, Merrick WC, Sen GC. 2000. A new pathway of translational regulation mediated by eukaryotic initiation factor 3. *EMBO J* **19:** 6891–6899.

Harb M, Becker MM, Vitour D, Baron CH, Vende P, Brown SC, Bolte S, Arold ST, Poncet D. 2008. Nuclear localization of cytoplasmic poly(A)-binding protein upon rotavirus infection involves the interaction of NSP3 with eIF4G and RoXaN. *J Virol* **82:** 11283–11293.

He B, Gross M, Roizman B. 1997. The $\gamma_1$34.5 protein of herpes simplex virus 1 complexes with protein phosphatase 1α to dephosphorylate the α subunit of the eukaryotic translation initiation factor 2 and preclude the shutoff of protein synthesis by double-stranded RNA-activated protein kinase. *Proc Natl Acad Sci* **94:** 843–848.

Henke JI, Goergen D, Zheng J, Song Y, Schuttler CG, Fehr C, Junemann C, Niepmann M. 2008. microRNA-122 stimulates translation of hepatitis C virus RNA. *EMBO J* **27:** 3300–3310.

Herdy B, Jaramillo M, Svitkin YV, Rosenfeld AB, Kobayashi M, Walsh D, Alain T, Sean P, Robichaud N, Topisirovic I, et al. 2012. Translational control of the activation of transcription factor NF-κB and production of type I Interferon by phosphorylation of the translation factor eIF4E. *Nat Immunol* doi: 10.1038/ni.2291.

* Hershey WB, Sonenberg N, Mathews MB. 2012. Principles of translational control: An overview. *Cold Spring Harb Perspect Biol* doi: 10.1101/cshperspect.a011528.

* Hinnebusch AG, Lorsch JR. 2012. The mechanism of eukaryotic translation initiation: New insights and challenges. *Cold Spring Harb Perspect Biol* doi: cshperspect.a011528/cshperspect.a011544.

Ho BC, Yu SL, Chen JJ, Chang SY, Yan BS, Hong QS, Singh S, Kao CL, Chen HY, Su KY, et al. 2010. Enterovirus-induced miR-141 contributes to shutoff of host protein translation by targeting the translation initiation factor eIF4E. *Cell Host Microbe* **9:** 58–69.

Hogg JR, Goff SP. 2010. Upf1 senses 3′UTR length to potentiate mRNA decay. *Cell* **143:** 379–389.

Hoque M, Shamanna RA, Guan D, Pe'ery T, Mathews MB. 2011. HIV-1 replication and latency are regulated by translational control of cyclin T1. *J Mol Biol* **410:** 917–932.

Horvath CM, Williams MA, Lamb RA. 1990. Eukaryotic coupled translation of tandem cistrons: Identification of the influenza B virus BM2 polypeptide. *EMBO J* **9:** 2639–2647.

Houck-Loomis B, Durney MA, Salguero C, Shankar N, Nagle JM, Goff SP, D'Souza VM. 2011. An equilibrium-dependent retroviral mRNA switch regulates translational recoding. *Nature* **480:** 561–564.

Huang JT, Schneider RJ. 1991. Adenovirus inhibition of cellular protein synthesis involves inactivation of cap-binding protein. *Cell* **65:** 271–280.

Huang C, Lokugamage KG, Rozovics JM, Narayanan K, Semler BL, Makino S. 2011. SARS coronavirus nsp1 protein induces template-dependent endonucleolytic cleavage of mRNAs: Viral mRNAs are resistant to nsp1-induced RNA cleavage. *PLoS Pathog* **7:** e1002433.

Hui DJ, Bhasker CR, Merrick WC, Sen GC. 2003. Viral stress-inducible protein p56 inhibits translation by blocking the interaction of eIF3 with the ternary complex eIF2·GTP·Met-tRNAi. *J Biol Chem* **278:** 39477–39482.

Hui DJ, Terenzi F, Merrick WC, Sen GC. 2005. Mouse p56 blocks a distinct function of eukaryotic initiation factor 3 in translation initiation. *J Biol Chem* **280:** 3433–3440.

Ilkow CS, Mancinelli V, Beatch MD, Hobman TC. 2008. Rubella virus capsid protein interacts with poly(A)-binding protein and inhibits translation. *J Virol* **82:** 4284–4294.

Jack K, Bellodi C, Landry DM, Niederer RO, Meskauskas A, Musalgaonkar S, Kopmar N, Krasnykh O, Dean AM, Thompson SR, et al. 2011. rRNA pseudouridylation defects affect ribosomal ligand binding and translational fidelity from yeast to human cells. *Mol Cell* **44:** 660–666.

* Jackson RJ. 2012. The current status of vertebrate cellular mRNA IRESs. *Cold Spring Harb Perspect Biol* doi: cshperspect.a011528/cshperspect.a011569.

Jackson BR, Boyne JR, Noerenberg M, Taylor A, Hautbergue GM, Walsh MJ, Wheat R, Blackbourn DJ, Wilson SA, Whitehouse A. 2011. An interaction between KSHV ORF57 and UIF provides mRNA-adaptor redundancy in herpesvirus intronless mRNA export. *PLoS Pathog* **7:** e1002138.

Jan E, Sarnow P. 2002. Factorless ribosome assembly on the internal ribosome entry site of cricket paralysis virus. *J Mol Biol* **324:** 889–902.

Jang SK, Krausslich HG, Nicklin MJ, Duke GM, Palmenberg AC, Wimmer E. 1988. A segment of the 5′ nontranslated region of encephalomyocarditis virus RNA directs internal entry of ribosomes during in vitro translation. *J Virol* **62:** 2636–2643.

Jangra RK, Yi M, Lemon SM. 2010. Regulation of hepatitis C virus translation and infectious virus production by the microRNA miR-122. *J Virol* **84:** 6615–6625.

Janzen DM, Frolova L, Geballe AP. 2002. Inhibition of translation termination mediated by an interaction of eukaryotic release factor 1 with a nascent peptidyl-tRNA. *Mol Cell Biol* **22:** 8562–8570.

Ji H, Fraser CS, Yu Y, Leary J, Doudna JA. 2004. Coordinated assembly of human translation initiation complexes by the hepatitis C virus internal ribosome entry site RNA. *Proc Natl Acad Sci* **101:** 16990–16995.

Ji WT, Wang L, Lin RC, Huang WR, Liu HJ. 2009. Avian reovirus influences phosphorylation of several factors involved in host protein translation including eukaryotic translation elongation factor 2 (eEF2) in Vero cells. *Biochem Biophys Res Commun* **384:** 301–305.

Jopling CL, Yi M, Lancaster AM, Lemon SM, Sarnow P. 2005. Modulation of hepatitis C virus RNA abundance by a liver-specific microRNA. *Science* **309:** 1577–1581.

Jopling CL, Schutz S, Sarnow P. 2008. Position-dependent function for a tandem microRNA miR-122-binding site located in the hepatitis C virus RNA genome. *Cell Host Microbe* **4:** 77–85.

Jordan R, Wang L, Graczyk TM, Block TM, Romano PR. 2002. Replication of a cytopathic strain of bovine viral diarrhea virus activates PERK and induces endoplasmic reticulum stress-mediated apoptosis of MDBK cells. *J Virol* **76:** 9588–9599.

Kafasla P, Morgner N, Robinson CV, Jackson RJ. 2010. Polypyrimidine tract-binding protein stimulates the poliovirus IRES by modulating eIF4G binding. *EMBO J* **29:** 3710–3722.

Kamitani W, Narayanan K, Huang C, Lokugamage K, Ikegami T, Ito N, Kubo H, Makino S. 2006. Severe acute respiratory syndrome coronavirus nsp1 protein suppresses host gene expression by promoting host mRNA degradation. *Proc Natl Acad Sci* **103:** 12885–12890.

Kamitani W, Huang C, Narayanan K, Lokugamage KG, Makino S. 2009. A two-pronged strategy to suppress host protein synthesis by SARS coronavirus Nsp1 protein. *Nat Struct Mol Biol* **16:** 1134–1140.

Kamoshita N, Nomoto A, RajBhandary UL. 2009. Translation initiation from the ribosomal A site or the P site, dependent on the conformation of RNA pseudoknot I in dicistrovirus RNAs. *Mol Cell* **35:** 181–190.

Kang JI, Kwon SN, Park SH, Kim YK, Choi SY, Kim JP, Ahn BY. 2009. PKR protein kinase is activated by hepatitis C virus and inhibits viral replication through translational control. *Virus Res* **142:** 51–56.

Katsafanas GC, Moss B. 2007. Colocalization of transcription and translation within cytoplasmic poxvirus factories coordinates viral expression and subjugates host functions. *Cell Host Microbe* **2:** 221–228.

Kawagishi-Kobayashi M, Silverman JB, Ung TL, Dever TE. 1997. Regulation of the protein kinase PKR by the vaccinia virus pseudosubstrate inhibitor K3L is dependent on residues conserved between the K3L protein and the PKR substrate eIF2α. *Mol Cell Biol* **17:** 4146–4158.

Kawaguchi Y, Matsumura T, Roizman B, Hirai K. 1999. Cellular elongation factor 1δ is modified in cells infected with representative alpha-, beta-, or gammaherpesviruses. *J Virol* **73:** 4456–4460.

Kazemi S, Papadopoulou S, Li S, Su Q, Wang S, Yoshimura A, Matlashewski G, Dever TE, Koromilas AE. 2004. Control of α subunit of eukaryotic translation initiation factor 2 (eIF2α) phosphorylation by the human papillomavirus type 18 E6 oncoprotein: Implications for eIF2 α-dependent gene expression and cell death. *Mol Cell Biol* **24:** 3415–3429.

Kempf BJ, Barton DJ. 2008. Poliovirus 2A(Pro) increases viral mRNA and polysome stability coordinately in time with cleavage of eIF4G. *J Virol* **82:** 5847–5859.

Khan MA, Miyoshi H, Ray S, Natsuaki T, Suehiro N, Goss DJ. 2006. Interaction of genome-linked protein (VPg) of turnip mosaic virus with wheat germ translation initia-

tion factors eIFiso4E and eIFiso4F. *J Biol Chem* **281:** 28002–28010.

Khan MA, Miyoshi H, Gallie DR, Goss DJ. 2008. Potyvirus genome-linked protein, VPg, directly affects wheat germ in vitro translation: Interactions with translation initiation factors eIF4F and eIFiso4F. *J Biol Chem* **283:** 1340–1349.

Khong A, Jan E. 2011. Modulation of stress granules and P bodies during dicistrovirus infection. *J Virol* **85:** 1439–1451.

Kim JH, Park SM, Park JH, Keum SJ, Jang SK. 2011. eIF2A mediates translation of hepatitis C viral mRNA under stress conditions. *EMBO J* **30:** 2454–2464.

Kolupaeva VG, Hellen CU, Shatsky IN. 1996. Structural analysis of the interaction of the pyrimidine tract-binding protein with the internal ribosomal entry site of encephalomyocarditis virus and foot-and-mouth disease virus RNAs. *RNA* **2:** 1199–1212.

Kolupaeva VG, Pestova TV, Hellen CU, Shatsky IN. 1998. Translation eukaryotic initiation factor 4G recognizes a specific structural element within the internal ribosome entry site of encephalomyocarditis virus RNA. *J Biol Chem* **273:** 18599–18604.

Komarova AV, Real E, Borman AM, Brocard M, England P, Tordo N, Hershey JW, Kean KM, Jacob Y. 2007. Rabies virus matrix protein interplay with eIF3, new insights into rabies virus pathogenesis. *Nucleic Acids Res* **35:** 1522–1532.

Kratzer F, Rosorius O, Heger P, Hirschmann N, Dobner T, Hauber J, Stauber RH. 2000. The adenovirus type 5 E1B-55K oncoprotein is a highly active shuttle protein and shuttling is independent of E4orf6, p53 and Mdm2. *Oncogene* **19:** 850–857.

Krausslich HG, Nicklin MJ, Toyoda H, Etchison D, Wimmer E. 1987. Poliovirus proteinase 2A induces cleavage of eucaryotic initiation factor 4F polypeptide 220. *J Virol* **61:** 2711–2718.

Kudchodkar SB, Yu Y, Maguire TG, Alwine JC. 2004. Human cytomegalovirus infection induces rapamycin-insensitive phosphorylation of downstream effectors of mTOR kinase. *J Virol* **78:** 11030–11039.

Kudchodkar SB, Yu Y, Maguire TG, Alwine JC. 2006. Human cytomegalovirus infection alters the substrate specificities and rapamycin sensitivities of raptor- and rictor-containing complexes. *Proc Natl Acad Sci* **103:** 14182–14187.

Kumar GR, Glaunsinger BA. 2010. Nuclear import of cytoplasmic poly(A) binding protein restricts gene expression via hyperadenylation and nuclear retention of mRNA. *Mol Cell Biol* **30:** 4996–5008.

Kumar GR, Shum L, Glaunsinger BA. 2011. Importin α-mediated nuclear import of cytoplasmic poly(A) binding protein occurs as a direct consequence of cytoplasmic mRNA depletion. *Mol Cell Biol* **31:** 3113–3125.

Kuyumcu-Martinez M, Belliot G, Sosnovtsev SV, Chang KO, Green KY, Lloyd RE. 2004a. Calicivirus 3C-like proteinase inhibits cellular translation by cleavage of poly(A)-binding protein. *J Virol* **78:** 8172–8182.

Kuyumcu-Martinez NM, Van Eden ME, Younan P, Lloyd RE. 2004b. Cleavage of poly(A)-binding protein by poliovirus 3C protease inhibits host cell translation: A novel mechanism for host translation shutoff. *Mol Cell Biol* **24:** 1779–1790.

Lancaster AM, Jan E, Sarnow P. 2006. Initiation factor-independent translation mediated by the hepatitis C virus internal ribosome entry site. *RNA* **12:** 894–902.

Landry DM, Hertz MI, Thompson SR. 2009. RPS25 is essential for translation initiation by the Dicistroviridae and hepatitis C viral IRESs. *Genes Dev* **23:** 2753–2764.

Larralde O, Smith RW, Wilkie GS, Malik P, Gray NK, Clements JB. 2006. Direct stimulation of translation by the multifunctional herpesvirus ICP27 protein. *J Virol* **80:** 1588–1591.

Lee YJ, Glaunsinger BA. 2009. Aberrant herpesvirus-induced polyadenylation correlates with cellular messenger RNA destruction. *PLoS Biol* **7:** e1000107.

Li Y, Zhang C, Chen X, Yu J, Wang Y, Yang Y, Du M, Jin H, Ma Y, He B, et al. 2011. ICP34.5 protein of herpes simplex virus facilitates the initiation of protein translation by bridging eukaryotic initiation factor 2α (eIF2α) and protein phosphatase 1. *J Biol Chem* **286:** 24785–24792.

Lindquist ME, Lifland AW, Utley TJ, Santangelo PJ, Crowe JE Jr. 2010. Respiratory syncytial virus induces host RNA stress granules to facilitate viral replication. *J Virol* **84:** 12274–12284.

Lischka P, Toth Z, Thomas M, Mueller R, Stamminger T. 2006. The UL69 transactivator protein of human cytomegalovirus interacts with DEXD/H-Box RNA helicase UAP56 to promote cytoplasmic accumulation of unspliced RNA. *Mol Cell Biol* **26:** 1631–1643.

Locker N, Easton LE, Lukavsky PJ. 2007. HCV and CSFV IRES domain II mediate eIF2 release during 80S ribosome assembly. *EMBO J* **26:** 795–805.

Locker N, Chamond N, Sargueil B. 2010. A conserved structure within the HIV gag open reading frame that controls translation initiation directly recruits the 40S subunit and eIF3. *Nucleic Acids Res* **39:** 2367–2377.

Low W, Harries M, Ye H, Du MQ, Boshoff C, Collins M. 2001. Internal ribosome entry site regulates translation of Kaposi's sarcoma-associated herpesvirus FLICE inhibitory protein. *J Virol* **75:** 2938–2945.

Lu C, Bablanian R. 1996. Characterization of small non-translated polyadenylylated RNAs in vaccinia virus-infected cells. *Proc Natl Acad Sci* **93:** 2037–2042.

Lu Z, Hu X, Li Y, Zheng L, Zhou Y, Jiang H, Ning T, Basang Z, Zhang C, Ke Y. 2004. Human papillomavirus 16 E6 oncoprotein interferences with insulin signaling pathway by binding to tuberin. *J Biol Chem* **279:** 35664–35670.

Luttermann C, Meyers G. 2009. The importance of inter- and intramolecular base pairing for translation reinitiation on a eukaryotic bicistronic mRNA. *Genes Dev* **23:** 331–344.

McInerney GM, Kedersha NL, Kaufman RJ, Anderson P, Liljestrom P. 2005. Importance of eIF2α phosphorylation and stress granule assembly in alphavirus translation regulation. *Mol Biol Cell* **16:** 3753–3763.

McKinney C, Perez C, Mohr I. 2012. Poly(A) binding protein abundance regulates eukaryotic translation initiation factor 4F assembly in human cytomegalovirus-infected cells. *Proc Natl Acad Sci* **109:** 5627–5632.

McMahon R, Zaborowska I, Walsh D. 2011. Noncytotoxic inhibition of viral infection through eIF4F-independent suppression of translation by 4EGi-1. *J Virol* **85:** 853–864.

Michel YM, Borman AM, Paulous S, Kean KM. 2001. Eukaryotic initiation factor 4G-poly(A) binding protein interaction is required for poly(A) tail-mediated stimulation of picornavirus internal ribosome entry segment-driven translation but not for X-mediated stimulation of hepatitis C virus translation. *Mol Cell Biol* **21:** 4097–4109.

Miller WA, Wang Z, Treder K. 2007. The amazing diversity of cap-independent translation elements in the 3′-untranslated regions of plant viral RNAs. *Biochem Soc Trans* **35:** 1629–1633.

Mir MA, Panganiban AT. 2008. A protein that replaces the entire cellular eIF4F complex. *EMBO J* **27:** 3129–3139.

Mir MA, Panganiban AT. 2010. The triplet repeats of the Sin Nombre hantavirus 5′ untranslated region are sufficient in *cis* for nucleocapsid-mediated translation initiation. *J Virol* **84:** 8937–8944.

Mir MA, Duran WA, Hjelle BL, Ye C, Panganiban AT. 2008. Storage of cellular 5′ mRNA caps in P bodies for viral cap-snatching. *Proc Natl Acad Sci* **105:** 19294–19299.

Mizutani T, Fukushi S, Saijo M, Kurane I, Morikawa S. 2004. Phosphorylation of p38 MAPK and its downstream targets in SARS coronavirus-infected cells. *Biochem Biophys Res Commun* **319:** 1228–1234.

Mohankumar V, Dhanushkodi NR, Raju R. 2011. Sindbis virus replication, is insensitive to rapamycin and torin1, and suppresses Akt/mTOR pathway late during infection in HEK cells. *Biochem Biophys Res Commun* **406:** 262–267.

Mohr I. 2005. To replicate or not to replicate: Achieving selective oncolytic virus replication in cancer cells through translational control. *Oncogene* **24:** 7697–7709.

Montero H, Arias CF, Lopez S. 2006. Rotavirus nonstructural protein NSP3 is not required for viral protein synthesis. *J Virol* **80:** 9031–9038.

Montero H, Rojas M, Arias CF, Lopez S. 2008. Rotavirus infection induces the phosphorylation of eIF2α but prevents the formation of stress granules. *J Virol* **82:** 1496–1504.

Moody CA, Scott RS, Amirghahari N, Nathan CA, Young LS, Dawson CW, Sixbey JW. 2005. Modulation of the cell growth regulator mTOR by Epstein-Barr virus-encoded LMP2A. *J Virol* **79:** 5499–5506.

Moorman NJ, Shenk T. 2010. Rapamycin-resistant mTORC1 activity is required for herpesvirus replication. *J Virol* **84:** 5260–5269.

Moorman NJ, Cristea IM, Terhune SS, Rout MP, Chait BT, Shenk T. 2008. Human cytomegalovirus protein UL38 inhibits host cell stress responses by antagonizing the tuberous sclerosis protein complex. *Cell Host Microbe* **3:** 253–262.

Mulvey M, Poppers J, Ladd A, Mohr I. 1999. A herpesvirus ribosome-associated, RNA-binding protein confers a growth advantage upon mutants deficient in a GADD34-related function. *J Virol* **73:** 3375–3385.

Mulvey M, Poppers J, Sternberg D, Mohr I. 2003. Regulation of eIF2α phosphorylation by different functions that act during discrete phases in the herpes simplex virus type 1 life cycle. *J Virol* **77:** 10917–10928.

Mulvey M, Arias C, Mohr I. 2007. Maintenance of endoplasmic reticulum (ER) homeostasis in herpes simplex virus type 1-infected cells through the association of a viral glycoprotein with PERK, a cellular ER stress sensor. *J Virol* **81:** 3377–3390.

Murphy E, Vanicek J, Robins H, Shenk T, Levine AJ. 2008. Suppression of immediate-early viral gene expression by herpesvirus-coded microRNAs: Implications for latency. *Proc Natl Acad Sci* **105:** 5453–5458.

Muthukrishnan S, Morgan M, Banerjee AK, Shatkin AJ. 1976. Influence of 5′-terminal m7G and 2′-O-methylated residues on messenger ribonucleic acid binding to ribosomes. *Biochemistry* **15:** 5761–5768.

Nachmani D, Stern-Ginossar N, Sarid R, Mandelboim O. 2009. Diverse herpesvirus microRNAs target the stress-induced immune ligand MICB to escape recognition by natural killer cells. *Cell Host Microbe* **5:** 376–385.

Napthine S, Lever RA, Powell ML, Jackson RJ, Brown TD, Brierley I. 2009. Expression of the VP2 protein of murine norovirus by a translation termination-reinitiation strategy. *PLoS ONE* **4:** e8390.

Nathans R, Chu CY, Serquina AK, Lu CC, Cao H, Rana TM. 2009. Cellular microRNA and P bodies modulate host-HIV-1 interactions. *Mol Cell* **34:** 696–709.

Nichols LA, Adang LA, Kedes DH. 2011. Rapamycin blocks production of KSHV/HHV8: Insights into the anti-tumor activity of an immunosuppressant drug. *PLoS ONE* **6:** e14535.

Nieto C, Piron F, Dalmais M, Marco CF, Moriones E, Gomez-Guillamon ML, Truniger V, Gomez P, Garcia-Mas J, Aranda MA, et al. 2007. EcoTILLING for the identification of allelic variants of melon eIF4E, a factor that controls virus susceptibility. *BMC Plant Biol* **7:** 34.

Novac O, Guenier AS, Pelletier J. 2004. Inhibitors of protein synthesis identified by a high throughput multiplexed translation screen. *Nucleic Acids Res* **32:** 902–915.

Oliere S, Arguello M, Mesplede T, Tumilasci V, Nakhaei P, Stojdl D, Sonenberg N, Bell J, Hiscott J. 2008. Vesicular stomatitis virus oncolysis of T lymphocytes requires cell cycle entry and translation initiation. *J Virol* **82:** 5735–5749.

O'Malley RP, Duncan RF, Hershey JW, Mathews MB. 1989. Modification of protein synthesis initiation factors and the shut-off of host protein synthesis in adenovirus-infected cells. *Virology* **168:** 112–118.

O'Neill RE, Racaniello VR. 1989. Inhibition of translation in cells infected with a poliovirus 2Apro mutant correlates with phosphorylation of the α subunit of eucaryotic initiation factor 2. *J Virol* **63:** 5069–5075.

Orlova M, Yueh A, Leung J, Goff SP. 2003. Reverse transcriptase of Moloney murine leukemia virus binds to eukaryotic release factor 1 to modulate suppression of translational termination. *Cell* **115:** 319–331.

O'Shea C, Klupsch K, Choi S, Bagus B, Soria C, Shen J, McCormick F, Stokoe D. 2005. Adenoviral proteins mimic nutrient/growth signals to activate the mTOR pathway for viral replication. *EMBO J* **24:** 1211–1221.

Otto GA, Puglisi JD. 2004. The pathway of HCV IRES-mediated translation initiation. *Cell* **119:** 369–380.

Parrish S, Moss B. 2007. Characterization of a second vaccinia virus mRNA-decapping enzyme conserved in poxviruses. *J Virol* **81:** 12973–12978.

Parrish S, Resch W, Moss B. 2007. Vaccinia virus D10 protein has mRNA decapping activity, providing a mechanism for control of host and viral gene expression. *Proc Natl Acad Sci* **104:** 2139–2144.

Parsons J, Castaldi MP, Dutta S, Dibrov SM, Wyles DL, Hermann T. 2009. Conformational inhibition of the hepatitis C virus internal ribosome entry site RNA. *Nat Chem Biol* **5:** 823–825.

Paulsen RB, Seth PP, Swayze EE, Griffey RH, Skalicky JJ, Cheatham TE III, Davis DR. 2010. Inhibitor-induced structural change in the HCV IRES domain IIa RNA. *Proc Natl Acad Sci* **107:** 7263–7268.

Pavio N, Romano PR, Graczyk TM, Feinstone SM, Taylor DR. 2003. Protein synthesis and endoplasmic reticulum stress can be modulated by the hepatitis C virus envelope protein E2 through the eukaryotic initiation factor 2α kinase PERK. *J Virol* **77:** 3578–3585.

* Pavitt GD, Ron D. 2012. New insights into translational regulation in the endoplasmic reticulum unfolded protein response. *Cold Spring Harb Perspect Biol* doi: cshperspect.a011528/cshperspect.a012278.

Pe'ery T, Mathews MB. 2007. Viral conquest of the host cell. In *Fields virology* (ed. Knipe DM, Howley PM), pp. 168–208. Lippincott Williams and Wilkins, Philadelphia.

Pelletier J, Sonenberg N. 1988. Internal initiation of translation of eukaryotic mRNA directed by a sequence derived from poliovirus RNA. *Nature* **334:** 320–325.

Perez C, McKinney C, Chulunbaatar U, Mohr I. 2011. Translational control of the abundance of cytoplasmic poly(A) binding protein in human cytomegalovirus-infected cells. *J Virol* **85:** 156–164.

Pestova TV, Shatsky IN, Fletcher SP, Jackson RJ, Hellen CU. 1998. A prokaryotic-like mode of cytoplasmic eukaryotic ribosome binding to the initiation codon during internal translation initiation of hepatitis C and classical swine fever virus RNAs. *Genes Dev* **12:** 67–83.

Pestova TV, de Breyne S, Pisarev AV, Abaeva IS, Hellen CU. 2008. eIF2-dependent and eIF2-independent modes of initiation on the CSFV IRES: A common role of domain II. *EMBO J* **27:** 1060–1072.

Pfeffer S, Zavolan M, Grasser FA, Chien M, Russo JJ, Ju J, John B, Enright AJ, Marks D, Sander C, et al. 2004. Identification of virus-encoded microRNAs. *Science* **304:** 734–736.

Piotrowska J, Hansen SJ, Park N, Jamka K, Sarnow P, Gustin KE. 2010. Stable formation of compositionally unique stress granules in virus-infected cells. *J Virol* **84:** 3654–3665.

Piron M, Vende P, Cohen J, Poncet D. 1998. Rotavirus RNA-binding protein NSP3 interacts with eIF4GI and evicts the poly(A) binding protein from eIF4F. *EMBO J* **17:** 5811–5821.

Pisarev AV, Chard LS, Kaku Y, Johns HL, Shatsky IN, Belsham GJ. 2004. Functional and structural similarities between the internal ribosome entry sites of hepatitis C virus and porcine teschovirus, a picornavirus. *J Virol* **78:** 4487–4497.

Plotch SJ, Bouloy M, Ulmanen I, Krug RM. 1981. A unique cap(m7GpppXm)-dependent influenza virion endonuclease cleaves capped RNAs to generate the primers that initiate viral RNA transcription. *Cell* **23:** 847–858.

Cite this article as *Cold Spring Harb Perspect Biol* doi: 10.1101/cshperspect.a012351

Polacek C, Friebe P, Harris E. 2009. Poly(A)-binding protein binds to the non-polyadenylated 3′ untranslated region of dengue virus and modulates translation efficiency. *J Gen Virol* **90:** 687–692.

Powell ML, Leigh KE, Poyry TA, Jackson RJ, Brown TD, Brierley I. 2011. Further characterisation of the translational termination-reinitiation signal of the influenza B virus segment 7 RNA. *PLoS ONE* **6:** e16822.

Poyry TA, Kaminski A, Connell EJ, Fraser CS, Jackson RJ. 2007. The mechanism of an exceptional case of reinitiation after translation of a long ORF reveals why such events do not generally occur in mammalian mRNA translation. *Genes Dev* **21:** 3149–3162.

Qin Q, Hastings C, Miller CL. 2009. Mammalian orthoreovirus particles induce and are recruited into stress granules at early times postinfection. *J Virol* **83:** 11090–11101.

Qin Q, Carroll K, Hastings C, Miller CL. 2011. Mammalian orthoreovirus escape from host translational shutoff correlates with stress granule disruption and is independent of eIF2α phosphorylation and PKR. *J Virol* **85:** 8798–8810.

Raaben M, Groot Koerkamp MJ, Rottier PJ, de Haan CA. 2007. Mouse hepatitis coronavirus replication induces host translational shutoff and mRNA decay, with concomitant formation of stress granules and processing bodies. *Cell Microbiol* **9:** 2218–2229.

Ramelot TA, Cort JR, Yee AA, Liu F, Goshe MB, Edwards AM, Smith RD, Arrowsmith CH, Dever TE, Kennedy MA. 2002. Myxoma virus immunomodulatory protein M156R is a structural mimic of eukaryotic translation initiation factor eIF2α. *J Mol Biol* **322:** 943–954.

Raoult D, Audic S, Robert C, Abergel C, Renesto P, Ogata H, La Scola B, Suzan M, Claverie JM. 2004. The 1.2-megabase genome sequence of Mimivirus. *Science* **306:** 1344–1350.

Redondo N, Sanz MA, Welnowska E, Carrasco L. 2011. Translation without eIF2 promoted by poliovirus 2A protease. *PLoS ONE* **6:** e25699.

Reguera J, Weber F, Cusack S. 2010. *Bunyaviridae* RNA polymerases (L-protein) have an N-terminal, influenza-like endonuclease domain, essential for viral cap-dependent transcription. *PLoS Pathog* **6:** e1001101.

Remm M, Remm A, Ustav M. 1999. Human papillomavirus type 18 E1 protein is translated from polycistronic mRNA by a discontinuous scanning mechanism. *J Virol* **73:** 3062–3070.

Richner JM, Clyde K, Pezda AC, Cheng BY, Wang T, Kumar GR, Covarrubias S, Coscoy L, Glaunsinger B. 2011. Global mRNA degradation during lytic gammaherpesvirus infection contributes to establishment of viral latency. *PLoS Pathog* **7:** e1002150.

Robert F, Kapp LD, Khan SN, Acker MG, Kolitz S, Kazemi S, Kaufman RJ, Merrick WC, Koromilas AE, Lorsch JR, et al. 2006. Initiation of protein synthesis by hepatitis C virus is refractory to reduced eIF2•GTP•Met-tRNA$_i^{Met}$ ternary complex availability. *Mol Biol Cell* **17:** 4632–4644.

Rodriguez Pulido M, Serrano P, Saiz M, Martinez-Salas E. 2007. Foot-and-mouth disease virus infection induces proteolytic cleavage of PTB, eIF3a,b, and PABP RNA-binding proteins. *Virology* **364:** 466–474.

Rothenburg S, Seo EJ, Gibbs JS, Dever TE, Dittmar K. 2009. Rapid evolution of protein kinase PKR alters sensitivity to viral inhibitors. *Nat Struct Mol Biol* **16:** 63–70.

Rothenburg S, Chinchar VG, Dever TE. 2011. Characterization of a ranavirus inhibitor of the antiviral protein kinase PKR. *BMC Microbiol* **11:** 56.

* Roux PP, Topisirovic I. 2012. Regulation of mRNA translation by signaling pathways. *Cold Spring Harb Perspect Biol* doi: cshperspect.a011528/cshperspect.a012252.

Saini HK, Fischer D. 2007. Structural and functional insights into Mimivirus ORFans. *BMC Genomics* **8:** 115.

Salaun C, Macdonald AI, Larralde O, Howard L, Lochtie K, Burgess HM, Brook M, Malik P, Gray NK, Graham SV. 2010. Poly(A)-binding protein 1 (PABP1) partially relocalises to the nucleus during HSV-1 infection in an ICP27-independent manner and does not inhibit virus replication. *J Virol* **84:** 8539–8548.

Salvatore M, Basler CF, Parisien JP, Horvath CM, Bourmakina S, Zheng H, Muster T, Palese P, Garcia-Sastre A. 2002. Effects of influenza A virus NS1 protein on protein expression: The NS1 protein enhances translation and is not required for shutoff of host protein synthesis. *J Virol* **76:** 1206–1212.

Sanchez R, Mohr I. 2007. Inhibition of cellular 2′–5′ oligoadenylate synthetase by the herpes simplex virus type 1 Us11 protein. *J Virol* **81:** 3455–3464.

Sandri-Goldin RM. 2011. The function and activities of HSV-1 ICP27, a multifunctional regulator of gene expression. In *Alphaherpesviruses: Molecular virology* (ed. Weller S), pp. 39–50. Caister Academic Press, Norfolk, UK.

Santhakumar D, Forster T, Laqtom NN, Fragkoudis R, Dickinson P, Abreu-Goodger C, Manakov SA, Choudhury NR, Griffiths SJ, Vermeulen A, et al. 2010. Combined agonist–antagonist genome-wide functional screening identifies broadly active antiviral microRNAs. *Proc Natl Acad Sci* **107:** 13830–13835.

Sanz MA, Castello A, Ventoso I, Berlanga JJ, Carrasco L. 2009. Dual mechanism for the translation of subgenomic mRNA from Sindbis virus in infected and uninfected cells. *PLoS ONE* **4:** e4772.

Sato H, Masuda M, Kanai M, Tsukiyama-Kohara K, Yoneda M, Kai C. 2007. Measles virus N protein inhibits host translation by binding to eIF3-p40. *J Virol* **81:** 11569–11576.

Schepetilnikov M, Kobayashi K, Geldreich A, Caranta C, Robaglia C, Keller M, Ryabova LA. 2011. Viral factor TAV recruits TOR/S6K1 signalling to activate reinitiation after long ORF translation. *EMBO J* **30:** 1343–1356.

Seo EJ, Liu F, Kawagishi-Kobayashi M, Ung TL, Cao C, Dar AC, Sicheri F, Dever TE. 2008. Protein kinase PKR mutants resistant to the poxvirus pseudosubstrate K3L protein. *Proc Natl Acad Sci* **105:** 16894–16899.

Shih SR, Krug RM. 1996. Surprising function of the three influenza viral polymerase proteins: Selective protection of viral mRNAs against the cap-snatching reaction catalyzed by the same polymerase proteins. *Virology* **226:** 430–435.

Shuda M, Kwun HJ, Feng H, Chang Y, Moore PS. 2011. Human Merkel cell polyomavirus small T antigen is an oncoprotein targeting the 4E-BP1 translation regulator. *J Clin Invest* **121:** 3623–3634.

Silverman RH. 2007. Viral encounters with 2′,5′-oligo-adenylate synthetase and RNase L during the interferon antiviral response. *J Virol* **81:** 12720–12729.

Simpson-Holley M, Kedersha N, Dower K, Rubins KH, Anderson P, Hensley LE, Connor JH. 2011. Formation of antiviral cytoplasmic granules during orthopoxvirus infection. *J Virol* **85:** 1581–1593.

Sizova DV, Kolupaeva VG, Pestova TV, Shatsky IN, Hellen CU. 1998. Specific interaction of eukaryotic translation initiation factor 3 with the 5′ nontranslated regions of hepatitis C virus and classical swine fever virus RNAs. *J Virol* **72:** 4775–4782.

Skabkin MA, Skabkina OV, Dhote V, Komar AA, Hellen CU, Pestova TV. 2010. Activities of Ligatin and MCT-1/DENR in eukaryotic translation initiation and ribosomal recycling. *Genes Dev* **24:** 1787–1801.

Smith JA, Schmechel SC, Raghavan A, Abelson M, Reilly C, Katze MG, Kaufman RJ, Bohjanen PR, Schiff LA. 2006. Reovirus induces and benefits from an integrated cellular stress response. *J Virol* **80:** 2019–2033.

Sodhi A, Chaisuparat R, Hu J, Ramsdell AK, Manning BD, Sausville EA, Sawai ET, Molinolo A, Gutkind JS, Montaner S. 2006. The TSC2/mTOR pathway drives endothelial cell transformation induced by the Kaposi's sarcoma-associated herpesvirus G protein-coupled receptor. *Cancer Cell* **10:** 133–143.

Sonenberg N, Rupprecht KM, Hecht SM, Shatkin AJ. 1979. Eukaryotic mRNA cap binding protein: Purification by affinity chromatography on Sepharose-coupled m7GDP. *Proc Natl Acad Sci* **76:** 4345–4349.

Sood R, Porter AC, Ma K, Quilliam LA, Wek RC. 2000. Pancreatic eukaryotic initiation factor-2α kinase (PEK) homologues in humans, *Drosophila melanogaster* and *Caenorhabditis elegans* that mediate translational control in response to endoplasmic reticulum stress. *Biochem J* **346:** 281–293.

Spahn CM, Jan E, Mulder A, Grassucci RA, Sarnow P, Frank J. 2004. Cryo-EM visualization of a viral internal ribosome entry site bound to human ribosomes: The IRES functions as an RNA-based translation factor. *Cell* **118:** 465–475.

Spangle JM, Munger K. 2010. The human papillomavirus type 16 E6 oncoprotein activates mTORC1 signaling and increases protein synthesis. *J Virol* **84:** 9398–9407.

Stanford MM, Barrett JW, Nazarian SH, Werden S, McFadden G. 2007. Oncolytic virotherapy synergism with signaling inhibitors: Rapamycin increases myxoma virus tropism for human tumor cells. *J Virol* **81:** 1251–1260.

Sukarieh R, Sonenberg N, Pelletier J. 2010. Nuclear assortment of eIF4E coincides with shut-off of host protein synthesis upon poliovirus infection. *J Gen Virol* **91:** 1224–1228.

Sweeney TR, Dhote V, Yu Y, Hellen CU. 2011. A distinct class of internal ribosomal entry site (IRES) in members of the *Kobuvirus* and proposed *Salivirus* and *Paraturdivirus* genera of *Picornaviridae*. *J Virol* doi: 10.1128/JVI.05862-11.

Talloczy Z, Jiang W, Virgin HWt, Leib DA, Scheuner D, Kaufman RJ, Eskelinen EL, Levine B. 2002. Regulation of starvation- and virus-induced autophagy by the eIF2α kinase signaling pathway. *Proc Natl Acad Sci* **99:** 190–195.

Taneja S, MacGregor J, Markus S, Ha S, Mohr I. 2001. Enhanced antitumor efficacy of a herpes simplex virus mu-tant isolated by genetic selection in cancer cells. *Proc Natl Acad Sci* **98:** 8804–8808.

Tang S, Bertke AS, Patel A, Wang K, Cohen JI, Krause PR. 2008. An acutely and latently expressed herpes simplex virus 2 viral microRNA inhibits expression of ICP34.5, a viral neurovirulence factor. *Proc Natl Acad Sci* **105:** 10931–10936.

Taylor DR, Shi ST, Romano PR, Barber GN, Lai MM. 1999. Inhibition of the interferon-inducible protein kinase PKR by HCV E2 protein. *Science* **285:** 107–110.

Taylor DR, Tian B, Romano PR, Hinnebusch AG, Lai MM, Mathews MB. 2001. Hepatitis C virus envelope protein E2 does not inhibit PKR by simple competition with autophosphorylation sites in the RNA-binding domain. *J Virol* **75:** 1265–1273.

Terenzi F, Hui DJ, Merrick WC, Sen GC. 2006. Distinct induction patterns and functions of two closely related interferon-inducible human genes, ISG54 and ISG56. *J Biol Chem* **281:** 34064–34071.

Thiebeauld O, Schepetilnikov M, Park HS, Geldreich A, Kobayashi K, Keller M, Hohn T, Ryabova LA. 2009. A new plant protein interacts with eIF3 and 60S to enhance virus-activated translation re-initiation. *EMBO J* **28:** 3171–3184.

Thivierge K, Cotton S, Dufresne PJ, Mathieu I, Beauchemin C, Ide C, Fortin MG, Laliberte JF. 2008. Eukaryotic elongation factor 1A interacts with Turnip mosaic virus RNA-dependent RNA polymerase and VPg-Pro in virus-induced vesicles. *Virology* **377:** 216–225.

Thomas MA, Broughton RS, Goodrum FD, Ornelles DA. 2009. E4orf1 limits the oncolytic potential of the E1B-55K deletion mutant adenovirus. *J Virol* **83:** 2406–2416.

Thoreen CC, Kang SA, Chang JW, Liu Q, Zhang J, Gao Y, Reichling LJ, Sim T, Sabatini DM, Gray NS. 2009. An ATP-competitive mammalian target of rapamycin inhibitor reveals rapamycin-resistant functions of mTORC1. *J Biol Chem* **284:** 8023–8032.

Treder K, Kneller EL, Allen EM, Wang Z, Browning KS, Miller WA. 2008. The 3′ cap-independent translation element of Barley yellow dwarf virus binds eIF4F via the eIF4G subunit to initiate translation. *RNA* **14:** 134–147.

Trono D, Pelletier J, Sonenberg N, Baltimore D. 1988. Translation in mammalian cells of a gene linked to the poliovirus 5′ noncoding region. *Science* **241:** 445–448.

Truniger V, Aranda MA. 2009. Recessive resistance to plant viruses. *Adv Virus Res* **75:** 119–159.

Umbach JL, Kramer MF, Jurak I, Karnowski HW, Coen DM, Cullen BR. 2008. MicroRNAs expressed by herpes simplex virus 1 during latent infection regulate viral mRNAs. *Nature* **454:** 780–783.

Ventoso I, Blanco R, Perales C, Carrasco L. 2001. HIV-1 protease cleaves eukaryotic initiation factor 4G and inhibits cap-dependent translation. *Proc Natl Acad Sci* **98:** 12966–12971.

Ventoso I, Sanz MA, Molina S, Berlanga JJ, Carrasco L, Esteban M. 2006. Translational resistance of late alpha-virus mRNA to eIF2α phosphorylation: A strategy to overcome the antiviral effect of protein kinase PKR. *Genes Dev* **20:** 87–100.

Villas-Boas CS, Conceicao TM, Ramirez J, Santoro AB, Da Poian AT, Montero-Lomeli M. 2009. Dengue virus-induced regulation of the host cell translational machinery. *Braz J Med Biol Res* **42**: 1020–1026.

Vyas J, Elia A, Clemens MJ. 2003. Inhibition of the protein kinase PKR by the internal ribosome entry site of hepatitis C virus genomic RNA. *RNA* **9**: 858–870.

Walsh D, Mohr I. 2004. Phosphorylation of eIF4E by Mnk-1 enhances HSV-1 translation and replication in quiescent cells. *Genes Dev* **18**: 660–672.

Walsh D, Mohr I. 2006. Assembly of an active translation initiation factor complex by a viral protein. *Genes Dev* **20**: 461–472.

Walsh D, Mohr I. 2011. Viral subversion of the host protein synthesis machinery. *Nat Rev Microbiol* **9**: 860–875.

Walsh D, Perez C, Notary J, Mohr I. 2005. Regulation of the translation initiation factor eIF4F by multiple mechanisms in human cytomegalovirus-infected cells. *J Virol* **79**: 8057–8064.

Walsh D, Arias C, Perez C, Halladin D, Escandon M, Ueda T, Watanabe-Fukunaga R, Fukunaga R, Mohr I. 2008. Eukaryotic translation initiation factor 4F architectural alterations accompany translation initiation factor redistribution in poxvirus-infected cells. *Mol Cell Biol* **28**: 2648–2658.

Wang M, Hudak KA. 2006. A novel interaction of pokeweed antiviral protein with translation initiation factors 4G and iso4G: A potential indirect mechanism to access viral RNAs. *Nucleic Acids Res* **34**: 1174–1181.

Wang FZ, Weber F, Croce C, Liu CG, Liao X, Pellett PE. 2008. Human cytomegalovirus infection alters the expression of cellular microRNA species that affect its replication. *J Virol* **82**: 9065–9074.

Wang Z, Parisien M, Scheets K, Miller WA. 2011. The cap-binding translation initiation factor, eIF4E, binds a pseudoknot in a viral cap-independent translation element. *Structure* **19**: 868–880.

Watanabe Y, Ohtaki N, Hayashi Y, Ikuta K, Tomonaga K. 2009. Autogenous translational regulation of the Borna disease virus negative control factor X from polycistronic mRNA using host RNA helicases. *PLoS Pathog* **5**: e1000654.

Welnowska E, Castelló A, Moral P, Carrasco L. 2009. Translation of mRNAs from vesicular stomatitis virus and vaccinia virus is differentially blocked in cells with depletion of eIF4GI and/or eIF4GII. *J Mol Biol* **394**: 506–521.

Welnowska E, Sanz MA, Redondo N, Carrasco L. 2011. Translation of viral mRNA without active eIF2: The case of picornaviruses. *PLoS ONE* **6**: e22230.

Werden SJ, Barrett JW, Wang G, Stanford MM, McFadden G. 2007. M-T5, the ankyrin repeat, host range protein of myxoma virus, activates Akt and can be functionally replaced by cellular PIKE-A. *J Virol* **81**: 2340–2348.

White JP, Cardenas AM, Marissen WE, Lloyd RE. 2007. Inhibition of cytoplasmic mRNA stress granule formation by a viral proteinase. *Cell Host Microbe* **2**: 295–305.

White LK, Sali T, Alvarado D, Gatti E, Pierre P, Streblow D, Defilippis VR. 2010. Chikungunya virus induces IPS-1-dependent innate immune activation and protein kinase R-independent translational shutoff. *J Virol* **85**: 606–620.

White JP, Reineke LC, Lloyd RE. 2011. Poliovirus switches to an eIF2-independent mode of translation during infection. *J Virol* **85**: 8884–8893.

Willcocks MM, Carter MJ, Roberts LO. 2004. Cleavage of eukaryotic initiation factor eIF4G and inhibition of host-cell protein synthesis during feline calicivirus infection. *J Gen Virol* **85**: 1125–1130.

Willcocks MM, Locker N, Gomwalk Z, Royall E, Bakhshesh M, Belsham GJ, Idamakanti N, Burroughs KD, Reddy PS, Hallenbeck PL, et al. 2011. Structural features of the Seneca Valley virus internal ribosome entry site element; a picornavirus with a pestivirus-like IRES. *J Virol* **85**: 4452–4461.

Wilson JE, Pestova TV, Hellen CU, Sarnow P. 2000. Initiation of protein synthesis from the A site of the ribosome. *Cell* **102**: 511–520.

Won S, Eidenschenk C, Arnold CN, Siggs OM, Sun L, Brandl K, Mullen TM, Nemerow GR, Moresco EM, Beutler B. 2011. Increased susceptibility to DNA virus infection in mice with a GCN2 mutation. *J Virol* **86**: 1802–1808.

Xi Q, Cuesta R, Schneider RJ. 2004. Tethering of eIF4G to adenoviral mRNAs by viral 100k protein drives ribosome shunting. *Genes Dev* **18**: 1997–2009.

Xi Q, Cuesta R, Schneider RJ. 2005. Regulation of translation by ribosome shunting through phosphotyrosine-dependent coupling of adenovirus protein 100k to viral mRNAs. *J Virol* **79**: 5676–5683.

Xiao H, Xu LH, Yamada Y, Liu DX. 2008. Coronavirus spike protein inhibits host cell translation by interaction with eIF3f. *PLoS ONE* **3**: e1494.

Yan W, Frank CL, Korth MJ, Sopher BL, Novoa I, Ron D, Katze MG. 2002. Control of PERK eIF2α kinase activity by the endoplasmic reticulum stress-induced molecular chaperone P58IPK. *Proc Natl Acad Sci* **99**: 15920–15925.

Yánguez E, Rodriguez P, Goodfellow I, Nieto A. 2012. Influenza virus polymerase confers independence of the cellular cap-binding factor eIF4E for viral mRNA translation. *Virology* **422**: 297–307.

Yatherajam G, Huang W, Flint SJ. 2011. Export of adenoviral late mRNA from the nucleus requires the Nxf1/Tap export receptor. *J Virol* **85**: 1429–1438.

Yeam I, Cavatorta JR, Ripoll DR, Kang BC, Jahn MM. 2007. Functional dissection of naturally occurring amino acid substitutions in eIF4E that confers recessive potyvirus resistance in plants. *Plant Cell* **19**: 2913–2928.

Yu Y, Alwine JC. 2006. 19S late mRNAs of simian virus 40 have an internal ribosome entry site upstream of the virion structural protein 3 coding sequence. *J Virol* **80**: 6553–6558.

Yu Y, Kudchodkar SB, Alwine JC. 2005. Effects of simian virus 40 large and small tumor antigens on mammalian target of rapamycin signaling: Small tumor antigen mediates hypophosphorylation of eIF4E-binding protein 1 late in infection. *J Virol* **79**: 6882–6889.

Yu Y, Abaeva IS, Marintchev A, Pestova TV, Hellen CU. 2011a. Common conformational changes induced in type 2 picornavirus IRESs by cognate trans-acting factors. *Nucleic Acids Res* **39**: 4851–4865.

Yu Y, Sweeney TR, Kafasla P, Jackson RJ, Pestova TV, Hellen CU. 2011b. The mechanism of translation initiation on Aichivirus RNA mediated by a novel type of picornavirus IRES. *EMBO J* **30**: 4423–4436.

Yueh A, Schneider RJ. 1996. Selective translation initiation by ribosome jumping in adenovirus-infected and heat-shocked cells. *Genes Dev* **10:** 1557–1567.

Yueh A, Schneider RJ. 2000. Translation by ribosome shunting on adenovirus and hsp70 mRNAs facilitated by complementarity to 18S rRNA. *Genes Dev* **14:** 414–421.

Zaborowska I, Walsh D. 2009. PI3K signaling regulates rapamycin-insensitive translation initiation complex formation in vaccinia virus-infected cells. *J Virol* **83:** 3988–3992.

Zaborowska I, Kellner K, Henry M, Meleady P, Walsh D. 2012. Recruitment of host translation initiation factor eIF4G by the vaccinia virus ssDNA binding protein I3. *Virology* **425:** 11–22.

Zhang B, Morace G, Gauss-Muller V, Kusov Y. 2007. Poly(A) binding protein, C-terminally truncated by the hepatitis A virus proteinase 3C, inhibits viral translation. *Nucleic Acids Res* **35:** 5975–5984.

Zhang F, Moon A, Childs K, Goodbourn S, Dixon LK. 2010. The African Swine Fever Virus DP71L protein recruits the protein phosphatase 1 catalytic subunit to dephosphorylate eIF2α and inhibits CHOP induction but is dispensable for these activities during virus infection. *J Virol* **84:** 10681–10689.

Zheng L, Ding H, Lu Z, Li Y, Pan Y, Ning T, Ke Y. 2008. E3 ubiquitin ligase E6AP-mediated TSC2 turnover in the presence and absence of HPV16 E6. *Genes Cells* **13:** 285–294.

Zuo X, Wang J, Yu P, Eyler D, Xu H, Starich MR, Tiede DM, Simon AE, Kasprzak W, Schwieters CD, et al. 2010. Solution structure of the cap-independent translational enhancer and ribosome-binding element in the 3′ UTR of turnip crinkle virus. *Proc Natl Acad Sci* **107:** 1385–1390.

Emerging Therapeutics Targeting mRNA Translation

Abba Malina[1], John R. Mills[1], and Jerry Pelletier[1,2]

[1]Department of Biochemistry, McGill University, Montréal, Québec H3G 1Y6, Canada

[2]Rosalind and Morris Goodman Cancer Center, McGill University, Montréal, Québec H3G 1Y6, Canada

Correspondence: jerry.pelletier@mcgill.ca

A defining feature of many cancers is deregulated translational control. Typically, this occurs at the level of recruitment of the 40S ribosomes to the 5′-cap of cellular messenger RNAs (mRNAs), the rate-limiting step of protein synthesis, which is controlled by the heterotrimeric eukaryotic initiation complex eIF4F. Thus, eIF4F in particular, and translation initiation in general, represent an exploitable vulnerability and unique opportunity for therapeutic intervention in many transformed cells. In this article, we discuss the development, mode of action and biological activity of a number of small-molecule inhibitors that interrupt PI3K/mTOR signaling control of eIF4F assembly, as well as compounds that more directly block eIF4F activity.

It would seem fitting to finish this collection with an article on the topic of "emerging therapeutics in mRNA translation" because much of our current thinking into the molecular biology of this rich field owes a great debt to inhibitors of protein synthesis. The experimental use of antibiotics that block the ribosome at key enzymatic steps has been instrumental in defining general features of translation and identifying paradigms of translational control. From a therapeutic perspective, however, it has only been those antibiotics that inhibit prokaryotic protein synthesis that have had any medical success, contributing significantly to the treatment of bacterial infectious disease (Chambers 2001a,b). It thus had seemed that the use and potential of eukaryotic protein synthesis inhibitors were limited to tools in basic research. This view has changed during the last 20 years of research. We now have a more profound understanding of the complex regulatory apparatus that the eukaryotic cell uses to control messenger RNA (mRNA) translation and, with the emergence of several novel compounds that target steps other than elongation, a greater appreciation of the rewiring of translation factors that occurs in human disease, and most notably, cancer (Silvera et al. 2010). This latter and exciting development forms the basis of this article. For a more complete description of other known protein synthesis inhibitors we refer the reader to Pelletier and Peltz (2007).

A. Malina et al.

TARGETING TRANSLATION INITIATION AS AN ANTINEOPLASTIC APPROACH

The earliest, most varied and most widely studied inhibitors of eukaryotic protein synthesis are those that target the elongation step of mRNA translation (Pestka 1977; Vazquez 1979). Most of them act either by impairing peptidyl transferase function, impeding translocation, or interfering with aminoacyl-tRNA (transfer RNA) binding or accommodation. Despite their structural diversity and nuanced specificities, most eukaryotic elongation inhibitors have shown limited therapeutic value especially when compared to their prokaryotic counterparts, most likely owing to nonspecific toxicity derived from blocking global protein synthesis in nontransformed cells at the doses tested. Even with these apparent limitations, there has been renewed interest in using translation elongation inhibitors as antineoplastics. For example, homoharringtonine (omacetaxine mepesuccinate or HHT), a tree-derived alkaloid from the conifer *Cephalotaxus harringtonia*, has shown promise in several clinical trials for myelodysplastic syndrome and in phase II clinical trials in patients with gleevec-resistant chronic myelogenous leukemia and in acute myeloid leukemia (AML) (Kantarjian et al. 2001; Kim et al. 2011). HHT is thought to inhibit peptide chain elongation by binding the A site of the ribosome, blocking aminoacyl-tRNA binding, and halting peptide chain elongation (Gurel et al. 2009). Recent experiments have provided some understanding of the possible mechanism of HHT's mode of action. On treatment of leukemic cells with HHT, the short-lived antiapoptotic protein Mcl-1 was observed to be rapidly degraded, an effect that was solely attributed to the acute loss in overall protein synthesis and that was reversed upon proteasome inhibition (Tang et al. 2006; Robert et al. 2009; Chen et al. 2011a). This, and perhaps the degradation of other short-lived prosurvival factors, explains some of the synergistic effects observed with different inhibitors of translation elongation in the Burkitt's-like Eμ-Myc lymphoma mouse model and would warrant greater study and further evaluation of these drugs in combination therapies (Robert et al. 2009).

Other elongation inhibitors have had less clinical success when applied toward the treatment of cancer owing to unacceptable toxicities or poor pharmacological properties limiting their therapeutic window (Dumez et al. 2009). Not surprisingly, the current trend in the development of therapeutic agents that disrupt translation has thus drifted away from inhibitors of elongation to those that target initiation, potentially shifting the pharmacological response from global protein synthesis to a more selective (and possibly more cancer cell-dependent) translationally regulated effect.

eIF4F and Tumorigenesis

The eIF4F translation complex is currently at the forefront of the development of pharmacological agents that block initiation. Briefly, eIF4F is a heterotrimeric complex composed of eIF4E, the m^7GpppN cap-binding protein that anchors the complex to the 5′-end of the mRNA; eIF4A, an RNA DEAD box helicase that is thought to unwind RNA secondary structure surrounding the cap and increase the efficiency of ribosome binding; and eIF4G, a large scaffolding protein that bridges eIF4F to the 43S ribosomal complex (see Lorsch et al. 2012). Given eIF4E's limited abundance, the recruitment of the 43S ribosomal complex to the mRNA by eIF4F is thought to be rate limiting for translation initiation (Duncan et al. 1987). The interaction of eIF4F with the cap structure is affected by mRNA proximal secondary structure with increased structure diminishing the efficiency of the eIF4E-cap interaction and/or imposing a structural barrier to the weak eIF4A helicase activity (Pelletier and Sonenberg 1985b; Lawson et al. 1986). Consequently, altering the levels of eIF4F can influence which mRNAs are more readily translated (Lawson et al. 1986, 1988), correlating with the degree of secondary structure in the 5′-untranslated regions (5′-UTRs) of the mRNA—the greater the thermal stability, the more poorly the transcript is translated, presumably by obstructing efficient ribosome loading and 48S complex

328   Cite this article as *Cold Spring Harb Perspect Biol* doi: 10.1101/cshperspect.a012377

formation (Pelletier and Sonenberg 1985a,b; Lawson et al. 1986; Babendure et al. 2006). Increasing the amounts of cellular eIF4E can thus disproportionately stimulate the expression of messages that were once outcompeted by unstructured mRNAs, encoding proteins that are often progrowth and prosurvival in nature. In fact, this is thought to be the underlying mechanism behind eIF4E's tumorigenic properties, the selective increase in translation of a limited set of oncogenic and metastatic transcripts (Lazaris-Karatzas et al. 1990; Ruggero et al. 2004; Wendel et al. 2004). The role that eIF4E plays in cancer has received much broader scientific and clinical interest of late since the discovery that its function is regulated by mTOR (mammalian target of rapamycin), a master regulator of cellular homeostasis and key signaling node often up-regulated in cancers, and that the chemotherapeutic rapamycin, its highly specific inhibitor, can block its activity (Kunz et al. 1993; Sabatini et al. 1994; see Dobson et al. 2012 for details).

Therapeutic Strategies Targeting Regulation of eIF4F Assembly

Rapalogs (Rapamycin Analogs)

mTOR is a member of the phosphoinositide 3-kinase (PI3K) -related protein kinase (PIKK) family and is the catalytic subunit of two functionally distinct complexes: mTORC1 and mTORC2—defined by the nature of interacting accessory proteins. mTORC2 integrates cell survival, proliferation, lipogenesis (as well as other catabolic processes), and cytoskeletal organization to growth factor signaling, primarily through phosphorylation of members of the AGC class of kinases upstream of mTORC1. mTORC1 serves as a regulator of eIF4F assembly and translation initiation, sensing the energy and nutrient status of a cell and transmitting either a growth or starvation response to the translation apparatus (Sengupta et al. 2010). Briefly, mTOR stimulates eIF4F formation through two mechanisms: (1) It phosphorylates eIF4E-binding proteins (4E-BPs; of which there are three, the most prominent being 4E-BP1

and 4E-BP2), which prevents them from disrupting the eIF4E/eIF4G interaction, and (2) mTOR signals the degradation of PDCD4, which interferes with the eIF4A/eIF4G interaction (Gingras et al. 1999b; Dorrello et al. 2006). Thus, active mTOR promotes eIF4F formation and translation initiation.

The identification of germline mutations in genes that encode negative regulators of mTOR signaling (e.g., *Pten* and *Tsc1/2*), the fact that rapamycin possesses antiproliferative activity against a number of cancer cell lines (Hidalgo and Rowinsky 2000), and the observation that a majority of human cancers arise owing to activated mTORC1 signaling (Yuan and Cantley 2008), emphasized the need to uncover the role of mTOR in human cancer. Importantly, it is well established that mTOR control of eIF4F assembly acts as a critical node for cancer cell survival and proliferation (Wendel et al. 2004; Mills et al. 2008).

The rapalogs temsirolimus and everolimus have been clinically approved for the treatment of metastatic renal cell carcinoma, whereas temsirolimus has also been approved for the treatment of mantle cell lymphoma (Table 1). Currently, rapalogs are under investigation as cancer therapies in numerous clinical trials (www.clinicaltrials.gov). There are, however, limitations to rapamycin-based therapies. Among these is the presence of a $p70^{S6K}$-IRS-1 (insulin receptor substrate-1) negative-feedback loop (Fig. 1A) (Harrington et al. 2005). The mTOR substrate $p70^{S6K}$ normally suppresses PI3K signaling by inactivating IRS-1 as well as platelet-derived growth factor receptor (PDGFR) and uncoupling PI3K from upstream growth factor signals (Harrington et al. 2005). Thus, exposure of tumor cells to rapamycin leads to inhibition of $p70^{S6K}$ activity and subsequent stimulation of upstream PI3K signaling—an unfavorable situation for cancer therapy because this activates several mTORC1-independent prosurvival and proliferative signals and is associated with treatment failure (O'Reilly et al. 2006). Additionally, the mitogen-activated protein kinase (MAPK) pathway can be activated upon mTORC1 inhibition via the $p70^{S6K}$/PI3K/Ras pathway and indeed MAPK activation has been noted in

Table 1. Compounds that affect eIF4F assembly or activity

Targets	Compound	Sponsor	Stage of development
mTORC1	Rapamycin (Sirolimus)	Pfizer	FDA approved
	RAD001 (Everolimus)	Novartis	FDA approved
	CCI-779 (Temsirolimus)	Pfizer	FDA approved
	AP23573 (Ridaforolimus)	Merck/ARIAD	Accepted for FDA approval
TOR-KI	Torin1	N/A	Ex vivo cell culture
	INK128	Intellikine	Phase I
	AZD8055	AstraZeneca	Phase I/II
	AZD2014		Phase I
	OSI027	OSI Pharmaceuticals	Phase I
PI3K/TOR-KI	PI-103	N/A	Ex vivo cell culture
	NVPBEZ235	Novartis	Phase I/II
	SF1126	Semafore	Phase I
	GSK2126458	Glaxo Smith Kline	Phase I
	XL765	Exelixis	Phase I/II
	BGT226	Novartis	Phase I/II
	GDC0980	Genetech	Phase I
	PF04691502	Pfizer	Phase I
	PKI587	Pfizer	Phase I
eIF4F	Cap analogs	N/A	In vitro studies
	4Ei-1	N/A	In vitro studies
	4EGI-1	N/A	Ex vivo cell culture
	4E1RCat	N/A	Active in preclinical models
	4E2RCat	N/A	Active in preclinical models
	ISIS-EIF4ERx	Eli Lilly/ISIS	Phase II
	Pateamine A	N/A	Ex vivo cell culture
	DMDA-Pat A	N/A	Active in preclinical models
	Hippuristanol	N/A	Active in preclinical models
	Rocaglamides (Silvestrol)	N/A	Active in preclinical models
Mnk	Cercosporamide	Eli Lilly	Active in preclinical models
	CGP57380	Novartis	Ex vivo cell culture

human tumors following RAD001 treatment (Fig. 1A) (Carracedo et al. 2008).

Rapamycin resistance can also be imparted by overexpression of eIF4E, which has been identified as a genetic modifier of the rapamycin response (Fig. 1B) (Wendel et al. 2004, 2006; Mills et al. 2008). Given that neither 4E-BP phosphorylation status nor $p70^{S6K}$ activity serve as reliable predictive markers of rapamycin sensitivity in human cancers (Noh et al. 2004; Satheesha et al. 2011), eIF4E expression levels should be assessed as a potential marker to inform on rapamycin resistance (Satheesha et al. 2011). These results suggest that direct inhibitors of eIF4E (and eIF4F) may synergize with rapamycin to diminish resistance.

TOR-Kinase Inhibitors (TOR-KI)

Rapamycin acts through an unusual allosteric mechanism wherein it binds FKBP12 and it is this dimer that then interacts with mTOR (Chen et al. 1995). Binding of FKBP12-rapamycin to mTOR induces a conformational change that specifically weakens the mTOR-Raptor interaction (Kim et al. 2002). In vitro kinase assays with mTORC1 have shown that prolonged incubation with rapamycin is necessary to inhibit 4E-BP1 phosphorylation (Burnett et al. 1998), which differs kinetically from what is observed with $p70^{S6K}$, where rapamycin rapidly blocks mTORC1 phosphorylation of $p70^{S6K}$. This is recapitulated in vivo, where rapamycin fully

Cite this article as *Cold Spring Harb Perspect Biol* doi: 10.1101/cshperspect.a012377

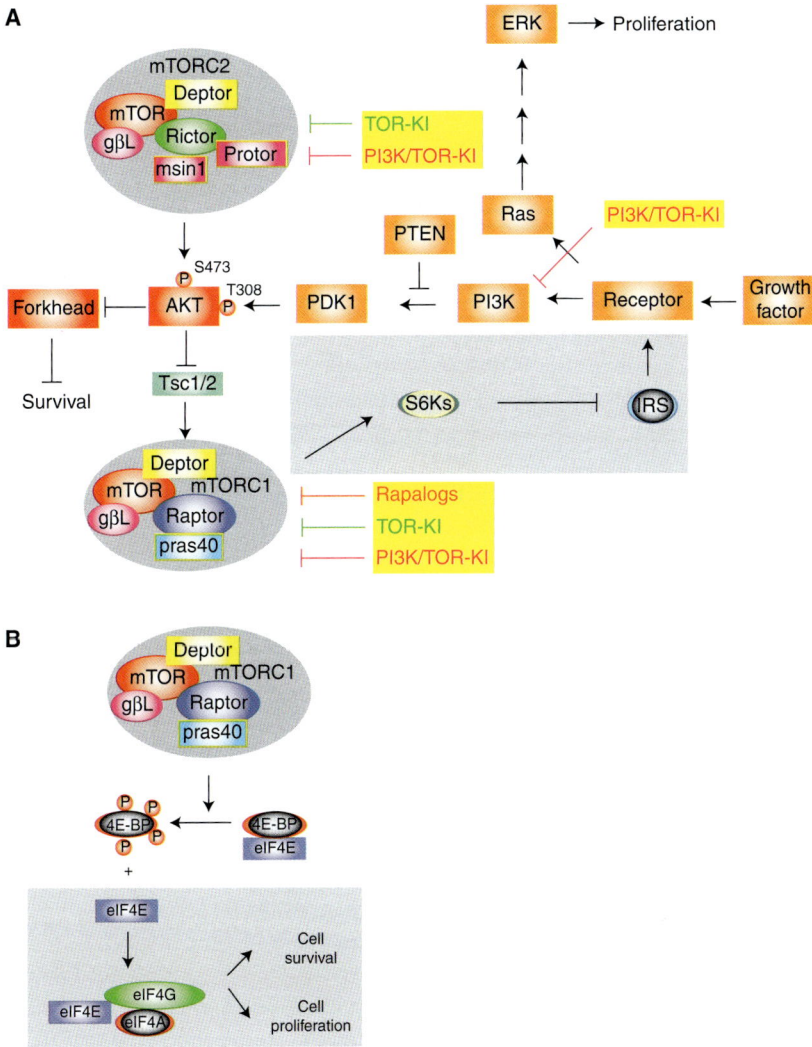

Figure 1. Diagram illustrating rapamycin-resistance pathways. (*A*) Inhibition of mTORC1 by rapalogs leads to dampening of the S6K/IRS feedback loop (in gray box) and increased signaling flux to PI3K and Ras. (*B*) Increased levels of eIF4E can circumvent effects on cell proliferation and cell survival (in gray box) mediated by inhibition of mTORC1 by rapamycin. See text for details.

suppresses p70^{S6K} phosphorylation but fails to completely dephosphorylate 4E-BP1 (Choo et al. 2008). The molecular basis responsible for these different phosphorylation kinetics is not well understood but may relate to rapamycin-induced structural changes in mTOR that exert more profound effects on mTOR-p70^{S6K} association than mTOR-4E-BP association, allowing uninterrupted signaling flux in the pres-

ence of rapamycin. Consistent with such a model is the finding that raptor binds 4E-BP1 more efficiently than p70^{S6K} (Hara et al. 2002). Furthermore, the FKBP12-rapamycin complex fails to effectively suppress mTORC2 kinase activity, limiting its use as a fully active inhibitor of mTOR (Sarbassov et al. 2004).

These issues prompted development of mTOR kinase inhibitors to block both mTOR complexes. Indeed several novel mTOR inhibitors were identified that specifically compete with ATP for access to the kinase active site and some of these compounds are currently under intensive clinical development (Table 1) (Wander et al. 2011). The TOR-KIs are superior to rapalogs in their ability to suppress growth, proliferation, and protein synthesis (Feldman et al. 2009; Thoreen et al. 2009) and they target mTOR irrespective of its protein-binding partners. One important and unanticipated result from these studies is that at least two TOR-KIs, Torin1 and PP242, more potently inhibit 4E-BP1 phosphorylation than do the rapalogs (Feldman et al. 2009; Thoreen et al. 2009). These results prompted a reexamination of how 4E-BP phosphorylation is regulated by mTOR. It is known that mTORC1 phosphorylates 4E-BP1 in a hierarchical manner, although there is some disagreement in the literature about how this occurs (Gingras et al. 1999a, 2001; Choo et al. 2008). In vitro, mTORC1 can phosphorylate 4E-BP1 on Thr37/46 (Burnett et al. 1998). Yet in some settings, rapamycin fails to block the phosphorylation of these sites in vivo (Choo et al. 2008). In contrast, rapamycin is very effective at allowing dephosphorylation of 4E-BP1 on Ser 65 and Thr 70 in vivo (Gingras et al. 1999a; Choo et al. 2008). The ability of rapamycin to suppress translation is variable between cell lines and correlates with differences in the extent of 4E-BP1 phosphorylation (Beretta et al. 1996; Choo et al. 2008).

In contrast, TOR-KIs rapidly and fully inhibit mTOR-mediated phosphorylation of both 4E-BP1 and p70^{S6K} and notably offer an improvement in their ability to suppress translation initiation (Feldman et al. 2009; Garcia-Martinez et al. 2009; Thoreen et al. 2009; Yu et al. 2010). Fully dephosphorylated 4E-BP1 is much better at inhibiting eIF4E function and translation initiation than its partially dephosphorylated variants (Mothe-Satney et al. 2000). Inhibition of mTORC1 with TOR-KIs suppresses cell proliferation in an mTORC1/4E-BP1/2-dependent manner, whereas cell growth inhibition is mediated by mTORC1/p70^{S6K} (Dowling et al. 2010). Oddly, however, in the absence of 4E-BP1/2, rapamycin can still mediate a partial block of cell proliferation suggesting that rapalogs influence mTOR activity through a more complex mechanism than is currently appreciated (Dowling et al. 2010).

Dual-Specificity PI3K/TOR-KI

A more recent class of mTOR inhibitors that also target PI3K has been characterized and entered into clinical development (Table 1) (Wander et al. 2011). The added advantage of these compounds over rapalogs is that they have the potential to block reactivated PI3K following p70^{S6K} inhibition. One limitation of these inhibitors is specificity. It turns out that some of these drugs have off-target inhibitory effects for related PIKK family members owing to shared structural overlap (Toledo et al. 2011). Although the effectiveness of this class of compounds at inhibiting eIF4F formation remains to be carefully assessed, an interesting link between the PI3K pathway and MYC has recently emerged. Elevated eIF4E and c-MYC levels can lead to PI3K/TOR-KI resistance (Ilic et al. 2011). In a breast cancer setting, MYC amplification leads to PI3K-independent tumor cell survival and resistance to PI3K inhibitors (Liu et al. 2011). A chemical genetics screen further validated c-MYC activation as a means to overcome the proliferative block induced by PI3K/TOR-KIs (Muellner et al. 2011). Elevated MYC and eIF4E levels are expected to increase eIF4F formation independent of mTOR (Jones et al. 1996; Rosenwald 1996; Lin et al. 2008) suggesting that inhibition of translation initiation may be a prominent mechanism by which PI3K/TOR-KIs exert their effects. Taken together, these results suggest that targeting components of the translation apparatus directly under mTOR control would be one avenue to overcome this resistance.

Therapeutic Strategies Directly Targeting eIF4F

There are several strategies under way to develop drugs that impede eIF4F activity, with efforts targeting different functions: (1) competing for eIF4E binding to the m^{7}GpppN cap structure,

(2) uncoupling the eIF4E-eIF4G interaction, (3) targeting eIF4E production, (4) blocking eIF4A activity, and (5) inhibiting eIF4E phosphorylation. All of these have diverse effects on cellular and therapeutic outcomes and all have shown some promising results.

Blocking eIF4E-Cap Interaction

The archetypal inhibitors of eIF4E activity in the field of mRNA translation are undoubtedly the cap analogs (Shatkin et al. 1982; Grudzien-Nogalska et al. 2007b). At the level of translation, these act by outcompeting nascent capped mRNA transcripts for eIF4E binding and preventing the preinitiation ribosomal complex from binding and commencing translation. The original competitors, first described over 30 years ago, were simple m^7GDP or m^7GTP derivatives. Now more than 75 analogs have been synthesized and extensively tested in vitro, with diverse chemical modifications from simple substitution of the methylated residue to bulkier aromatic groups and remodeling of the nucleotide backbone (Hickey et al. 1977; Adams et al. 1978; Darzynkiewicz et al. 1981, 1987, 1989; Cai et al. 1999; Grudzien-Nogalska et al. 2007a,b; Kowalska et al. 2008; Jemielity et al. 2010; Su et al. 2011). The best compounds that bind to eIF4E are, not surprisingly, those that model the natural mRNA cap residue—minimally, a m^7GpppN-like molecule—preserving the contacts between the π-π electron stacking of the eIF4E tryptophan residues (W56 and W102) and the m^7G ring, as well the electrostatic attractions between the basic residues of eIF4E (R122, R157, and K162) and the $5'$-$5'$ triphosphate bridge (Marcotrigiano et al. 1997; Niedzwiecka et al. 2002).

These compounds have been invaluable for the biophysical and biochemical study of eIF4E in vitro but their application in vivo has been limited owing to poor cross-membrane transport and low cellular stability arising from both intracellular and extracellular hydrolysis (Wagner et al. 2000; Jemielity et al. 2010). Two recent synthetic routes have been developed to circumvent such limitations, both involving modification of the phosphate backbone with nonnatural functional groups. One study focused primarily on altering the α, β, and γ phosphate groups to prevent hydrolysis by the decapping enzymes DcpS and/or Dcp1/2 (Grudzien-Nogalska et al. 2007a; Kowalska et al. 2008; Su et al. 2011). When attached at the $5'$ end of a luciferase-driven construct, a phosphoborate moiety introduced at the β position supported translation efficiencies similar to that of the parent $m_2^{7,2'}GpppG$ compound but imparted greater mRNA stability in vivo (Su et al. 2011)—the latter attributed to a combined effect of protection from decapping by association with eIF4E (Schwartz and Parker 2000) and increased resistance to endogenous Dcp1/2 (and probably DcpS as well) degradation.

Similarly, but with an emphasis toward more therapeutically viable eIF4E cap analog-based inhibitors, Ghosh et al. (2009) have published the synthesis of phosphoramidate derivates of m^7GTP. These "pronucleotides" harbor protecting groups that, on removal by endogenous phosphoramidases (histidine triad nucleotide-binding proteins), are converted to active derivatives. These cap analogs were able to inhibit cap-dependent translation in vitro in reticulocyte lysates, as well as in vivo when co-injected with reporter constructs into fertilized zebrafish eggs (Ghosh et al. 2009). Intriguingly, one of these (4Ei-1) was also able to block eIF4E-dependent epithelial-to-mesenchymal (EMT) transition of these eggs. Although promising, their therapeutic relevance remains unclear because both classes of cap derivatives appear unable to cross phospholipid bilayers (requiring either lipofection or ectopic injection), which precludes their use in more clinically relevant assays. Moreover, the use of cap analogs could influence cellular physiology more broadly than just via eIF4E-dependaent protein synthesis: For example, processes that rely on the nuclear cap-binding proteins CBP20 and CBP80, such as NMD, nuclear export, splicing, mRNA $3'$-end formation, and miRNA processing, may also be affected (Kim et al. 2008; Maquat et al. 2010). Alternate pharmacological routes may therefore be required to more directly inhibit eIF4E function.

Uncoupling eIF4E-eIF4G Interaction

As mentioned previously, a majority of the driver mutations in cancers up-regulate mTOR signaling, which ultimately promotes the eIF4E-eIF4G interaction. Reverting this dysregulation using nonphosphorylatable 4E-BP constructs often halts tumor cell proliferation and limits the extent of disease progression in mouse models, suggesting that inhibitors designed to prevent the binding of eIF4E to eIF4G would be therapeutically beneficial. Several modified peptides encompassing the eIF4G/4E-BP hydrophobic eIF4E-binding motif coupled to cell-penetrating moieties have been developed and most have proven successful in various model systems with few nonspecific effects (Herbert et al. 2000; Ko et al. 2009; Brown et al. 2011). Most of these peptides bind fairly efficiently to tagged recombinant eIF4E (in the low- to mid-nanomolar range), can inhibit cap-dependent translation in vivo and induce cancer cell lines to undergo apoptosis. In one impressive example, a 4E-BP-like peptide coupled to a gonadotropin-receptor agonist dramatically decreased tumor burden following intraperitoneal (i.p.) injection in an ovarian cell xenograft mouse model with seemingly no detectable toxicity (Ko et al. 2009).

In a search for small molecules that can disrupt the eIF4E-eIF4G interaction, the Wagner group has designed a high-throughput screen. To identify such compounds, chemical libraries were assayed that decreased the fluorescence polarization of a labeled eIF4G-like peptide when bound by recombinant eIF4E (Moerke et al. 2007). The most potent hit, called 4EGI-1, was evaluated further and found to inhibit cap-dependent translation in extracts, down-regulate known eIF4E-dependent transcripts in vivo, and trigger apoptosis in several tumor lines (Moerke et al. 2007). Based on nuclear magnetic resonance (NMR) spectra, 4EGI-1 was shown to directly bind eIF4E. Counterintuitively, 4EGI-1 did not appear to prevent 4E-BP1 from binding to eIF4E but rather promoted the interaction. Whether or not this "gain-of-function" activity contributes to some of the in vitro and in vivo effects observed remains to be tested. 4EGI-1

has biological properties associated with it that appear to be independent of its effects on cap-dependent translation (Fan et al. 2010; Pruvot et al. 2011) and medicinal chemistry efforts are now required to remove these features from 4EGI-1.

We used a similar strategy in which the eIF4E-eIF4G interaction was monitored using time-resolved fluorescence resonance energy transfer and identified several eIF4E-eIF4G inhibitors, among which the most potent were 4E1RCat and 4E2RCat (Cencic et al. 2011a,b). 4E1RCat blocked cap-dependent but not hepatitis C virus (HCV) internal ribosome entry site (IRES) -dependent translation of a bicistronic dual-luciferase reporter mRNA with an IC_{50} of $\sim$25 μM. Like 4EGI-1, 4E1RCat can prevent the association of eIF4G with m^7GTP-Sepharose-bound eIF4E but, unlike 4EGI-1, it also blocked 4E-BP1 binding with nearly the same efficacy (Cencic et al. 2011b). 4E1RCat was also pharmacologically active in cells and in mice, where it decreased the rate of overall protein synthesis by $\sim$30%. Most strikingly, 4E1RCat was able to improve the response to chemotherapy in the Eμ-myc mouse lymphoma model and thus prolong tumor-free survival. This synergy was probably owing to 4E1RCat's ability to lower Mcl-1 levels, an effect that it shares in common with other inhibitors of translation in vivo.

Targeting eIF4E Levels with Antisense Oligonucleotides (ASOs)

Early proof of principle using antisense-based approaches showed the feasibility of knocking down eIF4E levels to limit the tumorigenic potential of transformed cells, although these experiments often suffered from lack of specificity controls or used short-lived biomolecules (De Benedetti et al. 1991; Rinker-Schaeffer et al. 1993; DeFatta et al. 2000). Second generation ASOs against eIF4E have largely overcome such pitfalls by incorporating extensive phosphate backbone modifications imparting improved nuclease resistance and tissue stability to allow for effective systemic therapeutic delivery. Second generation ASOs targeting eIF4E have been made and results are encouraging (Graff et al.

2007). Despite decreasing overall protein synthesis only minimally (<20%), inhibition of known eIF4E-specific progrowth and prosurvival genes decreased in a dose-dependent manner, halting proliferation and eliciting apoptosis. Impressively, administration of the most potent eIF4E ASO (4E-ASO4, >80% reduction in eIF4E expression) showed pronounced in vivo activity in breast and prostate xenograft models, blunting tumor onset and growth and restricting endothelial cell tube formation, suggesting that eIF4E may also be a possible antiangiogenic target. Thus, targeting eIF4E expression directly might prove to be more sensitive and selective toward tumors through specific down-regulation of oncogenes under eIF4F translational control and afford a much wider therapeutic index and control.

Inhibiting eIF4A Helicase Activity

Another way to inhibit eIF4F-dependent translation initiation is to target eIF4A, the key enzymatic component of the complex. Although both eIF4A's helicase and ATPase functions are required for efficient ribosome loading and scanning, the consequences of eIF4A enzymatic activity in the mechanism of translation initiation is yet to be precisely defined. It has been suggested that eIF4A is required for unwinding 5′-proximal secondary structure to facilitate 40S ribosome recruitment; that eIF4A actively participates in the preinitiation complex scanning of the 5′UTR; and/or that hydrolysis of ATP may simply be necessary to rearrange protein–protein or protein–mRNA interactions (Rogers et al. 2002; Kapp and Lorsch 2004).

Three small-molecule inhibitors, pateamine A (Pat A), hippuristanol, and silvestrol (and related rocaglamide family members) have been characterized extensively as inhibitors of eIF4A and are currently being explored as potential chemotherapies. All of these compounds were originally discovered and purified from natural sources and all were initially characterized as having strong tumor cell-line-specific cytotoxic properties even though their shared molecular target was then unknown (Higa et al. 1981; Gonzalez et al. 2001; Hood et al. 2001; Hwang

et al. 2004). It was fortuitous that all three were identified in a screen designed to identify compounds that could differentially distinguish cap-driven versus IRES-driven translation in Krebs II ascites cell extracts (Novac et al. 2004; Bordeleau et al. 2005, 2006a, 2008). More specifically, they were all able to inhibit the expression of the first cap-dependent luciferase cistron (as part of a dual-luciferase bicistronic capped mRNA transcript) but not the eIF4A-independent HCV-IRES-driven downstream cistron, which provided the first clue that all three were specifically blocking an eIF4A-dependent process. NMR (Lindqvist et al. 2008), affinity selection using immobilized compounds (Bordeleau et al. 2005), and in vitro assays using purified eIF4A (Bordeleau et al. 2008) showed that each was able to bind and affect eIF4A activity directly. Even though all three compounds share a common target, they each have distinct mechanisms of action.

Hippuristanol is an allosteric inhibitor of eIF4A: it prevents both free eIF4 ($eIF4A_f$) and eIF4F complex bound eIF4A ($eIF4A_c$) from binding RNA, which in turn inhibits eIF4A helicase and ATPase activities (Bordeleau et al. 2006a). Consequently, ribosome loading on exogenous transcripts is prevented as eIF4F activity is greatly reduced. Pat A and silvestrol, on the other hand, promote the binding of recombinant eIF4A to RNA and stimulate by several fold the ATP hydrolysis activity of eIF4A (Bordeleau et al. 2005, 2006b, 2008; Low et al. 2005, 2007). Although perhaps perplexing at first glance, because both are as efficient in blocking ribosome loading as hippuristanol, it seems that either Pat A or silvestrol can only exert this effect on $eIF4A_f$, effectively limiting the amount of $eIF4A_c$, by sequestering $eIF4A_f$ nonspecifically on cellular RNAs (Bordeleau et al. 2005). Intriguingly, Pat A and hippuristanol bind to different regions on eIF4A (the binding of silvestrol has not yet been determined). Hippuristanol appears to make specific contacts in the carboxy-terminal domain (CTD) of eIF4A (Lindqvist et al. 2008), and whereas the exact binding site of Pat A is not well defined, Pat A does not appear to interact with the CTD of eIF4A (M Oberer, J Pelletier, and G Wager, unpubl.) and mutational analysis of eIF4A indicates that it requires the amino-terminal ATP-binding domain of eIF4A for its activity (Low et al. 2005). Similarly, a recently characterized bacterial toxin, BPSL1549, from *Burkholderia pseudomallei* was shown to deamidate Gln339 of eIF4A, a conserved residue in motifs V and VI within a loop flanking eIF4A's ATP- and RNA-binding sites (Cruz-Migoni et al. 2011). In fact, this maps precisely to the same region whose residues experienced the strongest chemical shifts in NMR on addition of hippuristanol. The Gln339 modification appeared to transform endogenous eIF4A into a dominant-negative mutant, robustly blocking cellular protein synthesis while inhibiting eIF4A helicase activity and preventing eIF4A recycling (Cruz-Migoni et al. 2011).

Hippuristanol suffers from poor solubility and low potency and thus in vivo studies are scarce, although novel synthetic routes will hopefully resolve these issues (Li et al. 2009; Ravindar et al. 2010, 2011). In one study, hippuristanol was able to selectively inhibit the proliferation of HTLV-1-infected T-cell leukemia cells, both in culture as well as when injected into immunodeficient mice (Tsumuraya et al. 2011). Pat A suffers from a different problem in that it seems to inhibit protein synthesis in an irreversible manner, both in vitro and in vivo, which might make Pat A more difficult to develop for further clinical use (Bordeleau et al. 2005). Nevertheless, in one report, a desmethyl desamino analog of Pat A (DMDA-Pat A) displayed single-agent activity against several human cancer cell xenografts in nude mice, although it appears to inhibit DNA synthesis as well, substantially differing in its mechanism of action from its precursor (Kuznetsov et al. 2009).

In terms of preclinical mouse models, silvestrol is at present the best-studied eIF4A inhibitor, mostly owing to good pharmacological tolerance and little apparent nonspecific toxicity in animals (Cencic et al. 2009a; Saradhi et al. 2011). Silvestrol has been tested in multiple settings, from cancer cell line xenografts to oncogene-derived genetic mouse models with encouraging results. It has been tested in xenografts of acute lymphoblastic leukemia (697 ALL), prostate (PC-3), and breast (MDA-MBA-231) cancer lines as well as the Eµ-Tcl-1 (CLL) and Eµ-Myc lymphoma mouse models (Bordeleau et al. 2008; Cencic et al. 2009a; Lucas et al. 2009). In almost all of these cases, administration of silvestrol as single-agent chemotherapy elicited apoptosis and caused tumor regression. In one study, it was observed that silvestrol appeared to be far more toxic toward B cells than T cells, an effect that was magnified in blood samples of CLL patients relative to normal healthy individuals, which suggests preferential killing of proliferating leukemic cells by the drug (Lucas et al. 2009). Likewise, silvestrol treatment alone showed nanomolar in vitro potency against $Tsc2^{-/-}$ Eµ-Myc lymphoma cultures, even in cells rendered rapamycin-resistant by overexpression of the PIM2 kinase (Schatz 2011). In contrast to these studies, silvestrol treatment of $Pten^{+/-}$ Eµ-Myc-derived tumors had minimal effect when used alone, perhaps as a result of the lower doses used (Bordeleau et al. 2008). When combined with doxorubicin, however, silvestrol greatly prolonged the tumor-free survival of mice, matching the synergy observed for combination rapamycin treatment (Bordeleau et al. 2008). Moreover, this combination therapy was even effective against eIF4E-driven Eµ-Myc lymphomas, which do not respond to monotherapy or combination chemotherapy treatment (Bordeleau et al. 2008). This is also what is seen for human AML cell lines, where silvestrol synergizes with the standard-of-care drugs daunorubicin and cytarabine (Cencic et al. 2009b). Finally, as with other inhibitors of translation, silvestrol's ability to chemosensitize tumors correlates with its ability to inhibit synthesis of Mcl-1 protein. Further improvement of silvestrol's pharmacology is becoming more feasible as new synthetic routes are helping uncover which structural features determine the activity of rocaglamides (El Sous et al. 2007; Gerard et al. 2007; Roche et al. 2010).

Preventing eIF4E Phosphorylation through Mnk Inhibition

Unlike the 4E-BP/mTOR axis of eIF4E regulation, the molecular and biological consequences of direct phosphorylation of eIF4E have only recently begun to be understood. eIF4E is phosphorylated on a single residue, Ser 209, by two kinases: the MAP kinaselike interacting kinases (Mnk) 1/2 (Pyronnet et al. 1999), which are both activated by the MAPK growth-induced and p38 stress-induced signaling pathways (Buxade et al. 2008). Exactly how the added phosphate residue alters eIF4E activity is not well understood. Theories ranging from increased "clamping" of eIF4E to the cap to greater stability in the eIF4F complex have thus far yielded few conclusive results. In lower organisms, eIF4E phosphorylation seems vital for organismal development, yet in mouse models loss of eIF4E phosphorylation has no noticeable phenotype (Ueda et al. 2004). What is emerging, however, is that eIF4E phosphorylation appears to be important in cancer—overexpression or knock-in of an S209A eIF4E mutant profoundly blunts tumor formation (Topisirovic et al. 2004; Wendel et al. 2007; Furic et al. 2010), which, in some circumstances, appears to rely on phospho-eIF4E's ability to regulate Mcl-1 expression levels (Wendel et al. 2007). This suggests that small-molecule inhibitors of the Mnk kinases might yield novel anti-cancer agents. Recently, a novel Mnk inhibitor, cercosporamide, was identified from a screen of 300,000 compounds using an in vitro Mnk1 kinase reaction (Konicek et al. 2011). Cercosporamide is more selective for Mnk2 than Mnk1 ($IC_{50} = 0.011$ μM versus $IC_{50} = 0.116$ μM), also potently inhibits Jak3 ($IC_{50} = 0.031$ μM), and is orally bioavailable, acting on tissues (e.g., liver) within 30 min after administration (Konicek et al. 2011). The compound was found to inhibit eIF4E phosphorylation in a dose-dependent manner, block cellular proliferation, and engage apoptosis (Konicek et al. 2010). In a xenograft colon (HCT116) and B16 lung metastasis model, treatment reduced tumor progression and incidence only modestly, probably reflecting the rebound in eIF4E phosphorylation seen after >4 h treatment in mouse tissues. Defining which cancers depend on eIF4E phosphorylation for tumor maintenance will be key in determining whether Mnk1/2 inhibition will become a therapeutically viable option.

Strategies Targeting Ternary Complex Formation

A second key point of regulation of translation initiation is through phosphorylation of eIF2α, an event that limits the availability of the ternary complex, (eIF2•GTP•tRNA$_i^{Met}$), integral for proper start codon selection and the formation of a competent 43S ribosomal scanning complex. eIF2α can be phosphorylated by four different intracellular kinases (PERK, PKR, GCN2, and HRI), which respond to a variety of cellular stress conditions (e.g., amino acid starvation, viral infection, oxidative stress, excess unfolded proteins, heme deficiency) to limit the protein synthetic load in a cell. Given the fact that cancer cells continuously grow, divide, and catabolize, they are generally under tremendous stress to maintain appropriate protein homeostatic load, a complex balancing act between protein biosynthesis, folding, translocation, assembly/disassembly, and clearance. Thus, relative to normal cells, tumors have constitutively higher levels of eIF2α phosphorylation (Rosenwald et al. 2003, 2008). Compounds designed to exploit this discrepancy between malignant and normal cells by tipping the scales of eIF2α phosphorylation could yield therapeutic benefit (and, arguably, may already form part of the mechanism of action of the approved drug bortezomib). Classic reagents used to induce the phosphorylation of eIF2α in cell culture are often too toxic for animal use (e.g., tunicamycin or thapsigargin) and thus finding novel compounds that can hinder ternary complex formation through alternative means is necessary. In a recent screen aimed to ferret out such chemicals, a group of novel activators of HRI was discovered (Chen et al. 2011b). These similarly structured N,N'-diarylurea compounds, the most potent being BTdCPU, were able to specifically activate eIF2α phosphorylation and downstream stress-associated pathways in a variety of conditions. Importantly, all appeared to cause eIF2α phosphorylation through direct activation of HRI and not through nonspecific oxidative damage, the usual route to HRI activation. Moreover, BTdCPU inhibited the growth of tumor cell lines, both in culture and in xenografts in nude mice, an effect that correlated with the induction of eIF2α phosphorylation. BTdCPU appeared to be well tolerated in vivo with little apparent weight loss and surprisingly minimal impact on the hematopoietic system. Whether BTdCPU will develop into an effective anticancer agent will depend crucially on the levels of expression of HRI in a given tumor cell. Still, its relatively potent in vivo activity means that it will prove a valuable tool in elucidating the impact of eIF2α phosphorylation and ternary complex formation on both normal cellular physiology and in various disease settings. It will also be interesting to see whether more direct and specific inhibitors of ternary complex formation (like NSC119889 [Robert et al. 2006]) will be able to target a more diverse set of tumor cell types.

FUTURE PERSPECTIVES

As the development of compounds targeting protein synthesis begins to transition away from preclinical models, the need for assays and biomarkers to reliably gauge drug efficacy and patient stratification becomes most pressing. Unlike the PI3K/TOR-KI inhibitors, where recent work on new pathway-based phosphoprofiling methodologies has proven quite successful in translation into the clinic (Andersen et al. 2010), biomarker readouts for eIF4F activity are less obvious. Polymerase chain reaction (PCR) -based methodologies coupled to next-generation sequencing are sensitive enough to measure polysome-bound mRNAs and may enable one to correlate drug treatment to tumor response on small biopsy specimens while informing on the relative tumorigenic contribution of cap-dependent versus cap-independent translation. In terms of patient stratification, clinical testing will be required to determine whether aberrant PI3K/mTOR activation in tumors can guide treatment modality involving translational inhibitors, particularly in cases of relapse, where increased eIF4F activity has been associated with chemoresistance.

We also anticipate continued screening for novel drug leads that can influence other translation initiation processes and factors, such as

(1) AUG selection, which potentially can alter the spectrum of therapeutically targeted mRNAs (Takacs et al. 2011); (2) translation reinitiation, which appears to contribute to the translation of a significant number of mRNA transcripts containing upstream open reading frames (ORFs), as recent deep sequencing data of ribosome-protected mRNA fragments suggests (Ingolia et al. 2011); (3) eIF3, a multisubunit complex that associates with the mTOR/raptor, eIF4G, and eIF4A (Korneeva et al. 2000; Holz et al. 2005) and has been implicated in the transformation process (Marchetti et al. 1995; Zhang et al. 2007, 2008); and (4) eIF4B, an essential factor whose regulation is under p70^{S6K} regulation (Raught et al. 2004; Shahbazian et al. 2006). Mammalian genetic screens using short hairpin RNA (shRNA) libraries will further tease out the unique oncogene and non-oncogene exploitable dependencies of mRNA translation factors and regulators (Solimini et al. 2007). Ultimately, these screens will need to be accomplished in more clinically relevant in vivo model systems that better replicate the natural tumor microenvironment and non-cell-autonomous interactions. Looking back at the evolution of this area of translation over the last 10 years and the remarkable progress that has been made, we are excited about the years to come and the potential that these inhibitors hold for both clinical use and in gaining further insight into the fundamental mechanisms of translational control.

ACKNOWLEDGMENTS

We sincerely apologize to those authors whose work is not cited herein owing to space constraints. Work in the authors' laboratory on this topic is supported by grants from the Canadian Institutes of Health Research and Canadian Cancer Society Research Institute.

REFERENCES

*Reference is also in this collection.

Adams BL, Morgan M, Muthukrishnan S, Hecht SM, Shatkin AJ. 1978. The effect of "cap" analogs on reovirus mRNA binding to wheat germ ribosomes. Evidence for

Cite this article as *Cold Spring Harb Perspect Biol* doi: 10.1101/cshperspect.a012377

enhancement of ribosomal binding via a preferred cap conformation. *J Biol Chem* **253:** 2589–2595.

Andersen JN, Sathyanarayanan S, Di Bacco A, Chi A, Zhang T, Chen AH, Dolinski B, Kraus M, Roberts B, Arthur W, et al. 2010. Pathway-based identification of biomarkers for targeted therapeutics: Personalized oncology with PI3K pathway inhibitors. *Sci Transl Med* **2:** 43ra55.

Babendure JR, Babendure JL, Ding JH, Tsien RY. 2006. Control of mammalian translation by mRNA structure near caps. *RNA* **12:** 851–861.

Beretta L, Gingras AC, Svitkin YV, Hall MN, Sonenberg N. 1996. Rapamycin blocks the phosphorylation of 4E-BP1 and inhibits cap-dependent initiation of translation. *EMBO J* **15:** 658–664.

Bordeleau ME, Matthews J, Wojnar JM, Lindqvist L, Novac O, Jankowsky E, Sonenberg N, Northcote P, Teesdale-Spittle P, Pelletier J. 2005. Stimulation of mammalian translation initiation factor eIF4A activity by a small molecule inhibitor of eukaryotic translation. *Proc Natl Acad Sci* **102:** 10460–10465.

Bordeleau ME, Mori A, Oberer M, Lindqvist L, Chard LS, Higa T, Belsham GJ, Wagner G, Tanaka J, Pelletier J. 2006a. Functional characterization of IRESes by an inhibitor of the RNA helicase eIF4A. *Nat Chem Biol* **2:** 213–220.

Bordeleau ME, Cencic R, Lindqvist L, Oberer M, Northcote P, Wagner G, Pelletier J. 2006b. RNA-mediated sequestration of the RNA helicase eIF4A by Pateamine A inhibits translation initiation. *Chem Biol* **13:** 1287–1295.

Bordeleau ME, Robert F, Gerard B, Lindqvist L, Chen SM, Wendel HG, Brem B, Greger H, Lowe SW, Porco JA Jr, et al. 2008. Therapeutic suppression of translation initiation modulates chemosensitivity in a mouse lymphoma model. *J Clin Invest* **118:** 2651–2660.

Brown CJ, Lim JJ, Leonard T, Lim HC, Chia CS, Verma CS, Lane DP. 2011. Stabilizing the eIF4G1 α-helix increases its binding affinity with eIF4E: Implications for peptidomimetic design strategies. *J Mol Biol* **405:** 736–753.

Burnett PE, Barrow RK, Cohen NA, Snyder SH, Sabatini DM. 1998. RAFT1 phosphorylation of the translational regulators p70 S6 kinase and 4E-BP1. *Proc Natl Acad Sci* **95:** 1432–1437.

Buxade M, Parra-Palau JL, Proud CG. 2008. The Mnks: MAP kinase-interacting kinases (MAP kinase signal-integrating kinases). *Front Biosci* **13:** 5359–5373.

Cai A, Jankowska-Anyszka M, Centers A, Chlebicka L, Stepinski J, Stolarski R, Darzynkiewicz E, Rhoads RE. 1999. Quantitative assessment of mRNA cap analogues as inhibitors of in vitro translation. *Biochemistry* **38:** 8538–8547.

Carracedo A, Ma L, Teruya-Feldstein J, Rojo F, Salmena L, Alimonti A, Egia A, Sasaki AT, Thomas G, Kozma SC, et al. 2008. Inhibition of mTORC1 leads to MAPK pathway activation through a PI3K-dependent feedback loop in human cancer. *J Clin Invest* **118:** 3065–3074.

Cencic R, Carrier M, Galicia-Vazquez G, Bordeleau ME, Sukarieh R, Bourdeau A, Brem B, Teodoro JG, Greger H, Tremblay ML, et al. 2009a. Antitumor activity and mechanism of action of the cyclopenta[b]benzofuran, silvestrol. *PLoS ONE* **4:** e5223.

Cencic R, Carrier M, Trnkus A, Porco JA Jr, Minden M, Pelletier J. 2009b. Synergistic effect of inhibiting transla-

tion initiation in combination with cytotoxic agents in acute myelogenous leukemia cells. *Leuk Res* **34:** 535–541.

Cencic R, Desforges M, Hall DR, Kozakov D, Du Y, Min J, Dingledine R, Fu H, Vajda S, Talbot PJ, et al. 2011a. Blocking eIF4E-eIF4G interaction as a strategy to impair coronavirus replication. *J Virol* **85:** 6381–6389.

Cencic R, Hall DR, Robert F, Du Y, Min J, Li L, Qui M, Lewis I, Kurtkaya S, Dingledine R, et al. 2011b. Reversing chemoresistance by small molecule inhibition of the translation initiation complex eIF4F. *Proc Natl Acad Sci* **108:** 1046–1051.

Chambers HF. 2001a. *Antimicrobial agents: Protein synthesis inhibitors and miscellaneous antibacterial agents*. McGraw-Hill, New York.

Chambers HF. 2001b. *Antimicrobial agents: The aminoglycosides*. McGraw-Hill, New York.

Chen J, Zheng XF, Brown EJ, Schreiber SL. 1995. Identification of an 11-kDa FKBP12-rapamycin-binding domain within the 289-kDa FKBP12-rapamycin-associated protein and characterization of a critical serine residue. *Proc Natl Acad Sci* **92:** 4947–4951.

Chen R, Guo L, Chen Y, Jiang Y, Wierda WG, Plunkett W. 2011a. Homoharringtonine reduced Mcl-1 expression and induced apoptosis in chronic lymphocytic leukemia. *Blood* **117:** 156–164.

Chen T, Ozel D, Qiao Y, Harbinski F, Chen L, Denoyelle S, He X, Zvereva N, Supko JG, Chorev M, et al. 2011b. Chemical genetics identify eIF2α kinase heme-regulated inhibitor as an anticancer target. *Nat Chem Biol* **7:** 610–616.

Choo AY, Yoon SO, Kim SG, Roux PP, Blenis J. 2008. Rapamycin differentially inhibits S6Ks and 4E-BP1 to mediate cell-type-specific repression of mRNA translation. *Proc Natl Acad Sci* **105:** 17414–17419.

Cruz-Migoni A, Hautbergue GM, Artymiuk PJ, Baker PJ, Bokori-Brown M, Chang CT, Dickman MJ, Essex-Lopresti A, Harding SV, Mahadi NM, et al. 2011. A Burkholderia pseudomallei toxin inhibits helicase activity of translation factor eIF4A. *Science* **334:** 821–824.

Darzynkiewicz E, Antosiewicz J, Ekiel I, Morgan MA, Tahara SM, Shatkin AJ. 1981. Methyl esterification of m7G5′p reversibly blocks its activity as an analog of eukaryotic mRNA 5′-caps. *J Mol Biol* **153:** 451–458.

Darzynkiewicz E, Ekiel I, Lassota P, Tahara SM. 1987. Inhibition of eukaryotic translation by analogues of messenger RNA 5′-cap: Chemical and biological consequences of 5′-phosphate modifications of 7-methylguanosine 5′-monophosphate. *Biochemistry* **26:** 4372–4380.

Darzynkiewicz E, Stepinski J, Ekiel I, Goyer C, Sonenberg N, Temeriusz A, Jin Y, Sijuwade T, Haber D, Tahara SM. 1989. Inhibition of eukaryotic translation by nucleoside 5′-monophosphate analogues of mRNA 5′-cap: Changes in N7 substituent affect analogue activity. *Biochemistry* **28:** 4771–4778.

De Benedetti A, Joshi-Barve S, Rinker-Schaeffer C, Rhoads RE. 1991. Expression of antisense RNA against initiation factor eIF-4E mRNA in HeLa cells results in lengthened cell division times, diminished translation rates, and reduced levels of both eIF-4E and the p220 component of eIF-4F. *Mol Cell Biol* **11:** 5435–5445.

DeFatta RJ, Nathan CA, De Benedetti A. 2000. Antisense RNA to eIF4E suppresses oncogenic properties of a head and neck squamous cell carcinoma cell line. *Laryngoscope* 110: 928–933.

* Dobson C, Christodoulou J, Cabrita L. 2012. *Cold Spring Harb Persepct Biol* doi: 10.1101/cshperspect.a012260.

Dorrello NV, Peschiaroli A, Guardavaccaro D, Colburn NH, Sherman NE, Pagano M. 2006. S6K1- and βTRCP-mediated degradation of PDCD4 promotes protein translation and cell growth. *Science* 314: 467–471.

Dowling RJ, Topisirovic I, Alain T, Bidinosti M, Fonseca BD, Petroulakis E, Wang X, Larsson O, Selvaraj A, Liu Y, et al. 2010. mTORC1-mediated cell proliferation, but not cell growth, controlled by the 4E-BPs. *Science* 328: 1172–1176.

Dumez H, Gallardo E, Culine S, Galceran JC, Schoffski P, Droz JP, Extremera S, Szyldergemajn S, Flechon A. 2009. Phase II study of biweekly plitidepsin as second-line therapy for advanced or metastatic transitional cell carcinoma of the urothelium. *Mar Drugs* 7: 451–463.

Duncan R, Milburn SC, Hershey JW. 1987. Regulated phosphorylation and low abundance of HeLa cell initiation factor eIF-4F suggest a role in translational control. Heat shock effects on eIF-4F. *J Biol Chem* 262: 380–388.

El Sous M, Khoo ML, Holloway G, Owen D, Scammells PJ, Rizzacasa MA. 2007. Total synthesis of (−)-episilvestrol and (−)-silvestrol. *Angew Chem Int Ed Engl* 46: 7835–7838.

Fan S, Li Y, Yue P, Khuri FR, Sun SY. 2010. The eIF4E/eIF4G interaction inhibitor 4EGI-1 augments TRAIL-mediated apoptosis through c-FLIP down-regulation and DR5 induction independent of inhibition of cap-dependent protein translation. *Neoplasia* 12: 346–356.

Feldman ME, Apsel B, Uotila A, Loewith R, Knight ZA, Ruggero D, Shokat KM. 2009. Active-site inhibitors of mTOR target rapamycin-resistant outputs of mTORC1 and mTORC2. *PLoS Biol* 7: e38.

Furic L, Rong L, Larsson O, Koumakpayi IH, Yoshida K, Brueschke A, Petroulakis E, Robichaud N, Pollak M, Gaboury LA, et al. 2010. eIF4E phosphorylation promotes tumorigenesis and is associated with prostate cancer progression. *Proc Natl Acad Sci* 107: 14134–14139.

Garcia-Martinez JM, Moran J, Clarke RG, Gray A, Cosulich SC, Chresta CM, Alessi DR. 2009. Ku-0063794 is a specific inhibitor of the mammalian target of rapamycin (mTOR). *Biochem J* 421: 29–42.

Gerard B, Cencic R, Pelletier J, Porco JA Jr. 2007. Enantioselective synthesis of the complex rocaglate (−)-silvestrol. *Angew Chem Int Ed Engl* 46: 7831–7834.

Ghosh B, Benyumov AO, Ghosh P, Jia Y, Avdulov S, Dahlberg PS, Peterson M, Smith K, Polunovsky VA, Bitterman PB, et al. 2009. Nontoxic chemical interdiction of the epithelial-to-mesenchymal transition by targeting cap-dependent translation. *ACS Chem Biol* 4: 367–377.

Gingras AC, Gygi SP, Raught B, Polakiewicz RD, Abraham RT, Hoekstra MF, Aebersold R, Sonenberg N. 1999a. Regulation of 4E-BP1 phosphorylation: A novel two-step mechanism. *Genes Dev* 13: 1422–1437.

Gingras AC, Raught B, Sonenberg N. 1999b. eIF4 initiation factors: Effectors of mRNA recruitment to ribosomes and regulators of translation. *Annu Rev Biochem* 68: 913–963.

Gingras AC, Raught B, Gygi SP, Niedzwiecka A, Miron M, Burley SK, Polakiewicz RD, Wyslouch-Cieszynska A,

Aebersold R, Sonenberg N. 2001. Hierarchical phosphorylation of the translation inhibitor 4E-BP1. *Genes Dev* 15: 2852–2864.

Gonzalez N, Barral MA, Rodriguez J, Jimenez C. 2001. New cytotoxic steroids from the gorgonian Isis hippuris. Structure-activity studies. *Tetrahedron* 57: 3487–3497.

Graff JR, Konicek BW, Vincent TM, Lynch RL, Monteith D, Weir SN, Schwier P, Capen A, Goode RL, Dowless MS, et al. 2007. Therapeutic suppression of translation initiation factor eIF4E expression reduces tumor growth without toxicity. *J Clin Invest* 117: 2638–2648.

Grudzien-Nogalska E, Jemielity J, Kowalska J, Darzynkiewicz E, Rhoads RE. 2007a. Phosphorothioate cap analogs stabilize mRNA and increase translational efficiency in mammalian cells. *RNA* 13: 1745–1755.

Grudzien-Nogalska E, Stepinski J, Jemielity J, Zuberek J, Stolarski R, Rhoads RE, Darzynkiewicz E. 2007b. Synthesis of anti-reverse cap analogs (ARCAs) and their applications in mRNA translation and stability. *Methods Enzymol* 431: 203–227.

Gurel G, Blaha G, Moore PB, Steitz TA. 2009. U2504 determines the species specificity of the A-site cleft antibiotics: The structures of tiamulin, homoharringtonine, and bruceantin bound to the ribosome. *J Mol Biol* 389: 146–156.

Hara K, Maruki Y, Long X, Yoshino K, Oshiro N, Hidayat S, Tokunaga C, Avruch J, Yonezawa K. 2002. Raptor, a binding partner of target of rapamycin (TOR), mediates TOR action. *Cell* 110: 177–189.

Harrington LS, Findlay GM, Lamb RF. 2005. Restraining PI3K: mTOR signalling goes back to the membrane. *Trends Biochem Sci* 30: 35–42.

Herbert TP, Fahraeus R, Prescott A, Lane DP, Proud CG. 2000. Rapid induction of apoptosis mediated by peptides that bind initiation factor eIF4E. *Curr Biol* 10: 793–796.

Hickey ED, Weber LA, Baglioni C, Kim CH, Sarma RH. 1977. A relation between inhibition of protein synthesis and conformation of 5′-phosphorylated 7-methylguanosine derivatives. *J Mol Biol* 109: 173–183.

Hidalgo M, Rowinsky EK. 2000. The rapamycin-sensitive signal transduction pathway as a target for cancer therapy. *Oncogene* 19: 6680–6686.

Higa T, Tanaka J, Yasumasa T, Hiroyuki K. 1981. Hippuristanols, cytotoxic polyoxygenated steroids from the gorgonian Isis hippuris. *Chem Lett* 11: 1647–1650.

Holz MK, Ballif BA, Gygi SP, Blenis J. 2005. mTOR and S6K1 mediate assembly of the translation preinitiation complex through dynamic protein interchange and ordered phosphorylation events. *Cell* 123: 569–580.

Hood KA, West LM, Northcote PT, Berridge MV, Miller JH. 2001. Induction of apoptosis by the marine sponge (Mycale) metabolites, mycalamide A and pateamine. *Apoptosis* 6: 207–219.

Hwang BY, Su BN, Chai H, Mi Q, Kardono LB, Afriastini JJ, Riswan S, Santarsiero BD, Mesecar AD, Wild R, et al. 2004. Silvestrol and episilvestrol, potential anticancer rocaglate derivatives from Aglaia silvestris. *J Org Chem* 69: 3350–3358.

Ilic N, Utermark T, Widlund HR, Roberts TM. 2011. PI3K-targeted therapy can be evaded by gene amplification along the MYC-eukaryotic translation initiation factor 4E (eIF4E) axis. *Proc Natl Acad Sci* 108: E699–E708.

Ingolia NT, Lareau LF, Weissman JS. 2011. Ribosome profiling of mouse embryonic stem cells reveals the complexity and dynamics of mammalian proteomes. *Cell* **147:** 789–802.

Jemielity J, Kowalska J, Rydzik AM, Darzynkiewicz E. 2010. Synthetic mRNA cap analogs with a modified triphosphate bridge—Synthesis, applications and prospects. *New J Chem* **34:** 829–844.

Jones RM, Branda J, Johnston KA, Polymenis M, Gadd M, Rustgi A, Callanan L, Schmidt EV. 1996. An essential E box in the promoter of the gene encoding the mRNA cap-binding protein (eukaryotic initiation factor 4E) is a target for activation by c-myc. *Mol Cell Biol* **16:** 4754–4764.

Kantarjian HM, Talpaz M, Santini V, Murgo A, Cheson B, O'Brien SM. 2001. Homoharringtonine: History, current research, and future direction. *Cancer* **92:** 1591–1605.

Kapp LD, Lorsch JR. 2004. The molecular mechanics of eukaryotic translation. *Annu Rev Biochem* **73:** 657–704.

Kim DH, Sarbassov DD, Ali SM, King JE, Latek RR, Erdjument-Bromage H, Tempst P, Sabatini DM. 2002. mTOR interacts with raptor to form a nutrient-sensitive complex that signals to the cell growth machinery. *Cell* **110:** 163–175.

Kim S, Yang JY, Xu J, Jang IC, Prigge MJ, Chua NH. 2008. Two cap-binding proteins CBP20 and CBP80 are involved in processing primary MicroRNAs. *Plant Cell Physiol* **49:** 1634–1644.

Kim TD, Frick M, le Coutre P. 2011. Omacetaxine mepesuccinate for the treatment of leukemia. *Expert Opin Pharmacother* **12:** 2381–2392.

Ko SY, Guo H, Barengo N, Naora H. 2009. Inhibition of ovarian cancer growth by a tumor-targeting peptide that binds eukaryotic translation initiation factor 4E. *Clin Cancer Res* **15:** 4336–4347.

Konicek BW, Stephens JR, McNulty AM, Robichaud N, Peery RB, Dumstorf CA, Dowless MS, Iversen P, Parsons SH, Ellis KE, et al. 2011. Therapeutic inhibition of MAP kinase interacting kinase blocks eukaryotic initiation factor 4E phosphorylation and suppresses outgrowth of experimental lung metastases. *Cancer Res* **71:** 1849–1857.

Korneeva NL, Lamphear BJ, Hennigan FL, Rhoads RE. 2000. Mutually cooperative binding of eukaryotic translation initiation factor (eIF) 3 and eIF4A to human eIF4G-1. *J Biol Chem* **275:** 41369–41376.

Kowalska J, Lewdorowicz M, Zuberek J, Grudzien-Nogalska E, Bojarska E, Stepinski J, Rhoads RE, Darzynkiewicz E, Davis RE, Jemielity J. 2008. Synthesis and characterization of mRNA cap analogs containing phosphorothioate substitutions that bind tightly to eIF4E and are resistant to the decapping pyrophosphatase DcpS. *RNA* **14:** 1119–1131.

Kunz J, Henriquez R, Schneider U, Deuter-Reinhard M, Movva NR, Hall MN. 1993. Target of rapamycin in yeast, TOR2, is an essential phosphatidylinositol kinase homolog required for G1 progression. *Cell* **73:** 585–596.

Kuznetsov G, Xu Q, Rudolph-Owen L, Tendyke K, Liu J, Towle M, Zhao N, Marsh J, Agoulnik S, Twine N, et al. 2009. Potent in vitro and in vivo anticancer activities of des-methyl, des-amino pateamine A, a synthetic analogue of marine natural product pateamine A. *Mol Cancer Ther* **8:** 1250–1260.

Lawson TG, Ray BK, Dodds JT, Grifo JA, Abramson RD, Merrick WC, Betsch DF, Weith HL, Thach RE. 1986. Influence of 5′ proximal secondary structure on the translational efficiency of eukaryotic mRNAs and on their interaction with initiation factors. *J Biol Chem* **261:** 13979–13989.

Lawson TG, Cladaras MH, Ray BK, Lee KA, Abramson RD, Merrick WC, Thach RE. 1988. Discriminatory interaction of purified eukaryotic initiation factors 4F plus 4A with the 5′ ends of reovirus messenger RNAs. *J Biol Chem* **263:** 7266–7276.

Lazaris-Karatzas A, Montine KS, Sonenberg N. 1990. Malignant transformation by a eukaryotic initiation factor subunit that binds mRNA 5′ cap. *Nature* **345:** 544–547.

Li W, Dang Y, Liu JO, Yu B. 2009. Expeditious synthesis of hippuristanol and congeners with potent antiproliferative activities. *Chemistry* **15:** 10356–10359.

Lin CJ, Cencic R, Mills JR, Robert F, Pelletier J. 2008. c-Myc and eIF4F are components of a feedforward loop that links transcription and translation. *Cancer Res* **68:** 5326–5334.

Lindqvist L, Oberer M, Reibarkh M, Cencic R, Bordeleau ME, Vogt E, Marintchev A, Tanaka J, Fagotto F, Altmann M, et al. 2008. Selective pharmacological targeting of a DEAD box RNA helicase. *PLoS ONE* **3:** e1583.

Liu P, Cheng H, Santiago S, Raeder M, Zhang F, Isabella A, Yang J, Semaan DJ, Chen C, Fox EA, et al. 2011. Oncogenic PIK3CA-driven mammary tumors frequently recur via PI3K pathway-dependent and PI3K pathway-independent mechanisms. *Nat Med* **17:** 1116–1120.

* Lorsch J, Dever T, Hinnebusch A, Green R. 2012. *Cold Spring Harb Perspect Biol* doi: 10.1101/cshperspect.a011544.

Low WK, Dang Y, Schneider-Poetsch T, Shi Z, Choi NS, Merrick WC, Romo D, Liu JO. 2005. Inhibition of eukaryotic translation initiation by the marine natural product pateamine A. *Mol Cell* **20:** 709–722.

Low WK, Dang Y, Bhat S, Romo D, Liu JO. 2007. Substrate-dependent targeting of eukaryotic translation initiation factor 4A by Pateamine A: Negation of domain-linker regulation of activity. *Chem Biol* **14:** 715–727.

Lucas DM, Edwards RB, Lozanski G, West DA, Shin JD, Vargo MA, Davis ME, Rozewski DM, Johnson AJ, Su BN, et al. 2009. The novel plant-derived agent silvestrol has B-cell selective activity in chronic lymphocytic leukemia and acute lymphoblastic leukemia in vitro and in vivo. *Blood* **113:** 4656–4666.

Maquat LE, Hwang J, Sato H, Tang Y. 2010. CBP80-promoted mRNP rearrangements during the pioneer round of translation, nonsense-mediated mRNA decay, and thereafter. *Cold Spring Harb Symp Quant Biol* **75:** 127–134.

Marchetti A, Buttitta F, Miyazaki S, Gallahan D, Smith GH, Callahan R. 1995. Int-6, a highly conserved, widely expressed gene, is mutated by mouse mammary tumor virus in mammary preneoplasia. *J Virol* **69:** 1932–1938.

Marcotrigiano J, Gingras AC, Sonenberg N, Burley SK. 1997. Cocrystal structure of the messenger RNA 5′ cap-binding protein (eIF4E) bound to 7-methyl-GDP. *Cell* **89:** 951–961.

Mills JR, Hippo Y, Robert F, Chen SM, Malina A, Lin CJ, Trojahn U, Wendel HG, Charest A, Bronson RT, et al.

2008. mTORC1 promotes survival through translational control of Mcl-1. *Proc Natl Acad Sci* **105:** 10853–10858.

Moerke NJ, Aktas H, Chen H, Cantel S, Reibarkh MY, Fahmy A, Gross JD, Degterev A, Yuan J, Chorev M, et al. 2007. Small-molecule inhibition of the interaction between the translation initiation factors eIF4E and eIF4G. *Cell* **128:** 257–267.

Mothe-Satney I, Yang D, Fadden P, Haystead TA, Lawrence JC Jr. 2000. Multiple mechanisms control phosphorylation of PHAS-I in five (S/T)P sites that govern translational repression. *Mol Cell Biol* **20:** 3558–3567.

Muellner MK, Uras IZ, Gapp BV, Kerzendorfer C, Smida M, Lechtermann H, Craig-Mueller N, Colinge J, Duernberger G, Nijman SM. 2011. A chemical-genetic screen reveals a mechanism of resistance to PI3K inhibitors in cancer. *Nat Chem Biol* **7:** 787–793.

Niedzwiecka A, Marcotrigiano J, Stepinski J, Jankowska-Anyszka M, Wyslouch-Cieszynska A, Dadlez M, Gingras AC, Mak P, Darzynkiewicz E, Sonenberg N, et al. 2002. Biophysical studies of eIF4E cap-binding protein: Recognition of mRNA 5′ cap structure and synthetic fragments of eIF4G and 4E-BP1 proteins. *J Mol Biol* **319:** 615–635.

Noh WC, Mondesire WH, Peng J, Jian W, Zhang H, Dong J, Mills GB, Hung MC, Meric-Bernstam F. 2004. Determinants of rapamycin sensitivity in breast cancer cells. *Clin Cancer Res* **10:** 1013–1023.

Novac O, Guenier AS, Pelletier J. 2004. Inhibitors of protein synthesis identified by a high throughput multiplexed translation screen. *Nucleic Acids Res* **32:** 902–915.

O'Reilly KE, Rojo F, She QB, Solit D, Mills GB, Smith D, Lane H, Hofmann F, Hicklin DJ, Ludwig DL, et al. 2006. mTOR inhibition induces upstream receptor tyrosine kinase signaling and activates Akt. *Cancer Res* **66:** 1500–1508.

Pelletier J, Peltz SW. 2007. *Therapeutic opportunities in translation.* Cold Spring Harbor Laboratory Press, Cold Spring Harbor, NY.

Pelletier J, Sonenberg N. 1985a. Insertion mutagenesis to increase secondary structure within the 5′ noncoding region of a eukaryotic mRNA reduces translational efficiency. *Cell* **40:** 515–526.

Pelletier J, Sonenberg N. 1985b. Photochemical cross-linking of cap binding proteins to eucaryotic mRNAs: Effect of mRNA 5′ secondary structure. *Mol Cell Biol* **5:** 3222–3230.

Pestka S. 1977. *Inhibitors of protein synthesis.* Academic Press, New York.

Pruvot B, Jacquel A, Droin N, Auberger P, Bouscary D, Tamburini J, Muller M, Fontenay M, Chluba J, Solary E. 2011. Leukemic cell xenograft in zebrafish embryo for investigating drug efficacy. *Haematologica* **96:** 612–616.

Pyronnet S, Imataka H, Gingras AC, Fukunaga R, Hunter T, Sonenberg N. 1999. Human eukaryotic translation initiation factor 4G (eIF4G) recruits mnk1 to phosphorylate eIF4E. *EMBO J* **18:** 270–279.

Raught B, Peiretti F, Gingras AC, Livingstone M, Shahbazian D, Mayeur GL, Polakiewicz RD, Sonenberg N, Hershey JW. 2004. Phosphorylation of eucaryotic translation initiation factor 4B Ser422 is modulated by S6 kinases. *EMBO J* **23:** 1761–1769.

Ravindar K, Reddy MS, Lindqvist L, Pelletier J, Deslongchamps P. 2010. Efficient synthetic approach to potent antiproliferative agent hippuristanol via Hg(II)-catalyzed spiroketalization. *Org Lett* **12:** 4420–4423.

Ravindar K, Reddy MS, Lindqvist L, Pelletier J, Deslongchamps P. 2011. Synthesis of the antiproliferative agent hippuristanol and its analogues via Suarez cyclizations and Hg(II)-catalyzed spiroketalizations. *J Org Chem* **76:** 1269–1284.

Rinker-Schaeffer CW, Graff JR, De Benedetti A, Zimmer SG, Rhoads RE. 1993. Decreasing the level of translation initiation factor 4E with antisense RNA causes reversal of ras-mediated transformation and tumorigenesis of cloned rat embryo fibroblasts. *Int J Cancer* **55:** 841–847.

Robert F, Kapp LD, Khan SN, Acker MG, Kolitz S, Kazemi S, Kaufman RJ, Merrick WC, Koromilas AE, Lorsch JR, et al. 2006. Initiation of protein synthesis by hepatitis C virus is refractory to reduced eIF2.GTP.Met-tRNAiMet ternary complex availability. *Mol Biol Cell* **17:** 4632–4644.

Robert F, Carrier M, Rawe S, Chen S, Lowe S, Pelletier J. 2009. Altering chemosensitivity by modulating translation elongation. *PLoS ONE* **4:** e5428.

Roche SP, Cencic R, Pelletier J, Porco JA Jr. 2010. Biomimetic photocycloaddition of 3-hydroxyflavones: Synthesis and evaluation of rocaglate derivatives as inhibitors of eukaryotic translation. *Angew Chem Int Ed Engl* **49:** 6533–6538.

Rogers GW Jr, Komar AA, Merrick WC. 2002. eIF4A: The godfather of the DEAD box helicases. *Prog Nucleic Acid Res Mol Biol* **72:** 307–331.

Rosenwald IB. 1996. Upregulated expression of the genes encoding translation initiation factors eIF-4E and eIF-2α in transformed cells. *Cancer Lett* **102:** 113–123.

Rosenwald IB, Wang S, Savas L, Woda B, Pullman J. 2003. Expression of translation initiation factor eIF-2α is increased in benign and malignant melanocytic and colonic epithelial neoplasms. *Cancer* **98:** 1080–1088.

Rosenwald IB, Koifman L, Savas L, Chen JJ, Woda BA, Kadin ME. 2008. Expression of the translation initiation factors eIF-4E and eIF-2* is frequently increased in neoplastic cells of Hodgkin lymphoma. *Hum Pathol* **39:** 910–916.

Ruggero D, Montanaro L, Ma L, Xu W, Londei P, Cordon-Cardo C, Pandolfi PP. 2004. The translation factor eIF-4E promotes tumor formation and cooperates with c-Myc in lymphomagenesis. *Nat Med* **10:** 484–486.

Sabatini DM, Erdjument-Bromage H, Lui M, Tempst P, Snyder SH. 1994. RAFT1: A mammalian protein that binds to FKBP12 in a rapamycin-dependent fashion and is homologous to yeast TORs. *Cell* **78:** 35–43.

Saradhi UV, Gupta SV, Chiu M, Wang J, Ling Y, Liu Z, Newman DJ, Covey JM, Kinghorn AD, Marcucci G, et al. 2011. Characterization of silvestrol pharmacokinetics in mice using liquid chromatography-tandem mass spectrometry. *AAPS J* **13:** 347–356.

Sarbassov DD, Ali SM, Kim DH, Guertin DA, Latek RR, Erdjument-Bromage H, Tempst P, Sabatini DM. 2004. Rictor, a novel binding partner of mTOR, defines a rapamycin-insensitive and raptor-independent pathway that regulates the cytoskeleton. *Curr Biol* **14:** 1296–1302.

Satheesha S, Cookson VJ, Coleman LJ, Ingram N, Madhok B, Hanby AM, Suleman CA, Sabine VS, Macaskill EJ,

Bartlett JM, et al. 2011. Response to mTOR inhibition: Activity of eIF4E predicts sensitivity in cell lines and acquired changes in eIF4E regulation in breast cancer. *Mol Cancer* **10:** 19.

Schatz JH, Oricchio E, Wolfe AL, Jiang M, Linkov I, Maragulia J, Shi W, Zhang Z, Rajasekhar VK, Pagano NC, et al. 2011. Targeting cap-dependent translation blocks converging survival signals by AKT and PIM kinases in lymphoma. *J Exp Med* **208:** 1799–1807.

Schwartz DC, Parker R. 2000. mRNA decapping in yeast requires dissociation of the cap binding protein, eukaryotic translation initiation factor 4E. *Mol Cell Biol* **20:** 7933–7942.

Sengupta S, Peterson TR, Sabatini DM. 2010. Regulation of the mTOR complex 1 pathway by nutrients, growth factors, and stress. *Mol Cell* **40:** 310–322.

Shahbazian D, Roux PP, Mieulet V, Cohen MS, Raught B, Taunton J, Hershey JW, Blenis J, Pende M, Sonenberg N. 2006. The mTOR/PI3K and MAPK pathways converge on eIF4B to control its phosphorylation and activity. *EMBO J* **25:** 2781–2791.

Shatkin AJ, Darzynkiewicz E, Furuichi Y, Kroath H, Morgan MA, Tahara SM, Yamakawa M. 1982. 5′-Terminal caps, cap-binding proteins and eukaryotic mRNA function. *Biochem Soc Symp* **47:** 129–143.

Silvera D, Formenti SC, Schneider RJ. 2010. Translational control in cancer. *Nat Rev Cancer* **10:** 254–266.

Solimini NL, Luo J, Elledge SJ. 2007. Non-oncogene addiction and the stress phenotype of cancer cells. *Cell* **130:** 986–988.

Su W, Slepenkov S, Grudzien-Nogalska E, Kowalska J, Kulis M, Zuberek J, Lukaszewicz M, Darzynkiewicz E, Jemielity J, Rhoads RE. 2011. Translation, stability, and resistance to decapping of mRNAs containing caps substituted in the triphosphate chain with BH3, Se, and NH. *RNA* **17:** 978–988.

Takacs JE, Neary TB, Ingolia NT, Saini AK, Martin-Marcos P, Pelletier J, Hinnebusch AG, Lorsch JR. 2011. Identification of compounds that decrease the fidelity of start codon recognition by the eukaryotic translational machinery. *RNA* **17:** 439–452.

Tang R, Faussat AM, Majdak P, Marzac C, Dubrulle S, Marjanovic Z, Legrand O, Marie JP. 2006. Semisynthetic homoharringtonine induces apoptosis via inhibition of protein synthesis and triggers rapid myeloid cell leukemia-1 down-regulation in myeloid leukemia cells. *Mol Cancer Ther* **5:** 723–731.

Thoreen CC, Kang SA, Chang JW, Liu Q, Zhang J, Gao Y, Reichling LJ, Sim T, Sabatini DM, Gray NS. 2009. An ATP-competitive mammalian target of rapamycin inhibitor reveals rapamycin-resistant functions of mTORC1. *J Biol Chem* **284:** 8023–8032.

Toledo LI, Murga M, Zur R, Soria R, Rodriguez A, Martinez S, Oyarzabal J, Pastor J, Bischoff JR, Fernandez-Capetillo O. 2011. A cell-based screen identifies ATR inhibitors with synthetic lethal properties for cancer-associated mutations. *Nat Struct Mol Biol* **18:** 721–727.

Topisirovic I, Ruiz-Gutierrez M, Borden KL. 2004. Phosphorylation of the eukaryotic translation initiation factor eIF4E contributes to its transformation and mRNA transport activities. *Cancer Res* **64:** 8639–8642.

Tsumuraya T, Ishikawa C, Machijima Y, Nakachi S, Senba M, Tanaka J, Mori N. 2011. Effects of hippuristanol, an inhibitor of eIF4A, on adult T-cell leukemia. *Biochem Pharmacol* **81:** 713–722.

Ueda T, Watanabe-Fukunaga R, Fukuyama H, Nagata S, Fukunaga R. 2004. Mnk2 and Mnk1 are essential for constitutive and inducible phosphorylation of eukaryotic initiation factor 4E but not for cell growth or development. *Mol Cell Biol* **24:** 6539–6549.

Vazquez D. 1979. Inhibitors of protein biosynthesis. *Mol Biol Biochem Biophys* **30:** 1–312.

Wagner CR, Iyer VV, McIntee EJ. 2000. Pronucleotides: Toward the in vivo delivery of antiviral and anticancer nucleotides. *Med Res Rev* **20:** 417–451.

Wander SA, Hennessy BT, Slingerland JM. 2011. Next-generation mTOR inhibitors in clinical oncology: How pathway complexity informs therapeutic strategy. *J Clin Invest* **121:** 1231–1241.

Wendel HG, De Stanchina E, Fridman JS, Malina A, Ray S, Kogan S, Cordon-Cardo C, Pelletier J, Lowe SW. 2004. Survival signalling by Akt and eIF4E in oncogenesis and cancer therapy. *Nature* **428:** 332–337.

Wendel HG, Malina A, Zhao Z, Zender L, Kogan SC, Cordon-Cardo C, Pelletier J, Lowe SW. 2006. Determinants of sensitivity and resistance to rapamycin-chemotherapy drug combinations in vivo. *Cancer Res* **66:** 7639–7646.

Wendel HG, Silva RL, Malina A, Mills JR, Zhu H, Ueda T, Watanabe-Fukunaga R, Fukunaga R, Teruya-Feldstein J, Pelletier J, et al. 2007. Dissecting eIF4E action in tumorigenesis. *Genes Dev* **21:** 3232–3237.

Yu K, Shi C, Toral-Barza L, Lucas J, Shor B, Kim JE, Zhang WG, Mahoney R, Gaydos C, Tardio L, et al. 2010. Beyond rapalog therapy: Preclinical pharmacology and antitumor activity of WYE-125132, an ATP-competitive and specific inhibitor of mTORC1 and mTORC2. *Cancer Res* **70:** 621–631.

Yuan TL, Cantley LC. 2008. PI3K pathway alterations in cancer: Variations on a theme. *Oncogene* **27:** 5497–5510.

Zhang L, Pan X, Hershey JW. 2007. Individual overexpression of five subunits of human translation initiation factor eIF3 promotes malignant transformation of immortal fibroblast cells. *J Biol Chem* **282:** 5790–5800.

Zhang L, Smit-McBride Z, Pan X, Rheinhardt J, Hershey JW. 2008. An oncogenic role for the phosphorylated h-subunit of human translation initiation factor eIF3. *J Biol Chem* **283:** 24047–24060.

Index